Cutting-Edge Analysis of Aqueous Suspension Filtration

Vadym Poliakov
Institute of Hydromechanics NAS, Ukraine, & Institute of Water Problems and Land Reclamation NAAS, Ukraine

IGI Global
Scientific Publishing
Publishing Tomorrow's Research Today

Vice President of Editorial Melissa Wagner
Director of Acquisitions Mikaela Felty
Director of Book Development Jocelynn Hessler
Production Manager Mike Brehm
Cover Design Phillip Shickler

Published in the United States of America by
IGI Global Scientific Publishing
701 East Chocolate Avenue
Hershey, PA, 17033, USA
Tel: 717-533-8845
Fax: 717-533-7115
Website: https://www.igi-global.com E-mail: cust@igi-global.com

Library of Congress Cataloging-in-Publication Data

Names: Poli a kov, V. L. (Vadym Leont′evich), author.
Title: Cutting-edge analysis of aqueous suspension filtration / edited by Vadym Poliakov.
Description: Hershey, PA : IGI Global Scientific Publishing, [2026] | Includes bibliographical references and index. | Summary: "The monograph summarizes the accumulated extensive knowledge on treatment of contaminated water in porous media (filters), develops and brings it to detailed recommendations to production. It pursues two main goals: creation of a methodological basis for a prediction of changes in the operational characteristics and development of the universal optimization methodology for technological and design parameters"-- Provided by publisher.
Identifiers: LCCN 2024058272 (print) | LCCN 2024058273 (ebook) | ISBN 9798369382110 (hardcover) | ISBN 9798369382127 (paperback) | ISBN 9798369382134 (ebook)
Subjects: LCSH: Water--Purification--Filtration. | Filters and filtration.
Classification: LCC TD441 .P65 2025 (print) | LCC TD441 (ebook) | DDC 628.1/64--dc23/eng/20250207
LC record available at https://lccn.loc.gov/2024058272
LC ebook record available at https://lccn.loc.gov/2024058273

British Cataloguing in Publication Data
A Cataloguing in Publication record for this book is available from the British Library.

Table of Contents

Detailed Table of Contents

The level of modern knowledge about filtration, that directly ensures the removal of the dispersed contamination from natural and waste waters, was analyzed in Chapter 1. A detailed description of the packed porous media saturated with contaminated water as physical two-phase system is given in Section 1. It is under the conditions of progressive reduction of the pore space. Section 2 includes the description of general approaches to mathematical modeling of processes of different nature, which together promote the productive operation of water treatment units with a porous functional layer. The principles of developing mathematical models, setting appropriate problems, methods of solving them using exact and approximate methods are discussed consistently and specifically for the various filtration conditions. Section 3 provides a base for the technological analysis, including five criteria parameters and equations. Using simple models, the significance of two mass transfer mechanisms is analyzed.

Chapter 2 highlights the basic provisions of modeling the detachment filtration of aqueous suspensions at a constant rate with linear mass exchange kinetics under the operating conditions similar to reality. Thus, the initial contamination of the packed bed of the rapid filter, significant changes in the initial content of the suspended solids over time are taken into account. Thanks to the use of exact and approximate analytical methods, formulas suitable for predicting an increase in the impurity concentrations and mechanical energy expenditure are derived, firstly, in a simple integral form, and secondly, in an even simpler form of combinations of elementary functions. In addition, similar solutions to the problem of clarification of the suspension in the packed bed of a special layered structure, as well as the calculation expressions, are obtained. There are comments on the key mass exchange coefficient.

The thorough technological and also design analysis of the rapid filters operation with a constant rate is carried out in Chapter 3. It is based on the solutions of the mathematical problems from the previous chapter, as well as sanitary and hydraulic criteria. The procedure for determining the duration of the filter run (operation period) is shown for similar conditions. This procedure is implemented for the purpose of applied optimization of the filter bed height, the most important design parameter. The consequences of the unavoidable sorting of the granular filtering material due to intensive washing are studied. Filtration is considered with a strict restriction of the impurity accumulation in the pores and, accordingly, the formation of two zones with complete and incomplete deposit saturation in the clogged bed. The technique of the filtration intensification by splitting a single suspension flow into two or more smaller flows is analyzed. The evaluation of the sensitivity of the technological parameters relatively temperature and deposit composition has been carried out.

Rigorous results of a theoretical study of the non-detachment filtration with nonlinear kinetics of suspension absorption and the formation of a gel-like deposit are obtained in Chapter 4. Much attention is paid to the technological aspects of filters functioning under the conditions of intensive accumulation of the non-consolidated and consolidated deposit during the sequence of filter runs to substantiate the service life of the successive replacement of the filter medium material. The technique of the filtration intensification due to a change in the filtration flow during the filter run is proposed and substantiated. The patterns of the suspension filtration with the formation of a surface layer on the packed bed surface under constant rate or pressure drop are established. Theoretical estimation of the effect of the filtration rate (varying many times) on mass transfer and hydraulic resistance of the clogged bed is performed as applied to detachment (Section 3.4) and non-detachment (Section 4.4) filtration.

The depth detachment filtration with linear kinetics and an unregulated rate is considered in detail. A significant filtration rate is initially provided due to the suspension accumulation in the storage tank. The early and preparation stages are specially studied. The main interest is the final stage, during which the suspension is processed with a declining filtration flow rate. The technological analysis is carried out considering the limitation on the decrease in the filter performance. The hydraulic properties of the filter communications and variable head losses in them are taken into account. Mathematical modeling is significantly complicated due to the close relationship between the kinetic parameters and the filtration rate. An approximate solution is obtained with minimal errors for the general form of such relationship. An exact solution was obtained and a reliable calculation method developed based on it. The limiting storage tank height was evaluated, calculations of the filter run duration and applied optimization of the bed absorption properties were performed.

Since the absorption rate of the suspended solids and the corresponding kinetic coefficient often decrease or increase as the deposit accumulates in the bed, Chapter 6 raises the issues of the detachment filtration with nonlinear kinetics. At the same time, the exact solutions of the suitable nonlinear problems were obtained in the presence of an initial deposit and a two-layer structure of the filter media, taking into consideration the autocatalysis phenomenon, progressive decrease in the absorptive capacity. Based on them the significance of the nonlinear effect was estimated, numerous technological calculations were performed. In view of the cumbersome calculation formalisms and the unambiguous choice of kinetics, an approximate solution of a similar problem with a kinetic equation of general form is obtained, which allows to calculate the characteristics of nonlinear filtration with minimal errors. The accuracy of the above solution is evaluated in case of linear filtration for the head losses due to severe clogging of the packed bed.

Preface

It is with great pleasure that I present this monograph, *Cutting-Edge Analysis of Aqueous Suspension Filtration*, a culmination of years of rigorous study, analytical exploration, and technological refinement in the field of water treatment through porous media. The need for efficient, adaptable, and economically sustainable filtration technologies methods has never been more pressing. Water, the essence of human life, is increasingly threatened by complex pollution patterns, evolving industrial activities, and unpredictable emergencies—natural and anthropogenic alike.

This work was driven by two core objectives: the creation of a robust methodological foundation for predicting changes in filtration facilities systems and the development of a universal optimization framework for both technological and design parameters. Through the formulation and solution resolution of a class of mathematical problems—interpretable as nonlinear filtration theory—this monograph bridges the gap between theoretical modeling and practical implementation in water treatment processes.

One of the distinctive features of this book is its comprehensive treatment of both the physical and chemical underpinnings that govern filter performance. It delves into the operational dynamics of various filtration facilities systems, encompassing depth and surface filters, linear and non-linear types of filtration, radial and plane planar designs, and systems filters with uniform or stratified media. The methodologies proposed enable a detailed evaluation and optimization of real-world filtration facilities configurations, including rapid, stepped, precoat, and soil-bed filters, across a wide spectrum of flow conditions.

Equally important is the monograph's focus on the economic and ecological imperatives of modern water treatment. By offering predictive models grounded in nonlinear mathematics and supported by systematic design analysis, this book provides engineers and researchers with a set of powerful tools for enhancing performance while reducing operational costs. The approach described herein offers clarity on the scale, structure, and cost-efficiency of experimental studies, promoting more targeted and less resource-intensive research. –

Of particular relevance is the emphasis on dispersed (granular) materials, whose mechanical and physical characteristics make them indispensable in filtration technologies. These materials, despite their cost, remain central to effective water treatment. This monograph offers insights into how to prolong their lifecycle, maximize efficiency, and integrate biological processes such as microbiota activity for even greater pollutant removal. –

The methodology elaborated in this work not only supports optimal design and operation of filtration facilities but also provides a replicable model for applying theoretical findings across engineering disciplines. The ability to forecast complex filtration behavior and translate it into actionable design and operational improvements holds profound implications for both current practices and future innovations.

This book is primarily tailored to professionals in water treatment and supply, yet I am confident it will also resonate with application engineers, researchers, and students in related domains. The interdisciplinary nature of the topic—and the clarity with which it is addressed—makes it a valuable reference for anyone interested in the advancement of environmental engineering.

I extend my sincere gratitude to the readers who engage with this work. May it serve as a practical guide, a source of inspiration, and a catalyst for continued innovation in the essential field of water purification.

ORGANIZATION OF THE BOOK

- **Chapter 1: Modern Theoretical Concepts of Water Treatment and Role of Porous Media in It**

In the opening chapter, I provide a foundational exploration of modern filtration science, with an emphasis on the role of porous media in the removal of dispersed contaminants from natural and wastewater sources. A comprehensive physical representation of saturated porous media as a two-phase system is developed under conditions of decreasing pore space. Building upon this, I delve into the general principles and techniques of mathematical modeling applicable to diverse filtration scenarios. These models are instrumental in enhancing the operational efficiency of water treatment systems. The chapter culminates in a technologically oriented analysis using a set of defined criteria and simplified models, which underscore the importance of key mass transfer mechanisms.

- **Chapter 2: Linear Theory of Detachment Filtration With Constant Rate**

Here, I lay out a linear theoretical framework for modeling detachment filtration processes operating under constant flow conditions and realistic assumptions of mass exchange kinetics. The analysis accommodates initial bed contamination and fluctuating concentrations of suspended solids, using both exact and approximate analytical solutions. These lead to user-friendly formulas that enable predictive assessments of contaminant buildup and energy expenditure. I also extend this approach to specialized filter beds with layered structures and discuss relevant mass exchange coefficients, setting the stage for applied filtration analysis.

- **Chapter 3: Applied Aspects of Constant Rate Filtration**

This chapter translates theoretical principles into practice through a detailed technological and design analysis evaluation of rapid filters functioning at a constant rate. Drawing on previously established mathematical solutions, I introduce a methodology to determine optimal filter run durations and assess the influence of media height, a key design parameter. The chapter also tackles the real-world challenges of granular media sorting due to backwashing and models scenarios with pore clogging-induced dual-zone formation. Additionally, I propose an intensification strategy based on flow division and conduct a sensitivity analysis of process variables such as temperature and deposit composition.

- **Chapter 4: Non-Detachment Filtration**

A shift to non-detachment filtration theory is made in this chapter, where I rigorously explore filtration with nonlinear absorption kinetics and gel-like deposit formation. Technological implications are investigated, particularly concerning filter longevity and the evolution of unconsolidated and consolidated deposits. I propose a novel intensification approach via filtration flow adjustments during operation. Both surface layer formation and internal pore saturation are considered under varying operational constraints. The chapter closes by analyzing the influence of filtration rate variations on mass transfer and resistance, offering insights into both detachment and non-detachment types of filtration regimes.

- **Chapter 5: Detachment Filtration with Unregulated Rate**

Chapter 5 is devoted to filtration facilities systems where the flow rate is unregulated and diminishes over time due to gravity-driven storage tanks. Here, I focus on the distinct filtration phases, especially the critical final stage characterized by declining flow. I incorporate real-world variables such as hydraulic losses and the dynamic behavior of kinetic parameters. A reliable approximate model is presented to manage the mathematical complexity, alongside a verified exact solution. This allows for precise calculation of filter run durations and supports optimization efforts, particularly concerning the storage tank design and media absorption efficiency.

- **Chapter 6: Nonlinear Detachment Filtration**

Recognizing the dynamic nature of filtration kinetics, I delve into nonlinear detachment scenarios in Chapter 6. I present exact solutions for filtration with variable absorption rates, initial deposits, and dual-layer filter media structures configurations. The analysis includes phenomena like autocatalysis and progressive flow capacity loss, leading to valuable conclusions about the impact of nonlinearity on process performance. I also develop an approximate solution based on generalized kinetics, offering a practical tool for engineers to calculate nonlinear filtration characteristics with minimal error. The chapter concludes by validating this approach under conditions of severe pore clogging and increasing head losses.

- **Chapter 7: Plane-Radial and Radial Detachment Filtration**

In this chapter, I explore a less conventional but highly promising filtration flow configuration: axisymmetric suspension flow within cylindrical filter media. Through both exact and approximate solutions, I analyze facilities systems where contaminated water enters through either the inner or outer surfaces of a cylindrical filter bed. Technological insights reveal that such configurations can significantly extend filter run due to enhanced absorption dynamics. I also perform a detailed design analysis, identifying optimal bed radii and ending with a study of radial filtration in spherical geometries. This investigation opens new pathways for efficient filtration unit design.

- **Chapter 8: Infiltration of Aqueous Suspension into Uniform Ground Bed**

Turning to environmental applications, I address the pressing issue of groundwater replenishment through infiltration of aqueous suspensions into uniform ground beds. Using analytical modeling, I examine how surface-derived water—often laden with dispersed impurities—interacts with and saturates hydrogeological systems. I conceptualize this as a three-phase technological process that captures joint fluid infiltration and solid absorption, offering vital insights for designing effective systems to restore depleted aquifers and safeguard water quality.

- **Chapter 9: Concluding Remarks and Future Trends**

The concluding chapter reflects on the motivations behind this monograph and synthesizes its key theoretical contributions. I highlight how the findings support the improvement of both filtration technologies and facility design. Emphasis is placed on future directions, especially the application of porous and pseudo-porous media properties in biological and membrane-based treatments. These trends offer fertile ground for research aimed at purifying waters from diverse origins more efficiently and sustainably.

CONCLUSION

In bringing this reference book to its completion, it is my sincere hope that the reader finds not only a structured exploration of filtration theory but also an integrated vision of how porous media science can elevate water treatment technologies to new standards of efficiency and adaptability. From the theoretical underpinnings of filtration kinetics to applied design insights and environmental implications, each chapter contributes a piece to the broader puzzle of sustainable and precise water purification.

The path of inquiry laid out in this volume is both grounded in rigorous mathematical modeling and guided by practical challenges faced in engineering and environmental sciences. Emphasis has been placed on presenting clear analytical frameworks and solution methods, while consistently reflecting on their technological relevance—from improving rapid filter runs to managing non-detachment clogging dynamics, and from layered filtration strategies to the replenishment of aquifers.

While we have achieved a level of clarity in understanding and simulating key filtration mechanisms, many questions remain open—particularly in the realm of nonlinear processes, hybrid systems, and biologically active filtration media. These areas invite further research and innovation. Future progress will undoubtedly rest on interdisciplinary collaboration, where theoretical precision meets real-world complexity.

I extend my gratitude to all researchers and practitioners committed to advancing water treatment technologies. May this work serve as both a foundation and a catalyst—for continued exploration, smarter designs, and ultimately, cleaner water for communities around the world.

Vadym Leontievich Poliakov
National Academy of Sciences of Ukraine, Ukraine

Chapter 1
Modern Theoretical Concepts of Water Treatment and Role of Porous Media in It

ABSTRACT

The level of modern knowledge about filtration, that directly ensures the removal of the dispersed contamination from natural and waste waters, was analyzed in Chapter 1. A detailed description of the packed porous media saturated with contaminated water as physical two-phase system is given in Section 1. It is under the conditions of progressive reduction of the pore space. Section 2 includes the description of general approaches to mathematical modeling of processes of different nature, which together promote the productive operation of water treatment units with a porous functional layer. The principles of developing mathematical models, setting appropriate problems, methods of solving them using exact and approximate methods are discussed consistently and specifically for the various filtration conditions. Section 3 provides a base for the technological analysis, including five criteria parameters and equations. Using simple models, the significance of two mass transfer mechanisms is analyzed.

1.1. POROUS MEDIA IN MODERN WATER TREATMENT TECHNOLOGIES

Porous media belong to one of the main components of the natural environment (biosphere) and are an important link in the natural cycles of water, gases, organic substances and biogenic elements. They are of exceptional socio-economic importance, providing not only conditions for human creative activity, but also for human existence. In addition to participating in the formation and functioning of the human environment (atmosphere; vegetation cover; hydrosphere, including surface and underground sources), porous media are used by humans for practical solution of numerous everyday issues. Thus, they have many functions, including those of vital importance to humans.

From the ecological point of view, the porous component in the biosphere plays a key role, because unlike the water and air components, it prevents the widespread of anthropogenic pollutants by trapping and accumulating them. It is capable of holding a large amount of water and harmful substances. However, it can become a source of secondary pollution in case of excessive content of the indicated substances and unfavorable change of conditions in adjacent components of ecosystems. Therefore, the forecast of physicochemical state of porous media in dynamics, also development of microbiocenosis, phytocenosis

DOI: 10.4018/979-8-3693-8211-0.ch001

in them and on this basis regulation of various conditions, primarily water-physical and closely related to it hydrochemical conditions and hydrobiological, are of great applied importance.

The monograph has a practical orientation, although it often touches upon issues of fundamental importance. The main attention is paid to various aspects of rational application of porous materials in modern (water) treatment technologies. It is here that they prove to be particularly effective due to the huge contact surface of the solid phase with the liquid (gas) phase. At the same time, such a surface and the accompanying extensive network of interconnected small voids (pores) negatively affect the intensity of the water flow process and the closely related performance of treatment plants and complexes. Thus, the area of application of porous media in practice is significantly limited and the search for ways to expand it is stimulated.

Both natural and artificial porous media most often represent a dispersed structured system, the elements (grains, granules) of which interact (cohesive media) or practically do not interact (non-cohesive, weakly cohesive) and are packed randomly. Nevertheless, simplified notions of porous media as a regular structure with simple geometry have been of great use in various applications. Their illustrative examples are, for example, the consideration of this structure as an ensemble of elements of the same size and regular (spherical, cylindrical, etc.) shape; pore space as a bundle of capillaries (Kozeny, 1931; MacDonald et al., 1991; Henderson et al., 2010; Tan et al., 2017), saturated dispersed medium as a set of spherical particles with a layer of liquid of appropriate thickness on their surface, etc. (Happel, 1958; Kuvabara, 1959; Payatakes et al., 1973; Reddy et al., 2005).

Mechanical, physical, chemical and, in the presence of organic substances, also biological properties of porous media are characterized by great diversity. It is natural that there are many classifications according to different features (by mechanical, chemical composition, flow and absorption capacity, origin, etc.). The mentioned properties and at the same time mechanical, hydrodynamic, physicochemical, biochemical, microbiological processes in the pore space and at interfacial boundaries have been studied for many decades both within separate sections of fundamental sciences (mechanics, hydraulics, chemistry, biology, geology) with the prefix "hydro" and at their junctions (geochemistry, biochemistry, etc.), as well as applied sciences (agrochemistry, hydrology, land reclamation, soil science, contaminant transport and phase transfer, microbiology, etc.).

Of primary importance for the characterization of non-cohesive and cohesive porous media, especially those of natural origin, is their mechanical composition according to (Polubarinova-Kochina, 1977; Sokovnin & Zagoskina, 2002; Vadyunina & Korchagina, 1961; Voznyuk et al., 1984). A complete representation of the number (volumetric content) of spherical particles of any size, from the smallest to the largest, is given by their diameter distribution function, which is described by the grain-size composition curve and cumulative curve (Rastvorova, 1983). The values of the particles diameter d_g may differ in soils by orders of magnitude.

In separation and, in particular, water treatment technologies, special filtering materials are used to enhance interfacial interaction, intensify the filtration flow process and, in general, the technological process of filtration (Kulikov et al., 2009; Fedorov et al., 1984). To make the work of filtration facilities effective, such material should be cheap, inert with respect to aggressive liquids, have high absorption and throughput capacities, large solids take-up capacity, strength, easy to regenerate. Obviously, a material that would satisfy all these requirements at once is unlikely to exist. In fact, it is rare for one material to combine cost-effectiveness and manufacturability. Synthetic fiber materials seem promising due to their very high porosity and solids take-up capacity, resistance to mechanical and chemical influences exposure (Bugai et al., 2007; Lebedev et al., 2006; Rahli et al., 1996; Stechkina & Kirsch, 2003; Tarnapolskaya, 2005). At the same time, they have a noticeably smaller interfacial surface area, although this disadvantage can be partially compensated for by the catalytic action of the accumulated deposit in large quantities. Also quite often there are serious difficulties in restoring the absorbing resource, which forces to apply expensive methods of the regeneration of filtering materials.

Wastes from local production are constantly attracting attention and becoming the subject of experimental studies due to their low cost (Aksogan et al., 2003; Kim et al., 2006). However, their absorption properties often do not meet practical needs. A detailed review and characterization of filter media used in (water)treatment technologies can be found in special reference literature and many publications (Ahsan et al., 2001; Ayukaev & Meltser, 1985; Girol et al., 1998; Han et al., 2024; Horan & Lowe, 2007; Shevchuk et al., 2005; Yaroshevskaya & Muravyev, 2002; Zosin et al., 2003).

Non-cohesive medium-dispersed materials composed of closely sized granular particles are usually used in water treatment filters. For their preparation on the basis of sandy soils these materials are sieved through the sets of sieves and thus particles of approximately the same size are selected. Some properties of a number of common heavy granular materials are reflected in Table 1.1 (Kulsky & Strokacz, 1986). Synthetic material, for example, polystyrene foam granules, may not need sorting if its particles are mass produced of certain sizes according to the established technology.

Table 1. Physical properties of heavy granular materials

Material name	Bulk density, *tonnes/m³*	Porosity %	"Critical" rate, *cm/s*
Quartz sand	1.7	30 - 40	16 - 25
Glass chips	1.3	35 - 48	14 - 17
Porcelain chips	1.4	35 - 42	15 - 18
Ceramic chips	1.6	36 - 46	14 - 15
Sand			
Of expanded clay	0.5	49 - 64	11 - 12
Forest agloporite	1.0	48 - 54	13 - 17
Blast furnace slag	1.5	43 - 44	12 - 14
Burnt rocks	1.6	38 - 41	13 - 15
Crushed anthracite	0.9	37 - 41	13 - 14

Homogeneous on mechanical composition, granular media have somewhat smaller total and specific surfaces in comparison with polydisperse media. However, the volume available to the liquid phase and especially the permeability of monodisperse media turn out to be clearly larger. Thus, applying filtering materials from sorted particles instead of ready-made natural materials with heterogeneous composition, it is possible to appreciably increase the productivity of filtration facilities, despite some weakening of the interphase mass exchange (Adin & Elimelech, 1989; Darby et al., 1991; Xu et al., 2006). However, there is a real possibility to compensate such weakening with excess due to pretreatment (modification) of the surface of these particles (Benkli et al., 2005; Chen et al., 1998; Foppen et al., 2008; Lau et al., 2004; Richard, 2004; Yin et al., 2007).

The main reason for low flow velocities of viscous fluid and large hydraulic resistances in porous media is the counteraction of the moving liquid phase by the immobile solid phase. It is realized through the vast surface of their interface. Therefore, it is important to reliably know the area of such a surface to estimate the resistance (Samuel et al., 2012; Trussell & Chang, 1999; Zhurba, 1980). It is easy to determine its specific value s_g (in a unit volume) in suspension flow through granular materials, if the distribution function of spherical particles by diameter d_g is known, namely, $f_d(d_g)$ is such that

$$\int_0^\infty f_d(\varsigma)d\varsigma = 1 - n_0,$$

where n_0 is the porosity of a clean porous medium. Then the desired function is defined by the formula

$$s_g(d_g) = 6\int_0^{d_g} \frac{f_d(\varsigma)}{\varsigma} d\varsigma. \tag{1.1}$$

The total area is formally equal to $s_g(\infty)$. In the simplest case of homogeneous composition of solid phase with particles of the diameter d_g the distribution function takes the form of the generalized δ - function $f_d(d_g)= (1-n_0)\delta(d_g)$.

Then the number of particles N_g and the area s_g per unit volume are

$$N_g = \frac{6(1 - n_0)}{\pi d_g^3}, \quad s_g = \frac{6(1 - n_0)}{d_g}. \tag{1.2}$$

After fractionation of natural granular material, particle sizes within each fraction noticeably differ due to the heterogeneity of its composition and discreteness of standard sieve mesh sizes. Therefore, in practice, no more than 2, at most, 3 times difference is allowed between the minimum and the maximum grain sizes (Kulikov et al. 2009; Orlov, 2005; Poliakov, 2009b; Reddy et al., 2005; Stevenson, 1997). Accordingly, the function $f_d(d_g)$ has a specific character – its values are noticeably greater than 0 on the interval $[d_{g1},d_{g2}=3d_{g1}]$ and close to 0 outside this interval, so that

$$\int_0^\infty f_d(x)dx \approx \int_{d_{g1}}^{3d_{g1}} f_d(x)dx.$$

Let the filter media be composed of particles of two grades with (equivalent) diameters d_{g1}, d_{g2}. Then the value s_g will depend on them as well as on the number of particles, for example, of the first grade N_{g1}

$$s_g(d_{g1},d_{g2},N_{g1}) = 6\frac{1 - n_0}{d_{g1}} + \pi d_{g1}^2 N_{g1}\left(1 - \frac{d_{g1}}{d_{g2}}\right).$$

If we take the maximum allowable ratio $d_{g2}=3d_{g1}$ for d_{g1}, d_{g2}, then

$$s_g(d_{g1},N_{g1}) = 2\frac{1 - n_0}{d_{g1}} + \frac{2\pi}{3} d_{g1}^2 N_{g1}(d_{g1}). \tag{1.3}$$

Expression (1.3) is considerably simplified by introducing the scales $s_m= 6(1-n_0)/d_{g1}$ and $N_{gm} = 6(1 - n_0)/(\pi d_{g1}^3)$, which correspond to the values s_g and N_{g1} when only small particles are present in the packed filter bed. Then the relation between $\bar{s}_g = s_g/s_m$ and $\bar{N}_{g1} = N_{g1}/N_{gm}$ is represented in the following form

$$\bar{s}_g(\bar{N}_{g1}) = \frac{1}{3} + \frac{2}{3}\bar{N}_{g1}.$$

Since $\bar{N}_{g1}$ can vary maximally from 0 to 1, the values of $\bar{s}_g$ will respectively be in the range from 0 to 1. Then the expenditure of mechanical energy, and with it the intensity of interfacial mass exchange in porous medium, in this case at fixed porosity, can vary depending on the ratio between the content of particles of both grades accordingly. Although it is often preferable to use in the calculations of the indicated both expenditure and intensity a refined (effective) quantity $\bar{s}_g$, which should reflect the influence of other factors, for example, the kinematic structure of a suspension flow (Last & Stichimair, 2002).

Due to the close connection between the technological process and the specific area of the interfacial surface s_g, it is reasonable to operate in parallel with two characteristic values of the equivalent diameter d_e. The first value (d_{ev}) is established on the basis of the equality of the volumes of actual and fictitious particles (v) or their ensembles in a unit volume of porous medium, and the second value (d_{es}) is established on the basis of the equality of areas of their surfaces (s_g). In general, it is sometimes necessary to use more than two dozen of special coefficients for a detailed description of the surface of solid particles of essentially irregular shape (Tanaka et al., 1995; Zielina, 2011). When modeling and calculating the operation of water treatment filters, the elements of the operating medium of which are often nonspherical, it is customary to limit ourselves to the shape factor κ_f (Mintz, 1964; Orlov, 2005; Reznikov, 1999; Sokovnin & Zagoskina, 2002). If we introduce the specified coefficient as a multiplier in the expression for s_g (1.2) at the specific volume of the solid phase $1-n_0$

$$s_g = \frac{6(1-n_0)}{\kappa_f d_e} \tag{1.4}$$

and identify d_e with d_{ev}, then the following expression is true for κ_f

$$\kappa_f = \left(\frac{d_{ev}}{d_{es}}\right)^2 \tag{1.5}$$

Identifying d_e as d_{es}, we obtain

$$\kappa_f = \left(\frac{d_{ev}}{d_{es}}\right)^3 \tag{1.6}$$

It should be noted that expression (1.4) reduces to (1.2) at $d_e = d$, $\kappa_f = 1$ and thus gives the exact value of s_g in the case of a porous medium of spherical particles of the same diameter d.

It is especially appropriate to note the multivaluedness of porosity n_0 in relation to mechanical composition. Indeed, the value of n_0 depends both on the size and shape of the medium particles, and on the way of their arrangement. In practice, n_0 varies within small limits for each type of porous (granular) material, which gives us the right to consider it as a conservative parameter in our reasoning and test calculations (Ison & Ives, 1969). Even in the case of world-wide fine-dispersed soils and a large difference in the content of clay, silt and sand fractions, their porosities differ by a maximum of 15% (FAO

UNESCO, 1974; FAO, 2006). Therefore, when performing quantitative analysis on test examples, it is justified to set a characteristic value of n_0. Moreover, if this parameter is used as a scale, it can be excluded from the calculations of dimensionless characteristics. It should be taken into account that the porosity of organic soils as well as clean fibrous filtering materials used in physical-chemical (rapid and ultra-rapid filters) and biological (bioconveyor facilities) water treatment is 2-3 times higher (Bugai et al., 2004; Dmitrenko & Gvozdyak, 2002; Rode, 1965).

The evaluation of possible values of n_0 in the case of a monodisperse structure formed of spherical non-deformable particles (Yu & Rowe, 2011) is indicative for the analysis of this quantity, as a basic generalized physical and mechanical characteristic of a clean granular medium. At minimum packing density (cubic packing), when each incompressible particle has a point contact with 6 neighboring ones, $n_0=\pi/6$. Such a value is actually an upper limit for n_0 in ideal granular media. When the packing is as dense as possible (hexagonal packing), each internal particle is in contact with 12 neighboring ones, and n_0 decreases to 0.2595. Naturally the way, in which the volume of the granular material is arranged, has marked effect on n_0. The value n_0 is in a narrow range around the value 0.39 in the case of ordinary, haphazard packing. Denser packing (n_0 = 0.33-0.36) is achieved with additional regular mechanical influences.

Porosity n_0 varies in even wider limits at heterogeneous mechanical composition of a granular material. Even an elementary calculation indicates a sharp (by several orders of magnitude) increase in the total surface area of a physical body (particles) when it is crushed with a decrease in the characteristic size by an order of magnitude. Small particles can fill voids between large particles and then n_0 decreases to 0.271 at packing of the former close to the latter. Formal continuation of such a procedure of additional packing leads to a reduction of n_0 by several times more. And it is natural that porosity of natural non-cohesive porous media is often almost twice less than the limiting value of n_0. A detailed review of works on the geometry of granular media, reflecting the influence of the effects of compressibility, shielding, nonsphericity of shape, and polydispersity of composition on the characteristics n_0, s_g is given in the monograph by Aerov et al. (1979). Accumulated extensive data on the value n_0 for granular media used in water treatment technologies indicate its fluctuations within a narrow range, despite possible serious differences in their properties, composition, packing method of structure-forming particles.

Porosity n_0 determines, although ambiguously, the maximum permeability and absorption capacity of both deformable and non-deformable porous media. Therefore, a detailed information on the pore size distribution, which is usually formalized in the form of a continuous function, is of interest in an in-depth study of filtration flow and contaminant spreading and phase exchange in polydisperse porous media. In such cases, it is convenient to use an analog of the function $f_d(d_g)$, namely, the pore size distribution function. Some experimental methods and results of its determination are discussed in (Kryuchkov, 1998; Pivovar et al., 2000). The relationship between both functions was established in (Aberg, 1996).

At the same time, special dispersed media with double porosity are often found in nature (fractured rocks, aggregated soils) and are increasingly used in water treatment technologies (activated carbon) (Corapciooglu & Wang, 1999; Crittenden et al., 1986; Ibaraki & Sudicky, 1995; Limbach & Wei, 1990). The free space in them is formed by pores of two types, which differ significantly in characteristic sizes. Macropores are much larger and differ relatively little between them. The volume of the space available to liquid in porous filtering materials (e.g., bulk activated carbon), more often than not, far exceeds the similar volume in traditional materials with solid structural elements due to the presence of the systems of branched micropores inside porous elements (Kryuchkov, 1998; Whitaker, 1988). Accordingly, the absorption capacity of biporous media of water treatment filters is much higher. Such materials are especially effective for removal of toxic dissolved pollutants. However, their application is constrained by high cost and difficulties with regeneration (Frank et. al., 2015).

Also, the pore space even in granular materials often turns out to be significantly heterogeneous in hydrodynamic respect and can be conditionally divided into flowing and non-flowing (dead-end) pores (Gremyachkin & Dementiev, 1994; Last & Stichimair, 2002; Molnar et al., 2015; Senyavin et al., 1977; Venitsianov & Rubinstein, 1983;). Moreover, stagnant zones may be present in the filtration flow, which complicates interfacial mass exchange (Corapciooglu & Wang, 1999, Shandyba & Weinstein, 1988).

When modeling flow of liquids pure or with inert impurity through uniform porous media under the conditions of a small pressure drop, it is justified to consider the quantity n_0 as a constant. It is determined relatively simply experimentally (Rastvorova, 1983; Shein, 2001; Sherard et al., 1984). In some cases, theoretical calculation of n_0 is possible (Duck et al., 2000; Zou et al., 2011).

Often, when implementing (bio)filtration in natural or artificial porous media, the pore space gradually reduced over time (Mays, 2010). Pore shrinkage can, firstly, be of mechanical nature, secondly, be a consequence of immobilization (sticking) of dispersed impurity, thirdly, be explained by the development of (micro)biocenosis (DeJong & Christoph, 2009; Glatstein & Francisca, 2014; Mitchell & Santamarina, 2005; Reddi et al., 2000; Seifert & Engesgaard, 2012). In the first case, it is caused by various deformations of the medium skeleton and, mainly, by repacking of its elements. Their direct cause is a significant difference in pressure (head) at the boundaries of the filtration flow region (Brook, 1989; Jönsson & Jönsson, 1992; Poliakov, 2014; Vorobiev & Tarasenko, 1987). The filtering material is practically incompressible and, consequently, the total volume of the intergranular space will be constant in the absence of such deformations, which is typical for conditions of aqueous suspensions clarification by rapid filtration with the usual relatively narrow ranges of low operating pressures (Ison & Ives, 1969). It is only at large pressure drops and loose structure of (organic) deposit, which is characteristic of dewatering, film filtration, membrane technology (ultrafiltration), compressibility in porous media starts to have a noticeable effect on the technological process (Abboud & Corapcioglu, 1993; Caliskaner & Tchobanoglous, 2005; Carrier, 2000; Diliunas et al., 1984; Landman & White, 1994).

The pores gradually clog due to intensive interphase mass exchange with prevailing deposition of suspended solids. The rate of clogging during the depletion of the absorbing resource can significantly slow down, becoming asymptotic with time. Deposit (and biomass) formed in the pores of the packed bed is similar to solid matter in many respects, in particular, it can be considered impermeable to liquid. Therefore, it is justified to nominally refer the deposit to the solid phase of the medium at least from the hydraulic point of view. At the same time, the physical (density, strength, sometimes compressibility), as well as geometric, physicochemical properties of the newly formed matter differ significantly from similar properties of the primary solid phase due to the predominant content of bound water in it, and possibly amorphous structure (Noskov et al., 2008; Paal et al., 1994; Plusnin & Golovanov, 1983). For example, the density of conditionally solid products of the filtration process in physicochemical (and biological) water treatment is close to the water density (Bugai et al., 2004). Some deposits can lose a significant part of bound water over time (dehydration). Thereby their volume decreases and the flow capacity (permeability) of the medium increases. Nevertheless, the dehydration-induced increase in the deposit density is usually small. The weakness of the impurity fixation on the filter bed elements as well as the weak structure of the deposit (and biomass) under intensive fluid flow conditions on a rapid filter often contributes to the detachment of its fragments. As a consequence, the pore space is reduced more slowly and the water being clarified turns out to be additionally contaminated (secondary contamination) (MacDonald et al., 1999; Mintz, 1964; Moran et al., 1993). However, sometimes structural bonds in a deposit, e.g. based on iron compounds, can gradually become noticeably stronger. In such situations, consolidation of the gel-like formation, similar to polymerization involving coagulants and flocculants, above all, can seriously complicate the regeneration (washing) of a volume of heavily clogged filter media. Also, the ability of the deposit to resist hydrodynamic effects (mechanical resistance) deserves special attention, since interphase mass exchange, algorithms of operational and long-term control of the technological process of clarification are closely connected with it (Hapun & Miltreiger, 1981; Aravin & Kochina, 1969). Thus, the emerging and rapidly increasing adhesion between deposit elements, on the one hand, hinders mobilization of the accumulated deposit, and, on the other hand, leads to the formation of a non-washable component (Ayukaev & Meltzer, 1985; Orlov, 2005; Reznikov, 1999).

Therefore, if mass transfer is accompanied by the deposition of the suspended solids with formation of conditionally solid (gel-like) matter in the pore space (Bugai et al., 2004; Noskov et al., 2008; Oiha & Graham, 1992), then the fraction of space remaining free and available to fluid usually already changes significantly with time and in the direction of a filtration flow.

The accumulation of filtration products and eventually of a deposit depends significantly on the kinematic structure of a suspension flow and should differ significantly in the transit (flowing pore system) and stagnant (non-flowing pore system) zones. The fact of gradual reduction of the pore space is of particular importance from the hydraulic point of view and has serious consequences for the movement of fluids and impurities in it (Ives, 1970; Mays & Hunt, 2006; Verigin, 1971). Formally, it is reflected, first of all, in the current porosity n, which is related to the initial porosity n_0 by the following relation (vertical filtration)

$$n(z,t) = n_0 - S_S(z,t) \tag{1.7}$$

where S_S is the volumetric concentration of newly formed solid matter (deposit, biomass). Individual values of n are difficult but realistic to determine experimentally. Full information about the quantity $n(z,t)$ can be obtained only by theoretical methods.

It should be noted that the function $n(z,t)$, as a rule, is monotonic and rapidly decreasing in the direction of suspension motion. When a zone with the maximum deposit content is formed, the quantity n here becomes constant after reaching the minimum value. (Kulikov, 2009; Poliakov, 2012b). This function acquires a bell-shaped form with the maximum value inside the bed if the filtration flow direction changes during the operating period.

The configuration of the free space becomes increasingly irregular as a deposit accumulates or biofilms develop. The trajectories of liquid and impurity particles can become noticeably and unpredictably straightened or curved. In such situations, estimation of water flow properties and determination of corresponding parameters on the basis of the detailed modeling of hydrodynamics, contaminant transport and phase transfer in the intergranular space are unpromising, and the only real way is to conduct experimental studies at the macro level under the conditions of controlled contamination of filtering material.

Progressive reduction of the pore space due to its clogging and biocenosis development, formally described by the function decreasing over time $n(z,t)$, inevitably leads to a significant deterioration of suspension flow conditions. As a result, the expenditure of mechanical energy increases unreasonably during the prolonged filtration of an aqueous suspension and the filter capacity decreases. Resulting clogging is slowed down if deposition is accompanied by the deposit mobilization. The detachment of the deposit fragments is inhibited due to the presence of sites in the pore space that are difficult for the moving fluid to reach (stagnant zones, dead-end pores, micropores), as well as the deposit physical and chemical structurization (Ayukaev & Meltser, 1985; Guo et al., 2015 Indraratna et al., 2013; Molnar et al., 2015; Orlov, 2005). However, the detachment increases as the deposit, biomass accumulates due to a decrease in their strength and increase in the fluid flow rate.

For long continuous operation of a filter, it is important to periodically increase the value of n substantially, thereby restoring the original absorption capacity of a filtering material as far as possible. In practice, it is not possible to completely remove the deposit accumulated in the bed, even if the filter media is suspended and in an unstable state. And, despite the numerous collisions of the bed elements (grains), a small part of the deposit, which is firmly fixed on them, is bound to remain. Therefore, it is necessary to arrange the filter operation in such a way as to prevent the gradual accumulation of non-washable deposit and, as a consequence, the shortening of the service life of a packed bed and an increase in the cost of treated water. Sometimes even at traditional depth filtration with solid bed elements, for example, in case of fibrous filtering materials, it is necessary to resort to physical (assuming mechanical, acoustic, thermal treatment), chemical (treatment with solvents, alkali, oxidizing agents) methods of regeneration, the efficiency and cost of which are noticeably higher than those of standard pure water washing at controlled fluidization of the operating layer (without its elements being carried away) (Arun et al., 2013; Davidson & Harrison, 1977; Sakkas & Lekkas, 1989; Stevenson, 1995).

Thus, filter regeneration (washing) causes a jump-like increase in porosity to values close to the initial n_0 (Cheng, 2003; Girol et al., 1998; Grabovsky et al. 2012; Hall &Fitzpatrick, 1998; Shandyba, 1988). The distribution of the residual deposit along a bed height is formally described in the general case by the function $S_S^0(z)$ (in engineering calculations due to intensive mixing of the filtering material during (back)washings, it is often and quite reasonably assumed $S_S^0 = const$(Grabovskiy & Gurinchik, 2005; Oliynik & Kalugin, 2005; Poliakov, 2006). Then the porosity value at the beginning of a new filter run n^0 by analogy with (1.7) will be as follows

$$n^0(z) = n_0 - S_S^0(z).$$

If the operating medium consists of 2 or more layers (Emelko, 2003; Girol et al. 1998; Ives, 1979; Karagyaur et al., 2013; Mohanka, 1969; Odegaard et al., 2003; Rodjers et al., 2005; Shevchuk et al., 2005; Williams et al., 2007; Yaroshevskaya & Muravyev, 2002; Zhurba & Myakishev, 1992), and the mechanical composition of the filtering material of each of them is stable, then the generalized porosity function for the entire clean bed is piecewise constant. The above function becomes significantly variable with prolonged filtration, but all its discontinuities of the first kind at the interlayer boundaries are still present. Similarly, porosity varies in stepped filters, which are a series of successively connected rapid filters, starting from a prefilter (Dantas & Bernardo, 2006; Kawamura, 1985; Kulsky et al., 1991; Poliakov & Shevchuk, 2007; Zhurba & Myakishev, 1992).

Fluid movement, suspended matter transfer and contamination removal are indirectly determined in filtration by the flow capacity of porous media, under which it is reasonable to generalize the filtration flow rate at a unit head gradient. The mentioned reduced flow rate depends mainly on the fluid viscosity and geometrical properties of a such medium, which, first of all, are characterized by two specific parameters changing in time – the area s_g and porosity n. As a rule, these parameters are insufficient for reliable accounting of the influence of solid phase even on the flow process in clean porous media due to the diversity of their composition and structure in practice. Therefore, it is usually necessary to introduce additional correction factors into the corresponding phenomenological expressions, which have a clear physical sense, for example, the previously mentioned shape factor. Empirical coefficients are used along with them, which have no such sense and are introduced to maximize the approximation of the calculated values to the experimental values. The permeability k_{p0} without taking into account the viscous properties of a fluid, and the hydraulic conductivity k_0 taking them into account can serve as a measure of the flow capacity and hydraulic resistance of a clean granular medium.

It should be noted that relatively reliable theoretical determination of the saturated hydraulic conductivity k_0 is possible only for porous media with regular structure and, first of all, for a clean granular media formed of sorted particles. But only some of the many existing experimental methods (e.g., express methods allow to estimate k_0 roughly) can provide the reliable determination of k_0 in other cases, and especially for natural media (Aravin & Nosova, 1969).

Water flow process in porous medium has linear nature, i.e. Darcy's law is valid, only in the range of Reynolds number values limited from below and above. According to different data $Re_g = Vd_g/\nu < 3–10$ (V is the water flow rate or filtration rate; ν is the kinematic viscosity). The specified range for water and the quantity Vd_g is equivalent to the range 0.014–0.047m^2/h at temperature 10 , and 0.011–0.036m^2/h at 20 . If we now take the value of 0.5 mm for d_g as a reference point, it becomes obvious that Darcy's law may be violated when operating filters with a large hydraulic load (rapid and especially ultra-rapid). Therefore, it is necessary to make sure that suspension flow process will be linear at the planned flow rate mode.

In general, the known empirical and semi-empirical formulas for the saturated hydraulic conductivity k and permeability k_p of various porous media are united by the proportionality of the calculation quantity to the square of the equivalent diameter of structure-forming particles d_e (in fact, the specific area s_g) (Aravin & Kochina, 1969; Bear, 1972; Hazen, 1911; Rahli et al., 1996). The proportionality coefficient depends on many factors and, first of all, on physical and geometrical properties of solid and liquid phases of the operating medium. For example, following the multiplicative principle, a correction multiplier is introduced in some formulas for k (Hazen, 1911; Poiseuille, 1869, etc.) and the basic hydrophysical characteristic to reflect the influence of unstable temperature conditions on water-physical properties of the medium, especially due to seasonal temperature fluctuations. This multiplier depends only on temperature and characterizes the correlation between it and the above-mentioned coefficient (or permeability), hydraulic resistance, and actually fluid viscosity (Grant & Salehzadeh, 1996; Hazen, 1911; van Genuchten, 1980). In part, the structure of the coefficient k can be revealed by involving simplified models of special porous media. A successful example of such an approach is the classical Kozeny-Karman formula, which allows us to reliably determine the values k_0 and k_{p0} in the case of a granular medium composed of spherical particles of approximately the same size (Kozeny, 1931)

$$k_0 = \frac{\nu}{g}k_{p0} = \frac{\nu}{g}\frac{n_0^3 d_e^2}{5 s_g^2} = \frac{\nu}{g}\frac{n_0^3 d_e^2}{180(1-n_0)^2}, \qquad (1.8)$$

where g is the gravity acceleration. When applying formula (1.8) in specific conditions, in particular, for granular beds with heterogeneous mechanical composition, non-Newtonian fluids, cohesive (sandy) and non-cohesive (sandy loam, loam, clay) soils, non-spherical structural elements, as well as fouled membranes, it is necessary to introduce corrections for heterogeneity of mechanical composition, aggregation and non-sphericity of medium particles, etc. (MacDonald et al., 1991; Rahli et al., 1996; Tan 2017 et al.; Zhurba, 2011). It should be emphasized that there is no practical use to seek an accurate determination of the coefficient k_0, since the head losses in a clean bed, as a rule, are only a small part of similar losses after its clogging, when the question arises about the reasonability or inexpediency of continuing the filter operation.

At especially high flow rates, which are characteristic for operation of rapid filters in a forced rate mode, but mainly for ultra-rapid filters, the contribution of the inertia forces to the hydrodynamic pattern is often commensurate with the contribution of the pressure, gravity, and drag forces (Hutchison, 1976; Lerch et al., 2016; Niu et al., 2016). Then nonlinear filtration flow takes place and it is necessary to use generalizations of Darcy's law for its description (Panfilov & Fourar, 2006). As a consequence, the expression for the hydraulic resistance takes a nonlinear form, namely, a two-term form (Barenblatt, 1987; Gal et al., 1985; Pivovar et al., 2000; Venetsianov & Rubinstein, 1983) or a power form (Aravin & Kochina, 1969; Polubarinova-Kochina, 1977; Zhernov, 1982). In both cases it is necessary to find two empirical coefficients instead of one coefficient, which greatly complicates experimental studies and processing of their data. In principle, both forms of the generalized equation are also suitable for describing the turbulent flow of liquid in pores, which, incidentally, is rarely observed under the conditions of water treatment by filtration. Sometimes similar flow conditions occur at the operation of ultra-rapid filters (approximately at $V= 40–100 m/h$) (Rudnitsky, 2008; Shtykov et al., 1987).

In the presence of particularly large suspended solids and significant constrictions in the pore chains (throats), their blockage (sieve effect) begins to play an important role in suspension clarification (Bradford & Bettahar, 2005; Iritani et al., 2007; Sharma & Yortos, 1987; Tufenkji et al., 2006; Wang et al., 2012). This effect can lead to a sharp decrease in suspension flow rate or an increase in mechanical energy losses and thus seriously complicate the realization of the technological process. It is mainly manifested immediately after the filter is turned on and becomes weaker later on as the pores shrink. The significance of the sieve effect is usually estimated based on the ratio between the characteristic sizes of the bed elements (grains) and contaminants. In the case of their spherical shape, the critical value of the ratio between the corresponding diameters is approximately equal 12 (Herzig et al., 1979; Stevik et al., 2004). At the same time, flow pore blockage can be appropriate if due to it a surface (dynamic) layer is formed at proper selection of impurity composition (in the presence of two or more fractions). And then surface (film) filtration either supplements the deep bed filtration or can be applied independently (precoat filter, secondary membrane) (Kuberkar & Davis, 2000; Leichkis, 1975; Liu et al., 2024). Incidentally, the liquid flow properties of the newly formed layer (of fine-dispersed impurity) and the basic layer (of filtering material) differ significantly – to a lesser extent their porosity and to a greater extent their permeability.

The increase in the (volumetric) concentration S_S entails a multiple decrease in the permeability of the medium over time, which has a negative effect directly on the filtration process (commensurate increase in the hydraulic resistance of the operating layer, head losses in it) and indirectly on the (bio)filter performance in general (Mays, 2010; Mitchell & Santamarina, 2005; Vandevivere & Haveye, 1992). It is relatively easy to estimate the effects of such an increase theoretically if we know the dependence of the hydraulic conductivity k on S_S. Usually, the following generalized expression is accepted for the dependence (Alekseev & Kommunar, 1974; Kiselev, 1997; Saatci, 1990; Venetsianov & Rubinstein, 1983)

$$k(S_S) = k_0 f_k(S_S), \tag{1.9}$$

so that $f_k(0)=1$, $k(0)=k_0$. Proposals concerning the form of the normalized function $f_k(S_S)$ are reduced mainly to its two forms, namely, the two-parameter form based on the power function (Mintz & Meltzer, 1970; Shechtman, 1961; Zhurba, 1993)

$$f_k(S_S) = \left[1 - \left(\frac{S_S}{n_0}\right)^{m_{k1}}\right]^{m_{k2}} \tag{1.10}$$

and one-parameter exponential function (Chiang & Tien, 1985; Khilar & Fogler, 1998; Nivala et al., 2012; Sen et al., 2004)

$$f_k(S_s) = e^{-m_{k3}S_s}, \tag{1.11}$$

where m_{k1}, m_{k2}, m_{k3} are the empirical coefficients (often taken m_{k1}=1, m_{k2}=3). The form of $f_k(S_S)$ becomes more complicated in the case of fibrous filtering material (Thomasa et al., 2001).

The equivalent form of the law of the hydraulic resistance for the clogged layer of filtering material is the empirical expression for the head (pressure) gradient (McDowell et al., 1986; Mitin, 1969; Senyavin et al., 1977) $i = i_0 f_k^{-1}(S_S)$, where i, i_0 are the current and initial concentration gradients.

Significantly polluted surface water (rivers, reservoirs, lakes, surface runoff (Molchak, 1975) can become a reliable source of quality water at minimal financial costs, if the surrounding easily permeable (sandy) soils are used as natural filters. When arranging water intake drains near the river (Bradford et al., 2002; Gupta et al., 2009), infiltration systems for groundwater recharge and rainwater drainage (basins) (Altshul et al, 1977; Burchak, 1978; Hutt et al., 2007), wetlands (Anderson et al., 2024; Jia, 2005; Kadlec & Knight, 1996), bioconveyor facilities (Gvozdyak, 2003; Sabliy, 2013) are quite feasible to achieve drinking water standards at the output. Such filters are able to retain a variety of contaminants in significant amounts and for a long time. At the same time, treatment of wastewater and surface runoff by the infiltration system allows us to use them in water supply, irrigated agriculture (Adin & Elimelech, 1987; Gordin & Markov, 1991; Jimenez et al., 2000; Maslov et al., 11989; Plotnikov 1983; Samuel et al., 2012; Shevtsov, 1988; Shtykov et al., 1987; Zakariae & Nore, 2024). In the latter case, serious complications often arise due to the need for selective removal of contaminants while retaining sufficient amounts of nutrients, easily degradable organic matter. Presence and systematic renewal of suspended solids in groundwater inevitably leads to the clogging of a porous medium and, what is especially undesirable, of the pre-drain zone (Tugay et al., 2004; Verigin & Kommunar, 1978). Thus, water flow conditions control, and indirectly closely related to it regulation of contaminant mass transfer between solid and liquid phases can be significantly impeded. However, restoration of the original properties of natural filters is extremely problematic.

Therefore, porous media have a huge productive potential. They are widely used in many sectors of the country's economy, including water management, chemical and water treatment technologies. However, their resource is still used incompletely and irrationally. As a consequence, scientific developments are still relevant at present, the ultimate aim of which are practical recommendations on optimization of technological processes occurring in natural and artificial porous media; improvement of purification plants design implementing such processes on an industrial scale and for personal needs. To achieve maximum efficiency of porous media in water treatment technologies is realistic only by combining experimental and theoretical methods of their research taking into account the influence of adjacent media. Methods of mathematical modeling play the decisive role at obtainment, processing and generalization of the results of such researches. They are able to give a reliable forecast of the development of the technological process of (bio)filtration and thus, significantly facilitate the search for ways of its intensification and optimization, more perfect designs of filtration facilities. Principles of their application in water treatment problems using porous materials for absorption of contaminants are covered in the next section.

1.2. METHODS OF MATHEMATICAL MODELING IN WATER TREATMENT TECHNOLOGIES

The logic of applied research requires the application of an integrated approach in the cognitive process, in which experimental and theoretical methods are rationally combined and complement each other. Mathematical modeling methods play an extremely important role in this process. In general, their main purpose is to identify significant effects by theoretical means, generalization of knowledge about the peculiarities of the course of certain operating processes, which are acquired most often in strictly controlled laboratory conditions, as well as in full-scale conditions at permanently operating engineering facilities. As a result, these methods contribute to the creation of a theoretical basis for the applied analysis of practical aspects of (bio)chemical technologies. The ultimate aim of such analysis should be the development of economically justified and, if possible, optimal technical and technological decisions. As applied to the problems of water treatment, the methods of mathematical modeling allow:

- to forecast the development of physical, chemical and hydrodynamic conditions in the operating volume of purification facilities, units and their transportation system;
- to analyze the contribution of individual processes and effects to the technological process and its sensitivity with respect to the model parameters;
- to estimate computational errors due, firstly, to the use of various simplifications in the formalization of the technological process and, secondly, to its analytical analysis by approximate methods;
- to improve the planning of experimental studies;
- to optimize the technological and design parameters.

Since the monograph is primarily aimed at engineering specialists, there is no practical sense in increased requirements to the formalization of the technological process of filtration. Most often both designers and operators of filtration facilities have incomplete and not quite reliable initial information. In addition, the conditions of (bio)filtration during long-term operation of a facility can change appreciably, in particular, the degree and nature of contamination of raw and waste waters; geometric, hydraulic, absorption and biological properties of individual elements of solid, biological phase and in general the operating medium (packed bed), and inevitably the temperature of the liquid in different periods of the year (Hopper-Jones et al., 2010; Shen et al., 2018; Vyakhirev et al., 2002). Therefore, compromises with respect to both the adequacy of the initial mathematical model to natural conditions and the rigor of mathematical solutions and also the accuracy of calculation dependencies are not only acceptable but also justified. Nevertheless, it is necessary to strive to minimize additional computational errors caused by simplifications of the basic model and the use of approximate methods in its solution.

The mathematical modeling was performed in the monograph sequentially in four or five steps, viz,

- correct formulation of the mathematical problem of the filtration species for the technological conditions similar to the original conditions;

- solving the problem by exact or approximate analytical methods and estimating the accuracy of the approximate solution;
- prediction of regular changes in the key filtration characteristics on test examples in order to illustrate the derived calculation dependencies and formulas;
- technological analysis based on the system of the special criteria and analytical dependencies adapted to it (Mintz, 1964; Poliakov, 2009a);
- design analysis using the results of technological modeling to improve the technical support of filtration (Poliakov, 2010d).

The emphasis was placed on the fourth and fifth steps. In addition to the highlighted steps, when planning experimental studies, it is advisable to perform the sensitivity analysis of the required characteristics with respect to the most important model parameters (Crittenden et al., 1986; Kim & Tobiasson, 2004a).

Figure 1 clearly presents the opportunities of mathematical modeling methods in relation to the contaminated water treatment by means of engineering units, facilities, systems with a porous operating layer. In general, structuring of the accumulated empirical knowledge about patterns of a complex of processes and effects of different nature and importance as well as the experience of long-term operation of filtration facilities the assessment of actual capabilities of mathematical modeling and engineering calculation methods in relation to practical filtration. As a result of structuring, it is preferable to distinguish and present in a visual form a set of interconnected compartments performing different (informative, representative, executive, advisory) functions. Thus, it is advisable to precede mathematical modeling by drawing up an auxiliary block diagram, which conceptually reflects both the prerequisites for it, and the most important cause-and-effect relations within and between the compartments with specifying of the operational information to be transmitted therewith and, at last, the final results in the form of the forecasts and the recommendations for improving the design and technology.

Exactly such block diagram is schematically depicted in Figure 1. In accordance with the conceptual model given in fact, the theoretical studies of filtration facilities operation were carried out sequentially, starting from the preparatory stage (compartment I), followed by the executive (compartments II-IV) and, lastly, the final (compartment VI) stages. The feedback between compartments I and VI, manifested in the applied optimization procedure, should be especially noted. Its essence consists in a purposeful search of realistic options (iterations) with the establishment of the most economical option in the end. The main functional purpose of the compartments in this diagram is as follows:

- input information compartment (I) provides initial data in full regarding geometric, mechanical, physical and chemical properties of the filtering material layer and impurities, hydraulic properties of support equipment, also formalized requirements to the quality of purification, energy status and productivity of the filtration facility, which are necessary for mathematical, technological and design modeling of separation of aqueous suspensions by rapid filtration;
- clarification compartment (II, key in the complex of functional compartments) directly reflects the regular character of suspensions clarification by means of interphase exchange of particles of (polydisperse) impurity at its arranged stabilized transport in the operating layer, which is composed of sorted particles (grains) of natural or artificial origin, also formed in the process of surface filtration from the coarse-grained fraction of suspension;
- filtration flow compartment (III) deals with saturated fluid flow in primary and secondary (newly formed), artificial and natural porous media under the conditions of significantly shrinking pore

space and accelerated increase of their hydraulic resistance, regulated and unregulated filtration flow rate;

- structural compartment (IV) emphasizes the role of individual elements of filtering material in removal of pollution, and its absorption is localized inside porous and on the surface of solid grains and is directly carried out due to sorption (in the broad sense) process, which is preceded and promoted by hydrodynamics of the flow around a grain, mechanisms of the particulates delivery, pollutants diffusion and adsorption, contaminants capture;
- hydraulic compartment (V) is used for the technical support of the stable functioning of the operating layer, the rate regulation and the control of operational indicators; it ensures uninterrupted supply of contaminated water and discharge of treated water, stability of the technological conditions;
- output information compartment (VI) is designed to forecast changes in the protective and throughput capacity of a filter bed, its energy resource, as well as applied optimization of technological and design parameters, and the results of applied modeling are the basis for the development of practical recommendations to designers and operators of filtration facilities in order to improve the facility design, technology of clarification of aqueous suspensions.

Figure 1. Block diagram to mathematical modeling of filtration

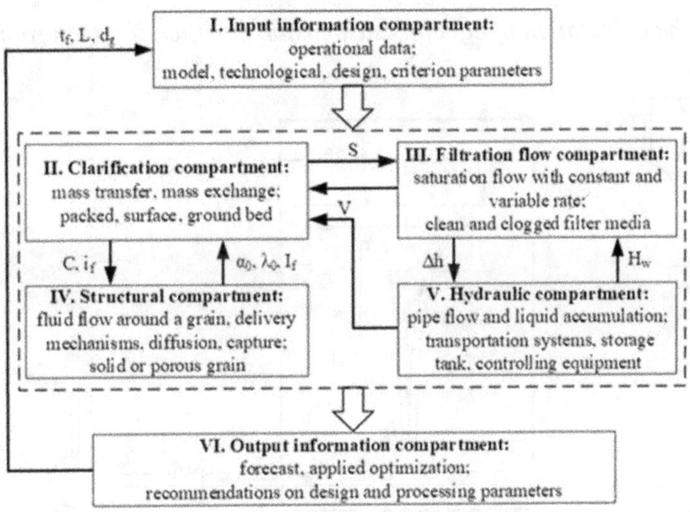

Note. C – (volumetric) concentration of pollutants (suspended solids); d_g – (equivalent) diameter of a particle of porous medium (filter bed, ground bed); H_w – depth of the suspension layer above the filter bed (surface layer, layer in the storage tank); I_f – absorption rate of pollutants distributed along the height of the operating layer; i_f – microflow of pollutants to a (porous) structural element (grain, biofilm); L – packed bed height; l_d – dynamic layer thickness; S – (volumetric) concentration of the deposited suspension particles; t_f – filter run duration; V – filtration (flow) rate; α_0, λ_0 – reduced rate coefficients of pollutant absorption (capture, sorption) by (clean) filtering material; Δh – head losses in the operating layer;

The mentioned, and at the same time auxiliary functions, which are not discussed here, are performed by means of appropriate complexes of interrelated processes of mechanical, physical, chemical, and possibly biological nature. They take place in the filtration facilities, the design of which allows flexible control of these processes and thus, minimize the cost of treated water. Long-term and effective functioning of the facilities of different design, purpose and field of application is facilitated by the coordinated action of its structural components, such as: the operating layer of different structure (packed bed, surface porous layer, their combination); supply communications, including the initial water collector, pipelines, pressure channel (Berman, 1953; Grabovsky et al, 2012; Lee & Clark, 1998); offtake communications including filtrate collector, pipeline, and drainage (Epoyan et al., 2016; Progulny & Ryabkov, 2019); storage tank (cartridge) (Grabovsky et al, 2012; Cleasby, 1993); support layer (Kumari & Kazmi, 2014; Young & Dahab, 1983); regulating and controlling devices (rate controllers, aerators, etc.); measuring equipment (Orlov & Martynov, 2011; Pittoors et al., 2014; Popkovic & Ropin,1986); pump etc. The scheme illustrating the design of the rapid filtration facility in operation is shown in **Figure 2**.

The technological aspects of the aqueous suspension clarification by filtration (compartment I) play a key role in prolonging the operation of a filtration facility and increasing its economic efficiency. Here it is relevant to mention special measures that precede the complex filtration process and contribute to its intensification. They are of physical (radiation, electric and magnetic fields) or chemical (reagents, strong oxidizing agents) nature (Babenkov, 1977; Kulkarni et al., 2005; Yakovlev et al., 1987; Yang & Langdon, 2004; Yuksel et al., 2003; Zapolsky & Baran, 1987). From a practical point of view, it is especially important to optimize the control algorithm of the filtration facility. An unsuccessful choice of such an algorithm inevitably leads either to an unacceptable decrease in treatment quality and facility productivity, or to an increase in the cost of water treatment and may even cause an emergency situation. The above-mentioned aspects are discussed in detail in the next Section.

Figure 2. Scheme for the calculation of aqueous suspension clarification on a rapid filter

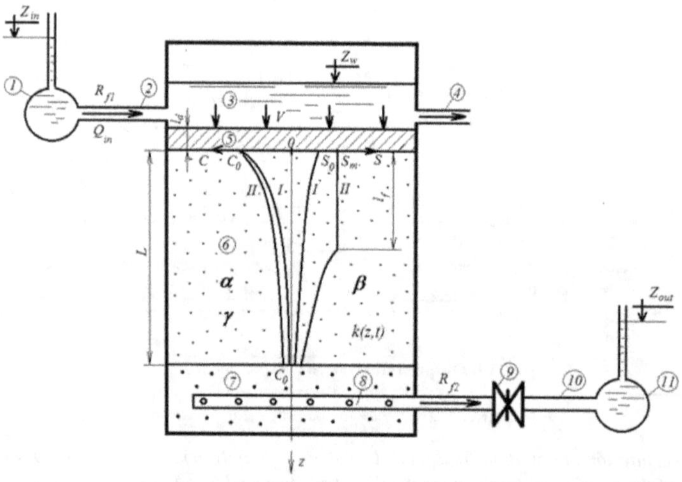

Note. 1 – initial (raw) water collector, 2 – supply communications, 3 – surface water layer in the filter, 4 – washing water discharge pipe, 5 – dynamic layer, 6 – layer of filtering material (packed bed), 7 – supporting layer, 8 – drain, 9 – rate regulator, 10 – offtake communications, 11 – filtrate collector

The following notations are taken here: Z_{in} is the piezometric mark of the initial suspension (raw water) collector; Z_{out} is the piezometric mark of the filtrate collector; Z_w is the piezometric mark of the suspension surface above the filter bed; R_{f1} is the resistance of the supply communications; R_{f2} is the resistance of the offtake communications; l_d is the thickness of the dynamic layer; l_f is the depth of the zone of maximum saturation with a deposit; I, II are the stages of saturation of the bed with a deposit

Mathematical modeling makes practical sense only if there is reliable baseline information in full (compartment I). Its composition is dictated by the basic mathematical model chosen and should include the reliable data on, first, model coefficients, second, design parameters, and third, criterion parameters. Let us especially note the technological parameters, some of which are adopted on the basis of accumulated experience, and the rest should be selected by means of applied optimization procedure, namely, calculation of a set of possible options and selection of the best one from the economic point of view.

Design materials, regulatory documents, and the body of literature gathered over several decades should be involved (Foppen & Schijven, 2006; Saiers et al., 1994). Although it is clearly preferable to design and conduct specific experimental studies under controlled and similar to actual conditions. This is especially important for the model coefficients that have a significant influence on the desired characteristics, but for the establishment of which, in principle, the direct measurements cannot be directly used. As a consequence, we are compelled to actually solve inverse problems that are mathematically incorrect (Mokrik & Strimaitite, 1984; Samani et al., 2007; Yeh, 1986). In fact, the procedure for finding such coefficients consists in fitting their values using special methods and preferably on the basis of already known analytical solutions in order to bring the calculated date as much close as possible to the experimental date (Alonso et al., 2000; Bai & Tien, 2000; Dharmappa et al., 1997; Pachepsky, 2000; Yenikeeva, 1985;). Full-scale experimental studies with the participation of the author were carried out for subsequent diverse analysis of rapid filtration for clarification of aqueous suspensions at different impurity concentrations (KNUCA, Kyiv) (Poliakov et al., 2019) and physicochemical iron removal from deep groundwater (NUWEE, Rivne) (Martynov & Poliakov, 2022a, b). In parallel, data on changes in the main characteristics of filtration were collected, which were then used to verify the basic models of water clarification with high suspended solids and iron content at constant and variable filtration rates.

The random nature of the packing of bed elements, especially if they differ appreciably in size and shape, generates significant uncertainty in the initial data and, as a consequence, in the operating processes. It is possible to take into account the results of experimental studies in full only by using methods of probability theory, which operate with random variables and processes (Dagan, 1984; Fiori & Dagan, 1999; Hassan et al., 1998; Miralles-Wilhelm et al., 1996; Polubarinova-Kochina et al., 1969; Schlze-Makuch et al., 1999; Shvidler, 1977; Sposito et al., 1986). However, such an approach turns out to be of little use for the design and operation of filtration facilities due to both the complexity of the corresponding mathematical apparatus and the need for a statistical set of initial information. Therefore, the deterministic approach is clearly preferable for engineering purposes. In this case we have to deal only with mathematical expectation of model coefficients and filtration characteristics at equal probability of all events (results of measurements). Thus, this monograph actually only carefully reveals trends in a change of the desired characteristics, which is quite sufficient for rational design and control of a filtration facility after the technological analysis.

Obviously, the choice of an initial mathematical model is crucial for adequate modeling of any technological process. In the case of treatment technologies such a model should provide reliable forecast of spatial and temporal changes of all calculated quantities contained in it (concentrations, heads, levels, rate), and under various conditions of filtration (from normal to extreme) and in all phases of continuous operation of the filtration facility (in special cases also its service life). The final phase deserves special attention, since it is during this phase that the responsible decision to turn off the filter must be taken.

Removal of contaminants is directly caused by mass exchange between the solid and the liquid phases of water treatment filter media with prevalence of their absorption. If the intensity of such exchange can be expressed only through macrocharacteristics, there is no need in its in-depth study (Jegatheesan &

Vigneswaran, 2005). In such cases, it is sufficient for the initial model to reflect only regular changes in the desired characteristics within the limits and on the scale of a volume of filtering material. And it is logical that the central place in the structure of the mathematical models of (bio)filtration is occupied by the compartment II (Figure 1), which describes the transfer, interphase exchange and transformation (in a broad sense) of contaminants at the macrolevel and in a complex, and if necessary also the balance of substances providing transformation (Fayollea et al., 2007; Ives, 1981; Mintz, 1964; Myint et al., 2007; Widdowson et al., 1988). The special significance of the mentioned compartment is that it directly provides, firstly, the prediction of changes in the physical and mechanical state of the filter bed, and secondly, the operational control over the filtrate quality. The compartment is based on the equations of contaminant transport and phase transfer, i.e. convection-diffusion transfer and kinetics of interphase exchange of contaminants. The first of them, relating to the mobile liquid phase in a uniform operating layer and at different configuration of the directed suspension flow, can be written in the following one-dimensional form

$$\frac{1}{x^{\chi_D}}\frac{\partial}{\partial x}\left(x^{\chi_D} D\frac{\partial C}{\partial x}\right) - V\frac{\partial C}{\partial x} + \frac{\partial S}{\partial t} = n_e\frac{\partial C}{\partial t} \qquad (1.12)$$

where C,S are the (volumetric) concentrations of contaminants in liquid and solid phases; D is the convective diffusion coefficient; values χ_D=1,2,3 correspond consistently to plane, plane-radial (axisymmetric) and radial liquid flow; n_e is the effective porosity. Thus, it is taken into account in (1.12) that the transfer of contaminants within the operating layer is carried out by jointly diffusive and convective mechanisms, and contaminated water is purified directly due to the inorganic impurity transition from liquid phase to solid phase (capture, adsorption). In principle, it is not difficult to take into account the spatial nature of a liquid flow in the operating volume formally, if we use three-dimensional material balance equations and supplement the operator of boundary and initial conditions with conditions on the side surface of the flow region. However, the evaluation of the near-wall effects indicates their low significance, so that, taking into consideration the applied orientation of the monograph, it is justified to limit the basic models to one-dimensional equation (1.12) and its simplified versions (Gursch et al., 2016; Kwapinski et al., 2004; Quintana-Solorzano et al., 2018).

The contaminants removal is directly described by the kinetic equation, the generalized form of which, allowing for the two-sided nature of interfacial mass exchange, can be represented in the following way (Mints, 1964; Oliynyk & Kalugin, 2005; Venitsianov & Rubinstein, 1983)

$$\frac{\partial S}{\partial t} = f_\alpha(S,V)C - f_\beta(S,V). \qquad (1.13)$$

Here the empirical functions in generalized form $f_\alpha(S,V), f_\beta(S,V)$ characterize the rates of contaminants exchange between liquid and solid (or biological) phases in both directions.

For several decades, considerable efforts have been made to identify and formalize the patterns of removal of the suspended and dissolved matter by (bio)filtration, and in fact to reveal the expression for the functional coefficient f_α (function of the rate of the suspended solids absorption). The fundamental importance is the kinetic model of I.Iwasaki for the modern theory of interphase mass exchange in relation to the problems of the clarification of aqueous suspensions (Abramson & Brown, 2007; Iwasaki, 1937; Mackie et al., 1987), which in general correctly reflects the experimentally established facts of relative stability of distribution and exponential nature of a decrease in the suspended solids content downstream (along the height of a filter bed at vertical filtration). Therefore, afterwards the above model served as a basis for theoretical studies of filtration and, essentially, only finalized and generalized, although often its specifications were fundamental. The corresponding activity was systematically and intensively carried out in two directions.

When developing the first direction, the attention was emphasized on deepening of ideas about the deposition of the suspended particles in artificial and natural porous media and subsequent correction of the above-mentioned model by accounting for the dependence of its rate on operating, design factors. In general, the transition of dispersed impurity from the liquid phase to the solid phase according to equation (1.9) is directly controlled by the rate coefficient f_α, which is actually a function of the main parameters of filtration (d_g, equivalent diameter of the suspended particles d_p, V, medium temperature T_t), and quite often of the deposit concentration S_S (or S). Since S_S among the listed quantities, and also V in the case of filtration with unregulated rate are variables (incidentally, they should be found in the course of solving the corresponding mathematical problem), it is justified to consider their contribution to the absorption process separately. And then, following the multiplicative principle and allocating a constant coefficient α_0, it is reasonable to take for f_α the following form

$$f_\alpha(d_g, d_p, T, V, S_S) = \alpha_0(d_g, d_p, T) \cdot \chi_V(V) \cdot \chi_S(S_S). \tag{1.14}$$

In the case of a constant filtration rate, it makes sense to combine the first two multipliers in the right part of equality (1.14), so that

$$\alpha_0 = \alpha_0(d_g, d_p, V, T). \tag{1.15}$$

A large number of studies by experimental and theoretical methods were dedicated to the establishment of analytical expressions for the constant α_0, normalized functional χ_V, χ_S coefficients and their approbation. The theoretical approach proved to be effective, first of all, with respect to the reduced rate coefficient of suspended solids attachment α_0 (1.15). And here it is appropriate to note that, according to the modern concepts three phases in the absorption of the suspended particles is reasonable to distinguish (Elimelech & O'Melia, 1990; Reznikov, 1999). In the first phase, these particles come maximally close to the bed elements (grains), and in the second phase they are fixed on their surface at contact with them. In the third phase the delayed particles chemically (firmly) and physically (weakly) bind with free water, thus, forming a deposit with specific properties. It has to be stated that meaningful theoretical results have so far been obtained for the first phase only. Moreover, they are valid only for its beginning, when the collectors-particles (grains) are still in the initial state and their entire surface is accessible to the approaching small suspension particles. Then formally the following equations are valid S_S=S=0, $\chi_S(0)$=1, f_α= α_0.

Now, to reflect the peculiarities of the initial immobilization of contaminant particles when setting α_0, first of all, a representative suspended solid was isolated and the mechanisms were analyzed by which it was delivered to the collector. Six such mechanisms were identified and evaluated in terms of their importance, and finally three main ones, namely, diffusion, sedimentation and interception were distinguished (Jegatheesan & Vigneswaran, 2005; Reznikov, 1999; Venitsianov & Rubinstein, 1983). All of them are involved in the movement of any suspension particle in the direction of a collector, but to different degrees. The determining factor here is its size (d_p). It has been found that the first mechanism dominates in the delivery of Brownian particles, while the second and third mechanisms are more important in the case of non-Brownian particles (Ben Aim et al., 1997; Kau & Lawler, 1995; Poliakov, 2007). At the same time, the parameters d_g, V, T_t (V=*const*) are relevant and present in the expressions characterizing the contribution of the mentioned mechanisms to the absorption process. In order to take into account the total effect from the simultaneous realization of the three mechanisms, these expressions are added up in accordance with the additive principle. Thus, a number of the generalized theoretical formulas similar in approach and formal features were derived for a single impurity particle (Radjagopalan & Tien, 1976; Ren et al., 2000; Tufenkji & Elimelech, 2004; Veerapaneni & Wiesner, 1993; Vigneswaran & Chang, 1986). This result (initial collector removal efficiency) was then extended to the association of such particles within a filter bed. As a result, a formula for calculating the desired α_0, but only under particularly favorable physicochemical conditions for the particle-collector interactions, such as high ionic strength, low pH, etc., was obtained. In practice, such conditions often turn out to be unfavorable and then to take them into account it is necessary to introduce into the above formulas a correction factor (the sticking or attachment efficiency factor α_c), which is equal to 1 in ideal cases (at complete destabilization of the operating medium), but more often it is less and sometimes, for example, at immobilization of microorganisms, significantly (Foppen & Schijven, 2005; Harvey & Garabedlan, 1991; Kuhnen et al., 2000; Ren et al., 2000; Shani et al., 2008; Shaoging, 1998). In order to make such conditions more suitable for suspended solids fixation and thus, increase the specified coefficient, reagents are mainly used (Bugai et al., 2007; Chuang & Li, 1997; Girol et al., 1998; Pin Gao et al., 2012; Shevchuk et al., 2005; Shirasaki et al., 2000; Zhu et al., 1996; Zosin et al., 2003). The most authoritative formula relative to α_0 is given and discussed with specific examples in Section 2.6. It is obvious that only calculation of α_0 is not enough for modeling the absorpsion process in depth filtration and even more so the operation of water treatment filters due to heavy clogging of their bed especially at the end of the successive operating period and multiple increase in head losses, as well as possible changes in V. As the experience of rapid filters practical application testifies, the reliability of the results obtained in this case is often too low. Nevertheless, such formulas are useful from the methodological point of view, as they give a clear idea of the nature of the relationship between the absorption process and the most important parameters of filtration.

Experimental substantiation of the functional coefficients $\chi_{V, S}$ allows to closely resemble the natural dynamics of contamination within the filter bed limits when modeling mass exchange, since theoretical approaches are ineffective here. According to the results of few experiments it is established that the coefficient χ_V can be expressed by a power function (Adin & Rebhun, 1987; Alekseev & Kommunar, 1974; Aronino et al., 2003; Kretzschmaa et al., 1997; Senyavin et al., 1977), namely,

$$\chi_V(V) = V^l \tag{1.16}$$

where the power exponent l varies from 0 to 1, although it can take small negative values according to the theoretical developments for α_0 in case of the rate mode V=*const*. However, the value of l can be reliably determined only by experiment especially for polydisperse impurities.

But the most problematic is the choice of the adhesion function $\chi_S(S)$ (Bai & Tien, 2000; Boller & Kavanauch, 1995; Ives, 1969; Ojha & Graham, 1992). The reason for serious difficulties is the complex nature of the influence of the accumulated deposit on the suspended solids removal. It is well established that the clarification process goes through four unequal stages in its development (Kau & Lawler, 1995; Kim & Tobiasson, 2004; Pin Gao et al., 2012; Vigneswaran & Tulachan, 1988). They are distinguished based on the characteristics of the variation of the outlet suspended solids concentration with time. During a short period of time, the first portion of suspension is processed and the primary contamination of the filter media is formed, but in fact, only its inlet section. Only at this stage it is acceptable to use the coefficient $_0$ for the filtration calculations and neglect the coefficient χ_S. The newly formed deposit is actively involved in the suspension separation at the second stage. The ripening of a filter bed or, in fact, autocatalysis, when the role of an additional collector for contaminants is played by a deposit, causes even a short-term improvement of the filtrate quality (Adin & Rebhun, 1977; Darby & Lawler, 1990). The exhaustion of the initial absorbing resource of the bed is compensated by the autocatalytic effect and the level of filtrate contamination stabilizes for some time at the third stage. The impurity content at the filter outlet sharply increases at the final, fourth stage, which is crucial for technological analysis and control of the filter operation. At the same time, there is an accelerated increase in the hydraulic resistance of the bed. It is monitoring the dynamics of the outlet suspended solids concentration that often makes it possible to maximize the time of continuous filter operation at the required filtrate quality.

Throughout each operating period, concentration S undergoes significant changes that have fundamental differences at the distinguished stages. As a consequence, the behavior of the concentration function $S(z,t)$ is much more complex than behavior of $C(z,t)$. The content of the deposited particles changes significantly by depth and over time, necessarily (except in special cases) decreasing in the direction of fluid flow. The (volumetric) concentration of the deposit S_S closely related to S is even more uncertain in a physical and formal sense due to the poorly understood of the third phase of the absorption of a dispersed impurity. It is important to note that the determination of the value S is not the ultimate goal of even predictive calculations and even more so of technological calculations, but plays a subordinate role. Just on the basis of complete information about deposition of the suspended solids and a pattern of deposit formation, then the distribution of deposit in the volume of filtering material is established in dynamics, which is necessary for subsequent hydraulic calculations. On the contrary, a deposit is first taken out in experimental studies and then all the bound water is removed at high temperature

Due to the limited pore space, the increase in (specific) deposit volume is possible only up to some limiting value S_{Sm}, which is actually much smaller than the initial porosity n_0 (Adin & Rebhun, 1987; Biserov, 1967; Bochever, 1977; Burchak, 1978; Kommunar & Bogoliubov, 1977; Ojha & Graham, 1994). Realistically, the solids take-up capacity (true with respect to a deposit) S_{Sm} in granular media is half or somewhat more of n_0, which is explained not only by the specific configuration of free space and the presence of non-flowing pores, but also, perhaps, by the preferential accumulation of deposit in narrowed places. However, since most of the deposit is a bound water, the maximum specific volume of impurity S_m that a filter media can capable of holding is many times smaller. Thus, the effective solids take-up capacity (true with respect to the deposited particles) S_m is necessarily only a fraction of the true capacity S_{Sm} (Adin & Rebhun, 1987; Saiers et al., 1994; Tobiason & Vigneswaran, 1994). The fundamental difference between the two is that the latter is essentially a physical constant, while the former already depends on the properties of an impurity, the age of a deposit. Nevertheless, due to the scarcity of information on S_m, such a detailed characterization is hardly meaningful in the calculation practice.

Due to both smallness of S_m and rapid accumulation of a deposit at the inlet section of a packed bed, the question arises about the peculiarities of mass exchange kinetics and the rate of generalized absorption at the final stage of filtration. Two characteristic situations are possible at high saturation of the medium with a deposit. In the first situation, the true deposition rate of the impurity depends on the deposit concentration and gradually decreases as S approaches S_m (Alekseev & Kommunar, 1974; Corapcioglu & Jiang, 1993; Jegatheesan & Vigneswaran, 2000; Mackie & Zhao, 1999; Ornatsky et al.,

1955; Tobiason & Vigneswaran, 1994; Vigneswaran & Tulachan, 1988). Due to the asymptotic nature of the S behavior here, the bed saturation limit is in principle not reached. Therefore, there is no need to introduce an additional stage and a moving boundary in the modeling. However, the mass exchange kinetics will inevitably be nonlinear even with a linear form of the functional coefficient $\chi_S(S)$ in (1.10). In the second situation, we have to consider filtration in the presence of an inactive area of maximum saturation with deposit, which expands with time, but the above kinetics can remain linear (Kulikov et al., 2009; Poliakov, 2012b).

It can be argued that a fresh deposit plays an important and positive role in clarifying the aqueous suspension in the early stages of filtration. (Calvin, 1992; Venitsianov & Rubinstein, 1983; Weber-Shirk & Chan, 2007). However, there is still no clear understanding of the significance of autocatalysis at the final stages, when the reduction of the absorption resource of the bed becomes crucial for the filter performance (Bai & Mackie, 1995; Darby et al., 1991; Moran et al., 1993). Focusing on the original dynamics of a deposit in the entire volume of a filter media, it is often accepted as a working hypothesis that autocatalysis is significant at first, and then the dynamics of the clogging of filtering material depends entirely on the reduction of its absorption capacity. However, physical preconditions become irrelevant if the expression for f_α and its components is established on the basis of experimental or operational data on concentrations. It is appropriate to notice that, apparently, the first effect should play an appreciable role until the end of filtration process in view of weak contamination of the peripheral part of a bed during the whole operating period. Summarizing the above mentioned with respect to χ_S, it is legitimate to take the following expression according to the multiplicative principle

$$\chi_S(S) = \chi_{Sa}(S)\, \chi_{Sm}(S). \tag{1.17}$$

Here it is justified to take a linear form for χ_{Sa} (autocatalysis), χ_{Sm} (absorbing resource depletion) for the investigations by analytical methods and test calculations, and at the same time for the generalization of experimental data. Quite often in modeling the removal of impurities in two-phase media it is necessary to rely on nonlinear mass exchange kinetics, in particular, classical Mono kinetics, its modifications, and to solve the corresponding mathematical problems using numerical methods and computers, thereby significantly narrowing the range of users of theoretical developments (Hekmat et al., 2006; Liu & Zachara, 2001; Lu et al., 2004; Meric et al., 2002; Mitchell & McNevin, 2001). Due to the significant capabilities of modern computing facilities, tabulation, interpolation and extrapolation procedures are no longer necessary when calculating special functions usually contained in the final dependencies.

Essentially heterogeneous mechanical composition of a dispersed impurity promotes the blockage of smaller pores, interstitial channels and, as a consequence, flow capacity of the filtering material can appreciably decrease. Formally, the sieve effect can be taken into account by an appropriate correction of f_α, namely, either by introducing a special summand or by refining the known rate coefficient of physicochemical deposition (Bradford & Bettahar, 2005; Tufenkji et al., 2006; Verigin, 1971; Wang et al., 2012). Then, when conducting experimental studies to obtain initial information for mathematical modeling, the values of the required model coefficients should be found by processing experimental data, taking into consideration the already given simple form of functional coefficients from (1.13).

It has been experimentally proved that the filtrate often contains suspension particles, which were already part of the solid phase of a filter bed (Kim & Tobiasson, 2004b; Moran et al., 1993). This fact means that there is also mobilization of these particles and fragments of the deposit formed on their basis (with possible re-deposition) in the operating layer along with the attachment (adhesion) process

(Bai & Tien, 1997; Bergendahi & Grasso, 2003; Corapcioglu & Jiang, 1993; Sharma et al., 1992). The direct cause of the poorly understood phenomenon of deposit detachment is the hydrodynamic force generated by suspension flow (not washing water flow) (Bergendahi & Grasso, 2003; Kim & Lawler, 2012). Obviously, the rate of mobilization of the deposited particles is closely related (usually assumed to be proportional) to their concentration. Therefore, it is justified to consider the detachment as a first-order reaction with a coefficient of its rate requiring predetermination. We can only state so far as for the functional coefficient f_β, that it does not depend on the content of suspension particles in both phases and, on the contrary, it depends essentially on the hydraulic load and the particle sizes. And it is natural that a power form of the dependence on V is recommended for f_β, as well as for f_α (Adin & Rebhun, 1987; Demchin et al., 1994; Kau & Lawler, 1995; Kim & Lawler, 2012). Incidentally, the physicochemical pattern in a packed bed becomes even more complicated due to the re-deposition of a part of the mobilized particles (secondary contamination) (Bai & Tien, 1997; Poliakov, 2010c). This phenomenon should be reflected in the coefficient f_α, which once again confirms the expediency of finding the latter precisely experimentally.

This phenomenon becomes particularly significant in the final fourth stage of filtration due to the large amount of a deposit, which is also concentrated near the bed inlet. This is where intensive mobilization can contribute to additional technological process limitation. During long clarification of a suspension, the state of filtering material in the inlet section asymptotically tends to a quasi-steady state, which is characterized by a dynamic equilibrium between the constituents of the impurities in the liquid and solid phases. In other words, the content of both suspended and deposited suspension particles here is already practically constant, but mass exchange and possibly intensive continues and is accompanied by the deposit renewal. The limiting value of the ratio of both concentrations under such conditions makes sense of solids take-up capacity, but now already dynamic S_d, because it depends not only on α_0, β_0, but also on V. It is often possible to increase S_d precisely by decreasing V towards the end of the operating period (Biserov, 1967; Dzyubo & Alferova, 2007; Lozada & Cruz-Velez, 2014; Yaroshevskaya et al., 1989).

The correctness of the formulation of the compartment II, and hence the uniqueness of its subsequent solution, is ensured by attaching to the system of equations (1.12), (1.13) a boundary operator in which the conditions at the inlet to the operating layer are determinant (Grabovskiy & Gurinchik, 2005; Oliynyk & Kalugin, 2005; Poliakov, 2006)

$$x=x_0,\ C=C_0(t), \tag{1.18}$$

and the inhomogeneous initial condition

$$t=0,\ S=S^0(x), \tag{1.19}$$

where C_0 is the (variable) inlet concentration of the impurity, S^0 is the initial concentration of the deposited suspension particles. It should be noted that conditions (1.18), (1.19), first of all, reflect, respectively, the magnitude and nature of the solids load on a filter, i.e. by contaminants as opposed to the hydraulic load (by V), as well as its initial operating condition. Indeed, noticeable changes in the composition and content of contamination are possible during the day at long continuous operation of the filter. In addition, a small part of the deposit accumulated by the beginning of each washing will be inevitably retained in the bed after it. More often than not, the deposit is distributed almost uniformly in the operating volume due to the intensive agitation of the granular material during washing. The residual deposit may even be necessary to start the technological process, as in the case of physicochemical iron removal.

In general, when modeling the operation of filters with a layered bed and stepped filters, respectively, the boundary conditions of the fourth kind must be strictly fulfilled on the interlayer boundaries and internal adjacent surfaces of the filters sequence. The influence of contaminant transport and phase transfer in any layer (filter) spreads on all subsequent layers and practically does not appear in previous layers at small diffusion transfer of contaminants (in comparison with convection transfer). As a consequence, the conditions of equality of contaminant fluxes in the absence of the storage tank are met automatically. Thus, the sequential calculation of the clarifying effect of layers (filters) on the basis of a unified mathematical model of the depth filtration through a uniform bed becomes possible. In this case, the appropriate model coefficients and the variable inlet concentration $C_0(t)$ are used at every stage. The indicated concentration is already the subject of preliminary calculations (Li & Cleall, 2008; Poliakov, 2008a; 2011; Tang & Aral, 1992). Consequently, the inlet concentration of contaminants for the second layer, step of a stepped filter (C_{02}) and subsequent layers, steps depends on the initial concentration C_0 and physicochemical properties of the previous layers, steps. The specified concentration is found as a result of a solution of the corresponding mathematical problem, which is a part of the general problem of suspension filtration in the combined (layered) operating medium. Since the contaminants do not accumulate on interlayer boundaries, the value C_{02} is equal to the contaminants concentration at the outlet of the first step (layer) C_{1e} and therefore increases with time. In this case preliminary averaging of $C_{02}(t)$ has no sense. It is much more difficult to determine the initial value $S^0(z)$ in case of the progressive reinforcement of a deposit (consolidation).

In the case of a consolidatable deposit, in order to calculate a constant initial concentration of the suspension particles in non-washable deposit S^0 for a representative filter run, it is necessary to pre-specify the empirical function of its resistance to the hydrodynamic effect $f_s(t)$, which is represented by the following relation

$$f_s(t) = \frac{s_e(t)}{s_0} \tag{1.20}$$

where S_0 and s_e are the concentrations of deposited particles (deposit) in the experimental sample of porous medium respectively initial and after its standard washing. If the deposited particles accumulate in the initially clean sample according to the dependence $s(t)$, their residual concentration $s_e(t)$ (after washing at the moment of time t_*) will be as follows

$$s_e(t_*) = \int_0^{t_*} f_s(t) \frac{\partial}{\partial t} s(t)\, dt \tag{1.21}$$

Let on the water with the concentration of contaminants C_0 was pre-fed to a uniform bed with the height L and bulk density ρ_s during time t_1 before the main calculation period (filter run). At smallness of diffusive transfer of the suspended matter, the expression for the balance of contaminants follows after double integration of equation (1.12) on z from 0 to L and on t from 0 to t in the integral form

$$V \int_0^t [C_0 - C_e(\xi)]\, d\xi = \rho_s \int_0^L [S(z,t) - S^0(z)]\, dz \tag{1.22}$$

where S, S^0 are the current and initial (t=0) concentrations of the deposited particles. High filtrate quality formally means that $C_0 >> C_e$ over any calculation period. As a rule, after each washing the residual deposit distributed approximately uniformly along a bed height, so that $S^0 = const$. Then it follows from (1.22)

$$\rho_s L S_a(t) = \rho_s L S^0 + V C_0 t \tag{1.23}$$

where S_a is the average concentration of the deposited particles over the bed height at the time point t. If $S_a(t)$ is identified as $s(t)$, then according to (1.21) and (1.23) the constant initial concentration of the deposited particles in the non-washable deposit S^0 for a representative operating period approximately is

$$S^0(t_1) = S_0 + \frac{VC_0}{\rho_s L}\int_0^{t_1} f_s(t)\, dt \tag{1.24}$$

The clarification compartment (II) is supplemented by the filtration flow compartment (III), which allows to monitor changes in the hydrodynamic pattern within the entire volume of the filtering material at the macrolevel. Depending on the filtration mode, the relation between the compartments II and III is two-sided ($V=V(t)$) or practically one-sided ($V=const$). The great importance of the filtration flow compartment for the modeling of filtration is explained by the fact that the moving fluid is actually a carrier of contaminants and delivers them to the places of their transformation and removal by means of two transport mechanisms.

From the technological point of view, this compartment is valuable because it allows to obtain objective information about local and general changes in the piezometric head, regular increasing the hydraulic resistance of the porous layer and decreasing the filtration rate. The basis of the compartment II is the equation of motion, which expresses the balance between the forces of pressure, gravity, resistance, inertia and can be presented in the following form (Noskov et al., 2008; Panfilov & Fourar, 2006; Romankov & Kurochkina, 1982; Trussell & Chang, 1999)

$$\frac{\partial h}{\partial x} + a_h(S_S)\, V + b_h(S_S)\, V^2 = 0 \tag{1.25}$$

Thus, the empirical functions $a_h(S_S)$ and $b_h(S_S)$ at a known rate V characterize, respectively, the resistance and inertia forces at any degree of pore space clogging. If linear filtration takes place (Gray & Miller, 2004; Vigneswaran et al., 1985), the equation of motion takes the traditional form

$$\frac{\partial h}{\partial x} + \frac{V}{k(S_S)} = 0 \tag{1.26}$$

In fact, the basic filtration flow law is generalized by taking into account the natural changes in the saturated hydraulic conductivity. Consequently, the function $a_h(S_S)$ is the inverse of the function of the current hydraulic conductivity $k(S_S)$ at $b_h=0$. It should be noted that experimental investigations of porous medium permeability under conditions of its progressive clogging were often focused on the establishment of the dependence $k(S_S)$ and the relationships between a_h, b_h and S_S have hardly been studied. And it is acceptable to use the well-known Ergun formula only in the case of a clean bed (Cheng, 2003; Ergun,

1952; Grabovsky & Gurinchik, 2008; Plessis & Woudberg, 2008; Williams et al., 2007; Zhurba, 1993), which is designed to calculate its reduced hydraulic resistance. Incidentally, sometimes it was also used for a clogged medium without proof. According to this formula

$$a_{h0} = a_h(0) = \frac{150\nu(1-n_0)^2}{g d_e^2 n_0^3}, \quad b_{h0} = b_h(0) = \frac{1.75\nu(1-n_0)}{g d_e n_0^3}. \tag{1.27}$$

It is pertinent to point out that the above formula can be formally generalized even to turbulent filtration flow (ultra-rapid filters) if one manages to empirically establish the functions $a_h(S_S)$, $b_h(S_S)$.

The head losses in the entire layer of a filtering material $\Delta h(t)$ and in its arbitrary section are established depending on time by integration of equation (1.26) at given outlet head and hydraulic conductivity as a function of the concentration of deposited particles S (Akgiray & Saatci, 1997; Saatci, 1990). The second approach to finding Δh is realized at unknown dependence $k(S)$ by using empirical expressions, which formally can be interpreted as an integral of equation (1.26) in the case of linear filtration, and as an integral of equation (1.25) in the case of nonlinear filtration. Thus, in the first case the hydraulic resistance $\Delta h/L$ is the following integral function of time

$$\frac{\Delta h(t)}{L} = \frac{V}{L}\int_0^L \frac{dz}{k(S(z,t))} \tag{1.28}$$

and in the second case it has to be determined directly from experimental data (piezometer readings).

Still, the first approach is preferable despite the need for preliminary studies of the pattern of a decrease in the hydraulic conductivity (permeability) of a filtering material $k(S)$ as it clogged up (or $a_h(S)$, $b_h(S)$). Really, it allows to calculate more justified from the hydraulic point of view the losses Δh in a bed of arbitrary thickness although requires additional integration according to (1.28). It should be noted that attempts to formally link the calculation formulas for current conditions of filtration with similar formulas for standard conditions (in fact, to use simple models of porous medium) are unjustified here because of uncertainty of changes in the geometry of pore space at intensive accumulation of a deposit in it.

Experimental methods proved that the gel-like matter accumulated in the pores of a bed contains mainly bound water and only small amounts of the contaminant itself (or products of its transformation). The ratio of the corresponding concentrations (bulk factor) $\gamma = S_S/S$ generalizes its composition and plays a key role in filtration flow calculations and at the same time depends significantly on the quantity S. Therefore, the following equality is true

$$S_S = \gamma(S)S. \tag{1.29}$$

Due to the extremely limited empirical information on the functional coefficient $\gamma(S)$, at best, its average (representative) value is taken in quantitative analysis, which is not always justified. Taking into account (1.8) and (1.29), equation (1.26) is supplemented at linear filtration, firstly, by the expression for the dependence $k(S)$ in the given form (Akgiray & Saatci, 1997; Saatci, 1990).

$$k(S) = k_0 f_k(S), \tag{1.30}$$

and, more often, the generalized approximation representation (1.6) is used for the normalized permeability function $f_k(S)$, secondly, one or two boundary conditions. Thus, it is enough to specify the head in the outlet cross-section of a bed in the rate mode V=*const*. Then the variable inlet head and total head losses in the filter media are determined at a given rate by calculation. If $V=V(t)$, then it is necessary to specify the head in both mentioned cross-sections, and as a result of a solution of the filtration flow compartment the pattern of decreasing the filtration rate as the layer of filtering material is being clogged will be found. The relation of this compartment with the hydraulic compartment is formally established by means of the conditions of heads equality at the boundaries of this layer, which can be calculated only after the final solution of the initial mathematical problem.

In fact, the compartments IV and V are not mandatory components of the general model. Their inclusion in it is justified in certain cases and then allows to bring the calculated pattern considerably closer to the natural one. The hydraulic compartment (V) makes it possible to predict the consequences for the technological process of liquid flow in the transport systems of a filtration facility, its accumulation in the storage tank above a bed. Such consequences are expressed in additional head losses in the supply and offtake communications, which are often commensurate with similar losses in the clogged bed, as well as in an increase in head at the inlet to it at the regulated (constant) and unregulated rate of filtration. In the latter case, suspension filtration and contaminant removal are intensified. In fact, the negative effect of the progressive clogging of the pore space can only be partially compensated in this way, so that the decrease in the filtration rate is only slightly slowed down. The compartment V, first of all, includes the differential equation for the balance of the initial water in the storage tank

$$\omega\frac{dH_W}{dt} = Q_{in} - \omega V \tag{1.31}$$

where ω is the inlet surface area of the filter bed, H_W is the suspension level in the storage tank, Q is the volumetric flow rate of suspension in the supply pipeline. The uniqueness of the solution of equation (1.31) is provided by setting the initial level H_W^0, so that

$$t = 0,\ H_W = H_W^0. \tag{1.32}$$

As a rule, it is possible to consider $H_W^0 = 0$. Finally, to determine the head losses in the supply and offtake pipelines (communications), it is reasonable to use the basic calculation formula for long pipelines (Bolshakov et al.,1979; Grabovsky, 2016; Naumenko, 2005; Tkachuk, 2022), which takes for the offtake pipeline the following form

$$H_d = H_{out} + R_{f2}\omega^2 V^2 \tag{1.33}$$

where H_d, H_{out} are the heads at the filter bed outlet and filtrate collector (mechanical energy expenditure in the supporting layer are negligible); R_{f2} is the resistance of the offtake pipeline (communications). Incidentally, acceptance of a single value of the head along the collectors of filtrate and raw water means that the head losses along their length are ignored. Such coarsening of the hydraulic pattern in the transportation system of the filtration facility is justified due to their usually short length and large diameter.

There are basically two ways of supplying water on a filter. Firstly, it is realistic to keep the flow discharge Q_{in} constant and then there is no point in finding the head losses in the supply pipeline when calculating the clarification process. If the head at the collector of initial suspension (raw water) H_{in} is constant, then it is connected with the head at the inlet to the filter (level in the storage tank) H_W by the equation

$$H_{in} = H_W + R_{f1} Q_{in}^2 \tag{1.34}$$

where R_{f1} is the resistance of the supply pipeline (communications).

It should be noted that under the filtration mode $V=const$ it is not reasonable to involve the hydraulic compartment for the water treatment calculations at all. In this case, the head losses in both communications are stable and are of interest only for the calculation of their hydraulic performance (diameter). Then the diameters of the supply and offtake pipelines are easy to calculate from (1.33), (1.34) at given H_{in}, H_{out} and $Q=\omega_p V$ (ω_p is the cross-sectional area of a pipeline). The compartment V must be included in the basic model in its entirety at variable rate V, and the subject of calculations based on the joint solution of the compartments II, III, V are the functions $V(t)$, $Q(t)$, $H_W(t)$, $H_d(t)$.

Quite often modeling of mass exchange only at the macro level has low reliability. Significant difficulties, first of all, may arise when giving the variable mass exchange (kinetic) coefficients due to their unpredictability (impossibility to set them in advance in the form of already known functions of time and position). It is possible to substantially approximate the calculated data to the experimental data due to the joint theoretical studies at two levels (macro and micro). The primary subject of the theoretical study of contaminant behavior at the micro level is its local removal by fixation on the surface and within individual (porous) bed elements (Poliakov, 2012c; Pritzker, 2003), as well as inside aggregated macroparticles (Venitsianov, 1998).

In principle, the structural compartment (IV) allows to considerably deepen the understanding of the contaminant removal patterns due to the formalization of operating processes in a small neighborhood and inside each particle-collector. Two approaches are possible in predicting the removal of impurities on the basis of the underlying two-level model. If from the practical point of view it is justified to carry out theoretical analysis and applied calculations on the ground of the specified model, then all compartments are considered as a single complex. The corresponding correctly formulated mathematical problem can be solved only by numerical methods, which significantly narrows the range of users of its solution.

It is possible to significantly simplify simulation, and sometimes even apply analytical methods, if the second approach is implemented. In this case, a structural compartment is isolated and solved separately. Its connection with other compartments is reflected by setting the input information about the current state of the filter media in a generalized form. Such an approach enables to perform modeling step-by-step, first to formally evaluate the mass exchange coefficients and the distributed absorption function of a contaminant, and later to analyze the technological process exclusively at the macro level.

Hence, the structural compartment should be included in the basic model where possible and linked with its basic compartments (II) and (III) corresponding to the macro level. In the case of in-depth modeling of contaminant transport and phase transfer, even its greater elaboration is possible, in particular, the consideration of hydrodynamics of individual particles and their groups, taking into account the action of a complex of mass and surface forces (Jegatheesan & Vigneswaran, 2005), transfer, exchange, transformation processes within individual pores (Gal et al., 1985; James & Chrysikopoulos, 2003; Kessler & Hunt, 1994; Knapp et al., 2000; Tung et al., 2004). The conceptual role in filtration of localized removal of contaminants by absorption is more clearly illustrated by the structural compartment (Fig. 1).

In the case of the contaminants absorption by the entire volume of each porous element, and, in fact, physical adsorption of contaminants on the micropore walls, internal mass transfer is of great importance. It is realized in a dense, branched network of micropores due only to the mechanisms of molecular and surface diffusion (Hu et al., 1994; Leyva-Ramos & Geankoplis, 1985; Riekert, 1985). Direct removal of dissolved contaminants from the liquid phase is provided by physical adsorption (Koganovsky et al., 1981; Venitsianov & Rubinstein, 1983). Over time, for at least partial restoration of adsorbent capacity, part of the accumulated adsorbate should be extracted using regeneration methods that initiate its desorption, oxidation, and destruction (Kulikov et al., 2009; Zapolsky et al., 2000). For practical implementation of operative desorption of contaminants it is necessary to spend a lot of resources, and the corresponding theoretical basis is still insufficiently developed (Chern & Wu, 2001; Mollah & Robinson, 1996).

It is usually assumed in structural models that a representative porous element has the strictly spherical shape and is characterized by an (equivalent) diameter. The amount of contaminants being trapped within any such element steadily increases over time. Therefore, the differential equation of its micro-balance is nonstationary and of parabolic type. Local absorption of contaminants is characterized by the intensity of contaminants inflow to an individual element i_f. The value of the corresponding microflow i_f depends on the position of the element under consideration and serves as a basis for the derivation of the continuous absorption function $I_f(z,t)$, describing the rate (distributed along the filter media height) of the contaminants removal.

At the same time, the collection of such elements can be considered as a non-cohesive secondary porous medium, which has specific physical and chemical properties, e.g., the secondary porosity and permeability (Hønning et al., 2007; Lewandowska et al., 2008; Liu et al., 2016; Šimůnek et al., 2003; Zhang & Li, 2010). Then the mentioned fictitious continuous medium is linked by means of the volumetric coefficients of the mass transfer in a two-phase contaminated system to the primary medium, whose porosity and permeability are entirely due to the macropore system.

Another illustrative example of a secondary medium active in relation to pollutants can be the microbial biomass (biofilms) fixed on packed bed elements. Extremely extensive literature has been devoted to the study of composition, development patterns of aerobic and anaerobic biofilms, peculiarities of their functioning under various conditions, for example (Henze et al. 2006; Rittmann, 1993; Swanson & Loeh, 1997). Mathematical modeling of dissolved pollutant removal due to the activity of microbiocenosis fixed on the solid phase surface is significantly complicated due to the need to control its quantity, composition and activity.

Biota is especially sensitive to heat, all phases of its development being closely related to the ambient temperature. There are optimal temperatures or even an interval of such temperatures at which living organisms feel most comfortable. When describing biomass growth, decomposition of organic matter, and the rate of chemical reactions under conditions of active heat exchange with the environment, the Van-Goff and Arrhenius equations are widely used, which formally reflect regular changes in the intensity of chemical and biological processes when the background temperature increases by a certain number of degrees (for water treatment, the intensity doubles when the temperature in the operating range increases by 10 degrees) (Andersson et al., 2001; Crohn & Valenzuela-Solano, 2003; Dalias et al., 2001). Also, the temperature factor has an appreciable effect on the viscosity properties of a liquid, and therefore internal friction and ultimately the hydraulic conductivity and suspension flow rate in porous components of filtration facilities, head losses (Daly & Harleman, 1971; Sterenlicht, 1984; Trussell & Chang, 1999). Finally, the liquid temperature has a measurable effect on the removal of contaminant particles, primarily by enhancing their delivery mechanisms to the bed elements (McDowell et al., 1986; Rajagopalan & Tien, 1976).

Informative reviews on adsorption, biological, membrane water treatment with the use of special porous media together with the results of our own theoretical and experimental studies in these areas will be presented in the next monograph.

As a rule, the complexity of an initial mathematical model, its analytical solution, and the cumbersome calculation dependencies resulting from it are noticeably simplified, and the simulation results are significantly generalized, without compromising their accuracy. If a standard procedure is used for reducing the initial model to a dimensionless form. first of all, the number of the model parameters can be significantly reduced thanks to the introduction of the scales for the dimensional variables. For example, this leaves 7 dimensionless complexes out of the original 15 dimensional parameters in the stationary two-level biofiltration model. Incidentally, an important result of data processing of systematized experiments is the identification of regular relationships between the selected complexes, in particular, the numbers Re, *Pe, Sh, Nu, Sc*, etc. (Badriyha et al., 2003; Ko et al., 2001). However, the number of calculations in conducting a thorough quantitative analysis on the basis of dimensionless mathematical models increases due to the significant (especially in multivariate problems) expansion of the limits of variation of complex parameters. Indeed, the minimum and maximum values of the model complexes are set as corresponding combinations of the limiting values of the initial parameters. At the same time, the relative dependent and independent variables are already usually changed within computationally convenient limits. Finally, the introduction of dimensionless complexes facilitates the search for and substantiation of possible simplifications of the underlying model. As a rule, such models include a number of parameters (C_0, L, etc.), that are directly suitable for the use as scales. Quite often some of the necessary parameter-scales are initially missing, in particular, a scale for the concentration S in the case of linear mass exchange kinetics. Then it is necessary to combine several given constant model coefficients. If a model parameter has a functional form, it is reasonable to choose its characteristic value as a scale, for example, the maximum or minimum possible value. Then the reference value for such a relative variable will be one. When performing a special analysis, for example, in order to assess the significance and sensitivity of the technological, design parameters that were initially used as scales and now become independent variables, it is justified to select appropriate scales for them in the same way.

When developing an applied calculation base at the final stage, analytical exact and approximate as well as numerical methods of solving mathematical problems in partial derivatives are involved. Exact solutions are unconditionally reliable from the mathematical point of view (Alekseev et al., 1989; Corapcioglu & Haridas, 1985; Huang et al., 1984; Ivanchenko et al., 2012; Philip, 1994). Due to the considerable capabilities of modern computational tools, tabulation, interpolation and extrapolation procedures are no longer necessary when calculating special functions, usually contained in final dependencies. However, in order to find an exact solution, it is usually necessary to further simplify the initial model, to impose more severe limitations on the operating processes, which narrow and sometimes significantly the field of its application. Approximate solutions, which can be obtained due to the use of special techniques that simplify the formulation of a mathematical problem or the course of its solution (different ways of averaging variable coefficients and characteristics, successive changes of stationary states, integral moments, non-standard techniques), can usually be used much more widely (Barenblatt et al, 1972; Evtushenko & Kotelkin, 1991; Ornatsky et al., 1955; Pang et al., 2003; Polubarinova-Kochina et al., 1969; Polyakov & Dilman, 2006; Serrano, 2001). But at the same time, they need an estimation of the errors possible in doing so, which is best done by comparison with a special exact solution.

Approximate and even more exact analytical methods are effective only if initial models are sufficiently simple in mathematical terms. The use of analytical methods becomes especially problematic in the presence of nonlinear effects. In those widespread situations when multifactor complex processes and phenomena in a multiphase medium should be calculated, the only way out is to use numerical methods, e.g. (Arora et al., 2007; Caputo et al., 2001; Dutournie et al., 2011; Grabovsky & Gurinchik, 2007; Haarstrick et al., 2004). Formally the validity of numerical calculations is verified by comparison with an exact solution of the original problem, which can be obtained, as a rule, only after its substantial simplification. A software program providing numerical solution of a complex problem of (bio)filtration is expedient to include in a common software product designed for operational control of the filtration unit as part of a large water treatment structure. However, it is usually difficult for a real user to directly apply numerical solutions in engineering practice, even in the dialog mode, due to the conservativeness of the programs implementing them and the diversity and variability of filtration conditions.

By applying analytical methods to the basic problems of (bio)filtration of low-concentrated suspensions and solutions, the dependencies linking the desired macro- and micro-characteristics (concentrations, flow rate, head, level) with coordinates (height), time and a set of model coefficients are found. We have to resort to complex transformations due to the nonlinearity usual for these problems as well as changes in the model coefficients in the course of their solution. As a result, the final expressions contain computationally problematic integrals (improper integral, with oscillating integrands, etc.), special functions and have a form, which is cumbersome and inconvenient for engineering calculations (Polyanin & Zaitsev, 2002; Verigin & Kommunar, 1978). In rare cases, and especially when using approximate methods, it is possible to derive simple formulas. Much more often the above-mentioned dependencies have an implicit form. For this reason it is necessary to determine the values of the desired characteristic by purposeful enumeration of options. If modern applied software packages oriented to engineers and solving practical problems, technical calculations using mathematical analysis tools (Mathcad, Matlab, Maple, Mathematica, etc.) are involved, the calculation procedure does not become more complicated, but only slightly increases its body. The analytical solutions often have a parametric form at the filtration with a variable rate. Then the calculation characteristic and its argument are linked with the newly introduced parameter separately, and calculations are carried out in parallel by two implicit dependencies. And here there is no fundamental difficulty.

If the theoretical analysis of the suspension clarification by filtration is performed with the purpose of revealing its general patterns, estimation of sensitivity of the calculation characteristics to individual indicators, effects, it should be based on dimensionless formalisms and completed with generalized conclusions. Naturally, when calculating functioning filtration facilities and making recommendations for rational control or design improvements, it is preferable to present the obtained results in dimensional form with indication of specific values in the generally accepted system of units.

Thereby, the primary goal of mathematical modeling of filtration of natural and waste waters systematically carried out in the monograph was the development of an effective computational tools for predicting the changes in physical and chemical conditions in the layer of filtering material with possible consideration of the hydraulics of auxiliary design elements (transportation systems, storage tank). It is obvious that the analytical form of such tools is clearly preferable for engineering practice, i.e. in the form of preferably simple formulas, dependencies, which would allow to perform calculations and estimations promptly and in large numbers. Another, extremely important field of practical application of these tools is the technological and design calculations. If the maximum permissible physical and mechanical bed state and filtrate quality are established in advance or, in other words, the system of the restrictions, within which the effective functioning of a filtration facility is guaranteed, is specified, then the basic analytical solution together with the mentioned criteria can serve as a reliable basis for improving the design, elaboration of the rational algorithm of the technological process control. Principle provisions of the technological analysis are presented in the following Section.

1.3. TECHNOLOGICAL CALCULATIONS IN DEEP BED FILTRATION PRACTICE

The most important application of the theoretical solutions of problems of aqueous suspension clarification is the technological analysis of the filtration facility operation. Since this analysis involves a large number of various calculations, analytical dependencies are clearly preferable. In this case the main purpose of the technological modeling is the development of an optimal control algorithm of the mentioned facility. In general, such an algorithm is a sequence of alternate activations and shutdowns of a water treatment filter. However, since washing of excessively clogged packed bed of a rapid filter is thoroughly developed technically and methodologically (Amburgey & Amirtharajah, 2005; Colton et al., 1996; Frank et al, 2015; Grabovsky et al., 2012), and the time spent on it is two orders of magnitude less than that spent directly on water treatment, the subsequent numerous technological calculations focus on the time intervals between activation and thereafter shutdown (operating periods). Afterwards, they are formally identified with the duration of filter runs $t_{fi}(i=1,2,\ldots)$. Then it is generally justified to interpret the service life of a filter media as the total duration of the corresponding series of filter runs, and the sequence of the filter activations (control algorithm) to be determined by a set of the values t_{fi}. The simplest algorithm takes place in the absence of the accumulation of the non-washable deposit in a sequence of filter runs and, therefore, $t_{fi}=t_f$ (except for the first filter run). In this case, the development of the control algorithm is reduced to the establishment of a single duration t_f. The determination of t_{fi}, t_f is preceded by the calculation of a number of technological (criterion) times. They constitute the content of technological calculations together with the subsequent comparative analysis (Huisman, 1980; Mints, 1964; Poliakov, 2009a; Zhurba, 2011).

It should be emphasized that quite often the establishment of a filter run duration is not a self-sufficient goal, but helps in solving purely practical issues on improvement of the filter design and intensification of its operation. The subject of the applied optimization here can be the height of a packed bed and filter body, grain-size composition of filtering material and an equivalent diameter of its elements, doses of coagulants and flocculants, and possibly hydraulic and dirt loads.

As a rule, additional information is required for the special technological calculations, e.g. on the relationship between the mass exchange coefficients and the design and the technological parameters. The most important result of such calculations should be the only values of these parameters at which the longest duration of continuous filtration is achieved under the considered conditions.

Historically, the theory of suspensions filtration was not developing fast enough for a long time and the needs in engineering calculations clearly exceeded the possibilities it provided. Therefore, it was necessary to search for an alternative way to substantiate technological parameters, which allowed to do without conservation laws and mathematical models of filtration based on them. And here, as often happened at impossibility to fully use the methods of mathematical modeling, the theory of dimensions was involved (Ives, 1970; Romankov & Kurochkina, 1982; Yavorsky & Detlaf, 1977). During a number of decades such an approach has been actively developed from the positions of the mentioned theory and general ideas about filtration of polluted water as a complex of physicochemical, hydromechanical processes. In fact, an independent direction has emerged in it, namely, the technological modeling (Ison & Ives, 1969; Yaroshevskaya & Kulsky, 1985; Zhurba, 1993). At the same time, it was possible to achieve certain successes especially with respect to the time of intensive deposition of dispersed impurity and, as a consequence, high quality of filtrate due to this time stability to changes in influence factors. Each of the calculation dependence derived using the analysis of dimensions contains purely empirical coefficients that cannot be interpreted from physical considerations. Such coefficients are characterized by a significant variability, which makes their generalization very difficult. Nowadays, when solutions of many important filtration problems have already been obtained and there are incomparably greater possibilities in calculations of any complexity, the technological analysis seems too unreliable and unpromising due to its isolation from the underlying models.

In contrast to the described simplified approach to the substantiation of aqueous suspension filtration technology, the monograph develops a methodology of technological analysis, which is just based on the above-mentioned solutions. Indeed, the dependencies derived from them and intended for the calculations of the key characteristics of filtration, are the basis of criterion equations after their adaptation to the criteria of effective operation of water treatment filters. It is the collection of the latter with respect to the technological times, covering a variety of conditions and modes of filtration in practice, has the right to be considered as a method of technological analysis. First of all, it provides rational choice of the control algorithm of a filtration facility. At the same time, this method can be successfully used as a reliable tool for theoretical substantiation of the facility design and its individual elements as well as the conditions of its effective operation.

Considering a filter operation during many filter runs, we have to face two characteristic situations. In the first situation, it is possible to restore almost completely the hydraulic performance and the absorption capacity of the filter bed after each its washing. Then the duration t_f under stable filtration conditions during the entire service life of the bed should be constant, which is not true. In practice the granular material is gradually abraded and crushed due to the regular forced turbulization of a suspension flow, fluidization of medium and accompanying intensive mixing with frequent collisions of chaotically moving elements of the bed. In the end, the mentioned deformations of the filtering material become the reason for a noticeable reduction in the equivalent diameter of its particles, first of all, due to smoothing of irregularities on their surface. Crushing and denser packing of the deformed particles cause noticeable changes in the filtration flow and absorption properties of a packed bed. In particular, annually the hydraulic conductivity of a washed claydite bed and t_f actually decrease by 5-6% (Ayukaev & Meltser, 1985; Tsitsishvili & Andronikashvili, 1985; Zhurba, 1980; 2011). Such an effect provides an increase in the equivalent grain diameter by 2.5-3% at unchanged initial porosity.

The deviations from normal filter operation caused by instability of filtration conditions, for example, the temperature fluctuations throughout a year are even more significant (Hopper-Jones et al., 2010; Shen et al., 2018). Still, in this situation, it is acceptable to assume the duration of a filter run t_f conditionally constant in the first approximation and to limit ourselves to the calculation of its single value (Ives, 1981; Krivonog & Krivonog, 1997; Mintz, 1964; Shevchuk et al., 2005; Tanaka et al., 1995; Yaroshevskaya & Sotkova, 1996; Zhurba, 1980). Then the behavior of filtration characteristics during all filter runs should be identical. However, the first filter run from the beginning of operation of a new filter bed deserves a special attention even with such a unified approach. In this case, the filtering material is initially clean, and in the future will necessarily contain some relatively small amount of retained contaminants even directly after each washing. It should be added that fundamental difficulties do not arise in the calculations of t_f both at a smooth and at a sharp change in filtration conditions (filtration rate, temperature, etc.), if we can adapt the model coefficients to them.

In the second situation, the technological calculations for suspended solids removal should be carried out from the beginning of operation of a new batch (change) of filtering material, consistently monitoring the dynamics of a deposit accumulation both over one filter run and in a sequence of filter runs up to the end of its service life (Lukinykh et al., 1978; Martynov & Poliakov, 2022b). Two trends are clearly observed here, namely, the gradual decrease in the absorption capacity of a given material and the increase in the hydraulic resistance of a bed. As a consequence, the time of effective continuous operation of the filter steadily decreases.

A number of requirements of different degrees of importance are made to the operation of water treatment filters. Regardless of the design and operating conditions of a filtration facility, the quality of filtrate must certainly be provided in accordance with the regulatory documents (Bespamyatov & Krotov, 1985; Isaji, 2003; Kulsky & Strokach, 1986; StSanRandN, 2010; American Assosiation, 1999). It is obvious that the content of dispersed impurity in filtrate is closely related to its amount in the suspension being clarified, characterized by the parameter C_0. In some cases, the initial concentration of the specified impurity can change significantly during one filter run. Then C_0 is acceptable to interpret as its average value for the corresponding period. Taking into account the high cost of water treatment in the modern technological schemes and also essentially limited opportunities in the associated removal of other types of pollution, primarily organic, firstly, the concentration C_0 is sought to be substantially reduced at the stages preceding rapid filtration preliminarily removing larger suspended particles by the sedimentation in settling tanks. (Belyaev & Kozachina, 2014; Xue & Sun, 2003), and secondly, to minimize the content of dissolved organic matter using biological methods and bioreactors. At extremely high initial content of dispersed and dissolved impurities, the necessary degree of purification can be ensured through the use of (bio)filters connected in series (Goncharuk et al., 2001; Gvozdyak, 1989; Komarova et al., 1982; Kulikov et al., 2009; Kulkarni et al., 2005; Poliakov, 2023).

Actually, water preparation for filtration and its final purification (secondary purification) are combined in a similar way in one technological scheme using units of the same type. Then, there is a practical question of rational arrangement of such a scheme, i.e. optimization of the number of the units, or possibly of the structure and shape of their operating volume, the parameters and operating conditions, which should be solved also based on the results of the technological modeling. Then it becomes possible to theoretically substantiate, for example, clarification by filtration of turbid surface waters, for example, surface runoff from the urbanized surface and, in particular, storm drains (Hutt et al., 2007; Zhurba & Myakishev, 1992), whose suspended solids concentration is one or two orders of magnitude higher than that of raw water or pretreated wastewater. At the same time, it is realistic to reach the drinking water standard for this indicator. Sometimes separately prefilters can be an acceptable alternative to primary sedimentation tanks (Stechkina & Kirsch, 2003; Tanaka et al., 1995; Wolska & Smolinska–Kempisty, 2017).

Other requirements are not so rigid and, in fact, are intended for the control of the hydraulic parameters (head losses, flow rate, level in the storage tank). At the same time, it should be kept in mind that in case of large and regularly replenished quantities of contaminated water, it is advisable to maintain high productivity of a filtration facility at all times, based on economic and environmental considerations. A number of requirements with regard to the hydraulic aspects of the filtration facility are closely related to the rate mode.

Filtration at a constant rate has become widespread in engineering practice of water treatment (Brook, 1992; Kim & Lawler, 2012; Reddy et al., 2005). Thus, due to the use of rate controllers the stability of filtration flow, mass exchange processes in a filter bed and productivity of a filtration facility are provided for a long period of time. Here it is appropriate to note that the constancy of filtration flow rate at progressive clogging of filtering material causes an increase in the true velocity of liquid flow in the pore space, and thus, the hydrodynamic impact of the flow on the solid phase and, as a consequence, increased detachment of the formed deposit (Reddy et al., 2005; Shechtman, 1961).

The second filtration (rate) mode is implemented in the absence of the filtration flow process regulation (Akgiray & Saatci,1998; Bass, 1978; Cleasby, 1993; Dzyubo & Alferova, 2007; Grabovskiy & Gurinchik, 2005; Ison & Ives, 1969; Jegatheesan & Vigneswaran, 2005; Lozada & Cruz-Velez, 2014; Martynov, 2014; Poliakov, 2010b; 2012a; Senyavin et al., 1977; Yaroshevskaya et al., 1989). In this case, the behavior of the flow characteristic V over time can change significantly. Currently, two variations of it are practiced with different ways of supplying contaminated water on a rapid filter. In the first case, water in all design elements of a filtration facility moves under the action of constant head differential at the raw water and filtrate collectors. The head losses inside the collectors are not taken into account due to their smallness (large diameter, low velocity). Thus, the rate V monotonically decreases because of the increasing clogging of a filter bed during each filter run (Grabovsky & Gurinchik, 2008; Oleynik & Tugay, 2001; Verigin, 1971). Also, the piezometric head gradually decreases along the entire path of liquid movement from the first collector to the second collector.

At another variety of the second mode (split filter flow) the head at the outlet of a filtration facility is also kept constant (Grabovsky, 2016; Saatci, 1990). But in this case, the suspension is fed on the filter with a constant flow rate, which, as a rule, exceeds the flow capacity of a clean bed and the more so of a clogged bed. The excessive amount of initial water is collected in the storage tank. This creates an increased pressure at the inlet to the bed, which increases over time. Thus, it is possible to compensate almost completely for the additional mechanical energy expenditure due to the clogging of the bed and minimize the reduction of the filtration rate for a certain period of time. However, the filter productivity eventually starts to drop even in this case. Fundamentally, a significant decrease in the specified productivity is actual in both varieties of the filtration mode with unregulated rate, due to which its further operation becomes unprofitable from the economic point of view. Therefore, the corresponding critical filtration flow rate should be determined on the basis of a technical and economic analysis.

When implementing the first two rate modes, filtration of a suspension takes place in the stabilized conditions. At the same time, sometimes a deliberate abrupt change in the filtration conditions is justified, because the filter operation is prolonged in this way (Dzyubo & Alferova, 2007; Saatci, 1990). An example of such, in fact, combined mode can be the clarification of suspension on a filter with the storage tank in two stages. In the first stage, the filter operates in the second variety of the second mode up to the point of overfilling its tank. If the filter performance and filtrate quality are still high enough by this moment of time, then it makes direct sense to continue its operation now at a constant water level above a bed surface, i.e. at the second stage. Calculations of the nonstationary filtration characteristics are also performed in two stages – sequentially according to different dependences before and after the moment of full filling of the tank. And if the bed is assumed to be (almost) clean before the first stage, then it most often turns out to be seriously contaminated by the beginning of the second stage. As a consequence, subsequent calculations of concentrations are much more complicated, and at the same time it is necessary to determine the already variable filtration rate.

Since the operation of rapid filters is always limited in time due to a progressive decrease in the absorption and flow capacities of their filtering material, the question inevitably arises of time limitations imposed on the nonstationary technological process of filtration. The composition and quantitative indicators of such restrictions were the subject of special experimental studies, technical and economic calculations. The extensive experience of the operation of filtration facilities on industrial scale and cottage units was attracted for their establishment. At present, the relevant information, which serves as a basis for the technological calculations, is contained in the reference literature on water treatment and filtration (Ayukaev & Meltser, 1985; Degremont, 2007; Dickinson, 1992; Kulsky et al., 1980; Twort et al., 2006). The complex of the above-mentioned constraints is formalized in the form of a system of the criteria for the efficiency of a filtration facility.

When preparing drinking water (to a lesser extent, technical water) it is essential to monitor the concentration of contaminants in the liquid phase at the filter outlet C_e with special attention, since human health is an absolute priority here. In this case it is necessary that the specified concentration does not exceed a given threshold value (a reserve is allowed for reliability), and in essence the maximum permissible value C_* (Isaji, 2003; Jimenez et al., 2000). The corresponding criterion of filtrate quality is symbolically expressed in the form of the relation

$$C_e(t) \leq C_*. \tag{1.35}$$

Obviously, its observance guarantees sanitary safety of filtrate in terms of suspended solids content and is mandatory.

The second, hydraulic criterion is relevant and often turns out to be decisive in the filter control if the first filtration mode ($V=const$) is implemented. Then it is undesirable that the current head losses Δh in a bed become greater than their maximum permissible value Δh_* (if the head losses in the filter transportation system are also taken into account, then Δh_* should be corrected accordingly) (Ayukaev, 2003; Jimenez et al., 2000; Lukinykh et al., 1978; Mays & Hunt, 2005). Thus, the formal representation for the second criterion will be

$$\Delta h(t) \leq \Delta h_*. \tag{1.36}$$

Ignoring (1.36) may cause excessive reduction of the filter performance, serious difficulties in its regeneration.

If no measures are taken to stabilize the filtration flow rate (the second filtration mode), the rate of water treatment is acceleratedly declining. In order to avoid a significant reduction of the suspension clarification rate with time, and thus, an increase in the filtrate cost, which is unacceptable for economic reasons, a new limitation is imposed on the rate V (Poliakov, 2022; Zhurba, 1980)

$$V(t) \leq V_*, \tag{1.37}$$

where V_* denotes the minimum allowable filtration rate by analogy with C_*, Δh_*.

The fourth, technical criterion is attracted for the technological calculations at the operation of a filter equipped with the storage tank in the second mode (second variety). In practice, such a filter is turned off every time its storage tank is filled with initial water up to the maximum mark (Akgiray & Saatci, 1998; Cleasby, 1993). Similar control method is convenient from the practical point of view due to the simplicity of monitoring the operating state of the filter. The conclusion about its (non)operability and the necessity of the next washing is made on the basis of the following inequality

$$H_W(t) \leq H_\bullet, \tag{1.38}$$

where $H_\bullet$ is the mark of the edge of the chute discharging the washing water (or its height above the bed surface). As already mentioned, it often happens that a filter is far from exhausting its clarifying resource by the time of overfilling its storage tank, so that the suspended solids content in the filtrate is still low, and the filtration flow rate is still high. Then the use of this criterion makes no practical sense, and it is reasonable to continue filter operation until the first or third criteria are violated.

Finally, the fifth, consolidation criterion should only be used in situations where fresh deposit can "age" relatively quickly (Ayukaev & Meltzer, 1985; Ishiyama, 2010; 2011; Shandyba & Weinstein, 1988). Generally, the term «aging» is often used in a broad sense to state the irreversible change in properties of substances of different nature and operating media over time (Gao et al., 2017; Schuberth et al., 2011). Here, aging is understood in a mechanical sense and is expressed as an increase in the strength of the deposit due to its partial dehydration, strengthening structural bonds in it. Obviously, it makes sense to consider the age of the deposit in engineering calculations only if the characteristic time of its consolidation is comparable to the time of continuous productive operation of the filter with high filtrate quality and flow head losses (duration t_f). Such a phenomenon, for example, is observed when the deposit with a high content of trivalent iron is formed (Orlov, 2008). In such cases, the progressive consolidation of a deposit with iron hydroxide as a structure-forming component leads to a noticeable increase in the capital and operating costs. As a consequence, the consolidation factor becomes a decisive factor in the development of the filter control algorithm. Since at mathematical and technological modeling of filtration facilities operation to solve the question of deposit ability to withstand the force impact of the filtering and washing water flows is of fundamental importance, and reliable consolidation models for gel-like deposits have not been developed yet, it is currently sufficient for considering the above-mentioned factor to establish the empirical function of hydrodynamic stability (mechanical resistance of a deposit in relation to the hydrodynamic force) $f_s(t, V)$ (Khapun & Miltreiger, 1981; Martynov & Poliakov, 2022b). Such a function should characterize the relative amount of the fresh deposit that can be removed from a unit volume of a clogged filter media by washing at standard duration and intensity, depending on the time of its formation. Naturally, the age of the deposit is closely related to the duration of filtration. It builds up in a bed practically uniformly over time and therefore its age varies from 0 to the calculation point in time. The exception is a small and, after many filter runs, possibly significant amount of the deposit that has remained after all washings (total non-washable deposit). To determine the threshold value of the above-mentioned function f_{s*}, starting from which the filter washing is delayed and does not allow to achieve the maximum effect due to filtration, it is possible either to rely on extensive experience of the operation of rapid filters under similar conditions, or to carry out special technical and economic calculations with full consideration of the capital and operating costs during the entire service life of one change of filter media. Then the fifth criterion is formalized in the following form

$$f_s(t, V) \leq f_{s*}. \tag{1.39}$$

It is not difficult to establish a suitable number of the control time moments (technological or criterion times) if there are reliable information on the relevant criterion parameters (C_*, Δh_*, V_*, H_*) and an analytical solution of the basic problem of suspension filtration. The results of their comparative analysis serve as a basis for continuation or termination of the filter operation. These times are mainly found by fitting as roots of certain (transcendental with integral coefficients) equations, which follow from the above-mentioned solution and can be interpreted as implicit functions. In other words, one should compute the values of a number of implicit functions from one or more variables. Exceptionally, when a suitable formula exists, the desired value is found immediately. Much more often, such a single value is simply fitted from also a single computation equation. Sometimes such a fitting is significantly, though not essentially, complicated, namely, when it is necessary to simultaneously find two or more characteristic times from a system of equations of similar order.

The primary goal of the technological analysis of a filtration facility is to calculate the duration of the time periods during which the relevant efficiency criteria are met, in fact, the technological times (Bass et al., 1978; Fominykh, 1980; Zhurba, 2011). It is natural that, first of all, we have to set the system of the specified criteria in accordance with the implemented filtration mode and only then proceed to the calculation of the required times. The first and basic of them, the so-called time of protective effect of a filter bed t_p (Ayukaev & Meltser, 1985; Demchin et al., 1984; Dzyubo & Alferova, 2007; Fominykh, 1980; James, 1981; Lukinykh et al., 1978; Senyavin et al., 1977; Yaroshevskaya & Muravyev, 2002; Zhurba, 1993), should be determined according to (1.35) from the following equality

$$C_e(t_p) = C_*. \tag{1.40}$$

From here it is obvious that the content of suspended solids in the filtrate will not exceed the standard value at $t \leq t_p$ due to the monotonicity of the function $C_e(t)$. And it should not be allowed that a filter continues to operate at $t > t_p$.

At a stable filtration flow rate (the first filtration mode) the filter operation has to be suspended also taking into account the time t_h of reaching the maximum allowable head losses in a packed bed (Ayukaev & Meltser, 1985; Demchin et al., 1984; Fominykh, 1980; James, 1981; Lukinykh et al., 1978; Senyavin et al., 1977; Yaroshevskaya & Muravyev, 2002; Zhurba, 1993). This time is calculated in accordance with (1.36) from the equation

$$\Delta h(t_h) = \Delta h_*. \tag{1.41}$$

Filtration of aqueous suspension with an unregulated rate inevitably leads to an accelerated decrease in the filter productivity. At its long continuous operation the current filtration rate will decrease after a strictly explicit time t_V (the time to reduce the filtration rate to an economically feasible minimum value) to the justified minimum V_* irrespective of the method of suspension feeding (Bass et al., 1978; Poliakov, 2010a; Wakeman, 2007). When the theoretical dependence $V(t)$ is known and V_* is specified, the time of the filter shutdown due to its low productivity is calculated by fitting from the equation

$$V(t_V) = V_*. \tag{1.42}$$

If the filtration of a suspension through a packed bed, that is being gradually clogged, is accompanied by filling of the storage tank with it (the second variety of the second rate mode), then, in accordance with the popular assessment of its operability by the level (mark) position H_W, the time t_H (time of the rise of the suspension level in the storage tank to the maximum mark H_m in fact) becomes important to control the technological process (Akgiray & Saatci, 1998). At the theoretically established pattern $H(t)$, the time t_H should satisfy according to (1.38) the equation

$$H(t_H) = H_*. \tag{1.43}$$

The greatest difficulties arise with the determination of the characteristic consolidation time of a gel-like deposit as a specific product of water filtration with dispersed impurity. So far, mathematical models and, moreover, the methods of calculating changes in mechanical and structural properties of such a specific conditionally solid matter under dynamic conditions have not been developed. Therefore, at present the only real possibility to physically justify the fifth criterion time t_c is to use experimentally determined function $f_s(t)$ and the parameter of the hydrodynamic resistance of a deposit f_{s*}. Then the required time is elementary found from the following equation

$$f_s(t_c) = f_{s*}. \tag{1.44}$$

Actually, f_{s*} means the maximum allowable increment of a deposit that cannot be removed by the washing water from its total amount accumulated during the representative filter run. The fundamental difference between equation (1.44) and previous equations (1.35) to (1.38) is that the functions $C_e(t)$, $\Delta h(t)$, $V(t)$, $H_W(t)$ are determined as a result of solving a specific mathematical problem, and the function $f_s(t)$ is purely empirical. Until recently, due to the lack of systematic data on the relationship between deposit strength and its age, one had to rely solely on extensive experience in filtration of anomalous suspensions that form a dynamically structured deposit after separation. For example, it has been established based on many years of practice, that at physicochemical iron removal from deep groundwater, backwashing of the filter bed contained the deposited and hydrated iron hydroxide gives a tangible economic effect under usual filtration conditions only if it is carried out not later than in two days. In recent years, however, unique data have been obtained in the NUWEE laboratory, which now allow a technological analysis to be performed over the lifetime of the filter media change in the case of a consolidatable deposit. These are outlined in detail by Martynov & Poliakov (2022b).

Obviously, the suspension filtration should be stopped and the next filter washing start immediately after violation of any of the above efficiency criteria. This technological requirement is formalized as follows (Bass et al., 1978; Gurinchik, 2010; Poliakov, 2009a; 2022).

$$t_f = \min(t_p, t_h, t_V, t_H, t_c). \tag{1.45}$$

The expression for the duration of a filter run (1.45) has a generalized form, because it reflects a variety of filtration conditions possible in practice. However, in each specific case, this expression is necessarily simplified depending on the filtration mode, the properties of impurities and filtering material due to a reduction of the number of criteria involved and, accordingly, the technological times. For example, the procedure for setting the duration t_f at constant filtration rate and unconsolidated deposit, and in fact the technological optimization, should be based on the following expression (Ayukaev & Meltser, 1985; Dharmappa et al., 1992; Krivonog & Krivonog, 1997; James, 1981; Lukinykh et al., 1978; Orlov, 2005; Saltnes et al., 2003; Zhurba, 1980).

$$t_f = \min(t_p, t_h). \tag{1.46}$$

Important to note that it is much easier to substantiate t_f on the basis of (1.45), because in this case it is possible to avoid a labor-intensive technical and economic analysis. At the same time, the required initial information is often outdated or not available in full. However, the economic indicators (cost varieties) are reflected in all the criteria parameters except C_*. If reliable data on the capital and operating costs can be obtained, it is preferable to optimize the filter control algorithm (values t_f, t_{fi}) on the basis of technical and economic calculations. This approach will be implemented for iron removal filters subsequently in order to select a rational control algorithm for them.

Technical support of the technological process of clarification should guarantee stable, long-term, effective functioning of filtration facilities. From the ecological and sanitary point of view their equipment, except for the filter media, is of auxiliary importance, however, significantly affects the efficiency of water treatment. Therefore, of great practical importance is the design analysis of filtration plant, which should be carried out in conjunction with technological modeling. A significant number of filter designs are proposed and widely used in water treatment structures due to the diversity of filtration conditions, which is caused by differences in the rate modes, configuration of suspension flow, the level and nature of contamination of initial water and the way of its supply to a filter, the peculiarities of interphase mass exchange. The monograph focused on predicting the removal of harmful impurities in special porous media (packed bed) as well as on taking into account the results of the mathematical modeling in the technological analysis of rapid filtration, and only the height of a bed and the size of its structural elements (grains) were selected from the design parameters for the corresponding analysis (as an exception, the contribution to the filtration facility hydraulics of its transport systems was estimated). Therefore, the author limited himself to mentioning some official documents, textbooks, reference books, which contain description of common designs and recommendations on their practical application (Epoyan et al., 2016; Kovalchuk, 2002; Kozhinov, 2008; Martinov & Orlova, 2022; SBC, 2013a; SBC, 2013b; Tugay & Orlov, 2009; Zhurba et al., 2010) and the scheme as applied to the rapid filtration (Figure 2).

It should be emphasized that analytical solutions of (bio)filtration problems can be especially useful in the design of rapid filters and bioreactor-filters. On their basis and with the involvement of the efficiency criteria, in principle, it is relatively easy to substantiate the design parameters of a filter and, first of all, the height of its operating layer L (Chaudry, 1987; Mintz & Melzer, 1973; Ncube et al, 2018; Orlov, 2005; Poliakov, 2007) as well as the size of filtering material elements (equivalent diameter d_g) (Adin & Elimelech, 1989; Farooq & Al-Yousel,1993; Kim & Tobiasson, 2004a; 2004b; Mintz & Meltzer, 1973; Odegaard, 2003; Saltnes et al., 2003). However, the rational choice of the d_g value is significantly complicated due to the high sensitivity of the complex process of the aqueous suspension clarification with respect to it (Chuang & Li, 1997; Tobiason et al., 1993). Indeed, it has long been known that even small changes in the grain diameter d_g cause major changes in all the most important characteristics of filtration. The existing proposals on selection of the (equivalent) size of particles-collectors of a packed bed are based on the results of practical application of grains of significantly different coarseness. Diameter d_g varies in rapid filters in a relatively wide limits while observing approximate homogeneity of mechanical composition in each case. Nevertheless, the indicated recommendations are excessively or insufficiently specific (Chuang & Li., 1997; Zhurba, 1980). In order to avoid a noticeable decrease in the efficiency of such filters at the stage of their design, it is necessary to refine these recommendations in relation to the realistic conditions, based on the results of the technological calculations. Such analysis requires reliable information, first of all, on the dependence of the absorption rate of the suspended matter by (mono)dispersed filter media on the diameter d_g. Here, the theoretical dependences of the rate coefficient of suspension particles attachment on the representative value of d_g are appropriate for approximate estimates (Rajagopalan & Tien, 1976; Tufenkji & Elimelech, 2004).

In reality, the choice of filtering materials and their particles size is usually very limited. In these situations, if deposit is unconsolidatable, it is expedient to carry out the procedure of applied optimization. Then the mentioned procedure consists in the technological calculations and comparative analysis of a small number of options (according to available varieties of filtering materials and their mechanical composition). From the set of representative values of the duration t_f calculated according to (1.45) the largest value is taken and the bed option is executed in practice that meets it and is thus the most economical.

The procedure for optimizing the bed height is fundamentally different, since it is technically feasible to arrange a column of filtering material at any economically substantiated value of L. In general, conducting a filter design analysis by theoretical means, which is based on the principle provisions of the technological analysis, it is reasonable to consider a selected parameter, in particular, the height L as a continuous function of the duration t_f (Demchin, 1994 et al.; Kim & Tobiasson, 2004a,b; Poliakov, 2007; Saltnes et al., 2003). The formal search for the optimal value of L is facilitated if we operate with its relative analog $\bar{L} = L/L_0$, and use the dependencies properly transformed of the dimensionless technological times on $\bar{L}$. In the course of such transformations, the independent variable $\bar{L}$ is isolated, and the coefficients contained in the previous calculation dependencies, which are dimensionless complexes of several primary model coefficients, now have a new length scale L_0 instead L. If we understand L_0 as the smallest possible value of the truly possible values of L, then the continuous function $\bar{t}_f(\bar{L})$ must be found as a result of a large number of calculations of the required technological times. The derivative $\partial \bar{t}_f / \partial \bar{L}$ can also be continuous. Then, due to the monotonicity of this function, its maximum value will be at one of the boundary values of $\bar{L}$. However, more often this derivative has a finite break at an intermediate value of $\bar{L}$, and its sign changes here. This is exactly $\bar{L}$ what corresponds to the maximum value of $\bar{t}_f$.

Often groups of rapid filters assembled in a special way (blocks) are used to operatively treat large volumes of water with high content of dispersed contaminants. The consistency of their operation is ensured due to sequential (stepped filters), parallel or combined connection (Akgiray & Saatci, 1998; Grabovsky et al., 2012; Yaroshevskaya & Sotkova, 1996). At the same time, significant difficulties arise in modeling the operation of the filter block due to the inevitable non-synchrony of their functioning. Filters are turned off at different times, as a rule, at excessive reduction of the filtration efficiency (controlled by the system of criteria (1.35) - (1.37). At stable water supply to the filter block, the hydraulic load on any filter after the start of its washing is redistributed among all operating filters. As a consequence, it is necessary to correct the set of the initial values C_{0i} and to take into account additional non-stationarity caused by their sudden change (operation of such blocks is usually arranged in the second mode of the rate).

The detailed described procedure of applied optimization of the deep bed filtration technology as applied to rapid filters at three rate modes and possible deposit aging served as a methodological basis for a versatile theoretical analysis. The technological modeling is systematically performed in the following chapters of the monograph and the results obtained are discussed. At the same time, great possibilities and perspectives of the technological modeling will be demonstrated with respect to, firstly, essential intensification of the clarification process; secondly, improvement of the filters design.

1.4. COMPARATIVE ANALYSIS OF IMPORTANCE OF MASS TRANSFER MECHANISMS IN FILTER MEDIA

Mathematical modeling of the operation of wastewater treatment structures is considerably complicated, mainly due to the nonlinearity of interphase mass exchange, the non-equilibrium of physical and chemical state of an operating medium, the unevenness of contaminants distribution in it. When developing the engineering calculation methods, nonlinearity is often eliminated by linearization of the corresponding equations and boundary conditions, non-stationarity is conventionally treated as quasi-stationarity. More formal complications arise at the consideration of the gradient mass transfer (diffusion mechanisms), which causes an increase in the order of the differential equations of contaminant transport and phase transfer. Then, it is necessary to set an additional boundary condition to ensure the correctness of formulation of corresponding mathematical problems. Obviously, in cases of essentially heterogeneous composition of contaminants and presence of three, four interacting phases, the difficulties in conducting theoretical studies are much more significant, although their nature does not change.

Depending on the relationship between the contributions of the convection and diffusion mechanisms to the total transfer of contaminants, three characteristic situations are possible. Sometimes the diffusion transport can be equal in importance to the convection transport, for example, in membranes during cross-flow filtration (Song, 1998). Moreover, it is diffusion that provides delivery of pollutants to active microorganisms, vacant places within biofilms, permeable filter bed elements, (Pritzker, 2003; Rittmann & Stiwell, 2002). When both contributions are commensurate, it is important to consider them equally carefully. In such cases, a number of well-known exact solutions of unsteady contaminant spreading and phase exchange problems containing a variety of source (or sink) functions as free terms in the linear formulation may be useful, e.g. (Alekseev et al., 1989; Barry & Sposito,1989; Cameron & Klute, 1977; Gao et al., 2013; Li & Cleall, 2010; Philip, 1994; Tom et al., 2008; Verigin, 1971). The expressions derived from such solutions are very inconvenient for applied computations because they are either represented as slowly converging series or contain special integral functions (Crank, 1975; Fogden et al., 1988).

Modern filtration facilities are designed for the accelerated (secondary) treatment of large amounts of contaminated water. And it is natural that the filtration process in their operating volume (packed bed), carried out and controlled by special technical means, proceeds much more intensively than in natural porous media (soils, rocks). As a consequence, here the convection mechanism of the transfer of dissolved and suspended substances dominates at their migration, and the diffusion mechanism plays a minor role. The contribution of the diffusion to the movement of pollutants within a packed bed of usually small thickness is noticeably increased at their high initial content and enhanced absorption by the solid phase, which is typical, in particular, for the water treatment in adsorbers (Kravchuk, 2000; Poliakov, 2012d; Sherwood et al., 1982). Then it makes practical sense to consider the redistribution of an impurity in the specified porous layer under the action of the concentration gradient (as a driving force of the process), as a correction. As a consequence, there is actually a small parameter at the highest order derivative in the mathematical problems of contaminant mass transfer between solid and liquid phases in filtration facilities, namely, in the equation of the contaminant balance. In such cases, it is justified to use asymptotic methods, assuming the refinement of the basic degenerate solution by taking into account the diffusion component of the mass transfer in the second approximation (Oyanader & Arce, 2007).

The third situation is encountered much more often at (bio)filtration of aqueous suspension (solution). Then the mass transfer occurs, in fact, only by means of the convection mechanism due to relatively small concentration gradients and filtration flow rate. It is appropriate to notice that the factors of nonlinearity and non-equilibrium at prolonged filtration comparatively weakly influence the above-mentioned relationship. Therefore, when estimating the significance of the diffusion as a component of mass transfer, it is inexpedient to consider the above factors in detail, to involve numerical solutions or cumbersome analytical expressions. Thus, to reveal the contribution of the diffusion flow it is quite enough to rely on simple mathematical models, which adequately, first of all, describe the process of contaminants transfer with the filtration flow through the layer of filtering material, and also reflect in general the effect of contaminant absorption by the solid phase changing along its height. Just such an approach is realized in this subsection, and the conclusions obtained in this case are further used in the choice of the underlying mathematical models of filtration.

Below we consider a test problem of the depth non-detachment direct filtration of contaminated water, analogs of which (differences only in values of the model coefficients) have found wide application in justification of soil conditions regulation (water, biogenic elements), (deep) groundwater resources management. Presented already in dimensionless form, this problem can be considered as representative one in relation to the technological process of filtration in spite of significant limitations of its formulation. On the other hand, its solution makes it possible to simply predict the deterioration or stabilization of the filtrate quality and to carry out an applied analysis on this basis. The following data of the author's theoretical studies are presented in a form that maximally facilitates the prediction of the reaction of the outlet concentration of contaminants to the changes in the filtration conditions.

The deep bed filtration with a constant rate is analyzed exclusively by analytical methods, and contaminants are removed according to linear kinetics (Iwasaki, 1937). Interpretation of the interphase mass exchange as a result of two divergent exchange processes and the choice of a hydraulically simpler filtration mode are quite justified in this case. Indeed, the focus here is exactly the mass transfer, directly providing active and prolonged involvement of more sorbent material in water treatment. Really, if the filtration coefficient λ and rate V are variable due to, for example, a deposit (or biomass) accumulation, unstable hydraulic load, it is sufficient to average them beforehand. Thus, the non-stationary one-dimensional problem describing the current balance of the free impurity in an initially clean semi-confined column of a filtering material has the following form

$$\frac{1}{Pe}\frac{\partial^2 \overline{C}}{\partial \overline{x}^2} - \frac{\partial \overline{C}}{\partial \overline{x}} - \overline{\lambda}\,\overline{C} = \overline{n}_e \frac{\partial \overline{C}}{\partial \overline{t}} \; ; \tag{1.47}$$

$$\overline{t} = 0, \quad \overline{C} = 0 \; ; \tag{1.48}$$

$$\overline{x} = 0, \quad \overline{C} = 1; \qquad \overline{x} \to \infty, \quad \overline{C} \to 0. \tag{1.49}$$

Here

$$\overline{C} = \frac{C}{C_0}, \; \overline{x} = \frac{x}{L}, \; \overline{t} = \frac{Vt}{n_0 L}; \; \overline{\lambda} = \frac{L\lambda}{V}, \; \overline{n}_e = \frac{n_e}{n_0}, \; Pe = \frac{VL}{D}, \; D = n_e \xi_w D_m + \lambda_h V^{m_h} \; ;$$

L is the length scale (height of the filter bed); Pe is the Peclet number; ξ_w is the tortuosity coefficient ($\xi_w \leq 1$) (Chou et al., 2012; Gao et al., 2014; Shena & Chen, 2007); D_m is the molecular diffusion coefficient in clean water; λ_h is the hydrodispersion coefficient (Bear, 1988; Chrysikopoulos & Katzourakis,

2015; Fried & Combarnous, 1971; Lee et. al., 2009; Nielsen et al., 1986), m_h is the empirical coefficient (often taken equal to 1) (Bear, 1988).

It should be noted that the diffusion transfer in both natural and artificial porous media is usually understood in a broad sense as a transport process, the rate of which is proportional to the concentration gradient of a free impurity. At direct flow filtration it is characterized by the first term in one-dimensional equation (1.47). In fact, the mentioned coefficient has several components, the physical prerequisites for which are fundamentally different (Alekseev et al., 1989; Crittenden et al., 1986; Piron et al., 1995). The first component, molecular diffusion, is generated by the thermal motion of the impurity particles in the liquid, and the second, dispersion component, is produced by a nonuniformity of the actual velocity field. Obviously, its deviation from the average macroscopic velocity depends significantly on the size of the pores and suspended particles, as well as on the macroscale (height of the operating layer). Moreover, due to the smallness of the first component in a variety of porous media, including granular ones, in fact, the primary cause of mass transfer in the case of intensive movement of suspensions and solutions through them is the filtration flow process, which controls both convection and diffusion (dispersion) mechanisms (Bear, 1972; Perkins & Johnston, 1963; Venitsianov & Rubinstein, 1983). It is appropriate to note that a form of the coefficient D can become complicated, for example, due to the interaction of the charged impurity particles and bed elements (Bresler et al., 1987), the spatial character of fluid motion (Lopato et al., 2011), the mobility and unevenness of the distribution of the specified particles along the particles-collectors active surface (Lyn, 1996).

The fact of the gradual accumulation of a substance (inorganic deposit, biomass) newly formed in the pore space is reflected in equation (1.47) by the coefficient $\overline{n}_e$, which actually changes from the initial value $\overline{n}_{eo} \leq 1$ (less than 1 at initial space contamination) to a value commensurate with the relative solids take-up capacity. However, having in mind the insignificance of the last term in equation (1.47), it is sufficient for estimation calculations to limit ourselves to the maximum (1) and minimum (0) values of $\overline{n}_e$. In practice, the progressive reduction of the pore volume causes a gradual change in the filtration characteristics over time. To make the calculated pattern as close as possible to the realistic pattern, it is necessary to involve also the kinetic equation characterizing the dynamics of contamination already in the solid phase. However, the mentioned changes occur so slowly that they cannot noticeably affect the estimations of the significance of the diffusion transfer, dynamics of contamination in the liquid phase of a filter bed for water treatment.

The exact solution of problem (1.47) - (1.49) is described by the known dependence

$$\overline{C}(\overline{x},\overline{t}) = e^{0.5Pe\,\overline{x}}\left[e^{-\delta\overline{t}}\,erfc\sqrt{\frac{\overline{n}_e Pe\;\overline{x}^2}{4\overline{t}}} + \delta\int_0^{\overline{t}} e^{-\delta\xi}\,erfc\sqrt{\frac{\overline{n}_e Pe\;\overline{x}^2}{4\xi}}\,d\xi\right], \tag{1.50}$$

where $\delta = \frac{Pe + 4\overline{\lambda}}{4\overline{n}_e}$, $erfc\;y = 1 - \frac{2}{\sqrt{\pi}}\int_0^y e^{-u^2}du$. Expression (1.50) can be represented in a more compact form, viz,

$$\overline{C}(\overline{x},\tau) = \sqrt{\frac{Pe}{4\pi}}\,e^{0.5Pe\,\overline{x}}\int_0^{\tau}\frac{e^{-\tilde{\delta}\xi-\frac{Pe\,\overline{x}^2}{4\xi}}}{\xi^{\frac{3}{2}}}\,d\xi, \tag{1.51}$$

where $\tau = \frac{\overline{t}}{\overline{n}_e}$, $\tilde{\delta} = \frac{\delta}{Pe} = \frac{Pe + 4\overline{\lambda}}{4\overline{n}_e}$. It follows from (1.51) that at $\tau \to \infty$ $(\overline{t} \to \infty \;\; or \;\; \overline{n}_e \to 0)$ the steady-state distributed quantity $\overline{C}^e_\infty$ will be

$$\overline{C}_{\infty}(\overline{x}) = \exp\frac{-2\overline{\lambda} Pe\, \overline{x}}{Pe + \sqrt{Pe^2 + 4\overline{\lambda} Pe}}. \tag{1.52}$$

It should be emphasized that the layer of a filtering material has a small thickness ($0 \le x \le L << \infty$) in actual water treatment filters due to its usually high absorption capacity and cost. Then the second condition (1.49) in the operator of boundary conditions (1.48), (1.49) is necessary to replace by the homogeneous condition

$$\overline{x} = 1, \quad \frac{\partial \overline{C}}{\partial \overline{x}} = 0. \tag{1.53}$$

The solution of the initial non-stationary problem is not difficult to obtain in a bounded filtration flow region as a series converging in $\overline{t}$, and its convergence is significantly worse at small values of $\overline{t}$. However, keeping in mind the goal here, it is easier to use in calculations the solution of the analogous stationary problem, which is expressed by the dependence

$$\overline{C}_{\infty}(\overline{x}) = \frac{(1-\delta)e^{0.5Pe[(1-\delta)\overline{x}-\delta]} - (1+\delta)e^{0.5Pe[(1-\delta)\overline{x}+\delta]}}{(1-\delta)e^{-0.5Pe\,\delta} - (1+\delta)e^{0.5Pe\,\delta}}, \tag{1.54}$$

where $\delta = \sqrt{1 + 4\overline{\lambda}/Pe}$.

To isolate the effect of the diffusion process in the broad sense it is necessary to have an asymptotic solution ($Pe \to \infty$) of problem (1.47) - (1.49). The corresponding expression for the desired function $\overline{C}(\overline{x}, \overline{t})$ can be obtained either from (1.50), (1.51) by passing to the Pe limit or directly solving equation (1.47) without the first term, i.e. the equation of the first order at the only boundary condition. It is represented in the following exact form

$$\overline{C}(\overline{x}, \overline{t}) = \begin{cases} e^{-\overline{\lambda}\overline{x}}, & at \quad 0 \le \overline{x} \le \frac{\overline{t}}{\overline{n}_e} \quad or \quad \overline{t} \ge n_e \overline{x}\ ; \\ 0, & at \quad \overline{x} > \frac{\overline{t}}{n_e} \quad or \quad 0 \le \overline{t} \le \overline{n}_e \overline{x}. \end{cases} \tag{1.55}$$

Thus, the non-stationarity in the behavior of the free impurity concentration function is manifested in the considered case only through the displacement of a stable pollution front. The distribution of an impurity along the height (length) of the layer $0 \le \overline{x} \le \overline{t}/\overline{n}_e$ does not change in time. Thus, the profile $\overline{C}(\overline{z}) = e^{-\overline{\lambda}\overline{x}}$ will be finally formed here after the time $\overline{n}_e$ at the bed height L and hence the relative height value 1. In fact, the content of the impurity in the solid phase in the entire operating layer except for its inlet cross-section will gradually increase and the pore space shrinks accordingly. It is established that almost all the matter coming into each elementary volume of a bed either passes into the solid phase or transits through it to the next similar volume. Therefore, when assessing the significance of the components of mass transfer, it is justified to take as a basis for comparison the outlet value of $\overline{C}$ at $\overline{t} > \overline{n}_e$

$$\overline{C}_e = e^{-\overline{\lambda}}. \tag{1.56}$$

The subject of the estimation calculations in this section is the relative outlet concentration of contaminants in the liquid phase $\overline{C}_e$ and ultimately the difference between its reference and current values for the steady-state and unsteady filtration process $\Delta\overline{C}_e$. All the above formulas are involved in the quantitative analysis. It is important to note their mathematical rigor, which ensures high reliability of the purposefully conducted comparison within the framework of the adopted model of the mass transfer in a two-phase contaminated system. The four figures show the dependences of $\Delta \overline{C}_e$ on $\bar{t}$ and *Pe*, $\bar{\lambda}$, with these coefficients using alternately as an argument or a discretely varying parameter. At the same time *Pe* changed by two orders of magnitude, and $\bar{\lambda}$ changed by one order of magnitude.

First of all, we analyzed the legitimacy of the formal replacement of the filtration flow region, bounded in the direction of fluid motion ($0 \leq \bar{x} \leq 1$), by an abstract semi-bounded region ($0 \leq \bar{x} < \infty$) or vice versa, which quite often practiced in applied problems. **Figure 3** shows the graphs of the dependence $\Delta \overline{C}_{e\infty}(\lg Pe)$, calculated at two values of $\bar{\lambda}$. The first value corresponds to a moderate and more realistic intensity of contamination removal and the second value meets a somewhat overestimated intensity. Here we compared the steady-state values of $\Delta \overline{C}_{e\infty}$ corresponding to $\bar{L} = 1$ and $\bar{L} \to \infty$, as well as the values calculated according to formulas (1.45) and (1.43) respectively. Thus, $\Delta \overline{C}_{e\infty}$ is expressed by the ratio

$$\Delta \overline{C}_{e\infty} = \frac{\overline{C}_{e\infty}(\bar{L} = 1) - \overline{C}_{e\infty}(\bar{L} \to \infty)}{\overline{C}_{e\infty}(\bar{L} = 1)}$$

Figure 3 demonstrates that a noticeable (up to 7%) computational error due to the extension of the actual operating layer of rectangular shape to a semi-bounded strip (in two-dimensional interpretation) takes place if $Pe<100$, $\bar{\lambda} = 8$. However, high quality of the filtrate is maintained under such conditions in practice for a long time and, therefore, does not limit the technological process. Consequently, the discussed replacement of the filtration flow region and the use of formula (1.50) are valid for predicting the increase in $\overline{C}_e$ over time under typical filtration conditions.

Sharp changes in the concentration $\overline{C}_e$ over time and, as a consequence, large values of the derivative $\partial\overline{C}/\partial\bar{t}$ are observed only at the short-term early stage of filtration, as evidenced by **Figure 4**. Here the curves are depicted of a decrease in the relative quantity $\Delta \overline{C}_e$ representing the following ratio in the interval $0 \leq \bar{t} \leq 1.5$

$$\Delta \overline{C}_e(\bar{t}) = \frac{\overline{C}_{e\infty} - \overline{C}_e(\bar{t})}{\overline{C}_{e\infty}}$$

Three values of $\overline{C}_{e\infty}$, corresponding to three values of *Pe*, were calculated by formula (1.52). The dynamics of $\overline{C}_e$ was calculated by (1.51). The specific form of curves 1-3 with two inflections is due to accelerated movement of contamination within the actual operating layer ($0 \leq \bar{x} \leq 1$) by the diffusion mechanism. The significance of such mechanism increases at lower values of *Pe* (curve 1). However, its contribution to the total transport turns out to be negligible in the entire considered range of *Pe* values already at $\bar{t} > 1.5$. Thus, Figure 4 clearly confirms the validity of the earlier conclusion about the smallness of the derivative of the concentration *C* with respect to time. Nevertheless, the technological process of water treatment by the depth filtration as a whole and especially the mass exchange between the solid and liquid phases have a pronounced unsteady nature, that is reflected at their mathematical modeling in a special kinetic equation (in biofiltration - biomass growth equation) supplementing (1.47).

It should be emphasized that in this section the errors in the calculations of the key (from the sanitary point of view) relative characteristic $\overline{C}_e$ due to the neglect of both molecular and dispersive components of the convection diffusion of an impurity are of the greatest interest for the subsequent theoretical studies. They were estimated from the data of calculating the quantity $\Delta \overline{C}_e$, which now has the following form

$$\Delta \overline{C}_e = \frac{\overline{C}_e(Pe) - \overline{C}_e(Pe \rightarrow \infty)}{\overline{C}_e(Pe \rightarrow \infty)}.$$

The results of calculating $\Delta \overline{C}_e$ as a function of lg Pe are shown in **Figure 5**, and as a function of $\overline{\lambda}$ presented in **Figure 6**. The large differences in the sensitivity of $\overline{C}_e$ and $\Delta \overline{C}_e$ with respect to the indicated coefficients (Pe and $\overline{\lambda}$) deserve special attention. In particular, their response nonlinearly increases with decreasing Pe (Figure 5) and increasing $\overline{\lambda}$ (Figure 6). The degree of such changes in turn depends significantly on the values of the second coefficient-parameter. For example, the deviation of approximate ($Pe \rightarrow \infty$) values of $\overline{C}_e$ from the actual ($Pe < \infty$) values turns out to be so significant in the range $1 \leq \lg Pe \leq 2$ and at large values of $\overline{\lambda}$ that its relative measure $\Delta \overline{C}_e$ far exceeds 1. And here it is important to note that especially severe errors in calculations should be expected in the case of a well-absorbing filtering material and significant molecular diffusion. However, such conditions are uncharacteristic for actual filtration facilities. Thus, the increased absorption of an impurity, on the one hand, causes the increase in protective capability of a filter bed. But on the other hand, it leads to a sharp increase in the hydraulic resistance of the clogged bed and, as a consequence, forced early washing of the filter. And then hydraulic aspects and economic indicators come to the forefront in the technological analysis. The quantity $\Delta \overline{C}_e$ is so small under normal conditions of filtration, characterized by small and moderate values of $\overline{\lambda}$, moderate and large values of Pe, that the convective diffusion can be ignored at all. Thus, it is justified to discard the term containing the second derivative in equation (1.47) and similar equations.

Figure 3. Dependence $\Delta \overline{C}_{e\infty}(Pe)$

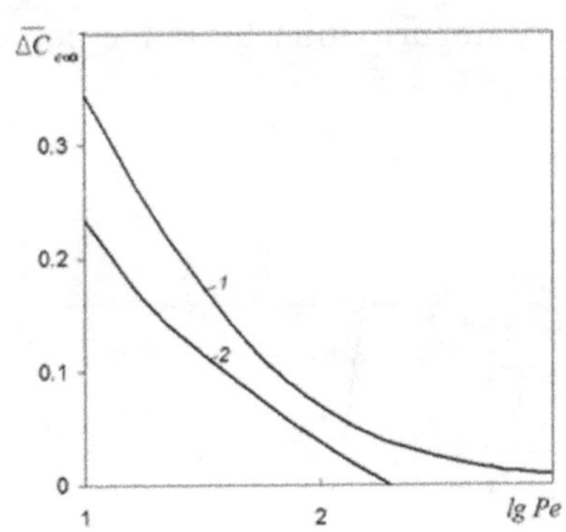

Note. $1 - \overline{\lambda} = 8$, $2 - \overline{\lambda} = 4$

Figure 4. Dependence $\Delta \overline{C}_e(\bar{t})$

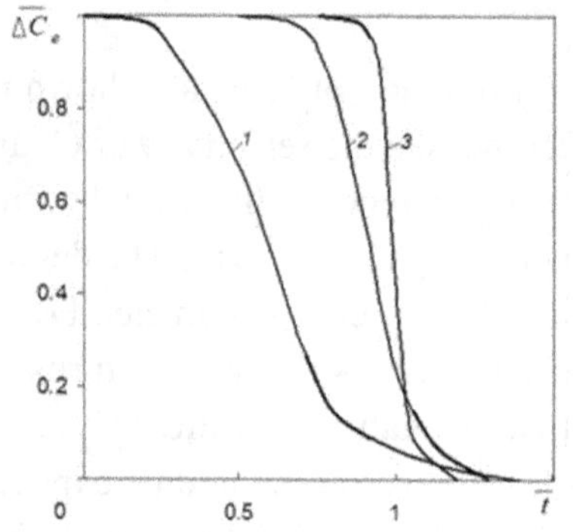

Note. 1– Pe=10, 2– Pe=100, 3– Pe=1000

The results of calculations of the outlet concentration $\overline{C}_e$ with and without considering diffusive impurity transfer presented in Figures 5 and 6 can serve as a guide in determining its role in the total combined mass transfer and, most importantly, in the impurity removal from a filter bed. However, having in mind the specificity of interrelation of the calculation characteristics and the model coefficients Pe, $\bar{\lambda}$, and also essentially simplified form of the accepted model and the test dependencies (1.50) - (1.52), (1.54), (1.55) resulting from its solution, it is preferable from the methodological point of view to carry out such comparative analysis for specific conditions of filtration, taking into account the requirements to the quality of water treatment. And it is possible to finally ascertain the validity (or otherwise) of neglecting the diffusion component of impurity transfer in the porous layer only on the results of the above-mentioned analysis.

Figure 5. Dependence $\Delta \overline{C}_{e\infty}(lg Pe)$

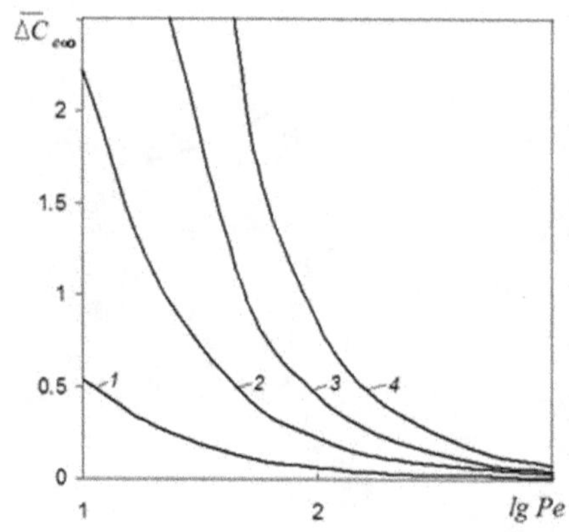

Note. 1 – $\bar{\lambda}$ = 2, 2 – $\bar{\lambda}$ = 4, 3 – $\bar{\lambda}$ = 6, 4 – $\bar{\lambda}$ = 8

Figure 6. Dependence $\Delta \overline{C}_{e\infty}(\overline{\lambda})$

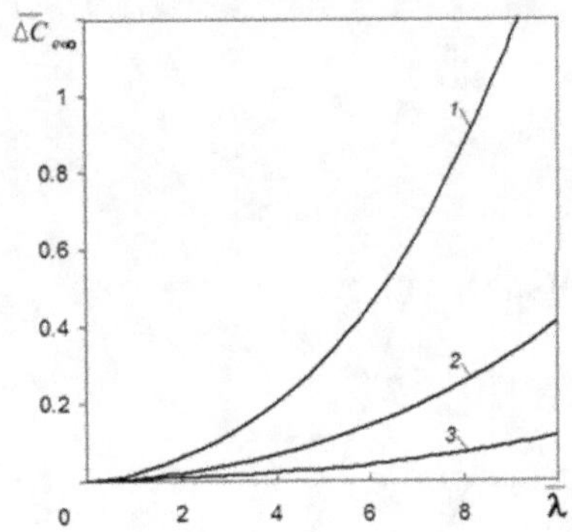

Note. 1 – Pe=100, 2 – Pe=300, 3 – Pe=1000

It is worth noting that the above analysis can in principle be easily extended to the filtration flow process and the key hydraulic characteristic, namely, head losses in a filter bed. It is necessary to involve a suitable kinetic equation and take into consideration the dependence of the coefficient λ on the content of the immobilized matter. However, the results of in-depth theoretical studies are unlikely to essentially affect the estimates obtained above.

The filtration of low concentration suspensions will be considered in the monograph mainly by analytical methods and primarily in relation to the normal and extreme conditions characterized by large values of *Pe*. Then, in accordance with the results obtained above, it is justified to neglect the diffusion (dispersion) transfer of impurity and thus, to simplify considerably the governing equation describing the general mass transfer. As a consequence, it is possible to significantly expand the application fields of the expressions recommended for engineering calculations.

If the diffusion mechanism still plays a significant role in the impurity transfer, which is characteristic, for example, for the infiltration of contaminated surface water into a ground bed, the removal of dissolved impurities in adsorbers, spreading of volley contaminants in natural porous medium, it is often enough to take it into account in the engineering calculations of contaminant transport and phase transfer, having adjusted properly the theoretical developments at $Pe \rightarrow \infty$ (actually the first approximation). Moreover, it is usually sufficient to limit ourselves to the second approximation with respect to the problems of (bio)filtration of suspensions (solutions) under conditions where the values of *Pe* are in the range from 10 - 25 to 100. Justification of such an approach is simply confirmed by the simulation of the steady convection-diffusion transfer of an impurity, which is removed according to the linear kinetics in the layer of a filtering material. Then the first approximation has form (1.56), and the second approximation, reflecting the main contribution to the mass transfer of diffusion transport, is represented by the expression

$$\overline{C}_{2\infty}(\overline{x}) = e^{-\overline{\lambda}\overline{x}}\left(1 + \frac{\overline{\lambda}^2}{Pe}\overline{x}\right) \tag{1.57}$$

The function $\overline{C}_{\infty}(\overline{x})$ is exactly calculated using formula (1.54). The errors in calculating the steady-state outlet concentration $\overline{C}_{e\infty}$ in the first and second approximations were estimated as follows

$$\Delta \overline{C}_{e\infty} = \frac{\overline{C}_{e\infty,точн} - \overline{C}_{e\infty,приб}}{\overline{C}_{e\infty,точн}}$$

Figure 7. Dependence $\Delta \overline{C}_e(lgPe)$:

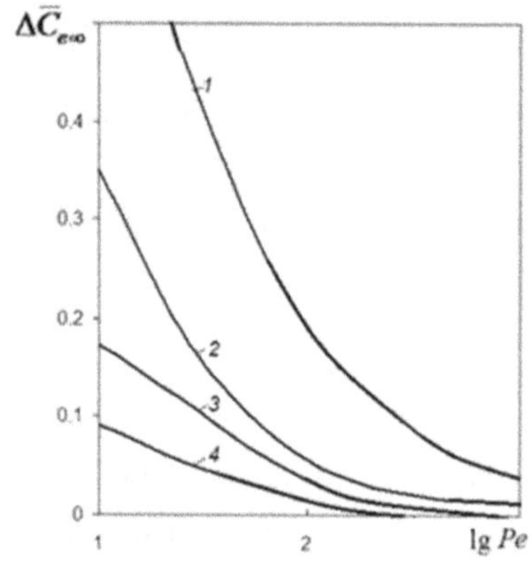

Note. 1, 2 - first approximation; 3, 4 - second approximation; 1, 3 – $\overline{\lambda}$ = 4; 2, 4 – $\overline{\lambda}$ = 2

The data of the calculations of the quantity $\Delta \overline{C}_{e\infty}$ as a function of *Pe* at two values of $\overline{\lambda}$ are shown in **Figure 7**. Obvious from the comparison of curves 1 and 3, 2 and 4 that it is possible to significantly expand the field of the basic solution application due to its slight modification. In particular, $\overline{C}_{e\infty}$ is now found with 10% error already at *Pe*=10, $\overline{\lambda} = 2$ or at *Pe*=25, $\overline{\lambda} = 4$.

1.5. ON INFORMATION SUPPORT OF MATHEMATICAL MODELING OF DEEP BED FILTRATION

All meaningful theoretical developments in the monograph are accompanied by the detailed calculations of illustrative examples, which allow to reveal the regular nature of behavior of the most important characteristics of filtration and rationally choose the technological and design parameters. A lot of attention was paid to the information support of the mathematical modeling. Its composition was dictated by the mathematical model adopted, and initial data were selected so as to cover the optimal, favorable and extreme conditions of deep bed filtration up to the critical conditions, when the productivity of a filtration facility is in question.

The data on numerous model coefficients (parameters) necessary for theoretical studies can be obtained less often by their direct measurements and much more often after special mathematical processing of an array of experimental points based on the solution of a basic mathematical problem or on a priori approximation expressions. First of all, characteristic (representative) values of the sought coefficients can be found in a similar way (Alonso et al., 2000; Beuling et al., 2000; De Beer et al., 1997; Martynov & Poliakov, 2022a, b). If a set of experiments involves the input parameters varying within certain limits, the ranges of the model coefficients corresponding to them are established as a result of regression analysis, solving inverse problems (Lee & Clark, 1998; Kiranmai et al., 2005). However, the greatest value is represented by generalizations of the results of experimental studies in the form of the dependencies that establish a unique correspondence between the specified coefficients (criterion equations) and also the desired quantities (Dytnersky, 1986; Ko et al., 2001; Ojha & Graham, 1992). If it is possible to obtain only a part of the initial information necessary for modeling from our own experimental studies, the missing data have to be borrowed from literature sources. Quite often the only source of suitable input information for the test calculations are the publications that are devoted to the mathematical modeling of a set of processes on natural or engineering objects for similar conditions (Benefield & Molz, 1984; Chang & Rittmann, 1987; Droste & Kennedy, 1988; Lindstrem, 1992; Mayer et al., 2001). The results of experimental studies contained in numerous publications were widely used in the monograph along with our data in preparation for the quantitative analysis of the filtration of aqueous suspensions. These data were obtained at the pilot unit at Kyiv National University of Civil Engineering and Architecture during 2016-2018 years. The specified unit was developed, planned and experiments were conducted (A.M.Kravchuk, A.A.Kravchuk) with the participation of the author based on the mathematical model of the deep bed filtration with unregulated rate, which is presented and implemented in Chapter 5.

In general, the content and main objectives of the conducted experimental studies on the deep bed filtration were as follows:

- collection and systematization of the regular observation (operational) data for spatial and temporal changes of the most important filtration characteristics at the operating pilot unit;
- selection of individual components and the basic mathematical model;
- development of the information support for mathematical modeling with attraction of special techniques for processing empirical data and obtaining the key modeling coefficients;
- verification of adequacy of the adopted model to the natural conditions of the depth filtration, the correction of theoretical and experimental base.

Due to the multifactor nature of the studied processes and effects, the composition of experimental studies of the clarification of aqueous suspensions with the use of reagents was initially planned based on the set of scientific and practical tasks. And since the main task was the applied optimization of the separation of low concentration suspensions on rapid filters at an unregulated filtration rate, the above studies were primarily aimed at collecting and preparing in full the initial information necessary for modeling of joint clarification and filtration flow processes in the second main rate mode of filtration. Therefore, the authors considered it possible to limit themselves to the only two natural test materials, which were selected on the basis of the economic, technological considerations, the technical capabilities for the arrangement of the operating layer of a filter unit (quartz sand) and the preparation of a model suspension (spondyl clay). Considerable efforts were made to obtain the reliable information on absorption and filtration flow properties of these materials in pure and transformed states.

We have chosen the sand as a granular filler for the pilot unit, which is used in the Dnipro water supply system of Kjiv. The true and bulk densities of the sand in the unconsolidated state were 1.412 and 2.65 g/cm^3, so that the porosity of the clean bed was equal to 0.47. The equivalent grain diameter and shape factor were respectively $1.15 \cdot 10^{-3} m$ and 1.19. Fractionation of the filter media was performed on a standard set of sieves. The graph describing the grain-size composition of the experimental sand is shown in **Figure 8**. It follows that the mechanical composition is dominated by the grains with the diameter from $3.15 \cdot 10^{-4}$ up to $6.3 \cdot 10^{-4} m$ (59.03% by mass). The mass of the coarser fraction is also very significant, namely, the grains with diameters from $6.3 \cdot 10^{-4}$ up to $1.25 \cdot 10^{-3} m$ (33.13%). The total mass of even larger and smaller grains amounted to only 7.84%. The equivalent particle diameter of the spondyl clay was measured after it had been thoroughly grinding and came to $7.7 \bullet 10^{-6} m$. To effectively delay them, the coagulant $Al_2(SO_4)_3$ was used at the doses of 20, 30*mg/l*.

Figure 8. Curve of the sand grain-size composition

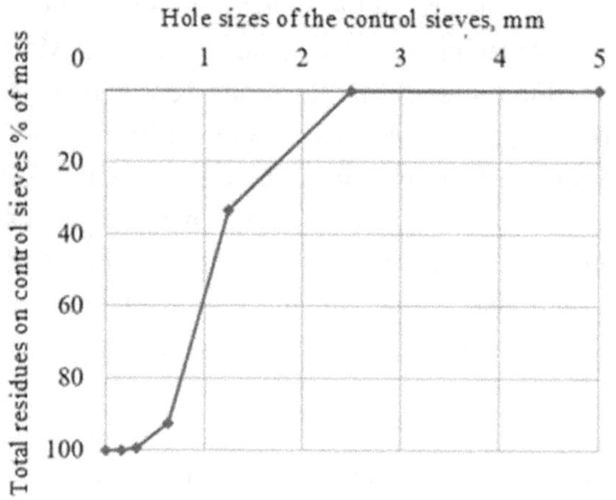

Note. Hole sizes of the control sieves, mm. Total residues on control sieves % of mass.

Also, the compliance with the linear filtration law (Darcy's law) was preliminarily checked on the clean bed, and then constantly controlled on the clogged bed. The validity of the above law was confirmed by the straight-line nature of the graphs of the dependence of the filtration flow rate on the hydraulic gradient and head distribution along the height of the operating layer throughout all experiments. In addition, the fact of significant detachment and downward movement of the deposit fragments with filtration flow was established and then repeatedly confirmed already in the first series of experiments with suspension being clarified. The noted phenomenon served as a basis for using mainly models of the detachment filtration at the simulation of the rapid filter operation. At the same time, Chapter 4 is entirely and reasonably dedicated to the non-detachment filtration.

Much attention in the laboratory studies of the suspension separation has been paid to the establishment of the model coefficients, which play a key role in modeling clarification and filtration flow processes under the conditions of continuously and unpredictably changing hydraulic load. In such cases, these coefficients actually turn out to be functions of the desired and therefore unknown characteristics – the variable filtration rate, the concentrations of deposit and deposited particles. Thus, simulation of the technological process is considerably complicated due to the close interconnection between the compartments of the original model. These functional coefficients could not be measured directly. Therefore, it was necessary, first, to obtain preliminary analytical solutions of the test problems in a general form, and second, to verify the desired coefficients in a special way to ensure sufficient proximity of experimental and theoretical results. The result of such activity was the following expressions for the above mentioned four variable coefficients. First of all, two mass exchange coefficients, namely, the rate coefficients of the suspended particles sticking (attachment) (α) and deposited particles detachment (β), were initially taken in the generalized power form corresponding, in particular, to (1.16)

$$\alpha(V)=\alpha_V V^l,\ \beta(V)=\beta_V V^q, \tag{1.58}$$

where α_V, β_V, l, q are the empirical constants under consideration. The legitimacy of the choice of this form follows from the results of statistical processing of the obtained experimental data. The representative values of the constant coefficients according to **Fig. 9, 10**, on which numerous experimental points are depicted and actually straight approximation lines generalizing them are plotted, were as follows:

$$\alpha_V = 6.65\, m^{-0.8} s^{-0.2},\ \beta_V = 0.025 m^{-0.9} s^{-0.1},\ l = 0.8,\ q = 0.9.$$

It should be emphasized that the theoretical developments in the monograph are finally presented exclusively in dimensionless form and all the test calculations were performed on their basis. Therefore, it is of a practical interest to estimate the characteristic values and intervals for the relative model coefficients. In the experiments with a fixed hydraulic load, its values discretely varied from 3.97 to 13.04*m/h*, which made it possible to cover practically the entire actual range of the rate values implemented on rapid filters. Taking into account the specified range for the filtration rate, and at the same time the proximity of the values of the bed thickness at the operated filters (they fluctuate near 1*m* in practice, and in addition, the height of the operating column in the pilot unit was also equal to 1*m*), the approximate intervals of the values of the dimensionless coefficients $\bar{\alpha}$, $\bar{\beta}$, corresponding to all the observational data, were:

$$\bar{\alpha} = \frac{\alpha L}{V} = \alpha_V L V^{l-1} \approx 3.98 - 5.047$$

$$\bar{\beta} = \frac{n_0 \beta L}{V} = n_0 \beta_V L V^{q-1} \approx 9.089 \cdot 10^{-3} - 1.024 \cdot 10^{-2}$$

Figure 9. Dependence of the rate coefficient of the suspended solids attachment on the filtration rate

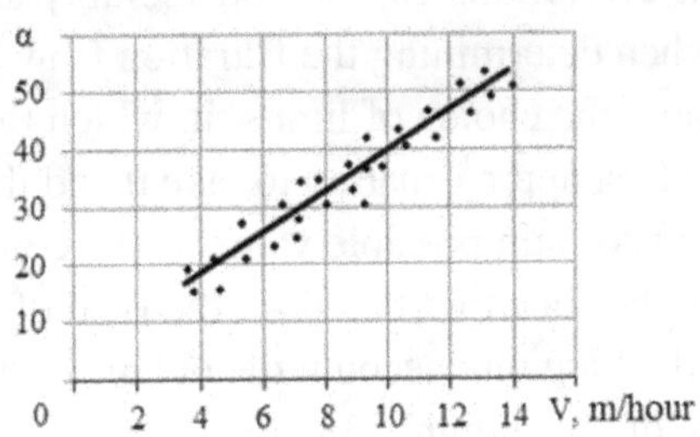

Figure 10. Dependence of the rate coefficient of the deposited particles detachment on the filtration rate

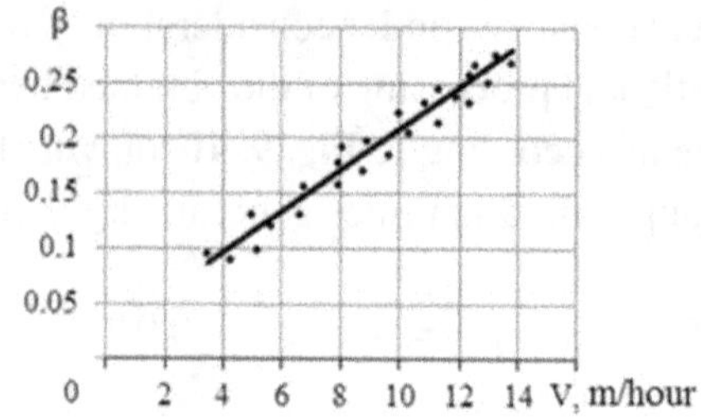

The flow capacity of a clogged bed and a deposit composition remain the least studied so far. Therefore, experimental studies have focused on the determination of the bulk factor γ and the saturated hydraulic conductivity k as the functions of the deposited particles (S) and deposit (S_S) concentrations. A set of empirical data completely characterizing the relationship between the quantity γ and the relative concentration S/n_0 under the considered conditions as well as the continuous approximation line describing them with minimum dispersion and up to the maximum possible values of the argument, is shown in **Figure 11**. The analytical expression for this graph has the following linear form

$$\gamma(S) = 57 - 67\frac{S_S}{n_0} \tag{1.59}$$

From (1.59) follows also the dependence of γ on S/n_0 and S

$$\gamma(S) = = 57 - \frac{3819S}{n_0 + 67S} \tag{1.60}$$

Calculations and mathematical derivations can be considerably simplified if, the function $\gamma(S)$ is replaced by its average value γ_a, when determining the filtration flow characteristics and, in particular, the variable flow rate. However, here the choice of limits, in which the quantity γ should be averaged, is problematic. It is justified in the first approximation to take 0 and the value of the concentration S at the inlet to the bed S_0 or even the maximum possible value as the concentration boundary values. The second boundary value of S should be specified based on the type of a kinetic equation. According to the obtained experimental data and taking into account the obtained relative value of the solids take-up capacity (S_{sm}/n_0), the average value of γ_a is roughly 35.

Figure 11. Dependence of the deposit composition factor on the relative deposit specific volume

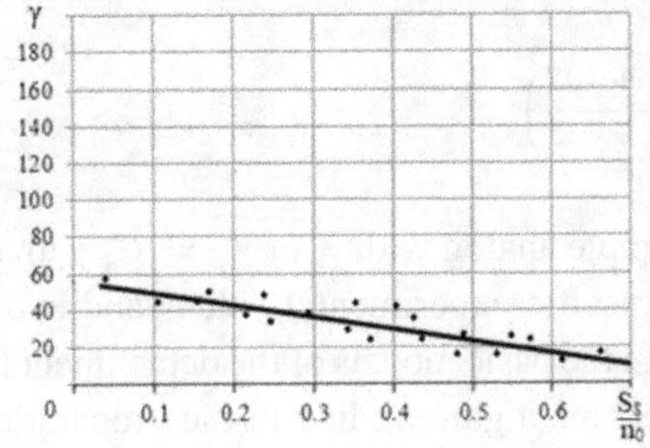

Figure 12. Dependence of the relative deposit specific volume on the relative hydraulic conductivity

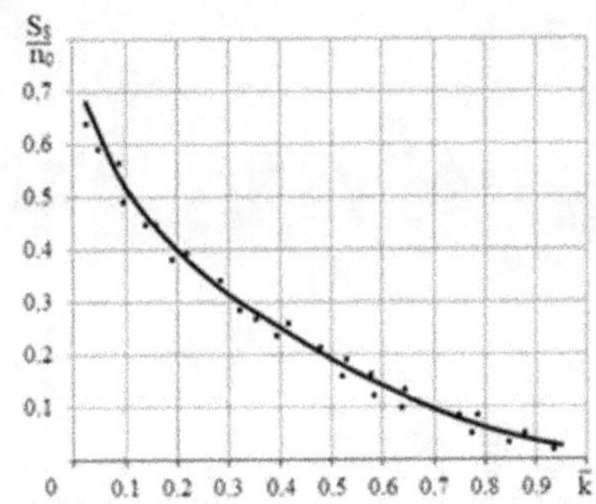

Finally, the pattern of the hydraulic conductivity (permeability) decrease of filtering material at its increasing clogging is of great importance, as well as $\gamma(S)$ for the modeling of the filtration flow process in the porous medium with a decreasing porosity. The corresponding set of the experimental points and the best mathematically approximating curve are shown in **Figure 12**. The mentioned curve allows to visualize an increase in the hydraulic resistance of the experimental sand column in the process of its progressive clogging characterized by the ratio S_S/n_0. The value of the hydraulic conductivity for the clean bed k_0, which is the maximum possible value for it and was used as a scale in the filtration flow compartment, was determined beforehand and equaled to 49*m/h*. The curve in Figure 12 corresponds to the following analytical representation

$$\frac{S_S}{n_0} = 1 - \sqrt[3]{\bar{k}_0} \tag{1.61}$$

Its equivalent form

$$\bar{k}(S_S) = \left(1 - \frac{S_S}{n_0}\right)^3 \tag{1.62}$$

is exactly as recommended in the monograph by D.M. Mintz (1964). Taking into account the initial expression for γ and (1.53), the dependence of $\bar{k}$ on S/n_0 or S is represented in the following form

$$\bar{k}(S) = \left(1 - \frac{57\, S/n_0}{1 + 67S/n_0}\right)^3 = \left(\frac{n_0 + 10S}{n_0 + 67S}\right)^3 \tag{1.63}$$

It is expression (1.60) or its simple analog with γ_a or $\bar{\gamma}_a = C_0\, \gamma_a$ that will be mainly used instead of $\gamma(S)$ hereafter. Just given above specific experimental values and expressions for the most important functional coefficients contained in the basic models of the detachment filtration of aqueous suspensions with the use of a reagent have become a guiding line in the preparation of a baseline information for the predictive and technological calculations with respect to the normal and extreme conditions of the clarification and filtration flow.

REFERENCES

Abboud, N. M., & Corapcioglu, M. Y. (1993). Modeling of compressible cake filtration. *J. Colloid. Int. Sci., 160*, 304–316.

Aberg, B. (1996). Void sizes in granular soils. *Journal of Geotechnical Engineering, 122*(3), 236–239. DOI: 10.1061/(ASCE)0733-9410(1996)122:3(236)

Abramson, A., & Brown, D. B. (2007). Influence of solution ionic strength on the collision efficiency distribution and predicted transport distance of a Sphingomonas sp. flowing through porous media. *Water Research, 41*(19), 4435–4445. DOI: 10.1016/j.watres.2007.06.005 PMID: 17644150

Adin, A., & Elimelech, M. (1989). Particle filtration for wastewater irrigation. *Journal of Irrigation and Drainage Engineering, 115*(3), 474–487. DOI: 10.1061/(ASCE)0733-9437(1989)115:3(474)

Adin, A., & Rebhun, M. (1977). A model to predict concentration and head loss profiles in granular bed filtration. *Journal - American Water Works Association, 69*(8), 444–453. DOI: 10.1002/j.1551-8833.1977.tb06787.x

Adin, A., & Rebhun, M. (1987). Deep-bed filtration: Accumulation-detachment model parameters. *Chemical Engineering Science, 42*(5), 1213–1219. DOI: 10.1016/0009-2509(87)80072-6

Aerov, M. E., Todes, O. M., & Narinsky, D. A. (1979). *Apparatuses with stationary granular layer*. Khimiya. Leningrad Branch. (in Russian)

Ahsan, S., Kaneco, S., Ohta, K., Mizuno, T., & Kani, K. (2001). Use of some natural and waste materials for waste water treatment. *Water Research, 35*(15), 3738–3742. DOI: 10.1016/S0043-1354(01)00047-1 PMID: 11561637

Akgiray, O., & Saatci, A. M. (1998). An algorithm for bank operation of declining rate filters. *Water Research, 32*(7), 2095–2105. DOI: 10.1016/S0043-1354(97)00459-4

Aksogan, S., Basturk, A., Yuksel, E., & Akgiray, O. (2003). On the use of crushed shells of apricot stones as the upper layer in duel media filters. *Water Science and Technology, 48*(11-12), 497–503. DOI: 10.2166/wst.2004.0899 PMID: 14753573

Alekseev, V. S., & Kommunar, G. M. (1974). Clogging of filters and prefilter zones of water intake wells. [in Russian]. *Water Resources*, (5), 170–177.

Alekseev, V.S., Kommunar, G.M., & Sherzhukov, B.S. (1989). *Mass transfer in water-saturated rocks. Results of Science and Technology. V.11*. Hydrogeology, Engineering Geology. (in Russian)

Alonso, C., Zhu, X., Suidan, M. T., Kim, B. R., & Kim, B. J. (2000). Parameter estimation in biofilter systems. *Environmental Science & Technology, 34*(11), 2318–2323. DOI: 10.1021/es990329o

Altshul, A. H., Usenko, V. S., & Chaban, M. O. (1977). *Regulation of groundwater reserves*. Kolos. (in Russian)

Amburgey, J. E., & Amirtharajah, A. (2005). Strategic filter backwashing techniques and resulting particle passage. *Journal of Environmental Engineering, 131*(4), 535–547. DOI: 10.1061/(ASCE)0733-9372(2005)131:4(535)

American Water Works Assosiation. (1999). *Water quality and treatment, a handbook of community water supply* (5th ed.). Mc Graw–Hill.

Anderson, A., Laurent, P., Kihn, A., Prévost, M., & Servais, P. (2001). Impact of temperature on nitrification in biological activated carbon (BAC) filters used for drinking water treatment. *Water Research, 35*(12), 2923–2934. DOI: 10.1016/S0043-1354(00)00579-0 PMID: 11471692

Anderson, E., Schilling, K. E., Just, C. L., & Seo, B. C. (2024). Quantifying the impact of a constructed wetland on downstream nitrate concentrations and loads in the US Midwest. *Frontiers in Environmental Science, 12*, 1416018. DOI: 10.3389/fenvs.2024.1416018

Aravin, V. I., & Kochina, P. Ya. (Eds.). (1969). *Development of research on the theory of filtration flow in the USSR (1917–1967)*. Nauka. (in Russian)

Aravin, V. I., & Nosova, O. N. (1969). *Natural studies of water flow (theoretical bases)*. Energia. (in Russian)

Aronino, R., Dludy, C., Arkhangelsky, E., Shandalov, S., Oron, G., Brenner, A., & Gitis, V. (2009). Removal of viruses from surface water and secondary effluents by sand filtration. *Water Research, 43*(1), 87–96. DOI: 10.1016/j.watres.2008.10.036 PMID: 19013631

Arora, S., Dhaliwal, S. S., & Kukreja, V. K. (2006). Simulation of washing of packed bed of porous particles by orthogonal collocation on finite elements. *Computers & Chemical Engineering, 30*(6-7), 1054–1060. DOI: 10.1016/j.compchemeng.2006.02.002

Arun, N., Abdul, A., Razack, B., & Sivasubramanian, V. (2013). Recent progress in hydrodynamics of inverse fluidized bed reactors: A review. *Chemical Engineering Communications, 200*(9), 1260–1277. DOI: 10.1080/00986445.2012.744747

Ayukaev, R. I., & Meltser, V. Z. (1985). *Production and application of filtering materials for water purification. Reference manual*. Stroyizdat. (in Russian)

Ayukaev, R. R. (2003). Slow filters in household and drinking water supply of small water consumers in Karelia. [in Russian]. *Water and Ecology*, (3), 16–28.

Babenkov, E. D. (1977). *Water purification by coagulants*. Nauka. (in Russian)

Badriyha, B. N., Ravindran, V., Den, W., & Pirbazari, M. (2003). Bioadsorber efficiency, design and performance forecasting for alachlor removal. *Water Research, 37*(17), 4051–4072. DOI: 10.1016/S0043-1354(03)00266-5 PMID: 12946887

Bai, R., & Mackie, R. I. (1995). Modeling the transition between deposition modes in deep bed filtration. *Water Research, 29*(1), 2601–2604. DOI: 10.1016/0043-1354(95)00078-Y

Bai, R., & Tien, C. (1997). Particle detachment in deep bed filtration. *Journal of Colloid and Interface Science, 186*(2), 307–317. DOI: 10.1006/jcis.1996.4663 PMID: 9056359

Bai, R., & Tien, C. (2000). Effect of deposition in deep-bed filtration determination and search of rate parameters. *Journal of Colloid and Interface Science*, *231*(2), 299–311. DOI: 10.1006/jcis.2000.7130 PMID: 11049680

Barenblatt, G. I. (1987). Nonlinear flow filtration – past, present and future. In *Problems of flow filtration theory and mechanics of oil recovery enhancement processes* (pp. 15–27). Nauka. (in Russian)

Barenblatt, G. I., Yentov, V. M., & Ryzhik, V. M. (1972). *Theory of non-stationary flow and gas*. Nedra. (in Russian)

Barry, D. A., & Sposito, P. (1989). Analytical solution of a convection-dispersion model with time-dependent transport coefficients. *Water Resources Research*, *25*(12), 2407–2416. DOI: 10.1029/WR025i012p02407

Bass, G.M., Grabovskiy, P.P., & Larkina, G.M. (1978). Optimal duration of the filter run of rapid filters operating with falling rate. *Science and Technology in Municipal Economy. Collected papers*, 38, 37–39. (in Russian)

Bear, J. (1988). *Dynamics of fluids in porous media*. Mineola.

Belyaev, N. N., & Kozachina, V. A. (2014). *Mathematical modeling of mass transfer in horizontal settling tanks*. Accent PP. (in Russian)

Ben Aim, R., Vigneswaran, S., Prasanthi, H., & Jegatheesan, V. (1997). Influence of particle size and size distribution in granular bed filtration and dynamic microfiltration. *Water Science and Technology*, *36*(4), 207–215. DOI: 10.2166/wst.1997.0121

Benefield, L., & Molz, F. (1985). Mathematical simulation of a biofilm process. *Biotechnology and Bioengineering*, *27*(7), 921–931. DOI: 10.1002/bit.260270702 PMID: 18553761

Benkli, Y. E., Can, M. F., Turan, M., & Celik, M. S. (2005). Modification of organo-zeolite surface for the removal of reactive azo dyes in fixed-bed reactors. *Water Research*, *39*(2-3), 487–493. DOI: 10.1016/j.watres.2004.10.008 PMID: 15644257

Bergendahi, J. A., & Grasso, D. (2003). Mechanistic basis for particle detachment from granular media. *Environmental Science & Technology*, *37*(10), 2317–2322. DOI: 10.1021/es0209316 PMID: 12785543

Berman, A. S. (1953). Laminar flow in channels with porous walls. *Journal of Applied Physics*, *24*(9), 1232–1235. DOI: 10.1063/1.1721476

Bespamyatnov, G. P., & Krotov, Y. A. (1985). *Maximum permissible concentrations of chemical substances in the environment*. Khimiya. (in Russian)

Beuling, E. E., van den Heuvel, J. C., & Ottengraf, S. P. P. (2000). Diffusion coefficients of metabolites in active biofilms. *Biotechnology and Bioengineering*, *67*(1), 53–60. DOI: 10.1002/(SICI)1097-0290(20000105)67:1<53::AID-BIT6>3.0.CO;2-H PMID: 10581435

Biserov, R. A. (1967). *Research of mud capacity of sand beds at clarification of preliminary coagulated water by the method of superrapid filtration*. Ph.D. dissertation abstract. LECI Press. (in Russian)

Bochever, F. M. (1977). About estimation of filtration intensity from basins in systems of artificial groundwater recharge. *Proceedings of VNII VODGEO*, 63, 3–10. (in Russian)

Boller, M. A., & Kavanauch, M. C. (1995). Particle characteristics and head loss increase in granular media filtration. *Water Research, 29*(4), 1139–1149. DOI: 10.1016/0043-1354(94)00256-7

Bolshakov, V. A., Konstantinov, Y. M., Popov, V. N., & Dadenkov, V. Y. (1979). *Handbook on Hydraulics*. Vishcha Shkola. (in Russian)

Bradford, S. A., & Bettahar, M. (2005). Straining, attachment and detachment of Cryptosporidium Oocysts in saturated porous media. *Journal of Environmental Quality, 34*(2), 469–478. DOI: 10.2134/jeq2005.0469 PMID: 15758099

Bradford, S. A., Yates, S. R., Bettahar, M., & Simunek, J. (2005). Physical factors affecting the transport and fate of colloids in saturated porous media. *Water Resources Research, 38*(12), 1327–1339.

Bresler, E., McNeil, B. L., & Carter, D. L. (1987). *Solonchaks and salt marshes. Principles. Dynamics. Modeling*. Gidrometeoizdat. (in Russian)

Brook, O. L. (1989). Theory and calculation of industrial filtration of suspensions with formation of highly compressible precipitates. [in Russian]. *Theoretical Foundations of Chemical Engineering, 23*(4), 495–498.

Brook, O. L. (1992). To calculation of parameters of industrial filtration processes. [in Russian]. *Theoretical Foundations of Chemical Engineering, 26*(6), 867–871.

Bugay, N.G., Krivonog, A.I., & Krivonog, V.V. (2004). Reagentless water purification by filters from fiber-porous polyethylene. *Applied Hydromechanics*, 6(78), 2, 27–43. (in Russian)

Bugay, N. G., Krivonog, A. I., & Krivonog, V. V. (2007). Contact reagent treatment of filtration water from fiber-porous material. [in Russian]. *Applied Hydromechanics, 9*(1), 8–22.

Burchak, T. V. (1978). *Infiltration basins*. Budivelnik. (in Russian)

Caliskaner, O., & Tchobanoglous, G. (2005). Modeling depth filtration of activated sludge effluent using a compressible medium filter water. *Environmental Research, 77*(7), 3080–3091. PMID: 16381157

Calvin, R. M. (1992). Ripening of silica sand used for filtration. *Water Research, 26*(5), 683–688. DOI: 10.1016/0043-1354(92)90245-Y

Cameron, D. A., & Klute, A. (1977). Convective-dispersive solute transport with a combined equilibrium and kinetic adsorption model. *Water Resources Research, 13*(1), 183–188. DOI: 10.1029/WR013i001p00183

Caputo, A. C., Pelagagge, P. M., & Scacchia, F. (2001). Software tool for environmental plant modeling. *Environmental Progress, 20*(4), 229–239. DOI: 10.1002/ep.670200411

Carrier, W. D. (2000). III. Compressibility of a compacted sand. *Journal of Geotechnical and Geoenvironmental Engineering, 126*(3), 273–275. DOI: 10.1061/(ASCE)1090-0241(2000)126:3(273)

Chang, H. T., & Rittmann, B. E. (1987). Verification of the model of biofilm on activated carbon. *Environmental Science & Technology, 21*(3), 280–288. DOI: 10.1021/es00157a009 PMID: 22185107

Chaundry, F.H. (1987), Theory of declining rate filtration. I. Continuous operation. *J. Environ. Eng. Div., ASCE*, 113(4), 834–851.

Chen, J., Truesdall, S., Zu, F., Zhan, G., Belvin, C., Koopmah, B., Farrah, S., & Shan, D. (1998). Long-term evaluation of aluminium hydroxide-coated sand for removal of bacteria from wastewater. *Water Research, 32*(7), 2171–2179. DOI: 10.1016/S0043-1354(97)00427-2

Cheng, N.-S. (2006). Application of Ergun equation to computation of critical shear velocity subject to seepage. *Journal of Irrigation and Drainage Engineering, 129*(4), 278–283. DOI: 10.1061/(ASCE)0733-9437(2003)129:4(278)

Chern, J.-M., & Wu, C.-Y. (2001). Desorption of dye from activated carbon beds: Effects of temperature, pH and alcohol. *Water Research, 35*(17), 4159–4165. DOI: 10.1016/S0043-1354(01)00127-0 PMID: 11791845

Chiang, H.-W., & Tien, C. (1985). Dynamics of deep-bed filtration. I. Analysis of two limiting situations. II. Experiment. *AIChE Journal. American Institute of Chemical Engineers, 31*(8), 1349–1371. DOI: 10.1002/aic.690310814

Chou, H., Wu, L., Zeng, L., & Chang, A. (2012). Evaluation of solute diffusion tortuosity factor models for variously saturated soils. Water Resour. Res., 48, 1–11. W10539, .DOI: 10.1029/2011WR011653

Chrysikopoulos, C. V., & Katzourakis, V. E. (2015). Colloid particle size-dependent dispersivity. *Water Resources Research, 51*(6), 4668–4683. DOI: 10.1002/2014WR016094

Chuang, C.-J., & Li, K.-Y. (1997). Effect of coagluant dosage and grain size on the performance of direct filtration. *Separation and Purification Technology, 12*(3), 229–241. DOI: 10.1016/S1383-5866(97)00048-8

Cleasby, J. L. (1993). Status of declining rate filtration design. *Water Science and Technology, 27*(10), 151–164. DOI: 10.2166/wst.1993.0224

Colton, J. F., Hillis, P., & Fitpatrick, C. S. B. (1996). Filter backwash and start-up strategies for enhanced particulate removal. *Water Research, 30*(10), 2502–2507. DOI: 10.1016/0043-1354(96)00085-1

Corapcioglu, M. Y., & Haridas, U. (1985). Microbial transport in soils and ground water: A numerical model. *Advances in Water Resources, 8*(4), 188–200. DOI: 10.1016/0309-1708(85)90063-6

Corapciooglu, M. Y., & Jiang, S. (1993). Colloid-facilitated groundwater contaminant transport. *Water Resources Research, 29*(7), 2215–2226. DOI: 10.1029/93WR00404

Corapciooglu, M. Y., & Wang, S. (1999). Dual porosity groundwater contaminant transport in the presence of colloids. *Water Resources Research, 35*(11), 3261–3273. DOI: 10.1029/1999WR900122

Crank, J. (1975). *The mathematics of diffusion* (2nd ed.). Clarendon Press.

Crittenden, J. C., Hutzler, N. J., Geyer, D. G., Oravitz, J. L., & Friedman, G. (1986). Transport of organic compounds with saturated groundwater flow: Model development and parameter sensitivity. *Water Resources Research, 22*(3), 271–284. DOI: 10.1029/WR022i003p00271

Crohn, D. M., & Valenzuela-Solano, C. (2003). Modeling temperature effects on decomposition. *Journal of Environmental Engineering, 12*(12), 1149–1156. DOI: 10.1061/(ASCE)0733-9372(2003)129:12(1149)

Dagan, G. (1984). Solute transport in heterogeneous porous formations. *Journal of Fluid Mechanics, 145*(-1), 151–177. DOI: 10.1017/S0022112084002858

Daley, J., & Harleman, D. (1971). Fluid Mechanics. [in Russian]. *Energy*.

Dalias, P., Anderson, J. M., Bottner, P., & Couteaux, M.-M. (2001). Long-term effects of temperature on carbon mineralization processes. *Soil Biology & Biochemistry, 33*(7-8), 1049–1057. DOI: 10.1016/S0038-0717(01)00009-8

Dantas, A. D. B., & Bernardo, L. D. (2006). Performance of two-stage filtration system for treating high turbidity water. *J. Wat. Suppl.: Res. and Technol. Aqua (Neu-Isenburg, Germany), 55*(7-8), 499–515.

Darby, J. L., & Lawler, D. F. (1990). Ripening in depth filtration: Effect of particle size on removal and head loss. *Environmental Science & Technology, 24*(7), 1069–1079. DOI: 10.1021/es00077a018

Darby, J. L., Lawler, D. F., & Wilshusen, T. P. (1991). Depth filtration of wastewater: Particle size and ripening. *Research Journal of the Water Pollution Control Federation, 63*(3), 228–238.

Davidson, I., & Harrison, D. (Eds.). (1974). *Fluidization*. Khimiya. (in Russian)

De Beer, D., Stoodley, P., & Lewandowski, Z. (1997). Measurement of local diffusion coefficients in biofilms by microinjection and confocal microscopy. *Biotechnology and Bioengineering, 53*(2), 151–158. DOI: 10.1002/(SICI)1097-0290(19970120)53:2<151::AID-BIT4>3.0.CO;2-N PMID: 18633959

De Jong, J. T., & Christoph, G. C. (2009). Influence of particle properties and initial specimen state on one-dimensional compression and hydraulic conductivity. *Journal of Geotechnical and Geoenvironmental Engineering, 135*(3), 449–454. DOI: 10.1061/(ASCE)1090-0241(2009)135:3(449)

Degremont (Ed,). (2007). *Water Treatment Handbook* (7th ed.). 2 Volume Set. Editions Tec & Doc (Lavoisier).

Demchin, I. I., Orlov, V. O., & Venitsianov, E. V. (1994). Calculation of technological parameters of filters with floating filter beds. [in Russian]. *Journal of Water Chemistry and Technology, 16*(2), 115–118.

Dharmappa, H. B., Prasanthi, H., Krishna, M. U., & Xiao, Y. (1997). Estimation of empirical coefficients of a granular filtration model. *Water Research, 31*(5), 1083–1091. DOI: 10.1016/S0043-1354(96)00329-6

Dharmappa, H. B., Verink, J., Fujiwara, O., & Vigneswaran, S. (1992). Optimization of granular bed filtration treating polydispersed suspension. *Water Research, 26*(10), 1307–1318. DOI: 10.1016/0043-1354(92)90125-N

Dickinson, C. (1992). *Filters and filtration handbook*. Elsevier Advanced Technology.

Diliunas, I. P., Kaminskas, M. M., & Yagminas, E. A. (1984). Evaluation of hydrodynamic parameters for artificial groundwater recharge from open flow structures. In *Methods of hydrogeological data analysis and processing for groundwater resources forecasting* (pp. 190–202). Academy of Sciences of the ESSR. (in Russian)

Dmitrenko, G. N., & Gvozdyak, P. I. (2002). Biotechnology of treatment of highly concentrated wastewater from organic solvents. [in Russian]. *Journal of Water Chemistry and Technology, 24*(2), 185–190.

Droste, R. L., & Kennedy, K. J. (1988). Dynamic anaerobic fixed-film reactor. *Journal of Environmental Engineering, 114*(3), 606–620. DOI: 10.1061/(ASCE)0733-9372(1988)114:3(606)

du Plessis, J. P., & Woudberg, S. (2008). Pore-scale derivation of the Ergun equation to enhance its adaptability and generalization. *Chemical Engineering Science, 63*(9), 2576–2586. DOI: 10.1016/j.ces.2008.02.017

Duck, J., Zvetanov, E., & Neebe, T. (2000). Porosity prediction for fine-grained filter cake. *Chemical Engineering & Technology, 22*(1), 18–22. DOI: 10.1002/(SICI)1521-4125(200001)23:1<18::AID-CEAT18>3.0.CO;2-P

Dutournie, P., Limousy, L., Deon, S., & Bourseau, P. (2011). Unsteady transport of divalent salt through a mineral membrane of ultrafiltration: Numeracal estimation of physical parameters. *Desalination, 265*(1-3), 184–189. DOI: 10.1016/j.desal.2010.07.050

Dytnersky, Y. I. (1986). *Baromembrane processes. Theory and calculation.* Khimiya. (in Russian)

Dzyubo, V. V., & Alferova, L. I. (2007). Natural water filtration in the mode of non-uniform rates. [in Russian]. *Bulletin of TSASU*, (2), 180–190.

Emelko, M. B. (2003). Removal of viable and inactivated Cryptosporidium by dual and tri-media filtration. *Water Research, 37*(12), 2998–3008. DOI: 10.1016/S0043-1354(03)00113-1 PMID: 12767303

Epoyan, S., Karagiaur, A., Volkov, V., & Yarkin, V. (2016). Improvement of filtration facilities operation efficiency at application of drainage systems from porous polymeric materials. *MOTROL. Commision of Motorization and Energetic on Agriculture, 18*(6), 102–109.

Ergun, S. (1952). Fluid flow through packed columns. *Chemical Engineering Progress, 48*(2), 89–94.

Evtushenko, A. I., & Kotelkin, V. D. (1991). Modeling of transfer processes in a fixed granular layer. [in Russian]. *Theoretical Foundations of Chemical Engineering, 25*(4), 511–523.

FAO. (2006). *Guidelines for soil description* (4th ed.).

FAO-UNESCO. (1974). *Soil map of the world* (Vol. 1). Legend.

Faroog, S., & Al-Yousel, A. K. (1993). Slow sand filtration of secondary effluent. *Journal of Environmental Engineering, 119*(4), 615–630. DOI: 10.1061/(ASCE)0733-9372(1993)119:4(615)

Fayollea, Y., Cockxb, A., Gillota, S., Roustanb, M., & Héduit, A. (2007). Oxygen transfer prediction in aeration tanks using CFD. *Chemical Engineering Science, 62*(24), 7163–7171. DOI: 10.1016/j.ces.2007.08.082

Fedorov, A.A., Larionov, A.I., & Ushakov, V.A. (1984). *Porous filtering materials.* SRITEhim. (in Russian)

Fiori, A., & Dagan, G. (1999). Concentration fluctuations in transport by groundwater. Comparison between theory and field experiments. *Water Resources Research, 35*(1), 105–112. DOI: 10.1029/98WR01862

Fogden, A., Landman, K. A., & White, L. R. (1988). Contaminant transport in fractured porous media steady state solutions by a boundary integral method. *Water Resources Research, 24*(8), 1384–1396. DOI: 10.1029/WR024i008p01384

Fominykh, A. M. (1980). Mathematical model of water purification process by filtration and its practical application. [in Russian]. *Water Resources*, (3), 194–199.

Foppen, J. W., Liem, Y., & Schijven, I. (2008). Effect of humic acid on the attachment of Escherichia coli in columns of goethite-water sand. *Water Research, 42*(1-2), 211–219. DOI: 10.1016/j.watres.2007.06.064 PMID: 17825871

Foppen, J. W., & Schijven, J. F. (2005). Transport of E. coli in columns of geochemically heterogeneous sediment. *Water Research, 39*(13), 3082–3088. DOI: 10.1016/j.watres.2005.05.023 PMID: 15996706

Foppen, J. W., & Schijven, J. F. (2006). Evaluation of data from the literature on the transport and survival of Escherichia coli and thermotolerant coliforms in aquifers under saturated conditions. [Review]. *Water Research, 40*(4), 401–426. DOI: 10.1016/j.watres.2005.11.018 PMID: 16434075

Frank, J., Ruhl, A. S., & Jekel, M. (2015). Impacts of backwashing on granular activated carbon filters for advanced wastewater treatment. *Water Research, 87*, 166–174. DOI: 10.1016/j.watres.2015.09.020 PMID: 26405842

Fried, J. J., & Combarnous, M. A. (1971). Dispersion in porous media. *Adv. Hydrosci, 7*, 169–282. DOI: 10.1016/B978-0-12-021807-3.50008-4

Gal, E., Tardos, G., & Pfeffer, R. (1971). A study of inertial effects in granular bed filtration. *AIChE Journal. American Institute of Chemical Engineers, 31*(7), 1093–1104. DOI: 10.1002/aic.690310707

Gao, M., Lv, M., Gong, X., Song, W., Qi, Y., & Zhen, L. (2017). Effect of temperature and aging on the adsorption and desorption of menadione in soil. *Environmental Progress & Sustainable Energy, 36*(4), 997–1004. DOI: 10.1002/ep.12540

Gao, P., Xue, G., Song, X., & Lin, Z. (2012). Depth filtration using novel fiber-ball filter media for the treatment of high-turbicity surface water. *Separation and Purification Technology, 95*, 32–38. DOI: 10.1016/j.seppur.2012.04.010

Gao, X., da Costa, J. C. D., & Bhatia, S. K. (2014). Understanding the diffusional tortuosity of porous materials: An effective medium theory perspective. *Chemical Engineering Science, 110*, 55–71. DOI: 10.1016/j.ces.2013.09.050

Girole, N. N., Zhurba, M. G., Semchuk, G. M., & Yakimchuk, B. N. (1998). Secondary treatment of waste water on granular filters (special ed.) SPOOO, "Levoberezhnaya". (in Russian)

Glatstein, D. A., & Francisca, F. M. (2014). Hydraulic conductivity of compacted soils controlled by microbial activity. *Environmental Technology, 35*(15), 1886–1892. DOI: 10.1080/09593330.2014.885583 PMID: 24956782

Goncharuk, V. V., Shevchuk, E. A., & Galinker, E. V. (2001). Theoretical studies of regularities of the process of filtration flow of low-concentration suspensions through granular beds. [in Russian]. *Journal of Water Chemistry and Technology*, (6), 159–161.

Gordin, I.V., & Markov, P.P. (1991). *Closed systems of agrarian-industrial water use.* VO "Agropromizdat". (in Russian)

Grabovsky, P. A. (2016). Water filtration through a granular layer with decreasing rate. [in Russian]. *Rep. NAS of Ukraine*, (8), 40–45.

Grabovsky, P. A., Larkina, G. M., & Progulny, V. I. (2012). Washing of water treatment filters. [in Russian]. *Optimum (Paris)*.

Grabovsky, P.O., & Gurinchyk, N.A. (2005). Numerical realization of the mathematical model of filtration. *Problems of Water Supply,* Sewage *and Hydraulics. Collected papers*, 6, 4–13. (in Russian)

Grabovsky, P.O., & Gurinchyk, N.A. (2007). Numerical modeling of filtration structures operation. *Bulletin of NUWEE*, 4(40), part 2, 429–434. (in Russian)

Grabovsky, P. O., & Gurinchyk, N. A. (2008). Water filtration through a granular layer with decreasing rate in time. *Water and Ecology.* [in Russian]. *Problems and Solutions*, *37*(4), 3–11.

Grant, S., & Salehzadeh, A. (1996). Calculation of temperature effects on wetting coefficients of porous solids and their capillary pressure functions. *Water Resources Research*, *32*(2), 261–270. DOI: 10.1029/95WR02915

Gray, W. G., & Miller, C. T. (2004). Examination of Darcy's law for flow in porous media with variable porosity. *Environmental Science & Technology*, *38*(22), 5895–5901. DOI: 10.1021/es049728w PMID: 15573587

Gremyachkin, V. M., & Dementiev, V. V. (1994). Inhomogeneities of flow rate distribution in a model reactor with a fixed granular bed. [in Russian]. *Theoretical Foundations of Chemical Engineering*, *28*(2), 99–103.

Guo, S., Zhang, F., Song, X., & Wang, B. (2015). Deposited sediment settlement and consolidation mechanisms. *Water Science and Engineering*, *8*(4), 335–344. DOI: 10.1016/j.wse.2015.10.002

Gupta, V., Johnson, W. P., Shafieian, P., Ryu, H., Alum, A., Abbaszadegan, M., Hubbs, S. A., & Rauch-Williams, T. (2009). Riverbank filtration: Comparison of pilot scale transport with theory. *Environmental Science & Technology*, *43*(3), 669–676. DOI: 10.1021/es8016396 PMID: 19245000

Gurinchik, N. A. (2010). Determination of the duration of operation of filters for different modes of filtration. [in Russian]. *Bulletin of ONACA*, *42*, 56–62.

Gursch, J., Giner, R., & Krammer, G. (2016). Influence of wall effects on filter upscaling during solid-liquid filter tests. *Chemical Engineering & Technology*, *39*(3), 417–424. DOI: 10.1002/ceat.201500312

Gvozdyak, P. I. (2003). On the principle of bioprocessing (Biotechnology of environmental protection). [in Russian]. *Bulletin of NAS of Ukraine*, (3), 29–36.

Haarstrick, A., Mora-Naranjo, N., Meima, J., & Hempel, D. C. (2004). Modeling anaerobic degradation in municipal landfills. *Environmental Engineering Science*, *21*(4), 471–484. DOI: 10.1089/1092875041358494

Hall, D., & Fitzpatrick, C. S. B. (1998). A mathematical filter backwash model. *Water Science and Technology*, *37*(12), 371–379. DOI: 10.2166/wst.1998.0563

Han, C., Yang, J., Dong, S., Ma, L., Dai, Q., & Guo, J. (2024). *Zeolite preparation from industrial solid waste: Current status, applications, and prospects*. Sep. and Purif. Tech.

Happel, J. (1958). Viscous flow in multiparticle systems: Slow motion of fluids relative to beds of spherical particles. *AIChE Journal. American Institute of Chemical Engineers*, *4*(2), 197–201. DOI: 10.1002/aic.690040214

Hassan, A. E., Cushman, J. H., & Delleur, J. W. (1998). Significance of porosity variability to transport in heterogeneous porous media. *Water Resources Research*, *34*(9), 2249–2259. DOI: 10.1029/98WR01470

Hazen, H. (1911). Dams on sand formation [discussion]. *Trans. ASCE*, *73*, 199–203.

Hekmat, D., Stephan, M., Bauer, R., Feuchtinger, A., & Vortmeyer, D. (2006). Modelling of multispecies biofilm population dynamics in a trickle-bed bioreactor used for waste gas treatment. *Proc. Biochem.*, 41, 1409–1416. Doi:, 2006.02.003.DOI: 10.1016/j.procbio

Henderson, N., Brettas, J. C., & Wagner, F. S. (2010). A three-parameter Kozeny-Carman generalized equation for fractal porous media. *Chemical Engineering Science*, *65*(15), 4432–4442. DOI: 10.1016/j.ces.2010.04.006

Henze, M., Armoes, P., La-Coeur-Jansen, I., & Arvan, E. (2006). *Treatment of sewage waters.* Mir. (in Russian)

Herzig, J. P., Leclerc, D. M., & Le Goff, P. (1979). Flow of suspension through porous media. Application to deep filtration. *Industrial & Engineering Chemistry*, *62*(5), 8–35. DOI: 10.1021/ie50725a003

Hønning, J., Broholm, M. M., & Bjerg, P. L. (2007). Role of diffusion in chemical oxidation of PCE in a dual permeability system. *Environmental Science & Technology*, *41*(24), 8426–8432. DOI: 10.1021/es0708417 PMID: 18200874

Hopper-Jones, C., Oldham, G., & Drewes, J. E. (2010). Attenuation of total organic carbon and unregulated trace organic chemicals in US riverbank filtration systems. *Water Research*, *44*(15), 4643–4659. DOI: 10.1016/j.watres.2010.06.022 PMID: 20599244

Horan, N. I., & Lowe, M. (2007). Full-scale trials of recycled glass as tertiary filter medium for wastewater treatment. *Water Research*, *41*(1), 253–259. DOI: 10.1016/j.watres.2006.08.028 PMID: 17070893

Hu, X., Do, D. D., & Rao, G. N. (1994). Experimental cjncentration dependence of surface diffusivity of hydrocarbons in activated carbon. *Chemical Engineering Science*, *49*(13), 2145–2152. DOI: 10.1016/S0009-2509(94)85059-3

Huang, J. C., Rothstein, D., & Madey, R. (1984). Analytical solution for a first-order reaction considered in a packed bed with diffusion. *AIChE Journal. American Institute of Chemical Engineers*, *30*(4), 660–662. DOI: 10.1002/aic.690300423

Huang, Q.-Z., Hsu, S.-Y., & Syu, C.-H. (2024). Empirical formulas for dynamic effective suction head in the modified Green-Ampt model. *Journal of Hydrology (Amsterdam)*, *642*(131820), 131820. Advance online publication. DOI: 10.1016/j.jhydrol.2024.131820

Huisman, L. (1979). *Rapid filtration. Part 1.* Delft University of Technology Press.

Hutchison, W. R. (1976). High-rate direct filtration. *Journal - American Water Works Association*, *68*(6), 292–298. DOI: 10.1002/j.1551-8833.1976.tb02412.x

Hutt, B. E., Fletcher, T. D., & Deletic, A. (2007). Treatment performance of gravel filter media: Implications for design and application of storm water infiltration system. *Water Research*, *41*(12), 2513–2524. DOI: 10.1016/j.watres.2007.03.014 PMID: 17475303

Ibaraki, M., & Sudicky, E. A. (1995). Colloid-facilitated contaminant transport in discretely fractured porous media. 1. Numerical formulation and sensitivity analysis. *Water Resources Research*, *31*(12), 2945–2960. DOI: 10.1029/95WR02180

Indraratna, B. F., Basack, S., & Rujikiatkamjorn, C. (2013). Numerical solution of stone column–improved soft soil considering arching, clogging, and smear effects. *Journal of Geotechnical and Geoenvironmental Engineering*, *139*(3), 377–394. DOI: 10.1061/(ASCE)GT.1943-5606.0000789

Iritani, E., Katagiri, N., Sugiyama, Y., & Yagishita, K. (2007). Analysis of flux decline behaviors in filtration of very dilute suspensions. *AIChE Journal. American Institute of Chemical Engineers*, *53*(9), 2275–2283. DOI: 10.1002/aic.11271

Isaji, C. R. (2003). Integrated water quality management for drinking water of good quality. *Water Science and Technology*, *47*(9), 15–23. DOI: 10.2166/wst.2003.0482 PMID: 12830935

Ishiyama, E. M., Coletti, F., Macchietto, S., Paterson, W. R., & Wilson, D. I. (2010). Impact of deposit ageing on thermal fouling: Lumped parameter model. *AIChE Journal. American Institute of Chemical Engineers*, *56*(2), 531–545. DOI: 10.1002/aic.11978

Ishiyama, E. M., Paterson, W. R., & Wilson, D. I. (2011). Exploration of alternative models for the ageing of fouling deposits. *AIChE Journal. American Institute of Chemical Engineers*, *57*(11), 3199–3209. DOI: 10.1002/aic.12514

Ison, C. R., & Ives, K. J. (1969). Removal mechanism in deep bed filtration. *Chemical Engineering Science*, *24*(4), 717–729. DOI: 10.1016/0009-2509(69)80064-3

Ivanchenko, O., Sindhwani, N., & Linninger, A. A. (2012). Exact solution of the diffusion-convection equation in cylindrical geometry. *AIChE Journal. American Institute of Chemical Engineers*, *58*(4), 1299–1302. DOI: 10.1002/aic.12663

Ives, K. J. (1970). Rapid filtration. *Water Research*, *4*(3), 201–223. DOI: 10.1016/0043-1354(70)90068-0

Ives, K.J. (1979). The basis for the application of multiple layer filters to water treatment. *Wasser und Abwasser – Forchung*, 12 (3), 106–110.

Ives, K. J. (1981). Optimization model of rapid filtration at tertiary treatment. In A. James (Ed.), *Mathematical models of water pollution control.* (pp.160 – 205). Mir. (in Russian)

Iwasaki, I. (1937). Some notes on sand filtration. *Journal - American Water Works Association*, *29*(10), 1591–1602. DOI: 10.1002/j.1551-8833.1937.tb14014.x

James, A. (Ed.). (1981). *Mathematical models of water pollution control.* Mir. (in Russian)

James, S. C., & Chrysikopoulos, C. (2003). Effective velocity and effective dispersion coefficient for finite-sized particles flowing in a uniform fracture. *Journal of Colloid and Interface Science*, *263*(1), 288–295. DOI: 10.1016/S0021-9797(03)00254-6 PMID: 12804914

Jegatheesan, V., & Vigneswaran, S. (2000). Transient stage deposition of submicron particles in deep bed filtration under unfavorable conditions. *Water Research, 34*(7), 2119–2131. DOI: 10.1016/S0043-1354(99)00381-4

Jegatheesan, V., & Vigneswaran, S. (2005). Deep bed filtration: Mathematical models and observations. *Critical Reviews in Environmental Science and Technology, 35*(6), 515–569. DOI: 10.1080/10643380500326432

Jia, H.-Y. (2005). Bioplato for wastewater treatment. *Problems of Water Supply,* Sewage *and Hydraulics. Collected papers*, 4, 114–124. (in Russian)

Jimenez, B., Chavez, A., Leyva, A., & Tchobanoglous, G. (2000). Sand and synthetic medium filtration of advanced primary treatment effluent from Mexico-city. *Water Research, 34*(2), 473–480. DOI: 10.1016/S0043-1354(99)00180-3

Jönsson, K. A.-S., & Jönsson, B. T. L. (1992). Fluid flow in compressible porous media. I. Steady-state conditions. II. Dynamic behavior. *AIChE Journal. American Institute of Chemical Engineers, 38*(9), 1340–1356. DOI: 10.1002/aic.690380904

Kadlec, R. H., & Knight, R. L. (1996). *Treatment wetlands*. Boca Ration. FL.: CRC Press.

Karagyaur, A. S., Epoyan, S. M., & Pabolkov, V. V. (2013). Choice of filter media for multilayer filters with horizontal filtration direction. [in Russian]. *MOTROL Commission of Motorization and Energetics in Agricultural, 15*(6), 11–18.

Kau, S. M., & Lawler, D. F. (1995). Dynamics of deep-bed filtration: Velocity, depth and media. *Journal of Environmental Engineering, 121*(12), 850–859. DOI: 10.1061/(ASCE)0733-9372(1995)121:12(850)

Kawamura, S. (1985). Two-stage filtration. *Journal - American Water Works Association, 77*(12), 42–47. DOI: 10.1002/j.1551-8833.1985.tb05656.x

Kessler, J. H., & Hunt, J. R. (1994). Dissolved and colloidal contaminant transport. *Water Resources Research, 30*(4), 1195–1206. DOI: 10.1029/93WR03555

Khapun, T. F., & Miltreiger, I. M. (1981). About mathematical models and engineering methods of calculating the process of natural water purification by filtration. [in Russian]. *Water Resources*, (3), 189–195.

Khilar, K. C., & Fogler, H. S. (1998). *Migration of bines in porous media*. Kluwer Academic Publishers. DOI: 10.1007/978-94-015-9074-7

Kim, A. N., Kolodkin, I. V., & Sharavii, Ch. O. (2006). Study of physicochemical and technological properties of filtering and sorption material of zeolite of Kholinsky deposit of Buryatia. [in Russian]. *Water and Ecology*, (1), 9–15.

Kim, J., & Lawler, D. F. (2012). The influence of hydraulic loads on depth filtration. *Water Research, 46*(2), 433–441. DOI: 10.1016/j.watres.2011.10.059 PMID: 22112925

Kim, J., & Tobiasson, J. E. (2004). Particles in filter effluent: The roles of deposition and detachment. *Environmental Science & Technology, 38*(22), 6132–6138. DOI: 10.1021/es0352698 PMID: 15573617

Kim, J., & Tobiasson, J. E. (2004). Origins of filter effluent particles: Experimental study of particle deposition and detachment. *Water Science and Technology, 50*(12), 215–222. DOI: 10.2166/wst.2004.0716 PMID: 15686024

Kiranmai, D., Jyothirmai, A., & Murty, C. V. S. (2005). Determination of kinetic parameters in fixed-film bio-reactors: An inverse problem approach. *Biochemical Engineering Journal, 23*(1), 73–83. DOI: 10.1016/j.bej.2004.10.005

Kiselev, S.K. (1997). Change of filtration properties of filter bed at water deferrization. *Land Reclamation and Water Management. Collected papers*, 84, 190–194. (in Russian)

Knapp, R. B., Chiaropa, M. L., & Durham, W. B. (2000). An experimental exploration of the transport and capture of abiotic colloids in a single fracture. *Water Resources Research, 36*(11), 3139–3149. DOI: 10.1029/2000WR900145

Ko, D. C. K., Porter, J. F. P., & McKay, G. (2001). Film-pore diffusion model for the fixed-bed sorption of copper and cadmium ions onto bone char. *Water Research, 35*(16), 3876–3886. DOI: 10.1016/S0043-1354(01)00114-2 PMID: 12230170

Koganovskiy, A. M., Levchenko, T. M., Roda, I. G., & Marotovskiy, R. M. (1981). *Adsorption technology of wastewater treatment*. Technics. (in Russian)

Komarova, E. A., Nakorchevskaya, V. F., & Kulsky, L. A. (•••). Application of stepped filtration in water treatment technology. [in Russian]. *Journal of Water Chemistry and Technology, 4*(3), 240–244.

Kommunar, G. M., & Bogolyubov, K. S. (1977). Some features of calculation of infiltration basins under conditions of colmatization and clogging. *Proceedings of VNII VODGEO*, 63, 10–16. (in Russian)

Kovalchuk, V.A. (2002). *Wastewater treatment*. OJSC "Rivne Printing House". (in Ukrainian)

Kozeny, J. (1931). Grundwasserbewegung bei freiem Spiegel, Fluss- und Kanal – versickerung. *Wasserkraft und Wasserwirts chaft*, 4.3, 28–31.

Kozhinov, V. F. (2008). *Purification of drinking and technical water. Textbook for universities* (4th ed.). BASTET. (in Russian)

Kravchuk, A.M. (2000). *Contaminant transport and phase transfer in the structures of water supply and sewage systems*. KNUCA Press. (in Russian)

Kretzsehmar, R., Barmettler, K., & Grolimund, D. (1997). Experimental determination of colloid deposition rates and collision efficiencies in natural porous media. *Water Resources Research, 33*(5), 1129–1137. DOI: 10.1029/97WR00298

Kryuchkov, Yu. N. (1998). Determination of porous structure parameters and permeability of porous materials. [in Russian]. *Theoretical Foundations of Chemical Engineering, 32*(5), 515–523.

Kryvonoh, O.I., & Kryvonoh, V.V. (1997). Study of the process of water clarification by filters made of fiber-porous polyethylene. *Land Reclamation and Water Management. Collected papers*, 84, 170–174. (in Russian)

Kuberkar, V. T., & Davis, R. H. (2000). Modeling of fouling reduction by secondary membranes. *Journal of Membrane Science*, *168*(1-2), 243–258. DOI: 10.1016/S0376-7388(99)00324-5

Kulikov, N.I., Naimanov, A.Y., Omelchenko, N.P., & Chernyshev, V.N. (2009). *Theoretical bases of water purification*. Publishing House "Znanie", (Donetsk branch). (in Russian)

Kulkarni, P., Dutari, G., Weingeis, D., Adin, A., Haught, R., & Biswas, P. (2005). Capture of water-borne colloids in granular beds using external electric fields: Improving removal of Cryptosporidium parvum. *Water Research*, *39*(6), 1047–1060. DOI: 10.1016/j.watres.2004.12.026 PMID: 15766959

Kulskiy, L. A., Goronovskiy, I. T., Koganovskiy, A. M., & Shevchenko, M. A. (1980). *Reference book on properties, methods of analysis and purification of water: in 2 parts*. Naukova dumka. (in Russian)

Kulskiy, L. A., & Strokach, P. P. (1986). *Technology of natural water purification*. Vishcha Shkola. (in Russian)

Kulsky, L. A., Yaroshevskaya, N. V., & Shevchuk, E. A. (1991). Increasing the retention capacity of filters in two-stage filtration schemes. [in Russian]. *Journal of Water Chemistry and Technology*, *13*(3), 239–241.

Kumari, M., & Kazmi, A. A. (2014). Effect of physical property of supporting media and variable hydraulic loading on hydraulic characteristics of advanced onsite wastewater treatment system. *Environmental Technology*, *36*(11), 1414–1422. DOI: 10.1080/09593330.2014.992480 PMID: 25428652

Kuwabara, S. (1959). The forces experienced by randomly distributed parallel circular cylinders or spheres in a viscous flow at small Reynolds numbers. *Journal of the Physical Society of Japan*, *14*(4), 527–532. DOI: 10.1143/JPSJ.14.527

Kwapinski, W., Winterberg, M., Tsotsas, E., & Mewes, D. (2004). Modeling of the wall effect in packed bed adsorption. *Chemical Engineering & Technology*, *27*(11), 1176–1186. DOI: 10.1002/ceat.200407001

Landman, K., & White, L. (1994). Solid/liquid separation of flocculated suspensions. *Advances in Colloid and Interface Science*, *51*, 175–246. DOI: 10.1016/0001-8686(94)80036-7

Last, W., & Stichimair, J. (2002). Influence of stagnant liquid elements on the effective interfacial area during chemical adsorption in packed columns. *Chemical Engineering & Technology*, *25*(11), 1075–1079. DOI: 10.1002/1521-4125(20021105)25:11<1075::AID-CEAT1075>3.0.CO;2-C

Lau, B. L., Harrington, G. W., Anderson, M. A., & Tejedor, I. (2004). Removal of nano and microparticles by granular dilter media coated with nanoporous aluminium oxide. *Water Science and Technology*, *50*(12), 223–228. DOI: 10.2166/wst.2004.0717 PMID: 15686025

Lebedev, I. A., Somin, V. A., Kondratyuk, E. V., & Komarova, L. F. (2006). Ecological technologies: Intensification of water purification from suspended solids by filtration method. [in Russian]. *Engineering Ecology*, (2), 20–27.

Lee, J., Fox, P. J., & Lenhart, J. J. (2009). Investigation of consolidation-Induced solute transport. I. Effect of consolidation on transport parameters. *Journal of Geotechnical and Geoenvironmental Engineering*, *135*(9), 1228–1238. DOI: 10.1061/(ASCE)GT.1943-5606.0000047

Lee, Y., & Clark, M. M. (1998). Modeling of flux decline during crossflow ultrafiltration of colloidal suspension. *Journal of Membrane Science*, *149*(2), 181–202. DOI: 10.1016/S0376-7388(98)00177-X

Leichkis, I. M. (1975). *Filtering with the use of auxiliary substances*. Tehnika. (in Russian)

Lerch, A., Nahrstedt, A., Gimbel, R., Gray, S., & Booker, N. (2002). On the development of a high-rate deep bed filtration unit. *Water Science and Technology: Water Supply*, *2*(1), 259–269. DOI: 10.2166/ws.2002.0031

Lewandowska, J., Ngoc, T. D. T., Vauclin, M., & Bertin, H. (2008). Water drainage in double-porosity soils: Experiments and micro–macro modeling. *Journal of Geotechnical and Geoenvironmental Engineering*, *134*(2), 231–243. DOI: 10.1061/(ASCE)1090-0241(2008)134:2(231)

Leyva-Ramos, R., & Geankoplis, C. J. (1985). Model simulation and analysis of surface diffusion of liquids in porous solids. *Chemical Engineering Science*, *40*(5), 799–807. DOI: 10.1016/0009-2509(85)85032-6

Li, Y.-C., & Cleall, P. J. (2010). Analytical solutions for contaminant diffusion in double-layered porous media. *Journal of Geotechnical and Geoenvironmental Engineering*, *136*(11), 1542–1554. DOI: 10.1061/(ASCE)GT.1943-5606.0000365

Limbach, K. W., & Wei, J. (1990). Restricted diffusion through granular materials. *AIChE Journal. American Institute of Chemical Engineers*, *36*(2), 242–248. DOI: 10.1002/aic.690360210

Lindstrem, F. T. (1992). A mathematical model for the one-dimensional transport and fate of oxygen and substrate in a water-saturated sorbing homogeneous porous medium. *Water Resources Research*, *28*(9), 2499–2511. DOI: 10.1029/92WR00898

Linninger, A. A. (2012). Exact solution of the diffusion-convection equation in cylindrical geometry. *AIChE Journal. American Institute of Chemical Engineers*, *58*(4), 1299–1302. DOI: 10.1002/aic.12663

Liu, C., & Zachara, J. M. (2001). Uncertainties of Monod kinetic parameters nonlinearly estimated from batch experiments. *Environmental Science & Technology*, *35*(1), 133–141. DOI: 10.1021/es001261b PMID: 11351998

Liu, L., Xue, Q., Wan, Y., & Tian, Y. (2016). Evaluation of dual permeability of gas flow in municipal solid waste: Experiment and modeling. *Environmental Progress & Sustainable Energy*, *35*(1), 1–7. DOI: 10.1002/ep.12184

Liu, Q., Chen, L., Minglei, Z., Ying, L., Yangyang, Y., Yuxuan, L., & Siyuan, M. (2024). Performance of a double-filter-medium tandem membrane bioreactor with low operating costs in domestic wastewater treatment. *Water (Basel)*, *16*(2), 361. DOI: 10.3390/w16020361

Lopato, L., Galaj, Z., Delpont, S., Binning, P. J., & Arvin, E. (2011). Heterogeneity of rapid sand filters and its effect on contaminant transport and nitrification performance. *Journal of Environmental Engineering*, *137*(4), 248–257. DOI: 10.1061/(ASCE)EE.1943-7870.0000321

Lozada, P. T., & Cruz-Velez, C. H. (2014). Influence of the flow control (declining – constant rate) on the filtration process in drinking water treatment. *Inginieria y Universidad*, (July), 217–234.

Lu, C., Chang, K., & Hsu, S. (2004). A model for treating isopropyl alcohol and acetone mixtures in a trickle-bed air biofilter. *Process Biochemistry (Barking, London, England), 39*(12), 1849–1858. DOI: 10.1016/j.procbio.2003.09.019

Lukinykh, N. A., Lipman, B. L., & Krishtul, V. P. (1978). *Methods of additional treatment of sewage waters*. Stroyizdat. (in Russian)

Lyn, D. A. (1996). Asymptotic analysis of intraparticle diffusion in GAC batch reactors. *Journal of Environmental Engineering, 122*(11), 1013–1022. DOI: 10.1061/(ASCE)0733-9372(1996)122:11(1013)

MacDonald, M. J., Chu, C.-F., Guilloit, P. P., & Ng Ka, M. (1991). A generalized Blake-Kozeny equation for multisized spherical particles. *AIChE Journal. American Institute of Chemical Engineers, 37*(10), 1583–1588. DOI: 10.1002/aic.690371016

MacDonald, T. R., Kitanidis, P. K., McCarty, P. L., & Roberts, P. V. (1999). Effects of shear detachment on biomass growth and in situ bioremediation. *Groundwater, 37*(4), 555–562. DOI: 10.1111/j.1745-6584.1999.tb01142.x

Mackie, R. I., Horne, R. M. W., & Jarvis, R. J. (1987). Dynamic modeling of deep-bed filtration. *AIChE Journal. American Institute of Chemical Engineers, 33*(11), 1761–1775. DOI: 10.1002/aic.690331102

Mackie, R. I., & Zhao, Q. (1999). A framework for modeling removal in the filtration of polydisperse suspensions. *Water Research, 33*(3), 794–806. DOI: 10.1016/S0043-1354(98)00267-X

Martynov, S. Yu. (2014). Modeling the operation of polystyrene filters with variable filtration rate. [in Ukrainian]. *Bulletin of NUWEE, 2*(66), 262–269.

Martynov, S. Yu., & Orlova, A. M. (2022). *Design of water clarification and decolorization stations. Study guide*. NUWEE Press. (in Ukrainian)

Martynov, S. Yu., & Poliakov, V. L. (2022). Experimental studies of iron transformations kinetics and autocatalysis during its physicochemical removal from underground water. *Water Science and Technology: Water Supply, 22*(3), 2883–2895. DOI: 10.2166/ws.2021.428

Martynov, S. Yu., & Poliakov, V. L. (2022). Experimental studies on the hydrodynamic properties of a deposit in rapid filters during physicochemical removal of iron from groundwater. *Water Science and Technology: Water Supply, 22*(11), 7603–7617. DOI: 10.2166/ws.2022.305

Maslov, B. S., Minaev, I. V., & Guber, K. V. (1989). *Reference book on land reclamation*. Rosagropromizdat. (in Russian)

Mayer, K. U., Frind, E. O., & Blowes, D. W. (2001). Multicomponent reactive transport modeling in variably saturated porous media using a generalized formulation for kinetically controlled reactions. *Water Resources Research, 38*(9), 1174–1183. DOI: 10.1029/2001WR000862

Mays, D. C. (2010). Contrasting clogging in granular media filters, soils, and dead-end membranes. *Journal of Environmental Engineering, 136*(5), 475–480. DOI: 10.1061/(ASCE)EE.1943-7870.0000173

Mays, D. C., & Hunt, J. R. (2005). Hydrodynamic aspects of particle clogging in porous media. *Environmental Science & Technology, 39*(2), 577–584. DOI: 10.1021/es049367k PMID: 15707058

McDowell, L. M., Hunt, J. R., & Sitar, N. (1986). Particle transport through porous media. *Water Resources Research, 22*(13), 1901–1921. DOI: 10.1029/WR022i013p01901

Meltser, V.Z. (1995). *Filtration structures in municipal water supply*. Stroyizdat.

Meric, S., Tunay, O., & Ali San, H. (2002). A new approach to modelling substrate inhibition. *Environmental Technology, 23*(2), 163–177. DOI: 10.1080/09593332508618428 PMID: 11950069

Mints, D. M. (1964). *Theoretical bases of water purification technology*. Publishing House on Construction. (in Russian)

Mints, D. M., & Meltser, V. Z. (1970). Hydraulic resistance of granular porous medium in the process of clogging. *Rep.* [in Russian]. *USSR Academy of Sciences, 192*(2), 304–306.

Mintz, D. M., & Meltser, V. Z. (1973). Simplified method of technological modeling of the filtration process. *Scientific Proceedings of AKH,* 98, 57–62. (in Russian)

Miralles-Wilhelm, F., & Gelhar, L. W. (1996). Stohastic analysis of sorption macrokinetics in heterogeneous aquifers. *Water Resources Research, 32*(6), 1541–1549. DOI: 10.1029/96WR00791

Mitchell, C., & McNevin, D. (2001). Alternative analysis of BOD removal in subsurface flow constructed wetlands employing Monod kinetics. *Water Research, 35*(5), 1295–1303. DOI: 10.1016/S0043-1354(00)00373-0 PMID: 11268849

Mitchell, J. K., & Santamarina, J. C. (2005). Biological considerations in geotechnical engineering. *Journal of Geotechnical and Geoenvironmental Engineering, 131*(10), 1222–1233. DOI: 10.1061/(ASCE)1090-0241(2005)131:10(1222)

Mitin, B. A. (1969). Research of influence of structural and mechanical properties of deposit on operation of granular clarifying filters. Ph.D. dissertation abstract. LECI Press. (in Russian)

Mohanka, S. S. (1969). Theory of multilayer filtration. *Journal of the Sanitary Engineering Division, 98*(6), 1079–1095. DOI: 10.1061/JSEDAI.0001033

Mokrik, R. V., & Strimaitite, R. I. (1984). About numerical solution of the inverse problem of stationary flow in multilayer strata. In *Methods of analysis and processing of hydrogeological data for groundwater resources forecasting* (pp. 175–178). Academy of Sciences of the ESSR. (in Russian)

Molchak, Y. A. (1984). *Rainfall runoff of the over-humidification zone of Ukraine (studies and calculations).* "Vishcha Shkola", Lviv State University Press. (in Russian)

Mollah, A. H., & Robinson, C. W. (1996). Pentachlorophenol adsorption and desorption characteristics of granular activated carbon. I. Isotherms. II. Kinetics. *Water Research, 30*(2), 2901–2913. DOI: 10.1016/S0043-1354(96)00131-5

Molnar, I. L., Gerhard, J. I., Willson, C. S., & O'Carroll, D. M. (2015). The impact of immobile zones on the transport and retention of nanoparticles in porous media. *Water Resources Research, 51*(11), 8973–8994. DOI: 10.1002/2015WR017167

Moran, M. C., Moran, D. C., Cushing, R. S., & Lawler, D. F. (1993). Particle behavior in deep bed filtration. 2. Particle detachment. *Journal - American Water Works Association, 85*(12), 82–93. DOI: 10.1002/j.1551-8833.1993.tb06121.x

Myint, M., Nirmalakhandan, N., & Speece, R. E. (2007). Anaerobic fermentation of cattle manure: Modeling of hydrolysis and acidogenesis. *Water Research, 41*(2), 323–332. DOI: 10.1016/j.watres.2006.10.026 PMID: 17141822

Naumenko, I. I. (2005). *Hydraulics*. NUWEE Press. (in Russian)

Ncube, P., Pidou, M., & Jarvis, P. (2018). The impact of filter bed depth and solids loading using a multimedia filter. *Separation Science and Technology, 53*(14), 2249–2258. DOI: 10.1080/01496395.2018.1439961

Nielsen, D. R., van Genuchten, M. T., & Biggar, J. W. (1986). Water flow and solute transport processes in the unsaturated zone. *Water Resources Research, 22*, 89–108. DOI: 10.1029/WR022i09Sp0089S

Niu, S., Park, K., Yu, J., & Kim, Y. (2016). Operation and performance evaluation of high-speed filter using porous non-woven filamentous fibre for the treatment of turbid water. *Environmental Technology, 37*(5), 577–589. DOI: 10.1080/09593330.2015.1075596 PMID: 26209268

Nivala, J., Knowles, P., Dotro, G., Garcı, J., & Wallace, S. (2012). Clogging in subsurface-flow treatment wetlands: Measurement, modeling and management. [Review]. *Water Research, 46*(6), 1625–1640. DOI: 10.1016/j.watres.2011.12.051 PMID: 22284912

Noskov, M. D., Zaitseva, M. S., Istomin, A. D., & Lukashevich, O. D. (2008). Mathematical modeling of the rapid filters operation. [in Russian]. *Bulletin of TSASU*, (2), 126–137.

Odegaard, H., Liao, Z., & Hansen, A. T. (2003). Coarse media filtration – an alternative to settling in wastewater treatment. *Water Science and Technology, 47*(12), 81–88. DOI: 10.2166/wst.2003.0631 PMID: 12926673

Ojha, C. S., & Graham, N. J. D. (1994). Computer-aided simulation of slow sand filter performance. *Water Research, 28*(5), 1025–1030. DOI: 10.1016/0043-1354(94)90187-2

Ojha, C. S. P., & Graham, N. J. D. (1992). Appropriate use of deep-bed filtration models. *Journal of Environmental Engineering, 118*(6), 964–980. DOI: 10.1061/(ASCE)0733-9372(1992)118:6(964)

Oleynik, A. Ya., & Tugay, A. M. (2001). Modeling of the processes of clogging and suffusion in the prefilter zone of the well. [in Russian]. *Rep. NAS of Ukraine*, (9), 190–194.

Oliynyk, O.Ya., & Kalugin, Y.I. (2005). Some results of theoretical studies of mass transfer processes in porous media. *Problems of Water Supply, Sewage and Hydraulic. Collected papers,* 5, 100–112. (in Ukrainian)

Orlov, V. O. (2005). *Water purification filters with granular filter media.* NUWEE Press. (in Russian)

Orlov, V. O. (2008). *Deferrization of groundwater by simplified aeration and filtration.* NUWEE Press. (in Ukrainian)

Orlov, V. O., & Martynov, S. Y. (2011). Aeration methods of water deferrization. [in Ukrainian]. *Water and Water Treatment Technologies, 2*(4), 42–52.

Ornatsky, N. S., Sergeev, E. M., & Shehtman, Y. M. (1955). *Investigation of the process of sand clogging*. Moscow State University Press. (in Russian)

Oyanader, M., & Arce, P. (2007). An effective analytical–algorithmic approach to solve diffusion-type differential equations with non-linear sources: Unbounded domains. *Computers & Chemical Engineering*, *31*(9), 1174–1180. DOI: 10.1016/j.compchemeng.2006.10.021

Paal, L. L., Karu, Y. A., Melder, H. A., & Repin, B. N. (1994). *Reference book on purification of natural and sewage waters*. Vishchya Shkola. (in Russian)

Pachepsky, J. O. (1990). *Mathematical models of physicochemical processes in soils*. Nauka. (in Russian)

Panfilov, M., & Fourar, M. (2006). Physical splitting of non-linear effects in high-velocity stable flow through porous media. *Advances in Water Resources*, *29*(1), 30–41. DOI: 10.1016/j.advwatres.2005.05.002

Pang, L., Goltz, M., & Close, M. (2003). Application of the method of temporal moments to interpret solute transport with sorption and degradation. *Journal of Contaminant Hydrology*, *60*(1-2), 123–134. DOI: 10.1016/S0169-7722(02)00061-X PMID: 12498577

Payatakes, A. C., Tien, C., & Turian, R. M. (1973). A new model for granular porous media. I. Model formulation. *AIChE Journal. American Institute of Chemical Engineers*, *19*(1), 58–67. DOI: 10.1002/aic.690190110

Perkins, T. K., & Johnston, O. C. (1963). A review of diffusion and dispersion in porous media. *Society of Petroleum Engineers Journal*, *3*(1), 70–84. DOI: 10.2118/480-PA

Philip, J. R. (1994). Some exact solutions of convection-diffusion and diffusion equations. *Water Resources Research*, *30*(12), 3545–3551. DOI: 10.1029/94WR01329

Piron, E., Rene, F., & Latrille, E. (1995). A cross-flow microfiltration model based on integration of the mass transport equation. *Journal of Membrane Science*, *108*(1-2), 57–70. DOI: 10.1016/0376-7388(95)00141-4

Pittoors, E., Guo, Y., & Van Hulle, S. W. H. (2014). Modeling dissolved oxygen concentration for optimizing aeration systems and reducing oxygen consumption in activated sludge processes: A review. *Chemical Engineering Communications*, *201*, 983–1002. DOI: 10.1080/00986445.2014.883974

Pivovar, N.G., Bugay, N.G., Friedrichson, V.L., Krivonog, A.I., & Krivonog, V.V. (2000). *Drainage with fiber filters for protection of territories from waterlogging*. NASU. Institute of Hydromechanics. (in Russian)

Plotnikov, N.A. (1983). *Design of systems of artificial replenishment of underground waters for water supply*. Stroyizdat. (in Russian)

Plusnin, I. I., & Golovanov, A. I. (1983). *Reclamation soil science*. Kolos. (in Russian)

Poliakov, V., Kravchuk, A., Kochetov, G., & Kravchuk, O. (2019). Clarification of aqueous suspensions with a high content of suspended solids in rapid sand filters. *Eureka: Physics and Engineering*, *1*(1), 28–45. DOI: 10.21303/2461-4262.2019.00827

Poliakov, V. L. (2006). About filtration of suspension at initial contamination of filter bed (linear kinetics of mass transfer). [in Russian]. *Rep. NAS of Ukraine, 10*, 65–71.

Poliakov, V.L. (2007). On the choice of the rate coefficient of suspended solids removal from suspensions during filtration. *Problems of Water Supply, Sewage and Hydraulics. Collected papers*, 8, 120–131. (in Russian)

Poliakov, V. L. (2008). Calculation of suspension filtration through multilayer filter bed under linear mass transfer kinetics. [in Russian]. *Journal of Water Chemistry and Technology, 30*(1), 3–31.

Poliakov, V.L. (2008). Filtration of suspensions through granular beds at variable suspended solids concentration. *Problems of Water Supply, Sewage and Hydraulics. Collected papers*, 11, 103–111. (in Russian)

Poliakov, V. L. (2009). Theoretical analysis of filter run duration. [in Russian]. *Journal of Water Chemistry and Technology, 31*(6), 605–618.

Poliakov, V.L. (2009). On the influence of grain sorting during filter bed operation on its efficiency. *Problems of Water Supply, Sewage and Hydraulics. Collected papers*, 12, 103–118. (in Russian)

Poliakov, V. L. (2010). On the forecast of head losses in the filter bed. [in Russian]. *Rep. NAS of Ukraine*, (3), 70–76.

Poliakov, V. L. (2010). Clarification of suspension by filtration with declining flow rate. [in Russian]. *Rep. NAS of Ukraine*, (5), 57–63.

Poliakov, V.L. (2010). Modeling the dynamics of secondary pollution in bioreactors-filters. *Problems of Water Supply, Sewage and Hydraulics. Collected papers*, 14, 152–162. (in Russian)

Poliakov, V.L. (2010). About influence of bed height and filtration rate on suspension clarification. *Problems of Water Supply, Sewage and Hydraulics. Collected papers*, 15, 131–141. (in Russian)

Poliakov, V. L. (2011). Filtration of suspensions through multilayer filter bed at nonlinear mass exchange kinetics. 1. Theory. [in Russian]. *Journal of Water Chemistry and Technology, 33*(1), 3–14.

Poliakov, V. L. (2012). Filtration of suspensions with decreasing rate at linear kinetics of mass transfer. [in Russian]. *Journal of Water Chemistry and Technology, 34*(2), 107–130.

Poliakov, V.L. (2012). Filtration of suspension through a granular bed at limiting saturation of its part with depositt. *Applied hydromechanics*, 14(86), 3, 55–63. (in Russian)

Poliakov, V. L. (2012). Theoretical analysis of the adsorbent grain action. [in Russian]. *Rep. NAS of Ukraine*, (4), 64–69.

Poliakov, V. L. (2012). Modeling of the water treatment process in the adsorber bed. [in Russian]. *Rep. NAS of Ukraine*, (5), 63–71.

Poliakov, V. L. (2014). *Filtration flow deformations in drained soils: theory and applications*. Agrar Media Group. (in Russian)

Poliakov, V. L. (2022). Theoretical analysis of reagent filtration of aqueous suspensions with unregulated filtration rate. *Theoretical Foundations of Chemical Engineering*, *56*(4), 477–496. DOI: 10.1134/S0040579522030101

Poliakov, V.L., & Shevchuk, E.A. (2007). Calculation of stepped filtration of suspensions through granular filter beds. Bulletin of NUWEE, 4(40), part 2, 544–553. (in Russian)

Polubarinova-Kochina, P. Ya. (1977). *Theory of groundwater movement*. Nauka. (in Russian)

Polubarinova-Kochina, P. Ya., Pryazhinskaya, V. G., & Emikh, V. N. (1969). *Mathematical methods in irrigation issues*. Nauka. (in Russian)

Polyakov, Y. S., & Dilman, V. V. (2006). Approximate method for nonlinear differential and integro-differential equations. *AIChE Journal. American Institute of Chemical Engineers*, *52*(11), 3813–3824. DOI: 10.1002/aic.10995

Polyanin, A. D., & Zaitsev, V. F. (2002). *Reference book on nonlinear equations of mathematical physics: exact solutions*. Fizmatlit. (in Russian)

Popkovich, G. S., & Ropin, B. N. (1986). *Systems of aeration of sewage waters*. Stroyizdat. (in Russian)

Pritzker, M. D. (2003). Model for parallel surface and pore diffusion of an adsorbate in a spherical adsorbent particle. *Chemical Engineering Science*, *58*(2), 473–478. DOI: 10.1016/S0009-2509(02)00554-7

Progulny, V. I., & Ryabkov, M. V. (2019). Porous polymer concrete in drains of pressure polystyrene foam filters. [in Russian]. *Bulletin of ONACA*, *74*, 147–153.

Quintana-Solorzano, R., Che-Galicia, G., Trejo-Reyes, M., & Armendariz-Herrera, H. (2018). On the influence of particle shape and process conditions in the pressure drop and hydrodynamics in a wall-effect fixed bed. *Chemical Engineering Communications*, *205*(10), 1323–1341. DOI: 10.1080/00986445.2018.1448274

Rahli, O., Tadrist, L., & Miscevic, M. (1996). Experimental analysis of fibrous porous media permeability. *AIChE Journal. American Institute of Chemical Engineers*, *42*(12), 3547–3549. DOI: 10.1002/aic.690421224

Rajagopalan, R., & Tien, C. (1976). Trajectory analysis of deep bed filtration with sphere in cell porous media model. *AIChE Journal. American Institute of Chemical Engineers*, *22*(3), 523–533. DOI: 10.1002/aic.690220316

Rastvorova, O. G. (1983). *Soil Physics. Practical guide*. Leningrad State University Press. (in Russian)

Reddy, L. N., Ming, X., Hajra, M. G., & Lee, I. M. (2000). Permeability reduction of soil filters due to physical clogging. *Journal of Geotechnical and Geoenvironmental Engineering*, *126*(3), 236–246. DOI: 10.1061/(ASCE)1090-0241(2000)126:3(236)

Reddy, L. N., Xiao, M., Hajra, M. C., & Lee, I. M. (2005). Physical clogging of soil filters under constant flow rate versus constant head. *Canadian Geotechnical Journal*, *42*(3), 804–811. DOI: 10.1139/t05-018

Ren, J., Packman, A. J., & Welty, C. (2000). Correlation of colloid efficiency with hydraulic conductivity of silica sands. *Water Resources Research*, *36*(9), 2493–2500. DOI: 10.1029/2000WR900163

Reznikov, G. D. (1999). About the particle absorption in filters of water treatment facilities. [in Russian]. *Engineering and Physical Journal*, *72*(1), 38–43.

Richard, L. (2004). Filter media surface modification technology: State of art. *Filtration & Separation*, *41*(9), 20–21. DOI: 10.1016/S0015-1882(04)00343-X

Riekert, L. (1985). The relative contribution of pore volume diffusion and surface diffusion to mass transfer in capillaries and porous media. *AIChE Journal. American Institute of Chemical Engineers*, *31*(5), 863–864. DOI: 10.1002/aic.690310525

Rittmann, B. E. (1993). The significance of biofilms in porous media. *Water Resources Research*, *27*(7), 2195–2202. DOI: 10.1029/93WR00611

Rittmann, B. E., & Stiwell, D. S. (2002). Modeling biological processes in drinking water treatment: An integral biofiltration model. [in Russian]. *Water and Ecology*, *3*, 13–29.

Rode, A. A. (1965). *Fundamentals of the doctrine of soil moisture* (Vol. 1). Gidrometeoizdat. (in Russian)

Rodgers, M., Healy, M. G., & Mulqueen, J. (2005). Organic removal and nitrification of high strength wastewater using stratified sand filters. *Water Research*, *39*(14), 3279–3286. DOI: 10.1016/j.watres.2005.05.035 PMID: 16005491

Romankov, P. G., & Kurochkina, M. I. (1982). *Hydromechanical processes of chemical technology*. Khimiya. (in Russian)

Rudnytsky, E.A. (2008). Hydraulic characteristics of the upward flow through the floating filter bed. *Land Reclamation and Water Management. Collected papers*, 96, 213–222. (in Russian)

Saatci, A.M. (1990). Application of the declining rate filtration theory. – Continuous operation. *J. Environ. Eng. Div., ASCE*, 116(1), 87–105.

Sabliy, L. A. (2013). *Physical, chemical and biological treatment of highly concentrated wastewater*. NUWEE Press. (in Russian)

Saiers, J. E., Hornberger, G. M., & Liang, Z. (1994). First- and second-order kinetics approaches for modeling the transport of colloidal particles in porous media. *Water Resources Research*, *30*(9), 2499–2506. DOI: 10.1029/94WR01046

Sakkas, N. D., & Lekkas, T. D. (1989). Hydrodynamic characteristics of the solid-liquid fluidized bed developing during the backwash of filter media. *Environmental Technology Letters*, *10*(2), 151–166. DOI: 10.1080/09593338909384728

Salthes, T., Eikenbrokk, B., & Odegard, H. (2003). Effect of filtration rate in a filter with two-layer expanded clay (filtralite) in comparison with a sand-anthracite filter. [in Russian]. *Water and Ecology*, (2), 23–32.

Samani, J. M. V., Fathi, P., & Homaee, M. (2007). Simultaneous prediction of saturated hydraulic conductivity and drainable porosity using the inverse problem technique. *Journal of Irrigation and Drainage Engineering*, *133*(2), 110–115. DOI: 10.1061/(ASCE)0733-9437(2007)133:2(110)

Samuel, M. P., Senthilvel, S., Tamilmani, D., & Mathew, A. C. (2012). Performance evaluation and modeling studies of gravel-coir fibre-sand multimedia stormwater filter. *Environmental Technology, 33*(17), 2057–2069. DOI: 10.1080/09593330.2012.660642 PMID: 23240200

SBC V.2.5-74:2013. (2013). *Water supply. External networks and structures. Basic design provisions.* MRRBZHKG of Ukraine. (in Ukainian)

SBC V.2.5-75:2013. (2013). *Sewerage. Design of external networks and structures. Basic provisions of design.* UkrNDIvodokanalproekt. (in Ukainian)

Schuberth, J., Mauschitz, G., & Höflinger, W. (2011). Clogging mechanisms involved in the aging process of cleanable filter media. *Separation and Purification Technology, 77*(2), 196–201. DOI: 10.1016/j.seppur.2010.12.004

Schulze-Makuch, D., Carlson, D. A., Cherkauer, D. S., & Malik, P. (1999). Scale dependency of hydraulic conductivity in heterogeneous media. *Groundwater, 37*(6), 904–919. DOI: 10.1111/j.1745-6584.1999.tb01190.x

Seifert, D., & Engesgaard, P. (2012). Sand box experiments with bioclogging of porous media: Hydraulic conductivity reductions. *Journal of Contaminant Hydrology, 136–137*, 1–9. DOI: 10.1016/j.jconhyd.2012.04.007 PMID: 22647500

Sen, T. K., Shanbhag, S., & Khilar, K. C. (2004). Subsurface colloids in groundwater contamination: A mathematical model. *Colloids and Surfaces. A, Physicochemical and Engineering Aspects, 232*(1), 29–38. DOI: 10.1016/j.colsurfa.2003.10.004

Senyavin, M. M., Venitsianov, E. V., & Ayukaev, R. I. (1997). About mathematical models and engineering methods for calculating the process of natural water purification by filtration. [in Russian]. *Water Resources*, (2), 157–170.

Serrano, S. E. (2001). Solute transport under non-linear sorption and decay. *Water Research, 35*(6), 1525–1533. DOI: 10.1016/S0043-1354(00)00390-0 PMID: 11317900

Shandyba, A. B., & Vainshtein, I. A. (1988). To calculation of washing of dispersed materials. [in Russian]. *Hydromechanics, 57*, 72–76.

Shani, C., Weisbrod, N., & Yakirevich, A. (2008). Colloid transport through saturated sand columns: Influence of physical and chemical properties on deposition. *Colloids and Surfaces. A, Physicochemical and Engineering Aspects, 316*(1-3), 142–150. DOI: 10.1016/j.colsurfa.2007.08.047

Shaoging, Q. (1998). Expression for critical suspended particle size in granular media filtration processes. *Water Research, 32*(9), 2856–2859. DOI: 10.1016/S0043-1354(98)00028-1

Sharma, M. M., Chaumon, H., Sarma, D. S., & Schecher, R. S. (1992). Factors controlling the hydrodynamic detachment of particles from surfaces. *Journal of Colloid and Interface Science, 149*(1), 121–134. DOI: 10.1016/0021-9797(92)90398-6

Sharma, M. M., & Yortos, Y. C. (1987). Transport of particulate suspensions in porous media: Model formulation. *AIChE Journal. American Institute of Chemical Engineers, 33*(10), 1636–1643. DOI: 10.1002/aic.690331007

Shehtman, Yu. M. (1961). *Filtration of low–concentration suspensions.* AS USSR Publishing House. (in Russian)

Shein, E. V. (Ed.). (2001). *Field and laboratory methods of research of physical properties and soil regimes: Methodical guide.* Moscow State University Press. (in Russian)

Shen, W., Yang, G.-H., Liu, P., Yao, Z., Shi, H., Cui, Y., & Li, Z. (2018). Effects of temperature on the filtration characteristics of a dual–layer granular bed filter. *Chemical Engineering & Technology, 41*(9), 1759–1766. DOI: 10.1002/ceat.201800071

Shena, L., & Chen, Z. (2007). Critical review of the impact of tortuosity on diffusion. *Chemical Engineering Science, 62*(14), 3748–3755. DOI: 10.1016/j.ces.2007.03.041

Sherard, J. L., Dunnigan, L. P., & Talbot, J. R. (1984). Basic properties of sand and gravel filters. *Journal of Geotechnical Engineering, 110*(6), 684–700. DOI: 10.1061/(ASCE)0733-9410(1984)110:6(684)

Sherwood, T., Pigford, R., & Wilkie, C. (1982). *Mass Transfer.* Khimiya. (in Russian)

Shevchuk, E. A., Mamchenko, A. V., & Goncharuk, V. V. (2005). Technology of direct flow filtration of natural and waste water through granular loads. [in Russian]. *Journal of Water Chemistry and Technology, 27*(4), 369–383.

Shevtsov, N. M. (1988). *Intra-soil purification and utilization of sewage waters.* Agropromizdat. (in Russian)

Shirasaki, N., Matsushita, T., Matsui, Y., Oshiba, A., & Ohno, K. (2010). Estimation of norovirus removal performance in a coagulation-rapid sand filtration process by using recombinant norovirus VLPs. *Water Research, 44*(5), 1307–1316. DOI: 10.1016/j.watres.2009.10.038 PMID: 19922975

Shterenlicht, D. V. (1984). *Hydraulics.* Energoatomizdat. (in Russian)

Shtykov, V.I., Shevelev, Ya, Z., & Koshevoy, O.Yu. (1987). *Use of livestock waste complexes on specialized systems.* Rosselhozizizdat. (in Russian)

Shvidler, M. I. (1987). Stochastic filtration flow transfer. In *Problems of filtration flow theory and mechanics of oil recovery enhancement* (pp. 203–211). Nauka. (in Russian)

Šimůnek, J., Jarvisb, N. J., van Genuchten, M. Th., & Gärdenäs, A. (2003). Review and comparison of models for describing non-equilibrium and preferential flow and transport in the vadose zone. *Journal of Hydrology (Amsterdam), 272*(1-4), 14–35. DOI: 10.1016/S0022-1694(02)00252-4

Sokovnin, O. T., & Zagoskina, N. V. (2002). Selection and determination of characteristic particle size. [in Russian]. *Theoretical Foundations of Chemical Engineering, 36*(6), 658–669. DOI: 10.1023/A:1021274019713

Song, L. (1998). Flux decline in crossflow microfiltration and ultrafiltration: Mechanisms and modeling of membrane fouling. *Journal of Membrane Science, 139*(2), 183–200. DOI: 10.1016/S0376-7388(97)00263-9

Sposito, G., Jury, W. A., & Gupta, V. K. (1986). Fundamental problems in the stochastic convection-dispersion model of solute transport in aquifers and field soils. *Water Resources Research, 22*(1), 77–88. DOI: 10.1029/WR022i001p00077

Stechkina, I. B., & Kirsh, V. A. (2003). Optimization of filter parameters in a multistage system of fine gas purification. [in Russian]. *Theoretical Foundations of Chemical Engineering, 37*(3), 236–245. DOI: 10.1023/A:1024071202844

Stevenson, D. G. (1995). Process conditions for the backwashing of filter with simultaneous air and water. *Water Research, 29*(11), 2594–2597. DOI: 10.1016/0043-1354(95)00041-I

Stevenson, D. G. (1997). Flow and filtration through granular media. The effect of grain and particle size dispersion. *Water Research, 31*(2), 310–322. DOI: 10.1016/S0043-1354(96)00271-0

Stevik, T. K., Aa, K., Ausland, G., & Hansen, J. F. (2004). Retention and removal of pathogenic bacteria in wastewater percolating through porous media: A review. *Water Research, 38*(6), 1355–1367. DOI: 10.1016/j.watres.2003.12.024 PMID: 15016513

StSanRandN 2.2.4-171-10. (2010). *Hygienic requirements for drinking water intended for human consumption*. Ministry of Health of Ukraine. (in Ukrainian)

Sutherland, K. (2008). Filtration overview: A closer look at depth filtration. *Filtration & Separation, 45*(October), 25–27. DOI: 10.1016/S0015-1882(08)70296-9

Swanson, W. J., & Loeh, R. C. (1997). Biofiltration: Fundamentals, design and operations principles, and applications. *Journal of Environmental Engineering, 123*(6), 538–546. DOI: 10.1061/(ASCE)0733-9372(1997)123:6(538)

Tan, X.-H., Jianh, L., Li, X.-P., Li, Y.-Y., & Zhang, K. (2017). A complex model for the permeability and porosity of porous media. *Chemical Engineering Science, 172*, 230–238. DOI: 10.1016/j.ces.2017.06.041

Tanaka, Y., Miyajuma, K., Funakosi, T., & Chida, S. (1995). Filtration of municipal sewage by ring shaped floating plastic net media. *Water Research, 29*(5), 1387–1392. DOI: 10.1016/0043-1354(94)00288-I

Tang, Y., & Aral, M. M. (1992). Contaminant transport in layered porous media. 1. General solution. *Water Resources Research, 28*(5), 1389–1397. DOI: 10.1029/92WR00292

Tarnapolskaya, M. G. (2005). Filtering materials for water purification from oil products and criterion of their choice. [in Russian]. *Water and Ecology*, (3), 74–79.

Thomasa, D., Penicota, P., Contala, P., Leclerca, D., & Vende, J. (2001). Clogging of fibrous filters by solid aerosol particles. Experimental and modelling study. *Chemical Engineering Science, 56*(11), 3549–3561. DOI: 10.1016/S0009-2509(01)00041-0

Tkachuk, O. A. (2022). *Hydraulic calculations of pipeline systems of water supply and drainage: monograph*. NUWEE Press. (in Russian)

Tobiason, J. E., Johnson, G. S., Westerhoff, P. K., & Vigneswaran, B. (1993). Particle size and chemical effects on contact filtration performance. *Journal of Environmental Engineering, 119*(3), 520–533. DOI: 10.1061/(ASCE)0733-9372(1993)119:3(520)

Tobiason, J. E., & Vigneswaran, B. (1999). Evaluation of a modified model for deep bed filtration. *Water Research, 28*(2), 335–342. DOI: 10.1016/0043-1354(94)90271-2

Tom, C. S., Zimbron, J. A., & Dandy, D. S. (2008). Effects of reduced contaminant loading on downgradient water quality in an idealized two-layer granular porous media. *Journal of Contaminant Hydrology, 102*(1-2), 72–85. DOI: 10.1016/j.jconhyd.2008.08.002 PMID: 18930336

Trussell, R. R., & Chang, M. (1999). Review of flow through porous media as applied to head loss in water filters. *Journal of Environmental Engineering, 125*(11), 998–1006. DOI: 10.1061/(ASCE)0733-9372(1999)125:11(998)

Tsitsishvili, G. V., & Andronikashvili, T. G. (1985). *Natural zeolites*. Khimiya. (in Russian)

Tufenkji, N., Dixon, D. R., Considine, R., & Drummond, C. J. (2006). Multi-scale Cryptosporidium sand interactions in water treatment. [Review]. *Water Research, 40*(18), 3315–3330. DOI: 10.1016/j.watres.2006.07.036 PMID: 16979211

Tufenkji, N., & Elimelech, M. (2004). Correlation equation for predicting single-collector efficiency in physicochemical filtration in saturated porous media. *Environmental Science & Technology, 38*(2), 529–536. DOI: 10.1021/es034049r PMID: 14750730

Tugay, A. M., Oleynyk, A. Ya., & Tugay, Y. A. (2004). *Productivity of water intake wells under conditions of clogging*. KHNAME Press. (in Russian)

Tugay, A.M., & Orlov, V.O. (2009). *Water supply*. Znannya. (in Ukrainian)

Twort, A. C., Ratnayaka, D. D., & Brandt, M. J. (2006). *Water Supply* (5th ed.). IWA Publishing.

Vadyunina, A.F., & Korchagina, Z.A. (1961). *Methods of determination of physical properties of soils and grounds in the field and laboratory*. State Publishing House "Higher School". (in Russian)

Van Genuchten, M. T. (1980). A closed-form equation for predicting the hydraulic conductivity of unsaturated soils. *Soil Science Society of America Journal, 44*(5), 892–898. DOI: 10.2136/sssaj1980.03615995004400050002x

Vandevivere, P., & Haveye, P. (1992). Saturated hydraulic conductivity reduction caused by aerobic bacteria in sand columns. *Soil Science Society of America Journal, 56*(1), 1–13. DOI: 10.2136/sssaj1992.03615995005600010001x

Veerapaneni, S., & Wiesner, M. R. (1993). Role of suspension polydispersivity in granular media filtration. *Journal of Environmental Engineering, 119*(1), 172–190. DOI: 10.1061/(ASCE)0733-9372(1993)119:1(172)

Venitsianov, E. V. (1998). Some features of heavy metals sorption by a layer of bottom sediments and soil. [in Russian]. *Water Resources, 25*(4), 462–466.

Venitsianov, E. V., & Rubinstein, R. N. (1983). *Dynamics of sorption from liquid media*. Nauka. (in Russian)

Verigin, N. N. (1971). *Clogging of radial filters with activated carbon*. Stroyizdat. (in Russian)

Verigin, N. N., & Kommunar, G. M. (1978). Filtration of suspensions through porous media. *Proceedings of VNII VODGEO*, 74, 120–124. (in Russian)

Vigneswaran, S., & Chang, J. S. (1986). Mathematical modeling of the entire cycle of deep bed filtration. *Water, Air, and Soil Pollution*, *29*(2), 155–164. DOI: 10.1007/BF00208406

Vigneswaran, S., Thiyagaram, M., & Ben Ami, R. (1985). Mathematical modeling of radial flow filtration and its application to groundwater recharge problems. *AIChE Journal. American Institute of Chemical Engineers*, *31*(11), 1929–1933. DOI: 10.1002/aic.690311121

Vigneswaran, S., & Tulachan, R. K. (1988). Mathematical modeling of transient behavior of deep bed filtration. *Water Research*, *22*(9), 1093–1100. DOI: 10.1016/0043-1354(88)90003-6

Vorobyev, E. I., & Tarasenko, A. P. (1987). Influence of sludge compressibility on the regularities of the process of suspension filtration. [in Russian]. *Theoretical Foundations of Chemical Engineering*, *21*(4), 507–514.

Voznyuk, S. T., Kuzmich, P. K., Krysztof, V. G., Kirienko, T. N., & Olinevich, V. A. (1984). *Reclamation soil science with the basics of hydrology*. "Vishcha Shkola", Lviv State University Press. (in Russian)

Vyakhirev, G. I., Zagnitiko, A. V., Burbasov, A. N., Pushko, A. I., Pushko, G. I., & Chaplygin, Yu. O. (2002). Gas-dynamic resistance at high-speed purification of gases by multilayer metal-ceramic filters. [in Russian]. *Theoretical Foundations of Chemical Engineering*, *36*(4), 353–357. DOI: 10.1023/A:1019887330128

Wakeman, R. (2007). The influence of particles properties on filtration. *Separation and Purification Technology*, *58*(2), 234–241. DOI: 10.1016/j.seppur.2007.03.018

Wang, Y., Kim, J.-H., Baek, J.-B., Miller, G. W., & Pennell, K. D. (2012). Transport behavior of functionalized multi-wall carbon nanotubes in water-saturated quartz sand as a function of tube length. *Water Research*, *46*(14), 4521–4531. DOI: 10.1016/j.watres.2012.05.036 PMID: 22704927

Weber-Shirk, M. L., & Chan, K. L. (2007). The role of aluminum in slow sand filtration. *Water Research*, *41*(6), 1350–1354. DOI: 10.1016/j.watres.2006.12.002 PMID: 17276479

Whitaker, S. (1988). Diffusion in packed beds of porous particles. *AIChE Journal. American Institute of Chemical Engineers*, *34*(4), 679–683. DOI: 10.1002/aic.690340419

Widdowson, M. A., Molz, F. J., & Benefield, L. D. (1988). A numerical transport model for oxygen- and nitrate-based respiration linked to substrate and nutrient availability in porous media. *Water Resources Research*, *24*(9), 1553–1565. DOI: 10.1029/WR024i009p01553

Williams, G. J., Sheikh, B., Holden, R. B., Kouretas, T. J., & Nelson, K. L. (2007). The impact of increased loading rate on granular media, rapid depth filtration of wastewater. *Water Research*, *41*(19), 4535–4545. DOI: 10.1016/j.watres.2007.06.018 PMID: 17632202

Wolska, J., & Smolinska–Kempisty, K. (2017). Polypropylene prefilters with surface imprinted layer. *Separation and Purification Technology*, *174*, 89–96. DOI: 10.1016/j.seppur.2016.10.007

Xu, S., Gao, B., & Saiers, J. E. (2006). Straining of colloidal particles in saturated porous media. *Water Resources Research*, *42*(12), 2006WR004948. Advance online publication. DOI: 10.1029/2006WR004948

Xue, B., & Sun, Y. (2003). Modeling of sedimentation of polydisperse spherical beads with a broadsize distribution. *Chemical Engineering Science*, *58*(8), 1531–1543. DOI: 10.1016/S0009-2509(02)00656-5

Yakovlev, S. V., Krasnoborodko, I. G., & Rogov, V. M. (1987). *Technology of electrochemical water purification*. Stroyizdat. (in Russian)

Yang, Z., & Langdon, A. G. (2004). The use of magnetic bed conditioning and pH control to enhance filtration by natural titanomagnetite. *Water Research*, *38*(14-15), 3304–3312. DOI: 10.1016/j.watres.2004.04.017 PMID: 15276747

Yaroshevskaya, N. V., Andrievskaya, M. D., Shvidenko, O. G., & Kulskiy, L. A. (1989). Water purification by filtration with non-stationary change in the rate. [in Russian]. *Journal of Water Chemistry and Technology*, *11*(3), 251–253.

Yaroshevskaya, N. V., & Kulskiy, L. A. (1985). Calculation method for filtration facilities operating with decreasing rate. [in Russian]. *Journal of Water Chemistry and Technology*, *7*(2), 3–5.

Yaroshevskaya, N. V., & Muravyev, V. R. (2002). Comparative evaluation of water purification filters loading. [in Russian]. *Journal of Water Chemistry and Technology*, *24*(1), 71–79.

Yaroshevskaya, N. V., & Sotkova, G. Z. (1996). Calculation of parameters of a block of filters with a drainage of a large resistance, operating with decreasing rate. [in Russian]. *Journal of Water Chemistry and Technology*, *18*(4), 356–366.

Yavorsky, B. M., & Detlaf, A. A. (1977). *Reference book on physics*. Nauka. (in Russian)

Yeh, W. W.-G. (1986). Review of parameter identification procedures in groundwater hydrology. The inverse problem. *Water Resources Research*, *22*(2), 95–108. DOI: 10.1029/WR022i002p00095

Yenikeeva, S. N. (1985). Application of nonlinear optimization method for determination of formation migration parameters from field experiments. In *Modeling of hydrogeochemical processes and scientific bases of hydrogeochemical forecasts* (pp. 110–117). Nauka. (in Russian)

Yin, C. Y., Aroua, M. K., & Daud, W. M. A. W. (2007). Review of modifications of activated carbon for enhancing contaminant uptakes from aqueous solutions. [Review]. *Separation and Purification Technology*, *52*(3), 403–415. DOI: 10.1016/j.seppur.2006.06.009

Young, J. C., & Dahab, M. F. (1983). Effect of media design on the performance of fixed–bed anaerobic filters. *Water Science and Technology*, *15*(8-9), 369–383. DOI: 10.2166/wst.1983.0179

Yu, Y., & Rowe, R. K. (2011). Improved solutions for porosity and specific surface of a uniform porous medium with attached film. *J. Environ. Eng*, *137*, 248–257.

Yuksel, E., Akgiray, O., & Soyer, E. (2003). Direct filtration with preozonation for small water treatment systems. *Water Science and Technology*, *48*(11-12), 473–479. DOI: 10.2166/wst.2004.0896 PMID: 14753570

Zakariae, N., & Nore, S. M. M. (2024). Harvesting solutions: Discover the evolution o" agriculture wastewater treatment through comprehensive bibliometric analysis using scopus database 1971-2023. *Desalination and Water Treatment*, *317*, 100291. DOI: 10.1016/j.dwt.2024.100291

Zapolsky, A. K., & Baran, A. A. (1987). *Coagulants and flocculants in water purification processes*. Khimiya. (in Russian)

Zapolsky, A. K., Mishkova-Klimenko, N. A., Astrelin, I. M., Bryk, M. G., Gvozdyak, P. I., & Knyazkova, T. V. (2000). *Physical and chemical bases of wastewater treatment technologies: Textbook*. Libra. (in Ukrainian)

Zhang, L. M., & Li, X. (2010). Microporosity structure of coarse granular soils. *Journal of Geotechnical and Geoenvironmental Engineering*, *136*(10), 1425–1436. DOI: 10.1061/(ASCE)GT.1943-5606.0000348

Zhernov, I. E. (1982). *Dynamics of underground waters*. Vishcha Shkola. (in Russian)

Zhurba, M. G. (1980). *Water purification on granular filters*. Vyshchaya shkola, Lviv State University Press. (in Russian)

Zhurba, M.G. (1993). Application of filtration theory in engineering calculations. *WSE. Water supply and sanitary engineering*, N°7, 2–6. (in Russian)

Zhurba, M.G. (2011). *Water treatment filters with floating filtering media. Scientific edition*. RNO VoSTU. (in Russian)

Zhurba, M.G., & Myakishev, V.A. (1992). Technology of water purification subjected to anthropogenic impact. *WSE. Water supply and sanitary engineering*, N°8, 3–5. (in Russian)

Zhurba, M.G., Sokolov, L.I., & Govorova, J.M. (2010). *Water Supply. Design of systems and structures: Textbook in 3 vol.* ASV Publishing House. (in Russian)

Zielina, M. (2011). Particle shapes in the drinking water filtration process. *Clean (Weinheim)*, *39*(11), 941–946. DOI: 10.1002/clen.201000432

Zosin, A. L., Priymak, T. I., Schall, E. E., Aleev, N. G., & Sulimenko, L. P. (2003). Filtering systems based on complex mineral loadings for treatment of mine water. [in Russian]. *Water and Ecology*, *3*(16), 58–65.

Zou, R. P., Gan, M. L., & Yu, A. B. (2011). Prediction of the porosity of multi-component mixtures of cohesive and non-cohesive particles. *Chemical Engineering Science*, *66*(20), 4711–4717. DOI: 10.1016/j.ces.2011.06.037

Chapter 2
Linear Theory of Detachment Filtration With Constant Rate

ABSTRACT

Chapter 2 highlights the basic provisions of modeling the detachment filtration of aqueous suspensions at a constant rate with linear mass exchange kinetics under the operating conditions similar to reality. Thus, the initial contamination of the packed bed of the rapid filter, significant changes in the initial content of the suspended solids over time are taken into account. Thanks to the use of exact and approximate analytical methods, formulas suitable for predicting an increase in the impurity concentrations and mechanical energy expenditure are derived, firstly, in a simple integral form, and secondly, in an even simpler form of combinations of elementary functions. In addition, similar solutions to the problem of clarification of the suspension in the packed bed of a special layered structure, as well as the calculation expressions, are obtained. There are comments on the key mass exchange coefficient.

The materials contained in this chapter deserve special attention because they are the basis for further theoretical studies of aqueous suspension filtration on rapid filters, as well as for the development of practical recommendations for their rational operation. First of all, it is necessary to note an extensive area of practical application of the new developments, which in the remaining chapters of the monograph is significantly expanded due to the correction of the basic model and the system of assumptions. The subsequent regular technological and design analysis is partly based on the results obtained in this chapter. The found and tested exact and approximate analytical solutions are of the methodological value. Indeed, the procedures of the derivation of the calculation dependences are practiced on their example, which are then used for more complicated from the formal point of view filtration conditions.

The subject of an in-depth study by analytical methods was the detachment direct flow deep bed filtration of aqueous suspension with a constant rate. The linear filtration and kinetics of exchange of dispersed impurity particles between liquid and solid phases of a special porous (granular) medium, which has high absorption and flow capacity, were assumed. A mathematical problem of clarification of a low-concentration suspension with stable composition and impurity content in an initially partially clogged uniform packed bed of a rapid filter is formulated and solved rigorously (Section 2.1) and approximated (Section 2.4). In addition, an exact solution of a similar problem is obtained, but without taking into account the initial contamination of the filtering material and with a significant change with time in the suspended solids concentration at the inlet to the filter media (Section 2.2). As a generalization of the first two problems, we can consider the problem of suspension separation in a layered bed (two-

DOI: 10.4018/979-8-3693-8211-0.ch002

layer, three-layer) with a special structure and properties that ensure active participation of all layers in the removal of suspended solids (Section 2.3). In the case of relatively high initial suspended solids concentration and low absorption resource of a uniform packed bed, the second stage of filtration was distinguished when the zone of complete saturation with deposit occurs (Section 2.5). Here modeling of the clarification of an aqueous suspension was performed using approximate analytical methods on the basis of the corrected formulation of the first problem. The main calculation expressions from the many ones presented in the first five sections of the chapter are illustrated by a large number of examples with initial data corresponding to normal and extreme filtration conditions. Finally, considering the key role of the sorption (in the wide sense) component of interfacial mass transfer in suspension separation, attention is emphasized on the mechanisms of delivery of its particles to the collectors-particles (grains) and the relationship between the most important mass exchange coefficient, which directly controls the clarification process, and the sizes of the specified particles (both equivalent diameters), as well as the filtration rate.

2.1. SUSPENSION FILTRATION THROUGH INITIALLY CLOGGED UNIFORM PACKED BED (EXACT SOLUTION)

The filtration of a low-concentration suspension (natural water or partially treated wastewater) through the granular medium of a rapid filter, which initially contains insignificant amounts of a solid contaminant, is investigated using analytical methods (Poliakov, 2006a). It is preliminarily assumed that

- the packed bed is made of highly permeable, well-absorbing filtering (granular) material;
- the operating volume (active medium) is a uniform, isotropic porous medium;
- constant filtration rate;
- composition and content of a dispersed impurity at the filter inlet do not change during the calculation period;
- convective mechanism of the dispersed impurity transport prevails (relevant estimates are made in Section 1.4);
- mass exchange between the liquid and solid phases of the bed is caused by the absorption of the suspended solids and the hydrodynamic detachment of the deposited suspension particles;
- linear mass exchange kinetics;
- suspension motion in the bed obeys the linear filtration flow law;
- vertical filtration;
- the bed contains a deposit except the start of operation;
- the filter transportation system causes additional head losses determined by suspension and filtrate flow rates.

Within the framework of this monograph, the above system of assumptions can be considered as a conceptual one, since it is used except for two or three positions, which are removed, replaced or supplemented when formulating most of the mathematical problems. The specified system is necessary for adequate reflection of filtration conditions and correct formulation, in essence, of the basic problem of a rapid filtration. In the future, the system will be expanded when specifying phenomenological relations contained in the initial mathematical models. And only analytical solutions of the classes of problems of detachment and non-detachment filtration formulated in the final form (assume their existence and uniqueness) will allow to perform various applied analysis of technological process necessary for engineers.

Thus, the generalized formulation of the first filtration problem is represented in the form of two fundamental interrelated compartments – the clarification and filtration flow ones (Figure 1 from Chapter 1). Since the studies in the monograph are carried out exclusively at the macro level, the structural compartment (at micro level) is not included in the general model. The participation of the hydraulic compartment at V=*const* is expressed in the establishment of stable corrections, caused by constant hydraulic resistance of the transportation system, to the total head losses. The first compartment provides for accounting of local changes (in time) of the suspended solids content in the packed bed, which also allows analyzing the peculiarities of the primary clarification of the suspension. First of all, it contains a system of the equations of the suspended solids balance and the mass exchange kinetics between the liquid and solid phases in the following form

$$V\frac{\partial C}{\partial z} + \frac{\partial S}{\partial t} + \frac{\partial (nC)}{\partial t} = 0, \tag{2.1}$$

$$\frac{\partial S}{\partial t} = \alpha C - \beta S. \tag{2.2}$$

The vertical axis is directed from the inlet cross-section (z=0) towards the outlet cross-section (z=L), so that V>0. To ensure the uniqueness of the solution of the clarification compartment, the boundary condition at the inlet of a bed is added to system (2.1), (2.2)

$$z = 0, \quad C = C_0 = const \tag{2.3}$$

and initial conditions

$$t = 0, \quad S = S^0(z); \quad C = C^0(z). \tag{2.4}$$

The function of the initial distribution of the deposited particles $S^0(z)$, as an exception, can be calculated but is usually empirical. In a sequence of filter runs it is either stable or regularly increases markedly (after each new backwashing). In general, each new change of clean filtering material in a filter is used for a long time due to periodic regeneration. The service life of such medium is most often limited due to the gradual abrasion and crushing of its elements. At the same time, it can be significantly reduced due to progressive fouling of pore space from run to run and aging of the deposit. At present, it is practically unrealistic to extract the accumulated deposit completely even from non-cohesive porous media.

Therefore, with the exception of the first filter run (immediately after replacing the filtering material that exhausted its capacity with a new one), there will always be a residual amount of immobilized suspension particles together with the bound water in the operating volume by the beginning of the successive filter run. The distribution of a deposit over the height of the packed bed due to intensive mixing of the elements of the non-cohesive granular material in fluidized bed conditions is often close to even distribution. Then, it is justified to consider the function $S^0(z)$ as a constant quantity. In the general case, when developing the methods of calculation of suspension clarification by filtration, it is still advisable to formally specify a function $S^0(z)$ of arbitrary type and, thus, the inhomogeneous initial condition (2.4) or, at least, a homogeneous analog after obligatory correction of the absorption capacity. Characteristic special cases of such a function, as confirmed by experimental studies, can be a linear and, more often, a noted constant function (Orlov et al., 2017).

It is important to note that the mass transfer equation (2.1) is nonlinear due to close relationship between γ_a and (C_0=5•10^{-5}), so that exact analytical methods are ineffective here. However, due to the relative smallness of $\partial(nC)/\partial t$ (except for the short initial filtration period, when the above term is commensurate with $V\partial C/\partial z$), it is justified to linearize this equation by replacing the variable n by its effective (constant) value n_e. Further it will be shown that the error in calculations introduced due to linearization of the last summand in (2.1) and, moreover, due to neglecting it, turns out to be negligible.

Traditionally, the filtration flow compartment is formed, first of all, by the equation of laminar inertia-free fluid motion (Darcy's law) (1.26)

$$V = -k(S_S)\frac{\partial h}{\partial z}, \tag{2.5}$$

as well as equation (1.30), which characterizes the decrease in the flow capacity of the filter media as it becomes clogged, and the deposit composition equation (1.29), which establishes the relationship between the concentrations S_S and S. It should be noted that the filtration flow is often inertial and sometimes turbulent in highly loaded filters. Then, instead of (2.5), it is necessary to rely on the more general (1.25).

The boundary condition necessary for the uniqueness of the solution of the filtration flow compartment has the form

$$z = L, \quad h = H_d \; ; \tag{2.6}$$

where H_d is the constant head at the packed bed outlet (set as the head in the filtrate collector minus the head losses in the support layer and offtake communications).

Dimensionless variables and parameters are introduced to represent the obtained results of theoretical studies in a generalized form:

$$\overline{C} = C/C_0, \; \overline{S} = S/(n_0 C_0), \; \overline{S}^0 = S^0/n_0 C_0, \; \tilde{h} = k_0(h - H_d)/(LV),$$

$$\overline{t} = Vt/(n_0 L), \overline{z} = z/L, \; \overline{\alpha} = L\alpha/V, \; \overline{\beta} = n_0 L\beta/V, \; \overline{n}_e = n_e/n_0.$$

Then the clarification compartment takes the form

$$\frac{\partial \overline{C}}{\partial \overline{z}} + \frac{\partial \overline{S}}{\partial \overline{t}} + \overline{n}_e \frac{\partial \overline{C}}{\partial \overline{t}} = 0 \; ; \tag{2.7}$$

$$\frac{\partial \overline{S}}{\partial \overline{t}} = \overline{\alpha}\,\overline{C} - \overline{\beta}\,\overline{S} \; ; \tag{2.8}$$

$$\overline{z} = 0, \quad \overline{C} = 1 \; ; \tag{2.9}$$

$$\overline{t} = 0, \quad \overline{S} = \overline{S}^0(\overline{z}). \tag{2.10}$$

The filtration flow compartment taking into account (1.29) is simplified as follows

$$f_k(\overline{S}_S)\frac{\partial \overline{h}}{\partial \overline{z}} = -1 \; ; \tag{2.11}$$

$$\overline{S}_S(\overline{S}) = \gamma(\overline{S})\overline{S} \; ; \tag{2.12}$$

$$\overline{z} = 1, \quad \overline{h} = 0. \tag{2.13}$$

Due to the constancy of the filtration rate, as well as the complete averaging of the variable n (by $\overline{z}$ and $\overline{t}$), the first compartment is now independent of the second compartment (the feedback is preserved and is essential). The exact solution of problem (2.7) - (2.10) is found using the Laplace transform and is expressed in the images $\overline{C}_L(\overline{z},p)$, $\overline{S}_L(\overline{z},p)$ by the dependencies

$$\begin{aligned} \overline{C}_L(\overline{z},p) &= e^{-\overline{\alpha}\overline{z}}\frac{e^{-\overline{n}_e\overline{z}p+\frac{\overline{\alpha}\overline{\beta}\overline{z}}{p+\overline{\beta}}}}{p}+ \\ &+\int_0^{\overline{z}} e^{-\overline{\alpha}(\overline{z}-\eta)}\left[\overline{n}_e\overline{C}^0(\eta)+\frac{\overline{\beta}}{p+\overline{\beta}}\overline{S}^0(\eta)\right]e^{-\overline{n}_e(\overline{z}-\eta)p+\frac{\overline{\alpha}\overline{\beta}}{p+\overline{\beta}}(\overline{z}-\eta)}\,d\eta \end{aligned} \tag{2.14}$$

$$\begin{aligned} \overline{S}_L(\overline{z},p) &= \overline{\alpha}\,e^{-\overline{\alpha}\overline{z}}\frac{e^{-\overline{n}_e\overline{z}p+\frac{\overline{\alpha}\overline{\beta}\overline{z}}{p+\overline{\beta}}}}{p(p+\overline{\beta})}+\frac{\overline{S}^0(\overline{z})}{p+\overline{\beta}}+ \\ &+\overline{\alpha}\int_0^{\overline{z}} e^{-\overline{\alpha}(\overline{z}-\eta)}\left[\overline{n}_e\overline{C}^0(\eta)+\frac{\overline{\beta}}{p+\overline{\beta}}\overline{S}^0(\eta)\right]\frac{e^{-\overline{n}_e(\overline{z}-\eta)p+\frac{\overline{\alpha}\overline{\beta}}{p+\overline{\beta}}(\overline{z}-\eta)}}{p+\overline{\beta}}\,d\eta. \end{aligned} \tag{2.15}$$

The inversion of expression (2.15) gives:

at $\overline{t} \le \overline{n}_e\overline{z}$

$$\overline{S}(\overline{z},\overline{t}) = \overline{S}^0(\overline{z})e^{-\overline{\beta}\overline{t}}, \tag{2.16}$$

at $\overline{t} > \overline{n}_e\overline{z}$

$$\begin{aligned} \overline{S}(\overline{z},\overline{t}) &= \overline{S}^0(\overline{z})e^{-\overline{\beta}\overline{t}}+\overline{\alpha}\,e^{-\overline{\alpha}\overline{z}}\int_0^{\overline{t}-\overline{n}_e\overline{z}} e^{-\overline{\beta}\xi}I_0\left(2\sqrt{\overline{\alpha}\overline{\beta}\overline{z}\xi}\right)d\xi+ \\ &+\overline{\alpha}\,\overline{n}_e\,e^{-\overline{\beta}\overline{t}}\int_0^{\overline{z}} e^{-(\overline{\alpha}-\overline{\beta}\overline{n}_e)(\overline{z}-\eta)}\cdot\overline{C}^0(\eta)I_0\left(2\sqrt{\overline{\alpha}\overline{\beta}(\overline{z}-\eta)(\overline{t}-\overline{n}_e\overline{z}+n_e\eta)}\right)d\eta+ \\ &+\overline{\alpha}\overline{\beta}\,e^{-\overline{\beta}\overline{t}}\int_0^{\overline{z}} e^{-(\overline{\alpha}-\overline{n}_e\overline{\beta})(\overline{z}-\eta)}\cdot\overline{S}^0(\eta)\int_{\overline{n}_e(\overline{z}-\eta)}^{\overline{t}} I_0\left(2\sqrt{\overline{\alpha}\overline{\beta}(\overline{z}-\eta)\varsigma}\right)d\varsigma\,d\eta. \end{aligned} \tag{2.17}$$

Here and in the following I_j is the symbol of the Bessel function of the imaginary argument of the first kind of the j order.

The concentration $\overline{C}$ depends weakly on $\overline{C}^0(\overline{z})$ during prolonged filtration. Therefore, the following formula is proposed for its practical calculations (at $\overline{t} > \overline{n}_e\overline{z}$)

$$\overline{C}(\overline{z},\overline{t}) = e^{-\overline{\alpha}\overline{z}}\left[e^{-\overline{\beta}(\overline{t}-\overline{n}_e\overline{z})}I_0\left(2\sqrt{\overline{\alpha}\overline{\beta}\overline{z}(\overline{t}-\overline{n}_e\overline{z})}\right)+\overline{\beta}\int_0^{\overline{t}-\overline{n}_e\overline{z}} e^{-\overline{\beta}\xi}I_0\left(2\sqrt{\overline{\alpha}\overline{\beta}\overline{z}\xi}\right)d\xi\right]+$$
$$+\overline{\beta}\int_0^{\overline{z}} e^{-\overline{\alpha}(\overline{z}-\eta)}\cdot\overline{S}^0(\eta)e^{-\overline{\beta}(\overline{t}-\overline{n}_e\overline{z}+\overline{n}_e\eta)}I_0\left(2\sqrt{\overline{\alpha}\overline{\beta}(\overline{z}-\eta)(\overline{t}-\overline{n}_e\overline{z}+\overline{n}_e\eta)}\right)d\eta. \tag{2.18}$$

The consequences from (2.17), (2.18) the most important for practice are formulas for calculating the relative concentrations of the deposited particles of suspension at the inlet to the bed and the suspended solids at the outlet from it. In particular, the content of the immobilized particles in the inlet cross-section varies with time according to the expression

$$\overline{S}_0(\overline{t}) = \overline{S}(0,\overline{t}) = \overline{S}^0(0)e^{-\overline{\beta}\overline{t}}+\frac{\overline{\alpha}}{\overline{\beta}}(1-e^{-\overline{\beta}\overline{t}}). \tag{2.19}$$

In the initial period of the clarification, a contamination front is formed, which moves within the bed with an average relative rate $\overline{n}_e$. It reaches its lower boundary at about time $\overline{t}_e = \overline{n}_e$. On the outside of the front, suspended solids appear solely due to the detachment of a deposit and suspension particles contained in it.

It is especially important for timely termination of filtration to monitor the dynamics of the outlet concentration of the suspended solids. Dependence (2.18) allows to predict it reliably, if to assume $\overline{z} = 1$ in it. Also, it is necessary to control in parallel the head at the upper boundary of the bed. And since the head at the outlet of the bed is usually constant and set at filtration with constant rate, the knowledge of the head at the inlet of the bed means that the head losses are known in it. It is these losses that tend to receive the most attention during the filter operation. The solution of the filtration flow compartment (2.11) - (2.13) makes it possible to predict the increase in the inlet head and mechanical energy losses in the bed with time. It is expressed in the general case by the following system of the dependences with respect to the reduced head $\overline{h}$:

at $\overline{t} \le \overline{n}_e$,
if $\overline{z} \le \overline{n}_e\overline{t}$, then

$$\overline{h}(\overline{z},\overline{t}) = \int_{\overline{z}}^{\overline{n}_e\overline{t}}\frac{d\eta}{f_k(\overline{S}(\eta,\overline{t}))}+\int_{\overline{n}_e\overline{t}}^{1}\frac{d\overline{z}}{f_k(\overline{S}^0(\overline{z}))}, \tag{2.20}$$

if $\overline{z} > \overline{n}_e\overline{t}$,then

$$\overline{h}(\overline{z})=\int_{\overline{z}}^{1}\frac{d\eta}{f_k(\overline{S}^0(\eta))}; \tag{2.21}$$

at $\overline{t} > \overline{n}_e$

$$\overline{h}(\overline{z},\overline{t}) = \int_{\overline{z}}^{1}\frac{d\eta}{f_k(\overline{S}(\eta,\overline{t}))}. \tag{2.22}$$

It follows from (2.20) - (2.22) that the relative head losses in the packed bed $\Delta\overline{h}$ are as follows:

at $\bar{t} \leq \bar{n}_e$

$$\Delta\bar{h}(\bar{t}) = \bar{h}(0,\bar{t}) = \int_0^{\bar{n}_e\bar{t}} \frac{d\bar{z}}{f_k(\bar{S}(\bar{z},\bar{t}))} + \int_{\bar{n}_e\bar{t}}^{1} \frac{d\bar{z}}{f_k(\bar{S}^0(\bar{z}))}, \tag{2.23}$$

at $\bar{t} > \bar{n}_e$

$$\Delta\bar{h}(\bar{t}) = \int_0^1 \frac{d\bar{z}}{f_k(\bar{S}(\bar{z},\bar{t}))}. \tag{2.24}$$

The exact solution of problem (1.29), (1.30), (2.1) - (2.5) obtained above is illustrated by the calculation of a number of examples with typical initial data. Their main purpose is to evaluate the significance of the initial contamination of a filter media for its efficient operation. The subject of calculations are relative quantities – the volumetric concentrations of the suspended and deposited particles of suspension in characteristic sections of filtration flow, as well as the total head losses in the bed. Special attention is paid to the outlet concentration $\bar{C}_e$, because it is an indicator of filtrate quality by dispersed contaminants. And it is natural that in practice a particularly strict requirement is imposed on the quantity $\bar{C}_e$, namely, not to exceed a rigorously defined (normative) value. Also of practical interest is the dynamics of accumulating deposited particles, and in fact of the deposit in the inlet cross-section of the bed, where especially unfavorable filtration conditions are formed, i.e. $\bar{S}_0(\bar{t}) = \bar{S}(0,\bar{t})$. Besides, for estimation of the filter operability, it is important to correctly determine the head losses in the bed in dynamics, as they characterize in general the hydraulic resistance of the filtering material. Here and in the following, when solving most filtration problems and then in specific calculations of the current head, its losses and variable filtration rate, we will use the approximation expression for the (normalized) permeability function (Mintz & Meltzer, 1970; Poliakov et al., 2019), which, allowing for (1.29), takes the following form

$$f_k(S) = \left(1 - \gamma(S)\frac{S}{n_0}\right)^3 \tag{2.25}$$

and is in fact both a generalization and a consequence of (1.10) at m_{k1}=1, m_{k2}=3. In some cases, an approximation expression in exponential form will be used for $f_k(S)$. Representation (2.25) can also be written in the equivalent form $[1 - \bar{\gamma}(\bar{S})\bar{S}]^3$. In the engineering calculations $\bar{\gamma}(\bar{S})$ is replaced, as a rule, by its average value $\bar{\gamma}_a$ due to the lack of detailed data on the content of suspension particles in a deposit.

The initial information for calculations included, firstly, fixed values of the relative mass exchange coefficients $\bar{\alpha}$ (8), $\bar{\beta}$ (0.02), and the value of the averaged functional coefficient $\gamma(\bar{S})$ (γ_a=0.003). Secondly, the initial content of the immobilized particles in the bed was varied in the examples. At the same time, firstly, a number of constant values starting from 0 (initially clean bed) was taken for the $\bar{S}^0$, and then the limiting linear functions $\bar{S}^0(\bar{z})$ corresponding to the initial total content of such particles were taken. In addition, the required characteristics were calculated when varying the values of $\bar{n}_e$ within the maximum limits.

First of all, the increase in the concentration $\bar{C}_e$ over time at different amounts of deposit, not washed after the successive washing, was calculated using formula (2.18). Since it was assumed here that the initial deposit is evenly distributed over the height of the bed, then four characteristic values of $\bar{S}^0$ (0, 20, 40, 80) were taken, which correspond to four increasing fractions of $\bar{\alpha}/\bar{\beta}$ (0, 0.05, 0.1, 0.2). The series of curves of the $\overline{C}_e$ increase is shown in **Figure 1**. It is clearly seen from here that the initial contamination of a bed at the detachment filtration has a very significant influence on the removal of the suspended matter from it during the entire calculation period. Although the specified effect is somewhat weaker during prolonged filtration than at its beginning. In general, Figure 1 already convincingly demonstrates the importance of careful consideration of the initial decrease in the absorption capacity of the packed bed due to the presence of the non-washable deposit in it when predicting the content of dispersed impurities in the filtrate.

It is relevant to note that the deposit retained after washing can also intensify the deposition of the suspended solids, initiating the beginning of filter ripening. In order to adequately account for this effect in modeling of a rapid filtration, it is necessary to make corrections to kinetic equation (2.2). In fact, we have to return to its generalized form (1.13) and consider the basic mass exchange coefficient as an appropriate function of S. However, taking into account a considerable duration of continuous filtration in four stages (Jegatheesan & Vigneswaran, 2005) and specificity of the verification of the mentioned coefficient on the basis of filtrate quality data, a constant value of α, at least for the fourth stage, which is the most important for the filter control, should plausibly reflect the joint participation of both filter media and fresh deposit in the removal of dispersed contaminants. The forecast of the suspension clarification after a sufficiently long time based on the solution of problem (1.29), (1.30), (2.1) - (2.6) can serve as a basis for the subsequent technological analysis.

Figure 2 allows us to clearly evaluate possible consequences of neglecting the initial filter media contamination for both concentrations being considered. The data are presented here of calculating the special parameters

$$\Delta_C(\bar{t}) = \frac{\overline{C}_e(\bar{t};\ \overline{S}^0) - \overline{C}_e(\bar{t};\ \overline{S}^0 = 0)}{\overline{C}_e(\bar{t};\ \overline{S}^0 = 0)}, \qquad \Delta_S(\bar{t}) = \frac{\overline{S}_0(\bar{t};\ \overline{S}^0) - \overline{S}_0(\bar{t};\ \overline{S}^0 = 0)}{\overline{S}_0(\bar{t};\ \overline{S}^0 = 0)},$$

which characterize the deviation of the true values of $\overline{S}^0$ (at $\overline{S}^0 > 0$) from the reference values for comparison (at $\overline{S}^0 = 0$) at prolonged continuous filtration. The quantity $\overline{S}^0$ was found by formula (2.19). The behavior of the graphs $\Delta_S(\bar{t})$ (curves 1-3) and Δ_C (curves 4-6) is fundamentally different. The first ones have significant values only at the initial stage of filtration. Then the difference between the actual and similar approximate ($\overline{S}^0 = 0$) values is quickly leveled off with time. The quantity Δ_C turns out to be several orders of magnitude greater than Δ_S, during the entire calculation period. In general, the results of calculations of Δ_C and $\overline{C}_e$ (Figure 1) agree and testify to the importance of thorough washing, although the final conclusion about its duration and intensity can be made only on the basis of the technical and economic analysis of the filter operation.

Figure 1. Increase in the relative volumetric concentration of suspended solids in the filtrate over time

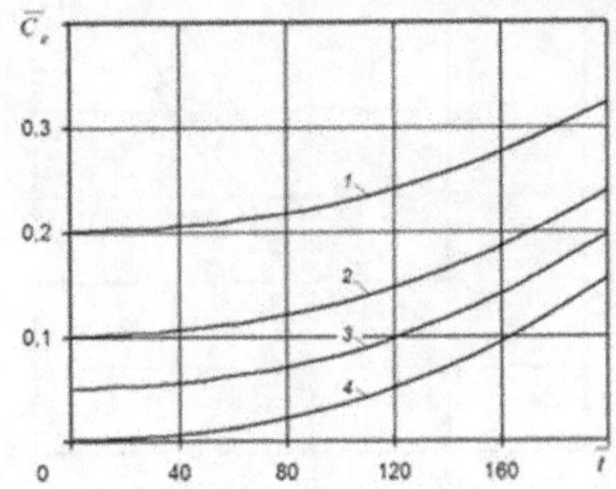

Note. 1 – $\bar{S}^0$ = 80, 2 – $\bar{S}^0$ = 40, 3 – $\bar{S}^0$ = 20, 4 – $\bar{S}^0$ = 0

Figure 2. Dependencies $\Delta_C(\bar{t})$, $\Delta_S(\bar{t})$

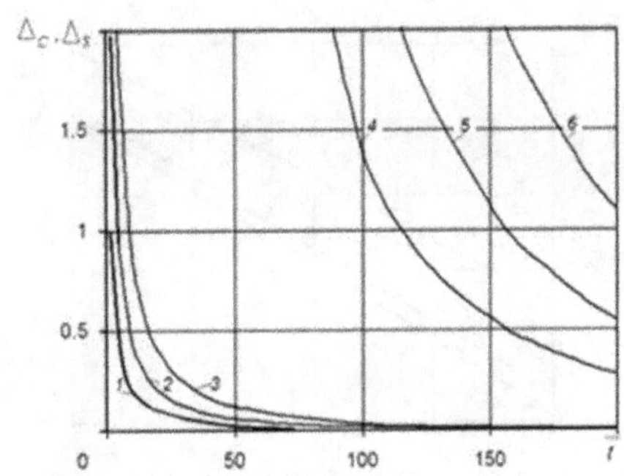

Note. 1-3 – Δ_S, 4-6 – Δ_C; 1, 4 – $\bar{S}^0$ = 20, 2, 5 – $\bar{S}^0$ = 40, 3, 6 – $\bar{S}^0$ = 80

To a lesser extent, the initial even clogging of a packed bed ($\bar{S}^0$ = *const*) is reflected in the head losses within it, as evidenced by **Figure 3**. Here are the increase curves of the indicated losses $\Delta\bar{h}$, calculated by formulas (2.17), (2.24) at three constant values of $\bar{S}^0$, including 0. Increase in losses due to the primary reduction in the flow capacity of the bed was 0.13 at $\bar{S}^0$ = 20 and 0.687 at $\bar{S}^0$ = 80, i.e. 13% and 68.7% of their value in a clean filter media. As can be seen from Fig. 3, initial water contamination significantly affects the value of $\Delta\bar{h}$ during the entire calculation period.

A better understanding of the importance of a constant value of $\bar{S}^0$ for the clarification process and its consequences can be obtained from **Figure 4**. It shows graphs of the dependence of $\bar{C}_e$ on $\bar{S}^0$ at a continuous variation of the argument over a wide range (its maximum value is 160 or $0.4\bar{\alpha}/\bar{\beta}$). Three representative moments in time are chosen corresponding to filtration under conditions of weak ($\bar{t}$ = 50), moderate ($\bar{t}$ = 100) and severe ($\bar{t}$ = 200) clogging of the bed. It is natural that the quality of filtrate steadily deteriorates with the increase in $\bar{S}^0$ and the calculation point of time, and this dependence is almost linear. Thus, in order to avoid too early and frequent filter shutdowns due to excessive weakening of its protective capability, it is necessary to thoroughly wash out the clogged bed.

Figure 3. Increase in the relative head losses in the packed bed over time

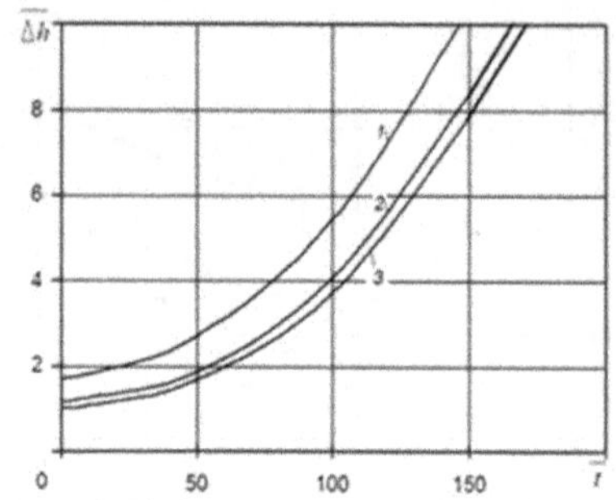

Note. 1 – $\overline{S}^0$ = 80, 2 – $\overline{S}^0$ = 20, 3 – $\overline{S}^0$ = 0

Figure 4. Dependence $\overline{C}_e(\overline{S}^0)$

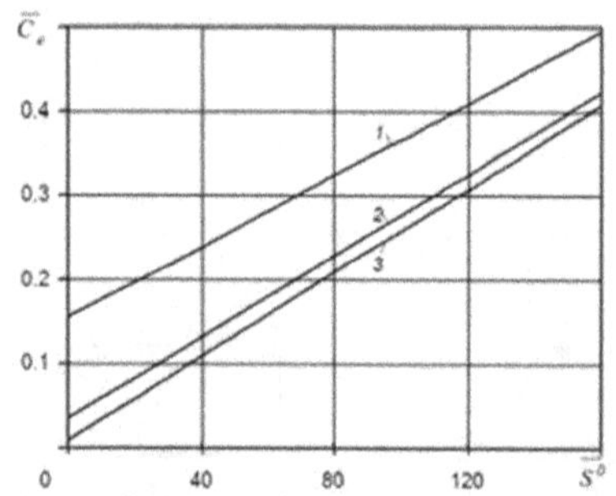

Note. 1 – $\overline{t}_*$ = 200, 2 – $\overline{t}_*$ = 100, 3 – $\overline{t}_*$ = 50

Finally, the filtrate contamination dynamics was analyzed again, but now with different nature of the deposit distribution in the bed immediately after its washing. The total primary volume of the deposited particles was assumed to be unified and equal (in relative units) to 20. Further calculations of $\overline{C}_e$ were carried out for three options of setting the quantity $\overline{S}^0$. First of all, the concentration $\overline{S}^0$ was considered constant and the corresponding curve was obtained earlier (Figure 1). The other two options were a linear function of the initial distribution of the deposited particles and a limit function in the sense that $\overline{S}^0$ takes the minimum value (0) at one bed boundary and the maximum value (40) at the other boundary. Thus, two more such expressions were used for the function $\overline{S}^0(\overline{z})$

$$\overline{S}^0_1(\overline{z}) = 40\overline{z}, \qquad \overline{S}^0_2(\overline{z}) = 40 - 40\overline{z}.$$

Consequently, the deposit was initially located at $\overline{S}^0_1(\overline{z})$ predominantly in the lower part of the bed, while at $\overline{S}^0_2(\overline{z})$, on the contrary, in the upper part. The data of calculations of $\overline{C}_e$ are shown in **Figure 5**. The curve $\overline{C}_e(\overline{t})$ occupies an intermediate position at $\overline{S}^0$ = *const*. The curve corresponding $\overline{S}^0_1$ up to approximately $\overline{t}$ = 100 is located noticeably higher than the other two curves. And all curves get closer with increasing time and $\overline{C}_e$ takes practically the same value (0.196) at $\overline{t}$ = 200 for all options of $\overline{S}_0$. It is obvious that the unevenness factor can seriously affect the quality of filtrate in the considered conditions of the detachment filtration at duration of filter operation before the next washing less than 100. But this factor becomes insignificant at longer duration.

It should be emphasized that the main filtration characteristics were calculated in parallel for two limiting values of $\overline{n}_e$ (0 and 1). This allowed to use dependences (2.17), (2.18), (2.24) in both cases. In fact, in a similar way the limit values (majorants) of the mentioned characteristics were found, which limit the entire array of their possible values from above and below. And since the difference between the majorants at $\overline{t} > 1$ turned out to be negligible (much less than 1%), the term with local derivative $\partial C/\partial t$ was discarded in the subsequent theoretical studies of the deep bed filtration.

Figure 5. Increase in the relative volumetric concentration of suspended solids in the filtrate over time

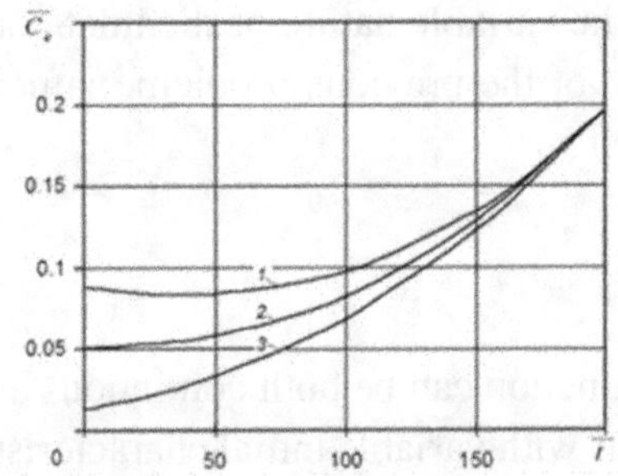

Note. $1 - \overline{S}^0 = \overline{S}^0_1(\overline{z})$, $2 - \overline{S}^0 = 20$, $3 - \overline{S}^0 = \overline{S}^0_2(\overline{z})$

At the end of this Section it is expedient to emphasize the formal existence of a limit value S_d for the concentration S within the adopted model, which according to (1.13), (2.19) is reached at $\overline{t} \rightarrow \infty$ and is simply expressed through α, β, C_0. In fact, the given value characterizes the (situational limit) absorption capacity of the filter media in the emerging hydrodynamic conditions. When the hydraulic load changes, the dynamic solids take-up capacity with respect to the deposited suspension particles S_d also changes accordingly. In contrast to S_d, the true solids take-up capacity S_m (or its analog S_{Sm} with respect to the deposit) depends mainly on the structural and physicochemical properties of the filtering material of the packed bed. Therefore, two characteristic situations can occur depending on the nature of a deposit accumulation, and in fact the relationship between S_d and S_m with linear mass exchange kinetics (linear filtration). The first situation ($S_m > S_d$) has been discussed above. When the second situation is realized ($S_d > S_m$), there should come a moment in time t_s, during the calculation period when the true solids take-up capacity is exhausted at the inlet cross-section of the bed. Then further use of the calculation dependences derived in this section will lead to errors increasing from this moment. They can be noticeably reduced via appropriately correcting the basic problem. Section 2.5 is dedicated to such adjustment and clarification of its consequences.

2.2. SUSPENSION FILTRATION WITH VARIABLE INITIAL CONCENTRATION OF SUSPENDED SOLIDS

Stability of the hydraulic load, content and composition of fine impurities in the suspension being clarified contributes to the stable operation of water treatment filters for a long time. Therefore, theoretical studies of the clarification process by filtration were previously carried out, as a rule, under the condition of constancy of the initial content of the suspended solids (C_0=*const*). The corresponding boundary condition (2.3) was included in the basic problem formulation from the previous section. However, in practice, the inlet concentration of impurity can sometimes change rapidly with time, which is conditioned by man-made and natural features of water pollution, functioning of the filtration facilities. Changes in the quantity C_0 can be smooth or abrupt and are related to the change of the contaminated water source, extraordinary events in the environment (heavy rains, floods), cyclic water management activities of industrial, agricultural, municipal facilities, etc. The situation is similar in the case of clarification of turbid water, e.g., storm drains on stepped filters and filters with nonuniform medium. The similar situation is also formed at the clarification of turbid waters, for example, storm drains, on step filters and filters with layered medium. It will be analyzed in detail in the next section.

Thus, to accurately account for the variable nature of the inlet concentration C_0, it is necessary and sufficient to correct the formulation of the previous problem, namely, to replace condition (2.3) with the following more general condition

$$z = 0, \qquad C = C_0(t), \tag{2.26}$$

where the initial concentration function can be both continuous and discontinuous.

Traditionally, in applied problems with variable initial characteristics and coefficients one resorts to a significant simplification of the latter, which consists in their averaging over the calculation period. However, this formal technique is difficult to apply in the considered filtration problem with condition (2.26) due to the uncertainty of the specified time interval – the duration of the working period is unknown in advance.

Other corrections, which are made in the formulation of the basic problem (2.7) -(2.13), allow us to formulate the new problem in a simpler way. Thus, taking into account the relative smallness, with few exceptions, of changes in the suspended solids concentration with time, $\partial(nC)/\partial t$ is discarded in equation (2.1). Also, the packed bed is assumed to be initially clean (S^0=0). Such simplifications are justified since the focus of this section is on the variable $C_0(t)$ and its role in the clarification process. The other components of the above-mentioned formulation are retained (Poliakov, 2008a). A peculiarity of the new problem is the absence in its conditions of a scale for the desired concentrations due to the now unstable nature of C_0. However, the choice of such a scale is not of fundamental importance and it is possible, for example, to correlate C and S with the initial value $C_{00}=C_0(0)$. Then the relative outlet concentration at the first calculation point in time, as in the problem from Section 2.1 at $\bar{S}^0 = 0$, will be $e^{-\bar{a}}$. In this case, the form of the adjusted boundary condition (2.26) after making it dimensionless, in fact, remains the same as before

$$\bar{z} = 0, \qquad \bar{C} = \bar{C}_0(\bar{t}). \tag{2.27}$$

Thus, equation (2.7) and the initial condition after the above simplifications will be

$$\frac{\partial \bar{C}}{\partial \bar{z}} + \frac{\partial \bar{S}}{\partial \bar{t}} = 0, \tag{2.28}$$

$$\bar{t} = 0, \quad \bar{S} = 0, \tag{2.29}$$

where $\overline{C} = C/C_{00}$, $\overline{S} = S/(n_0 C_{00})$.

To solve the clarification compartment in a new formulation (Poliakov, 2008a), the operational method is again applied, so that the initial concentrations in the images $\overline{C}_L$, $\overline{S}_L$ are of the form

$$\overline{C}_L(\overline{z},p) = \overline{C}_{0L}(p)e^{-\frac{\overline{\alpha}p}{p+\overline{\beta}}\overline{z}}, \quad \overline{S}_L(\overline{z},p) = \frac{\overline{\alpha}\,\overline{C}_{0L}(p)}{p+\overline{\beta}}e^{-\frac{\overline{\alpha}p}{p+\overline{\beta}}\overline{z}}. \tag{2.30}$$

The inversion of (2.30) gives for the original $\overline{S}$ this expression

$$\overline{S}(\overline{z},\overline{t}) = \overline{\alpha}\,e^{-\overline{\alpha}\overline{z}}\int_0^{\overline{t}} e^{-\overline{\beta}(\overline{t}-\xi)}I_0\left(2\sqrt{\overline{\alpha}\overline{\beta}\overline{z}(\overline{t}-\xi)}\right)\overline{C}_0(\xi)d\xi \tag{2.31}$$

The dependence $\overline{C}(\overline{z},\overline{t})$ is derived using the equation

$$\overline{C} = \frac{1}{\overline{\alpha}}\frac{\partial\overline{S}}{\partial\overline{t}} + \frac{\overline{\beta}}{\overline{\alpha}}\overline{S}. \tag{2.32}$$

First of all, by differentiating (2.31) on $\overline{t}$ obtain

$$\frac{\partial\overline{S}}{\partial\overline{t}} = \overline{\alpha}\,e^{-\overline{\alpha}\overline{z}}\left[\overline{C}_0(\overline{t}) - \overline{\beta}\overline{S} + \sqrt{\overline{\alpha}\overline{\beta}\overline{z}}\int_0^{\overline{t}}\frac{e^{-\overline{\beta}(\overline{t}-\xi)}I_1\left(2\sqrt{\overline{\alpha}\overline{\beta}\overline{z}(\overline{t}-\xi)}\right)}{\sqrt{\overline{t}-\xi}}\overline{C}_0(\xi)d\xi\right]. \tag{2.33}$$

Then it follows from (2.32) taking into account (2.33) that

$$\overline{C}(\overline{z},\overline{t}) = e^{-\overline{\alpha}\overline{z}}\left[\overline{C}_0(\overline{t}) + \sqrt{\overline{\alpha}\overline{\beta}\overline{z}}\int_0^{\overline{t}}\frac{e^{-\overline{\beta}(\overline{t}-\xi)}I_1\left(2\sqrt{\overline{\alpha}\overline{\beta}\overline{z}(\overline{t}-\xi)}\right)}{\sqrt{\overline{t}-\xi}}\overline{C}_0(\xi)d\xi\right]. \tag{2.34}$$

From (2.33), (2.36) we obtain the important in practical terms formulas for calculating the concentration of the deposited suspension particles increasing over time at the inlet to the filter media

$$\overline{S}_0(\overline{t}) = \overline{S}(0,\overline{t}) = \overline{\alpha}\int_0^{\overline{t}} e^{-\overline{\beta}(\overline{t}-\xi)}\overline{C}_0(\xi)d\xi \tag{2.35}$$

and the outlet concentration of the suspended solids

$$\overline{C}_e(\overline{t}) = \overline{C}(1,\overline{t}) = e^{-\overline{\alpha}}\left[\overline{C}_0(\overline{t}) + \sqrt{\overline{\alpha}\overline{\beta}}\int_0^{\overline{t}}\frac{e^{-\overline{\beta}(\overline{t}-\xi)}I_1\left(2\sqrt{\overline{\alpha}\overline{\beta}(\overline{t}-\xi)}\right)}{\sqrt{\overline{t}-\xi}}\overline{C}_0(\xi)d\xi\right]. \tag{2.36}$$

It is possible that the impurity content in the suspension may change abruptly in some periods of filter operation. Then the quantity $\overline{C}_0$ should be taken as the following piecewise constant function

$C_0 = C_{01} = const$, at $0 < t \leq t_1$;

$C_0 = C_{02} = const$, at $t > t_1$.

If we use the value of C_{01} to scale the concentrations, then the expression for $\overline{C}_0$ is

$$\begin{aligned} \overline{C}_0 &= 1, \text{ at } 0 \le \bar{t} \le \bar{t}_1 \,; \\ \overline{C}_0 &= \overline{C}_{02}, \text{ at } \bar{t} > t_1 \,; \end{aligned} \tag{2.37}$$

where $\overline{C} = C/C_{01}$, $\overline{C}_{02} = C_{02}/C_{01}$. In this case it is possible to use at the first stage of the calculations of both concentrations either formulas (2.17), (2.18), derived in the previous section, or equivalent formulas (2.31), (2.34) at $\overline{C}_0 = 1$ up to the moment of a jump in the inlet concentration of the impurity, when their distributions along the height of a bed will appear in the following form: $\overline{S}_1(\bar{z}) = \overline{S}(\bar{z}, \bar{t}_1)$, $\overline{C}_1(\bar{z}) = \overline{C}(\bar{z}, \bar{t}_1)$. At the second stage $(\bar{t} > \bar{t}_1)$ the following expressions will be valid

$$\overline{S}_2(\bar{z}, \bar{t}) = \overline{S}_1(\bar{z}) + \bar{\alpha}\, \overline{C}_{02} e^{-\bar{\alpha}\bar{z}} \int_{\bar{t}_1}^{\bar{t}} e^{-\bar{\beta}(\bar{t}-\xi)} I_0\left(2\sqrt{\bar{\alpha}\bar{\beta}\bar{z}(\bar{t}-\xi)}\right) d\xi, \tag{2.38}$$

$$\begin{aligned} \overline{C}_2(\bar{z}, \bar{t}) = e^{-\bar{\alpha}\bar{z}} \Bigg\{ \overline{C}_{02} + \sqrt{\bar{\alpha}\bar{\beta}\bar{z}} \Bigg[\int_0^{\bar{t}_1} \frac{e^{-\bar{\beta}(\bar{t}_1-\xi)} I_1\left(2\sqrt{\bar{\alpha}\bar{\beta}\bar{z}(\bar{t}_1-\xi)}\right)}{\sqrt{\bar{t}_1-\xi}} d\xi + \\ + \overline{C}_{02} \int_{\bar{t}_1}^{\bar{t}} \frac{e^{-\bar{\beta}(\bar{t}-\xi)} I_1\left(2\sqrt{\bar{\alpha}\bar{\beta}\bar{z}(\bar{t}-\xi)}\right)}{\sqrt{\bar{t}-\xi}} d\xi \Bigg] \Bigg\}. \end{aligned} \tag{2.39}$$

Therefore, in particular,

$$\overline{S}_{20}(\bar{t}) = \frac{\bar{\alpha}}{\bar{\beta}} \left[1 + \overline{C}_{02} - e^{-\bar{\beta}\bar{t}_1} - \overline{C}_{02} e^{-\bar{\beta}(\bar{t}-\bar{t}_1)} \right]. \tag{2.40}$$

Calculations of the impurity content in the deposited and suspended states are simplified if the quantity $\overline{C}_0(\bar{t})$ is averaged over the calculation period $[0,T]$, and then to use in the above formulas instead of the variable $\overline{C}_0(\bar{t})$ its average value

$$\overline{C}_{0a} = \frac{1}{T} \int_0^T \overline{C}_0(\bar{t})\, d\bar{t}$$

As a result, for instance,

$$\overline{S}_0(\bar{t}) = \frac{\bar{\alpha}\, \overline{C}_{0a}}{\bar{\beta}} \left(1 - e^{-\bar{\beta}\bar{t}}\right), \tag{2.41}$$

$$\overline{C}_e(\bar{t}) = \overline{C}_{0a} e^{-\bar{\alpha}} \left[1 + \sqrt{\bar{\alpha}\bar{\beta}} \int_0^{\bar{t}} \frac{e^{-\bar{\beta}(\bar{t}-\xi)} I_1\left(2\sqrt{\bar{\alpha}\bar{\beta}(\bar{t}-\xi)}\right)}{\sqrt{\bar{t}-\xi}} d\xi \right]. \tag{2.42}$$

Incidentally, at (2.37)

$$\overline{C}_{0a} = \frac{\bar{t}_1 + \overline{C}_{02}(T - \bar{t}_1)}{T}.$$

Obviously, the determination of $\overline{C}_{0a}$ becomes more complicated when there is a priori unknown T. The representation of the filtration flow compartment for the first problem and the problem considered here is identical. Naturally, the calculation formulas derived in the previous section concerning the reduced head $\tilde{h}$ and its relative losses $\Delta\bar{h}$ remain the same. However, if $\overline{C}_0(\bar{t})$ has the form (2.37), then the head losses by the moment of its sharp change will be as follows

$$\Delta\bar{h}(\bar{t}_1) = \int_0^1 \frac{d\bar{z}}{f_k(\overline{S}_1(\bar{z}))}.$$

Their further increase $(\bar{t} > \bar{t}_1)$ is described by the dependence

$$\Delta\bar{h}(\bar{t}) = \int_0^1 \frac{d\bar{z}}{f_k(\overline{S}_2(\bar{z},\bar{t}))}. \tag{2.43}$$

The quantitative analysis was performed in this Section to illustrate the calculation formulas derived therein, to assess the reality of the process intensification by purposefully changing the initial suspended solids content, and finally, to verify the validity of averaging the variable inlet concentration of the suspended solids. Attention was focused on the behavior of the key filtration characteristics, first of all, $\overline{C}_e$ and also $\overline{S}_0$ and $\Delta\bar{h}$, but now in connection with the quantity $\overline{C}_0(\bar{t})$. Four examples were calculated in parallel. Single values of $\bar{\alpha}$ (8), $\bar{\beta}$ (0.02) and $\bar{\gamma}_a$ (0.003) were assumed in them, as before. The differences between the examples are due to the different nature of the change in $\overline{C}_0$. Thus, a linear increasing function is given in the first example

$$\overline{C}_0(\bar{t}) = 1 + a_c\bar{t}, \tag{2.44}$$

with the value of 0.01 selected for a_c. Since the duration of the calculation period was taken equal to 200, the suspended solids content at the inlet to the bed increased by 3 times by its end. Such an extreme situation can occur in the surface water treatment after storm precipitation, intensive snowmelt in early spring. In the second example, the function $\overline{C}_0(\bar{t})$ was piecewise constant (12), with a jump at the time $\bar{t}_1 = 100$. In order to ensure that the same specific (through the inlet surface of a unit area) volume of impurity W_C, as in the first example, was delivered to the bed in the entire time T=200, exactly,

$$W_C = \int_0^T (1 + a_c\bar{t})d\bar{t} = T(1 + 0.5a_cT) = 400,$$

the value of $\overline{C}_{02}$ should be

$$\overline{C}_{02} = (W_C - \bar{t})/(T - \bar{t}_1) = 3.$$

In the third example, the value of the averaging constant $\overline{C}_{0a} = 2$ was used to preserve the value of W_C. Finally, the fourth example also involved for $\overline{C}_0$ a linear but decreasing function

$$\overline{C}_0(\bar{t}) = 3 - b_c\bar{t}, \tag{2.45}$$

and to ensure the same value W_C=400, it was assumed b_c=a_c=0.01.

The series of graphs of the dependence $\overline{C}_e(\bar{t})$, which were calculated for the selected examples by formulas (2.36), (2.39) and (2.42), are shown in **Figure 6**. Of particular interest is the mutual arrangement of curves 1-3, corresponding to three significantly different ways of smooth adjusting the initial suspended solids content. Hence, it is obvious that with regard to the quality of aqueous suspension clarification, the increase in its contamination (curve 3) is clearly preferable during the filter operation. The breakthrough of the suspension particles slightly increased at the stable $\overline{C}_0$ (curve 2). The worst result here was obtained when the suspension contamination dropped (curve 1). And the outlet suspended solids concentration increased most slowly in the case of abrupt change in contamination from low to high level (curve 4).

Figure 6. Increase in the relative volumetric concentration of suspended solids in filtrate over time

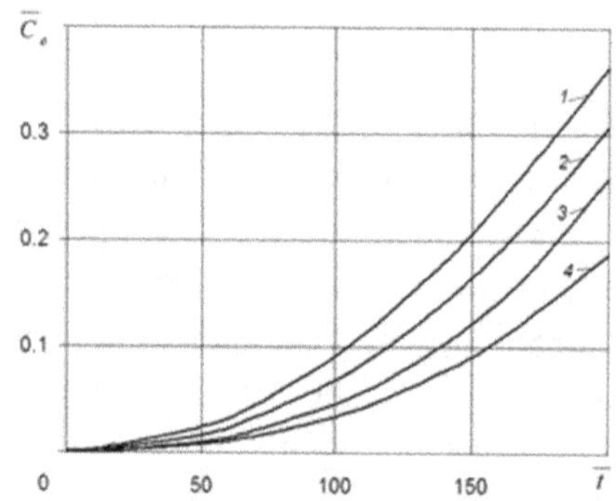

Note. 1 – example 4, 2 – example 3, 3 – example 1, 4 – example 2

Curves 1-3 also assist to assess the validity of formal averaging of the quantity $\overline{C}_0(\bar{t})$. Curve 2 was actually determined with the averaging constant ($\overline{C}_{0a} = 2$) single for the calculation period. That caused increasing divergence of curve 2 from the basic (accurate) curves 1, 3 with time, and hence an increase in errors when calculating $\overline{C}_e$. It is possible to reduce them noticeably if we resort to an unconventional adaptive averaging. Then the specified constant is found at each computational step ($\bar{t} = T$) anew, so that $\overline{C}_{0a} = \overline{C}_{0a}(T)$. Curves 1 – 3 in **Figure 7** are duplicated from Fig. 6 and supplemented with curves 2, 4, which were also determined by formula (2.42), but with the constant $\overline{C}_{0a}$ changed in accordance with the procedure of the adaptive averaging. Thus, a high degree of water purification (reduction of $\overline{C}_0$ by an order of magnitude or more) occurred at relatively small values of $\bar{t}$ ($\bar{t} < 100 - 140$). It is here that the new averaging procedure proved to be especially effective and gave the values of $\overline{C}_e$ which were very significantly different when using (2.44), (2.45) and approached the exact values.

Figure 7. Growth of the relative volumetric concentration of suspended solids in the filtrate over time

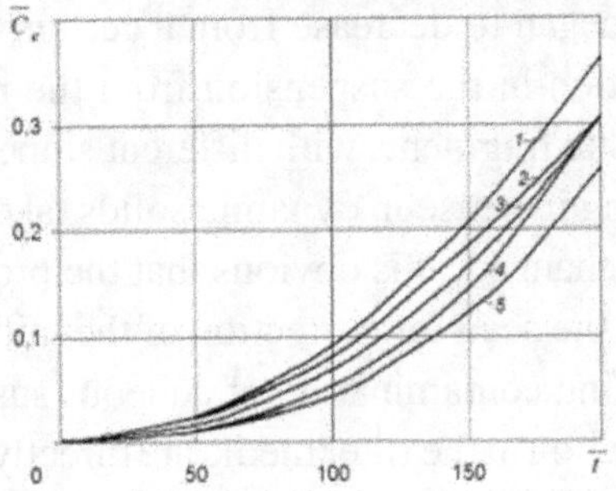

Note. 1 – example 4, 2 – example 3 at $\overline{C}_{0a}(T)$ and (b_c), 3 – example 3$(\overline{C}_{0a} = 2)$, 4 – example 3 at $\overline{C}_{0a}(T)$ and (a_c), 5 – example 1

Figure 8. Dependence $\overline{C}_e(\overline{C}_{0a})$

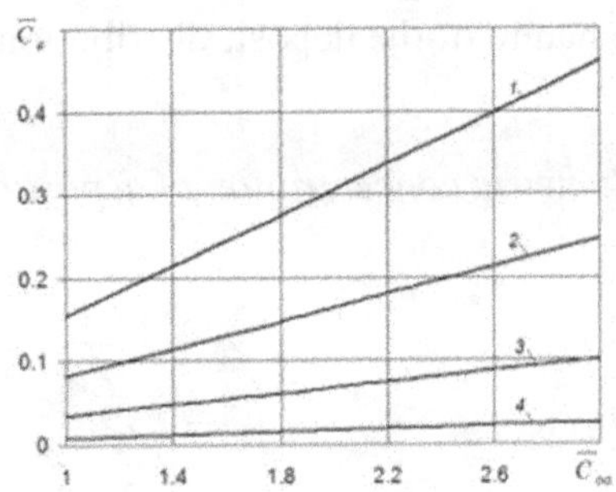

Note. 1 – $\bar{t}$ = 200, 2 – $\bar{t}$ = 150, 3 – $\bar{t}$ = 100, 4 – $\bar{t}$ = 50

The necessity to be cautious at formal simplification of the filtration conditions at the filter inlet is also evidenced by **Figure 8**, where curves of the dependence of the concentration $\overline{C}_e$ on the averaging constant $\overline{C}_{0a}$ are depicted. Calculations were carried out by formula (2.42) with continuous variation of $\overline{C}_{0a}$ at the entire interval of the values according to the accepted initial data (from 1 to 3) for four moments in time. These moments correspond to the following exact values of $\overline{C}_e$: using (2.44) and (2.45) respectively 0.01 and 0.024 at $\bar{t} = 50$; 0.046 and 0.09 at $\bar{t} = 100$; 0.259 and 0.363 at $\bar{t} = 200$. To make the approximate values of the desired $\overline{C}_e$ coincide with the exact values, we should take $\overline{C}_{0a} = 1.18$, 2.76 in the first case, $\overline{C}_{0a} = 1.335$, 2.64 – in the second case, $\overline{C}_{0a} = 1.685$, 2.365 – in the third case. Thus, it is very problematic to successfully select a single value of $\overline{C}_{0a}$, especially for a long calculation period and significant changes in $\overline{C}_0$ during it.

While the nature of the change in the key characteristic $\overline{C}_e$ from the sanitary point of view is similar for the various options selected for the function $\overline{C}_0(\bar{t})$, and only the rate of its increase differs, and to a small extent, then the effect of C_0 directly on the concentration S and indirectly on the head losses Δh is clearly stronger and can lead to fundamental differences in the form of the corresponding relationships. **Figure 9** shows the curves of the change in the inlet concentration $\overline{S}_0$ with time calculated for these four examples using formulas (2.35), (2.40) and (2.41). And if at constant and increasing smoothly or stepwisely concentration C_0 (examples 1-3) the required value increased monotonically, though with essentially different rate, then at decreasing concentration C_0 by the end of the calculation period $\overline{\alpha}\,\overline{C}$ there was less $\overline{\beta}\,\overline{S}$ at the upper site, and thus, detachment already prevailed over attachment. As a result, the amount of a deposit here even began to decrease from a certain point in time (curve 1). The sharp increase of the impurity concentration in the suspension from the moment $\overline{t}_1$ is reflected in curve 4, which clearly shows two characteristic fragments with different slopes. It should be emphasized that the increase in C_0 means a corresponding increase in dynamic solids take-up capacity, but not in true solids take-up capacity, which remains unchanged. It is obvious that the probability of premature depletion of the true solids take-up capacity and the need for correction of the initial model increases in this connection under the conditions of increasing contamination of aqueous suspension.

The peculiarities of deposit formation in the filter media are directly reflected in the head losses, which were determined here using formula (2.24). The calculated curves of the dependence $\Delta\overline{h}(\overline{t})$ for different options of $\overline{C}_0(\overline{t})$ are shown in **Figure 10**. Partial mobilization at the top of the medium of the deposit in case of gradual reduction of its initial content leads to some leveling of the profiles $\overline{S}(\overline{z})$, which in turn can cause such an anomalous for filtration consequence as a decrease in the value $\Delta\overline{h}$ over time (curve 1). And it is important to note that such a decrease is observed against the background of progressive medium clogging as a whole. This fact once again indicates the great importance for the head losses in the layer of filtering material of the nature of the deposit distribution in it.

Figure 9. Change in the relative volumetric concentration of deposited suspension particles at the inlet to the bed over time

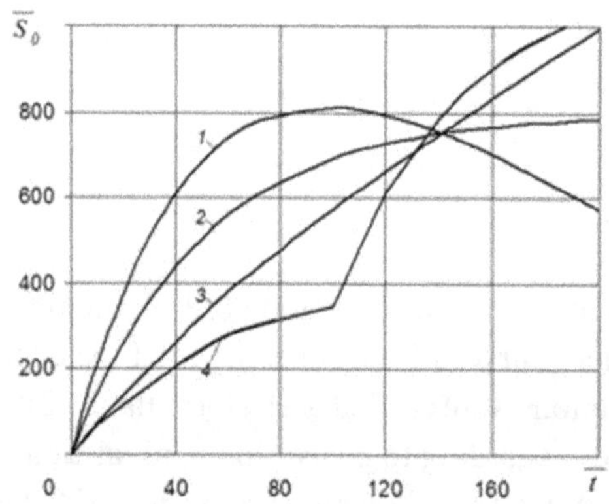

Note. 1 – example 4, 2 – example 3, 3 – example 1, 4 – example 2

Figure 10. Growth of the relative head losses in the bed over time

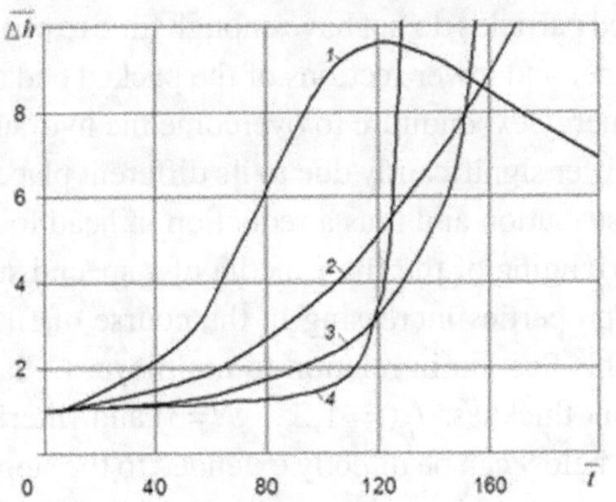

Note. 1 – example 4, 2 – example 3, 3 – example 1, 4 – example 2

Increased rise in the deposit volume since moment $\bar{t}_1 = 100$ (example 2) formally favors early unrestricted increase in $\Delta\bar{h}$ (curve 4). From a technological point of view, this means that in such situation (jump of C_0) a noticeable extension of high-quality clarification of the suspension is possible (curve 4 in Figure 6), which, however, is accompanied by premature excessive clogging of a packed bed. The latter circumstance may cause forced washing of the filter in a short time after the start of the next operating run. It is appropriate to note that such a rapid increase in $\Delta\bar{h}$ contradicts the physical essence of the filtration process, since, firstly, a very significant part of the pore space (sometimes up to half) always remains inaccessible to a deposit, and, therefore, the flow capacity of the bed is in principle unable to fall to 0. Secondly, a decrease in the functional coefficient $\gamma(S)$ is observed with increasing the concentration S (Ojha & Graham, 1993; Poliakov et al., 2019). There are two ways to avoid physically unjustified excessive rise of hydraulic resistance of the bed. First, while maintaining the linearity of the mass exchange kinetics, it is proposed to introduce a zone of maximum saturation with a deposit, expanding with time, and the second stage of filtration, which will be done in the fifth section of this chapter. The second way is to use a nonlinear kinetic equation, which contains a special parameter (true solids take-up capacity) that strictly limits the deposit volume increase. It will be realized in chapter 6. Finally, if curves 1, 3 are considered as the reference curves due to the strictness of accounting $C_0(t)$ and curve 2 as an approximate one due to the use of the constant $\bar{C}_{0a}$, the fact that they intersect at $\bar{t} \approx 156$ (curves 1 and 2) and at $\bar{t} \approx 150$ (curves 2 and 3), followed by a sharp divergence, indicates once again the possible serious complications associated with the averaging of the inlet concentration function.

In conclusion, it should be emphasized that for reliable calculation of the clarification of aqueous suspensions by filtration at extreme water contamination and, as a consequence, clogging of the filter bed, we must have accurate baseline information in its entirety for a wide range of physicochemical and water flow conditions.

2.3. SUSPENSION FILTRATION THROUGH MULTILAYER FILTER MEDIA

The efficiency of rapid filters with uniform medium is usually significantly reduced due to the preferential deposition of suspension particles at its inlet section. Obviously, the filtering material must have a high absorption capability to guarantee an almost complete removal of the dispersed impurity at a generally small thickness of the operating porous layer and the time of the presence of the suspension inside. However, most suspended particles do not have enough time to penetrate deeply into the indicated layer. As a consequence, the upper and lower sections of the packed bed are clogged quite differently. It is known that the mechanical energy expenditure to overcome the hydraulic resistance of the bed at the same amount of a deposit can differ significantly due to its different placement. There are many ways to achieve a more even deposit distribution and thus a reduction in head losses. One of the effective ways of filtration intensification is arranging of the filter media of a special structure, namely, composed of several layers with absorption properties increasing in the course of filtration. The theory of suspension filtration is developed in this Section in relation to multilayer (N layers), and actually three-layer bed formed by layers of different thickness l_i (i= 1,2,…,N=3) and filtering material (Poliakov, 2006b, 2008b). All the results obtained below can be directly extended to the stepped filters, since the operation of each of a number of filters connected in series can be formally identified with the operation of the corresponding layer of a nonuniform bed (Poliakov & Shevchuk, 2007).

It is logical to assume that mathematical model similar to (2.7) - (2.13) is justified to apply to each of the layers of a multilayer bed, but necessarily taking into account its individual absorption and filtration flow properties. Therefore, the general model of the depth filtration in this case is compiled as a set of models identical in structure and similar to those presented in Sections 2.1, 2.2 with indication by means of appropriate indexing of dependent variables and parameters (of the specific layer) to which they directly attribute. Thus, the clarification component (compartment) of the general model will be in dimensionless form

$$\frac{\partial \overline{C}_i}{\partial \overline{z}} + \frac{\partial \overline{S}_i}{\partial \overline{t}} = 0, \tag{2.46}$$

$$\frac{\partial \overline{S}_i}{\partial \overline{t}} = \overline{\alpha}_i \overline{C}_i - \overline{\beta}_i \overline{S}_i \ ; \tag{2.47}$$

$$\overline{z} = 0, \quad \overline{C}_1 = 1; \qquad \overline{t} = 0, \quad \overline{S}_i = 0; \qquad i = 1,\ 2, ...,\ N, \tag{2.48}$$

where $\overline{C}_i = C_i/C_0$; $\overline{S}_i = S_i/(n_{01} C_0)$; $\overline{t} = Vt/(n_{01} L)$; n_{01} is the porosity of the clean top layer (downward suspension flow); L is the thickness of the whole bed, so that $L = \sum_{i=1}^{N} l_i$. Due to the smallness of the diffusion transport of the suspended matter in the bed (fifth position from the conceptual system of assumptions), the connection between its layers has a one-way nature, namely, physical and chemical processes in the overlying layer actively influence similar processes in the underlying layer, and the reverse influence is practically absent. Therefore, N–1 boundary conditions expressing equality of the suspended matter concentrations on both sides of each interlayer boundary are accepted, i.e.

$$\overline{z} = \overline{L}_i = \sum_{j=1}^{i} \overline{l}_j, \qquad \overline{C}_i = \overline{C}_{i+1}; \quad (i = 1,\ 2, ...,\ N-1) \tag{2.49}$$

where $\overline{L}_i = L_i/L$, $\overline{l}_j = l_j/L$. The conditions of equality of mass flows on the same boundaries due to (2.49) and constancy of the filtration flow rate are automatically fulfilled.

The filtration flow component of the model includes a system of N nonlinear equations of linear (Darcy's law) water flow

$$\bar{k}_{0i} f_{ki}(\bar{S}_{si}) \frac{\partial \bar{h}_i}{\partial \bar{z}} = -1, \tag{2.50}$$

as well as similar systems of the expressions for the relative hydraulic conductivity of all layers

$$\bar{k}_i(\bar{S}_{si}) = \bar{k}_{0i} f_{ki}(\bar{S}_{si}) \tag{2.51}$$

and the equations of a deposit composition within each layer

$$\bar{S}_{si}(\bar{S}_i) = \gamma_i(\bar{S}_i)\, \bar{S}_i. \tag{2.52}$$

Here $\bar{h}_i = k_{01}(h_i - H_d)/(VL)$, $\bar{k}_i = k_i/k_{01}$, $\bar{k}_{0i} = k_{0i}/k_{01}$. The conditions at the bottom and interlayer bed boundaries are added to equations (2.50) - (2.52)

$$\bar{z} = 1, \quad \bar{h}_N = 0\ ; \tag{2.53}$$

$$\bar{z} = \bar{L}_i, \quad \bar{h}_i = \bar{h}_{i+1}; \quad i = 1, 2, ..., N-1. \tag{2.54}$$

Since the suspension in the upper layer is clarified at a constant filtration rate regardless of the structure and properties of the rest of the packed bed, the exact solution of the clarification compartment for it is actually already presented in Section 2.1. Below it is duplicated for the convenience of the calculations, but in a simplified form due to the homogeneity of conditions (2.48) and with indication of its belonging to the first (upper) layer. Thus, at $\bar{l}_1 \geq \bar{z} \geq 0$

$$\bar{C}_1(\bar{z},\bar{t}) = e^{-\bar{\alpha}_1 \bar{z}}\left[e^{-\bar{\beta}_1 \bar{t}} I_0\left(2\sqrt{\bar{\alpha}_1 \bar{\beta}_1 \bar{z}\bar{t}}\right) + \bar{\beta}_1 \int_0^{\bar{t}} e^{-\bar{\beta}_1 \xi} I_0\left(2\sqrt{\bar{\alpha}_1 \bar{\beta}_1 \bar{z}\xi}\right) d\xi\right], \tag{2.55}$$

$$\bar{S}_1(\bar{z},\bar{t}) = \bar{\alpha}_1 e^{-\bar{\alpha}_1 \bar{z}} \int_0^{\bar{t}} e^{-\bar{\beta}_1 \xi} I_0\left(2\sqrt{\bar{\alpha}_1 \bar{\beta}_1 \bar{z}\xi}\right) d\xi. \tag{2.56}$$

From (2.55), (2.56) there follow the consequences useful for the following analysis

$$\bar{C}_{1e}(\bar{t}) = \bar{C}_1(\bar{l}_1,\bar{t}) = e^{-\bar{\alpha}_1 \bar{l}_1}\left[e^{-\bar{\beta}_1 \bar{l}_1} I_0\left(2\sqrt{\bar{\alpha}_1 \bar{\beta}_1 \bar{l}_1 \bar{t}}\right) + \bar{\beta}_1 \int_0^{\bar{t}} e^{-\bar{\beta}_1 \xi} I_0\left(2\sqrt{\bar{\alpha}_1 \bar{\beta}_1 \bar{l}_1 \xi}\right) d\xi\right], \tag{2.57}$$

$$\bar{S}_{10}(\bar{t}) = S_1(0,\bar{t}) = \frac{\bar{\alpha}_1}{\bar{\beta}_1}(1 - e^{-\bar{\beta}_1 \bar{t}}). \tag{2.58}$$

For the calculations of the clarification process in the second (underlying) layer, the exact solution, which is obtained in the previous Section for uniform filter media at an essentially arbitrary function of the initial water contamination $C_0(t)$, is suitable. In this case, the role of such a function is played by the intermediate outlet quantity $\bar{C}_{1e}(\bar{t})$. Thus, the concentrations of the suspended matter at the lower boundary of the considered layer, as well as the deposited matter within its limits $(\bar{l}_1 + \bar{l}_2 \geq \bar{z} \geq \bar{l}_1)$ and at its upper boundary will be

$$\bar{C}_{2e}(\bar{t}) = e^{-\bar{\alpha}_2 \bar{l}_2}\left[\bar{C}_{1e}(\bar{t}) + \sqrt{\bar{\alpha}_2 \bar{\beta}_2 \bar{l}_2}\int_0^{\bar{t}} \frac{e^{-\bar{\beta}_2(\bar{t}-\xi)} I_1\left(2\sqrt{\bar{\alpha}_2 \bar{\beta}_2 \bar{l}_2 (\bar{t}-\xi)}\right)}{\sqrt{\bar{t}-\xi}} \bar{C}_{1e}(\xi)\, d\xi\right], \tag{2.59}$$

$$\bar{S}_2(\bar{z},\bar{t}) = \bar{\alpha}_2 e^{-\bar{\alpha}_2(\bar{z}-\bar{l}_1)} \int_0^{\bar{t}} e^{-\bar{\beta}_2(\bar{t}-\xi)} I_0\left(2\sqrt{\bar{\alpha}_2 \bar{\beta}_2 (\bar{z}-\bar{l}_1)(\bar{t}-\xi)}\right) \bar{C}_{1e}(\xi)\, d\xi, \tag{2.60}$$

$$\bar{S}_2(\bar{l}_1,\bar{t}) = \frac{\bar{\alpha}_2}{\bar{\beta}_2}\int_0^{\bar{t}} e^{-\bar{\beta}_2(\bar{t}-\xi)} \bar{C}_{1e}(\xi)\, d\xi. \tag{2.61}$$

Since a filter media of four and more layers are rarely used in the water treatment practice, it is enough to consider the separation and filtration of a suspension in a three-layer structure. Then the expressions for both concentrations by analogy with (2.59) - (2.61), but with respect to the third (lower) layer, will be as follows $(1 \geq \bar{z} \geq \bar{L}_2 = \bar{l}_1 + \bar{l}_2)$.

$$\bar{C}_{3e}(\bar{t}) = e^{-\bar{\alpha}_3 \bar{l}_3}\left[\bar{C}_{2e}(\bar{t}) + \sqrt{\bar{\alpha}_3 \bar{\beta}_3 \bar{l}_3}\int_0^{\bar{t}} \frac{e^{-\bar{\beta}_3(\bar{t}-\xi)} I_1\left(2\sqrt{\bar{\alpha}_3 \bar{\beta}_3 \bar{l}_3 (\bar{t}-\xi)}\right)}{\sqrt{\bar{t}-\xi}} \bar{C}_{2e}(\xi)\, d\xi\right], \tag{2.62}$$

$$\bar{S}_3(\bar{z},\bar{t}) = \bar{\alpha}_3 e^{-\bar{\alpha}_3(\bar{z}-\bar{L}_2)} \int_0^{\bar{t}} e^{-\bar{\beta}_3(\bar{t}-\xi)} I_0\left(2\sqrt{\bar{\alpha}_3 \bar{\beta}_3 (\bar{z}-\bar{L}_2)(\bar{t}-\xi)}\right) \bar{C}_{2e}(\xi)\, d\xi, \tag{2.63}$$

$$\bar{S}_3(\bar{L}_1,\bar{t}) = \frac{\bar{\alpha}_3}{\bar{\beta}_3}\int_0^{\bar{t}} e^{-\bar{\beta}_3(\bar{t}-\xi)} \bar{C}_{2e}(\xi)\, d\xi. \tag{2.64}$$

Formally, it is not difficult to continue the procedure of determining the dependences for $\bar{C}_i$, $\bar{S}_i$ if $i \geq 4$. However, the multiplicity of the integrals present in them increases accordingly in this case. Then the time of their computation increases sharply in the case of 6, 7 layers even using the modern software for mathematical analysis.

The set of the dependences $\bar{S}_i(\bar{z},\bar{t})$, three of which are presented above, allows us to get a complete pattern of the contribution of individual layers to the filter operation by comparing the relative current (over time $\bar{t}$) volumes of the suspended particles trapped in them. In particular, such volume W_{Si} for i-th layer will be

$$W_{Si}(\bar{t}) = \int_{\bar{L}_{i-1}}^{\bar{L}_i} \bar{S}_i(\bar{z},\bar{t})\, d\bar{z}. \tag{2.65}$$

Their maximum specific relative volume that the multilayer bed is able to absorb (at $t \to \infty$) in this case is equal to the sum of the dynamic solids take-up capacities of all layers (W_d), i.e.

$$W_d = \sum_{i=1}^{N} \bar{S}_{di} = \sum_{i=1}^{N} \frac{\bar{\alpha}_i}{\bar{\beta}_i} \bar{l}_i. \tag{2.66}$$

If the true solids take-up capacity in relation to the specified particles of at least one layer is less than the corresponding dynamic layer, the limiting value of W_{Si} will be less than that calculated by formula (2.66).

It is worth noting that the use of special layered media leads to a slight deterioration in the filtrate quality, which is usually accompanied by a noticeable reduction in the head losses. In order to fully appreciate the positive effect achieved as a result, it is necessary to carefully monitor the total head losses in dynamics. For this purpose, the filtration flow compartment of the general model including equations (2.50) - (2.52) and conditions (2.53), (2.54) is realized below. As a result, it is established that the distribution of the reduced head in the third (lower) $\bar{h}_3$, and then in the second $\bar{h}_2$ and the first $\bar{h}_1$ layers is described by the following system of expressions

$$\bar{h}_3(\bar{z},\bar{t}) = \frac{1}{\bar{k}_{03}} \int_{\bar{z}}^{1} \frac{d\eta}{f_{k3}(\bar{S}_3(\eta,\bar{t}))}, \tag{2.67}$$

$$\bar{h}_2(\bar{z},\bar{t}) = \frac{1}{\bar{k}_{02}} \int_{\bar{z}}^{\bar{l}_1+\bar{l}_2} \frac{d\eta}{f_{k2}(\bar{S}_2(\eta,\bar{t}))} + \frac{1}{\bar{k}_{03}} \int_{\bar{l}_1+\bar{l}_2}^{1} \frac{d\bar{z}}{f_{k3}(\bar{S}_3(\bar{z},\bar{t}))}, \tag{2.68}$$

$$\bar{h}_1(\bar{z},\bar{t}) = \int_{\bar{z}}^{\bar{l}_1} \frac{d\eta}{f_{k1}(\bar{S}_1(\eta,\bar{t}))} + \frac{1}{\bar{k}_{02}} \int_{\bar{l}_1}^{\bar{l}_1+\bar{l}_2} \frac{d\bar{z}}{f_{k2}(\bar{S}_2(\bar{z},\bar{t}))} + \frac{1}{\bar{k}_{03}} \int_{\bar{l}_1+\bar{l}_2}^{1} \frac{d\bar{z}}{f_{k3}(\bar{S}_3(\bar{z},\bar{t}))}. \tag{2.69}$$

The relative total head losses in the three-layer bed are equal in accordance with the representation adopted for dimensionless heads to the value of the reduced head $\bar{h}_1$ at the its surface (z=0). Then according to (2.69) the following formula is valid for $\Delta\bar{h}$

$$\Delta\bar{h}(\bar{t}) = \int_{0}^{\bar{l}_1} \frac{d\bar{z}}{f_{k1}(\bar{S}_1(\bar{z},\bar{t}))} + \frac{1}{\bar{k}_{02}} \int_{\bar{l}_1}^{\bar{l}_1+\bar{l}_2} \frac{d\bar{z}}{f_{k2}(\bar{S}_2(\bar{z},\bar{t}))} + \frac{1}{\bar{k}_{03}} \int_{\bar{l}_1+\bar{l}_2}^{1} \frac{d\bar{z}}{f_{k3}(\bar{S}_3(\bar{z},\bar{t}))}. \tag{2.70}$$

Therefore, the expressions for the concentrations $\bar{S}_1$, $\bar{S}_2$, $\bar{S}_3$ without taking into account the initial bed contamination are given above, that are (2.56), (2.59), (2.63), respectively. The set of N permeability functions $f_{ki}(\bar{S}_i)$ should be determined empirically for every layer. In this case, it is necessary to preliminarily establish the deposit composition (bulk) functions $\gamma_i(S_i)$ also by experimental methods.

The exact solution of the mathematical problem of suspension filtration through a three-layer packed bed, which can be easily generalized to beds of any number of layers, obtained above in the framework of linear mass exchange kinetics, served as a basis for the derivation of a set of the calculation dependencies. Collectively, they allow to predict the development of the mentioned technological process in a comprehensive and reliable manner. Insignificant difficulties arising at the applied calculations are easily overcome due to modern high-performance computing facilities, which gives the right to interpret the derived dependencies as an engineering method and recommend them for the application in the practice of water treatment.

In order to demonstrate the wide possibilities of the above-mentioned dependences for the intensification of filtration, numerous calculations of the suspension separation in three-layer, two-layer and uniform bed have been performed. The data of the calculations for three-layer bed are especially significant in this respect, because from their comparison with similar data for its other structural varieties the advantages of the layered structure of the operating medium over uniform structure, three-layer structure over two-layer one become obvious. The key parameters of the initial model, namely, the rate coefficient of suspended solids attachment, were specially selected in order to achieve a more even deposition of these particles over the entire height of the bed (Example 1: $\bar{\alpha}_1 = 2,\ \bar{\alpha}_2 = 5,\ \bar{\alpha}_3 = 10$). In three further examples, the filter media had a two-layer structure, with its top and bottom layers having the same absorption and flow capacity as the first and third layers of the three-layer analog (Example 2: $\bar{\alpha}_1 = 2,\ \bar{\alpha}_2 = 10$), the first and second layers (Example 3: $\bar{\alpha}_1 = 2,\ \bar{\alpha}_2 = 5$), and the second and third layers (Example 4: $\bar{\alpha}_1 = 5,\ \bar{\alpha}_2 = 10$). The thicknesses of the layers were assumed to be the same ($\bar{l}_1 = \bar{l}_2 = \bar{l}_3 = 0.333$, $\bar{l}_1 = \bar{l}_2 = 0.5$) except in the case of a focused study of the impact of individual thickness. Finally, the absorption properties of the uniform beds were matched to those of the top (Example 5: $\bar{\alpha} = 2$), middle (Example 6: $\bar{\alpha} = 5$) and bottom (Example 7: $\bar{\alpha} = 10$) layers of the three-layer medium as a basic medium for the comparison.

In carrying out a versatile quantitative analysis, several objectives were pursued, namely, it was necessary to verify the reality of a significant increase in the filter efficiency when replacing uniform media with layered media; to illustrate at least the main proposed calculation dependencies; to assess the significance of medium parameters and the contribution of its individual layers in the suspension clarification. In addition, the third chapter will discuss a step-by-step procedure for rational selection of the technological parameters for layered beds.

The initial model contains a large number of coefficients (kinetic, filtration flow, design, etc.) and it is not possible to study their significance in detail here. Therefore, first of all, we had to limit the number of basic parameters. Most of the model parameters had fixed values, although some of them were discretely and even continuously varied if necessary. It is particularly noteworthy that generally filtering materials with worse absorption capacity are weaker in retaining adhering particles. Consequently, the relationship $\bar{\beta}_1 < \bar{\beta}_2 < \bar{\beta}_3$ is valid for the three-layer bed. However, taking into account the formulated objectives of this study, it was assumed $\bar{\beta}_1 = \bar{\beta}_2 = \bar{\beta}_3 = 0.01$.

Figure 11. Increase in the relative volume of the deposited particles in individual layers and the whole bed over time

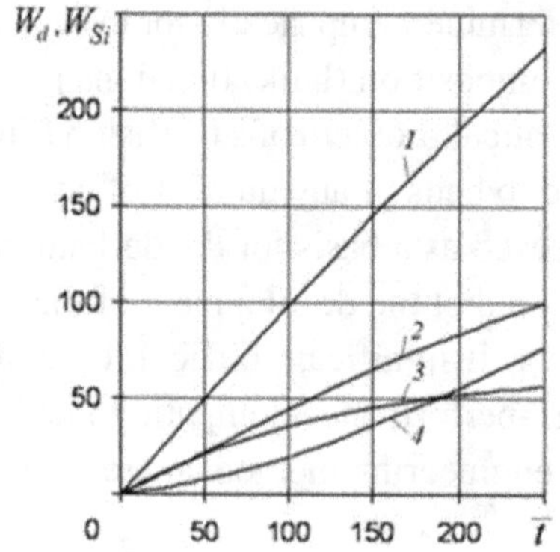

Note. 1 – W_d, 2 –W_{S2}, 3 –W_{S1}, 4 – W_{S3}

Figure 12. Increase in the relative volumetric concentration of the suspended solids in the filtrate

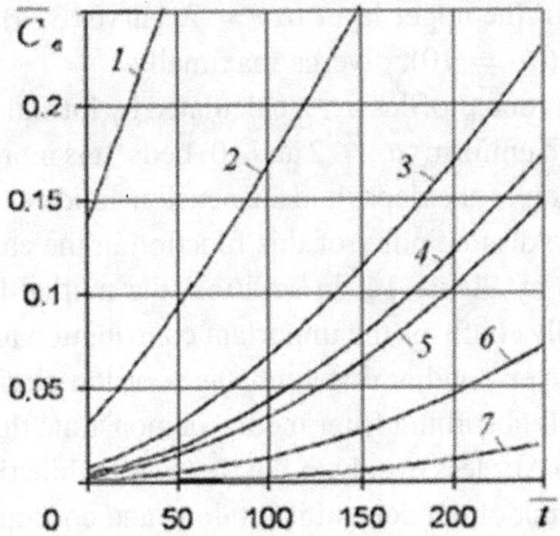

Note. 1 – example 5, 2 – example 3, 3 – example 6, 4 – example 1, 5 – example2, 6 – example 4, 7 – example 7

First of all, the dynamics of the accumulation of the deposited particles in three layers of the bed was calculated in order to confirm the correctness of the choice of the kinetic coefficients. The calculations were carried out by formulas (2.56), (2.60), (2.63) and (2.65), and their data are presented in the form of curves $W_{Si}(\bar{t})$ $(i = 1, 2, 3)$ in **Figure 11**. At the beginning of the operating period, the upper two layers retained a similar amount of the suspended matter, but the absorption resource of the first layer was almost exhausted by the time $\bar{t} = 200$. Further, particles were deposited mainly in the pores of the two lower layers. By the end of the calculation period the total dynamic solids take-up capacity, which according to (2.66) was 566.7 at the given values of $\bar{\alpha}_i$, $\bar{\beta}_i$, $\bar{l}_i$ $(i = 1, 2, 3)$, was exhausted by 41%, and 10% of this volume fell on the upper layer, 17.5% – on the middle layer, 13.5% – on the lower layer.

Further, the results of the calculations of the sanitary index of water treatment filter efficiency, which were carried out for seven examples by formulas (2.55), (2.59), (2.62), are shown in the form of the series of curves $\overline{C}_e(\bar{t})$ in **Figure 12**. First of all, it follows from them that at relatively small values of $\bar{\alpha}$ (here $\bar{\alpha} = 2$) even a clean uniform bed after processing of the first portion of suspension ($\bar{t} \approx 1$) passes too much suspended matter – in fact, the initial value for the subsequent calculations of $\overline{C}_e$ reaches 0.14 (curve 1). The two-layer bed with a weakly absorbing top layer (curve 2) also performs unsatisfactorily. Here the filtrate contains at once 3.18% of impurity from the initial volume, and its reduction, for example, by an order of magnitude or more will be provided for a short time (up to $\bar{t} = 61$). Naturally, the required quality of the filtrate for the longest time is also guaranteed by a uniform bed, but made of well-absorbing material ($\bar{\alpha} = 10$, curve 7). Curve 4, which corresponds exactly to the three-layer bed, occupies the intermediate position. It should be emphasized that the information obtained by the calculation regarding the behavior of $\overline{C}_e$ under various conditions of filtration is not enough for rational choice of the bed structure, because it is also necessary to study in detail the hydraulic aspects of the filtration process and, as a result, to establish the dynamics of layer-by-layer and total head losses. Such a study should start with a theoretical study of the process of the deposition of suspension particles in the pores of the bed and the search for the pattern of the deposit accumulation in different parts of the column. **Figure 13** gives a primary idea about the peculiarities of the three-layer bed clogging. It shows the increase in the concentrations of the deposited suspension particles with time at the upper $(\overline{S}_{i0})$

and lower $(\overline{S}_{ie})$ boundaries of all three layers (i= 1,2,3). The difference between the sets of two values of $\overline{S}_{i0}$ and $\overline{S}_{ie}$ depends on the absorption properties of the corresponding layer. And it is logical that this difference will be the smallest for the upper layer ($\overline{\alpha}_1 = 2$, curves 3, 4). On the contrary, curves 1, 6, corresponding to the third layer ($\overline{\alpha}_3 = 10$), diverge maximally.

Figure 14 shows the concentration profiles $\overline{S}(\overline{z})$, calculated by formulas (2.56), (2.60), (2.63) for two points in time, three-layer and two uniform ($\overline{\alpha} = 2$ and 10) beds. It is more illustrative for understanding the peculiarities of the distribution of the deposited suspension solids along the height of the bed. And first of all, it should be noted the discontinuity of this function in the case of the three-layer structure, which, incidentally, is indicated by Figure 14. In addition, the mutual location of three fragments of both discontinuous curves visually confirms the important contribution to the removal of the suspended solids of the second and third layers, and at different stages of the clarification process. Curves 3, 5, corresponding to weakly and well-absorbing filter media, demonstrate that a relatively small amount of the deposit is distributed in it more or less evenly in the first case, while the opposite pattern takes place in the second case – a large amount of the deposit is retained and concentrated at the top.

Figure 13. Increase in the relative volumetric concentration of the deposited particles at the inner and outer boundaries of layers of three-layer bed over time

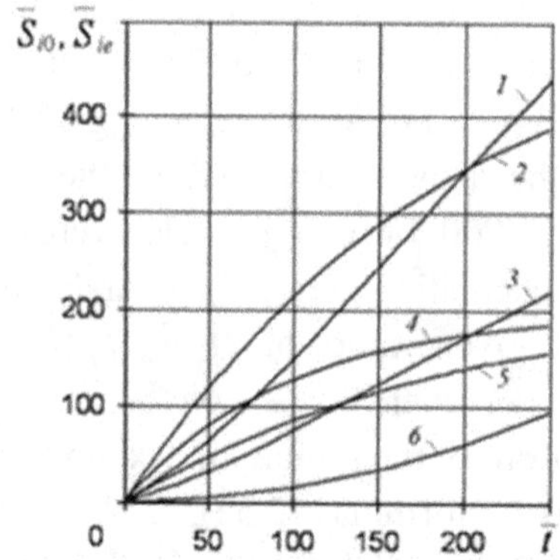

Note. 1 – $\overline{S}_{30}$, 2 – $\overline{S}_{20}$, 3 – $\overline{S}_{2e}$, 4 – $\overline{S}_{10}$, 5 – $\overline{S}_{1e}$, 6 – $\overline{S}_{3e}$

Figure 14. Profiles of the relative volumetric concentration of the deposited particles

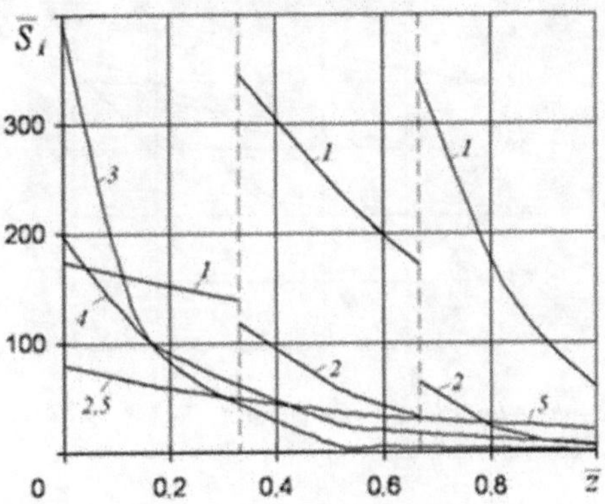

Note. $1 - \bar{t} = 200$, $2 - 5 - \bar{t} = 50$; *1, 2 – three-layer bed, 3 - 5 – uniform*

One of the direct consequences of the clogging of a bed is a decrease in its permeability. The hydraulic conductivity of a clean layered medium is a piecewise constant function of $\bar{z}$. After the filter is turned on, this coefficient decreases monotonically everywhere, but jumps of its values at the interlayer boundaries are maintained due to the existence of similar jumps in the concentration function of the deposited particles $\bar{S}(\bar{z})$ ($\bar{S}_{20} > \bar{S}_{1e}$, $\bar{S}_{30} > \bar{S}_{2e}$ – in case of a three-layer structure).The profiles of the relative hydraulic conductivity, as well as $\bar{S}(\bar{z})$, are plotted for the three-layer and three different uniform beds at two time points using the universal formula (2.25) at γ_a=0.003 (**Figure 15**). Significant decrease in the permeability is observed in the upper parts of a uniform media (examples 6, 7; curves 4, 5), which should greatly complicate the filtration flow in them.

Due to the active formation of a deposit in the filter bed pores, a corresponding increase in the inlet head is required to maintain the initial water flow rate. In other words, the increase in the head $\bar{h}_1(0, \bar{t})$ must exactly match the increase in the head losses $\Delta\bar{h}$ and thus compensate for it. It is quite easy to establish a suitable dependence $\bar{h}_1(0, \bar{t})$ (or $\Delta\bar{h}$) by using formula (2.69). The dependence $\Delta\bar{h}(\bar{t})$ was determined for all seven examples, and the data obtained are shown in **Figure 16**. As a rule, layers in multilayer beds are composed of materials that differ not only in the absorption but also in the filtration flow properties. Most often, the permeability of the lower layers is less than the permeability of the upper layers. Therefore, the values of 0.6 and 0.3 were chosen and fixed for the relative coefficients $\bar{k}_{20}$, $\bar{k}_{30}$ respectively. Thus, the relative initial head losses amounted to 2 according to (2.72). The operation of the uniform well-absorbing bed causes an extremely fast increase in the head losses (curve 1), so that the resource of controlling the process of suspension clarification (reserve of mechanical energy) must be very large in this situation, otherwise the duration of the operating period will be too short and economically unprofitable. The hydraulic resistance of the three-layer bed increases much slower, which contributes to a longer operation of the filter until the successive washing. The other extreme also occurs in the case of a uniform bed, but with a low value of the coefficient $\bar{\alpha}$ (curve 7). Here $\bar{h}_1(0, \bar{t})$ and $\Delta\bar{h}(\bar{t})$ change in time very slowly, which is explained not only by the almost even spatial distribution of a deposit, but also by a low quality of clarification. Indeed, Figure 12 shows the breakthrough of a significant amount of the suspended solids, and it is observed almost immediately after feeding the filter with contaminated water.

Figure 15. Profiles of the relative hydraulic conductivity

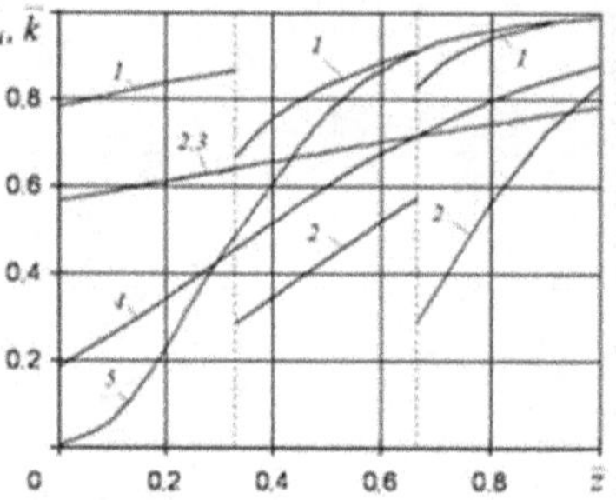

Note. 1 – $\bar{t}$ = 50, 2 - 5– $\bar{t}$ = 200; 1 - 2 – three-layer bed, 3 - 5 – uniform

Figure 16. Increase in the relative head losses in the bed over time

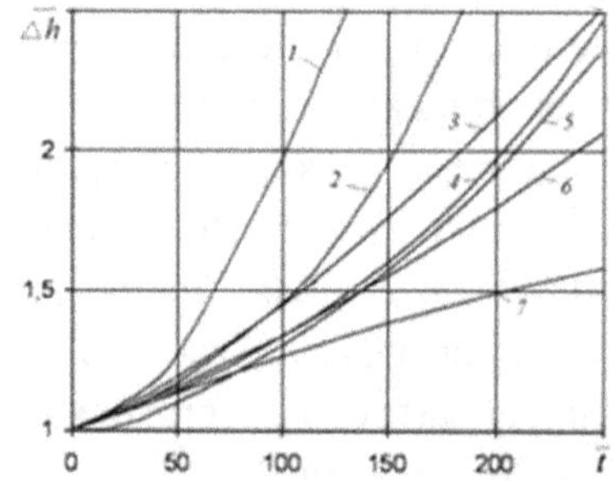

Note. 1 – example 7, 2 – example 2, 3 – example 6, 4 – example 1, 5 – example 4, 6 – example 3, 7 – example 5

At arrangement of a layered media, it is really appreciable to increase the efficiency of water treatment at rapid filters due to reasonable change in the thickness of their layers. **Figure 17**, which shows the curves of the outlet concentration $\overline{C}_e$ change depending on the relative thicknesses $\bar{l}_1$, $\bar{l}_2$, $\bar{l}_3$ (three-layer medium), allows to convince in it once again. The time points $\bar{t} = 50$ and 200 were chosen as before. The parameters $\bar{l}_i$ were changed alternately. At the same time, the two remaining thicknesses have already been determined accordingly. Since they were considered to be the same then, for example, at a given $\bar{l}_1$ they were equal to $\bar{l}_2 = \bar{l}_3 = 0.5(1 - \bar{l}_1)$. As the data of Figure 17 (curves 1 and 2) show, the thickness $\bar{l}_3$ has a very significant effect on the quality of the clarification. An equally significant but opposite effect is produced by varying the thickness $\bar{l}_1$. It should be noted that a variation of $\bar{l}_2$ even within full limits has little effect on the value of $\overline{C}_e$.

Figure 17. Dependence $\overline{C}_e(\overline{l}_i)$

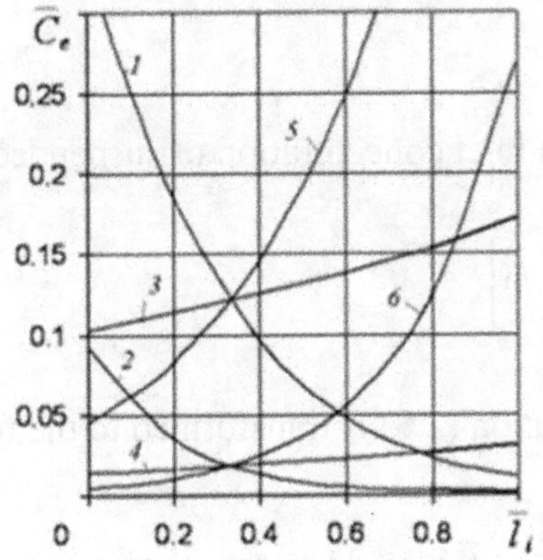

Note. 1, 4, 5 – $\bar{t}$ = 200; 2, 3, 6 – $\bar{t}$ = 50; 1, 2 – $\overline{l}_3$, 3, 4 – $\overline{l}_2$, 5, 6 – $\overline{l}_1$

2.4. SUSPENSION FILTRATION THROUGH UNIFORM AND TWO-LAYER FILTER BED (APPROXIMATE SOLUTION)

The possibilities of exact analytical methods for applied problems in general and nonlinear filtration problems in particular are extremely limited, and their successful application, as a rule, leads to cumbersome computational formalisms. Approximate analytical methods serve as a worthy alternative to them. As a rule, they allow to obtain the calculation dependences and equations, firstly, for more general conditions of filtration, and secondly, in essentially simpler form. At the same time, of primary importance is the evaluation of the accuracy of an approximate solution, which is best performed by comparing it with a particular exact solution. The first example of the effective use of approximate methods is the solution of the basic problem from Section 2.1, obtained and verified below. It is assumed that n_e=0 due to the transience of the initial filtration with a moving front of contamination. Thus, the formulation of the above problem, consisting of two compartments, in dimensionless form includes the system of equations (2.7), (2.8), (2.11), (2.12) and the boundary conditions operator (2.9), (2.10), (2.13).

For approximate solution of the set mathematical problem, averaging of the second term in the right part of equation (2.8) (characterizes the detachment of deposited particles under the action of the hydrodynamic force) is performed over the calculation period [0,*T*] (Poliakov, 2009, 2011). Here *T* plays the role of a parameter of the problem, which can take any current values of time up to the final moment in time and is considered to be a constant quantity in the operations of integration and differentiation. Therefore, the formal replacement of the independent variable $\bar{t}$ in the required concentrations by also variable duration of the calculation period can be appropriately interpreted as their dynamic averaging. Thus,

$$\overline{S}_a(\bar{z},T) = \frac{1}{T}\int_0^T \overline{S}(\bar{z},\bar{t})\,d\bar{t}. \tag{2.71}$$

As a result, system (2.7), (2.8) reduces to the equation

$$\frac{\partial \overline{C}}{\partial \overline{z}} + \overline{\alpha}\,\overline{C} = \overline{\beta}\,\overline{S}_a(\overline{z}, T). \tag{2.72}$$

The solution of (2.72) at a given inlet concentration of suspended solids (2.9) gives

$$\overline{C}(\overline{z}, T) = e^{-\overline{\alpha}\overline{z}}\left[1 + \overline{\beta}\int_0^{\overline{z}} e^{\overline{\alpha}\eta}\,\overline{S}_a(\eta, T)\,d\eta\right]. \tag{2.73}$$

Taking into account (2.73), equation (2.8) is transformed to the form

$$\frac{\partial \overline{S}}{\partial \overline{t}} = \overline{\alpha}\,e^{-\overline{\alpha}\overline{z}}\left[1 + \overline{\beta}\int_0^{\overline{z}} e^{\overline{\alpha}\eta}\,\overline{S}_a(\eta, T)d\eta\right] - \overline{\beta}\,\overline{S}_a(\overline{z}, T). \tag{2.74}$$

The following expression is obtained by integrating (2.74) under condition (2.10)

$$\overline{S}(\overline{z}, \overline{t}) = \overline{S}^0(\overline{z}) + \left[\overline{\alpha}\,e^{-\overline{\alpha}\overline{z}} + \overline{\alpha}\overline{\beta}\,e^{-\overline{\alpha}\overline{z}}\int_0^{\overline{z}} e^{\overline{\alpha}\eta}\,\overline{S}_a(\eta, T)d\eta - \overline{\beta}\,\overline{S}_a(\overline{z}, T)\right]\overline{t}. \tag{2.75}$$

The equation relatively $\overline{S}_a$ is obtained by averaging both parts of (2.75) over $\overline{t}$. Then

$$2e^{\overline{\alpha}\overline{z}}\,\overline{S}_a(\overline{z}, T) = 2e^{\overline{\alpha}\overline{z}}\,\overline{S}^0(\overline{z}) + T\left[\overline{\alpha} + \overline{\alpha}\overline{\beta}\int_0^{\overline{z}} e^{\overline{\alpha}\eta}\,\overline{S}_a(\eta, T)d\eta - \overline{\beta}\,e^{\overline{\alpha}\overline{z}}\,\overline{S}_a(\overline{z}, T)\right]. \tag{2.76}$$

Next, change of variables is performed

$$y = \int_0^{\overline{z}} e^{\overline{\alpha}\eta}\,\overline{S}_a(\eta, T)\,d\eta, \tag{2.77}$$

so that

$$\frac{dy}{d\overline{z}} = e^{\overline{\alpha}\overline{z}}\,\overline{S}_a(\overline{z}, T). \tag{2.78}$$

Taking into account (2.77), (2.78), equation (2.76) is transformed in the following way

$$(2 + \overline{\beta}T)\frac{dy}{d\overline{z}} = \overline{\alpha}\overline{\beta}Ty + \overline{\alpha}T + 2e^{\overline{\alpha}\overline{z}}\,\overline{S}^0(\overline{z}). \tag{2.79}$$

The function $y(\overline{z}; T)$ is determined by solving (2.79) under the homogeneous condition

$$\overline{z} = 0, \quad y = 0. \tag{2.80}$$

and will be

$$y(\bar{z},T) = \frac{1}{\bar{\beta}}(e^{\delta(T)\bar{z}} - 1) + \frac{2}{2+\bar{\beta}T}\int_0^{\bar{z}} e^{\bar{\alpha}\eta - \delta(T)(\bar{z}-\eta)}\,\bar{S}^0(\eta)d\eta, \tag{2.81}$$

where $\delta(T) = \bar{\alpha}\bar{\beta}T/(2+\bar{\beta}T)$. Keeping in mind (2.78), $\bar{S}_a$ takes the form

$$\begin{aligned}\bar{S}_a(\bar{z},T) &= \frac{\bar{\alpha}T}{2+\bar{\beta}T}e^{-\frac{2\bar{\alpha}\bar{z}}{2+\bar{\beta}T}} + \\ &+ \frac{2}{2+\bar{\beta}T}e^{\bar{\alpha}\bar{z}}\bar{S}^0(\bar{z}) - \frac{2\delta(T)}{2+\bar{\beta}T}\int_0^{\bar{z}} e^{-[\bar{\alpha}+\delta(T)](\bar{z}-\eta)}\,\bar{S}^0(\eta)d\eta\end{aligned} \tag{2.82}$$

The dependence $\bar{S}^0 = 0$ follows from expression (2.75) taking into account (2.82)

$$\begin{aligned}\bar{S}(\bar{z},\bar{t};T) &= [\bar{\alpha} - \delta(T)]\bar{t}\,e^{[\delta(T)-\bar{\alpha}]\bar{z}} + \bar{S}^0(\bar{z})\left(1 - \frac{2\bar{\beta}\bar{t}}{2+\bar{\beta}T}\right) + \\ &+ \frac{[\bar{\alpha}+\delta(T)]2\bar{\beta}\bar{t}}{2+\bar{\beta}T}\int_0^{\bar{z}} e^{-[\bar{\alpha}+\delta(T)](\bar{z}-\eta)}\,\bar{S}^0(\eta)d\eta.\end{aligned} \tag{2.83}$$

To obtain the calculation formula for the desired $\Delta_S(\bar{t})$, the parameter Δ_S is identified with the variable Δ_C and after simple transformations the following expression is finally obtained

$$\begin{aligned}\bar{S}(\bar{z},\bar{t}) &= \frac{1}{2+\bar{\beta}\bar{t}}\Big[2\bar{\alpha}\bar{t}\,e^{-\lambda_t\bar{z}} + (2-\bar{\beta}\bar{t})\bar{S}^0(\bar{z}) + \\ &+ 2\bar{\beta}\bar{t}\,\lambda_t(1+\bar{\beta}\bar{t})\int_0^{\bar{z}} e^{-\lambda_t(1+\bar{\beta}\bar{t})(\bar{z}-\eta)}\,\bar{S}^0(\eta)d\eta\Big],\end{aligned} \tag{2.84}$$

where $\lambda_t = 2\bar{\alpha}/(2+\bar{\beta}\bar{t})$.

Expression for the concentration $\bar{C}$ follows from (2.73) and, taking into consideration (2.81), will be

$$\bar{C}(\bar{z},\bar{t}) = 2e^{-\lambda_t\bar{z}} - e^{-\bar{\alpha}\bar{z}} + \frac{2\bar{\beta}}{2+\bar{\beta}\bar{t}}\int_0^{\bar{z}} e^{-\lambda_t(1+\bar{\beta}\bar{t})(\bar{z}-\eta)}\,\bar{S}^0(\eta)d\eta. \tag{2.85}$$

Important special cases of (2.84), (2.85) are similar simple dependencies for the initially clean bed (S^0=0). Then

$$\bar{S}(\bar{z},\bar{t}) = \lambda_t\bar{t}\,e^{-\lambda_t\bar{z}}, \tag{2.86}$$

$$\bar{C}(\bar{z},\bar{t}) = 2e^{-\lambda_t\bar{z}} - e^{-\bar{\alpha}\bar{z}}. \tag{2.87}$$

Useful for the practical calculation formulas follow from (2.87)

$$\bar{C}_e(\bar{t}) = \bar{C}(1,\bar{t}) = e^{-\bar{\alpha}}\left(2e^{\frac{\bar{\alpha}\bar{\beta}\bar{t}}{2+\bar{\beta}\bar{t}}} - 1\right), \tag{2.88}$$

$$\bar{S}_0(\bar{t}) = \bar{S}(0,\bar{t}) = \lambda_t\bar{t}. \tag{2.89}$$

Due to the simplicity of expression (2.86), it is possible to derive a formula for the relative head losses $\Delta\bar{h}$, containing only elementary functions. Then, using the formula for $\Delta\bar{h}$ (2.24) and the simplified approximate expression for f_k (2.25)

$$f_k(\bar{S}) = \left[1 - (\bar{\gamma}_a\bar{S})^{m_{k1}}\right]^3, \tag{2.90}$$

where $\bar{\gamma}_a = \gamma_a C_0$, m_{k1} is the empirical constant, it is easy to obtain

$$\begin{aligned}\Delta\bar{h}(\bar{t}) = {} & \frac{1}{m_{k1}\lambda_t}\ln\frac{e^{m_{k1}\lambda_t} - \phi_t}{1 - \phi_t} - \frac{2\phi_t(1 - e^{m_{k1}\lambda_t})}{m_{k1}\lambda_t(1 - \phi_t)(e^{m_{k1}\lambda_t} - \phi_t)} - \\ & -\frac{\phi_t^2}{2m_{k1}\lambda_t}\left[\frac{1}{(e^{m_{k1}\lambda_t} - \phi_t)^2} - \frac{1}{(1 - \phi_t)^2}\right].\end{aligned} \tag{2.91}$$

Here $\phi_t = (\bar{\gamma}_a\lambda_t t)^{m_{k1}}$.

The theoretical results obtained for the uniform bed are generalized to the two-layer bed, but now we assume $S^0=0$ to simplify the calculations and the final expressions. The formulation of the corresponding mathematical problem is actually given in Section 2.3, and its approximate solution is obtained in two stages. At the first stage, the depth filtration and its consequences are considered exclusively within the upper layer. And here the dependences derived for a uniform bed and close to the exact dependences are justified to apply without changes, only resorting to the appropriate indexing of characteristics and parameters due to the layered structure of the bed (index "1" indicates their belonging to the first layer). But the main thing is that the mentioned results are necessary, firstly, to determine the contribution of the hydraulic resistance of the upper layer to the total resistance, secondly, and mainly, to calculate the most important characteristics of the suspension filtration inside and at the boundaries of the second layer. The relative value of the head losses in the upper layer $\Delta\bar{h}$ is calculated according to formula (2.24) independently of mass transfer, mass exchange and suspension flow in the lower layer. For the theoretical analysis of the technological process in the lower part of the bed ($l_1+l_2 \geq z \geq l_1$), it is necessary to set the now known variable value for $\bar{C}_2$ at the upper boundary ($z=l_1$), namely, due to the equality here of the concentrations $\bar{C}_1$ and $\bar{C}_2$ the corresponding boundary condition will be

$$\bar{z} = \bar{l}_1, \quad \bar{C}_2 = \bar{C}_{1e}(\bar{t}) = e^{-\lambda_{1t}\bar{l}_1} - e^{-\bar{\alpha}_{1t}\bar{l}_1}, \tag{2.92}$$

where $\lambda_{1t} = 2\bar{\alpha}_1/(2 + \bar{\beta}_1\bar{t})$, $\bar{l}_1 = l_1/(l_1 + l_2)$. The subsequent procedure for solving problem (2.46) - (2.54) at the second stage, in fact, the main stage with respect to the evaluation of the filtrate quality, is similar to the procedure described above in case of a uniform bed.

First of all, $\bar{C}_2$ is expressed according to (2.46), (2.47) through $\bar{S}_2$ as follows

$$\bar{C}_2(\bar{z},\bar{t}) = \bar{C}_{1e}(\bar{t})e^{-\bar{\alpha}_2(\bar{z}-\bar{l}_1)} + \bar{\beta}_2\int_{\bar{l}_1}^{\bar{z}} e^{\bar{\alpha}_2(\eta-\bar{z})}\bar{S}_2(\eta,\bar{t})d\eta. \tag{2.93}$$

Then a partial averaging of the intensity of the mass exchange process, and in fact the desired concentration $\bar{S}_2$ is performed over the current calculation period $[0,T]$

$$\bar{S}_{2a}(\bar{z},T) = \frac{1}{T}\int_0^T \bar{S}_2(\bar{z},\bar{t})d\bar{t} \tag{2.94}$$

Thus, by analogy with (2.71), the parameter T is introduced in the problem, which will be treated as a variable and identified with $\bar{t}$ only in the final calculation formulas. As a result of integration (2.47), taking into account (2.93), (2.94) we obtain:

$$\bar{S}_2(\bar{z},\bar{t};T) = \left[\bar{\alpha}_2 e^{-\bar{\alpha}_2(\bar{z}-\bar{l}_1)} F_s(T) + \bar{\alpha}_2\bar{\beta}_2 \int_{\bar{l}_1}^{\bar{z}} e^{-\bar{\alpha}_2(\eta-\bar{z})} \bar{S}_{2a}(\eta,T)d\eta - \bar{\beta}_2\bar{S}_{2a}(\bar{z},T)\right]\bar{t}, \tag{2.95}$$

where

$$\begin{aligned} F_s(T) &= \frac{1}{T}\int_0^T \bar{C}_{1e}(\bar{t})d\bar{t} = \frac{2}{\bar{\beta}_1 T}\Big\{ (2+\bar{\beta}_1 T)\, e^{-\frac{2\bar{\alpha}_1\bar{l}_1}{2+\bar{\beta}_1 T}} - 2e^{-\bar{\alpha}_1\bar{l}_1} + \\ &+2\bar{\alpha}_1\bar{l}_1\left[Ei(-\bar{\alpha}_1\bar{l}_1) - Ei\left(-\frac{2\bar{\alpha}_1\bar{l}_1}{2+\bar{\beta}_1 T}\right)\right]\Big\} - e^{-\bar{\alpha}_1\bar{l}_1}. \end{aligned} \tag{2.96}$$

The integral exponent function $Ei(-x)$ was previously tabulated in detail (Vasiliev et al., 1970) and has the form

$$Ei(-x) = \int_x^\infty \frac{e^{-\xi}}{\xi}d\xi.$$

To derive the equation relative to $\bar{S}_{2a}$, both parts of equation (2.95) are averaged in a similar manner

$$\begin{aligned} &\bar{S}_{2a}(\bar{z},T) = \\ &= \left[\bar{\alpha}_2 e^{-\bar{\alpha}_2(\bar{z}-\bar{l}_1)} F_s(T) + \bar{\alpha}_2\bar{\beta}_2 \int_{\bar{l}_1}^{\bar{z}} e^{-\bar{\alpha}_2(\eta-\bar{z})} \bar{S}_{2a}(\eta,T)d\eta - \bar{\beta}_2\bar{S}_{2a}(\bar{z},T)\right]\frac{T}{2} \end{aligned} \tag{2.97}$$

In order to reduce the integral equation (2.97) to a differential equation, substitution of variables similar to (2.77) is carried out

$$y = \int_{\bar{l}_1}^{\bar{z}} e^{-\bar{\alpha}_2\eta}\bar{S}_{2a}(\eta,T)d\eta, \quad \frac{dy}{d\bar{z}} = e^{\bar{\alpha}_2\bar{z}}\bar{S}_{2a}(\bar{z},T). \tag{2.98}$$

Then equation (2.97) relative to $\bar{l}_1$ can be written in the form

$$(2+\bar{\beta}_2 T)\frac{dy}{d\bar{z}} - \bar{\alpha}_2\bar{\beta}_2 Ty = \bar{\alpha}_2 T e^{\bar{\alpha}_2\bar{l}_1} F_s(T) \tag{2.99}$$

To it in accordance with (2.98) the condition

$$\bar{z} = \bar{l}_1, \quad y = 0 \tag{2.100}$$

is added to (2.99). The solution of (2.99), (2.100) yields

$$y(\bar{z},T) = \frac{e^{\bar{\alpha}_2 \bar{l}_1} F_s(T)}{\bar{\beta}_2}\left[e^{\frac{\bar{\alpha}_2 \bar{\beta}_2 T}{2+\bar{\beta}_2 T}(\bar{z}-\bar{l}_1)} - 1\right], \tag{2.101}$$

and, therefore,

$$\bar{S}_{2a}(\bar{z},T) = \frac{\bar{\alpha}_2 T}{2+\bar{\beta}_2 T} F_s(T)\, e^{-\frac{2\bar{\alpha}_2}{2+\bar{\beta}_2 T}(\bar{z}-\bar{l}_1)}. \tag{2.102}$$

Then from (2.95) we find taking into account (2.102)

$$\bar{S}_2(\bar{z},\bar{t};T) = \frac{2\bar{\alpha}_2 \bar{t}}{2+\bar{\beta}_2 T} F_s(T) e^{-\frac{2\bar{\alpha}_2}{2+\bar{\beta}_2 T}(\bar{z}-\bar{l}_1)}. \tag{2.103}$$

As a result of the transition in (2.103) from T to $\bar{t}$, the formula for calculating $\bar{S}_2$ takes the form

$$\bar{S}_2(\bar{z},\bar{t}) = \lambda_{2t}\bar{t}\cdot F_s(\bar{t})\, e^{-\lambda_{2t}(\bar{z}-\bar{l}_1)}, \tag{2.104}$$

where $\lambda_{2t} = 2\bar{\alpha}_2/(2+\bar{\beta}_2\bar{t})$, $F_s(\bar{t})$ is represented by formula (2.96), in which T is formally replaced by $\bar{t}$. The calculation formula for $\bar{C}_2(\bar{z},\bar{t})$ can be derived from (2.93) allowing for (2.104)

$$\bar{C}_2(\bar{z},\bar{t}) = 2F_s(\bar{t})\, e^{-\lambda_{2t}(\bar{z}-\bar{l}_1)} + \left[\bar{C}_{1e}(\bar{t}) - 2F_s(\bar{t})\right] e^{-\bar{\alpha}_2(\bar{z}-\bar{l}_1)}. \tag{2.105}$$

The total head losses in both layers $\Delta\bar{h}$ in accordance with (2.70) are represented in a general form as follows

$$\Delta\bar{h}(\bar{t}) = \int_0^{\bar{l}_1} \frac{d\bar{z}}{f_{k1}(\bar{S}_1(\bar{z},\bar{t}))} + \frac{1}{\bar{k}_{02}}\int_{\bar{l}_1}^{1} \frac{d\bar{z}}{f_{k2}(\bar{S}_2(\bar{z},\bar{t}))}. \tag{2.106}$$

If the approximation representation (2.90) is valid for $f_{ki}(\bar{S}_i)$ (i= 1,2), the indicated losses in the first layer will be

$$\Delta\bar{h}_1(\bar{t}) = \int_0^{\bar{l}_1} \frac{d\bar{z}}{\left\{1-\left[\bar{\gamma}_{1a} S_1(\bar{z},\bar{t})\right]^{m_{k1}^{(1)}}\right\}^3} = \frac{1}{m_{k1}^{(1)}\lambda_t}\cdot\left\{\ln\frac{e^{m_{k1}^{(1)}\lambda_{1t}\bar{l}_1} - \phi_{1t}}{1-\phi_{1t}} - \right.$$
$$\left. -2\phi_{1t}\left(\frac{1}{e^{m_{k1}^{(1)}\lambda_{1t}\bar{l}_1} - \phi_{1t}} - \frac{1}{1-\phi_{1t}}\right) - \frac{\left(m_{k1}^{(1)}\lambda_t\right)^2}{2}\left[\frac{1}{\left(e^{m_{k1}^{(1)}\lambda_{1t}\bar{l}_1} - \phi_{1t}\right)^2} - \frac{1}{\left(1-\phi_{1t}\right)^2}\right]\right\}. \tag{2.107}$$

and the losses in the second layer are as follows

$$\Delta\bar{h}_2(\bar{t}) = \int_{\bar{l}_1}^{1} \frac{d\bar{z}}{\left\{1-\left[\bar{\gamma}_{2a}\bar{S}_2(\bar{z},\bar{t})/\bar{n}_{02}\right]^{m_{k1}^{(2)}}\right\}^3} = \frac{1}{\bar{k}_{02} m_{k1}^{(2)}\lambda_{2t}}\left\{\ln\frac{e^{m_{k1}^{(2)}\lambda_{2t}} - \phi_{2t}}{1-\phi_{2t}} - \right.$$
$$\left. -2\phi_{2t}\left(\frac{1}{e^{m_{k1}^{(2)}\lambda_{2t}} - \phi_{2t}} - \frac{1}{1-\phi_{2t}}\right) - \frac{\left(m_{k1}^{(2)}\lambda_{2t}\right)^2}{2}\left[\frac{1}{\left(e^{m_{k1}^{(2)}\lambda_{2t}} - \phi_{2t}\right)^2} - \frac{1}{\left(1-\phi_{2t}\right)^2}\right]\right\}. \tag{2.108}$$

Here

$$\phi_{1t} = \left(\bar{\gamma}_{1a}\lambda_{1t}\bar{t}\right)^{m_{k1}^{(1)}}, \quad \phi_{2t} = \left[\frac{\bar{\gamma}_{2a}}{\bar{n}_{02}}\lambda_{2t}\bar{t}\,F_S(\bar{t})\right]^{m_{k1}^{(2)}}, \quad \bar{n}_{02} = \frac{n_{02}}{n_{01}}.$$

The two solutions obtained above, although they are approximate due to partial averaging of the function $\bar{S}(\bar{z},\bar{t})$, but, nevertheless, provide high accuracy of the calculations of all required filtration characteristics. Comparison of the results of the parallel calculations by exact and approximate formulas of the relative values, namely, the outlet concentration of the suspended solids $\bar{C}_e$, the inlet concentration of the deposited particles $\bar{S}_0$ and the head losses $\Delta\bar{h}$ as functions of time convinces us of the validity of this statement. At the same time, the duration of the calculation periods significantly exceeded the time of the filter operation before the successive filter washing, which allowed to evaluate more fully the possibilities of the new solutions.

Figure 18. Increase in the relative volumetric concentration of the suspended solids in the filtrate over time

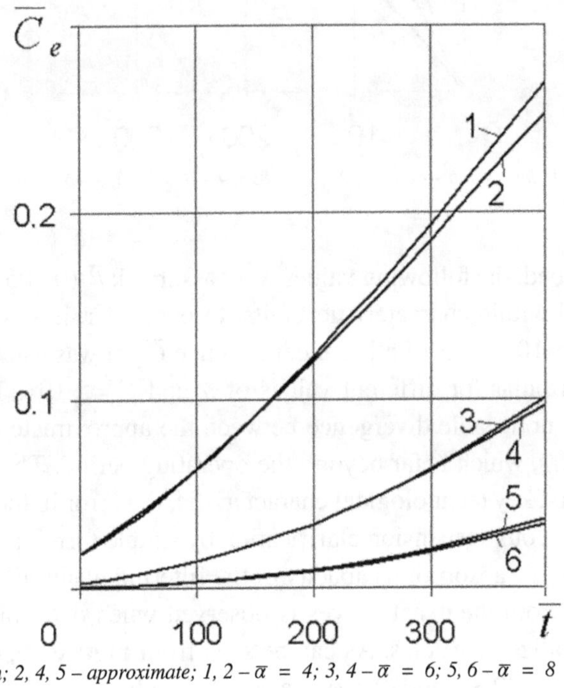

Note. 1, 3, 6 – exact calculation; 2, 4, 5 – approximate; 1, 2 – $\bar{\alpha}$ = 4; 3, 4 – $\bar{\alpha}$ = 6; 5, 6 – $\bar{\alpha}$ = 8

Figure 19. Increase in the relative volumetric concentration of the deposited suspension particles at the inlet to the bed over time

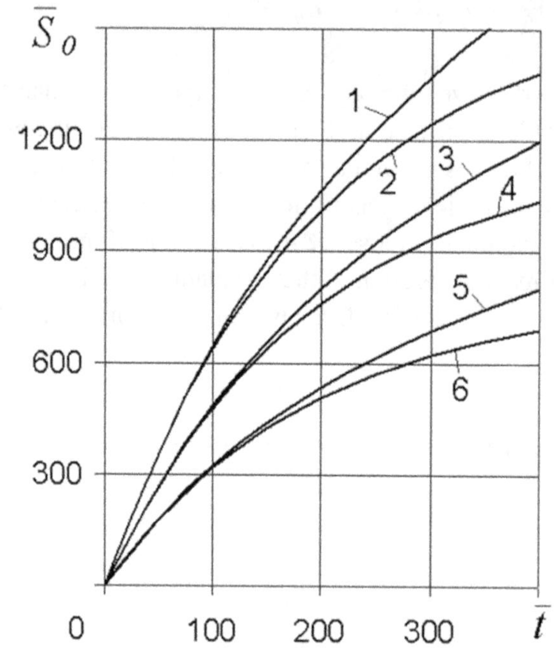

Note. 1, 3, 5 – approximate calculation; 2, 4, 6 – exact; 1, 2 – $\overline{\alpha}$ = 8; 3, 4 – $\overline{\alpha}$ = 6; 5, 6 – $\overline{\alpha}$ = 4

In the case of a uniform bed, the following values were assumed: $\overline{\beta}$ (0.005), γ_a(20). Other parameters of the problem were varied within characteristic limits. In particular, it was assumed that $\overline{\alpha} = 4, 6, 8$ and C_0=2.5•10^{-5}, 5•10^{-5}, 7.5•10^{-5}. First of all, the dependence $\overline{C}_e(\overline{t})$ was established using exact (2.18) and approximate (2.88) formulas for different values of $\overline{\alpha}$ and C_0=5•10^{-5}. The corresponding curves are shown in **Figure 18**. A noticeable divergence between the approximate and control curves begins approximately after $\overline{t}$ = 300, which is far beyond the operating period. Thus, according to the results of the calculations of the first key technological characteristic, the error in the approximate quantitative analysis of the quality of aqueous suspension clarification by simple formula (2.88) in practical terms is negligible, and regardless of the absorption capacity of the filtering material. A slightly larger deviation of the approximate curves from the exact curves is observed when determining the dynamics of the accumulation of the immobilized particles. As can be seen from **Figure 19**, obvious differences in the values of $\overline{S}_0$, found by formulas (2.89) and (2.19), are noted already at $\overline{t}$ > 200. It should be emphasized that the important value for the applied analysis here is not the noted differences, but their influence on the reliability of calculating the second key technological characteristic – head losses. **Figure 20** shows a possible appreciable advance of the indicated losses according to (2.24), (2.86) over the actual losses (within the framework of the accepted mathematical model) in accordance with (2.17), (2.24), but only at time points significantly exceeding the duration of the operating period.

Figure 20. Increase in the relative head losses in the bed over time

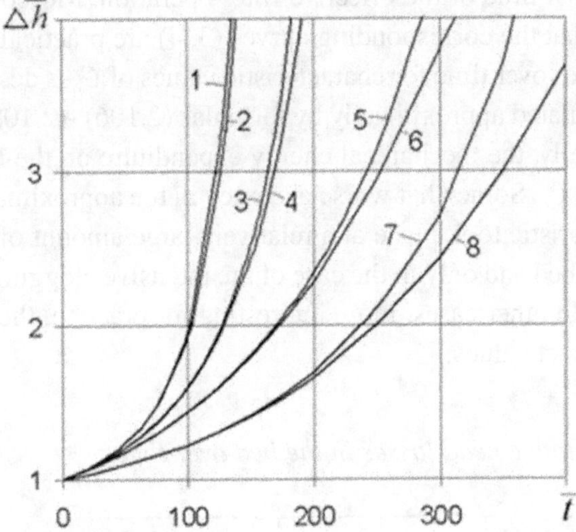

Note. 1, 3, 5, 7 – approximate calculation ;2, 4, 6, 8 – exact; 1, 2 – $\bar{\alpha}$ *= 10; 3, 4 –* $\bar{\alpha}$ *= 8; 5, 6 –* $\bar{\alpha}$ *= 6; 7, 8 –* $\bar{\alpha}$ *= 4*

Figure 21. Increase in the relative outlet volumetric concentration of the suspended solids over time

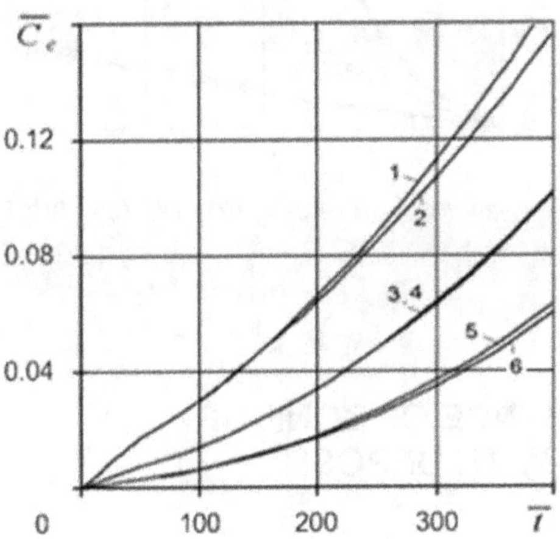

Note. 1, 3, 6 – exact calculation; 2, 4, 5 – approximate; 1, 2 – $\bar{l}_1$ *= 0.75; 3, 4 –* $\bar{l}_1$ *= 0.5; 5, 6 –* $\bar{l}_1$ *= 0.25*

In the case of two-layer bed, the initial information primarily included the single values $\bar{\beta}_1 = \bar{\beta}_2 = 0.005$, $\bar{\gamma}_{1a} = \bar{\gamma}_{2a} = 20$, as well as a set of two values $\bar{\alpha}_1 = 4$, $\bar{\alpha}_2 = 8$. The thicknesses of both layers were varied and the relative thickness of the upper layer was $\bar{l}_1 = 0.25$, 0.5 or 0.75. On the other hand, the thickness $\bar{l}_2$ complemented $\bar{l}_1$ to one. The former three values were taken for the initial suspended solids concentration. First of all, the main quality indicator of the suspension clarification at different thicknesses of the upper and respectively lower layers was calculated. Exact (2.59) and approximate (2.105) formulas were used, and the curves obtained in this way are shown in

Figure 21. The differences between the pairs of the curves corresponding to the same value of $\bar{l}_1$ turned out to be minimal even at significant deterioration of the filtrate quality, i.e. in the time interval, which significantly exceeds the usual time of the effective filter operation. Moreover, these differences were so small at $\bar{l}_1 = \bar{l}_2 = 0.5$ that the corresponding curves (3, 4) are practically coincided.

The increase in head losses over time for characteristic values of C_0 is described by curves in **Figure 22**. The losses $\Delta\bar{h}$ were calculated approximately by formulas (2.106) -(2.108), and exactly by formulas (2.56), (2.60), (2.70). Naturally, the mechanical energy expenditure on the filtration process increased significantly with increasing C_0. Somewhat worse accuracy in the approximate calculations of the most important hydraulic characteristic took place at a relatively large amount of impurities in the clarified suspension at the inlet to the bed and only in the case of the intensive clogging of the bed with a multiple increase in the head losses. In other cases, more interesting for practice, the approximate values of $\Delta\bar{h}$ almost coincided with the exact values.

Figure 22. Increase in the relative head losses in the bed over time

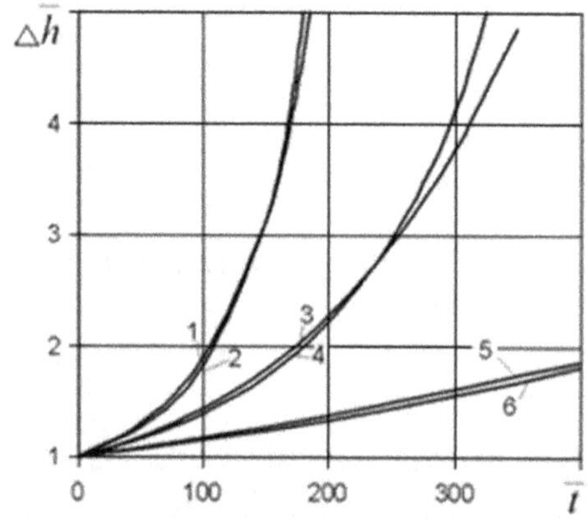

Note. 1, 3, 5 – exact calculation; 2, 4, 6 – approximate; 1, 2 – C_0=7.5·10⁻⁵; 3, 4 – C_0=5·10⁻⁵; 5, 6 – C_0=2..5·10⁻⁵

2.5. FILTRATION IN PRESENCE OF ZONE OF MAXIMUM SATURATION WITH DEPOSIT

The clarification of aqueous suspension on rapid filters is usually restrained by a number of factors. A substantial part of the pore space always remains inaccessible to a deposit (Kulikov et al., 2009; Poliakov, 2012). In addition, the force generated by the suspension flow partially detaches the deposit in case of its loose fixation by the solid phase, thus weakening the effect of the suspension separation due to the absorption of impurities and causing the secondary contamination of a packed bed. And here the physicochemical properties of filtering material and dispersed impurity become important. The first limiting factor is characterized by the parameters of the true solids take-up capacity S_m, S_{Sm}, which by analogy with (1.29) are related by the nonlinear relation

$$S_{Sm} = \gamma(S_m)S_m. \tag{2.109}$$

Quantitative indicators of the second factor are the dynamic solids take-up capacities S_d and S_{Sd} (with respect to the deposited suspension particles and deposit). The concentration S_d is expressed at the linear mass exchange kinetics through the constant mass exchange coefficients, the initial concentration of the suspended solids and according to (2.19) will be

$$S_d = \frac{\alpha}{\beta} C_0,\ \bar{S}_d = \frac{S_d}{n_0 C_0} = \frac{\bar{\alpha}}{\beta}.$$

In reality, only this constraint can be implemented, to which the smaller of the values of S_m, S_d correspond. It usually has to be taken into account in the technological process modeling because of the long continuous operation of a filtration facility and the considerable unevenness of the deposit distribution in the bed volume. The technological consequences of both limiting factors are similar and manifest themselves in two ways. First of all, they cause a deterioration of the filtrate quality. Their positive value, however, lies in the reduction of the head losses in case of a constant filtration rate and in the slowing down the decrease of the facility productivity in case of unregulated filtration rate.

The nature of pore space saturation with a deposit is significantly depends on the ratio between S_m and S_d. Below we analyze in detail the detachment linear filtration under strict limitation of the impurity absorption ($S_m < S_d$). Since a bed is minimally clogged after the successive washout ($S^0 << S_m$), a certain time (t_s) is required for this restriction to affect the accumulation and distribution of the deposit. Therefore, it is reasonable to divide the technological process in time into the first ($t_s \geq t \geq 0$) and second ($t_f \geq t > t_s$) stages. For the calculations of the detachment filtration at the first stage up to the moment in time t_s, it is necessary to use exact or approximate solutions from Sections 2.1, 2.4 (the system of initial assumptions does not change). Later on ($t > t_s$), the bed area with the maximum possible clogging of the pore space is formed and gradually expands.

The deposit here is constantly in a state of dynamic equilibrium, so that its loss is immediately compensated by the absorption component of the interfacial mass exchange due to the hydrodynamic detachment. Thus, the concentration of the suspended solids in the suspension flow does not change at this section, although now they are partially in a bound state (inside fragments of the immobilized deposit). The rest of the bed, i.e. the zone of active suspension clarification, is correspondingly shrunk. Mathematical modeling is significantly complicated at the second stage due to the need to additionally find the pattern of moving the boundary between the zones of maximum saturation and active clarification, which is initially unknown. As a consequence, exact analytical methods are ineffective here. Therefore, an efficient simplified approach is recommended and developed below, which relies on the balance of the suspended and deposited matter in the bed and the known exact solution of the problem of the linear detachment filtration with a constant rate.

Thereby, the basis for the theoretical studies of the operation of rapid filters with a stable high hydraulic load, and then for the technological calculations was the known mathematical model of the direct-flow depth detachment linear filtration. It consists in this case of two compartments and is presented in Section 2.1. Its clarification compartment includes the system of equations of mass transfer (at n_e=0) and kinetics of interphase mass exchange (2.1), (2.2), to which the boundary condition (2.3) and a special case of the initial condition (2.4) are added, namely,

$$t=0,\ S=S^0=const. \tag{2.110}$$

These conditions are characteristic for the practical separation of low-concentration aqueous suspensions. Condition (2.110) means that residual contamination is evenly distributed in the washed bed due to intensive mixing of its elements.

The filtration flow compartment contains the equation of motion (2.5), the dependencies characterizing the hydraulic resistance of a clogged bed (1.30) and the deposit state (1.29), as well as the condition at the lower boundary of the bed (downward flow) (2.6). After the introduction of dimensionless variables and parameters in the standard way (Section 2.1) and exclusion of S_S, mathematical problem (1.29), (1.30), (2.1) - (2.6) is expressed by the system of equations (2.7), (2.8), (2.11) and the boundary conditions operator (2.9), (2.10), (2.13).

The exact solution of this problem in general form is presented in Section 2.1. If the deposit is evenly distributed along the height in accordance with (2.110) by the beginning of the calculation period, then the following special dependencies in integral form are valid for both concentrations ($\overline{C}$, $\overline{S}$) and head losses $\Delta\overline{h}$

$$\overline{C}(\overline{z},\overline{t}) = e^{-\overline{\alpha}\overline{z}}\left[e^{-\overline{\beta}\overline{t}} I_0\left(2\sqrt{\overline{\alpha}\overline{\beta}\overline{z}\overline{t}}\right) + \overline{\beta}\int_0^{\overline{t}} e^{-\overline{\beta}\xi} I_0\left(2\sqrt{\overline{\alpha}\overline{\beta}\overline{z}\xi}\right) d\xi\right] + \\ + \overline{\beta}\,\overline{S}^0 e^{-\overline{\beta}\overline{t}} \int_0^{\overline{z}} e^{-\overline{\alpha}\eta} I_0\left(2\sqrt{\overline{\alpha}\overline{\beta}\overline{t}\eta}\right) d\eta, \tag{2.111}$$

$$\overline{S}(\overline{z},\overline{t}) = \overline{S}^0 e^{-\overline{\beta}\overline{t}} + \overline{\alpha}\, e^{-\overline{\alpha}\overline{z}} \int_0^{\overline{t}} e^{-\overline{\beta}\xi} I_0\left(2\sqrt{\overline{\alpha}\overline{\beta}\overline{z}\xi}\right) d\xi + \\ + \overline{\alpha}\overline{\beta}\,\overline{S}^0 e^{-\overline{\beta}\overline{t}} \int_0^{\overline{z}} e^{-\overline{\alpha}\eta} \int_0^{\overline{t}} I_0\left(2\sqrt{\overline{\alpha}\overline{\beta}\eta\xi}\right) d\xi d\eta, \tag{2.112}$$

$$\Delta\overline{h}(\overline{t}) = \int_0^1 \frac{d\overline{z}}{f_k(\overline{S}(\overline{z},\overline{t}))}. \tag{2.113}$$

This solution, if the formulated problem is adequate to the real conditions of filtration, allows us to reliably calculate its characteristics up to the exhaustion of the smaller of the above-mentioned solids take-up capacities. If such is the dynamic solids take-up capacity, which is characteristic of poorly absorbing filtering materials and unstable deposits, the basic model and dependences (2.111) - (2.113) are suitable for versatile analysis (forecasts, design and technological calculations) during the entire filter run of any duration. At limitation of the absorption process and deposit accumulation by the true solids take-up capacity with parameters S_{Sm} or S_m, the calculation expressions (2.111) - (2.113) is acceptable to use only up to the moment in time t_s, when the concentration of the absorbed particles in the inlet section of a bed reaches the maximum value S_m. The relative time $\overline{t}_s$ is calculated exactly by the formula

$$\overline{t}_s = \frac{1}{\overline{\beta}} \ln \frac{\overline{\beta}\,\overline{S}^0 - \overline{\alpha}}{\overline{\beta}\,\overline{S}_m - \overline{\alpha}} \tag{2.114}$$

Here $\overline{S}_m = S_m/(n_0 C_0)$; index 1, and further on index 2 mean the belonging of the characteristic containing them respectively to the previously thoroughly considered first and still unexplored second stages of the technological process.

An essentially inactive area (between 0 and $l_f(t)$, where l_f is the depth of the inactive area) is formed and expanded in the inlet section of a filter bed during the second stage ($t>t_s$). This is where the regular renewal of a part of the deposit takes place, while the total amount of the deposit remains the same. Thus, the suspended solids content in the filtration flow decreases only within the active area $\left(1 \geq \bar{z} > \bar{l}_f(\bar{t})\right)$, the extent of which steadily decreases over time. Naturally, when modeling linear filtration at the second stage under the conditions of the corrected mathematical problem, along with the required quantities C, S, h, there should be one more unknown variable ($\bar{l}_f$), which monotonically increases over time from 0 to 1. In fact, we have to formulate and solve the problem belonging to the class of problems with moving boundaries. Their complexity is significantly higher than that of the traditional problems with fixed boundaries. In this case, it is even more aggravated due to the unknown pattern of the boundary movement. And standard methods of solving linear problems in partial derivatives, for example, integral transformations, are not effective for this class. Therefore, it is necessary to search for original approaches and techniques based on the specific conditions at the engineering site.

For the detachment filtration with structural restriction of a deposit accumulation ($S_m<S_d$) considered in this chapter, an approximate approach has been developed based on the above-mentioned exact solution (Section 2.1). Dependencies (2.111) - (2.113) that follow from it, which are directly applicable to the calculations of the suspension filtration at the first and then second stages, can be used due to the following two assumptions. First, it is assumed that the amount and composition of the deposit produced per unit time with and without the severe limitation on the deposition are the same due to the smallness of S_m. Its validity is obvious from the technological considerations. Indeed, the filter operation will be effective only at the minimum (in accordance with the existing standards) removal of impurity from a bed. The past of suspended solids through the bed is taken into account very carefully in the balance of the deposited suspension particles. The inaccuracy allowed here is caused by neglecting the increase in the impurity in the filtrate directly due to the reduction of the active volume of the bed, which is only a small part of the small amount of the impurity carried away with filtrate.

Secondly, the deposited particles are distributed in the active area along the height at any moment of time ($t>t_s$) according to expression (2.112) with correction for the actual position of the interzone boundary. The second assumption is more debatable. Its application is justified by the mathematical accuracy of the concentration profile $S_1(z,\, t_1) \;=\; S_2^0(z)$ initial for the second stage (at $t=t_s$) and the slowness of its subsequent transformations, which weakly affect the second important technological characteristic, namely, the head losses in a uniform filter media.

The limitation ($S\leq S_{Sm}$) imposed on the deposit formation and accumulation in the modeling of the linear detachment filtration at the second stage is taken into account in the following way. Proceeding from the disperse impurity balance in differential form, a similar balance for the current point in time $t>t_s$ is presented in integral form. In this case, the mass transfer equation (2.1) is integrated twice – by relative time from $\bar{t}_s$ to $\bar{t}$ and by relative depth from 0 to 1. In this way, the following equality is obtained

$$\bar{t} - \bar{t}_s - \int_{\bar{t}_s}^{\bar{t}} \bar{C}_e(\xi)d\xi + \int_0^1 \bar{S}_2(\eta,\bar{t})\,d\eta - \int_0^1 \bar{S}_2(\eta,\bar{t}_s)\,d\eta \;=\; 0 \tag{2.115}$$

Then we consider two options of setting the function $\bar{S}_2(\bar{z},\bar{t})$. First, it is taken according to the above-mentioned exact solution of the problem at $S_m>S_d$, namely (2.112). Secondly, an expression for $\bar{S}_2$ is chosen to reflect the fact of the formation of two characteristic zones in a bed

$$\bar{S}_2(\bar{z},\bar{t}) = \begin{cases} \bar{S}_m, & at \ \ \bar{z} \le \bar{l}_f(\bar{t}) \ ; \\ \bar{S}_2(\bar{z},\bar{t}_s), & at \ \ \bar{z} > \bar{l}_f(\bar{t}); \end{cases} \tag{2.116}$$

where $\bar{S}_2(\bar{z},\bar{t}_s)$ formally coincides with $\bar{S}_2(\bar{z},\bar{t})$ according to (2.112) at $\bar{t}_s = \bar{t}$. Here $\bar{t}_s$ is a new parameter, which is larger $\bar{t}$ and uniquely related to it. It should be set by calculation and ensure that the material balance is maintained under the current spatial restriction on the deposit accumulation (with formation of the zone of maximum saturation that is inactive area). Using expressions (2.112) and (2.116) for $\bar{S}$, two forms of the integral balance of impurity are obtained. One of the corresponding equations is then subtracted from the other equation and the result is the following equality

$$\bar{S}_m \bar{l}_f \int_0^1 \bar{S}_2(\bar{z},\bar{t}_*)\,d\bar{z} = \int_{\bar{l}_f}^1 \bar{S}_2(\bar{z},\bar{t})d\bar{z} + \int_{\bar{t}_s}^{\bar{t}} \left[\bar{C}_e(\xi) - \bar{C}_e(\bar{t}_s(\xi))\right] d\xi \tag{2.117}$$

The second integral in the right part of equality (2.117) has small values during a filter run and; therefore, it is justified to neglect it in the first approximation. To specify (2.117), expressions (2.112) and (2.116) are used.

Thus, at the second stage, the equality of the volumes of the deposited particles, calculated at an arbitrary time point on the basis of (2.112) in the presence and in the absence of the constraint $S \le S_m$, is expressed by the following equation

$$\begin{aligned} &\bar{S}_m \bar{l}_f + \bar{\alpha} \int_{\bar{l}_f}^1 e^{-\bar{\alpha}\eta} \int_0^{\bar{t}_s} e^{-\bar{\beta}\xi} I_0\left(2\sqrt{\bar{\alpha}\bar{\beta}\eta\xi}\right) d\xi d\eta = \\ &= \bar{\alpha} \int_0^1 e^{-\bar{\alpha}\eta} \int_0^{\bar{t}_s} e^{-\bar{\beta}\xi} I_0\left(2\sqrt{\bar{\alpha}\bar{\beta}\eta\xi}\right) d\xi d\eta. \end{aligned} \tag{2.118}$$

Here and in the following, we assume $S^0=0$ for more compactness of the computational equations. If necessary, it is easy to generalize them to the case $S^0>0$. The integrand in the first integral describes the profile of the relative volumetric concentration of the solids trapped in the active clarification area $\left(1 \ge \bar{z} > \bar{l}_f\right)$ at a later time point (t_s) than the calculated time (t). In fact, the new relative quantities $\bar{t}_s$ and $\bar{l}_f$ are unknown functions of the real time and must be established by calculation. Therefore, equation (2.118) is supplemented by the equality, which formally expresses the restriction of the desired concentration $\bar{S}_2$ from above by the value $\bar{S}_m$

$$\bar{\alpha} e^{-\bar{\alpha}\bar{l}_f} \int_0^{\bar{t}_s} e^{-\bar{\beta}\xi} I_0\left(2\sqrt{\bar{\alpha}\bar{\beta}\,\bar{l}_f \xi}\right) d\xi = \bar{S}_m \tag{2.119}$$

For each given moment in time, the corresponding set of two values of $\bar{t}_s$ and $\bar{l}_f$ is found by the fitting procedure from the system of equations (2.118), (2.119). By doing a lot of similar calculations, the relationship between $\bar{t}$ and $\bar{t}_s$, $\bar{l}_f$ is concretized in the form of the dependencies $\bar{t}_s(\bar{t})$, $\bar{l}_f(\bar{t})$. Then the refined expression for the desired concentration $\bar{S}_2$ takes the following form

$$\overline{S}_2(\overline{z},\overline{t}_s(\overline{t})) = \overline{\alpha}\,e^{-\overline{\alpha}\overline{z}} \int_0^{\overline{t}_s(\overline{t})} e^{-\overline{\beta}\xi} I_0\left(2\sqrt{\overline{\alpha}\overline{\beta}\overline{z}\xi}\right) d\xi \tag{2.120}$$

Now, involving the empirical normalized permeability function $f_k(\overline{S})$ and representation (2.120), it is easy to determine the current (at a relative time point $\overline{t}$) relative head losses in a filter bed

$$\Delta\overline{h}(\overline{t}) = \frac{\overline{l}_f(\overline{t})}{f_k(\overline{S}_m)} + \int_{\overline{l}_f(\overline{t})}^{1} f_k\left(\overline{S}(\overline{z},\overline{t}_s(\overline{t}))\right) \tag{2.121}$$

The corrected relative suspended solids concentration within the saturation zone is found by integrating the mass transfer equation under the condition

$$\overline{z} = \overline{l}_f(\overline{t}),\ \overline{C} = 1 \tag{2.122}$$

and expressed in the following way

$$\overline{C}_2\left(\overline{z},\overline{l}_f(\overline{t}),\overline{t}_s(\overline{t})\right) = e^{-\overline{\alpha}[\overline{z}-\overline{l}_f(\overline{t})]} + \overline{\beta} \int_{\overline{l}_f(\overline{t})}^{\overline{z}} e^{-\overline{\alpha}(\overline{z}-\eta)}\,\overline{S}_2(\eta,\overline{t}_s(\overline{t}))\,d\eta \tag{2.123}$$

Then the relative outlet concentration $\overline{C}_e$ will be

$$\overline{C}_{2e}(\overline{t}) = e^{-\overline{\alpha}[1-\overline{l}_f(\overline{t})]} + \overline{\beta} \int_{\overline{l}_f(\overline{t})}^{1} e^{-\overline{\alpha}(1-\eta)}\,\overline{S}_2(\eta,\overline{t}_s(\overline{t}))\,d\eta \tag{2.124}$$

Since in developing the calculation procedures described above for the second stage of the linear filtration two assumptions were made, which inevitably reduce the accuracy of calculations of its characteristics, the errors introduced by them were evaluated first of all. An exact solution of the analogous problem with respect to the non-detachment filtration also in two stages served as a reference. The detailed analysis by analytical methods of the non-detachment filtration will be carried out in Chapter 4. The only difference in the formulations of the reference and the above problems is in the kinetic equation, which was simplified for the comparative analysis (assume $\overline{\beta} = 0$) and thus, reduced to the classical form (Iwasaki, 1937),

$$\frac{\partial \overline{S}_2}{\partial \overline{t}} = \overline{\alpha}\,\overline{C}_2 \tag{2.125}$$

The mass transfer equation (2.28) and boundary (2.9), (2.122) and initial (2.10) conditions were not changed. The solution of this problem with respect to the concentrations $\overline{C}_1$, $\overline{S}_1$ (first stage) is described by the simple dependences

$$\overline{C}_1(\overline{z}) = e^{-\overline{\alpha}\overline{z}} \tag{2.126}$$

$$\overline{S}_1(\overline{z},\overline{t}) = \overline{\alpha}\,e^{-\overline{\alpha}\overline{z}}\,\overline{t} + \overline{S}^0 \tag{2.127}$$

It follows that the first stage ends by the point of time

$$\bar{t}_s = \frac{\bar{S}_m - \bar{S}^0}{\bar{\alpha}} \tag{2.128}$$

and at that the final for it and the initial concentration profile of the deposited suspension particles will be for the second stage

$$\bar{S}_1(\bar{z}, \bar{t}_s) = \bar{S}_2^0(\bar{z}) = \bar{S}^0 + (\bar{S}_m - \bar{S})\, e^{-\bar{\alpha}\bar{z}} \tag{2.129}$$

At the second stage of filtration, as follows from equation (2.9) and condition (2.122), the suspended solids will be distributed within the bed according to the expression

$$\bar{C}_2(\bar{z}, \bar{t}) = e^{-\bar{\alpha}[\bar{z} - \bar{l}_f(\bar{t})]} \tag{2.130}$$

From equation (2.125) and expression (2.129) follows

$$\bar{S}_2(\bar{z}, \bar{t}) = \bar{S}^0 + (\bar{S}_m - \bar{S}^0)\, e^{-\bar{\alpha}\bar{z}} + \bar{\alpha} \int_{\bar{t}_s}^{\bar{t}} e^{-\bar{\alpha}[\bar{z} - \bar{l}_f(\xi)]} d\xi \tag{2.131}$$

The relative function $\bar{l}_f(\bar{t})$, describing the displacement of the interzone boundary, is found using the condition

$$\bar{z} = \bar{l}_f(\bar{t}),\ \bar{S}_2 = \bar{S}_m. \tag{2.132}$$

From (2.131), taking into account (2.132), we derive the equation with respect to $\bar{l}_f(\bar{t})$

$$\int_{\bar{t}_s}^{\bar{t}} e^{\bar{\alpha}\bar{l}_f(\xi)} d\xi = \frac{\bar{S}_m - \bar{S}^0}{\bar{\alpha}} (e^{\bar{\alpha}\bar{l}_f(\bar{t})} - 1) \tag{2.133}$$

which is transformed into the simple differential equation

$$\frac{d\bar{l}}{d\bar{t}} = \frac{1}{\bar{S}_m - \bar{S}^0} \tag{2.134}$$

Solution of equation (2.134) at the initial condition

$$\bar{t} = \bar{t}_s,\ \bar{l}_f = 0 \tag{2.135}$$

is the following linear function

$$\bar{l}_f(\bar{t}) = \frac{\bar{t} - \bar{t}_s}{\bar{S}_m - \bar{S}^0} \tag{2.136}$$

Thus, finally the exact representations for the required concentrations will be

$$\overline{C}_2(\overline{z},\overline{t}) = e^{-\overline{\alpha}\left(\overline{z}-\frac{\overline{t}-\overline{t}_s}{\overline{S}_m-\overline{S}^0}\right)} \tag{2.137}$$

$$\overline{S}_2(\overline{z},\overline{t}) = \overline{S}^0 + \left(\overline{S}_m - \overline{S}^0\right) \exp\left[\frac{\overline{\alpha}}{\overline{S}_m - \overline{S}^0}(\overline{t} - \overline{t}_s) - \overline{\alpha}\overline{z}\right] \tag{2.138}$$

When establishing the filtration characteristics, according to the above procedure, first of all, a system of equations is prepared with respect to $\overline{l}_f, \overline{t}_s$. Proceeding from the exact solution of this problem at the first stage of filtration, it has the following form

$$\overline{S}_m\overline{l}_f + \int_{\overline{l}_f}^{1} (\overline{S}^0 + \overline{\alpha}\, e^{-\overline{\alpha}\overline{z}}\, \overline{t}_s)\, d\overline{z} = \int_0^1 (\overline{S}^0 + \overline{\alpha}\, e^{-\overline{\alpha}\overline{z}}\overline{t})\, d\overline{z} \tag{2.139}$$

$$\overline{S}^0 + \overline{\alpha}\, e^{-\overline{\alpha}\overline{l}}\, \overline{t}_s = \overline{S}_m \tag{2.140}$$

It follows from (2.140)

$$\overline{t}_s = \frac{\overline{S}_m - \overline{S}^0}{\overline{\alpha}} e^{\overline{\alpha}\overline{l}_f} \tag{2.141}$$

Then equation (2.139) transforms as follows

$$\overline{l}_f - \frac{e^{\overline{\alpha}(1-\overline{l}_f)}}{\overline{\alpha}} = (1 - e^{-\overline{\alpha}})\frac{\overline{t}}{\overline{S}_m - \overline{S}^0} - \frac{1}{\overline{\alpha}} \tag{2.142}$$

Comparing equations (2.136) and (2.142), we obtain for the discrepancy $\Delta\overline{l}_f$ between the exact and approximate values of $\overline{l}_f$ the expression

$$\Delta\overline{l}_f = e^{-\overline{\alpha}}\left(\frac{e^{-\overline{\alpha}\overline{l}_f}}{\overline{\alpha}} - \frac{\overline{t}}{\overline{S}_m - \overline{S}^0}\right) \tag{2.143}$$

This discrepancy is a small quantity compared to $\overline{l}_f$ in the realistic filtration conditions. For example, the error will be about 1% at $\overline{\alpha} = 5, \overline{l}_f = 0.5$.

The relative concentration $\overline{C}_2$ is calculated using the approximate solution with somewhat larger errors. When deriving the expression with respect to the function $\overline{C}_2(\overline{z},\overline{t})$, first of all, an approximate representation for the concentration $\overline{S}_2$ was obtained according to (2.127), (2.141)

$$\overline{S}(\overline{z},\overline{t}) = \overline{S}^0 + \left(\overline{S}_m - \overline{S}^0\right) e^{-\overline{\alpha}(\overline{z}-\overline{l}_f)} \tag{2.144}$$

The formula for $\overline{C}_2$ follows from equation (2.125)

$$\overline{C}_2(\overline{z},\overline{t}) = \left(\overline{S}_m - \overline{S}^0\right) e^{-\overline{\alpha}[\overline{z}-\overline{l}_f(\overline{t})]} \frac{d\overline{l}_f}{d\overline{t}}$$

where according to (2.142)

$$(\bar{S}_m - \bar{S}^0)\,\frac{d\bar{l}_f}{d\bar{t}} = \frac{1 - e^{-\bar{\alpha}}}{1 - e^{-\bar{\alpha}-\bar{l}_f}}$$

Then

$$\bar{C}_2(\bar{z},\bar{t}) = \frac{1 - e^{-\bar{\alpha}}}{1 - e^{-\bar{\alpha}[1-\bar{l}_f(\bar{t})]}}\,e^{-\bar{\alpha}[\bar{z}-\bar{l}_f(\bar{t})]} \tag{2.145}$$

From (2.137) and (2.145) it follows that the relative error $\Delta\bar{C}_e = (\bar{C}_{e,exact} - \bar{C}_{e,appr})/C_{e,exact}$ depending on $\bar{l}_f$ will be

$$\Delta\bar{C}_e(\bar{l}_f) = \left|\frac{1 - e^{\bar{\alpha}\bar{l}_f}}{e^{\bar{\alpha}} - e^{\bar{\alpha}\bar{l}_f}}\right| \tag{2.146}$$

If we determine $\bar{C}_e$ on the basis of the unadjusted solution and without taking into account the solids take-up capacity constraint ($S \leq S_m$), then it is easy to obtain using (2.126)

$$\Delta\bar{C}_e(\bar{l}_f) = 1 - e^{\bar{\alpha}\bar{l}_f} \tag{2.147}$$

Figure 23 shows the graphs of the dependence $\Delta\bar{C}_e(\bar{l}_f)\Delta\bar{C}_e(\bar{l}_f)$, calculated by (2.146), (2.147) when changing $\bar{l}_f$ in practical limits and for two values of $\bar{\alpha}$. Hence, firstly, the possibility and expediency of careful consideration of the existence of a part of the pore space inaccessible to a deposit in the framework of the linear filtration models, secondly, the legitimacy of the proposed approximate approach to taking into account the noted factor are obvious. Indeed, if we model filtration with the linear mass exchange kinetics and uncontrolled rise of the quantity $S_s = \gamma(S)S$ up to the maximum possible value n_0, the true values of the outlet suspended solids concentration can be many times less than when such rise is restrained ($S \leq S_m$). At the same time, if limiting the deposition of impurities, based on the filter bed solids take-up capacity, is taken into account according to the above recommendations, the errors in determining the content of contamination in the filtrate will be minimal for a long time, and even more reduced with the improvement of the absorption properties of the filtering material. They reach 14.9% at $\bar{\alpha} = 5, \bar{l}_f = 0.6$ and also 64% at the same value of $\bar{l}_f$ and $\bar{\alpha} = 7$ in the situation being considered. And here it should be kept in mind that the bed by the end of the calculation period has long since exhausted its clarification resource and needs urgent and intensified washing. It is logical, that the results of the performed comparative analysis give the right to rely on the above developed approximate dependences and equations in future when conducting the forecasts and technological calculations in two stages (in accordance with two stages of the detachment linear filtration).

Figure 23. Dependence $\Delta \overline{C}_e(\overline{l}_f)$

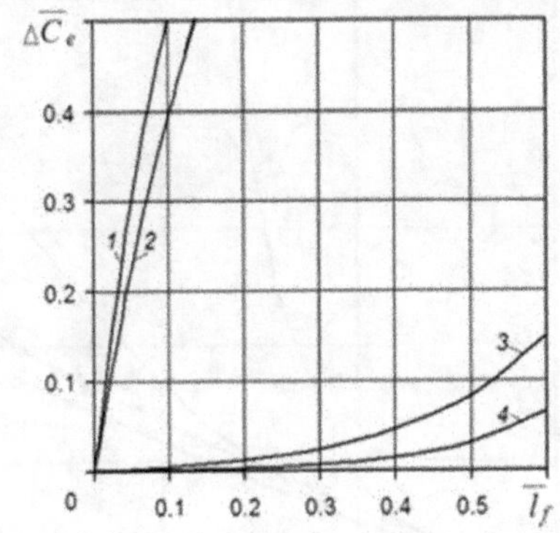

Note. 1 – by formula (3.118); 2 – by (3.119); 1, 4 – $\overline{\alpha}$ = 7; 2, 3 – $\overline{\alpha}$ = 5

Figure 24. Increase in the relative outlet volumetric concentration of the suspended matter over time

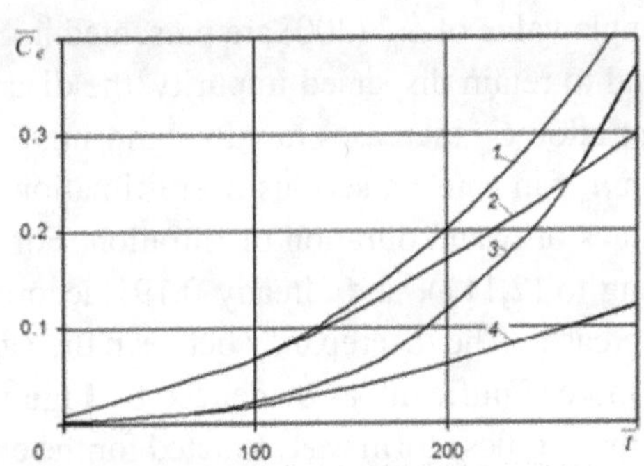

Note. 1, 3 – with correction; 2, 4 –without correction; 1, 2 – $\overline{\alpha}$ = 5; 3, 4 – $\overline{\alpha}$ = 7

Subsequent quantitative analysis was carried out on numerous examples to demonstrate possible significant errors in the calculation of the filtration characteristics due to neglecting the limitation of the absorption resource by volume ($S_{Sm} = \gamma(S_m)\ S_m < n_0$). The calculations were carried out in parallel using the unadjusted (2.111) - (2.113) and implicitly adjusted (2.118) - (2.121), (2.124) dependencies. The first system of the calculation dependencies follows from an exact solution of the problem of the linear detachment filtration and is mathematically rigorous. The second system reflects the fact of formation and development of the zone of maximum saturation with a deposit and, as a consequence, increased removal of the suspension particles from a filter bed, slowing down a rise of the head losses in it. Initial values of $\overline{\beta}(0.01)$, $\overline{\gamma}_a(0.002)$ were assumed; the bed was clean before the beginning of filtration ($S^0 = 0$) ($S^0 = 0$). The coefficient $\overline{\alpha}$ was varied discretely or continuously in wide ranges; two characteristic values were taken for the parameter $\overline{S}_m$.

Figure 25. Increase in the relative head losses in the bed over time

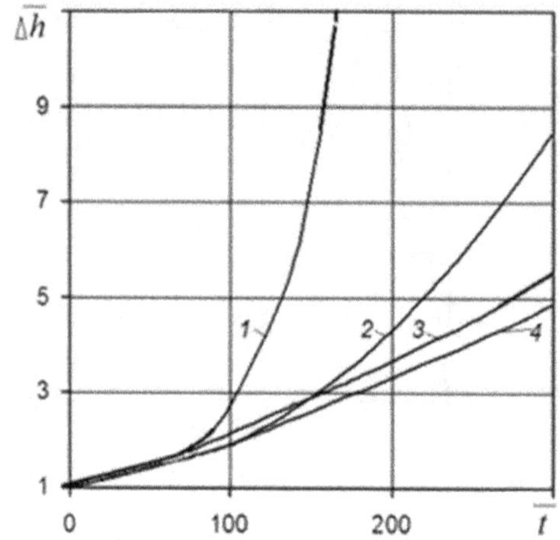

Note. 1, 2 – without correction; 3, 4 – with correction; 1, 3 – $\bar{\alpha}$ = 7; 2, 4 – $\bar{\alpha}$ = 5

The results of calculating the increase in the outlet concentration of the suspended solids with time at two values of $\bar{\alpha}$ (5, 7) and a single value of $\bar{S}_m$ (300) are presented in **Figure 24**. With increasing the capability of the filtering material to retain dispersed impurity, the discrepancy between the reference and refined values of the concentration $\bar{C}_e$ increases rapidly. Ignoring the well-known fact that $\bar{S}_m$ one and a half or two times less than n_0 can cause a serious overestimation of the protective capability of a bed, and especially at high values of $\bar{\alpha}$ and duration of filtration. For example, $\bar{C}_e$ is equal to 0.085 at $\bar{\alpha}$ = 7 and $\bar{t}$ = 240 according to (2.111), and already 0.191 according to the corrected data and further this difference rapidly increases. The discrepancy between the values of the head losses at their calculation by two ways is even more significant, as evidenced by **Figure 25**. Here the partially linearized representation (2.25) in a dimensionless form was attracted for the empirical dependence $f_k(\bar{S})\,f_k(\bar{S})$

$$f_k(\bar{S}) = (1 - \bar{\gamma}_a\bar{S})^3 \tag{2.148}$$

In the case of the detachment linear filtration with the dynamic limitation of the deposit formation ($S_m \geq S_d$), which was thoroughly discussed in the previous sections of this chapter, the head losses in the bed increase faster due to the intensive clogging of the inlet section and their sharp increase is possible by the end of the filter run (curve 1). The behavior of similar curves 3, 4 essentially differs at $\bar{S}_m$ = 300. They are characterized by a smooth rise during the entire calculation period, which is explained by a significant residual permeability of the bed. At the same time, the head losses are limited from above by the limiting value $\Delta\bar{h}_{\max} = (1 - \bar{\gamma}_a\bar{S}_m)^{-3}$, although their actual value should be much less due to the accelerated deterioration of the filtrate quality by the end of the filter run. A significant decrease in the sensitivity of $\Delta\bar{h}\Delta\bar{h}$ to the absorption properties of the filtering material due to the correction of the characteristics $\bar{S}_2$, $\bar{h}\,\bar{h}$ (curves 3, 4) is fundamental.

It is appropriate to note that the choice of an approximation expression for the dependence $f_k(\bar{S})\,f_k(\bar{S})$ (2.148) in combination with the values γ_a=0.002, S_m=500, when modeling the filtration flow process in the porous medium intensively being clogged, leads to unphysical results of the flow characteristic calculations. Indeed, in this case, at the maximum saturation of the filter media with a deposit, its permeability drops to 0, which does not correspond to reality. In practice, there will necessarily be free space from the deposit in a packed bed at any degree of clogging, including the maximum degree, and hence the residual permeability, which is usually much less than the initial permeability (by an order of magnitude or more). It is logical that the decrease in permeability of the inlet section of the bed (as it becomes progressively clogged) calculated according to (2.148) and (2.112) is much ahead of the actual decrease. Thus, the second criterion is violated too early. As a consequence, filter washing will be premature. However, the inactive area is not formed at all under such conditions, so that this development is not relevant. Asymptotic character of the deposit accumulation in the case $S_m \leq S_d$ and thus, a limited increase in the head losses in the bed can be taken into account due to the use of the nonlinear kinetics, which reflects the close relationship between the coefficient α and S. Just such kinetics will be the subject of the theoretical studies in Chapter 6 with linear and nonlinear relationships between them.

Application of the above presented methodology of the calculation of the detachment linear filtration at the rigidly limited absorption capacity of filtering material for the purpose of the theoretical substantiation of the technological parameters will be the subject of discussion in Section 3.1.

2.6. ABOUT CHOICE OF COEFFICIENT OF SUSPENDED SOLIDS ATTACHMENT RATE

The process of the attachment of the suspended matter to the structural elements of a packed bed (grains, granules) plays the key role in the clarification of aqueous suspensions by the deep bed filtration. As a rule, its rate is assumed to be proportional to the concentration of the suspended solids, so that the kinetics of the detachment nonlinear filtration is generally described by equation (1.13) (Ives, 1970; Iwasaki, 1937; Mintz, 1964). The reliability of the results of mathematical modeling of the (non)detachment filtration, and eventually the degree to which the calculated pattern corresponds to the actual pattern, is largely determined by the choice of the (functional) proportionality coefficient (in fact the function of dispersed contaminant absorption) f_α. This coefficient directly controls the rate of the suspended solids deposition and, in a first approximation, especially if the technological analysis is to be performed, can be considered constant at V=*const*, i.e. $f_\alpha(S, V) = \alpha = const$.

The only reliable way to establish an adequate value of the coefficient α, controlling the absorption of the suspended solids, in the case of the progressive clogging of a bed is to conduct serial experiments with samples of a filter media at different exposure times and identical technological conditions. However, the experimental determination of the mentioned important model coefficient, which essentially depends on many factors and can be unstable in reality, requires considerable costs and efforts. It is especially difficult to arrange and conduct appropriate experimental studies at the operating water treatment plants and filtration facilities. Therefore, it is natural that considerable attention has been paid to the theoretical substantiation of this coefficient for a number of decades. To understand more deeply the physical prerequisites for the attachment of small mobile particles to the surface of stationary large particles-collectors, it was necessary to involve the concepts and special models of the hydrodynamics of two-phase flows, molecular and statistical physics. As a result, two characteristic phases were distinguished in the absorption of dispersed contaminants, namely, delivery of the suspended solids to grains and then their firm fixation. Since from the suspension particles in contact with the grain only a part of them is actually permanently retained, a representation for α was adopted in the form of the product of two coefficients. The first of them (α_d) characterizes the delivery rate of the suspension particles to vacant places on the grain surface, and the sticking (or attachment) efficiency factor α_c is the fraction of the delivered particles that are fixed here. The coefficient α_c shows increased sensitivity to physicochemical conditions, water flow properties of the porous medium and is purely empirical, with its characteristic values ranging from 1 to values close to 0 (Landkamer et al., 2013; Parker et al., 2017; Ren et al., 2000; Shen et al., 2010; Yiantsios & Karabelas, 2003). The coefficient α_d, on the other hand, has been the subject of a number of the theoretical studies, during which three mechanisms were identified and formalized that ensure the convergence and contact of the suspended solids with regularly shaped particles-collectors. In general, the contribution of diffusion, interception and sedimentation to the deposition of the suspended solids in the pores of a filter media depends on many factors. But the size of the suspended particles (can vary by several orders of magnitude), packed bed elements (by an order of magnitude), as well as the rate mode have a particularly significant effect on the delivery and, in general, the absorption process. Without going into the specifics of the practiced approaches, especially since some remarks in this regard are made in Section 1.2, it still makes sense to highlight the main formulas for calculating the coefficient α_d and recommendations for its selection in the works (Elimelech & O'Melia, 1990; Martin et al., 1992; Nelson & Ginn, 2011; Rajagopalan & Tien, 1976; Tufenkji & Elimelech, 2004). In doing so, the authors relied on the results of modeling with the use of numerical methods of the suspended solids transport by laminar flow in the intergranular space under the action of forces of inertia, resistance and gravity. As it turned out over time, the above formulas allow to calculate the values of α_d, which are in satisfactory agreement with the corresponding experimental data. It is also important that they plausibly reflect the influence of those factors that mainly contribute to the removal of the suspended solids. The formulas are universal in the sense that they are applicable over a wide range of variations in the determining parameters (sizes of the suspended solids and particles-collectors, filtration rate). Finally, they have the convenient form for the quantitative analysis of interfacial mass exchange, generalizations and specification of the initial mathematical model of the deep bed filtration.

For example, the widely known formula (Rajagopalan & Tien, 1976) allows, in the absence of suitable experimental data, to provide mathematical modeling of the operation of rapid and ultra-rapid filters with fine-, medium- and coarse-grained filter bed with values of the coefficient α_d close to the realistic values. The frequency of possible contacts of the suspension particles with the grains of the bed is closely related to their equivalent diameters, respectively d_p, d_g and filtration flow rate V, which was the reason for the transformation of the above formula and the extraction of these parameters from the dimensionless complexes. Thus, the basic expression for the dimensionless coefficient $\overline{\alpha}_d$ (equal to $L\alpha_d/V$) contains three chief parameters (d_p, d_g, V) in an explicit form, which change significantly in practice (Poliakov, 2007), namely,

$$\overline{\alpha}_d = 1.347 \cdot F_1(n)\left(\frac{k_B T_t}{\mu}\right)^{\frac{2}{3}} \cdot d_g^{-\frac{5}{3}}\, d_p^{-\frac{2}{3}}\, V^{\frac{1}{3}} + 1.175 \cdot F_2(n)\left(\frac{H_G}{\mu}\right)^{\frac{1}{8}} d_g^{-\frac{23}{8}}\, d_p^{\frac{13}{8}}\, V^{\frac{7}{8}} + \\ +1.578 \cdot 10^{-4}\; F_2(n)\left[\frac{g(\rho_p - \rho_w)}{\mu}\right]^{\frac{6}{5}} d_g^{-\frac{3}{5}}\, d_p^{2}\, V^{-\frac{1}{5}}, \quad s^{-1}. \tag{2.149}$$

Here $F_1(n) = (1 - n)A_\alpha^{\frac{1}{3}}(n)$, $F_2(n) = (1 - n)^{\frac{5}{3}}A_\alpha(n)$,

$$A_\alpha(n) = \frac{1 - (1 - n)^{\frac{5}{3}}}{1 - 1.5\left(1 - n\right)^{\frac{1}{3}} + 1.5\left(1 - n\right)^{\frac{5}{3}} - \left(1 - n\right)^{2}}$$

$n=n_0-S_S$ is the current porosity of the filter medium; k_B is the Boltzmann constant, equal to $1.38\bullet10^{-23}J$; T_t is the temperature in Kelvin; μ is the dynamic viscosity, $kg/(m\bullet s)$; $\bar{\imath}_f$ is the Gamaker constant (typical values in the range from $10^{-19}$to 10^{-20}), J; g is the acceleration of gravity, m/s_2; ρ_p, ρ_w are the matter densities of solids and water, kg/m^3; d_p, d_g, V are measured in m and $m\bullet s$. It should be emphasized that theoretical formula (2.149) and its analogues are mainly of the methodological value and can only serve as a guide in selecting the value of the coefficient α. First, the value of α depends on the coefficient α_c, which can be noticeably less than 1. Incidentally, we can consider in the first approximation that α_c does not depend on d_g, d_p, V. Secondly, when deriving theoretical formulas for α_d, the configuration of the clogged pore space was significantly schematized. Nevertheless, it is justified to calculate α_d coarsely by formula (2.149) at carrying out the applied analysis and, first of all, for the estimation of the importance of the specified design and technological parameters in relation to the clarification process and its consequences, and to take α_c equal to 1, identifying, thus, α with α_d, in the case of a well-absorbing filtering material, the application of reagents.

The first summand in formula (2.149) characterizes the rate of the contaminant particles removal from a suspension due to the Brownian diffusion, the second summand – removal due to their interception, the third summand – removal due to the sedimentation. According to (2.149) the coefficient α should change in natural conditions even at the unchanged filtration rate and composition of contaminants, and these changes are not connected with the autocatalytic phenomenon. Indeed, as the deposit accumulates in the pores, the pore space declines, which causes a decrease in the functions $F_1(n)$, $F_2(n)$ and the required α_d, α with them. Such behavior of these functions means the absence of autocatalytic effect. Incidentally, if the suspension is filtered at a constant head differential, the decrease in α will be even faster. And nevertheless, such decrease is so slow that gives the right to assume α constant during the estimation calculations. And it is acceptable to be oriented on its initial value at the porosity of a clean filter media n_0. Thus, the application of the linear kinetics of the interphase mass exchange is partially justified even with increasing clogging of the porous medium.

Particularly important for the interaction of the mobile and immobile particles are their sizes. It was found that the diffusion transport of the suspended solids to grains dominates at $d_p\leq10^{-6}m$ (colloids); all three mechanisms of the suspension particle delivery to collectors give a comparable contribution in the interval $10^{-6}<d_p<5\bullet10^{-6}m$, and interception, sedimentation become the main mechanisms at $d_p>5\bullet10^{-6}m$. Here in the calculations of the coefficient α (or α_d) by formula (2.149), the sufficiently representative range (10^{-7},...,10^{-5}) was assumed for d_p, within which the ratios between the contributions of diffusion, interception and sedimentation to the removal of the suspended solids from the liquid phase are diametrically change. Although granular media can effectively retain particles with the effective diameter in the range from 10^{-8} to $10^{-4}m$, but the main part of the suspended matter in practical natural and waste waters is composed of particles of just these intermediate sizes. For convenience of presentation of the results of the subsequent theoretical analysis of the absorption capacity of the filtering material taking into account the properties of dispersed contaminants and water flow process, first of all, the basic formula (2.149) is presented in the additive form $\alpha= \alpha_B+ \alpha_I+ \alpha_G$, where the components α_B, α_I, α_G reflect the contribution of diffusion, interception and sedimentation, respectively, to the total value α. In addition, here only the relations α_j/α, α/α_{max}, α/α_{min} were the subject of calculations thus, avoiding complications related to the choice of the coefficient α_c.

The results of the calculations of $\bar{\alpha}_j$, $\bar{\alpha}$ are given in **Figure 26**, where the curves of the dependences of the ratio of individual components α_j to their sum α on the particles diameter of the monodisperse suspension are shown. Knowing the (equivalent) diameter d_p, it is simple to estimate the significance of different mechanisms of particles removal using Figure 26. The number of particles with diameter $2 \cdot 10^{-6} m$, transported to the grains by the Brownian diffusion and the interception is equalized, and slightly fewer such particles contact them due to the sedimentation. A kind of transition zone, within which the role of the above-mentioned mechanisms is polarized, is $d_p = 10^{-6}$-$5 \cdot 10^{-6} m$. Curves 1, 4, 6 are calculated at $d_g = 1mm$ and $V = 50 m/h$ (Example 2). However, here the choice of d_g and V is not of fundamental importance, because even at $d_g = 0.3mm$ and $V = 3m/h$ (example 1; curves 2, 3, 5) the values of $\bar{\alpha}_B/\bar{\alpha}$ will differ in both examples by at most 2%. Therefore, curves 3, 4 are practically coincided and shown as one curve. At the same time, the differences between α_I/α for the cases being considered rise with an increase in d_p and reach the maximum of 10% in the accepted range at $d_p = 10^{-5} m$. Similarly, the largest difference between the values of α_G/α is also at 19.2%. In general, the difference between the values α_j remains relatively small even when changing d_p, d_g, V within significant limits.

Of particular interest for mathematical modeling of the detachment filtration with variable rate is the nature of the relationship between α and V. The results of the calculations of the normalized dependence $\alpha/\alpha_{min}(V/V_{min})$ are shown in **Figure 27** in the form of a series of graphs. At the same time, d_g and d_p were varied discretely, (0.3, 1mm) and (10^{-7}, 10^{-6}, $10^{-5} m$) respectively; and the rate V continuously in the range from $3m/h$ to $60m/h$. The smallest of the calculated values of α, corresponding to $V_{min} = 3m/h$, was taken as α_{min}. The largest and the smallest rates of α rise were observed as an increase in V at $d_p = 10^{-5} m$, which indicates a significant effect of the grain size on the calculation dependence. But the main conclusions here, which will be taken into account in the subsequent theoretical studies of the second main mode of filtration, are as follows: first, the key mass exchange coefficient depends significantly on the filtration rate in a wide range of its values, typical for rapid and ultra-rapid filters; second, they are either related in an almost linear way, or, at least in the formalization of the mass exchange kinetics, it is justified to assume α directly proportional to V.

Figure 26. Dependences of α_j/α on lg d_p

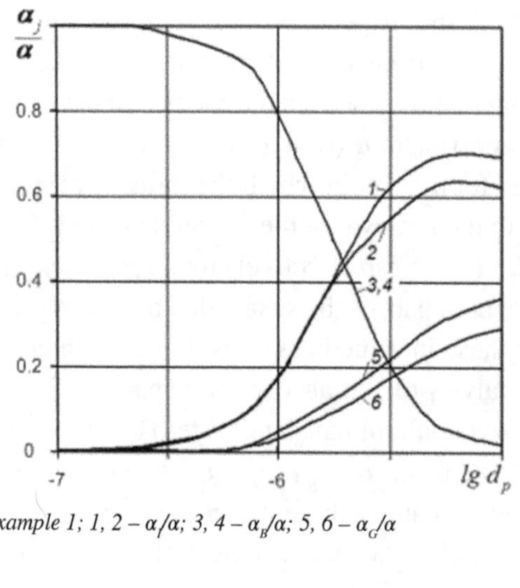

Note. 1, 4, 6 – example 2; 2, 3, 5 – example 1; 1, 2 – α_I/α; 3, 4 – α_B/α; 5, 6 – α_G/α

Figure 27. Dependence of α/α_{min} on (V/V_{min})

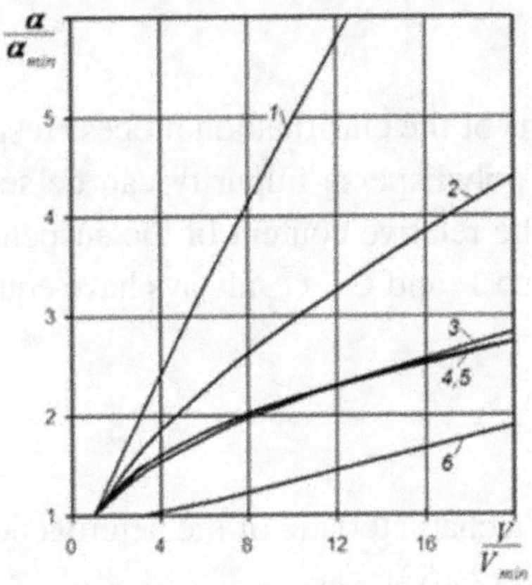

Note. 1, 2, 4 – d_g=0.3; 3, 5, 6 – d_g=1mm; 1, 6 – d_p=10^{-5}; 2, 3 – d_p=10^{-6}; 4, 5 – d_p=10^{-7}m

The rational choice of the grain size can be aided by the data from the calculations of the coefficient α at the continuously varying diameter d_g. To give them a more general form, the ratio $\alpha/\alpha_{\max}$ was defined as a function of $d_g/d_{g\,min}$. Since the range [0.3, 1.5] was adopted for d_g expressed in *mm*, then $d_{g\,min}$=0.3*mm*, $d_g/d_{g\,min}$ varied by a maximum of 5 times. The standard values 10^{-7}, 10^{-6}, $10^{-5}m$ were chosen for d_p, and the values 3 and 50*m/h* for the rate *V*. The curves of the dependence $\alpha/\alpha_{\max}(d_g/d_{g,\min})$ are shown in **Figure 28** and consist of two characteristic pieces. They sharply decrease at the first curve piece in the range of the values $(d_g/d_{g,\min})$ from 1 to about 2.2, which corresponds $d_{g,\min}$ and $d_g \approx 0.65mm$ at the given data. At greater values of the argument, the required α and $\alpha/\alpha_{\max}$ as it increases, continue to decrease, but much slower. However, according to Figure 28, from the practical point of view, it is not expedient to use granular materials with large elements (d_g>0.65*mm*) for water treatment filters due to their unsatisfactory absorption properties. In the case of especially fine suspension particles ($d_p = 10^{-7}m$) hydraulic load does not seriously affect the coefficient α, at least in the considered interval of d_g values (curves 2, 3 actually coincide). On the contrary, such influence is quite significant (curves 1, 5) in the case of large suspended solids ($d_p = 10^{-5}m$).

Finally, the possibility of a simplified theoretical calculation of the mass exchange coefficient α based on the equivalent diameter of the suspension particles d_{pe} was estimated. The impurity was assumed to be polydisperse and formed by fine particles of three grades (fractions). The total volumetric concentration of the impurity *C* included in different ratios: concentration C_1 of the particles with the (average) diameter $d_{p1}=10^{-7}m$, concentration C_2 of the particles with the diameter $10^{-6}m$, concentration C_3 of the particles with the diameter $d_{p3}=10^{-5}m$. Thus,

$$C_1 + C_2 + C_3 = C, \ \overline{C}_1 + \overline{C}_2 + \overline{C}_3 = 1, \tag{2.150}$$

where $\overline{C}_l = C_l/C, l=1,2,3$. Then the actual intensity I_s of the polydisperse impurity removal is equal to

$$I_s = \sum_{l=1}^{3} \alpha_l C_l \tag{2.151}$$

where α $\alpha_l = \alpha(d_{pl})$. The intensity of the removal of the equivalent monodisperse impurity is as follows

$$I_e = \alpha(d_{pe})C. \tag{2.152}$$

Possible errors in the calculations of the clarification process resulting from a formal simplification of the mechanical composition of polydisperse impurity can be seen in **Figure 29**. Here are shown graphs of the dependence I_e/I_s on the relative content of the suspension particles of three varieties. It was assumed that $\overline{C}_1$ varies from 0 to 1, and $\overline{C}_2$, $\overline{C}_3$ always have equal values and according to (2.150)

$$\overline{C}_2 = \overline{C}_3 = \frac{1 - \overline{C}_1}{2} \tag{2.153}$$

In fact, $\overline{C}_1$ can be considered as a characteristic of the heterogeneity of the aqueous suspension impurity. Taking into account (2.151) - (2.153),

$$\frac{I_e}{I_s} = \frac{2\alpha\left(d_{pe}\right)}{2\alpha_1\overline{C}_1 + (\alpha_2\alpha_3)\left(1 - \overline{C}_1\right)} \tag{2.154}$$

The value of d_{pe} was linked with $\overline{C}_1$ in two ways. In the first case, d_{pe} was determined from the equality of quantities and volumes of actual and fictitious equal-sized solids. Then

$$d_{pe}\left(\overline{C}_1\right) = \left(\frac{\overline{C}_1}{d_{p1}^3} + \frac{\overline{C}_2}{d_{p2}^3} + \frac{\overline{C}_3}{d_{p3}^3}\right)^{-\frac{1}{3}} = \left[\frac{\overline{C}_1}{d_{p1}^3} + \frac{1 - \overline{C}_1}{2}\left(\frac{1}{d_{p2}^3} + \frac{1}{d_{p3}^3}\right)\right]^{-\frac{1}{3}} \tag{2.155}$$

Figure 28. Dependence of α/α_{max} on $(d_g/d_{g,min})$

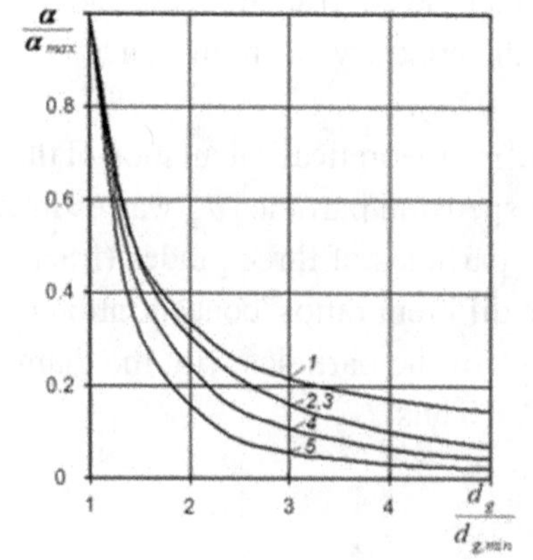

Note. 1, 3, 5 – V=3; 2, 4 – V=50m/h; 1, 5 – $d_p = 10^{-5}$; 2, 3 – $d_p = 10^{-7m}$; 4 – $d_p = 10^{-6}$m

Figure 29. Dependence of I_e/I_s on $\overline{C}_1$

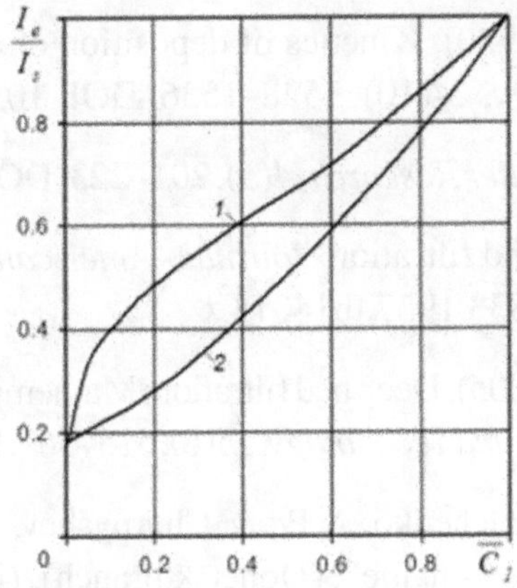

Note. 1 – d_{pe} according to formula (2.155), 2 – d_{pe} by (2.156)

In the second case, the total surface areas of the indicated particles were equated, so that

$$d_{pe}(\overline{C}_1) = \left(\frac{\overline{C}_1}{d_{p1}} + \frac{\overline{C}_2}{d_{p2}} + \frac{\overline{C}_3}{d_{p3}}\right)^{-1} = \left[\frac{\overline{C}_1}{d_{p1}} + \frac{1-\overline{C}_1}{2}\left(\frac{1}{d_{p2}} + \frac{1}{d_{p3}}\right)\right]^{-1} \quad (2.156)$$

The calculations were performed in parallel for the same two examples at $\overline{d}_{pe}(\overline{C}_1)d_{pe}(\overline{C}_1)$ according to (2.155) and (2.156). Their results were almost identical. The degree of the deviation of the calculated value I_e/I_s from 1 gives an idea of possible computational errors in the calculations of the removal rate and the amount of the trapped suspended matter due to the use of an inadequate value of the coefficient α, found for the monodisperse impurity with particles of the diameter d_{pe}. Maximum errors are observed in the absence of fine particles (C_1=0). At the same time, the true values of α and I are underestimated several times due to the formal averaging of the particle sizes even for the bidisperse impurities. Naturally, when $\overline{C}_1$ approaching the maximum value, I_e tends to I_s and they are equalized at $\overline{C}_1 = 1$ (monodisperse impurity). In general, Figure 29 clearly demonstrates the ineffectiveness of the introduction of an equivalent diameter by different ways in mathematical modeling of the clarification of polydisperse suspension.

Thus, the value of theoretical formula (2.149) and similar formulas is mainly in the fact that they allow us to quantify the significance of the properties of the suspension particles and bed elements, hydrodynamic factor for the clarification process. Moreover, involving them, in principle, it is possible to avoid time-consuming and expensive experimental studies of the mass exchange kinetics at moderate requirements to the reliability of modeling results. Still, the use of such formulas even in the engineering calculations of the designed or operated filtration facilities taking into account the progressive clogging of a packed bed is hardly justified because of the uncertainty of α_c and serious assumptions made in the theoretical determination of α_d.

REFERENCES

Elimelech, M., & O'Melia, C. R. (1990). Kinetics of deposition of colloidal particle in porous media. *Environmental Science & Technology*, *24*(10), 1528–1536. DOI: 10.1021/es00080a012

Ives, K. J. (1970). Rapid filtration. *Water Research*, *4*(3), 201–223. DOI: 10.1016/0043-1354(70)90068-0

Iwasaki, I. (1937). Some notes on sand filtration. *Journal - American Water Works Association*, *29*(10), 1591–1602. DOI: 10.1002/j.1551-8833.1937.tb14014.x

Jegatheesan, V., & Vigneswaran, S. (2005). Deep bed filtration: Mathematical models and observations. *Critical Reviews in Environmental Science and Technology*, *35*(6), 515–569. DOI: 10.1080/10643380500326432

Kulikov, N.I., Naimanov, A.Y., Omelchenko, N.P., & Chernyshev, V.N. (2009). *Theoretical bases of water purification*. Publishing House "Znanie", (Donetsk branch). (in Russian)

Landkamer, L. L., Harvey, R. W., Scheibe, T. D., & Ryan, J. N. (2013). Colloid transport in saturated porous media: Elimination of attachment efficiency in a new colloid transport model. *Water Resources Research*, *49*(5), 2952–2965. DOI: 10.1002/wrcr.20195

Martin, R.E., Bouwer, E.J., & Hanna, L.H. (1992). Application of clean-bed filtration theory to bacterial deposition in porous media. *Environ. Scl. Technol.*, 1053-1058.

Mints, D. M. (1964). *Theoretical bases of water purification technology*. Publishing House on Construction. (in Russian)

Mints, D. M., & Meltser, V. Z. (1970). Hydraulic resistance of granular porous medium in the process of clogging. *Rep.* [in Russian]. *Academy of Sciences of USSR*, *192*(2), 304–306.

Nelson, K. E., & Ginn, T.R. (2011). New collector efficiency equation for colloid filtration in both natural and engineered flow conditions. *Water Resour. Res.*, 47, 1-17. W05543, .DOI: 10.1029/2010WR009587

Ojha, C. S., & Graham, N. J. D. (1993). Theoretical estimates of bulk specific deposit in deep-bed filters. *Water Research*, *27*(3), 377–387. DOI: 10.1016/0043-1354(93)90037-I

Orlov, V. O., Martynov, S. Yu., Orlova, A. M., Zoshchuk, V. O., & Kunytskyi, S. O. (2017). *Water treatment on polystyrene foam filters*. NUWEE Press.

Parker, E. A., Rippy, M. A., Mehring, A. S., Winfrey, B. K., Ambrose, R. F., Levin, L. A., & Grant, S. B. (2017). Predictive power of clean bed filtration theory for fecal indicator bacteria removal in stormwater biofilters. *Environmental Science & Technology*, *51*(10), 5703–5712. DOI: 10.1021/acs.est.7b00752 PMID: 28445642

Poliakov, V., Kravchuk, A., Kochetov, G., & Kravchuk, O. (2019). Clarification of aqueous suspensions with a high content of suspended solids in rapid sand filters. *Eureka: Physics and Engineering*, *1*(1), 28–45. DOI: 10.21303/2461-4262.2019.00827

Poliakov, V. L. (2006). About filtration of suspension at initial contamination of filter bed (linear kinetics of mass exchange). [in Russian]. *Rep. NAS of Ukraine*, *10*, 65–71.

Poliakov, V.L. (2006). Calculation of suspension separation by filtration through a three-layer bed at linear mass transfer: *Problems of Water Supply, Sewage and Hydraulics. Collected papers*, 7, 122–131. (in Russian)

Poliakov, V.L. (2007). On the choice of the rate coefficient of suspended solids removal from suspensions during filtration. *Problems of Water Supply, Sewage and Hydraulics. Collected papers*, 8, 120–131. (in Russian).

Poliakov, V.L. (2008). Filtration of suspensions through granular beds at variable suspended solids concentration. *Problems of Water Supply, Sewage and Hydraulics. Collected papers*, 11, 103–111. (in Russian)

Poliakov, V. L. (2008). Calculation of suspension filtration through multilayer filter bed under linear mass transfer kinetics. [in Russian]. *Journal of Water Chemistry and Technology*, *30*(1), 3–31.

Poliakov, V.L. (2009). New analytical solutions of mathematical problems of suspension filtration at linear mass-exchange kinetics. *Bulletin of NUWEE*, 3(47), part 2, 264-271.

Poliakov, V. L. (2011). Engineering calculation of suspension filtration through a two-layer filter bed under linear mass transfer kinetics. [in Russian]. *Journal of Water Chemistry and Technology*, *33*(4), 367–380.

Poliakov, V.L. (2012). Filtration of suspension through a granular bed at limiting saturation of its part with deposit. *Applied hydromechanics*, 14(86), 3, 55–63. (in Russian)

Poliakov, V.L., & Shevchuk, E.A. (2007). Calculation of stepped filtration of suspensions through granular filter beds. Bulletin of NUWEE, 4(40), part 2, 544–553. (in Russian)

Rajagopalan, R., & Tien, C. (1976). Trajectory analysis of deep bed filtration with sphere in cell porous media model. *AIChE Journal. American Institute of Chemical Engineers*, *22*(3), 523–533. DOI: 10.1002/aic.690220316

Ren, J., Packman, A. J., & Welty, C. (2000). Correlation of colloid efficiency with hydraulic conductivity of silica sands. *Water Resources Research*, *36*(9), 2493–2500. DOI: 10.1029/2000WR900163

Shen, C., Huang, Y., Li B., & Yan Jin, Y. (2010). Predicting attachment efficiency of colloid deposition under unfavorable attachment conditions. *Water Resour. Res.*, 46, 1-12. W11526, .DOI: 10.1029/2010WR009218

Tufenkji, N., & Elimelech, M. (2004). Correlation equation for predicting single-collector efficiency in physicochemical filtration in saturated porous media. *Environmental Science & Technology*, *38*(2), 529–536. DOI: 10.1021/es034049r PMID: 14750730

Vasiliev, S. V., Verigin, N. V., & Glazer, B. A.. (1970). *Methods of water flow calculations of hydroreclamation systems*. Kolos.

Yiantsios, S. G., & Karabelas, A. J. (2003). Deposition of micron-sized particles on flat surfaces: Effects of hydrodynamic and physicochemical conditions on particle attachment efficiency. *Chemical Engineering Science*, *58*(14), 3105–3113. DOI: 10.1016/S0009-2509(03)00169-6

Chapter 3
Applied Aspects of Constant Rate Filtration

ABSTRACT

The thorough technological and also design analysis of the rapid filters operation with a constant rate is carried out in Chapter 3. It is based on the solutions of the mathematical problems from the previous chapter, as well as sanitary and hydraulic criteria. The procedure for determining the duration of the filter run (operation period) is shown for similar conditions. This procedure is implemented for the purpose of applied optimization of the filter bed height, the most important design parameter. The consequences of the unavoidable sorting of the granular filtering material due to intensive washing are studied. Filtration is considered with a strict restriction of the impurity accumulation in the pores and, accordingly, the formation of two zones with complete and incomplete deposit saturation in the clogged bed. The technique of the filtration intensification by splitting a single suspension flow into two or more smaller flows is analyzed. The evaluation of the sensitivity of the technological parameters relatively temperature and deposit composition has been carried out.

The chapter focuses on the applied aspects of the theory of the detachment deep bed filtration. In general, the development of the filtration theory, first of all, implies the study of new processes and effects in connection with the increasing requirements of water treatment practice. At the same time, fundamental knowledge about the patterns of changes in the filtration characteristic can be acquired through in-depth analysis of the fundamental knowledge about the patterns of changes in the filtration characteristic can be acquired through in-depth analysis of the validated mathematical models. Indeed, the analytical solutions and the numerous theoretical dependencies arising from them, which are contained in the previous chapter and will also be presented in most of the subsequent chapters of the monograph, can provide a reliable prediction of spatial and temporal changes in all the filtration characteristics involved in the basic models and initially unknown. And nevertheless, not a system solution of the classes of correct mathematical problems of various contaminants removal in special porous media became the main purpose of the monograph. From the practical point of view it is even more important to realize the full potential of the significant opportunities of new solutions as a basis for the technological and design calculations of filtration facilities. Therefore, the efficiency of the operation of the mentioned facilities (actually the economics of filtration) will constantly be in the focus of attention in the future, and the results of the mathematical modeling of the clarification process will logically culminate in the (applied) optimization of the technological and design parameters. In principle, analytical dependences for the

DOI: 10.4018/979-8-3693-8211-0.ch003

concentrations of dispersed impurity, head losses, derived in case of a constant filtration rate at linear mass exchange kinetics and detachment filtration (Chapter 2), at nonlinear kinetics and non-detachment filtration (Chapter 4), and in case of the controlled head differential for the impurity concentrations and filtration rate also at linear kinetics and detachment filtration (Chapter 5), etc. are suitable for realization of applied optimization procedure. At the same time, however, they need to be adapted using efficiency criteria to the technological requirements for the filtrate quality, the expenses of mechanical energy for filtration flow process, water flow rate.

Therefore, the analytical solutions of four problems of the detachment filtration of aqueous suspension with a constant rate obtained in Chapter 2 allow to predict reliably the increase in the outlet concentration of contaminants and head losses in the packed bed over time, and at given values of C_e, Δh to determine the time for which they are reached. The calculation dependences resulting from these solutions, in essence, serve as computational tools, which in combination with the principles of the applied optimization and technological modeling make it possible to theoretically substantiate rational design of a filtration facility and its control algorithm for the appropriate conditions of filtration.

If restrictions on the filtrate quality and mechanical energy expenditure are set in advance based on the sanitary and economic requirements (normative values C_*, Δh_*), then the expressions for $C_e(t)$, $S(z,t)$, $\Delta h(t)$ given in Chapter 2 are transformed into transcendental equations with respect to the required technological times. As a result, systems of such equations are formulated, which are not difficult to solve by fitting. The procedure for finding a unique solution, and in fact a set of suitable values of the technological times t_p, t_h, t_f is described in detail and implemented by characteristic examples in Section 3.1 and (Poliakov, 2009a). The filter run duration calculated in this way corresponds to a strictly explicit bed height and an equivalent grain diameter. If it is practically possible to vary them, the design analysis, which is the subject of Section 3.2, becomes important. At the same time, of particular interest for practice is the alternate establishment of the values of these design parameters (optimal L and d_g), at which the duration of the filter run is maximized. Possible consequences for a high-rate filtration of regular changes in the mechanical composition of granular material (grain grading) along the height of the filter media, caused by its intensive mixing during washing, are specially analyzed. The peculiarities of the suspension clarification at stable high-rate mode are considered in detail. The possibilities of the intensification of filtration facilities operation by increasing a hydraulic load are discussed on concrete examples. In individual cases, e.g. in case of the accelerated clogging of a filtering material, it is suggested to reduce the rate of medium fouling to supply the necessary amount of water into the filter at different levels, splitting into several flows. Section 3.5 develops a theoretical basis for the substantiation of such separation and establishment of rational technological parameters. The temperature of water being treated in warm and cold seasons differs significantly. As a consequence, physical, chemical, biological processes, which significantly depend on temperature, proceed in water treatment filters throughout the year at different rates. Obviously, the purification effect of (bio)filtration can also have a pronounced seasonal nature. Analytical methods have been used to evaluate the effect of temperature instability in the environment on suspension filtration, and at the same time the active water binding by the deposited contaminants, which significantly accelerates clogging of a bed and frequently noticeably reduces the time of continuous operation of the filter.

3.1. METHOD OF TECHNOLOGICAL CALCULATION OF DEEP BED FILTRATION

If systems of equations (transformed dependences for the calculation of concentrations, head, filtration rate) concerning the required technological times are grouped on the basis of the similarity of filtration conditions, then they are justified to consider as a number of the techniques of the engineering calculation of designed and operated filtration facilities, and together as an effective engineering method. This method significantly facilitates the search for rational solutions of practical issues of rapid filtration. The first techniques of the method are based on exact and approximate solutions of the problems of the detachment filtration of suspensions through a uniform and layered bed, presented in Sections 2.1, 2.3, 2.4. Below the calculation equations with respect to dimensionless time of the protective effect of the filter bed $\bar{t}_p$ and the time of reaching the maximum allowable head losses $\bar{t}_h$ are successively given and accompanied by the test examples and comments. These equations are presented for initially contaminated as well as clean uniform and three-layer, two-layer beds.

Thus, in the case of a uniform bed and the successive (representative) filter run, the time $\bar{t}_p$ is proposed to be found according to (2.18) by fitting from the following exact equation

$$e^{-\bar{\beta}\bar{t}_p} I_0\left(2\sqrt{\bar{\alpha}\bar{\beta}\,\bar{t}_p}\right) + \bar{\beta}\int_0^{\bar{t}_p} e^{-\bar{\beta}\xi} I_0\left(2\sqrt{\bar{\alpha}\bar{\beta}\xi}\right) d\xi + \\ +\bar{\beta}\, e^{-\bar{\beta}\bar{t}_p} \int_0^1 e^{-\bar{\alpha}(1-\eta)}\, \bar{S}^0(\eta) I_0\left(2\sqrt{\bar{\alpha}\bar{\beta}\,\bar{t}_p(1-\eta)}\right) d\eta = \bar{C}_* e^{\bar{\alpha}}. \tag{3.1}$$

The second technological time in accordance with (2.24) must satisfy the equation

$$\int_0^1 \frac{d\bar{z}}{f_k(\bar{S}(\bar{z},\bar{t}_h))} = \Delta\bar{h}_*, \tag{3.2}$$

where

$$\bar{S}(\bar{z},\bar{t}_h) = \bar{S}^0(\bar{z}) e^{-\bar{\beta}\bar{t}_h} + \bar{\alpha}\, e^{-\bar{\alpha}\bar{z}} \int_0^{\bar{t}_h} e^{-\bar{\beta}\xi} I_0\left(2\sqrt{\bar{\alpha}\bar{\beta}\bar{z}\xi}\right) d\xi + \\ +\bar{\alpha}\bar{\beta}\, e^{-\bar{\beta}\bar{t}_h} \int_0^{\bar{z}} e^{-\bar{\alpha}(\bar{z}-\eta)}\, \bar{S}^0(\eta) \int_0^{\bar{t}_p} I_0\left(2\sqrt{\bar{\alpha}\bar{\beta}(\bar{z}-\eta)\varsigma}\right) d\varsigma\, d\eta. \tag{3.3}$$

If a packed bed has a three-layer or, as a special case, two-layer structure and is initially clean, then the corresponding system of equations follows from dependences (2.62), (2.70). Then the required times $\bar{t}_p$, $\bar{t}_h$ are reliably calculated respectively, first, from the equation

$$\bar{C}_{2e}\left(\bar{t}_p\right) + \sqrt{\bar{\alpha}_3\bar{\beta}_3\bar{l}_3} \int_0^{\bar{t}_p} \frac{e^{-\bar{\beta}_3(\bar{t}_p-\xi)} I_1\left(2\sqrt{\bar{\alpha}_3\bar{\beta}_3\bar{l}_3(\bar{t}_p-\xi)}\right)}{\sqrt{\bar{t}_p-\xi}} \bar{C}_{2e}(\xi)\, d\xi = \bar{C}_* e^{\bar{\alpha}_3\bar{l}_3}, \tag{3.4}$$

where $\bar{C}_{2e}\left(\bar{t}_p\right)$ is determined from (2.59) and $\bar{C}_{1e}\left(\bar{t}_p\right)$ thus, from (2.55) at $\bar{t} = \bar{t}_p$. Second, the equation with respect to $\bar{t}_h$ will be

$$\int_0^{l_1}\frac{dz}{f_{k1}(\overline{S}_1(z,\overline{t}_h))}+\frac{1}{\overline{k}_{02}}\int_0^{l_2}\frac{dz}{f_{k2}(\overline{S}_2(z,\overline{t}_h))}+\frac{1}{\overline{k}_{03}}\int_0^{l_3}\frac{dz}{f_{k3}(\overline{S}_3(z,\overline{t}_h))}=\Delta\overline{h}_*, \tag{3.5}$$

where expressions (2.56), (2.60), (2.63) are valid for $\overline{S}_1(z,\overline{t}_h)$, $\overline{S}_2(z,\overline{t}_h)$, $\overline{S}_3(z,\overline{t}_h)$ respectively at $\overline{t}=\overline{t}_h$.

The times $\overline{t}_p$, $\overline{t}_h$ are determined easier and with the minimal errors for similar filtration conditions based on the approximate solutions from Section 2.4. Here it was even possible to derive a formula allowing to calculate directly the value of $\overline{t}_p$, as well as equations with respect to $\overline{t}_h$ containing only elementary functions. If a uniform bed is initially contaminated, the time $\overline{t}_p$ should be chosen in such a way that the following equality is satisfied

$$2e^{-\lambda_p}-e^{-\overline{\alpha}}+\frac{2\overline{\beta}}{2+\overline{\beta}\,\overline{t}_p}\int_0^1 e^{-\lambda_p(1+\overline{\beta}\overline{t}_p)(1-\eta)}\,\overline{S}^0(\eta)\,d\eta=\overline{C}_*, \tag{3.6}$$

where $\lambda_p=2\overline{\alpha}/(2+\overline{\beta}\,\overline{t}_p)$. In the special case $\overline{S}^0=0$ the simple formula follows from (3.6)

$$\overline{t}_p=\frac{2}{\overline{\beta}}\left[\frac{\overline{\alpha}}{\ln 2-\ln(\overline{C}_*+e^{-\overline{\alpha}})}-1\right]. \tag{3.7}$$

The time $\overline{t}_h$ should be determined from the general equation (3.2), in which

$$\overline{S}(z,\overline{t}_h)=\frac{1}{2+\overline{\beta}\,\overline{t}_h}\Big[2\overline{\alpha}\,\overline{t}_h e^{-\lambda_h z}+\left(2-\overline{\beta}\,\overline{t}_h\right)\overline{S}^0(z)+ \\ +2\overline{\beta}\,\overline{t}_h\lambda_h\left(1+\overline{\beta}\,\overline{t}_h\right)\int_0^z e^{-\lambda_h(1+\overline{\beta}\overline{t}_h)(z-\eta)}\,\overline{S}^0(\eta)d\eta\Big], \tag{3.8}$$

where $\lambda_h=2\overline{\alpha}/(2+\overline{\beta}\,\overline{t}_h)$. Again expression (3.8) is essentially simplified at $\overline{S}^0=0$ and the following equation was obtained relatively $\overline{t}_h$ in case $f_k(\overline{S})=\left[1-(\overline{\gamma}_a\overline{S})^{m_{k1}}\right]^3$

$$\frac{1}{m_{k1}\lambda_h}\ln\frac{e^{m_{k1}\lambda_h}-\phi_h}{1-\phi_h}-\frac{2\phi_h(1-e^{m_{k1}\lambda_h})}{m_{k1}\lambda_h(1-\phi_h)(e^{m_{k1}\lambda_h}-\phi_h)}- \\ -\frac{\phi_h^2}{2m_{k1}\lambda_h}\left[\frac{1}{(e^{m_{k1}\lambda_h}-\phi_h)^2}-\frac{1}{(1-\phi_h)^2}\right]=\Delta\overline{h}_*, \tag{3.9}$$

where $\phi_h=\left(\overline{\gamma}_a\lambda_h\overline{t}_h\right)^{m_{k1}}$.

Finally, the approximate equation with the unknown $\overline{t}_p$ for a two-layer bed in the absence of its initial contamination according to (2.107) will be

$$2F_S(\overline{t}_p)e^{-2\lambda_{2p}l_2}+\left[\overline{C}_{1e}(\overline{t}_p)-2F_S(\overline{t}_p)\right]e^{-\overline{\alpha}_2 l_2}=\overline{C}. \tag{3.10}$$

Here

$$F_S(\overline{t}_p)=\frac{2}{\overline{\beta}_1\overline{t}_p}\left\{\left(2+\overline{\beta}_1\overline{t}_p\right)e^{-\lambda_{1p}l_1}-2e^{-\overline{\alpha}_1 l_1}+2\overline{\alpha}_1 l_1\left[Ei(-\overline{\alpha}_1 l_1)-Ei(-\lambda_{1p}l_1)\right]\right\}-e^{-\overline{\alpha}_1 l_1}$$

$\lambda_{1p} = 2\overline{\alpha}_1/(2+\overline{\beta}_1\overline{t}_p)$; $\overline{C}_{1e}(\overline{t}_p)$ is calculated by formula (2.90) when $\overline{\alpha} = \overline{\alpha}_1$, $\overline{\beta} = \overline{\beta}_1$, $\overline{t} = \overline{t}_p$; $Ei(-x)$ is the integral exponential function. Then the time $\overline{t}_h$ should be fitted by solving the equation

$$\int_0^{\overline{l}_1} \frac{dz}{f_{k1}(\overline{S}_1(z,\overline{t}_h))} + \frac{1}{\overline{k}_{02}} \int_{\overline{l}_1}^{1} \frac{dz}{f_{k2}(\overline{S}_2(z,\overline{t}_h))} = \Delta\overline{h}_*, \tag{3.11}$$

where $\overline{S}_1(z,\overline{t}_h) = \lambda_{1h}\overline{t}_h e^{-\lambda_{1h}z}$, $\overline{S}_2(z,\overline{t}_h) = \lambda_{2h}\overline{t}_h e^{-\lambda_{2h}(z-\overline{l}_1)}$, $\lambda_{2h} = 2\overline{\alpha}_2/(2+\overline{\beta}_2\overline{t}_h)$.

If by analogy with (2.92) for both layers the following expressions are true

$$f_{k1}(\overline{S}_1) = \left[1-(\overline{\gamma}_{1a}\overline{S}_1)^{m_{k1}^{(1)}}\right]^3, \quad f_{k2}(\overline{S}_2) = \left[1-\left(\frac{\overline{\gamma}_{2a}}{\overline{n}_{02}}\overline{S}_2\right)^{m_{k1}^{(2)}}\right]^3,$$

then equation (3.11) reduces to a simple, though cumbersome equality

$$\Delta\overline{h}_1(\overline{t}_h) + \Delta\overline{h}_2(\overline{t}_h) = \Delta\overline{h}_*. \tag{3.12}$$

Here

$$\Delta\overline{h}_1(\overline{t}_h) = \frac{1}{m_{k1}^{(1)}\lambda_{1h}}\left\{\ln\frac{e^{m_{k1}^{(1)}\lambda_{1h}}-\phi_{1h}}{1-\phi_{1h}} - 2\phi_{1h}\left(\frac{1}{e^{m_{k1}^{(1)}\lambda_{1h}\overline{l}_1}-\phi_{1h}} - \frac{1}{1-\phi_{1h}}\right) - \frac{(m_{k1}^{(1)}\lambda_{1h})^2}{2}\left[\frac{1}{(e^{m_{k1}^{(1)}\lambda_{1h}\overline{l}_1}-\phi_{1h})^2} - \frac{1}{(1-\phi_{1h})^2}\right]\right\},$$

$$\Delta\overline{h}_2(\overline{t}_h) = \frac{1}{\overline{k}_{02}m_{k1}^{(2)}\lambda_{2h}}\left\{\ln\frac{e^{m_{k1}^{(2)}\lambda_{2h}}-\phi_{2h}}{1-\phi_{2h}} - 2\phi_{2h}\left(\frac{1}{e^{m_{k1}^{(2)}\lambda_{2h}}-\phi_{2h}} - \frac{1}{1-\phi_{2h}}\right) - \frac{(m_{k1}^{(2)}\lambda_{2h})^2}{2}\left[\frac{1}{(e^{m_{k1}^{(2)}\lambda_{2h}}-\phi_{2h})^2} - \frac{1}{(1-\phi_{2h})^2}\right]\right\},$$

$$\phi_{1h} = (\overline{\gamma}_{1a}\lambda_{1h}\overline{t}_h)^{m_{k1}^{(1)}}, \quad \phi_{2h} = \left(\frac{\overline{\gamma}_{2a}}{\overline{n}_{02}}\lambda_{2h}\overline{t}_h F_S(\overline{t}_h)\right)^{m_{k1}^{(2)}}.$$

The actual first part of the general method of the technological calculation of rapid filters presented in this Section is illustrated by the determination of three relative technological times $(\overline{t}_p,\ \overline{t}_h,\ \overline{t}_f)$ depending on the absorption capacity of a filter media in relation to a dispersed impurity, as well as on its initial content in the suspension. In this case, the coefficients $\overline{\alpha}$, $\overline{\alpha}_j (j = 1,\ 2)$ varied from the minimum values, at which the bed is not yet able to reduce the level of contamination to the standard level even of the first portion of the suspension, to the maximum value 10, which corresponds to enhanced absorption. The inlet concentration C_0 changed continuously by an order of magnitude in a number of examples. The required technological times were calculated sequentially on the basis of the corresponding exact and approximate equations with respect to them for initially clean uniform and two-layer beds. Since the attention was focused in the quantitative analysis on the absorption component of the interphase mass exchange, a deposit detachment and the relationship between the hydraulic resistance of any clogged filtering material and the deposit concentrations were characterized by unified dependencies with the same values of the model coefficients contained in them. Thus, we took $\bar{\beta} = \bar{\beta}_1 = \bar{\beta}_2 = 0.01$, and also $m_{k1} = m_{k1}^{(j)} = 1$, $m_{k2} = m_{k2}^{(j)} = 3$ for f_{k1}, f_{k2} in (3.11). In addition, the deposit composition was assumed to be stable with $\bar{\gamma}_a = \bar{\gamma}_{1a} = \bar{\gamma}_{2a} = 0.0015$ in all cases being considered. The exception was the case of calculating the dependences of $\bar{t}_{p,h}$ on C_0, where we had to use a new constant to scale the criterion parameter $\bar{C}_*$, concentration S and coefficient γ_a. The following characteristic values were chosen for the relative criterion parameters: values 0.05, 0.1 and 0.2 for $\bar{C}_*$, which approximately correspond to deep, normal and primary water treatment (in terms of the suspended solids content in the filtrate); values 6 and 8 for $\Delta\bar{h}_*$, corresponding to hard and soft requirements to the mechanical energy expenditure for the filtration flow process.

First of all, on the basis of strict equations (3.1), (3.2) we have performed the calculations of the times $\bar{t}_p$, $\bar{t}_h$ at hypothetical change in the absorption capacity of a uniform bed in wide ranges. The results thus obtained are presented in **Figure 1** in the form of five graphs describing a regular increase in $\bar{t}_p$ and a decrease in $\bar{t}_h$ with increasing $\bar{\alpha}$. The change in C_* affects the time $\bar{t}_p$ significantly, and the change in $\Delta\bar{h}_*$ affects $\bar{t}_h$ much weaker, which is explained by a sharp deterioration of the permeability of a filtering material at its enhanced clogging. From Fig. 1 shows an important practical conclusion about the necessity of a correct selection of the specified material. Indeed, if the curves of the dependencies $\bar{t}_p(\bar{\alpha})$ and $\bar{t}_h(\bar{\alpha})$ are combined in accordance with both the actual filtration conditions and the requirements of efficient filter operation, the curve $\bar{t}_f(\bar{\alpha})$ thus formed has a distinct peak. The maximum value $\bar{t}_f$ should correspond to a strictly explicit value of $\bar{\alpha}_{opt}$. And if the actual value of $\bar{\alpha}$ differs significantly from $\bar{\alpha}_{opt}$, then a significant reduction of the effective filter operation time is to be expected.

Figure 1. Dependencies $\bar{t}_p(\bar{\alpha})$, $\bar{t}_h(\bar{\alpha})$

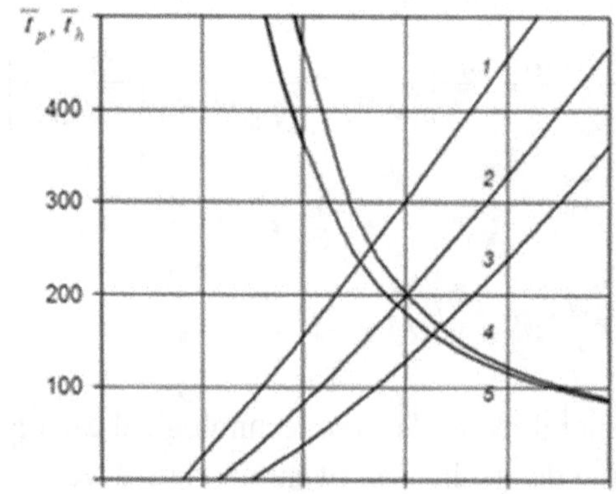

Note. 1-3 – $\bar{t}_p$; 4, 5 – $\bar{t}_h$; 1 – $\bar{C}_* = 0.2$, 2 – $\bar{C}_* = 0.1$, 3 – $\bar{C}_* = 0.05$; 4 – $\Delta\bar{h}_* = 8$, 5 – $\Delta\bar{h}_* = 6$

Figure 2. Dependencies $\bar{t}_p(\overline{C}_0)$, $\bar{t}_h(\overline{C}_0)$

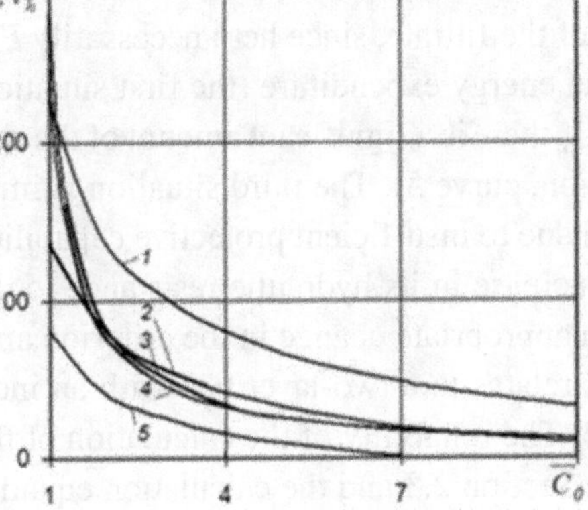

Note. 1, 2, 5 – $\bar{t}_p$; 3, 4 – $\bar{t}_h$; 1 – $\overline{C}_* = 0.2$, 2 – $\overline{C}_* = 0.1$, 3 – $\Delta\bar{h}_* = 8$, 4 – $\Delta\bar{h}_* = 6$, 5 – $\overline{C}_* = 0.05$

Further, the relationship between the technological times $\bar{t}_p$, $\bar{t}_h$ and the content of an impurity in the aqueous suspension was specified. Curves of the exact dependences of $\bar{t}_p$, $\bar{t}_h$ on the relative value $\overline{C}_0$ are shown in **Figure 2**. Here the variable C_0 was compared with its smallest value from the accepted interval [10^{-5}, 10^{-4}], namely $C_{00}=10^{-5}$. Thus, $\overline{C}_0 = C_0/C_{00}$ varied in the interval between 1 and 10. Since in the basic solution presented in dimensionless form in Section 2.1, the scale for both desired concentrations was a variable value C_0, in order to preserve the generalized (dimensionless) form, the calculation equations (3.1), (3.2) with respect to $\bar{t}_p$, $\bar{t}_h$ had to be transformed with explicit allocation of the relative variable $\overline{C}_0$. After simple transformations we obtained

$$\overline{C}_0\left[e^{-\bar{\beta}\bar{t}_p} I_0\left(2\sqrt{\bar{\alpha}\bar{\beta}\bar{t}_p}\right) + \bar{\beta}\int_0^{\bar{t}_p} e^{-\bar{\beta}\xi} I_0\left(2\sqrt{\bar{\alpha}\bar{\beta}\xi}\right) d\xi\right] + \\ +\bar{\beta}\, e^{-\bar{\beta}\bar{t}_p}\int_0^1 e^{-\bar{\alpha}(1-\eta)}\,\overline{S}^0(\eta) I_0\left(2\sqrt{\bar{\alpha}\bar{\beta}\bar{t}_p(1-\eta)}\right) d\eta = \overline{C}_* e^{\bar{\alpha}} ; \tag{3.13}$$

$$\int_0^1 \frac{d\bar{z}}{f_k(\overline{S}(\bar{z},\bar{t}_h))} = \Delta\bar{h}_* ; \tag{3.14}$$

$$\overline{S}(\bar{z},\bar{t}_h) = \overline{S}^0(\bar{z})e^{-\bar{\beta}\bar{t}_h} + \bar{\alpha}\,\overline{C}_0 e^{-\bar{\alpha}\bar{z}}\int_0^{\bar{t}_h} e^{-\bar{\beta}\xi} I_0\left(2\sqrt{\bar{\alpha}\bar{\beta}\bar{z}\xi}\right) d\xi + \\ +\bar{\alpha}\bar{\beta}\, e^{-\bar{\beta}\bar{t}_h}\int_0^{\bar{z}} e^{-\bar{\alpha}(\bar{z}-\eta)}\,\overline{S}^0(\eta)\int_0^{\bar{t}_p} I_0\left(2\sqrt{\bar{\alpha}\bar{\beta}(\bar{z}-\eta)\varsigma}\right) d\varsigma\, d\eta ; \tag{3.15}$$

where $\overline{C}_* = C_*/C_{00}$, $\overline{S} = S/(n_0 C_{00})$, $\overline{S}^0 = S^0/(n_0 C_{00})$. From Figure 2 it is obvious a sharp reduction of the effective filter operation time with increasing amount of the suspended solids in the supplied suspension. It should be especially mentioned that depending on the values of the criterion $\overline{C}_*$

, $\Delta \bar{h}_*$ and model parameters, different reasons for limiting the time of continuous filtration are possible here. Three characteristic situations take place in this series of examples. But, first of all, it should be noted that the technological process has to be interrupted at the values $\bar{C}_0$ close to 1 (low-concentration suspension) due to low quality of the filtrate, since here necessarily $\bar{t}_p$ less than $\bar{t}_h$. With increasing $\bar{C}_0$ or rapidly increasing mechanical energy expenditure (the first situation, curves 1, 2 and 3, 4) or in the entire accepted range of values $\bar{C}_0$ there is a significant amount of the suspended particles pasted through the filter bed (the second situation, curve 5). The third situation is implemented if at lower and higher values of $\bar{C}_0$ filtration is limited due to insufficient protective capability of the bed, and at intermediate values of $\bar{C}_0$ due to excessive increase in its hydraulic resistance (curves 2 and 3, 4). Certainly, other situations are also possible with appropriate change in the criterion and model parameters.

The next series of examples relates to a two-layer bed with an increase in the absorption capacity in the flow direction$(\bar{\alpha}_2 \geq \bar{\alpha}_1)$. The reliability of the calculation of the times $\bar{t}_p$, $\bar{t}_h$ is ensured by the rigor of the basic solution from Section 2.3 and the calculation equations derived from it. These times were consistently calculated by purposefully changing, firstly, both relative coefficients of the deposit attachment rate $(\bar{\alpha}_1, \ \bar{\alpha}_2)$, and, secondly, the ratio between the thicknesses of both bed layers. **Figure 3** shows the time $\bar{t}_p$ rise curves at a coordinated increase in $\bar{\alpha}_1$ and $\bar{\alpha}_2$ from 2 to 8, so that $\bar{\alpha}_2 > > \bar{\alpha}_1$. First, the value of $\bar{\alpha}_1$ (2) was fixed and the value of $\bar{\alpha}_2$ increased within the indicated limits (curves 4-6). Then $\bar{\alpha}_2$ was kept constant and equal to 8, and $\bar{\alpha}_1$ in turn varied from 2 to 8. All six graphs approach in shape straight lines. In case $\bar{\alpha}_1 = 2$ and $\bar{C}_* = 0.05, \ 0.1$ the clarification of a suspension in principle takes place at respectively $\bar{\alpha}_2 > 4$ and 2.65. Naturally, in this series of examples, the required degree of water purification was achieved for the longest time at $\bar{\alpha}_1 = \bar{\alpha}_2 = 8, \ \bar{C}_* = 0.2$ and the corresponding value of $\bar{t}_p$ was 454. However, it is only possible here to specify the time of the filter shutdown by comparing Figure 3 with **Figure 4**, which graphically describes the relationship between the second technological time and the absorption properties of both bed layers. The non-monotonic character of curves 3, 4 and the presence of their maxima inside the calculation interval should be especially pointed. It turned out that at a high absorption capacity of the lower layer and, on the contrary, at a low capacity of the upper layer as the coefficient $\bar{\alpha}_1$ increases, even a slight increase in $\bar{t}_h$ is observed at first due to a more even distribution of a deposit along the height, but further the above-mentioned curves steadily decrease. It is logical that in case of a weak absorption of the suspended solids in both layers (curves 1, 2), the values of $\bar{t}_h$ are much higher due to both more even distribution of the deposit along the height and significant breakthrough of the suspended matter through the bed layers.

When a layered bed is arranged, the choice of its layers thickness can be of essential importance for the intensification of the filter operation, which should be made on the basis of the specific filtration conditions and requirements to its operation. Possible type of curves of the dependences $\bar{t}_p(\bar{l}_1), \ \bar{t}_h(\bar{l}_1)$ is illustrated in **Figure 5**. It is obvious that the duration of the filter run will change with increasing $\bar{l}_1$. At first $\bar{t}_f$ increases and reaches a maximum at a strictly explicit value $\bar{l}_{1m}$, and then it decreases very quickly. For example, it is found at $\bar{C}_* = 0.1$ and $\Delta \bar{h}_* = 6$ that the maximum value of $\bar{t}_f$ will be about 190 at $\bar{l}_{1m} = 0.36$. For comparison, if in this case we take $\bar{l}_1 = \bar{l}_2 = 0.5$, then $\bar{t}_f$ is equal to $\bar{t}_h$ and will be only 135.

Figure 3. Dependencies $\bar{t}_p(\bar{\alpha}_1)$, $\bar{t}_p(\bar{\alpha}_2)$

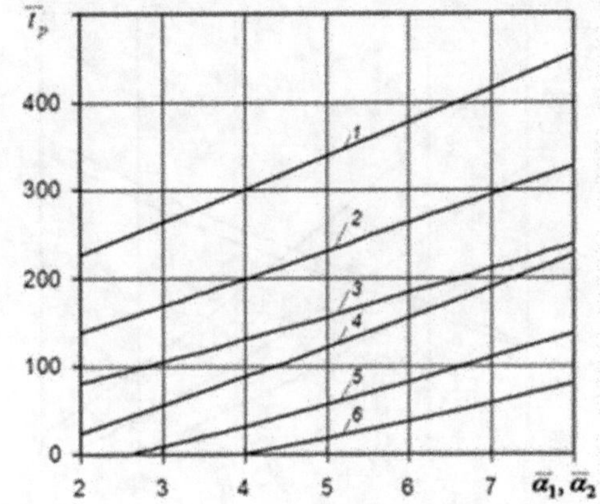

Note. 1-3 – $\bar{t}_p(\bar{\alpha}_1)$; 4-6 – $\bar{t}_p(\bar{\alpha}_2)$; 1, 4 – $\bar{C}_* = 0.2$; 2, 5 – $\bar{C}_* = 0.1$; 3, 6 – $\bar{C}_* = 0.05$

Figure 4. Dependencies $\bar{t}_h(\bar{\alpha}_1)$, $\bar{t}_h(\bar{\alpha}_2)$

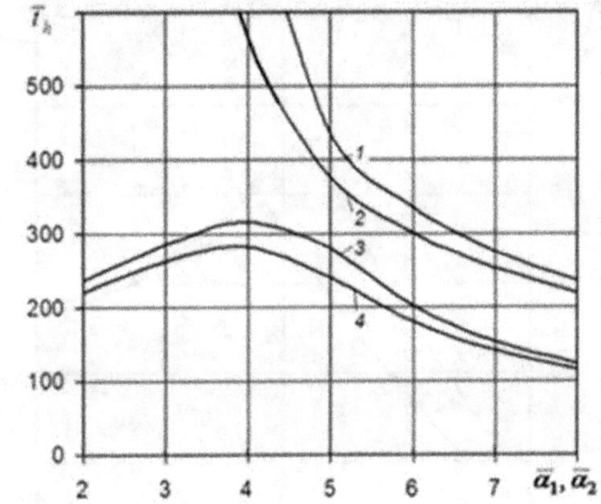

Note. 1, 2 – $\bar{t}_h(\bar{\alpha}_2)$; 3, 4 – $\bar{t}_h(\bar{\alpha}_1)$; 1, 3 – $\Delta\bar{h}_* = 8$; 2, 4 – $\Delta\bar{h}_* = 6$

Figure 5. Dependencies $\bar{t}_p(\bar{l}_1)$, $\bar{t}_h(\bar{l}_1)$

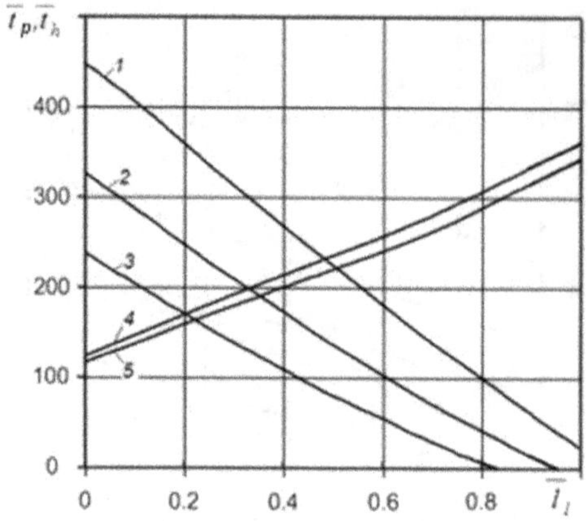

Note. 1-3 – $\bar{t}_p$; 4, 5 – $\bar{t}_h$; 1 – $\bar{C}_* = 0.2$, 2 – $\bar{C}_* = 0.1$, 3 – $\bar{C}_* = 0.05$, 4 – $\Delta\bar{h}_* = 8$, 5 – $\Delta\bar{h}_* = 6$

Figure 6. Dependences $\bar{t}_p(\bar{\alpha})$, $\bar{t}_h(\bar{\alpha})$, $\bar{t}_f(\bar{\alpha})$

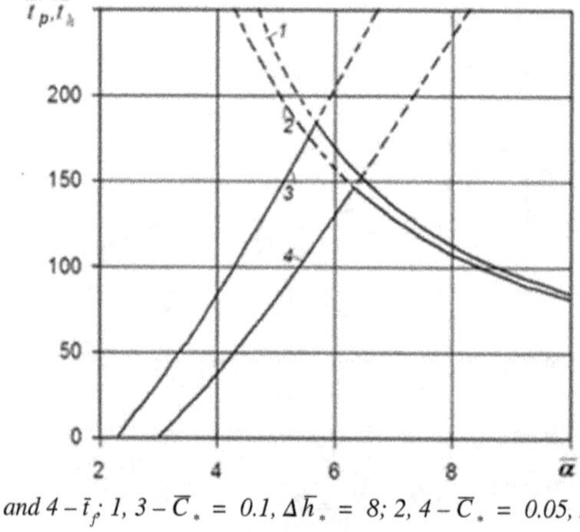

Note. 1, 2 – $\bar{t}_p$; 3, 4 – $\bar{t}_h$; 1 and 3, 2 and 4 – $\bar{t}_f$; 1, 3 – $\bar{C}_* = 0.1$, $\Delta\bar{h}_* = 8$; 2, 4 – $\bar{C}_* = 0.05$, $\Delta\bar{h}_* = 6$

Taking into account the importance of the technological calculations for the practice and the simple form of the approximate solutions from Section 2.4, it is reasonable to verify the validity of their use for calculating the times $\bar{t}_p$, $\bar{t}_h$, $\bar{t}_f$. It is especially easy to find the values of $\bar{t}_p$ at $\bar{S}^0 = 0$ due to simplicity of formula (3.7). First of all, **Figure 6** shows the plots of the dependencies $\bar{t}_p(\bar{\alpha})$, $\bar{t}_h(\bar{\alpha})$, calculated by the above formula and from equation (3.9) at $\bar{\alpha} = 5$, $\bar{\gamma}_a = 0.0015$ for sets of two values $\bar{C}_*$ (0.05, 0.1) and $\Delta\bar{h}_*$ (6, 8). For certainty, the values of $\bar{C}_*$ and $\Delta\bar{h}_*$, corresponding to harder $\left(\bar{C}_* = 0.05 \textit{ and } \Delta\bar{h}_* = 6\right)$ and softer $\left(\bar{C}_* = 0.1 \textit{ and } \Delta\bar{h}_* = 8\right)$ requirements to the filter operation, are selected and combined. In this case, the corresponding curves describing the relationship between $\bar{t}_f$ and $\bar{\alpha}$ turned out to be continuous lines with a single fracture at $\bar{\alpha} = \bar{\alpha}_{opt}$. They are composed of fragments of the curves describing the dependencies $\bar{t}_p(\bar{\alpha})$, $\bar{t}_h(\bar{\alpha})$ and are shown in Fig. 6 with solid lines. In the first and second cases, $\bar{\alpha}_{opt}$ is equal to 6.3 and 5.5 respectively. Then, the maximum values of $\bar{t}_f$ were 141 and 185. Special attention should be paid to the fact of a rapid increase in the time $\bar{t}_p$ and reduction of the time $\bar{t}_h$ as the absorption properties of the filtering material improve. Thus, the use of the material, the coefficient $\bar{\alpha}$ of which will differ significantly from the value $\bar{\alpha}_{opt}$ in one or another direction, will necessarily lead to a serious reduction in the duration of the filter run and, as a consequence, to a large overconsumption of clarified water for the filter washing. Similar exact calculations in both cases gave, respectively, the values $\bar{\alpha}_{opt} = 6.5$ and 6, also $\bar{t}_f = 128$ and 195. Consequently, the error in calculations of the key technological parameter was within 5 and 9%. Also, the approximate values of $\bar{\alpha}_{opt}$ were close to the exact values. It should be noted that with the deterioration of the filtration conditions, for example, due to the increase in the suspended solids content in a suspension, the accuracy of calculating $\bar{t}_h$ significantly increases. The time $\bar{t}_p$ was determined with the minimal errors in all considered examples.

The sharp reduction in all technological times due to an order of magnitude increase in the impurity content in the aqueous suspension is clearly shown in **Figure 7**. Here their relationship with the introduced variable parameter $\bar{C}_0$ is presented in the form of sharply descending curves 1-4, corresponding to three standard values of $\bar{C}_*$ and one value of $\Delta\bar{h}_*$ (8). It should be noted that the curve $\bar{t}_h(\bar{C}_0)$ was practically coincided with curve 3 at $\Delta\bar{h}_* = 6$. All curves were calculated using formula (3.7) and equation (3.9) transformed with the isolation of $\bar{C}_0$. Thus, the quantity $\bar{t}_p(\bar{C}_0)$ was determined as follows

$$\bar{t}_p(\bar{C}_0) = \frac{2}{\bar{\beta}}\left[\frac{\bar{\alpha}}{\ln(2\bar{C}_0) - \ln(\bar{C}_* + \bar{C}_0 e^{-\bar{\alpha}})} - 1\right], \tag{3.16}$$

and it was taken in equation (3.9)

$$\phi_h(\bar{t}_h, \bar{C}_0) = (\bar{C}_0\,\bar{\gamma}_a\,\lambda_h\,\bar{t}_h)^{m_{k1}}, \quad \bar{\gamma}_a = \bar{\gamma}/\bar{C}_0.$$

Fig. 7 shows that three different situations can be practically implemented depending on the relationship between the values $\bar{C}_*$ and $\Delta\bar{h}_*$. If a high suspended solids content is allowed $(\bar{C}_* = 0.2)$ in the filtrate, the filtration is time-limited over the entire calculation interval for $\bar{C}_0$ due to severe clogging of the bed (curve 3 is everywhere below curve 1). The opposite is the case when a high-quality clarification of a suspension is required, which is the main reason for early filter shutdown (curve 4 is below curve 3). An intermediate situation occurs if $\bar{C}_* = 0.1$. Then the duration of the technological process is limited due to the excessive content of suspended solids in the filtrate at first (at $\bar{C}_0 < 1.7$) and at the end ($\bar{C}_0 > 8.7$), and the increase in the hydraulic resistance of the bed is decisive for the control of the filter operation in the interval ($8.7 > \bar{C}_0 > 1.7$). The accuracy of calculating the technological times here improves with increasing $\bar{C}_0$.

The behavior of three relative technological times ($\bar{t}_p$, $\bar{t}_h$, $\bar{t}_f$) was also analyzed for a two-layer bed at continuous variation of the coefficients $\bar{\alpha}_1$, $\bar{\alpha}_2$, but now unlike the results in Figs. 3, 4, the calculations were performed on the basis of the approximate equations (3.10), (3.11). The calculated curves were obtained at a single value of the criterion parameters $(\bar{C}_* = 0.1,\ \Delta\bar{h}_* = 8)$ and are shown in **Figure 8**. Two situations typical for the dependences $\bar{t}_f(\bar{\alpha}_1)$, $\bar{t}_f(\bar{\alpha}_2)$ are actually reflected here. In the first situation, due to a low absorption capacity of the bottom layer $(\bar{\alpha}_1 = 2)$, the bed as a whole is either not able to immediately provide the required degree of water treatment $(\bar{\alpha}_2 < 2.8)$, or it copes properly with contamination for a short time $(\bar{\alpha}_2 > 2.8)$. Naturally, the time $\bar{t}_p$ will reach the maximum value 138 at $\bar{\alpha}_2 = 8$. At the same time, due to the active participation of the entire bed in the removal of the suspended matter, as well as the removal of its significant part, the time $\bar{t}_h$ is much longer than $\bar{t}_p$

(curves 1 and 4). Thus, the curve of the desired dependence $\bar{t}_f(\bar{\alpha}_2)$ should be identified with curve 1, representing initially the function $\bar{t}_p(\bar{\alpha}_2)$, which is shown, therefore, by a solid line. And only the improvement of the absorption properties of the upper layer at $\bar{\alpha}_2 = 8$ can guarantee a significant prolongation of the filter operation, despite a serious increase in the clogging of the bed. Indeed, curve 3, corresponding to the example with the maximum value of $\bar{\alpha}_2$ (8) and increasing values of $\bar{\alpha}_1$, is located clearly above the analogous curve 4. However, starting from the value $\bar{\alpha}_2 \approx 5$, the clarification process becomes increasingly limited due to excessive head losses in the bed. Therefore, in this case the solid continuous piecewise curve consists of two fragments of curves 2 and 3 and has a peak at the mentioned value of $\bar{\alpha}_2$. It is obvious that an unsuccessful choice of filter media (in this case $\bar{\alpha}_1$ noticeably more or less than 5) will necessarily cause a significant reduction in the filter run duration. Also of interest is the fact of the intersection of solid lines at $\bar{\alpha}_1 = \bar{\alpha}_2 \approx 7.4$, which in fact is another argument in favor of the use of layered filter media. Indeed, the piece of curve 2 corresponding to $\bar{\alpha}_1 > 7.4$, $\bar{\alpha}_2 = 8$, appears to be lower than the piece 4 corresponding to $\bar{\alpha}_1 = 2$, $\bar{\alpha}_2 > 7.4$. Finally, comparison of the data of approximate calculating $\bar{t}_f$ (Fig. 8) with similar exact data (Fig. 3, Fig. 4) allows us to estimate the errors in determining the technological times due to the use of the dynamic averaging technique (Section 2.4) when analytical solving the mathematical problem of the suspension filtration through a layered bed. Thus, approximate calculations of the maximum duration of the filter run in the option $\bar{\alpha}_2 = 8$ and variable $\bar{\alpha}_1$ gave the value $\bar{t}_f = 222$ at $\bar{\alpha}_{1opt} = 4.75$. According to equations (3.4), (3.5), the corresponding exact values are $\bar{t}_f = 230$ and $\bar{\alpha}_{1opt} = 5.17$. Thus, the error in determining the optimum value of $\bar{t}_f$ here is equal to 3.5%. It is even more reliably calculated from equations (3.10) and (3.11) in the option $\bar{\alpha}_1 = 2$ and variable $\bar{\alpha}_2$. Then the approximate value of $\bar{t}_f$, incidentally coinciding with $\bar{t}_p$, reaches the maximum value 138 at $\bar{\alpha}_2 = 8$. The exact value of $\bar{t}_f$ is 137.6 at the same value of $\bar{\alpha}_2$, so that the computational error is reduced by an order of magnitude.

Figure 7. Dependencies $\bar{t}_p(\bar{C}_0)$, $\bar{t}_h(\bar{C}_0)$

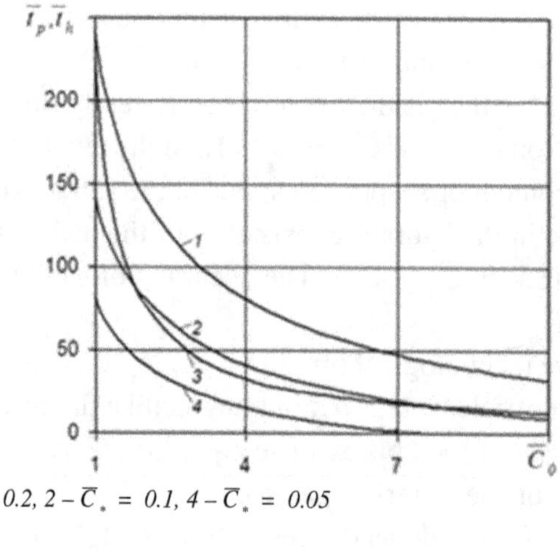

Note. 1, 2, 4 – $\bar{t}_p$; 3 – $\bar{t}_h$; 1 – $\bar{C}_* = 0.2$, 2 – $\bar{C}_* = 0.1$, 4 – $\bar{C}_* = 0.05$

Figure 8. Dependencies $\bar{t}_{p,h,f}(\bar{\alpha}_1)$, $\bar{t}_{p,h,f}(\bar{\alpha}_2)$

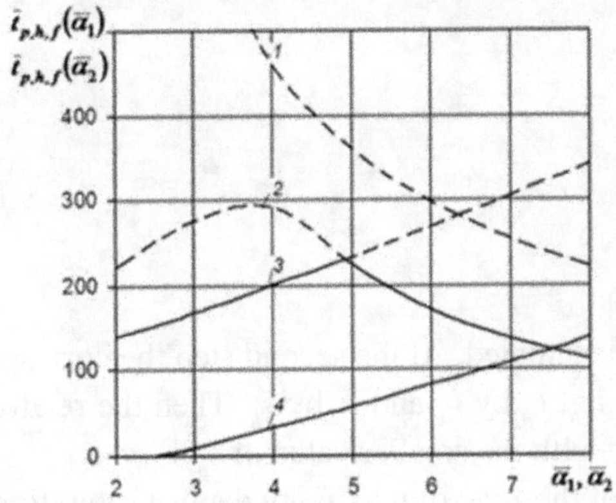

Note. $1 - \bar{t}_h(\bar{\alpha}_2)$; $2 - \bar{t}_h(\bar{\alpha}_1)$; $3 - \bar{t}_p(\bar{\alpha}_1)$, $4 - \bar{t}_p(\bar{\alpha}_2)$

Consequently, both approximate solutions from Section 2.4, related to uniform and layered bed, allow both to calculate the main technological parameter $\left(\bar{t}_f\right)$, and to specify the optimal values of the key mass exchange coefficients $\left(\bar{\alpha},\ \bar{\alpha}_1,\ \bar{\alpha}_2\right)$ with a high accuracy.

And finally, the technological calculations of the continuous prolonged filtration with a constant rate at the formation of the zone of maximum saturation with a deposit (relative duration of the filter run is more than $\bar{t}_s$) were carried out on the basis of equations (2.120), (2.121), (2.124). At the same time, the criterion parameters $\bar{C}_*$, $\Delta\bar{h}_*$ and equations (1.35), (1.36) were used. In this case, it is inexpedient to model filtration at the second stage for technological reasons, if both or at least one of the conditions

$$\bar{C}_e\left(\bar{t}_s\right) < \bar{C}_*,\ \Delta\bar{h}\left(\bar{t}_s\right) < \Delta\bar{h}_*$$

are met. The corresponding relative time of the protective capability of the bed $\bar{t}_p$ should be found at the maximum allowable relative concentration of the suspended solids in the filtrate $\bar{C}_*$ in this sequence. First, it is necessary to solve the system of equations concerning the parameters $\bar{t}_{*p}$, $\bar{l}_p$, which follows from equations (2.119) and (2.124) involving the quality criterion, namely,

$$\bar{\alpha}\, e^{-\bar{\alpha}\bar{l}_p} \int_0^{\bar{t}_{*p}} e^{-\bar{\beta}\xi} I_0\left(2\sqrt{\bar{\alpha}\bar{\beta}\,\bar{l}_p\xi}\right) d\xi = \bar{S}_m \tag{3.17}$$

$$\bar{\beta} \int_{\bar{l}_p}^{1} e^{\bar{\alpha}\eta}\, \bar{S}_2\left(\eta, \bar{t}_{*p}\right) d\eta = e^{\bar{\alpha}}\,\bar{C}_* - 1 \tag{3.18}$$

Then the relative first technological time is determined at the known values of $\bar{t}_{*p}$, $\bar{l}_p$ solving equation (2.118) which now takes the following form

$$\begin{aligned} &\bar{S}_m \bar{l}_p + \bar{\alpha} \int_{\bar{l}_p}^{1} e^{-\bar{\alpha}\eta} \int_0^{\bar{t}_{*p}} e^{-\bar{\beta}\xi} I_0\left(2\sqrt{\bar{\alpha}\bar{\beta}\eta\xi}\right) d\xi d\eta = \\ &= \bar{\alpha} \int_0^{1} e^{-\bar{\alpha}\eta} \int_0^{\bar{t}_p} e^{-\bar{\beta}\xi} I_0\left(2\sqrt{\bar{\alpha}\bar{\beta}\eta\xi}\right) d\xi d\eta. \end{aligned} \tag{3.19}$$

The second technological time $\left(\bar{t}_h\right)$ is established in the same sequence. First of all, from the system of equations containing the criterion parameter $\Delta\bar{h}_*$,

$$\bar{\alpha}\, e^{-\bar{\alpha}\bar{l}_h} \int_0^{\bar{t}_{*h}} e^{-\bar{\beta}\xi} I_0\left(2\sqrt{\bar{\alpha}\bar{\beta}\,\bar{l}_h\xi}\right) d\xi = \bar{S}_m \tag{3.20}$$

$$\frac{\bar{l}_h}{f_k(\bar{S}_m)} + \int_{\bar{l}_h}^{1} \frac{dz}{f_k(\bar{S}_2(z,\bar{t}_{*h}))} = \Delta\bar{h}_* \tag{3.21}$$

a set of two values $\bar{t}_{*h}$, $\bar{l}_h$ is calculated. At the second step the corresponding value of $\bar{t}_h$ is matched after formal substitution $\bar{t}_{*p}$ by $\bar{t}_{*h}$, $\bar{l}_p$ by $\bar{l}_h$ and $\bar{t}_p$ by $\bar{t}_h$. Then the relative duration of the filter run $\bar{t}_f$ according to (1.41) is identified with the smaller value of $\bar{t}_p$, $\bar{t}_h$.

The effect of the expansion of the zone of maximum deposit saturation on the technological times is illustrated in **Figure 9**. Here the relation of the relative technological times $\left(\bar{t}_p,\ \bar{t}_h\right)$ with the absorption properties of a filtering material ($\bar{\alpha}$) is shown in graphical form – without $\left(\bar{S}_m > \bar{S}_d\right)$ and with taking into account $\left(\bar{S}_m = 300,\ 500 \le S_d\right)$ the limitation of the absorption capacity for the deposited particles (deposit). Curves 1 and 5 are calculated exactly by formulas (2.111) and (2.113) taking into consideration (2.112). Curves 2, 3, which in essence refine them, were determined on the basis of equations (3.17) - (3.19), and curve 4 – on the basis of system (3.19) - (3.21). It follows from the comparison of curves 2, 3 with 1 and 4 with 5 that the protective capability of a filter bed and the mechanical energy expenditure for the filtration of a suspension being clarified can significantly decrease due to a limitation of the deposit accumulation. With an increase in the coefficient $\bar{\alpha}$, the effect due to limiting the absorption process by solids take-up capacity clearly increases. It especially affects the quantity $\bar{t}_h$. In particular, the specified time increases in 2.15 times at $\bar{\alpha} = 9$ because of the relative smallness of S_m. Nevertheless, the optimum values of $\bar{\alpha}_{opt}$, at which the relative duration of the filter run is the greatest, are approximately in the middle of the calculation interval. Thus, the optimum value is $\bar{\alpha}_{opt} = 5$ and accordingly $\bar{t}_f$ is about 140 in the reference case (curves 1 and 5). If a linear filtration takes place under the condition $\bar{S} \le \bar{S}_m = 300$, then already $\bar{\alpha}_{opt} = 5.8$ and $\bar{t}_f = 163$.

Figure 9. Dependences $\bar{t}_p(\bar{\alpha})$, $\bar{t}_h(\bar{\alpha})$:

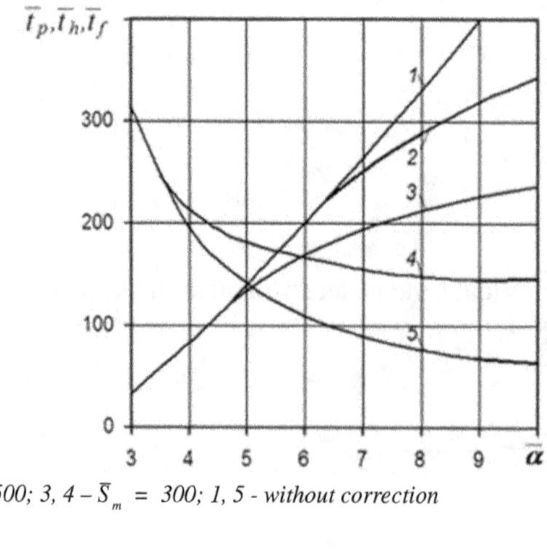

Note. 1-3 – $\bar{t}_p$; 4-5 – $\bar{t}_h$; 2 – $\bar{S}_m = 500$; 3, 4 – $\bar{S}_m = 300$; 1, 5 - without correction

3.2. METHOD OF DESIGN CALCULATION

Technological calculations are not necessarily the ultimate goal of the applied analysis of the aqueous suspension filtration. They can be successfully applied not only for the operational control of a filter, but also at the stage of its design for the theoretical substantiation of the rational design parameters. Then the corresponding determination of the technological times and, most importantly, of the filter run duration can be justified to be considered as the design calculations. The main design parameters for which such calculations should be carried out first of all are the total height of a packed bed L, the thickness of the l_i-th layer in case of a layered (i=1,2...) structure (vertical filtration), as well as the size (equivalent diameter) of the elements (grains) of the packed bed d_g. It is possible to significantly influence the future results of the clarification process by varying the quantity L in advance. A filter media with the thickness ranging from a few centimeters to several meters is practically feasible. And the actual question, which is important to get a clear answer in this case, is to establish, based on the expected conditions of the filtration process and the requirements for the operation of the filter, the optimal value of the desired height. It is logical to consider as optimal such value of L, at which the maximum duration of the filter run and, consequently, the minimum losses of clarified water are provided for washings.

In this Section and a number of other Sections of the monograph, a bed height will be consistently optimized for two characteristic options with respect to the availability of a filtering material (Poliakov, 2010). They are essentially limit and opposite in cost terms. In the first option it is assumed that locally produced solid wastes are used as such a material. Then, their use for the water treatment filter media can generally be construed as an efficient way of utilization. As a consequence, they do not increase the capital cost of filtration and, therefore, the bed height is not limited for economic reasons. The second option is opposite in terms of cost. Now a conditioned filtering material is acutely scarce and must be consumed as economically as possible. In this case, it is necessary to determine the optimal size of the operating medium at a fixed volume of the specified material and liquid flow rate on the basis of the design calculations. In fact, at the standard geometric form of the volume of a filtering material it is required to find such a ratio between the height and the area of the (inlet) surface of the bed, at which the terms of the next washing are maximally postponed.

Thus, the specified area and filtration rate do not change in the first option. Since the former linear scale L now becomes a variable quantity, we have to use some fixed value L_0 as a new length scale. For practical reasons it is convenient to take the smallest possible value of L. Then, in order to preserve the generalized form of the calculation equations, the dimensionless variables and model coefficients in the initial model and its solutions are transformed with explicit allocation of the new variable $\overline{L}$. Then, its minimum value will be 1 and the relation $\overline{L} \geq 1$is valid. Therefore,

$$\overline{z}(\overline{L}) = \overline{\overline{z}}/\overline{L}, \quad \overline{t}(\overline{L}) = \overline{\overline{t}}/\overline{L}, \quad \overline{\alpha}(\overline{L}) = \overline{\overline{\alpha}}\overline{L}, \quad \overline{\beta}(\overline{L}) = \overline{\overline{\beta}}\overline{L}, \quad \Delta\overline{h}_*(\overline{L}) = \Delta\overline{\overline{h}}_*/\overline{L},$$

where

$$\overline{\overline{t}} = Vt/(n_0L_0), \quad \overline{\overline{z}} = z/L_0, \quad \overline{\overline{\alpha}} = L_0\alpha/V, \quad \overline{\overline{\beta}} = n_0L_0\beta/V, \quad \Delta\overline{\overline{h}}_* = k_0\Delta h_*/(VL_0),$$

and all of them do not depend on $\overline{L}$. Then equations (3.1), (3.2) are transformed to the following form

$$e^{-\overline{\beta}\overline{t}_p} I_0\left(2\sqrt{\overline{\alpha}\overline{\beta}\overline{L}\overline{t}_p}\right) + \overline{\beta}\int_0^{\overline{t}_p} e^{-\overline{\beta}\xi} I_0\left(2\sqrt{\overline{\alpha}\overline{\beta}\overline{L}\xi}\right) d\xi +$$
$$+\overline{\beta}\, e^{-\overline{\beta}\overline{t}_p} \int_0^{\overline{L}} e^{-\overline{\alpha}(\overline{z}-\eta)} \overline{S}^0(\eta) I_0\left(2\sqrt{\overline{\alpha}\overline{\beta}\,\overline{t}_p(\overline{z}-\eta)}\right) d\eta = \overline{C}_* e^{\overline{\alpha}\overline{L}}, \tag{3.22}$$

$$\int_0^{\overline{L}} \frac{d\overline{z}}{f_k\left(\overline{S}(\overline{z},\overline{t}_h)\right)} = \Delta \overline{h}_*. \tag{3.23}$$

Here expression (3.3), in which $z,\ t_h,\ \alpha,\ \beta$ are formally replaced by $\overline{z},\ \overline{t}_h,\ \overline{\alpha},\ \overline{\beta}$, should be used for $\overline{S}(\overline{z},\overline{t}_h)$. Only the range of $\overline{z}$ values changes – now from 0 to $\overline{L}$.

If we use the approximate solution from Section 2.4 for a uniform bed in the calculations of $\overline{t}_{p,h,f}(\overline{L})$, the time $\overline{t}_p$ is related to $\overline{L}$ by the equation

$$2e^{-\overline{\lambda}_p\overline{L}} - e^{-\overline{\alpha}\overline{L}} + \frac{2\overline{\beta}}{2+\overline{\beta}\overline{t}_p}\int_0^{\overline{L}} e^{-\overline{\lambda}_p(1+\overline{\beta}\overline{t}_p)(\overline{z}-\eta)}\,\overline{S}^0(\eta)d\eta = \overline{C}_*, \tag{3.24}$$

where $\overline{\lambda}_p = 2\overline{\alpha}/(2+\overline{\beta}\overline{t}_p)$. The simple formula for $\overline{t}_p$ would be in the case of $\overline{S}^0 = 0$

$$\overline{t}_p(\overline{L}) = \frac{2}{\overline{\beta}}\left[\frac{\overline{\alpha}\overline{L}}{\ln 2 - \ln\left(\overline{C}_* + e^{-\overline{\alpha}\overline{L}}\right)} - 1\right]. \tag{3.25}$$

The form of the equation with respect to $\overline{t}_h$ (3.9) and expression (3.8) for $\overline{S}(\overline{z},\overline{t})$ is retained and the above substitution is also performed.

In the second option (expensive filtering material), the height of a bed L and its inlet surface area ω are varied in a coordinated manner so that the operating volume W_f remains constant. Hence, $\omega L = W_f = const.$

In addition, the filter performance W_w is also assumed to be stable when varying L. It means that the filtration rate should be corrected at each calculation step on the basis of the following equality

$$\omega V = W_w = const. \tag{3.26}$$

Thus, the rate V is related to L in a linear fashion, viz,

$$V(L) = \frac{W_w}{W_f}L = \frac{W_w}{W_f}L_0\overline{L} = V_0\overline{L}, \tag{3.27}$$

where V_0 is the rate that provides the performance W_w in the case of the bed height L_0.

Given the close relationship between V and the mass exchange coefficients α,β, it is necessary to involve appropriate dependencies for the subsequent analysis. As a first approximation, taking into account both the results of the studies in Section 2.6 and numerous literature data, e.g. (Caliskaner and Tchobanoglous, 2005; Logan et al., 1995), it is acceptable to assume that

$$\alpha = \alpha_V V, \quad \beta = \beta_V V. \tag{3.28}$$

Then the variables and model parameters here allowing for (3.27), (3.28) will be

$$\bar{t} = \bar{t}, \quad \Delta\bar{h}(\bar{L}) = \Delta\bar{h}/\bar{L}^2, \quad \bar{\alpha}(\bar{L}) = \bar{\alpha}_V\bar{L}, \quad \bar{\beta}(\bar{L}) = \bar{\beta}_V\bar{L}, \tag{3.29}$$

where $\bar{t} = V_0 t/(n_0 L_0)$, $\bar{\alpha}_V = L_0\alpha_V$, $\bar{\beta}_V = n_0 L_0 \beta_V$. Considering (3.29), the equations concerning the technological times (3.1) - (3.3) are transformed to the following form

$$\begin{aligned} & e^{-\bar{\beta}_V\bar{L}\bar{t}_p} I_0\left(2\sqrt{\bar{\alpha}_V\bar{\beta}_V\bar{L}^2\bar{t}_p}\right) + \bar{\beta}_V\bar{L}\int_0^{\bar{t}_p} e^{-\bar{\beta}_V\bar{L}\xi} I_0\left(2\sqrt{\bar{\alpha}_V\bar{\beta}_V\bar{L}^2\xi}\right) d\xi + \\ & + \bar{\beta}_V e^{-\bar{\beta}_V\bar{L}\bar{t}_p}\int_0^{\bar{L}} e^{-\bar{\alpha}_V(\bar{z}-\eta-\bar{L})}\bar{S}^0(\eta) I_0\left(2\sqrt{\bar{\alpha}_V\bar{\beta}_V\bar{L}\bar{t}_p(\bar{z}-\eta)}\right) d\eta = \bar{C}_* e^{\bar{\alpha}_V\bar{L}}, \end{aligned} \tag{3.30}$$

$$\bar{L}\int_0^{\bar{L}} \frac{d\bar{z}}{f_k(\bar{S}(\bar{z},\bar{t}_h;\bar{L}))} = \Delta\bar{h}_*, \tag{3.31}$$

Here

$$\begin{aligned} \bar{S}(\bar{z},\bar{t}_h;\bar{L}) &= \bar{S}^0(\bar{z})e^{-\bar{\beta}_V\bar{L}\bar{t}_h} + \bar{\alpha}_V\bar{L}e^{-\bar{\alpha}_V\bar{z}}\int_0^{\bar{t}_h} e^{-\bar{\beta}_V\bar{L}\xi} I_0\left(2\sqrt{\bar{\alpha}_V\bar{\beta}_V\bar{L}\bar{z}\xi}\right) d\xi + \\ & + \bar{\alpha}_V\bar{\beta}_V\bar{L}e^{-\bar{\beta}_V\bar{L}\bar{t}_h}\int_0^{\bar{z}} e^{-\bar{\alpha}_V(\bar{z}-\eta)}\bar{S}^0(\eta)\int_0^{\bar{t}_h} I_0\left(2\sqrt{\bar{\alpha}_V\bar{\beta}_V\bar{L}(\bar{z}-\eta)\varsigma}\right) d\varsigma d\eta. \end{aligned} \tag{3.32}$$

If we use the approximate solution of the same problem, then (3.30), (3.32) simplify to the following form

$$2e^{-\bar{\lambda}_p(\bar{L})} - e^{-\bar{\alpha}_V\bar{L}} + \frac{2\bar{\beta}_V}{2+\bar{\beta}_V\bar{t}_h}\int_0^{\bar{L}} e^{-\bar{\lambda}_p(\bar{L})(1+\bar{\beta}_V\bar{L}\bar{t}_h)(\bar{L}-\eta)}\bar{S}^0(\eta)d\eta = \bar{C}_*, \tag{3.33}$$

$$\begin{aligned} \bar{S}(\bar{z},\bar{t}_h;\bar{L}) &= \frac{1}{2+\bar{\beta}_V\bar{L}\bar{t}_h}\Big[2\bar{\alpha}_V\bar{L}\bar{t}_h e^{-\bar{\lambda}_h(\bar{L})\bar{z}} + \left(2-\bar{\beta}_V\bar{L}\bar{t}_h\right)\bar{S}^0(\bar{z}) + \\ & + 2\bar{\beta}_V\bar{t}_h\bar{\lambda}_h(\bar{L})\left(1+\bar{\beta}_V\bar{L}\bar{t}_h\right)\int_0^{\bar{z}} e^{-\bar{\lambda}_h(\bar{L})(1+\bar{\beta}_V\bar{L}\bar{t}_h)(\bar{z}-\eta)}\bar{S}^0(\eta)\,d\eta\Big], \end{aligned} \tag{3.34}$$

where $\bar{\lambda}_h(\bar{L}) = 2\bar{\alpha}_V\bar{L}/(2+\bar{\beta}_V\bar{L}\bar{t}_h)$. In the special case $\bar{S}^0 = 0$, there will be

$$\bar{t}_p(\bar{L}) = \frac{2}{\bar{\beta}_V\bar{L}}\left[\frac{\bar{\alpha}_V\bar{L}}{\ln 2 - \ln(\bar{C}_* + e^{-\bar{\alpha}_V\bar{L}})} - 1\right], \tag{3.35}$$

$$\bar{S}(\bar{z},\bar{t}_h;\bar{L}) = \bar{\lambda}_h(\bar{L})\,\bar{t}_h e^{-\bar{\lambda}_h(\bar{L})\bar{z}}. \tag{3.36}$$

Absorption capacity curves of the filtering (granular) material play the key role in studying the influence of its geometric properties (size and shape) on the technological process and parameters. Such curves describe the relationship between the ratios α/α_0 and d_g/d_{g0}. As scales d_{g0}, α_0, it is convenient to take the minimum equivalent diameter of a grain from the range considered for it $[d_{g,min}, d_{g,max}]$, as well as the maximum coefficient α_{max} corresponding to it. In view of the author's lack of reliable empirical information concerning regular and consistent changes in α and d_g, we use the theoretical formula for α (2.149) from Section 2.6. Examples of plots of the normalized dependence

$$\frac{\bar{\alpha}}{\bar{\alpha}_{max}} = f_g\left(\bar{d}_g\right), \tag{3.37}$$

where $\bar{d}_g = d_g/d_{g,min}$, are also given there for different combinations of the suspended solids size and filtration rate. Dependence (3.37) is calculated in general case after setting specific values of d_p, V, n_0. At the same time, $\bar{\alpha}_{max}$ and a number of values of $\bar{\alpha}$ are calculated sequentially at the values of $\bar{d}_g$ selected with a certain discreteness. When performing the technological and design calculations, it makes no practical sense to search for an analytical representation for the function $f_g\left(\bar{d}_g\right)$ found in discrete form. At the same time, such representation is easy to establish in those situations when one of the mechanisms of delivery of suspension particles to particles-collectors clearly prevails. Then, the specified function should be in accordance with (2.149) a power function, i.e.

$$f_g\left(\bar{d}_g\right) = \bar{d}_g^{\,m_g}, \tag{3.38}$$

where m_g= -5/3, -11/3, 3/5 at dominance of Brownian diffusion, interception and sedimentation mechanisms respectively.

The effect of a bed height on the relative technological parameters $\bar{t}_p$, $\bar{t}_h$, $\bar{t}_f$ was estimated in a series of examples with typical initial data on the basis of the strict calculation equations (3.1), (3.2), as well as approximate formula (3.7) and equation (3.9). The values $\bar{\alpha} = 5$, $\bar{\beta} = 0.01$, $\bar{\gamma}_a = 0.0015$ were fixed beforehand and the bed was considered initially clean ($\bar{S}^0 = 0$). The criterion parameters varied in a standardized manner: $\bar{C}_*$ (0.05, 0.1, 0.2) and $\Delta\bar{h}_*$ (6, 8). The results of the exact calculations of the two times when varied L by a factor of four for the first of the two options considered above are shown in **Figure 10**. It is natural that the time $\bar{t}_p$ increases rapidly with increasing $\bar{L}$ due to the high absorption properties of the filtering material. Therefore, the hypothetically incremental part of the bed can retain a larger part of the suspended solids passing through it at a lower L. At the same time, the small hydraulic resistance of the clean bed and the amount of the additionally retained suspended solids explain a slow decrease in the time $\bar{t}_h$ as it thickens, which in addition slows down even more at high values of $\bar{L}$.

The situation with the technological times fundamentally changes in the second option (**Figure 11**). Due to the increase in the bed thickness, its surface area decreases and the hydraulic load increases in a coordinated manner. As a consequence, the coefficient $\bar{\alpha}$ increases and the absorption of the suspended solids intensifies, and the inlet section of the bed is especially intensively clogged. As a result, curves 4, 5 decrease sharply in contrast to the similar curves in Figure 10. On the other hand, a much larger amount of the deposit is detached, which leads to a serious increase in the removal of the suspension particles from the filter and, thus, the rate of the rise of curves 1-3 is much lower than in Figure 10.

Figure 10. Dependencies $\bar{t}_p(\bar{L})$, $\bar{t}_h(\bar{L})$

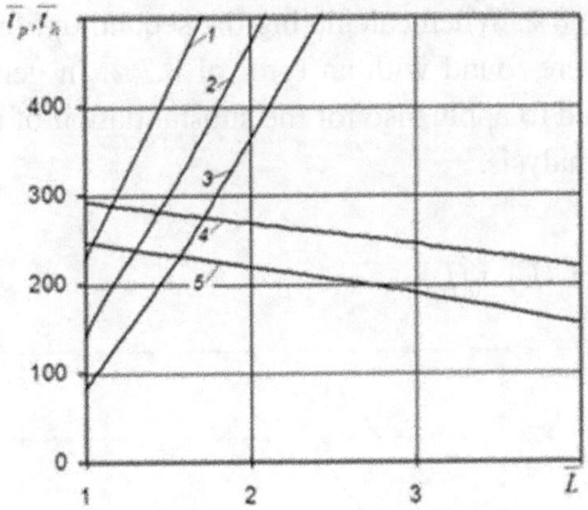

Note. 1-3 – $\bar{t}_p$; 4, 5 – $\bar{t}_h$; 1 – $\bar{C}_* = 0.2$, 2 – $\bar{C}_* = 0.1$, 3 – $\bar{C}_* = 0.05$, 4 – $\Delta\bar{h}_* = 8$, 5 – $\Delta\bar{h}_* = 6$

Figure 11. Dependencies of $\bar{t}_p(\bar{L})$, $\bar{t}_h(\bar{L})$

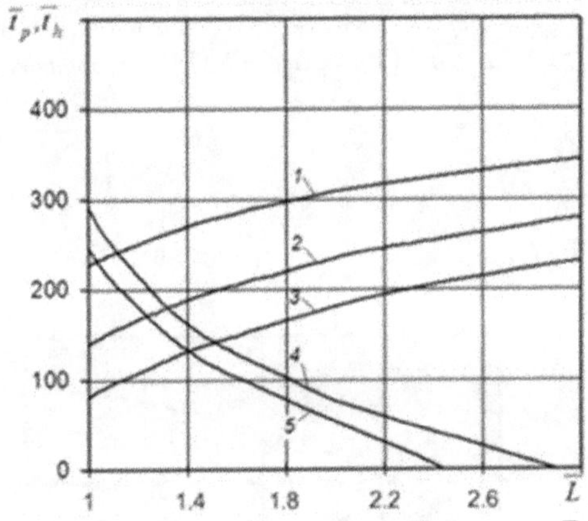

Note. 1-3 – $\bar{t}_p$; 4, 5 – $\bar{t}_h$; 1 – $\bar{C}_* = 0.2$, 2 – $\bar{C}_* = 0.1$, 3 – $\bar{C}_* = 0.05$, 4 – $\Delta\bar{h}_* = 8$, 5 – $\Delta\bar{h}_* = 6$

The same calculation outcomes pattern is observed when calculating the times $\bar{t}_p$, $\bar{t}_h$ using approximate expressions. Here, the values of two corresponding criterion parameters $\bar{C}_* = 0.05$, $\Delta\bar{h}_* = 6$ and $\bar{C}_* = 0.1$, $\Delta\bar{h}_* = 8$ were combined, which allowed us to clearly distinguish the curves of the time $\bar{t}_f$ variation when increasing $\bar{L}$ by factors of 3, 4. These curves together with the basic graphs of the dependences $\bar{t}_p(\bar{L})$, $\bar{t}_h(\bar{L})$ are shown in **Figure 12** for the first option, and for the second one – in **Figure 13**. They clearly demonstrate the real danger of a sharp reduction in the duration of continuous operation of the filter in both options. However, such a situation can occur even at a relatively small increment of L in the second option due to an excessive increase in the hydraulic resistance of the bed. In addition, the computational errors due to the use of the approximate solution were analyzed in relation to the technologically optimal situations when the longest filter run duration is ensured. The differences between the exact and approximate values of $\bar{t}_f$ and $\bar{L}_{opt}$ were estimated for four cases (in different combinations of two options and two sets of the values of $\overline{C}_*$ and $\Delta\overline{h}_*$). For example, the values $\overline{L}_{opt} = 1.48$ and 1.45 were obtained in the first option, respectively, at $\overline{C}_* = 0.05$, $\Delta\overline{h}_* = 6$ and the error in the determination of $\overline{t}_f$ amounted to 8%. When calculating the second option at $\overline{C}_* = 0.1$, $\Delta\overline{h}_* = 8$, the values $\overline{L}_{opt} = 1.325$ and 1.4 were found with an error of 8.3%. In general, approximate calculation expressions can be recommended to apply also for the substantiation of a bed height without resorting to the technical and economic analysis.

Figure 12. Dependencies $\overline{t}_p(\overline{L})$, $\overline{t}_h(\overline{L})$, $\overline{t}_f(\overline{L})$

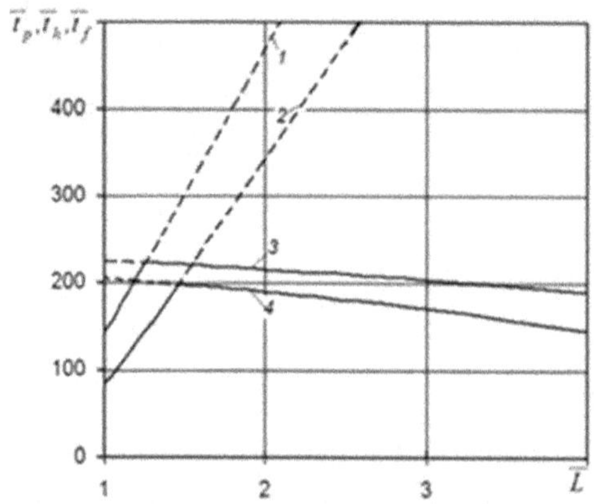

Note. 1, 2 – $\overline{t}_p$; 3, 4 – $\overline{t}_h$; 1 – $\overline{C}_* = 0.1$, 2 – $\overline{C}_* = 0.05$, 3 – $\Delta\overline{h}_* = 8$, 4 – $\Delta\overline{h}_* = 6$; solid lines – $\overline{t}_f$

Figure 13. Dependencies $\overline{t}_p(\overline{L})$, $\overline{t}_h(\overline{L})$

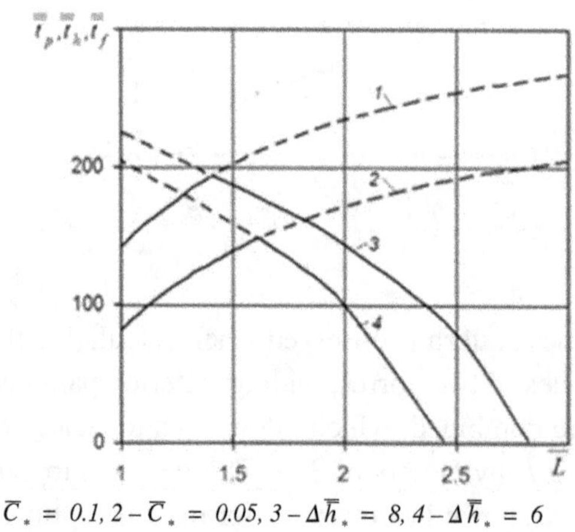

Note. $\overline{t}_f(\overline{L})$: 1, 2 – $\overline{t}_p$; 3, 4 – $\overline{t}_h$; 1 – $\overline{C}_* = 0.1$, 2 – $\overline{C}_* = 0.05$, 3 – $\Delta\overline{h}_* = 8$, 4 – $\Delta\overline{h}_* = 6$

Finally, the consequences of the hypothetical diameter d_g variation within the wide range ($\overline{d}_g$ increases from 1 to 5) were analyzed for the technological times. Their numerous calculations, since d_g was not used as a linear scale when introducing dimensionless variables and parameters, were performed directly by the equations with the criterion parameters (3.1), (3.2). The dimensionless coefficient $\overline{a}$

contains, along with the dimensional analog α, also the design L and mode V parameters, which are not initially fixed. Therefore, there is a possibility to choose the value $\overline{\alpha}_{max}$ based on common sense (the value $\overline{\alpha}_{max} = 12$ is accepted). Here, as well as in case of incomplete or unreliable initial information, special attention should be paid not to the absolute values of the calculation quantities, but to the trends in their changes. Thus, the dependencies $\overline{t}_p(\overline{d}_g)$ and $\overline{t}_h(\overline{d}_g)$ were calculated at various combinations of the values of V (3 and $50 m/h$), and d_p(10^{-6} and $10^{-5} m$), and the corresponding curves are presented in **Figure 14**. If six curves of the dependence $\overline{t}_f(\overline{d}_g)$ are identified in a standard way, the peaks are clearly visible on them. The dominance of these peaks means that even small deviations of the $\overline{d}_g$ value from the optimal value d_{opt}, and in any direction, will inevitably lead to a disproportionately large reduction in the duration $\overline{t}_f$. Thereby, the presence of a pronounced maximum at the functions $\overline{t}_f(\overline{d}_g)$ indicates the importance of the theoretical substantiation of the element size (grain diameter) of the filtering material from the technological point of view. Consequently, when designing a rapid filter, it is necessary to take into account the high sensitivity of the design parameters in relation to its operability.

Figure 14. Dependencies $\overline{t}_p(\overline{d}_g)$, $\overline{t}_h(\overline{d}_g)$

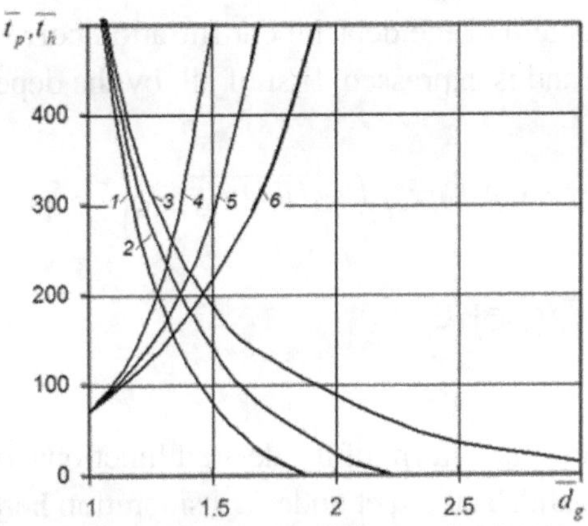

Note. 1-3 – $\overline{t}_p$; 4-6 – $\overline{t}_h$; 1, 6 – V=3, 2-5 – V=50m/h, 1, 3, 4, 6 – $d_p=10^{-5}$; 2, 5 – $d_p=10^{-6}$m

3.3. EFFECT OF GRAINS SORTING DURING PACKED BED OPERATION ON ITS EFFICIENCY

Natural filtering materials even after their special mechanical processing (grinding, screening, etc.) retain a noticeable heterogeneity of granulometric composition. At that, the sizes (equivalent diameters) of large and small structure-forming particles in a packed bed can differ three times, although the ratio of maximum and minimum diameters is mainly about two (Ayukaev & Meltser, 1985; Stevenson, 1995; Zapolsky et al., 2000). Due to the peculiarities of operation of the filters with downward and upward flow of aqueous suspension, in particular, their systematic regeneration by backflow of wash water, the initial homogeneous structure often undergoes significant changes, expressed in divergently directional redistribution of larger and smaller grains. As a result, deep bed filtration can be carried out in the direction of both increasing and decreasing grain size (Shevchuk et al., 2005; Yaroshevskaya

& Muravyev, 2002). Obviously, the absorption capacity of a packed bed must change, because small granular particles with their equal total volume with large particles have a larger surface area, and hence a larger number of vacancies for the suspended solids. Then, for example, at downward filtration, the absorption properties of a filter media will change after its washing according to the increase in the concentration of relatively small grains in the upper part of the bed and larger grains in the lower part. Thus, the unevenness of the deposition of the suspended solids by height and head losses will increase, which can lead to a noticeable reduction in the duration of a filter run. Below quantitative estimation is carried out of the effect of grain sorting during filter operation on the most important characteristics of filtration and technological parameter (Poliakov, 2009b). The mathematical model of filtration with a constant flow rate at linear kinetics of mass exchange between the liquid and solid phases of initially clean bed from Section 2.1 is used as a basis for the conducted theoretical studies. The only change in it concerns the kinetic equation, which now takes a more complex form, namely, after the introduction of dimensionless variables and coefficients

$$\frac{\partial \bar{S}}{\partial \bar{t}} = \bar{\alpha}(\bar{z})\bar{C} - \bar{\beta}\bar{S} \tag{3.39}$$

The solution of the basic problem of the detachment filtration corrected in this way is obtained by means of the Laplace transform and is expressed, first of all, by the dependencies

$$\bar{C}(\bar{z},\bar{t}) = e^{-A(\bar{z})}\left[e^{-\bar{\beta}\bar{t}} I_0\left(2\sqrt{\bar{\beta}A(\bar{z})\bar{t}}\right) + \bar{\beta}\int_0^{\bar{t}} e^{-\bar{\beta}\xi} I_0\left(2\sqrt{\bar{\beta}A(\bar{z})\xi}\right) d\xi\right] \tag{3.40}$$

$$\bar{S}(\bar{z},\bar{t}) = \bar{\alpha}(\bar{z}) e^{-A(\bar{z})} \int_0^{\bar{t}} e^{-\bar{\beta}\xi} I_0\left(2\sqrt{\bar{\beta}A(\bar{z})\xi}\right) d\xi \tag{3.41}$$

where $A(\bar{z}) = \int_0^{\bar{z}} \bar{\alpha}(\zeta)\, d\zeta$. The general form of the desired functions $\bar{h}(\bar{z}, \bar{t})$, $\Delta\bar{h}(\bar{t})$ remains the same, i.e., formulas (2.22), (2.24) are valid. The effect under consideration here affects only the quantity $\bar{S}$.

Due to the limited and incomplete data on the vertical distribution of grains of different sizes in actual conditions, we have to accept different plausible hypotheses. It is justified to dwell on cases of essentially limiting redistribution of grains – the largest and smallest grains are located at the opposite ends of a bed and, in addition, there are only structural particles of the same size at any height. In reality, such sorting is unlikely to be possible, but may be useful to estimate this effect in principle. Two cases that differ significantly in the nature of the grain distribution are discussed below. In the first case, the larger grains are concentrated near the inlet to the bed, while in the second case, on the contrary, the smaller grains are concentrated there. The ratio between the volumes of the filtering material particles of different grades (fractions) varies widely in practice, and here two extreme situations are chosen for examples. In the first situation, the volumes of all fractions are assumed to be equal, and in the second situation, the number of grains in each fraction is considered to be equal. Therefore, the larger grains will occupy a noticeably larger volume.

Let the distribution of the volumes of the grain sortament by their diameters d_g be described by the function $f_g(d_g)$, which is related to the distribution function $f_d(d_g)$ (Section 1.1) by the relation $f_g(d_g) = Lf_d(d_g)$. Then the volume of a small fraction of grains having a diameter in the range from d_g to $d_g + \Delta d_g$

is $\Delta\vartheta = f_\vartheta(d_g)\Delta d_g$. Obviously, when their diameter changes from the minimum d_{min} to the maximum value d_{max}, the equality takes place

$$\int_{d_{min}}^{d_{max}} f_\vartheta(d_g)\, dd_g = (1 - n_0)L \tag{3.42}$$

If the volume of grains of each variety is the same, then $f_\vartheta = f_{\vartheta 0}$ =*const*, and from (3.42) follows

$$f_{\vartheta 0} = \frac{(1 - n_0)L}{d_{max} - d_{min}} \tag{3.43}$$

The required dependence $d_g(z)$ is found at the displacement of large skeleton particles to the inlet cross-section of a packed bed from the condition

$$\int_{d_g}^{d_{max}} f_{\vartheta 0}\, dd_g = (1 - n_0)z \tag{3.44}$$

Integration of (3.44) taking into account (3.43) gives

$$\frac{d_{max} - d_g}{d_{max} - d_{min}} = \bar{z} \tag{3.45}$$

Then the inhomogeneity parameter $\bar{d}_{max} = d_{max}/d_{min}$ is introduced and (3.45) is transformed to the final form

$$\bar{d}_g = \frac{d_g}{d_{max}} = \bar{d}_{max} - (\bar{d}_{max} - 1)\bar{z} \tag{3.46}$$

At a different sorting nature (large grains move in the opposite direction and concentrate near the outlet cross-section instead of (3.44) will be

$$\int_{d_{min}}^{d_g} f_{\vartheta 0}\, dd_g = (1 - n_0)z$$

and, finally, we obtain the expression for $\bar{d}_g$

$$\bar{d}_g = 1 + (\bar{d}_{max} - 1)\bar{z} \tag{3.47}$$

In the second case, it is reasonable to introduce a distribution function of the number of grains also by their diameters $f_n(\bar{d}_g)$. If the number of grains of every sort coincides, then $f_n = f_{n0}$=*const*. To calculate f_{n0} we use the equality

$$\frac{\pi}{6}\int_{d_{min}}^{d_{max}} f_{n0} d_g^3 \, dd_g = (1 - n_0)L$$

from which follows

$$f_{n0} = \frac{24L(1 - n_0)}{\pi(d_{max}^4 - d_{min}^4)} \tag{3.48}$$

Let there are grains of diameter d_g at a distance z from the inlet to the bed, and larger grains are concentrated closer to it. Then the following equality is true

$$\frac{\pi}{6}\int_{d_g}^{d_{max}} f_{n0} d_g^3 \, dd_g = (1 - n_0) z$$

Hence it follows that

$$\frac{d_{max}^4 - d_g^4}{d_{max}^4 - d_{min}^4} = \overline{z}$$

or finally

$$\overline{d}_g = \left[\overline{d}_{max}^4 - (\overline{d}_{max}^4 - 1)\,\overline{z}\right]^{\frac{1}{4}} \tag{3.49}$$

At the concentration of coarse grains near the outlet cross-section there will be

$$\frac{\pi}{6}\int_{d_{min}}^{d_g} f_{n0} d_g^3 \, dd_g = (1 - n_0) z$$

so that the distribution of the relative grain diameter by height is described by the dependence

$$\overline{d}_g = \left[1 + (\overline{d}_{max}^4 - 1)\,\overline{z}\right]^{\frac{1}{4}} \tag{3.50}$$

Transformation of the initial homogeneous structure of a filter media into a clearly inhomogeneous media with a monotonically varying grain variety along its height can have a significant impact on the process of the clarification of aqueous suspension, its quality, as evidenced by the data of the calculations of numerous examples. The subject of the calculations here were the main relative characteristics of filtration (volumetric concentrations of the suspended solids at the outlet, deposited suspension particles, head losses), as well as the time of protective effect of the bed. The exact solution from Section 2.1 was used as a comparison standard for identifying the role of the grain redistribution effect at S^0=0, corresponding to $\overline{\alpha} = \overline{\alpha}_0 = const$. In order to evaluate the significance of this effect for filtration practice, four cases of the grain diameter distribution by height, the formulas for which are derived above, were analyzed quantitatively. In all examples, the only values of the coefficients $\overline{\beta}$ (0.005) and γ_a(20) were taken. The initial volumetric concentration C_0 was either assumed to be $2.5 \cdot 10^{-5}$, or continuously varied from 10^{-5} to 10^{-4}. Finally, the standard values of 1 and 3, respectively, were taken for the parameters m_{k1}, m_{k2} from (1.10).

The deposition rate of the suspended solids is related to the grain size according to formula (2.149). As previously discussed, the contribution of the delivery mechanisms of suspension particles to the bed elements (grains), and eventually the attachment of the former to the latter, depends significantly on d_g. If we follow formula (2.149), then the dependence $\alpha(d_g)$ has in general a power form at the dominance of any of the three main mechanisms

$$\alpha \approx \alpha_j = \tilde{\alpha} d_g^{m_g} \tag{3.51}$$

Next, the focus will be on the removal of the colloidal particles ($d_p \leq 10^{-5}M$), so that the Brownian diffusion mechanism should prevail and then $m_g = -5/3$.

For correctness of the comparative analysis a constant quantity α_0 for the initial granular medium with homogeneous mechanical composition was found by averaging the function $\alpha(d_g)$ over all grain diameters from $d_{\min}$ to $d_{\max}$. In this case

$$\alpha_0 = \tilde{\alpha} \frac{\int_{d_{\min}}^{d_{\max}} d_g^{m_g} f_{i0} \, d d_g}{\int_{d_{\min}}^{d_{\max}} f_{i0} \, d d_g} = \frac{\tilde{\alpha}}{m_g + 1} \frac{d_{\max}^{m_g+1} - d_{\min}^{m_g+1}}{d_{\max} - d_{\min}}, \quad i = \vartheta, n. \tag{3.52}$$

Then for the transformed medium it is true

$$\alpha(z) = \tilde{\alpha} d_g^{m_g}(z) \tag{3.53}$$

Based on (3.52), $\tilde{\alpha}$ was expressed through α_0, which is taken into account in (3.53). As a result, the following formula is obtained

$$\overline{\alpha}(\overline{z}) = \overline{\alpha}_0 (m_g + 1) \overline{d}_g^{m_g}(\overline{z}) \frac{\overline{d}_{\max} - 1}{\overline{d}_{\max}^{m_g} - 1}$$

which will be used in the following calculations. It was assumed $\overline{d}_{\max} = 2$ and the value 6 was chosen for $\overline{\alpha}_0$, and, consequently, the function $\overline{\alpha}(\overline{z})$ according to formulas (3.46), (3.47), (3.49), (3.50) took the following forms: $10.81 \ (2 - \overline{z})^{-5/3}$ (example 1), $10.81 \ (1 + \overline{z})^{-5/3}$ (example 2), $10.81 \ (16 - 15\overline{z})^{-5/12}$ (example 3), $10.81 \ (1 + 15\overline{z})^{-5/12}$ (example 4) and, finally, $\overline{\alpha}$ had a reference constant value (example 5).

First of all, the increase over time in the content of the suspended solids pasted through the bed was determined for all the listed options $\overline{d}_g(\overline{z})$, $\overline{\alpha}(\overline{z})$. The results of the calculations of the dependence $\overline{C}_e(\overline{t})$ are presented in **Figure 15**, and in fact there were only two of them. The first curve corresponds to three examples at once – the reference example and the two other examples with a linear change in $\overline{d}_g$ along height. This fact is explained by taking into account the absorption process in the dependence for $\overline{C}_e$ (3.40) by means of a single integral parameter $\int_0^1 \overline{\alpha}(\overline{z}) \, d\overline{z}$. In the case of a linear form of the function $\overline{d}_g(\overline{z})$, the values of this parameter coincide at $\overline{\alpha} = \overline{\alpha}(\overline{z})$ and $\overline{\alpha} = \overline{\alpha}_0$. In the case of a nonlinear form of $\overline{d}_g(\overline{z})$, the value of this parameter differs from $\overline{\alpha}_0$ and, as a consequence, curve 2 is located noticeably higher.

Figure 15. Change in the relative volumetric outlet concentration of the suspended solids over time

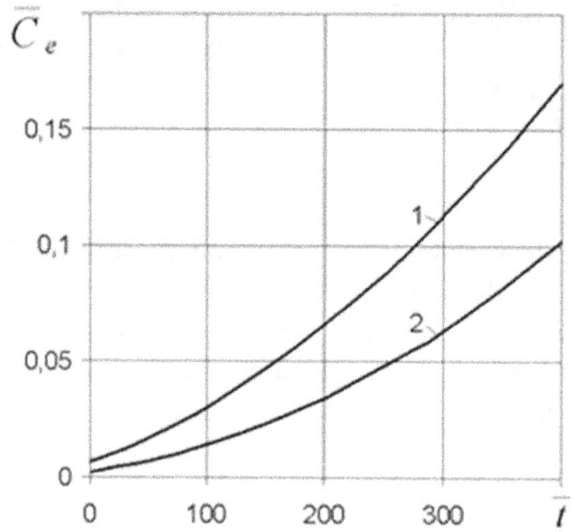

Note. 1 – examples 3,4; 2 – examples 1, 2, 5

Figure 16. The relative volumetric concentration profiles of the deposited particles

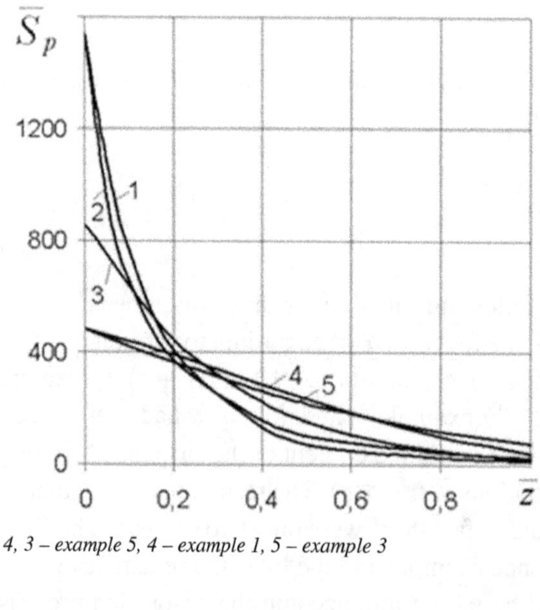

Note. 1 – example 2, 2–example 4, 3 – example 5, 4 – example 1, 5 – example 3

Depending on the direction of grain redistribution in relation to the suspension flow, the change in the absorption properties of the filter media dictated by it can significantly increase the unevenness of the deposit accumulation by height or, on the contrary, decrease it. **Figure 16** shows the concentration profiles $\overline{S}(\overline{z})$, which are plotted for all examples at the same time point during the final stage of filtration. The increase in the attachment activity at the inlet section of the bed (examples 1 and 3) leads here to a significant additional deposition of the suspended solids, which is expressed as an almost twofold increase in $\overline{S}$ (curves 1, 2). At the far section of the bed, the concentration $\overline{S}$ remains low and its changes due to the grain redistribution are small. Reduction of the absorption capacity of the granular medium near the inlet surface (examples 3, 4) leads to the opposite results – also almost twice as much as in example 5, but the deposit content near the inlet cross-section decreases and increases noticeably at the outlet section of the bed. According to Figure 15, the redistribution of grains according to their fineness can significantly affect the hydraulic aspects of the filtration process, which is confirmed in subsequent calculations, including technological calculations.

Hydraulic characteristics are particularly sensitive to the grain sorting, as indicated by **Figure 17**. According to the mutual arrangement of the curves of the dependence $\Delta\overline{h}(\overline{t})$, the ratio between the volumes of different fractions is not so important as the nature of redistribution of large and small grains. With filtration in the direction of decreasing their diameter there is a very dramatic reduction in the head losses (curves 4, 5) compared to the standard losses (curve 3). The opposite situation is observed for the other considered method of the grain redistribution, which leads to a rapid increase in the losses (curves 1, 2).

Figure 17. Change in the relative head losses in the bed

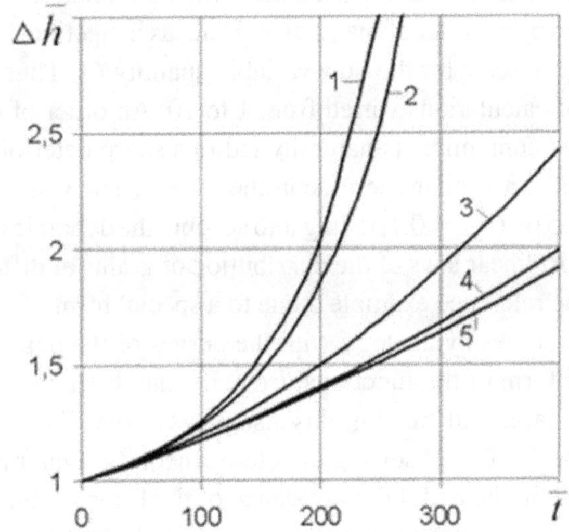

Note. 1 – example 2, 2 – example 4, 3 – example 5, 4 – example 1, 5 – example 3

Figure 18. Dependence $\bar{t}_p(\overline{C}_0)$

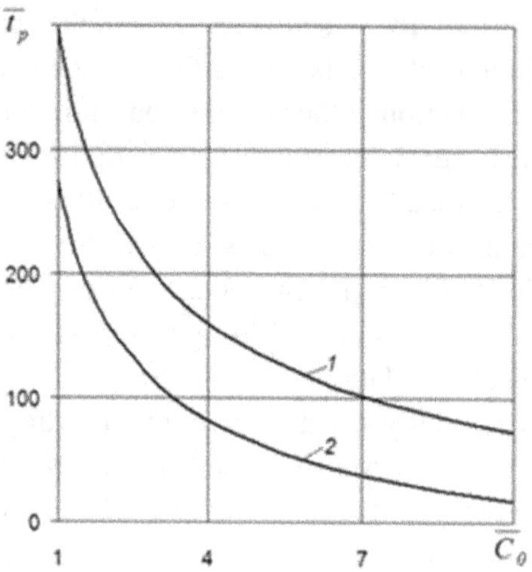

Note. 1 – examples 1, 2, 5; 2 – examples 3, 4

Finally, the effect of the vertical grain size redistribution on the technological parameters was analyzed. **Figures 18, 19** show the results of the technological analysis of the protective effect of the bed and the consequences of its clogging for the same five examples when the relative initial contamination $\overline{C}_0 = C_0/C_{00}$ is continuously changed by an order of magnitude. Here, as in Section 3.1, the fixed value $C_{00}(10^{-5})$ served in calculating $\bar{t}_p$, $\bar{t}_h$ as the scale for the now variable quantity C_0. Thus, the relative inlet concentration $\overline{C}_0$ in the last series of calculations varied from 1 to 10. An order of magnitude increase in the initial content of the dispersed contaminants naturally led to a sharp deterioration in the quality of the filtrate and, as a consequence, to a similar decrease in the time $\bar{t}_p$. The values of $\bar{t}_p$ were systematically determined from equation (1.35) at $\overline{C}_* = 0.1$, taking into account the dynamics of the outlet concentration increase according to (3.40). At linear laws of the distribution of grains of different grades (3.46), (3.47) (examples 1 and 2) and for the reference example 5 due to a special form of $A(\bar{z})$ and the coefficient α_0, the corresponding calculated curves by analogy with the curves of the dependence $\overline{C}_e(\bar{t})$ in Figure 15 coincided (1). At a nonlinear form of the functions $\bar{d}_g(\bar{z})$, i.e. the distribution laws (3.49), (3.50), only the ratio between the sizes of large and small grains also plays a role. That is why one graph (curve 2) corresponds to examples 3, 4. The forced sorting of heterogeneous granular material was more significantly reflected in the increase in the hydraulic resistance of the filter medium, as evidenced by Figure 19. Numerous calculations of $\bar{t}_h$ were performed at $\Delta\bar{h}_* = 6$ on the basis of equation (1.36) taking into account (3.2), (3.41). Here the reference curve 3 occupies an intermediate position and is clearly isolated. A noticeable positive effect of sorting is evident in the case of a relatively large number of fine grains (1, 2 compared to 3) and, on the contrary, a negative effect if the coarse-grained fraction predominates. It should also be noted that the way the grains are distributed (coarse grains predominantly in the near or distant sections of the bed) in the calculations, in contrast to $\overline{C}_e$, $\bar{t}_p$ still manifested, although weakly and only at the values of $\overline{C}_0$ close to 1 (curves 1 and 2, 4 and 5 in pairs).

Thus, a comparison of the times $\bar{t}_p$, $\bar{t}_h$, calculated at $\bar{\alpha} = \bar{\alpha}(\bar{z})$ and $\bar{\alpha} = const$, indicate possible very serious errors in the technological calculations due to the simplified representation of an actual bed, for example, as a monodisperse filter media. In general, it can be stated that the inevitable sorting of grains of inhomogeneous filtering (granular) material by size, which is consistently observed in the operation of a water treatment filter, can very actively affect the efficiency of its clarifying effect. Therefore, it is advisable already at designing and preparation of the filter for operation, firstly, to establish the best (from the technological point of view) equivalent diameter of particles-collectors by calculation, and secondly, to minimize the heterogeneity of the mechanical composition of the above-mentioned material.

Figure 19. Dependence $\bar{t}_h(\bar{C}_0)$

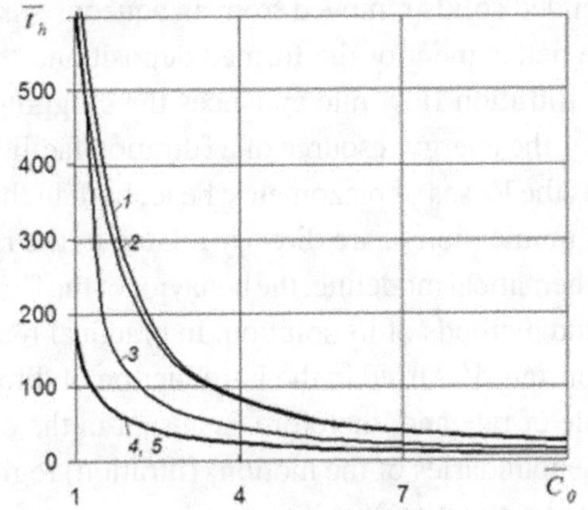

Note. 1– example 3, 2 – example 1, 3 – example 5, 4 – example 4, 5 – example 2

3.4. ESTIMATION OF FILTRATION RATE EFFECT ON TECHNOLOGICAL TIMES

A sharp increase in the performance of filtration facilities is currently a tempting prospect, especially in connection with the constantly and rapidly increasing volumes of the contaminated water. For its practical realization it is necessary to create an appropriate theoretical foundation and then on this basis to improve both the design of facilities and the algorithm of their control. From the physical point of view, rising the filtration rate means not only a proportional increase in the contamination load, but also the intensification of mass exchange processes. Thus, the unevenness of the suspended matter deposition inevitably increases. At the same time, inertia forces become significant, the turbulization of the filtration flow and the compaction of a porous medium are possible. As a result, we should expect an abnormally large increase in the head losses and, consequently, a corresponding reduction in the time t_h. In order to rationally select a forced rate mode, it is necessary to evaluate its feasibility from the economic point of view. Numerous attempts have been made for a number of decades to elucidate the implications for the clarification of suspensions by filtration of a substantial acceleration of liquid flow and can be highlighted, for example, the works (Hutchison, 1976; Jimenez & Buitron, 1990; Lerch et al., 2002; Rudzsky & Kim, 1983; Schneider et al., 2002). However, based on the results of the studies carried out by experimental and theoretical methods, it has not yet been possible to come up with concrete proposals for the application of the above-mentioned mode, although the reality of multiple increases in the filtration rate especially when special filtering materials are used has been constantly pointed out (Niu et al., 2016; Pin Gao et al., 2012). The theoretical results presented below contribute to an in-depth understanding of new opportunities in the filtration intensification due to the dramatic increase in a hydraulic load.

In general, the dynamism of the physicochemical pattern in the layer of a filtering material, hydraulics of all structural components of a filtration facility, providing its long-term and stable functioning, are determined by the filtration rate (hydraulic load) V. Concretizing the practical significance of the rate V, first of all, it should be noted that after a simple recalculation it is identified with the filter performance and, thus, is closely related to the volume of the clarified suspension. Also, this flow rate characteristic has a significant influence on both components of interfacial mass exchange, and thus on their joint result – the amount of the suspended solids removed from an aqueous suspension. Indeed, the filtration flow process promotes both the detachment of the formed deposit and the deposition of the dispersed impurity. In the latter case, the filtration flow rate enhances the clogging of the filter bed, although it can sometimes weaken it. Finally, the energy resource of a filtration facility as a whole and its individual structural components, as well as the losses of piezometric head, both in the bed and in the transportation system (supply and offtake communications), are directly related to the rate V.

From the point of view of mathematical modeling, the behavior of the flow characteristic predetermines the choice of the initial model and methods of its solution. In practical realization of the rate mode with a constant hydraulic load, the constant V is used in the introduction of dimensionless characteristics and parameters of filtration as a scale of rate and, therefore, is absent in the calculation dependencies in an explicit form. If the heads on the boundaries of the motion (filtration) region are controlled, the variable nature and initial uncertainty of the quantity $V(t)$ become the reason of continuous changes over time and in the operating volume, as well as the uncertainty of the mass-exchange coefficients. Therefore, one of the main goals of the filtration modeling should now be the determination of the function $V(t)$.

The filtration rate has a significant effect on the protective capability of a filter media, the mechanical energy expenditure. As a consequence, the technological times are significant functions of V. Thus, it is realistic to intensify the technological process by adjusting the filtration rate mode accordingly.

A more objective approach to the substantiation of the rational filter control algorithm requires a detailed technical and economic analysis of its operability, during which, first of all, losses of conditioned water for the preparation for the next filter run are taken into account, and in addition to them other operating costs. The theoretical studies carried out in this Section, in fact, of the sensitivity of the technological times in relation to the rate V under various conditions of filtration can be considered as the first approximation of such an analysis. In the presence of all necessary initial information, first of all, about economic indicators, in principle, it is not difficult to develop calculations on the basis of the balances of the consumed aqueous suspension and dispersed contamination and bring them to the logical end, namely, to establish the optimal value of V and corresponding value of t_f, at which the required volumes of a suspension will be clarified in a short time and as economically as possible.

The influence of the filtration rate on the technological process and its time characteristics is estimated there on the basis of exact and approximate solutions of the mathematical problem of the detachment filtration under linear mass exchange kinetics, presented in Sections 2.1, 2.4. However, the calculation expressions derived from them, due to their dimensionless form, do not explicitly contain a constant V (its value is used as one of the scales). Therefore, in order to operate only with relative quantities in the future, we have to isolate also the relative rate $\overline{V}$ by analogy with the procedure implemented for the variable L in Section 3.2. Thus, the rate V continuously varied within given limits is now related to its initially fixed value V_0. When examining the significance of V for the performance of the rapid and ultra-rapid filters, it is particularly important to consider the close relationship of both mass-exchange coefficients to it. Therefore, for the convenience of the theoretical analysis it is assumed that

$$\alpha = \alpha_V V^l, \beta = \beta_V V^q, \tag{3.54}$$

where α_V, β_V are the reduced analogs of the coefficients α, β, which do not depend on V; l, q are empirical constants. Then according to (3.54) it will be

$$\bar{\alpha}(\bar{V}) = \bar{\alpha}_V \bar{V}^{l-1}, \bar{\beta}(\bar{V}) = \bar{\beta}_V \bar{V}^{q-1}, \tag{3.55}$$

where $\bar{\alpha}_V = L V_0^{l-1} \alpha_V$, $\bar{\beta}_V = n_0 L V_0^{q-1} \beta_V$. In addition, new dimensionless variables are introduced

$$\bar{\bar{t}} = V_0 t/(n_0 L), \Delta\bar{\bar{h}} = k_0(h - H_d)/(L V_0), \tag{3.56}$$

which are related to $\bar{t}$ and $\Delta\bar{h}$ by the relations $\bar{\bar{t}} = \bar{t}/\bar{V}$, $\Delta\bar{\bar{h}} = \bar{V}\Delta\bar{h}$.

As a result of transition from the previous variables $\bar{t}$, $\Delta\bar{h}$, and coefficients $\bar{\alpha}$, $\bar{\beta}$ to new $\bar{\bar{t}}$, $\Delta\bar{\bar{h}}$, $\bar{\alpha}_V$, $\bar{\beta}_V$ in (3.54) - (3.56), the following expressions for the required filtration characteristics, now as functions of the rate $\bar{V}$, are obtained

$$\bar{C}_e(\bar{\bar{t}}; \bar{V}) = \tag{3.57}$$

$$= e^{-\bar{\alpha}_V \bar{V}^{l-1}} \left[e^{-\bar{\beta}_V \bar{V}^q \bar{\bar{t}}} I_0\left(2\sqrt{\bar{\alpha}_V \bar{\beta}_V \bar{V}^{l-q+1} \bar{\bar{t}}}\right) + \bar{\beta}_V \bar{V}^q \int_0^{\bar{\bar{t}}} \left(2\sqrt{\bar{\alpha}_V \bar{\beta}_V \bar{V}^{l-q+1}\xi}\right) d\xi \right]$$

$$\bar{S}(\bar{z}, \bar{\bar{t}}; \bar{V}) = \bar{\alpha}_V \bar{V}^l e^{-\bar{\alpha}_V \bar{V}^{l-1} \bar{z}} \int_0^{\bar{\bar{t}}} e^{-\bar{\beta}_V \bar{V}^q \xi} I_0\left(2\sqrt{\bar{\alpha}_V \bar{\beta}_V \bar{V}^{l-q+1}\xi}\right) d\xi \tag{3.58}$$

$$\Delta\bar{\bar{h}}(\bar{\bar{t}}; \bar{V}) = \bar{V} \int_0^1 \frac{d\bar{z}}{f_k(\bar{S}(\bar{z}, \bar{\bar{t}}; \bar{V}))}. \tag{3.59}$$

Then the technological times $\bar{\bar{t}}_p$, $\bar{\bar{t}}_h$, which are also functions of $\bar{V}$, should be found by fitting procedure from the equations

$$e^{-\bar{\beta}_V \bar{V}^q \bar{\bar{t}}} I_0\left(2\sqrt{\bar{\alpha}_V \bar{\beta}_V \bar{V}^{l-q+1} \bar{\bar{t}}_p}\right) + \tag{3.60}$$

$$+ \bar{\beta}_V \bar{V}^q \int_0^{\bar{\bar{t}}_p} e^{-\bar{\beta}_V \bar{V}^q \xi} I_0\left(2\sqrt{\bar{\alpha}_V \bar{\beta}_V \bar{V}^{l-q+1}\xi}\right) d\xi = \bar{C}_* e^{\bar{\alpha}_V \bar{V}^{l-1}}$$

$$\bar{V} \int_0^1 \frac{d\bar{z}}{f_k\left(\bar{S}(\bar{z}, \bar{\bar{t}}_h; \bar{V})\right)} = \Delta\bar{\bar{h}}_* \tag{3.61}$$

Proceeding from the above approximate solution, the equation similar to (3.60) has the form

$$2\exp\left(-\frac{2\overline{\alpha}_V\overline{V}^{l-1}}{2+\overline{\beta}_V\overline{V}^{q-1}\overline{t}_p}\right) - \exp\left(\overline{\alpha}_V\overline{V}^{l-1}\right) = \overline{C}_* \tag{3.62}$$

The formula for the calculation of $\overline{t}_p$ depending on $\overline{V}$ follows from (3.62)

$$\overline{t}_p(\overline{V}) = \frac{2}{\overline{\beta}_V\overline{V}^q}\left[\frac{\overline{\alpha}_V\overline{V}^{l-1}}{\ln 2 - \ln(\overline{C}_* + e^{-\overline{\alpha}_V\overline{V}^{l-1}})} - 1\right] \tag{3.63}$$

The approximate expression for the relative concentration $\overline{S}$ will be

$$\overline{S}(\overline{z}, \overline{t}_h; \overline{V}) = \frac{2\overline{\alpha}_V\overline{V}^l\overline{t}_h}{2+\overline{\beta}_V\overline{V}^q\overline{t}_h}\exp\left(\frac{2\overline{\alpha}_V\overline{V}^{l-1}\overline{z}}{2+\overline{\beta}_V\overline{V}^q\overline{t}_h}\right) \tag{3.64}$$

and then the time $\overline{t}_h$ is approximately determined from (3.61) after substituting (3.64) into it.

An identical result follows from (3.60), (3.62) concerning the limit value $\overline{V}_m$ for $\overline{V}$ in the sense that the filter does not provide any clarifying effect at $\overline{V} \geq \overline{V}_m$ at all, i.e. $\overline{t}_p \approx 0$. It is easy to see that according to (3.60) and (3.62)

$$\overline{V}_m = \left(-\frac{\ln\overline{C}_*}{\overline{\alpha}_V}\right)^{\frac{1}{l-1}} \tag{3.65}$$

From this we can see that $\overline{t}_p > 0$ is true at l=1 and any arbitrarily large value of $\overline{V}$.

Quantitative analysis of the dependences $\overline{t}_p(\overline{V})$, $\overline{t}_h(\overline{V})$, $\overline{t}_f(\overline{V})$ was performed on numerous examples when the rate $\overline{V}$ was changed several times. In fact, three series of calculations were carried out for different forms of the dependence of the coefficient α on V. The examples predominantly assumed the values $\overline{\alpha}_V$ (5), $\overline{\beta}_V$ (0.01), $\overline{\gamma}_a$ (0.0015) and the standard for the monograph sets of the values of $\overline{C}_*$ (0.05, 0.1, 0.2) and $\Delta\overline{h}_*$ (6, 8). In the first series of examples, a number of curves were obtained and shown in **Figures 20, 21** for the relative technological times corresponding to characteristic situations with moderate (l=1/3) and enhanced (l=1) influence of the hydraulic load on the deposition of the suspended solids. The first thing to note here is the monotonically descending nature of all the curves shown. There are obvious fundamental differences between both situations, which are expressed in the following. At weakened absorption (Fig. 20), firstly, the technological process at any values of $\overline{V}$ is limited by the quality of the aqueous suspension clarification. Secondly, the protective capability of the bed as a whole steadily decreases with increasing $\overline{V}$, and it is initially absent at $\overline{V} \geq \overline{V}_m$. The values $\overline{V}_m$ were found using formula (3.65) and were 2.156, 3.2 and 5.476, respectively, at $\overline{C}_* = 0.05, 0.1$, and 0.2. Hence it is obvious that an unsuccessful choice of the rate mode, dictated, for example, by the desire to maximize the filter productivity, can even lead to the opposite result, namely, from the very beginning of the successive filter run the outlet concentration of the suspended solids will exceed the maximum allowable value. It is logical that the practically realizable rate V must necessarily be somewhat less than V_m.The second situation is justified to consider as limiting in the sense that the time $\overline{t}_p$, and together with it $\overline{t}_h$ will be greater than 0 only at l=1 at any finite values of $\overline{V}$. But as can be seen from Fig. 21, with increasing $\overline{V}$ head losses are decisive for the decision to stop filtration. As a consequence, in contrast to the first situation, in view of the change of factors limiting the filter operation, the curves of the dependence $\overline{t}_f(\overline{V})$ become broken lines in the range accepted for V. It is important to note that the common value for $\overline{t}_p$ and

$\bar{t}_h$ at the intersection of the calculated curves is not an extremum for the function $\bar{t}_f(\bar{V})$. The quantity $\bar{t}_f$ has the highest value at the minimum of the considered rates.

Figure 20. Dependencies $\bar{t}_p(\bar{V})$, $\bar{t}_h(\bar{V})$

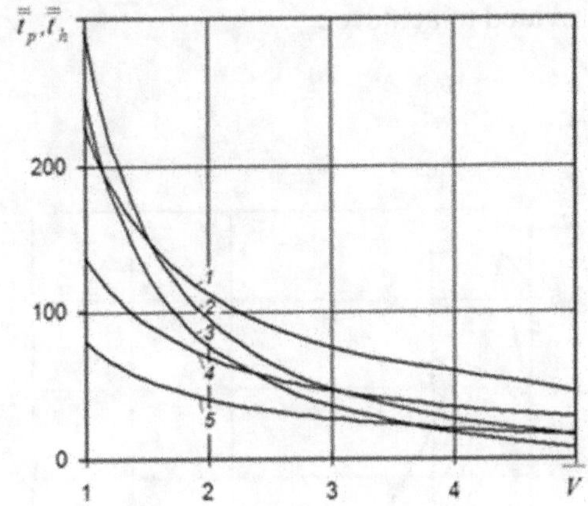

Note. 1, 4, 5 – $\bar{t}_p$; 2, 3 – $\bar{t}_h$; 1 – $\bar{C}_* = 0.2$, 2 – $\Delta\bar{h}_* = 8$, 3 – $\Delta\bar{h}_* = 6$, 4 – $\bar{C}_* = 0.1$, 5 – $\bar{C}_* = 0.05$

Figure 21. Dependencies $\bar{t}_p(\bar{V})$, $\bar{t}_h(\bar{V})$:

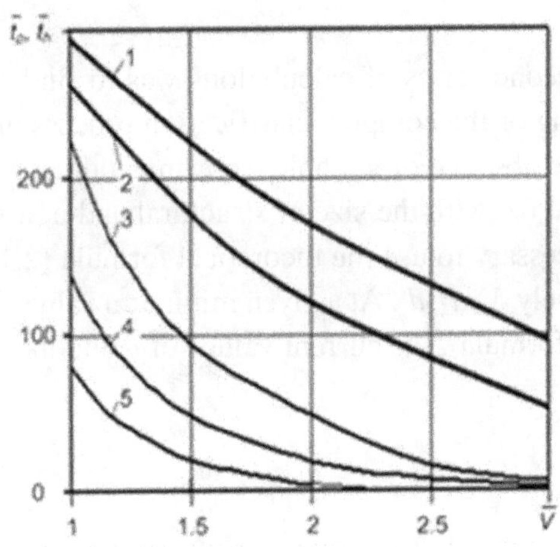

Note. 1, 2 – $\bar{t}_h$; 3-5 – $\bar{t}_p$; solid lines – $\bar{t}_f$; 1 – $\Delta\bar{h}_* = 8$, 2 – $\Delta\bar{h}_* = 6$, 3 – $\bar{C}_* = 0.2$, 4 – $\bar{C}_* = 0.1$, 5 – $\bar{C}_* = 0.05$

When using a well-absorbing filtering material, the nature of the limitation of the technological process can change significantly, as evidenced by **Figure 22**. Here the dependence $\bar{t}_f(\bar{V})$ is shown in graphic form at $\bar{C}_* = 0.1, \Delta\bar{h}_* = 6$ and different combinations of $\bar{\alpha}_V$ (9, 11) and l (-0.3, 1/3). At comparatively small values of the filtration rate ($\bar{V}$ less than 1.31–1.77) the technological process has to be interrupted because of rapidly rising expenses of mechanical energy, so that the corresponding time $\bar{t}_h$ is identical to the required time $\bar{t}_f$. If we continue to formally increase the rate $\bar{V}$, the increasing absorption will result in extreme clogging of the inlet section of the bed, which in turn, after a sufficiently long time will cause a sharp increase in the detachment of a deposit, and with it the removal of the suspension particles from the bed. Thus, the quality of the filtrate becomes the main limiting factor and $\bar{\bar{t}}_f$ is already equal to $\bar{\bar{t}}_p$. As a consequence, the continuous curves $\bar{t}_f(\bar{V})$ have a fracture, the identification of which is of particular interest from a practical point of view. Indeed, the peak formed at the fracture point indicates, firstly, the maximum possible filter run duration $\bar{t}_{fm}$ under the given conditions, and secondly, the (optimal) hydraulic load that should be provided to achieve $\bar{t}_{fm}$.

Figure 22. Dependence $\bar{t}_f(\bar{V})$

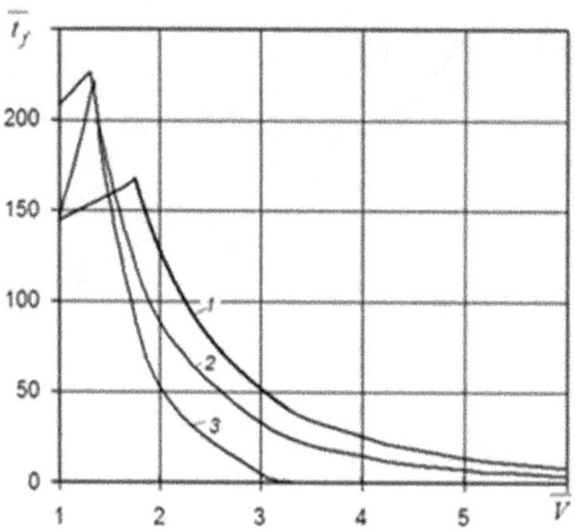

Note. 1 – $\bar{\alpha}_V$ = 11, l=1/3; 2 – $\bar{\alpha}_V$ = 9, l=1/3; 3 – $\bar{\alpha}_V$ = 11, l= -0.3

The main purpose of the second series of calculations was to find out the role of the geometrical parameters involved in modeling of the complex clarification process under the conditions of varying filtration rate mode. Due to the absence of suitable experimental data concerning the relation of the main mass-exchange coefficient (α) with the size of structural and non-structural particles (equivalent diameters d_g and d_p), it was necessary to use the theoretical formula (2.149), according to which α is a function of three variables, namely, V, d_g, d_p. At a given minimum value $V_{\min}$ from the range accepted for the rate V (based on the above formula), the current values of α and its reference value for calculations α_m are simply specified

$$\alpha = f_\alpha\left(V;\ d_g,\ d_p\right)$$

$$\alpha_m = f_\alpha\left(V_{\min};\ d_g,\ d_p\right)$$

Their ratio is represented in the functional form

$$\frac{\alpha}{\alpha_m} = F_V\left(V;\ d_g,\ d_p\right) \qquad (3.66)$$

Equation (3.66) is transformed in order to obtain a dimensionless representation convenient for the subsequent technological calculations, namely,

$$\overline{\alpha}\left(\overline{V};\, d_g,\, d_p\right) = \overline{\alpha}_m \frac{F_V\left(\overline{V};\, d_g,\, d_p\right)}{\overline{V}} \tag{3.67}$$

where $\overline{\alpha} = \alpha L/V$, $\overline{\alpha}_m = \alpha_m L/V_{\min}$ and, therefore, $\overline{\alpha}$ will vary, and $\overline{\alpha}_m$ will be constant when varying V. The value $\overline{\alpha}_m$ was selected based on the requirement for the filter to provide a given quality of the filtrate already at the beginning of the filter run (autocatalysis was not taken into account). It should be noted that the choice of $\overline{\alpha}_m$ in this case is not essentially important, since the focus is on the parameters d_p, d_g, $\overline{V}$ and their role in the suspension separation.

To calculate the times $\overline{t}_p$, $\overline{t}_h$, we used the exact solution of the problem of the detachment filtration with a constant rate and fixed coefficient $\overline{\alpha}$, obtained in Section 2.1. The values of $\overline{\alpha}$ were calculated according to (2.149) and (3.67) for varying discretely d_p, d_g, and continuously the relative rate $\overline{V}$ within the range from 1 to 5. Then they were substituted into the calculation equations with respect to $\overline{t}_p$, $\overline{t}_h$ resulting from the above solution, which are now conveniently represented as follows

$$e^{-\overline{\beta}\overline{t}} I_0\left(2\sqrt{\overline{\alpha}(\overline{V})\overline{\beta}\overline{z}\overline{t}}\right) + \overline{\beta}\int_0^{\overline{t}} e^{-\overline{\beta}\xi} I_0\left(2\sqrt{\overline{\alpha}(\overline{V})\overline{\beta}\overline{z}\xi}\right) d\xi = \overline{C}_* e^{\overline{\alpha}(\overline{V})} \tag{3.68}$$

$$\overline{S}(\overline{z}, \overline{t};\, \overline{V}) = \overline{\alpha}(\overline{V})\, e^{-\overline{\alpha}(\overline{V})\overline{z}}\, \overline{V} \int_0^{\overline{t}} e^{-\overline{\beta}\xi} I_0\left(2\sqrt{\overline{\alpha}(\overline{V})\overline{\beta}\overline{z}\xi}\right) d\xi \tag{3.69}$$

In addition, the general form of the equation relatively $\overline{t}_h$ remains the same as (3.59), but the expression for $\overline{S}(\overline{z}, \overline{t};\, \overline{V})$ should be taken according to (3.69). The data of calculating $\overline{t}_p$, $\overline{t}_h$ at the single values of $\overline{C}_*$ (0.01) and $\Delta\overline{h}_*$ (8), and also various combinations of the values of the diameters d_g($3{\cdot}10^{-4}$, $10^{-3}m$) and d_p(10^{-7}, 10^{-6}, $10^{-5}m$) are shown in **Figures 23, 24**. It was also preliminarily assumed that during primary clarification ($\overline{t} \approx 1$) the initial suspended solids concentration is reduced by a factor of 200, which gives the value 5.298 for $\overline{\alpha}_m$. All six curves calculated according to (3.68), (3.69) have a descending nature. Four of them, as more illustrative, are chosen and shown in the figures. The other two curves almost coincide with the depicted curves and, therefore, are omitted. The rate of the decline of the curves $\overline{t}_p(\overline{V})$ and $\overline{t}_h(\overline{V})$ is markedly different. The quantity $\overline{t}_p$ not only decreases faster with increasing $\overline{V}$, but reaches actually 0 at relatively small hydraulic loads with the exception of one case (curve 3 in Figure 23). The corresponding relative critical rate $\overline{V}_m$ is not difficult to calculate in this series from the equation

$$\overline{\alpha}_m F_V\left(\overline{V}_m;\, d_g,\, d_p\right) + \overline{V}_m \ln \overline{C}_* = 0$$

And only at smaller grains of a packed bed and larger particles of a suspension with increase in $\overline{V}$ in five times the time $\overline{t}_f$ is noticeably more than 0. At especially small solids the sizes of the bed elements almost do not affect its absorption properties, and thereby the values of $\overline{t}_p$, $\overline{t}_h$ at d_p= $10^{-7}m$, d_g= $3{\cdot}10^{-4}$ and $10^{-3}m$, and also at any of the accepted values of $\overline{V}$ practically coincided. The significant difference between Figures 23 and 24 is related to the mutual location of the corresponding curves $\overline{t}_{p,h}(\overline{V})$. The technological process was limited exclusively by the filtrate quality in three options and the curves $\overline{t}_{p,h}$ ($\overline{V}$) intersected only in one option $\left(d_g = 3 \cdot 10^{-4},\ d_p = 10^{-5}\, m\right)$, which indicates here the change of the limiting factor at about $\overline{V} = 2.5$.

Figure 23. Dependencies $\bar{t}_p(\bar{V})$, $\bar{t}_h(\bar{V})$

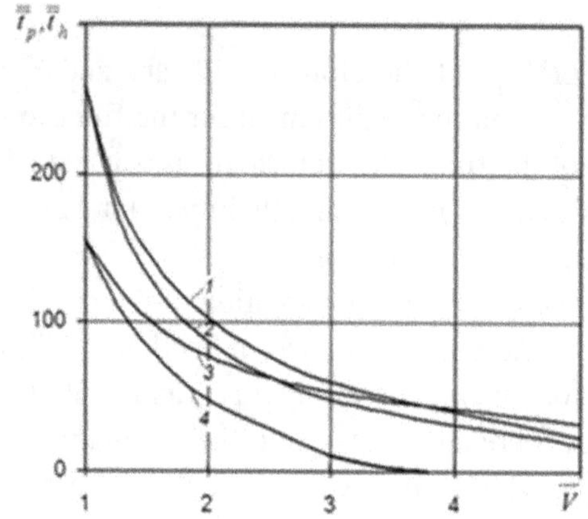

Note. 1, 2 – $\bar{t}_h$; 3, 4 – $\bar{t}_p$; 1, 4 – d_p=10^{-7}; 2, 3 – d_p=10^{-5}m

Figure 24. Dependencies $\bar{t}_p(\bar{V})$, $\bar{t}_h(\bar{V})$

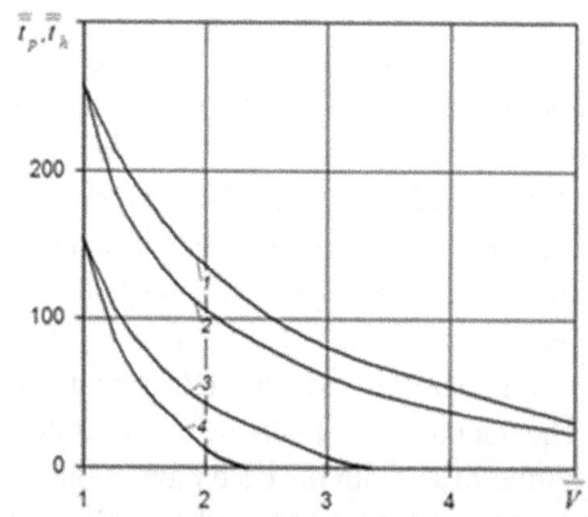

Note. 1, 2 – $\bar{t}_h$; 3, 4 – $\bar{t}_p$; 1, 4 – d_p=10^{-5}; 2, 3 – d_p=10^{-6}m

Three approaches can be recommended for the rational choice of the filtration rate, which differ in both the amount of baseline information and the degree of reliability of the results obtained during their implementation. The first in order and the simplest approach is based on purely technological calculations of a water treatment filter, the most important provisions of which are outlined in Section 3.1. To determine the limit value $\bar{V}_m$ as a reference point, limiting from above the range of the values of $\bar{V}$, at which in principle the clarifying effect is possible, it is proposed to use formula (3.65). The true optimal value $\bar{V}_{opt}$ will necessarily be less than $\bar{V}_m$. It is often possible to find it in the first approximation thanks to the approach following by order and essence. In this case, the unproductive expenditure of the clarified aqueous suspension for washing of a bed is already taken into account. Here two options of washing deserve attention. In the first option the same amount of pure water is spent for each washing, i.e. its specific (per unit inlet surface area) consumption is always the same. In the second option, however, the losses of water purified with time t_f are strictly linked to the total amount of water, so that $W_{tot}=\chi_w \omega V t_f$, where the water consumption factor χ_w<<1. For further analysis of the significance of $\bar{V}$ it is expedient to introduce the objective function of the useful output $F_V(\bar{V})$ or, in other words, the effective filtration rate V_e, which characterizes the actual (directly supplied to the consumer) volume of the conditioned water obtained per unit time. Thus, at $\bar{W}_{tot} = W_{tot}/(n_o \omega L) = const$

$$F_V(\bar{V}) = \bar{V}_e = \bar{V} - \frac{\bar{W}_{tot}}{\bar{t}_f(\bar{V})} \tag{3.70}$$

If $\bar{W}_{tot} = \bar{\chi}_w \bar{V} \bar{t}_f$, then

$$F_V(\bar{V}) = (1 - \bar{\chi}_w) \cdot \bar{V} \tag{3.71}$$

Only function (3.66) has an intermediate maximum, and its corresponding value $\bar{V}_{opt}$ should be found from the equation

$$\frac{dF_V}{d\bar{V}} = 1 + \frac{\bar{W}_{tot}}{\bar{t}_f^2(\bar{V}_{opt})} \cdot \frac{d\bar{t}_f}{d\bar{V}}(\bar{V}_{opt}) = 0 \tag{3.72}$$

Since, according to Figs. 20, 21 a filter has to be turned off because of excessive weakening of the protective capability of the bed, then $\bar{t}_f$ equates to $\bar{t}_p$. The strict expression for the derivative from equation (3.72) has the general form

$$\frac{d\bar{t}_p}{d\bar{V}} = -\frac{\partial F_c/\partial \bar{V}}{\partial F_c/\partial \bar{t}_p} \tag{3.73}$$

where according to (3.73)

$$F_c(\bar{t}_p, \bar{V}) = e^{-\bar{\beta}_V \bar{V}^q \bar{t}_p} I_0\left(2\sqrt{\bar{\alpha}_V \bar{\beta}_V \bar{V}^{l-q+1} \bar{t}_p}\right) +$$
$$+ \bar{\beta}_V \bar{V}^q \int_0^{\bar{t}_p} e^{-\bar{\beta}_V \bar{V}^q \xi} I_0\left(2\sqrt{\bar{\alpha}_V \bar{\beta}_V \bar{V}^{l-q+1} \xi}\right) d\xi - \bar{C}_* e^{\bar{\alpha}_V \bar{V}^{l-1}}.$$

As a result, expression (3.73) is cumbersome and, therefore, is not given. The approximate solution from Section 2.4 allows to obtain a representation for the derivative close to the exact expression (3.73). Then it follows from (3.63) that

$$\frac{d\bar{t}_p}{d\bar{V}}(\bar{V}) =$$
$$= \frac{2}{\bar{\beta}_V}\left\{ \frac{(l-q-1)\bar{\alpha}_V \bar{V}^{l-q-2}}{\ln 2 - \ln(\bar{C}_* + e^{-\bar{\alpha}_V \bar{V}^{l-1}})} - \frac{(l-1)\bar{\alpha}_V^2 \bar{V}^{2l-q-3} e^{-\bar{\alpha}_V \bar{V}^{l-1}}}{(\bar{C}_* + e^{-\bar{\alpha}_V \bar{V}^{l-1}})\left[\ln 2 - \ln(\bar{C}_* + e^{-\bar{\alpha}_V \bar{V}^{l-1}})\right]^2} + q\bar{V}^{-q-1} \right\}.$$

The useful relative filter productivity should be calculated by formula (3.70) at the value $\bar{V}_{opt}$ found from (3.72). Incidentally, $\bar{V}_{opt}$ can be established using (3.71) only on the basis of the technical and economic calculations. In general, such calculations are the most reliable way to substantiate V_{opt}. However, they require complete information on the capital and operating costs, which is often unavailable. For now, just note that in the third approach, due to frequent filter switching, progressive abrasion and crushing of the filtering material the value $\overline{V}_{opt}$ may be noticeably lower than in the second approach and even more so in the first approach. Further, the first two approaches are implemented on specific examples in view of the absence of the data on economic indicators.

The data of the calculations of $\overline{V}_{opt}$ from equation (3.72) and by formula (3.70) when changing the argument $\overline{w}$ from 0.01 to 10 are presented in **Figure 25**. It is logical that intensive washing of a filter media can significantly affect both the conditionally optimal (without taking into account the economic component of the technological process) filtration rate and the net output of clarified suspension (effective filtration rate). The difference between $\overline{V}_{opt}$ and $\overline{V}_e$ increases with rising $\overline{C}_*$ and $\overline{W}_{tot}$, reaching actually 40% in the case of $\overline{C}_* = 0.2$, $\overline{W}_{tot} = 10$. If a filter efficiency is estimated only on the basis of the clarification volumes and washing losses, the situation with its optimization changes dramatically at $l=q=1$. Then it turns out to be $\overline{V}_e \sim \overline{V}$, and it is reasonable to increase the hydraulic load as much as possible taking into account the technical possibilities. It is obvious that frequent manipulations with the filter are in reality undesirable because of considerable additional operating and eventually capital costs. Therefore, final and reliable recommendations regarding the choice of $\overline{V}$ in this and any other situation can only be made on the basis of the technical and economic calculations.

Figure 25. Dependencies $\overline{V}_{opt}(\overline{W}_{tot})$, $F_v(\overline{W}_{tot})$

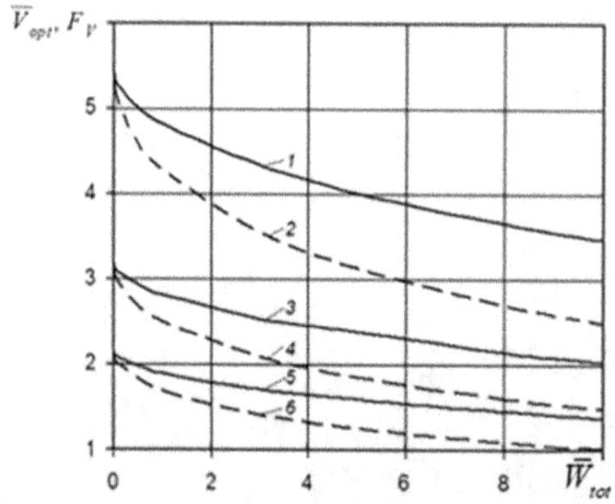

Note. 1, 3, 5 – $\overline{V}_{opt}$; 2, 4, 6 – F_V; 1, 2 $\overline{C}_* = 0.2$; 3, 4 – $\overline{C}_* = 0.1$; 5, 6 – $\overline{C}_* = 0.05$

Figure 26. Dependence $\overline{V}_{opt}(l)$

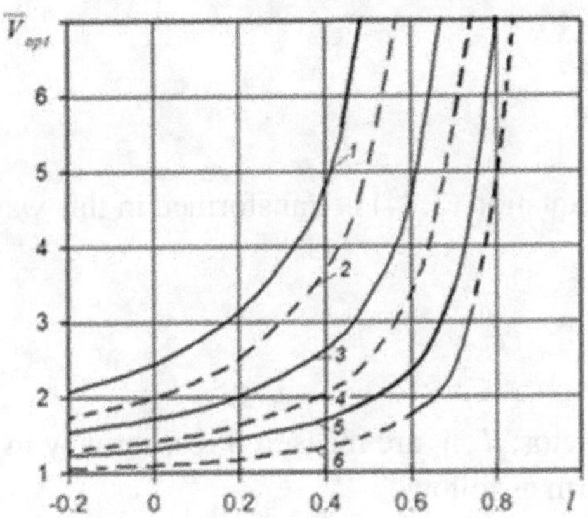

Note. 1, 3, 5 – $\overline{V}_{opt}$; 2, 4, 6 – F_V; 1, 2 – $\overline{C}_* = 0.2$; 3, 4 – $\overline{C}_* = 0.1$; 5, 6 – $\overline{C}_* = 0.05$

Further, the sensitivity of the parameters $\overline{V}_{opt}$, $F_V\left(\overline{V}_{opt}\right)$ was analyzed with respect to the absorption capacity of the filter media. In this case, formal changes in α were caused by the increase in the exponent l in formula (3.54) from the minimum possible value -0.2 to the maximum value 1. The curves $\overline{V}_{opt}(l)$, $F_V(l)$ were determined at $\overline{W}_{tot} = 5$ and a standard set of values of $\overline{C}_*$ from equation (3.72) and are shown in **Figure 26**. The desired quantities increase indefinitely at any actual values of $\overline{C}_*$ if $l \to 1$, which is contrary to reality. In this case, the rise of the calculated curves should be significantly slowed down, again for economic reasons. The significant difference between the corresponding pairs of curves $\overline{V}_{opt}(l)$, $F_V(l)$ is explained by a very large water expenditure for washing the bed.

The role of inertia forces in the filtration flow process sharply increases at high hydraulic loads, which are characteristic partly for rapid filters, but especially for ultra-rapid filters. Moreover, the turbulization of the suspension flow in pores is often observed at the intensive operation of ultra-rapid filters. In both cases, as noted in Section 1.2, there is essentially nonlinear filtration and, as a result, the hydraulic resistance of a filter media can increase significantly compared to linear filtration. To accurately account for additional mechanical energy expenditure, extensive experimental studies of the filtration in high-rate modes are required. Since the author is not aware of such studies that would provide mathematical modeling with the necessary initial information, he had to use the well-known Ergun formula (1.27) (Ergun, 1952) developed for a clean granular media. It is logical to generalize it and the two-term filtration flow law to the case of a clogged filter media, which consists in taking, by analogy with the hydraulic conductivity (1.16), the coefficients from the above law as functions of the deposit concentration S_S or deposited suspension solids concentration S (1.9). Then the corresponding equation can be written in the following form (Poliakov, 2007)

$$a_h\left(S_S\right)\left[V(t)+\frac{b_h(S_S)}{a_h(S_S)}V^2(t)\right] = -\frac{\partial h}{\partial z} \tag{3.74}$$

Two assumptions were made with respect to (3.74). First, the functional coefficient $a_h(S_S)$ was identified with the inverse quantity of $k(S_S)$. Second, it was assumed that the ratio b_h/a_h does not change at any values of S_S. Then, using the expressions for a_h, b_h from the above-mentioned formula, the following expression is obtained

$$\frac{b_h(n)}{a_h(n)} = \frac{1.167 \cdot 10^{-2}}{1-n} = \frac{1.167 \cdot 10^{-2}}{1-n_0+S_S} \tag{3.75}$$

Taking into account (3.75), equation (3.74) is transformed in this way

$$\frac{V(t)}{k(S_S)}\left[1+\frac{1.167 \cdot 10^{-2} \cdot \chi_t d_g}{1-n_0+S_S} V(t)\right] = -\frac{\partial h}{\partial z} \tag{3.76}$$

where χ_t is the conversion factor; d_g, V are measured respectively in m and m/h. Equation (3.76) is represented in dimensionless form as follows

$$\frac{\bar{V}(\bar{t})}{f_k(\bar{S})}\left[1+b_s(\bar{S}) \cdot \bar{V}(\bar{t})\right] = -\frac{\partial \bar{h}}{\partial \bar{z}} \tag{3.77}$$

where $b_s(\bar{S}) = \dfrac{42 d_g V_0}{1-n_0+n_0\bar{\gamma}(\bar{S}) \cdot \bar{S}}$.

Integration of (3.77) under condition (2.13) gives for the reduced head the expression

$$\bar{h}(\bar{z},\bar{t}) = \bar{V}(\bar{t})\left[\int_{\bar{z}}^{1}\frac{d\eta}{f_k(\bar{S}(\eta,\bar{t}))}+\bar{V}(\bar{t})\int_{\bar{z}}^{1}\frac{b_s(\bar{S}(\eta,\bar{t}))}{f_k(\bar{S}(\eta,\bar{t}))}d\eta\right] \tag{3.78}$$

Relative head losses in the packed bed by the point of time $\bar{t}$ at high-rate filtration modes are identical to the value $\bar{h}(0,\bar{t})$ according to (3.78). Thus, the addition $\Delta\bar{h}(0,\bar{t})$ due to nonlinearity of filtration to the specified losses at linear filtration will approximate as follows

$$\Delta\bar{h}(0,\bar{t}) = \bar{V}^2(\bar{t})\int_{0}^{1}\frac{b_s(\bar{S}(\eta,\bar{t}))}{f_k(\bar{S}(\eta,\bar{t}))}d\eta \tag{3.79}$$

Finally, the value of the time $\bar{t}_h$ in the constant filtration rate mode, refined by taking into account inertial forces and flow turbulence, should be found from the equation

$$\bar{V}\left[\int_{0}^{1}\frac{d\eta}{f_k(\bar{S}(\eta,\bar{t}_h;\bar{V}))}+\bar{V}\int_{0}^{1}\frac{b_s(\bar{S}(\eta,\bar{t}_h;\bar{V}))}{f_k(\bar{S}(\eta,\bar{t}_h;\bar{V}))}d\eta\right] = \Delta\bar{h}_* \tag{3.80}$$

In conclusion, possible consequences for the technological times $(\bar{t}_h, \bar{t}_f)$ of deviations of the actual filtration flow law (in the two-term form (1.27)) from the linear law were analyzed. The simplified approach to modeling of nonlinear suspension flow in porous media with varying porosity implemented above gives only a general idea of its significance for the operation of (ultra)rapid filters. However, it is not possible to apply it systematically in the monograph due to the lack of experimental information concerning the violations of linear (Darcy's) law in clogged porous (granular) media. Nevertheless, even the results obtained here after careful consideration, quite convincingly demonstrate the importance of taking into account in the predictive, technological calculations the nonlinear effects associated with inertia and turbulence of filtration flow when operating filters in the forced rate modes. Therefore, many calculations of the time $\bar{t}_h$ have been performed for the following initial data: $\bar{\alpha}_V = 5$, $\bar{\gamma}_a = 0.0015$, n_0=0.4, l=1/3 and at four values of the complex parameter $d_g V_0$, including 0. In the latter case, the set of the values of $\bar{t}_h$ can be considered as a basic set when estimating the contribution to the total head losses of their component which is caused by the above-mentioned effects. As can be seen from **Figure 27**, the increase of the specified parameter in actual limits can not only to lead to appreciable reductions of the time $\bar{t}_h$ and possibly $\bar{t}_f$, but also to narrow the interval of the filtration rates allowed in practice.

Figure 27. Dependence $\bar{t}_h(\bar{V})$

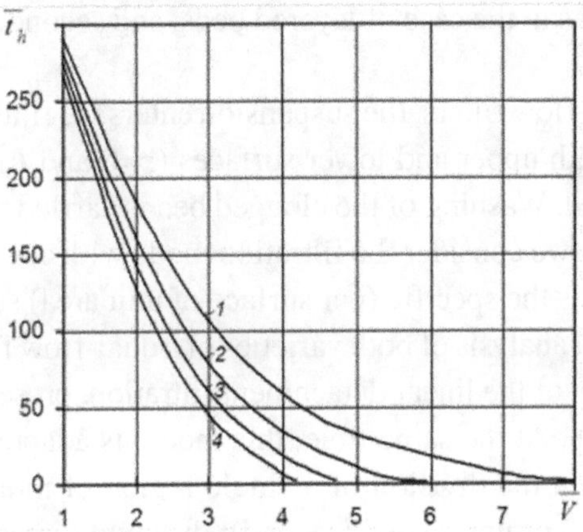

Note. 1 – $d_g V_0$=0, 2 – $d_g V_0$=0.001, 3 – $d_g V_0$=0.002, 4 – $d_g V_0$=0.003

It is obvious that the theoretical studies of water treatment filters operation under especially high hydraulic load should be developed in order to optimize the suspension clarification technology, but only after providing them with reliable baseline information in full.

3.5. ALTERNATIVE WAYS TO FEED SUSPENSION INTO BED (DUAL AND MULTI-FLOW FILTERS)

In case of the systematic and significant reduction of the filter efficiency due to rapidly increasing head losses, special measures should be taken in order to use mechanical energy more economically. Such measures are diverse and can be envisaged in the design of filtration facilities and implemented in their operation. Directly they should be aimed at ensuring a more even distribution of the deposit in the volume of a filter media and thus, at reducing its hydraulic resistance. In practice it is impossible to achieve the strictly even deposit accumulation. But it is quite realistic to redistribute part of the deposit along the course of the filtration flow. And it is especially important to at least partially equalize the deposit concentration profiles for a well-absorbing filter media, which guarantees a high filtrate quality over a long period of time. Incidentally, the use of such materials in modern depth filtration practice is just constrained by the excessive increase of the head losses in a packed bed, and mainly at its inlet section. It is possible to significantly reduce energy expenditure by using dual and multi-flow filters (Adelman et al., 2012; Poliakov & Kurganska, 2025). Here, a suspension enters the filter media also locally, but unlike traditional single-flow filters, it is fed simultaneously in two, three or more places. Obviously, most of the deposit will accumulate near them. Thus, it is possible, firstly, to sharply reduce the peak value of the deposit concentration S_0 in the inlet section of the bed; secondly, to ensure a generally deeper penetration of solid particles into the filter media. As a consequence, the rate of an increase in the head losses becomes noticeably lower.

According to the principle of the arrangement of the filtration flow there are two varieties of two-flow filters. The essence of the first type is a combination of the external (through the inlet cross-section) and internal (the suspension source is located at the depth $L_1<L$) suspension supply. Moreover, a suspension can be constructively fed in several places (internal cross-sections) at the same time. Thus, the hydraulic load is effectively split into two or more parts. Obviously, the deposit concentration profile firstly becomes discontinuous, as in the case of layered beds, and secondly, its fragments become flatter if their number increase.

In the second variety of dual-flow filters, the suspension enters the filters also simultaneously in equal or unequal amounts through both upper and lower surfaces (z=0 and L). The filtrate is taken away by drainage in the middle of the bed. Washing of the clogged bed is carried out by the flow of treated water in the opposite direction. Below we consider the filtration mode with constant rates (V, V_1=$const$) where V is the total hydraulic load, V_1 is the specific (per surface of unit area) suspension flow rate in the inlet cross-section. The technological analysis of both varieties of a dual-flow filter is performed in turn on the basis of the mathematical model of the linear detachment filtration, presented in Section 2.2, at variable initial concentration of impurity. At the same time, this model is adapted to new filtration conditions. Such conditions are expressed in the division of a single region of motion into two sub-regions, that are interrelated (the first variety) or not related (second); different directionality of the flows (second), uncertainty of boundary conditions at the boundary between the flows (first).

Thus, at the downward filtration and suspension supply through the upper surface of the bed (z=0) and its cross-section at the depth L_1, respectively, with the constant specific flow rates V_1 and $V_2= V-V_1$, the initial model taking into account (3.54) will include, with respect to the upper section of the bed ($0\leq z\leq L_1$), the system of the equations

$$V_1\frac{\partial C_1}{\partial z}+\frac{\partial S_1}{\partial t} = 0 \tag{3.81}$$

$$\frac{\partial S_1}{\partial t} = \alpha_V V_1^l C_1 - \beta_V V_1^q S_1 \tag{3.82}$$

$$V_1 = -k_0 f_k(S_1)\frac{\partial h_1}{\partial z} \tag{3.83}$$

$$f_k(S_i) = \left\{1 - \left[\gamma(S_i)S_i\right]^{m_{k1}}\right\}^{m_{k2}},\ (i = 1,\ 2), \tag{3.84}$$

and the boundary conditions operator

$$t = 0,\quad S_1 = 0;\quad z = 0,\quad C_1 = C_0;\quad z = L_1,\quad h_1 = h_2. \tag{3.85}$$

Similarly for the bottom section ($L_1\leq z\leq L$) will be

$$V\frac{\partial C_2}{\partial z} + \frac{\partial S_2}{\partial t} = 0, \tag{3.86}$$

$$\frac{\partial S_2}{\partial t} = \alpha_V V^l C_2 - \beta_V V^q S_2, \tag{3.87}$$

$$V = -k_0 f_k(S_2)\frac{\partial h_2}{\partial z}, \tag{3.88}$$

$$t = 0, \quad S_2 = 0; \quad z = L, \quad h_2 = H_d. \tag{3.89}$$

The missing condition, namely, the concentration C_2 at the inner boundary ($z=L_1$) is derived for the correct formulation of this mathematical problem from the balance of an impurity on it

$$V_1 C_{1e}(t) + (V - V_1)C_0 = VC_2, \tag{3.90}$$

where C_{1e} is the variable volumetric concentration of the suspended solids that has passed through the top section. In choosing the basic model, it is assumed that the same suspension enters both the bed sections. Since the relationship between the rates V_1, V_2, V can vary over a wide range, the relationship between them and the mass exchange coefficients α,β is revealed. Thus, α_V, ${}_V$ are the reduced coefficients of the attachment and detachment rates of suspension particles, which do not depend on the flow characteristics. It should also be noted that due to the difficulty of averaging the deposit composition function $\gamma(S_i)$, it is not applied further. After introducing the dimensionless variables and parameters in a standardized way, and at the same time

$$\bar V_1 = \frac{V_1}{V}, \quad \bar L_1 = \frac{L_1}{L}, \quad \bar\alpha_V = \alpha_V V^{l-1}, \quad \bar\beta_V = \beta_V V^{q-1} \tag{3.91}$$

problem (3.81) - (3.90) is transformed to the following form: for the upper section $(0 \le \bar z \le \bar L_1)$

$$\bar V_1\frac{\partial \bar C_1}{\partial \bar z} + \frac{\partial \bar S_1}{\partial \bar t} = 0, \tag{3.92}$$

$$\frac{\partial \bar S_1}{\partial \bar t} = \bar\alpha_V \bar V_1^l \bar C_1 - \bar\beta_V \bar V_1^q \bar S_1, \tag{3.93}$$

$$\bar V_1 = -f_k(\bar S_1)\frac{\partial \tilde h_1}{\partial \bar z}, \tag{3.94}$$

$$f_k(\bar S_i) = \left\{1 - \left[\bar\gamma(\bar S_i)\bar S_i\right]^{m_{k1}}\right\}^{m_{k2}}; \tag{3.95}$$

$$\bar t = 0, \quad \bar S_1 = 0; \quad \bar z = 0, \quad \bar C_1 = 1; \quad \bar z = \bar L_1, \quad \tilde h_1 = \tilde h_2; \tag{3.96}$$

for the lower section $(\bar L_1 \le \bar z \le 1)$

$$\frac{\partial \bar C_2}{\partial \bar z} + \frac{\partial \bar S_2}{\partial \bar t} = 0, \tag{3.97}$$

$$\frac{\partial \overline{S}_2}{\partial \overline{t}} = \overline{\alpha}_V \overline{C}_2 - \overline{\beta}_V \overline{S}_2, \tag{3.98}$$

$$1 = -f_k(\overline{S}_2)\frac{\partial \tilde{h}_2}{\partial \overline{z}} ; \tag{3.99}$$

$$\overline{t} = 0, \quad \overline{S}_2 = 0; \quad \overline{z} = \overline{L}_1, \quad \tilde{h}_2 = 0. \tag{3.100}$$

Moreover, the last boundary condition follows from (3.90)

$$\overline{z} = \overline{L}_1, \quad \overline{C}_2(\overline{t}) = 1 - \overline{V}_1 + \overline{V}_1 \overline{C}_{1e}(\overline{t}). \tag{3.101}$$

The solution of problem (3.92) - (3.101) is expressed by two sets of dependences for the upper and lower bed sections, respectively. Thus, the spatial and temporal variations of the relative concentrations and head losses at the upper section are described by the following dependences

$$\begin{aligned} \overline{C}_1(\overline{z},\overline{t}) = {} & e^{-\overline{\alpha}_V \overline{V}_1^{l-1} \overline{z}} \Big[e^{-\overline{\beta}_V \overline{V}_1^{q} \overline{t}} I_0\Big(2\sqrt{\overline{\alpha}_V \overline{\beta}_V \overline{V}_1^{l+q-1} \overline{z}\overline{t}}\Big) + \\ & + \overline{\beta}_V \overline{V}_1^{q} \int_0^{\overline{t}} e^{-\overline{\beta}_V \overline{V}_1^{q} \xi} I_0\Big(2\sqrt{\overline{\alpha}_V \overline{\beta}_V \overline{V}_1^{l+q-1} \overline{z}\xi}\Big)\, d\xi \Big], \end{aligned} \tag{3.102}$$

$$\overline{S}_1(\overline{z},\overline{t}) = \overline{\alpha}_V \overline{V}_1^{l} e^{-\overline{\alpha}_V \overline{V}_1^{l-1} \overline{z}} \int_0^{\overline{t}} e^{-\overline{\beta}_V \overline{V}_1^{q} \xi} I_0\Big(2\sqrt{\overline{\alpha}_V \overline{\beta}_V \overline{V}_1^{l+q-1} \overline{z}\xi}\Big)\, d\xi, \tag{3.103}$$

$$\tilde{h}_1(\overline{z},\overline{t}) = \overline{V}_1 \int_{\overline{z}}^{\overline{L}_1} \frac{d\overline{z}}{f_k(\overline{S}_1(\overline{z},\overline{t}))} + \int_{\overline{L}_1}^{1} \frac{d\overline{z}}{f_k(\overline{S}_2(\overline{z},\overline{t}))}. \tag{3.104}$$

As follows from (3.103), the relative inlet concentration of the deposited particles $\overline{S}_{10}$ will increase over time according to the formula

$$\overline{S}_{10}(\overline{t}) = \frac{\overline{\alpha}_V}{\overline{\beta}_V} \overline{V}_1^{l-q}\left(1 - e^{-\overline{\beta}_V \overline{V}_1^{q} \overline{t}}\right). \tag{3.105}$$

The relative head losses in the entire bed are numerically equal to $\tilde{h}_1(0,\overline{t})$ and according to (3.104) will be as follows

$$\Delta\tilde{h}(\overline{t}) = \overline{V}_1 \int_0^{\overline{L}_1} \frac{d\overline{z}}{f_k(\overline{S}_1(\overline{z},\overline{t}))} + \int_{\overline{L}_1}^{1} \frac{d\overline{z}}{f_k(\overline{S}_2(\overline{z},\overline{t}))}. \tag{3.106}$$

A similar set of the dependencies for the lower bed section includes

$$\begin{aligned} \overline{C}_2(\overline{z},\overline{t}) = {} & e^{-\overline{\alpha}_V(\overline{z}-\overline{L}_1)} \cdot \Big\{ 1 - \overline{V}_1 + \overline{V}_1 \overline{C}_{1e}(\overline{t}) + \\ & + \sqrt{\overline{\alpha}_V \overline{\beta}_V (\overline{z} - \overline{L}_1)} \int_0^{\overline{t}} \frac{e^{-\overline{\beta}_V(\overline{t}-\xi)} I_1\Big(2\sqrt{\overline{\alpha}_V \overline{\beta}_V (\overline{z}-\overline{L}_1)(\overline{t}-\xi)}\Big)}{\sqrt{\overline{t}-\xi}} \left[1 - \overline{V}_1 + \overline{V}_1 \overline{C}_{1e}(\xi)\right] d\xi \Big\}, \end{aligned} \tag{3.107}$$

$$\overline{S}_2(\overline{z},\overline{t}) = e^{-\overline{\alpha}_0(\overline{z}-\overline{L}_1)} \int_0^{\overline{t}} e^{-\overline{\beta}_0(\overline{t}-\xi)} I_0\Big(2\sqrt{\overline{\alpha}_0 \overline{\beta}_0 (\overline{z}-\overline{L}_1)(\overline{t}-\xi)}\Big) \left[1 - \overline{V}_1 + \overline{V}_1 \overline{C}_{1e}(\xi)\right] d\xi \tag{3.108}$$

$$\Delta\bar{h}_2(\bar{z},\bar{t}) = \int_{\bar{z}}^{1} \frac{d\bar{z}}{f_k(\bar{S}_2(\bar{z},\bar{t}))}. \tag{3.109}$$

Here $\bar{C}_{1e}(\bar{t})$ follows from (3.102)

$$\bar{C}_{1e}(\bar{t}) = e^{-\bar{\alpha}_V \bar{V}_1^{l-1} \bar{L}_1} \times$$
$$\times \left[e^{-\bar{\beta}_V \bar{V}_1^{q} \bar{t}} I_0\left(2\sqrt{\bar{\alpha}_V \bar{\beta}_V \bar{V}_1^{l+q-1} \bar{L}_1 \bar{t}}\right) + \bar{\beta}_V \bar{V}_1^{q} \int_0^{\bar{t}} e^{-\bar{\beta}_V \bar{V}_1^{q} \xi} I_0\left(2\sqrt{\bar{\alpha}_V \bar{\beta}_V \bar{V}_1^{l+q-1} \bar{L}_1 \xi}\right) d\xi \right]$$

The accumulation of the delayed suspension particles at the inlet to the second section $(\bar{z} = \bar{L}_1)$ is described by the exact dependence

$$\bar{S}_{20}(\bar{t}) = \bar{S}_2(\bar{L}_1,\bar{t}) = \bar{\alpha}_V \int_0^{\bar{t}} [1 - \bar{V}_1 + \bar{V}_1 \bar{C}_{1e}(\xi)] e^{-\bar{\beta}_V(\bar{t}-\xi)} d\xi. \tag{3.110}$$

The technological calculation of the deep bed filtration when arranging two counter flows through the upper and lower bed surfaces is simpler, because the processes of mass transfer, mass exchange and filtration flow on its upper and lower sections are modeled in a similar way, and the inlet suspended solids concentrations are constant and the same. At the same time, only some signs differ in the corresponding mathematical models due to different flow directions. The following are only the calculation formulae that are necessary to gain a full understanding of the results and consequences of the suspension clarification in a dual-flow filter with different flow directions.

Thereby, formulas (3.102), (3.103) should be used for the upper section in the calculations of the filtration process. Besides, the head losses $\Delta\bar{h}_1$ are accurately calculated by the formula

$$\Delta\bar{h}_1(\bar{t}) = \bar{V}_1 \int_0^{\bar{L}_1} \frac{d\bar{z}}{f_k(\bar{S}_1(\bar{z},\bar{t}))}. \tag{3.111}$$

The following dependencies are recommended for similar calculations at the lower section

$$\bar{C}_{2e}(\bar{t}) = e^{-\bar{\alpha}_V(1-\bar{V}_1)^{l-1}(1-\bar{L}_1)} \left[e^{-\bar{\beta}_V(1-\bar{V}_1)^{q}\bar{t}} I_0\left(2\sqrt{\bar{\alpha}_V \bar{\beta}_V (1-\bar{V}_1)^{l+q-1}(1-\bar{L}_1)\bar{t}}\right) + \right.$$
$$\left. + \bar{\beta}_V (1-\bar{V}_1)^{q} \int_0^{\bar{t}} e^{-\bar{\beta}_V(1-\bar{V}_1)^{q}} I_0\left(2\sqrt{\bar{\alpha}_V \bar{\beta}_V (1-\bar{V}_1)^{l+q-1}(1-\bar{L}_1)\xi}\right) d\xi \right], \tag{3.112}$$

$$\bar{S}_2(\bar{z},\bar{t}) =$$
$$= \bar{\alpha}_V (1-\bar{V}_1)^{l} e^{-\bar{\alpha}_V(1-\bar{V}_1)^{l-1}(1-\bar{z})} \int_0^{\bar{t}} e^{-\bar{\beta}_V(1-\bar{V}_1)^{q}\xi} I_0\left(2\sqrt{\bar{\alpha}_V \bar{\beta}_V (1-\bar{V}_1)^{l+q-1}(1-\bar{z})\xi}\right), \tag{3.113}$$

$$\Delta\bar{h}_2(\bar{t}) = (1-\bar{V}_1) \int_{\bar{L}_1}^{1} \frac{d\bar{z}}{f_k(\bar{S}_2(\bar{z},\bar{t}))}. \tag{3.114}$$

The relative impurity concentration in the filtrate and the total head losses in the bed are calculated as the following sums

$$\bar{C}_e(\bar{t}) = \bar{V}_1 \bar{C}_{1e}(\bar{t}) + (1 - \bar{V}_1)\bar{C}_{2e}(\bar{t}), \tag{3.115}$$

$$\Delta\bar{h}(\bar{t}) = \Delta\bar{h}_1(\bar{t}) + \Delta\bar{h}_2(\bar{t}) \tag{3.116}$$

Finally, the clogging of the lower bed surface is proposed to be calculated based on the formula

$$\bar{S}_{2e}(\bar{t}) = \bar{S}_2(1,\bar{t}) = \frac{\bar{\alpha}_V}{\bar{\beta}_V}(1 - \bar{V}_1)^{l-q}(1 - e^{-\bar{\beta}_V(1-\bar{V}_1)^q}).$$

The clarifying efficiency of both varieties of the dual-flow filter was evaluated on a number of test cases with the following fixed data:

$$\bar{\beta}_V = 0.01,\ q = 1,\ \bar{C}_* = 0.1,$$
$$\Delta\bar{h}_* = 8,\ \bar{\gamma}(\bar{S}) = 2.5 \cdot 10^{-3} - 2 \cdot 10^{-6}\bar{S},\ f_k(\bar{S}) = [1 - \bar{\gamma}(\bar{S})\,\bar{S}]^3.$$

In addition, two values of $l(1/3, 1)$, characteristic for Brownian ($d_p \leq 10^{-6}$m) and non-Brownian ($d_p \geq 10^{-6}$) particles of a suspension, respectively, were chosen. Finally, either their typical values or wide ranges of possible values are adopted for $\bar{V}_1$, $\bar{L}_1$, $\bar{\alpha}_V$. The main subject of the calculations was the relative duration of the filter run $\bar{t}_f$ as a key technological parameter. For its establishment, the relative times $\bar{t}_p$ and $\bar{t}_h$ were previously found with the involvement of the criterion equations (1.35), (1.36). At already known values of $\bar{t}_f$, the relative excesses (or reductions in unfavorable conditions) $\Delta\bar{t}_f$ corresponding to them were determined due to the separation of a hydraulic load. Thus, $\Delta\bar{t}_f = \bar{t}_f/\bar{t}_{f*}$, where $\bar{t}_{f*}$ is the relative duration of the filter run at downward filtration under similar conditions.

First of all, the consequences for the technological process were analyzed of the simultaneous aqueous suspension supply through the upper surface of the bed and inside it. It should be noted that the ratio between the corresponding suspension flow rates (V_1 and V), as well as the relative depth of the internal water source $\bar{L}_1$ varied continuously within the maximum limits – from 0 to 1. A large value was accepted for $\bar{\alpha}_0$ (8) so that the filter operation was limited precisely by the head losses. Then t_f=t_h. The calculations of the time $\bar{t}_p$ for the dual-flow filter were based on formulas (3.102) for the upper bed section and (3.107) for the lower section. The time $\bar{t}_h$ was calculated in the same order using formulas (3.103), (3.104) and (3.108), (3.109). In this case the value of $\bar{t}_{f*}$ ($\bar{V}_1 = 1$) is 123.6. The results of calculating $\Delta\bar{t}_f$ are presented in **Figure 28** in the form of two graphs of the dependence of the desired quantity on $\bar{V}_1$ at $\bar{L}_1 = 0.5$.The entire range of the values of $\bar{V}_1$ can be conditionally divided into two intervals and indicate the value $\bar{V}_{1*}$ that delimits them. The traditional single-flow filter is more productive ($\Delta\bar{t}_f < 1$) at $\bar{V}_1 \leq \bar{V}_{1*}$, and already dual-flow filter($\Delta\bar{t}_f > 1$) is better at $\bar{V}_1 > \bar{V}_{1*}$. The maximum effect from the separation of the hydraulic load is achieved at $\bar{V}_1 = 0.75 - 0.85$ and approximately the same (in percentage terms about 70%) at different grades of the impurity particles. However, an unsuccessful redistribution of the specified load (with a higher internal flow supply) can give a negative result at all, and the worst result will naturally be at $\bar{V}_1 = 0$. Then, in fact, the height of the operating bed is halved, while the time $\bar{t}_f$ is reduced by three times or more.

Figure 28. Dependence $\Delta \bar{t}_f(\bar{V}_1)$

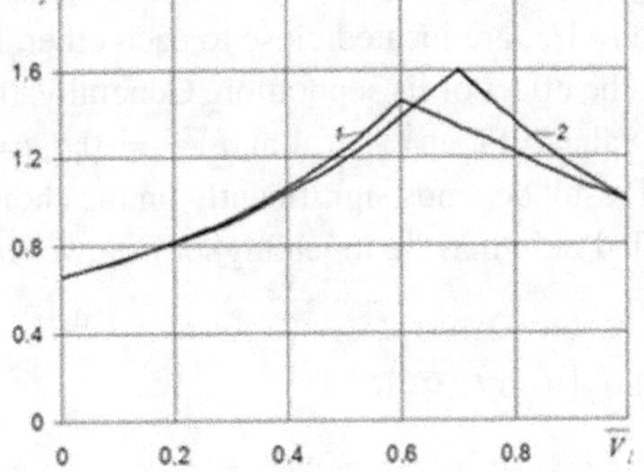

Note. 1 – l=1/3, 2 – l=1

Figure 29. Dependence $\Delta \bar{t}_f(\bar{L}_1)$

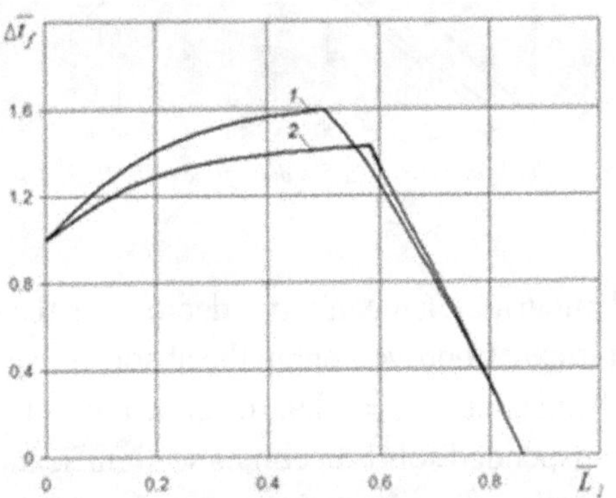

Note. 1 – l=1, 2 – l=1/3

The different, although partly similar, calculated pattern is observed in **Figure 29**, where the data are also given for $\Delta \bar{t}_f$, but now at the constant $\bar{V}_1$ (0.75) and continuously varying $\bar{L}_1$. Here the reference value $\bar{t}_{f*}$ corresponds to the value of $\bar{t}_f$ at $\bar{L}_1 = 0$ and is also equal to 123.6. The importance of making the right choice of $\bar{L}_1$ is evident from Fig. 29. By analogy with $\bar{V}_1$, there is also an interval of the values of $\bar{L}_1$, bounded from above by the special value $\bar{L}_{1*}$ (0.52 at l=1 and 0.58 at l=1/3), within which the dual-flow filter is preferable $\left(\Delta \bar{t}_f > 1\right)$. If, in fact, the residual amount of a suspension $\left(\bar{V} - \bar{V}_1 = 0.3\right)$ is injected too deeply into the bed, the filter may initially be incapable of operation due to poor filtrate quality immediately after the start of the process filtration. With the right choice of $\bar{L}_1$ it is also possible to extend the continuous operation of the filter by 45-60%.

Also advantages of using a dual-flow filter and suitable conditions for implementing the unconventional deep bed filtration clearly demonstrates **Figure 30**. This depicts the curves of the dependencies of the three relative technological times $\left(\bar{t}_p, \bar{t}_h, \bar{t}_f\right)$ on the absorption properties of the filter media $(\bar{\alpha}_V)$. The solid lines shows continuous broken curved lines describing the change in $\bar{t}_f$ with increasing $\bar{\alpha}_0$

from the minimum value $\bar{\alpha}_{\min}$ (when the bed is not able to properly clarify already the first portion of a suspension) to 12 (very high absorption capability). The solid curve formed by fragments of curves 3 and 4 and calculated at $\bar{V}_1 = 1$ (single flow filter) is the baseline for comparisons. Similar curves 2 and 5, 3 and 6, corresponding to l=1 and l=1/3, are located close to each other, indicating the insignificance of the suspension particles size for the effect of its separation. Generally, the absolute effect, measured by the ratio between the maximum values of $\bar{t}_f$ at $\bar{V}_1 = 1$ and $\bar{V}_1 = 0.5$, l=1/3 or $\bar{V}_1 = 0.5$, l=1, is minor (about 15%). However, the final result depends significantly on the choice of a filtering material and in a particular situation, the specified ratio may be much higher or lower in the absence of such a choice.

Figure 30. Dependencies $\bar{t}_p(\bar{\alpha}_V)$, $\bar{t}_h(\bar{\alpha}_V)$, $\bar{t}_f(\bar{\alpha}_V)$:

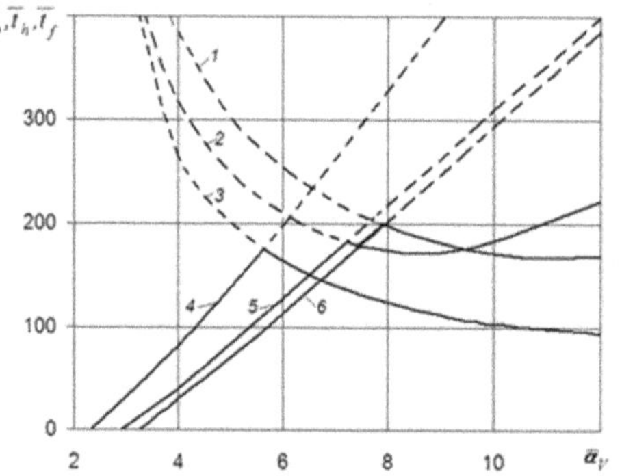

Note. 1-3 – $\bar{t}_h$, 4-6 – $\bar{t}_p$, $\bar{t}_f$ – solid lines; 1, 6 –dual-flow (l=1); 2, 5 – dual-flow (l=1/3); 3, 4 – single-flow

In the case of finely dispersed impurities, a meaningful decrease in the mechanical energy expenditure for filtration process due to a significant improvement in the absorption properties of a bed deserves special attention (curve 2). Starting from about $\bar{\alpha}_V = 8.9$, the contribution to the interfacial mass exchange of the detachment of the trapped suspended solids becomes so significant that the formal increase in this coefficient and the enhanced deposition of an impurity caused by it from the hydraulic point of view is more than compensated by the accompanying intensive deposit mobilization. As a result, the concentration profiles of the deposit and deposited particles are flatten out and, as a consequence, the resulting effect is expressed in a noticeable increase in the time $\bar{t}_h$, and thus, in the accompanying increase in the protective capability of the bed and $\bar{t}_f$ (curve 2).

A similar effect of dividing the hydraulic load V into two parts is also observed when a suspension enters the bed from both sides. Incidentally, it is technically easier to arrange the feeding from the second side than from the inside. The second series of examples was calculated with the same initial data as the first series. The dependences for the relative concentrations $\bar{C}$, $\bar{S}$ and head losses $\Delta\bar{h}$ were involved for the first section ($0 \le z \le L_1$) (3.102) - (3.104) and for the second section $(L_1 < \bar{z} \le L)$ (3.112) - (3.114) respectively. In order to establish the conditions that contribute to the productive operation of the second variety of dual-flow filter and to evaluate the gain therein due to the distribution of a hydraulic load between the upper and lower bed surfaces by analogy with the first series, the duration $\bar{t}_f$ was calculated depending on the position of the (drainage) filtrate outlet (depth of the treated water diversion L_1), as well as on the ratio between the upper (V_1) and lower (V–V_1) flow rates and, finally, on the absorption capability of the filter bed $(\overline{\alpha}_V)$. The mass exchange coefficients were assumed to be independent of the filtration flow direction.

First of all, the sensitivity of the relative duration $\overline{t}_f$ to the position of the filtrate diverted device (drain) was analyzed at equal upper and lower flow rates $(\overline{V}_1 = \overline{V}_2 = 0.5)$. It was found out that the technological process is limited in time only by the protective capability of a bed for coarse impurities, and it is reasonable to place the device in the middle of the filter media $(\overline{L}_1 = 0.5)$. Thus, it is possible to increase $\overline{t}_f$ by 32.7% in comparison with a traditional filtration. In case of finely dispersed impurity, firstly, the effect of splitting hydraulic load is even greater – the maximum value of $\Delta \overline{t}_f$ (1.643) is reached at $\overline{L}_1 = \overline{L}_* = 0.256$, and the filter operation has to be stopped at $\overline{L}_1 < \overline{L}_*$, because of the excessive deterioration of the filtrate quality, and due to the excessive head losses at $\overline{L}_1 > \overline{L}_*$. However, the calculated value of $\Delta \overline{t}_f$ is only slightly less than the maximum value (1.632) even at $\overline{L}_1 = 0.5$. Therefore, the value $\overline{L}_1 = 0.5$ is fixed in subsequent calculations. At the same time, it is inadmissible to set $\overline{L}_1$ too low, which may cause complete non-functionality of the filter. Indeed, it is not difficult to specify such a critical value L_{cr}, that the first portion of a suspension will be insufficiently clarified at $L_1 \le L_{cr}$. Proceeding from dependences (3.102), (3.112) the following equation concerning $\overline{L}_{cr}$ is derived

$$\overline{V}_1 e^{-\overline{\alpha}_V \overline{V}_1^{l-1} \overline{L}_{kp}} + \left(1 - \overline{V}_1\right) e^{-\overline{\alpha}_V (1-\overline{V}_1)^{l-1} (1-\overline{L}_{kp})} = \overline{C}_*$$

Its equivalent form will be

$$\overline{V}_1 e^{-\overline{\alpha}_V [\overline{V}_1^{l-1} + (1-\overline{V}_1)^{l-1}] \overline{L}_{cr}} - \overline{C}_* e^{-\overline{\alpha}_V (1-\overline{V}_1)^{l-1} \overline{L}_{cr}} + \left(1 - \overline{V}_1\right) e^{-\overline{\alpha}_V (1-\overline{V}_1)^{l-1}} = 0 \tag{3.117}$$

Equation (3.117) is easily solved in the general case by the fitting procedure, and the simple formula follows from it in the special case $\overline{V}_1 = \overline{V}_2 = 0.5$

$$\overline{L}_{cr} = -\frac{\ln\left(\overline{C}_* \pm \sqrt{\overline{C}_*^2 - e^{-0.5^{l-1}\overline{\alpha}_V}}\right)}{0.5^{l-1}\overline{\alpha}_V} \tag{3.118}$$

Both roots of equation (3.118) have physical meaning and correspond to two critical positions of the diverter, which are equidistant from the center of the bed and its nearby inlet. Specifically, in the second series of examples, $\overline{L}_{cr}$ equals to 0.1267 and 0.8733, respectively, at l=1/3 and already $\overline{L}_{cr} = 0.2022$ and 0.7978 at l=1. And further attention was emphasized on the relationship between $\overline{t}_f$ and $\overline{V}_1$ or $\overline{\alpha}_V$. Curves describing the excess of $\overline{t}_f$ over the basic value for comparison $\overline{t}_{f*}$ (single-flow filtration with parameters $\overline{V}_1 = \overline{L}_1 = 1$) at $\overline{L}_1 = 0.5$ and changing $\overline{V}_1$ from 0 (upward filtration through a bed with half the thickness) to 1 (downward filtration through a similar bed) are shown in **Figure 31**. Naturally, the reduction of the total bed thickness in both limiting situations $(\overline{V}_1 = 0$ and 1) halving causes a serious (one and a half times) reduction of $\overline{t}_f$ at the same total hydraulic load. If the suspension is fed through the surfaces with equal flow rates, the increment of $\overline{t}_f$ is maximum and will be 32.7% at l=1/3, and it doubles (63.2%) at l=1.

Finally, the technological times $\overline{t}_p$, $\overline{t}_h$, $\overline{t}_f$ were calculated as functions of $\overline{\alpha}_V$ at the fixed values of $\overline{V}_1$ (0.7), $\overline{L}_1$. The two series of plots obtained for the fine and coarse impurities based on (3.102) - (3.104) and (3.112) - (3.114) are shown in **Figure 32**. By analogy with Fig. 30 the solid lines highlight the most important for practice curves of the dependence $\overline{t}_f(\overline{\alpha}_V)$, which are also continuous and have one fracture each. Comparison of the optimum values of $\bar{t}_f$ in three considered cases of dual and single-flow filtration is indicative. A comparison of the peak values of $\bar{t}_f$ shows that the corresponding effect can be estimated at 30%. Also, as in the case of the combination of the external and internal feeding of a suspension into the bed, an increase in the times $\bar{t}_h$, $\bar{t}_f$ is observed here with an increase in $\bar{\alpha}_V$ from the value 7.9. However, the observed even more significant effect should not be overestimated due to the possible inadequacy of linear kinetics of mass exchange during prolonged filtration through a well-absorbing medium.

Figure 31. Dependence $\Delta\bar{t}_f(\bar{V}_l)$

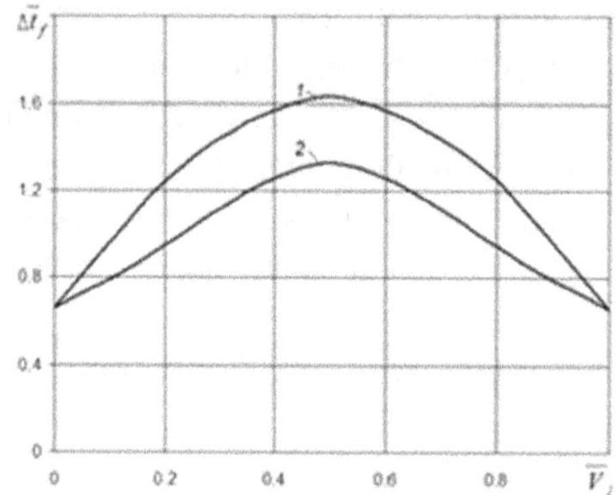

Note. 1 – l=1/3, 2 – l=1

Figure 32. Dependencies $\bar{t}_p(\bar{\alpha}_V)$, $\bar{t}_h(\bar{\alpha}_V)$, $\bar{t}_f(\bar{\alpha}_V)$:

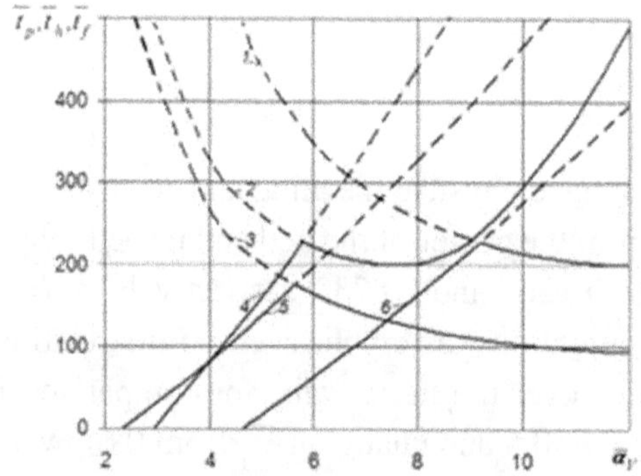

Note. $\bar{t}_f(\bar{\alpha}_V)$: 1-3 – $\bar{t}_h$, 4-6 – $\bar{t}_p$, $\bar{t}_f$ – solid lines; 1, 6 – dual-flow (l=1); 2, 4 – dual-flow (l=1/3); 3, 5 – single-flow

Thus, due to simultaneous feeding of a suspension into the bed in two places, it is possible to significantly intensify the clarification process also in the section of the bed, which in traditional (downward or upward) filtration practically did not take part in it. Thus, firstly, the absorption resource of the entire filter media is more fully implemented, and secondly, the concentration profiles of the deposit and deposited particles are flattened out and the mechanical energy expenditure is reduced accordingly. However, the quality of the filtrate deteriorates faster due to the reduction of the filtration rate and the weakening of the absorption (*l*>0). Therefore, it is necessary to preselect the method of feeding a suspension into the filter bed, based on the results of a joint technological analysis of dual-flow and single-flow filtration. And if a dual-flow filtration is exactly preferable at the available material, then further it is necessary to analyze the efficiency of both its varieties, varying at the same time V_1, L_1.

3.6. ESTIMATION OF SIGNIFICANCE OF FILTER MEDIA TEMPERATURE AND DEPOSIT COMPOSITION

The influence of the media temperature (T_t) and deposit composition (γ) on the clarification effect of water treatment filters deserves special attention. These factors differ significantly in their nature. Nevertheless, from the methodological point of view, they are united, firstly, by the previously cited solid argumentation in favor of the significance of such influence, and secondly, by the absence of even attempts to analyze the sensitivity of the most important characteristics of filtration, technological parameters to them (Gao, et al., 2017; Levenspiel, 1999; Ojha & Graham, 1993). Calculations of the detachment linear filtration with a constant rate on a rapid filter, performed sequentially for the warm or cold seasons, as well as with different techniques of the formalization of the deposit composition (Poliakov, 2009c; 2020), allow us to partially eliminate such an omission. The results of the technological analysis, which was carried out at varying T_t and γ in actual ranges, allow to objectively assess the consequences of inadequate consideration of the seasonal dynamics of the climatic conditions, peculiarities of a deposit formation directly for the control of the filtration facility. Thus, the final conclusions should follow from the established patterns of a change in the technological times t_p, t_h, and especially the duration of the filter run.

The operation of filters depends significantly on the temperature of the environment and, above all, of the suspension being clarified. First of all, the thermal mode in a bed is reflected on the filtration flow and the suspended solids absorption. But since the characteristic time of the thermal processes is much shorter than that of the clarification process, it makes no sense to involve complex heat transfer models. Therefore, the temperature of the operating medium T_t should be considered at the filtration with a constant rate along with d_g, d_p, V as another important (physical) parameter. Indeed, it does not change during many filter runs in practice. However, the suspension temperature has a pronounced seasonal dynamics. Naturally, the values of T_t in the warm season tend to be much higher than in the cold season. As a consequence, the liquid is filtered and the suspended matter is deposited in winter and summer with significantly different rates. And since filters operate all year round, there is a need for the mathematical modeling of the technological process taking into account climatic conditions in order to periodically make adjustments to the algorithm of their control. In this case, in order to avoid, for example, unjustified increase in an impurity content in the filtrate and head losses in a bed, it is advisable to emphasize the periods with extreme temperatures.

Theoretically, the processes of the dispersed contaminants deposition and filtration in the water-saturated destabilized porous medium can be connected with the temperature factor by means of the known formulas for the reduced coefficient of the suspended particles attachment rate, as well as the representation for the hydraulic conductivity as a ratio of the permeability of a medium (does not depend on T_t) and dynamic viscosity of the filtering fluid (significantly depends on T_t). Relying on them and methodological provisions from Section 3.1, in principle it is not difficult to conduct the technological analysis to study the response of the key parameter, namely, the duration of the filter run t_f, to hypothetical changes in the temperature background within a wide range.

The importance of T_t for the filter performance is obvious already by formal considerations due to the presence of the dynamic viscosity of a suspension in all the calculation dependences (in the coefficients α, α_0, k_0). Incidentally, its values for water and low-concentration suspension practically coincide. And it is logical that the determining value for changes in the above coefficients due to the temperature fluctuations of a filter media has the dependence $\mu(T_t)$, which is adequately described for water by the Poiseuille formula

$$\mu(T_t) = \mu_0 f_T(T_t) = \mu_0 (1 + 0.0337\, T_t + 0.00022\, T_t^2)^{-1}, \tag{3.119}$$

where μ_0 is the value of μ at $T_t = 0$ (measured in degrees Celsius).

Absorption of the suspension particles is considered as a result of the simultaneous implementation of three mechanisms of their delivery to the grains-collectors with subsequent obligatory fixation on the surface of the latter (α_c=1, reagent filtration). Then the coefficient α is equal to α_d and can be represented as a sum of the coefficients α_B, α_I, α_G, which correspond respectively to the delivery mechanisms due to Brownian diffusion, interception and sedimentation (Section 2.6). Consequently, according to (2.149) $\alpha_{B,I,G}$ will be directly proportional to the following functions of T_t

$$\alpha_B \sim \left[\frac{\mu(T_t)}{273 + T_t}\right]^{-\frac{2}{3}}, \ \alpha_I \sim [\mu(T_t)]^{-\frac{1}{8}}, \ \alpha_G \sim [\mu(T_t)]^{-\frac{5}{6}}. \tag{3.120}$$

Thus, the relative excess of the standard relative values of the functional coefficients $\bar{\alpha}_B(T_t)$, $\bar{\alpha}_I(T_t)$, $\bar{\alpha}_G(T_t)$ at the temperature increase by T_{t0} taking into account (3.160) will be as follows

$$\frac{\bar{\alpha}_B(T_t + T_{t0})}{\bar{\alpha}_B(T_t)} = \left[\frac{\mu(T_t + T_{t0}) \cdot (273 + T_t)}{\mu(T_t)(273 + T_t + T_{t0})}\right]^{-\frac{2}{3}},$$
$$\frac{\bar{\alpha}_I(T_t + T_{t0})}{\bar{\alpha}_I(T_t)} = \left[\frac{\mu(T_t + T_{t0})}{\mu(T_t)}\right]^{-\frac{1}{8}}, \ \frac{\bar{\alpha}_G(T_t + T_{t0})}{\bar{\alpha}_G(T_t)} = \left[\frac{\mu(T_t + T_{t0})}{\mu(T_t)}\right]^{-\frac{5}{6}}. \tag{3.121}$$

It should be emphasized that the form of the dependencies for both relative concentrations is independent of the thermal state of the filter bed. Therefore, before calculating the functions $\bar{C}_e(\bar{t})$, $\bar{S}(\bar{z}, \bar{t})$, it is sufficient to adapt the initial data to the specific temperature conditions, namely, to set the value of $\bar{\alpha}$ corresponding to them in the framework of linear or nonlinear filtration with a constant rate. Then the corresponding formulas from Chapters 2 or 6 should be used.

When introducing dimensionless variables and coefficients in the filtration flow compartment and deriving the formula for relative head losses (2.24), the coefficient k_0, which also significantly depends on the parameter T_t, was used as one of the scales. If the medium temperature deviates from the normal (standard) temperature, it is necessary to introduce into the above formula instead of the implicitly present coefficient k_0 a correction analog representing the ratio of its values in actual and standard conditions. Then the refined formula for the losses takes the following form

$$\Delta\bar{h}(\bar{t}) = \frac{f_T(T_t + T_{t0})}{f_T(T_t)} \int_0^1 \frac{d\bar{z}}{f_k(\bar{S}(\bar{z}, \bar{t};\, T_t + T_{t0})}. \tag{3.122}$$

To illustrate the above proposed approach to the temperature factor accounting and to assess its significance for the clarification of an aqueous suspension on the rapid filter, the technological analysis was performed using the developed method (Section 3.1). Its subject, first of all, was the relative times $\bar{t}_p$, $\bar{t}_h$ (constant filtration rate), but special attention was paid to the relative duration of the filter run $\bar{t}_f$. The only values of the criterion parameters were assumed to be $\bar{C}_* = 0.1$, $\Delta\bar{h}_* = 8$. The following values were also recorded $\bar{\beta} = 0.01$, $\bar{\gamma}_a = 0.002$. Two series of examples were calculated.

In the first series of examples, the performance of the rapid filter considering the medium temperature was analyzed in two stages. At the first stage suitable values of the relative coefficients $\bar{\alpha}_{B,I,G}$ were calculated at a high temperature (T_f+T_{f0}). The corresponding procedure was implemented step by step and repeated many times. Originally, the first specific value of $\bar{\alpha}_0$, which corresponded to a certain temperature T_f, was accepted from the range selected for it. Then, the values of the coefficients $\bar{\alpha}_{B,I,G}$ (denoted $\bar{\alpha}_{B0}$, $\bar{\alpha}_{I0}$, $\bar{\alpha}_{G0}$), also corresponding to T_f, were formally identified with the accepted value of $\bar{\alpha}_0$. Finally, the values of the same coefficients were found at the temperature T_f+T_{f0} according to (3.121).

The desired times $\bar{t}_{p,h,f}$ were directly determined at the second stage, using the known dependencies $\bar{C}_e(\bar{t})$, $\bar{S}(\bar{z},\bar{t})$, $\Delta\bar{h}(\bar{t})$ at the given $\bar{C}_*$, $\Delta\bar{h}_*$, $\bar{\alpha}_0$ and pre-calculated $\bar{\alpha}_{B,I,G}$ (T_f+T_{f0}). The calculation cycle was repeated at a new value of $\bar{\alpha}_0$, etc. As a result, single-valued correspondences were established between the values of $\bar{\alpha}_0$, running over the interval from 3 to 7, the values of the coefficients $\bar{\alpha}_{B,I,G}$ in the heated medium (T_f+T_{f0}) and the technological times $\bar{t}_{p,h,f}$. **Figure 33** presents in graphical form the dependencies of $\bar{t}_{p,h,f}$ on $\bar{\alpha}_0$ with regard to the different mechanisms of the contaminant delivery to grains. Curves 2 and 8, as well as the continuous broken curve made of their fragments and depicted by a solid line, became fundamental for the estimation of the effect of the medium temperature on the technological process. They were calculated at T_t=5 and did not depend on the physical prerequisites of absorption (delivery mechanisms) due to the systematic equalization of the values of $\bar{\alpha}_0$ and $\bar{\alpha}_{B0}$, $\bar{\alpha}_{I0}$, $\bar{\alpha}_{G0}$, and thus, on the mechanical composition of the impurity. Other curves are characteristic of a high temperature (T_{t0}=10) and differed significantly in the cases of fine and coarse dispersed impurities. All the resulting curves for $\bar{t}_f$ are composed of two fragments (1, 7; 2, 8; 3, 6; 4, 5) and have maxima in the form of a peak at very specific (optimal) values of the arguments $(\bar{\alpha}_0;\ \bar{\alpha}_{B0},\ \bar{\alpha}_{I0},\ \bar{\alpha}_{G0})$. Their deviation from the optimum in both directions inevitably leads to a noticeable reduction in the duration of continuous filtration.

Figure 33. Dependencies $\bar{t}_p(\bar{\alpha}_0)$, $\bar{t}_h(\bar{\alpha}_0)$

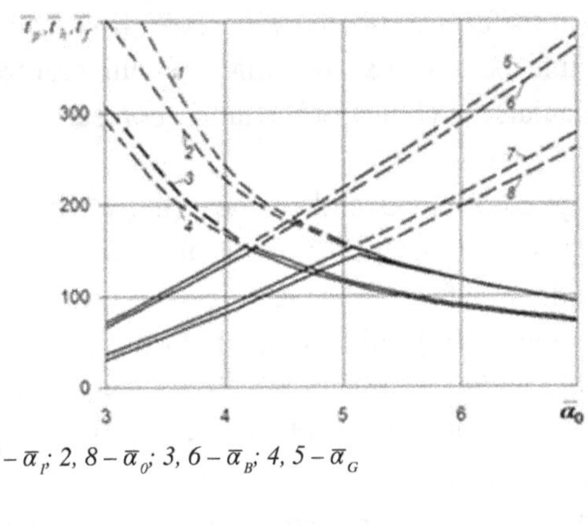

Note. $\bar{t}_f(\bar{\alpha}_0)$: 1-4 – $\bar{t}_h$; 5-8 – $\bar{t}_p$; 1, 7 – $\bar{\alpha}_I$; 2, 8 – $\bar{\alpha}_0$; 3, 6 – $\bar{\alpha}_B$; 4, 5 – $\bar{\alpha}_G$

Figure 34. Dependencies $\bar{t}_p(T_{t0})$, $\bar{t}_h(T_{t0})$, $\bar{t}_f(T_{t0})$

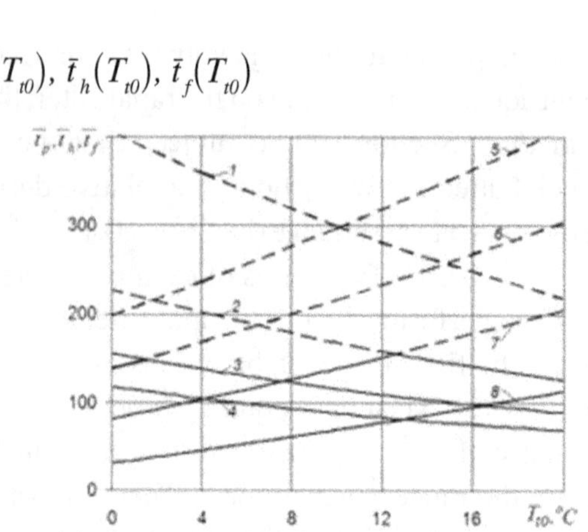

Note. 1-4 – $\bar{t}_h$; 5-8 – $\bar{t}_p$; 1, 8 – $\bar{\alpha}_{BO}$ = 3; 2, 7 – $\bar{\alpha}_{BO}$ = 4; 3, 6 – $\bar{\alpha}_{BO}$ = 5; 4, 5 – $\bar{\alpha}_{BO}$ = 6

Also indicative for analyzing the temperature effect is **Figure 34**. Here are the results of the calculations of the second series of examples, and the desired $\bar{t}_f$ already depends directly on the temperature difference in the warm and cold seasons T_{t0}. In contrast to the first series, now T_{t0} was the argument and varied continuously from 0 to 20 . The suspension contained exclusively Brownian particles. Four typical values were accepted for $\bar{\alpha}_{B0}$ between 3 and 6. The graphs of the dependence $\bar{t}_f(T_{t0})$, depicted by the solid lines, are the continuous broken (2, 7; 3, 6) or smooth (4; 8) curves. Their significant differences in shape and position are explained, first of all, by the change in the limiting factor (filtrate quality, mechanical energy expenditure) as $\bar{\alpha}_{B0}$ increases. In the extreme cases ($\bar{\alpha}_{B0}$ = 3 or 6) the corresponding calculated curves 8 and 4 changed smoothly. In the first case the technological process was limited in time by a low absorption capacity of the bed, and in the second case – by the rapidly increasing head losses. At intermediate values of $\bar{\alpha}_{B0}$ the required $\bar{t}_f$ reached a distinct maximum at specific values of T_{t0*}. Thus, T_{t0*}=1.6 at $\bar{\alpha}_{B0}$ = 4, and T_{t0*}=12.4 at $\bar{\alpha}_{B0}$ = 5. Excessive heating of the packed bed of well-absorbing material (curve 4) or cooling of poorly absorbing material (curve 8) is undesirable, since the duration $\bar{t}_f$ can be significantly reduced.

Summarizing the first part of this Section, it is important to note the significance of the temperature factor in the suspension clarification by filtration. Neglecting it can have serious negative consequences for the filter control during the calendar year or its separate periods such as a reduction in duration of the filter run duration, the increase in the cost of treated water, etc. It is realistic to increase the efficiency of rapid filters taking into account the seasonal dynamics of temperature and properties of impurity, filtering material due to, in particular, the choice of the latter, correction of the control algorithm using the above methodology.

Along with the mass exchange coefficients and permeability of the clogged bed, the choice of the (functional) coefficient of a deposit composition γ is problematic in the preparation of information support for the mathematical modeling of the aqueous suspension filtration. It is necessary to state that the modern level of knowledge about peculiarities of the deposit formation, its composition, properties and, the main thing for modeling, about the function $\gamma(S)$ does not meet the increased requirements, that have recently been placed on the prediction and technological calculations of high-rate filtration. It should be emphasized that the mentioned coefficient (bulk factor) is crucial for the calculation of the filtration flow process, since it determines the amount of a bound water and, in fact, the reduction of the porosity of a clogged bed over time. Indeed, the experimentally established fact of physical binding of a disproportionately large volume of water by the deposited suspension particles means that the actual volume of a deposit can be tens of times greater than the volume of the particles it contains. As a consequence, the deposit density is almost comparable to that of water (Poliakov et al., 2019). Thus, even when filtering low-concentration suspensions, large head losses are actual.

According to the few experimental data, this coefficient (function) depends primarily on the concentration of the deposited particles. Its decrease when increasing the concentration S is explained by the dehydration of an old deposit and weakening of the hydration of the deposited particles during the formation of a new deposit. Thus, the mathematical description of the filtration flow process and the determination of the head within the bed been clogging are complicated, but not fundamentally. However, considerable difficulties in obtaining experimental data on the relationship between γ and S and, as a consequence, the scarcity of already available information on the empirical dependence $\gamma(S)$ often force to introduce a constant γ_a instead of it. If the filtration flow calculations are performed for a given point in time, it is logical to interpret γ_a as an average value of the variable quantity γ during the corresponding time period. It is much more difficult to select the value γ_a in the technological calculations, since the calculation time and the corresponding physical and chemical state of the bed are not known in advance. In general, usually low reliability of the above-mentioned information and its formal coarsening in such cases especially negatively affect the accuracy of forecasts of the increase in the hydraulic resistance of the clogged bed, validity of the recommendations on rational control of a filtration facility.

Below, the estimation of errors in the calculations of the technological times t_h, t_f due to partial or complete averaging of the dependence $\bar{\gamma}(\bar{S})$ is performed with respect to the detachment linear filtration. At the same time, we used our own experimental data from Section 1.6. The results of calculating the relative head losses using exact formula (2.26) in view of (2.27), which allows an arbitrary form for the function $\bar{\gamma}(\bar{S})$, served as a standard for a comparative analysis. In parallel, the specified function was averaged in three ways. First, it was averaged over the argument in the maximum possible range, from 0 to the value $\bar{S}_d$. The relative dynamic solids take-up capacity $\bar{S}_d$ is formally achieved at $t\to\infty$ and is $\bar{\alpha}/\bar{\beta}$ for the detachment linear filtration. Thus,

$$\bar{\gamma}_{a1} = \frac{\bar{\beta}}{\bar{\alpha}}\int_0^{\bar{\alpha}/\bar{\beta}} \bar{\gamma}(\bar{S})d\bar{S} \tag{3.123}$$

Second, the quantity $\bar{\gamma}$ was averaged only over the height of a bed and hence the corresponding quantity $\bar{\gamma}_{a2}$ was already a function of time, i.e.

$$\bar{\gamma}_{a2}(\bar{t}) = \int_0^1 \bar{\gamma}(\bar{S}(\bar{z},\bar{t}))d\bar{z}. \tag{3.124}$$

Finally, the average value of the function $\bar{\gamma}(\bar{S})$ was found by integrating it over both height and time during the calculation period, the duration of which T is not specified. Thus,

$$\bar{\gamma}_{a3}(T) = \frac{1}{T}\int_0^T\int_0^1 \bar{\gamma}(\bar{S}(\bar{z},\bar{t}))d\bar{z}d\bar{t}. \tag{3.125}$$

Here T is a parameter. Since the calculation time (filter shutdown time) is initially unknown, we had to sequentially and alternately adjust $\overline{\gamma}_{a3}$ and $\overline{t}_h$ when calculating the technological times.

In order to evaluate the significance of the empirical information on a deposit composition for the technological analysis of a rapid filter operation, two series of examples were calculated with fixed values of the model coefficients $\overline{\beta}$ (0.01), $C_0(5 \cdot 10^{-5})$, $\overline{C}_*$ (0.1), $\Delta \overline{h}_*$ (6). The subject of many calculations in relation to the conditions of the detachment filtration with a constant rate has become the relative technological time $\overline{t}_h$. Therefore, the expressions were set for the functional coefficients f_k and $\overline{\gamma}$ contained in the filtration flow compartment. They were obtained experimentally and are given in Section 1.5. Thus, the dimensionless analogs of expressions (1.60), (1.63) are involved in the calculations

$$\overline{\gamma}(\overline{S}) = \frac{57\ C_0}{1 + 67\ C_0\ \overline{S}},\ f_k(\overline{S}) = \left(\frac{1 + 10\ C_0\ \overline{S}}{1 + 67\ C_0\ \overline{S}}\right)^3. \tag{3.126}$$

The time $\overline{t}_h$ variations are caused by varying the absorption properties of the filter media, namely, the range [3, 7] was chosen for the coefficient $\overline{\alpha}$ in the examples of the first series. The procedure for calculating the value of $\overline{t}_h$ was based on the expressions for $\overline{S}$ (2.17), $\Delta\overline{h}$ (2.24) and consisted in searching the values of $\overline{t}_h$ until the criterion equation (1.36) was satisfied. The time $\overline{t}_h$ was also determined at the maximum $\overline{\gamma}_{max} = \overline{\gamma}(0) = 2.85 \cdot 10^{-3}$ and minimum $\overline{\gamma}_{min} = \overline{\gamma}(\overline{S}_0)$ values of $\overline{\gamma}$ in addition to the calculations at $\overline{\gamma}(\overline{S})$ according to (3.126), at $\overline{\gamma}_{a1}$, $\overline{\gamma}_{a2}$, $\overline{\gamma}_{a3}$ according to (3.123), (3.124), (3.125). All curves describing the dependence $\overline{t}_h(\overline{\alpha})$ for the listed alternatives are presented in **Figure 35**. From the comparison of the calculated curves, we can immediately conclude that it is inadmissible to neglect the deposit dehydration and thus, reduce the deposit volume in the technological calculations. Indeed, the dehydration of a deposit occurs mainly in the inlet section of a bed, which is the most important for the filtration flow process. It is here that mechanical energy is mainly lost due to the significant clogging of the pore space. Due to intensive dewatering of the deposit, the increase in the head losses in this section and in the entire filter media can be considerably slowed down and the time $\overline{t}_h$ can be increased to the same extent. The values of $\overline{\gamma}_a$, as well as $\overline{\gamma}_{max}$, adopted as a result of averaging $\overline{\gamma}(\overline{S})$ by three techniques caused a significant downward deviation of the approximate calculated curves 3-6 from the reference curve 2. Thus, such simplifications of the quantity $\overline{\gamma}(\overline{S})$ led to a significant underestimation of the time $\overline{t}_h$. At the same time, there was already a noticeable overestimation of $\overline{t}_h$ (curve 1) due to the formal replacement of $\overline{S}$ in the expression for $\overline{\gamma}$ by the largest quantity under the considered conditions $\overline{S}_0$.

The second series of examples was calculated to evaluate the consequences of varying the deposit composition for the time $\overline{t}_h$. A special coefficient a_γ was introduced into the expression for $\overline{\gamma}$ (3.126) as follows

$$\overline{\gamma}\left(\overline{S}; a_\gamma\right) = \frac{57\ C_0}{1 + a_\gamma C_0 \overline{S}}. \tag{3.127}$$

It can be interpreted as the sensitivity of the ratio between the content of the bound water and the deposited solids in the deposit with respect to changes in the second component. If we formally consider $\overline{\gamma}$ as a function of a_γ and allow the argument to vary over a semi-constrained interval from 0 to ∞, then, firstly, it is not difficult to specify the first critical value $a_{\gamma*}$ such that the inlet cross-section is completely blocked at $a_{\gamma*} > a_\gamma \geq 0$ by the time $\overline{t}_b$ and, consequently, filtration stops. The specified time is exactly calculated by the formula

$$\bar{t}_b(a_\gamma) = -\frac{1}{\beta}\ln\left[1 - \frac{\beta}{\bar{\alpha}\, C_0(57 - a_\gamma)}\right]. \tag{3.128}$$

Figure 35. Dependence $\bar{t}_h(\bar{\alpha})$

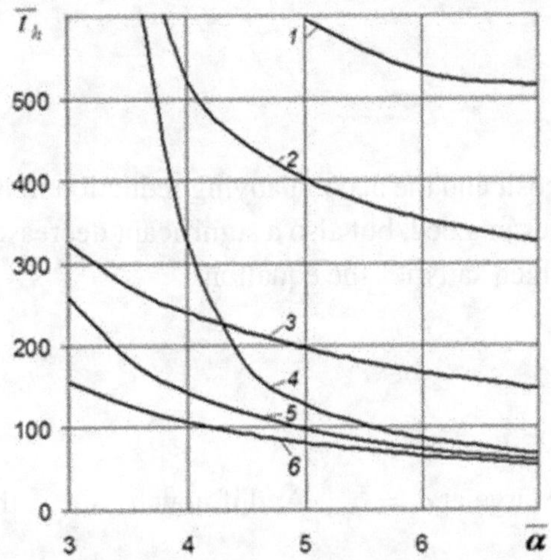

Note. 1 – $\bar{\gamma} = \bar{\gamma}_{min}$, 2 – reference, 3 – technique 1, 4 – technique 2, 5 – technique 3, 6 – $\bar{\gamma} = \bar{\gamma}_{max}$

Figure 36. Dependences $\bar{t}_h(a_\gamma)$, $\bar{t}_p(a_\gamma)$

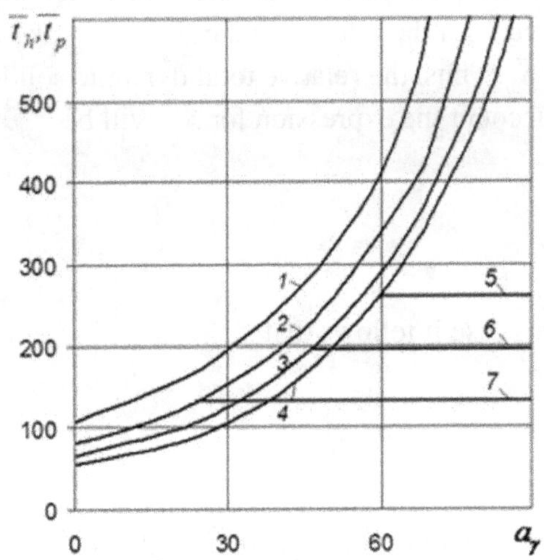

Note. 1-4 – $\bar{t}_h$, 5-7 – $\bar{t}_p$; 1 – $\bar{\alpha}$ = 4; 2, 7 – $\bar{\alpha}$ = 5; 3, 6 – $\bar{\alpha}$ = 6; 4, 5 – $\bar{\alpha}$ = 7

On the other hand, the corresponding value a_γ should be found for a given value $\bar{t}_b$ by the formula

$$a_\gamma(\bar{t}_b) = 57 - \frac{\bar{\beta}}{\bar{\alpha}\, C_0(57 - e^{-\bar{\beta}\bar{t}_b})}. \qquad (3.129)$$

Obviously, the required time $\bar{t}_h$ will be somewhat less than $\bar{t}_b$. In the case $a_\gamma \geq a_{\gamma*}$ the blockage of the inlet pores under the considered conditions is in principle impossible. Then the time $\bar{t}_h$ is found depending on a_γ by fitting procedure from the equation

$$\int_0^1 \frac{d\bar{z}}{f_k\left(\bar{S}(\bar{z}, \bar{t}_h), a_\gamma\right)} = \Delta \bar{h}_*. \qquad (3.130)$$

Severe dewatering of the deposit and the accompanying reduction in its volume can cause not only a slowed increase in the head losses in a bed, but also a significant decrease in their limiting value. There is a second critical value $a_{\gamma**}$, which satisfies the equation

$$\int_0^1 \frac{d\bar{z}}{f_k\left(\bar{S}(\bar{z}, \infty), a_{\gamma**}\right)} = \Delta \bar{h}_*. \qquad (3.131)$$

Hence, $\bar{t}_h$ becomes infinitely large at $a_\gamma = a_{\gamma**}$. And if now $a_\gamma > a_{\gamma**}$, then the inequality is true

$$\int_0^1 \frac{d\bar{z}}{f_k\left(\bar{S}(\bar{z}, \infty), a_{\gamma**}\right)} < \Delta \bar{h}_*, \qquad (3.132)$$

that is, the head losses in the bed will always be less than the maximum allowable head losses due to the rapid reduction of the deposit volume by the dehydration. The second limiting case is precisely the case if the absorption resource of the filter material is being exhausted infinitely long ($t \rightarrow \infty$). Then the steady-state value $\bar{S}_\infty$ is equalized with the relative dynamic solids take-up capacity (in relation to the deposited suspension particles) $\bar{S}_d$. Thus, the relative total dynamic solids take-up capacity (in relation to the deposit) $\bar{S}_{Sd}$, taking into account the expression for $\bar{S}_d$, will be

$$\bar{S}_{Sd} = \frac{\bar{\gamma}(\bar{S}_d)\bar{S}_d}{C_0} = \frac{\bar{\alpha}}{\bar{\beta} + \bar{\alpha}\, a_\gamma C_0} \qquad (3.133)$$

From equation (3.130) for this case it follows that

$$\bar{S}_{Sd}(\Delta \bar{h}_*) = \frac{1}{C_0}\left(1 - \frac{1}{\sqrt[3]{\Delta \bar{h}_*}}\right) \qquad (3.134)$$

Based on equations (3.133) and (3.134), the second critical value $a_{\gamma**}$ is related to $\Delta \bar{h}_*$ as follows

$$a_{\gamma**} = \frac{1}{C_0}\left(\frac{b_\gamma}{\bar{S}_{Sd}(\Delta \bar{h}_*)} - \frac{\bar{\beta}}{\bar{\alpha}}\right)$$

Therefore, the mechanical energy expenditure for deep bed filtration will always be within the allowable limits due to the intensive dehydration of the deposit with known $\Delta \bar{h}_*$ and at $a_\gamma > a_{\gamma**}$. Specifically, the values $a_{\gamma*}$=17, $a_{\gamma**}$=90, $\bar{t}_b(0) = 121$ are obtained for the given initial data and, for example, $\bar{\alpha} = 5$.

The results of the calculation of the relative time $\bar{t}_h$ by (3.126), (3.130) as a function of the coefficient a_γ at discrete change in $\bar{\alpha}$ are shown as a series of curves in **Figure 36**. It should be noted the pronounced nonlinearity of these calculated curves and the accelerated increase in the quantity $\bar{t}_h$ with increasing a_γ. The starting points for these results are the values of $\bar{t}_h$, corresponding to a deposit with stable composition, which means a_γ=0 and in this case also $\bar{\gamma} = \bar{\gamma}_{max} = 57\,C_0$. Naturally, these values appear to be minimal and larger at lower values of $\bar{\alpha}$. Increasing dehydration causes a nonlinear increase in the calculation time up to ∞ when $a_\gamma \to a_{**}$. Thus, a reasonable choice of this coefficient is particularly important when conditions are unfavorable for the water binding and favorable for the deposit dehydration. The influence of the absorption properties of the bed is not so significant, although it is clearly evident. Curves of the dependence $\bar{t}_h(a_\gamma)$ are supplemented with pieces of horizontal straight lines (5, 6, 7) due to the independence of the second technological time ($\bar{t}_h$) from the coefficient a_γ. The low position of the pieces relative to the curves indicates an excessively fast rise of the suspended solids content in the filtrate regardless of an amount of the bound water and its loss. As a consequence, it is the quality of the filtrate that limits the filter operation at weak absorption capacity of the filtering material. However, with increasing $\bar{\alpha}$ at lower values of a_γ the effects of the hydration of the matter being deposited and the dehydration of the deposit become important, and criterion (1.31) becomes determinant. Graphs of the dependence of the relative duration $\bar{t}_f$ on a_γ represent combinations of the straight line segments mentioned and located below the pieces of the corresponding curves. In fact, these graphs are represented by continuous broken curves, and the fractures on them indicate a change in the nature of the technological process limitation when varying the indicators of (de)hydration. In principle, three situations are possible with the limitation of the technological process at a continuous variation of the coefficient a_γ within the given limits. In the first situation, one limiting factor has time to be replaced by another at some intermediate value of the specified coefficient. Then the graph of the dependence of $\bar{t}_f$ on a_γ appears as a continuous line with one fracture just at the given value of the argument (2 and 7, 3 and 6, 4 and 5). In the other two situations, only one limiting factor is decisive for the filter control. Now the graph for $\bar{t}_f$ is identified as a smooth line, which will be based on the nature of the imposed constraint either horizontal or a curve of the dependence of $\bar{t}_h$ on a_γ.

In general, the condition of the inlet cross-section and the adjacent bed section deserves special attention, as it is here that the dispersed impurities are mainly trapped. Nevertheless, in-depth theoretical studies (at the microlevel) of the aqueous suspension filtration through heavily clogging filtering material are hardly expedient due to a complex and unpredictable configuration of significantly reduced pore space, hydrodynamic pattern in it. Part of it turns out to be difficult and even inaccessible for both liquid and deposit due to the formation of stagnant zones, dead-end pores, an increase in flow velocity and hydrodynamic force generated by it. The pore space has a regular structure in the clean granular bed and is formed by alternating and interconnected local expansions (voids) and constrictions (pore necks). To stop the suspension filtration through such a layer, in principle only a relatively small amount of a deposit is sufficient to completely block the pore necks of the inlet pores. However, it is in such places that the flow velocity and deposit detachment increase accordingly, thus preventing ultimate clogging (but not blockage) of the inlet cross-section of a filter bed. In engineering calculations, it is only realistic to take these factors into account through a physically based choice of the functions $f_k(S)$, $\gamma(S)$. Therefore, generalizations concerning their form, for example, representations (1.10), (3.127), should be treated critically. In each particular case, when establishing these functions, it is necessary to strive for the adequacy of their description of the experimental data. In each specific case when establishing these functions it is necessary to strive for the adequacy of their description of the experimental data. And at long continuous operation of filtration facilities of great importance acquire capacitive parameters, which characterize the potential capability of a filtering material in general or to accumulate the suspension particles (true solids take-up capacity S_m), or to hold a certain amount (volume) of them for a long time in the conditions of intensive filtration flow and the interphase mass transfer (dynamic solids take-up capacity S_d). Therefore, the first thing to be monitored is the accumulation of a deposit over time in the inlet cross-section of the bed. And then two characteristic cases are possible at long filtration, which became the subject of a detailed discussion in the monograph. In the first case, kinetics of mass exchange becomes nonlinear and we have to use the results of analytical studies presented in Chapter 6. In the second case, the linearity of filtration is preserved, but a zone of maximum saturation with deposit is formed. The results of such studies are described in Section 3.5. In reality, it is always S_s less than n_0, and the permeability of a packed bed cannot fall to 0 even under enhanced clogging (Mintz & Melzer, 1970). Therefore, the reasoning concerning critical values of a_γ and $\bar{t}_b$ is of interest only from the methodological point of view. In the presence of coarse-grained fraction in the impurity, blockage of the inlet small pores is also possible, which weakens the deep bed filtration and promotes the formation of a dynamic layer on the bed surface. Corresponding theoretical studies are carried out in Section 4.3.

REFERENCES

Adelman, M. J., Monroe, L., Weber-Shirk, M. L., Cordero, A. N., Coffey, S. L., Maher, W. J., Dylan Guelig, D., Jeffrey, C., Will, J. C., Stodter, S. C., Hurst, M. W., & Lion, L. W. (2012). Stacked filters: Novel approach to rapid sand filtration. *Journal of Environmental Engineering*, *138*(10), 999–1008. DOI: 10.1061/(ASCE)EE.1943-7870.0000562

Ayukaev, R. I., & Meltser, V. Z. (1985). *Production and application of filtering materials for water purification. Reference manual*. Stroyizdat. (in Russian)

Caliskaner, O., & Tchobanoglous, G. (2005). Modeling depth filtration of activated sludge effluent using a compressible medium filter water. *Environmental Research*, *77*(7), 3080–3091. PMID: 16381157

Ergun, S. (1952). Fluid flow through packed columns. *Chemical Engineering Progress*, *48*(2), 89–94.

Gao, M., Lv, M., Gong, X., Song, W., Qi, Y., & Zhen, L. (2017). Effect of temperature and aging on the adsorption and desorption of menadione in soil. *Environmental Progress & Sustainable Energy*, *36*(4), 997–1004. DOI: 10.1002/ep.12540

Gao, P., Xue, G., Song, X., & Lin, Z. (2012). Depth filtration using novel fiber-ball filter media for the treatment of high-turbicity surface water. *Separation and Purification Technology*, *95*, 32–38. DOI: 10.1016/j.seppur.2012.04.010

Hutchison, W. R. (1976). High-rate direct filtration. *Journal - American Water Works Association*, *68*(6), 292–298. DOI: 10.1002/j.1551-8833.1976.tb02412.x

Jimenez, B., & Buitron, G. (1990). Comparison between three secondary effluents in tertiary high rate filtration. *Environmental Technology*, *117*(9), 987–995. DOI: 10.1080/09593331708616467

Lerch, A., Nahrstedt, A., Gimbel, R., Gray, S., & Booker, N. (2002). On the development of a high-rate deep bed filtration unit. *Water Science and Technology: Water Supply*, *2*(1), 259–269. DOI: 10.2166/ws.2002.0031

Levenspiel, O. (1999). *Chemical reaction engineering* (3rd ed.). Wiley.

Logan, B. E., Jewett, D. G., Arnold, D. G., Bouwer, E. J., & O'Melia, C. R. (1995). Clarification of clean-bed filtration models. *Journal of Environmental Engineering*, *12*(12), 869–873. DOI: 10.1061/(ASCE)0733-9372(1995)121:12(869)

Mints, D. M., & Meltser, V. Z. (1970). Hydraulic resistance of granular porous medium in the process of clogging. *Rep.* [in Russian]. *USSR Academy of Sciences*, *192*(2), 304–306.

Mitin, B. A. *Research of influence of structural and mechanical properties of deposit on operation of granular clarifying filters*. Ph.D. dissertation abstract. LECI Press. (in Russian)

Niu, S., Park, K., Yu, J., & Kim, Y. (2016). Operation and performance evaluation of high-speed filter using porous non-woven filamentous fibre for the treatment of turbid water. *Environmental Technology*, *37*(5), 577–589. DOI: 10.1080/09593330.2015.1075596 PMID: 26209268

Ojha, C. S., & Graham, N. J. D. (1993). Theoretical estimates of bulk specific deposit in deep-bed filters. *Water Research*, *27*(3), 377–387. DOI: 10.1016/0043-1354(93)90037-I

Poliakov, V., Kravchuk, A., Kochetov, G., & Kravchuk, O. (2019). Clarification of aqueous suspensions with a high content of suspended solids in rapid sand filters. *Eureka: Physics and Engineering*, *1*(1), 28–45. DOI: 10.21303/2461-4262.2019.00827

Poliakov, V.L. (2007). Modeling of suspensions clarification in granular beds at nonlinear filtration flow. *Municipal economy of cities. Collected papers*, 74, 77– 86. (in Russian)

Poliakov, V. L. (2009). Theoretical analysis of filter run duration. [in Russian]. *Journal of Water Chemistry and Technology*, *31*(6), 605–618.

Poliakov, V.L. (2009). On the influence of grain sorting during filter bed operation on its efficiency. *Problems of Water Supply, Sewage and Hydraulics. Collected papers*, 12, 103–118. (in Russian)

Poliakov, V.L. (2009). About mathematical modeling and engineering calculations of suspension clarification by filtration taking into account the deposition structure. *Problems of Water Supply, Sewage and Hydraulics. Collected papers*, 13, 91–102. (in Russian)

Poliakov, V.L. (2010). About influence of bed height and filtration rate on suspension clarification. *Problems of Water Supply, Sewage and Hydraulics. Collected papers*, 15, 131–141. (in Russian)

Poliakov, V. L. (2020). Theoretical analysis of the influence of the seasonal temperature changes on aqueous suspension filtration. In: *Proceedings of International Scientific and Practical Conference "Energoeffektivne misto XXI stolittya"*, 42–47. (in Russian)

Rudzskiy, G. G., & Kim, A. N. (1983). Application of superfast filtration method for additional treatment of waste water. [in Russian]. *Chemistry and Water Technology*, *5*(1), 65–67.

Schneider, O. D., Nickols, D., VandeVenter, L., & Leggiero, S. (2002). An evaluation of high rate filtration using particle counting and turbidimetry for low turbidity waters. *Water Science and Technology: Water Supply*, *2*(1), 249–258. DOI: 10.2166/ws.2002.0030

Shevchuk, E. A., Mamchenko, A. V., & Goncharuk, V. V. (2005). Technology of direct flow filtration of natural and waste water through granular loads. [in Russian]. *Journal of Water Chemistry and Technology*, *27*(4), 369–383.

Stevenson, D. G. (1995). Process conditions for the backwashing of filter with simultaneous air and water. *Water Research*, *29*(11), 2594–2597. DOI: 10.1016/0043-1354(95)00041-I

Yaroshevskaya, N. V., & Muravyev, V. R. (2002). Comparative evaluation of water purification filters loading. [in Russian]. *Journal of Water Chemistry and Technology*, *24*(1), 71–79.

Zapolsky, A. K., Mishkova-Klimenko, N. A., Astrelin, I. M., Bryk, M. G., Gvozdyak, P. I., & Knyazkova, T. V. (2000). *Physical and chemical bases of wastewater treatment technologies: Textbook*. Libra. (in Ukrainian)

Chapter 4
Non-Detachment Filtration

ABSTRACT

Rigorous results of a theoretical study of the non-detachment filtration with nonlinear kinetics of suspension absorption and the formation of a gel-like deposit are obtained in Chapter 4. Much attention is paid to the technological aspects of filters functioning under the conditions of intensive accumulation of the non-consolidated and consolidated deposit during the sequence of filter runs to substantiate the service life of the successive replacement of the filter medium material. The technique of the filtration intensification due to a change in the filtration flow during the filter run is proposed and substantiated. The patterns of the suspension filtration with the formation of a surface layer on the packed bed surface under constant rate or pressure drop are established. Theoretical estimation of the effect of the filtration rate (varying many times) on mass transfer and hydraulic resistance of the clogged bed is performed as applied to detachment (Section 3.4) and non-detachment (Section 4.4) filtration.

Mass exchange during the aqueous suspensions filtration on rapid filters is often multidirectional due to the developed significant filtration hydrodynamic force, so that kinetic equations in general case have the form (1.13). At the same time, at a high absorption capacity of filtering material it can be (almost) one-sided and essentially nonlinear. Thus, the suspended solids are actively and firmly fixed on the elements of a packed bed due to the absorption (adhesion) process together with a large amount of bound water and further successfully resist the filtration flow. Then filtration is actually a non-detachable physicochemical process and it is necessary to assume $f_\beta=0$ in equation (1.13). Sometimes it is allowed to use kinetics and models of the non-detachment filtration and theoretical dependences obtained on their basis also in the presence of an insignificant deposit mobilization.

In the simplest case, the kinetics of impurity deposition is described by a linear equation (Iwasaki, 1937). However, the process of the firm fixation of the dispersed contaminants (irreversible physical sorption applied to the dissolved contaminants) depends on many factors and, first of all, on the concentration of a deposit. Therefore, in order to achieve a plausible description of the process of the contaminant absorption, it is necessary to adopt complex approximation expressions for the single mass exchange coefficient λ. As a rule, it is assumed that the filtration coefficient λ is a function of the concentrations S, S_S and its form should adequately reflect the specific impact of different factors during prolonged filtration.

An exact solution of the corresponding mathematical problem was obtained in Section 4.1 for an arbitrary form of the functional filtration coefficient $\lambda(S)$, which made it possible to take into account such nonlinear effects as the filter ripening (autocatalysis), the limited absorption resource of the filtering material and the bed as a whole. Specific patterns of a change in desired characteristics, inherent exactly

DOI: 10.4018/979-8-3693-8211-0.ch004

to the non-detachment type of the filtration process, are revealed at performance of a considerable body of the predictive calculations. The equations intended for the direct calculation of the technological times are presented in a generalized form and for some special cases. The sensitivity of these times is analyzed in relation to the model parameters characterizing the potential capability of a bed and deposit to retain the suspended matter, also the level of the initial contamination of an aqueous suspension.

In view of the regular use in practice only a small part of the absorption capacity of the filter media with predominant deposition of the suspended solids at its inlet section, we studied the non-detachment suspension filtration with a change in the direction of the filtration flow during the filter run. Thus, at the implementation of initially downward filtration, a suspension should be fed into the bed already through its lower surface at the second stage. Quantitative estimation of the results achieved due to a significantly more even distribution of the deposit in the operating volume of the filter has been carried out. They are expressed in a serious reduction in the rate of the head losses increase in the bed being clogged with minimal deterioration of the filtrate quality. In principle, it is possible to noticeably increase the time of continuous effective operation of the filter in this way.

The effectiveness of the surface filtration of essentially heterogeneous suspension was analyzed in detail, which promotes the formation and development of a dynamic filtering layer on the retention surface with a fine perforation. When formulating the corresponding mathematical problem, the mobility of the inlet surface of this layer was taken into account. Two main rate modes of filtration, namely, with a constant rate and at a constant pressure (head) differential at the boundaries of the newly formed incompressible layer, are considered sequentially. The derived dependences are used in the predictive and technological calculations for specific various conditions of filtration.

By analogy with Section 3.4, the role of a hydraulic load in the non-detachment filtration and the possibility to increase the filter productivity by increasing the load were studied. Based on a rigorous solution of the corresponding mathematical problem (Section 4.1), a number of formulas were obtained for calculating the critical parameters and, in particular, the rate V_m, when the filter is initially unable to separate the suspension in the required manner. It should be emphasized that the calculation equations are derived in which the variable filtration rate is already explicitly present and which allow to determine the technological times as functions of it. Using realistic examples, the significance of this rate and parameters characterizing its relationship with the rate of the irreversible immobilization of the suspended matter is studied for the technological process and its parameters.

Finally, the operation of a rapid filter was analyzed in detail for the duration of the service life of a filter media change under particularly unfavorable conditions, when the amount of a non-washable deposit gradually and significantly increases in the sequence of filter runs. Two control algorithms providing different approaches to the determination of the filter run durations with the consolidation criterion are consecutively considered. Suggestions are made for their rational application.

When using the equations of the kinetics of the non-detachment filtration, the choice of a generalized mass exchange rate coefficient (effective filtration coefficient) is problematic in the case of the detachment filtration. Then, in fact, a single coefficient should simultaneously control both components of the interfacial mass exchange, the peculiarities of which differ significantly. From the formal point of view, analytical methods applied to the models of the non-detachment filtration have essentially greater possibilities. And first of all, it was possible to obtain exact solutions for a class of the nonlinear filtration problems reflecting a variety of properties of the filtering material, dispersed impurity, and technological conditions.

4.1. CALCULATION OF FILTRATION WITH CONSTANT RATE

The formulation of the mathematical problem of the aqueous suspension clarification now in the case of the non-detachment filtration duplicates the formulation of the basic problem from Section 2.1 except for the kinetic equation. This equation, unlike the linear equation (2.2), is nonlinear and is initially taken in the following general form

$$\frac{\partial S}{\partial t} = \lambda(S) \cdot C \tag{4.1}$$

where the functional filtration coefficient $\lambda(S)$ can have a very complicated form due to complex consideration of the factors that limit and intensify the technological process. At first, the function $\lambda(S)$ and the calculation dependencies and equations resulting from the exact solution keep a general form in the theoretical part of this section. And they are specified and then used in numerous test examples only at the end of this part. A traditional representation is chosen for $\lambda(S)$, which formally reflects well-known facts about the deterioration of the suspended solids absorption due to a reduction of pore space and, at the same time, its intensification as the deposit accumulates. After introducing dimensionless variables and coefficiens, the form of equation (4.1) remains the same, namely,

$$\frac{\partial \bar{S}}{\partial \bar{t}} = \bar{\lambda}(\bar{S})\bar{C} \tag{4.2}$$

where $\bar{\lambda} = L\lambda/V$.

The system of equations (2.30), (4.2) reduces to one second-order partial derivative equation with respect to the concentration $\bar{S}$, integration of which within the range from 0 to $\bar{t}$ gives

$$\frac{1}{\bar{\lambda}(\bar{S})}\frac{\partial \bar{S}}{\partial \bar{z}} + \bar{S} = \frac{1}{\bar{\lambda}(\bar{S}^0)}\frac{d\bar{S}^0}{d\bar{z}} + \bar{S}^0 = G(\bar{z}). \tag{4.3}$$

Equation (4.3) should be solved under the boundary condition

$$\bar{z} = 0, \quad \bar{S} = \bar{S}_0(\bar{t}), \tag{4.4}$$

where the relative inlet concentration $\bar{S}_0$ must be determined in advance. From the kinetic equation (4.2), the general relationship between $\bar{t}$ and $\bar{S}_0$ is simply derived taking into account condition (2.11), viz,

$$\bar{t} = \int_{\bar{S}_0^0}^{\bar{S}_0} \frac{d\varsigma}{\bar{\lambda}(\varsigma)}, \tag{4.5}$$

where $\bar{S}_0^0$ is the initial value of the concentration of the deposited suspension particles at the inlet cross-section.

The analytical solution of problem (4.4), (4.5) can be obtained for an arbitrary function $\bar{\lambda}(\bar{S})$ only if $\bar{S}^0 = G = const$. Then it is represented as the inverse integral function

$$\bar{z} = \int_{\bar{S}}^{\bar{S}_0(\bar{t})} \frac{d\varsigma}{(\varsigma - \bar{S}^0) \cdot \bar{\lambda}(\varsigma)}. \tag{4.6}$$

The changes in the relative suspended solids concentration over time and along the bed height are obtained from the solution of the equation

$$\frac{1}{\bar{C}}\frac{\partial \bar{C}}{\partial \bar{z}} + \bar{\lambda}(\bar{S}) = 0$$

under condition (2.11). After integration of this equation and a number of transformations, the relation between $\bar{C}$ and $\bar{S}$ is obtained in the following form

$$\bar{C}(\bar{z},\bar{t}) = \frac{\bar{S}(\bar{z},\bar{t}) - \bar{S}^0}{\bar{S}_0(\bar{t}) - \bar{S}^0}. \tag{4.7}$$

It follows from (4.7) that after the contamination front passes through a bed at the initial stage of filtration, the suspended solids will be distributed exponentially along height, namely,

$$\bar{C}^0(\bar{z}) = e^{-\bar{\lambda}(\bar{S}^0)\bar{z}}, \tag{4.8}$$

so that its concentration in the first filtrate portion will be

$$\bar{C}^0(1) = e^{-\bar{\lambda}(\bar{S}^0)}. \tag{4.9}$$

Formula (4.9) is useful in that it allows us to estimate the operability of the filter in principle.
The relative head losses $\Delta\bar{h}$ should be determined according to (2.13) and (2.15) by the formula

$$\Delta\bar{h}(\bar{t}) = \int_0^1 \frac{d\bar{z}}{f_k(\bar{S}(\bar{z},\bar{t}))}. \tag{4.10}$$

However, due to the implicit form of the solution of the suspension clarification compartment (2.9), (2.11), (4.2), formula (4.10) is transformed to a more convenient for the engineering calculations form

$$\Delta\bar{h}(\bar{t}) = \int_{\bar{S}_e(\bar{t})}^{\bar{S}_0(\bar{t})} \frac{d\varsigma}{(\varsigma - \bar{S}^0)\bar{\lambda}(\varsigma) f_k(\varsigma)}, \tag{4.11}$$

where the relative outlet ($\bar{z} = 1$) concentration of the deposited particles $\bar{S}_e$ must be found as a function of $\bar{t}$ from the equation

$$\int_{\bar{S}_e}^{\bar{S}_0(\bar{t})} \frac{d\varsigma}{(\varsigma - \bar{S}^0)\bar{\lambda}(\varsigma)} = 1. \tag{4.12}$$

For specific calculations it is necessary to give the empirical functions $\bar{\lambda}(\bar{S})$ and $f_k(\bar{S})$.The reliability of the results of modeling of the non-detachment filtration of the aqueous suspensions largely depends on a choice of the approximation function $\lambda(S)$. Recommendations concerning its form are contained, for example, in the works (Alekseev & Kommunar, 1974; Chang et al., 2009; Skouras et al., 2011; Tan, 1994; Tufenkji & Elimelech, 2004). According to (Boller & Kavanauch, 1995; Ojha & Graham, 1992), the best approximation of the calculated data to the experimental data is achieved with the following complex form of $\lambda(S)$

$$\lambda(S) = \bar{\lambda}_0 (1 + a_{\lambda 1} S)^{m_{\lambda 1}} (1 + a_{\lambda 2} S)^{m_{\lambda 2}} (1 + a_{\lambda 3} S)^{m_{\lambda 3}} \tag{4.13}$$

where $a_{\lambda i}$, $m_{\lambda i}$ (i=1,2,3) are the empirical constants. At the same time, it is realistic to achieve an acceptable correspondence between theoretical and experimental data, if we take the simplified dimensionless expression for the specified function

$$\bar{\lambda}(\bar{S}) = \bar{\lambda}_0(1 + \theta \bar{S})(\bar{S}_m - \bar{S}) \tag{4.14}$$

and then to choose a suitable value for the relative reduced coefficient $\bar{\lambda}_0$. Here θ is the autocatalytic coefficient; $\bar{S}_m = S_m/(n_0 C_0)$, S_m is the solids take-up capacity of a filtering material relative to a dispersed impurity. In such a situation it is fundamental from the physical point of view to take into account two nonlinear effects caused by the autocatalysis and the reduction of the absorption resource of the filter bed. Such accounting formally provides both (4.13) and (4.14). However, in the second case it is possible to obtain a finite representations for the desired characteristics in the form of combinations of elementary functions. Thus, the forecasting and technological calculations are considerably facilitated and the body of calculations is considerably reduced. Therefore, further in this Section the calculation dependencies and equations are specified exactly at $\bar{\lambda}(\bar{S})$ according to (4.14). Naturally, in case of more complicated form of $\bar{\lambda}(\bar{S})$ and the initial deposit evenly distributed along the bed volume, we have to involve (4.6), (4.7), (4.11) and to carry out more laborious calculations. The integrals contained in them, as a rule, can be determined only numerically, which is easy to do using the software designed for the mathematical analysis (Mathcad, Matlab, etc.).

To describe the decrease in the flow capacity of a bed due to its progressive clogging, the power and less often exponential forms of the dependence $k(S_S)$ are used (Khilar &Fogler, 1998; Zhurba, 2011). In the first case, it is taken generalized (1.10). According to the available experimental data on the deposit composition, the relationship between the concentrations S_S and S is justified to characterize by a linear function (Poliakov et al., 2019), so that

$$\gamma(S_S) = \frac{S_S}{S} = a_s - b_s \frac{S_S}{n_0}. \tag{4.15}$$

Hence, $\bar{S}_S$ is expressed through $\bar{S}$ in the following way

$$\bar{S}_S = \frac{a_s \bar{S}}{1 + b_s C_0 \bar{S}}$$

Then (1.10) is transformed to a dimensionless form

$$f_k(\overline{S}) = \left[1 - \left(\frac{a_s C_0 \overline{S}}{1 + b_s C_0 \overline{S}}\right)^{m_{k1}}\right]^{m_{k2}}. \tag{4.16}$$

In accordance with the results of experimental studies (Mintz & Meltzer, 1970) m_{k1}=1, m_{k2}=3, and hence,

$$f_k(\overline{S}) = \left[\frac{1 + (b_s - a_s)\, C_0 \overline{S}}{1 + b_s C_0 \overline{S}}\right]^3. \tag{4.17}$$

It is pertinent to note here the difficulties in experimental establishing the function $\gamma(S)$ due to the smallness and variability of the solid component content in the gel-like deposit (Ojha & Graham, 1993).

The calculation dependences (4.5), (4.6), (4.11) can be given in a final form convenient for a quantitative analysis. In particular, the representations for the concentrations $\overline{S}_0$, $\overline{S}$ in explicit and implicit forms containing only elementary functions are the following

$$\overline{S}_0(\bar{t}) = \frac{\psi_0 \overline{S}_m e^{\bar{\lambda}_0(1+\theta \overline{S}_m)\bar{t}} - 1}{\psi_0 e^{\bar{\lambda}_0(1+\theta \overline{S}_m)\bar{t}} + \theta}, \tag{4.18}$$

$$F(\overline{S}, \overline{S}_0(\bar{t})) - \bar{\lambda}_0 \bar{z} = 0, \tag{4.19}$$

where

$$F(\overline{S}, \overline{S}_0(\bar{t})) = \frac{1}{(1 + \theta \overline{S}^0)(\overline{S}_m - \overline{S}^0)} \ln \frac{\overline{S}_0(\bar{t}) - \overline{S}^0}{\overline{S} - \overline{S}^0} + \frac{\theta}{(1 + \theta \overline{S}^0)(1 + \theta \overline{S}_m)} \ln \frac{1 + \theta S}{1 + \theta \overline{S}_0(\bar{t})} + $$
$$+ \frac{1}{(1 + \theta \overline{S}_m)(\overline{S}_m - \overline{S}^0)} \ln \frac{\overline{S}_m - \overline{S}}{\overline{S}_m - \overline{S}_0(\bar{t})},$$

$$\psi_0 = (1 + \theta \overline{S}_0^o)/(\overline{S}_m - \overline{S}_0^o).$$

The explicit form for the concentration $\overline{S}$ can be obtained only in special cases of the insignificant autocatalytic effect (θ=0) and initial contamination of the bed (S^0=0). If θ=0, then

$$\overline{C}(\bar{z}, \bar{t}) = \frac{1}{1 - e^{-\bar{\lambda}_0 \bar{t}} + e^{-\bar{\lambda}_0 \bar{t} + \bar{\lambda}_0(\overline{S}_m - \overline{S}^0)\bar{z}}}$$

$$\overline{S}(\bar{z}, \bar{t}) = \frac{\overline{S}_m \psi(\bar{t}) e^{-\bar{\lambda}_0(\overline{S}_m - \overline{S}^0)\bar{z}} + \overline{S}^0}{\psi(\bar{t}) e^{-\bar{\lambda}_0(\overline{S}_m - \overline{S}^0)\bar{z}} + 1} = \frac{\overline{S}_m (1 - e^{-\bar{\lambda}_0 \bar{t}}) + \overline{S}^0 e^{+\bar{\lambda}_0(\overline{S}_m - \overline{S}^0)\bar{z} - \bar{\lambda}_0 \bar{t}}}{1 + e^{+\bar{\lambda}_0(\overline{S}_m - \overline{S}^0)\bar{z} - \bar{\lambda}_0 \bar{t}} - e^{-\bar{\lambda}_0 \bar{t}}},$$

where

$$\psi(\bar{t}) = \left[\overline{S}_0(\bar{t}) - \overline{S}^0\right] / \left[\overline{S}_m - \overline{S}_0(\bar{t})\right].$$

If also $\overline{S}^0 = 0$, then

$$\overline{S}(\overline{z},\overline{t}) = \frac{\overline{S}_m\overline{S}_0(\overline{t})e^{-\overline{\lambda}_0\overline{S}_m\overline{z}}}{\overline{S}_0(\overline{t})e^{-\overline{\lambda}_0\overline{S}_m\overline{z}} - \overline{S}_0(\overline{t}) + \overline{S}_m}.$$

The implicit form is preserved in the presence of the specified effect and $\overline{S}^0 = 0$. Then

$$\overline{\lambda}_0\overline{z} = \frac{1}{\overline{S}_m}\ln\frac{\overline{S}_0(\overline{t})}{\overline{S}} + \frac{\theta}{1+\theta\overline{S}_m}\ln\frac{1+\theta\overline{S}}{1+\theta\overline{S}_0(\overline{t})} + \frac{1}{\overline{S}_m(1+\theta\overline{S}_m)}\ln\frac{\overline{S}_m - \overline{S}}{\overline{S}_m - \overline{S}_0(\overline{t})}.$$

The relative outlet suspended solids concentration at the values of $\overline{S}$ at the bed boundaries established from (4.17) - (4.19) is simply calculated by formula (4.7) at $\overline{z} = 1$.

Finally, the formula for the losses $\Delta\overline{h}$ (4.11) is transformed to the following form

$$\Delta\overline{h}(\overline{t}) = \frac{1}{\overline{\lambda}_0}\int_{\overline{S}_e(\overline{t})}^{\overline{S}_0(\overline{t})}\frac{d\varsigma}{(\varsigma-\overline{S}^0)(1+\theta\varsigma)(\overline{S}_m-\varsigma)f_k(\varsigma)}, \tag{4.20}$$

and taking into account (4.15) will be

$$\Delta\overline{h}(\overline{t}) = \frac{1}{\overline{\lambda}_0}\int_{\overline{S}_e(\overline{t})}^{\overline{S}_0(\overline{t})}\frac{(1+bC_0\varsigma)^3 d\varsigma}{(\varsigma-\overline{S}^0)(1+\theta\varsigma)(\overline{S}_m-\varsigma)\left[1+(b_s-a_s)C_0\varsigma\right]^3} \tag{4.21}$$

In principle, the integral in (4.21) can be calculated and then the right side of this equation will contain only elementary functions. However, the final expression is cumbersome and it is easier to perform applied calculations directly using (4.21).

The relative technological times $\overline{t}_p$, $\overline{t}_h$ were calculated on the basis of the exact solution obtained above using the method described in Section 3.1. The relative criterion parameters were involved $\overline{C}_*$ and also$\Delta\overline{h}_*$ due to the constancy of the filtration rate. Both of these times are found in two steps. First, the set of two parameters $\overline{S}_{0p}$, $\overline{S}_{ep}$ and $\overline{S}_{0h}$, $\overline{S}_{eh}$, respectively, are calculated at the times $\overline{t}_p$ and $\overline{t}_h$, and in fact the values of the relative concentration of the deposited particles at the inlet to the bed and at the outlet cross-section ($\overline{S}_{ep}$, $\overline{S}_{eh}$) are calculated . Thus, when calculating $\overline{t}_p$, first of all, the value $\overline{S}_{0p}$ is selected, which satisfies the equation

$$\int_{\overline{S}_{ep}(\overline{S}_{0p})}^{\overline{S}_{0p}}\frac{d\varsigma}{(\varsigma-\overline{S}^0)\overline{\lambda}(\varsigma)} = 1 \tag{4.22}$$

where $\overline{S}_{ep}$ is related to $\overline{S}_{0p}$ according to (4.7) as follows

$$\overline{S}_{ep}(\overline{S}_{0p}) = \overline{C}_*\overline{S}_{0p} + \overline{S}^0(1-\overline{C}_*). \tag{4.23}$$

Then the following formula should be used for $\overline{t}_p$ at the known value $\overline{S}_{0p}$

$$\overline{t}_p = \int_{\overline{S}^0}^{\overline{S}_{0p}}\frac{d\varsigma}{\overline{\lambda}(\varsigma)}. \tag{4.24}$$

More difficult to determine the time $\bar{t}_h$, since here it is necessary to solve in the first step the following system of equations with respect to $\bar{S}_{0h}$, $\bar{S}_{eh}$

$$\int_{\bar{S}_{eh}}^{\bar{S}_{0h}} \frac{d\varsigma}{(\varsigma - \bar{S}^0)\bar{\lambda}(\varsigma)} = 1 \tag{4.25}$$

$$\int_{\bar{S}_{eh}}^{\bar{S}_{0h}} \frac{d\varsigma}{(\varsigma - \bar{S}^0)\bar{\lambda}(\varsigma) f_k(\varsigma)} = \Delta \bar{h}_*, \tag{4.26}$$

First of all, we set the actual value of $\bar{S}_{0h}$ and find the value of $\bar{S}_{eh}$ satisfying equation (4.25) by fitting procedure. Then this set of two values is substituted into equation (4.26) and the discrepancy is determined as a measure of the deviation from the desired solution. Starting from it, we successively refine $\bar{S}_{0h}$ and$\bar{S}_{eh}$, which allows us to reduce the initial discrepancy. Following the described procedure, as a rule, it is possible to minimize this deviation in several steps so much that the last set of the values of $\bar{S}_{0h}$, $\bar{S}_{eh}$ can be justifiably used for calculating $\bar{t}_h$ by the formula

$$\bar{t}_h = \int_{\bar{S}^o}^{\bar{S}_{0h}} \frac{d\varsigma}{\bar{\lambda}(\varsigma)}. \tag{4.27}$$

This calculation scheme with respect to expression (4.13) is somewhat simplified due to the known dependence (4.17). In this case, the time $\bar{t}_p$ is immediately found by fitting from the equation

$$F\left(\bar{S}_e(\bar{t}_p, \bar{C}_*), \bar{S}_0(\bar{t}_p)\right) = \bar{\lambda}_0, \tag{4.28}$$

where again (4.23) is used. When determining $\bar{t}_h$ at the first stage it is also necessary to solve the system of the equations

$$F(\bar{S}_{eh}, \bar{S}_{0h}) = \bar{\lambda}_0, \tag{4.29}$$

$$\int_{\bar{S}_{eh}}^{\bar{S}_{0h}} \frac{(1 + b_s C_0 \varsigma)^3 d\varsigma}{(\varsigma - \bar{S}^0)(1 + \theta\varsigma)(\bar{S}_m - \varsigma)\left[1 + (b_s - a_s) C_0 \varsigma\right]^3} = \bar{\lambda}_0 \Delta \bar{h}_*. \tag{4.30}$$

If the value of $\bar{S}_{0h}$ is known, the corresponding value of $\bar{t}_h$ is calculated by the formula

$$\bar{t}_h = \frac{1}{\bar{\lambda}_0(1 + \theta \bar{S}_m)} \ln\left(\frac{1 + \theta \bar{S}_0}{1 + \theta \bar{S}_0^o} \cdot \frac{\bar{S}_m - \bar{S}_0^o}{\bar{S}_m - \bar{S}_0}\right). \tag{4.31}$$

The filter run time should be set according to (1.41).

A large body of calculations was performed on the test examples to illustrate, firstly, the above obtained exact solution of the nonlinear problem of the non-detachment filtration of a low-concentrated suspension with constant rate through a uniform bed of a rapid filter, and secondly, the method of calculating the technological times with emphasis on the time of the filter run.

In the first series of examples, the reference for a comparative analysis was the basic example with the typical input data: $\bar{\lambda}_0 = 0.02$, $\bar{S}_m = 200$, $\theta=0.01$, $\bar{S}^0 = 10$, $a_s=50$, $b_s=30$, $C_0=10^{-4}$. Three other examples were calculated in parallel to estimate the significance of the absorption capacity of the filter media ($\bar{\lambda}_0$), fresh and residual deposit (θ). An increase of 2.5 times was allowed in the second example. The autocatalysis ($\theta=0$) and initial bed contamination ($\bar{S}^0 = 0$) were considered insignificant, respectively, in the third and fourth examples. The subject of the predictive calculations in this series of examples were the relative characteristics: outlet suspended solids concentration $\bar{C}_e$, head losses in the bed $\Delta\bar{h}$, and concentration profiles of the deposited impurities at fixed time points $\bar{S}(\bar{z})$.

The deterioration of the filtrate quality over time was calculated by formula (4.7). The values of $\bar{S}_0$ were preliminarily calculated for the given time points, first, by formula (4.17) and then by fitting procedure from (4.18) at $\bar{z} = 1$. Four curves of the dependence $\bar{C}_e(\bar{t})$ obtained in this way are presented in **Figure 1** for the above examples. The reference for a comparison of the results here is curve **2.** The figure clearly shows the importance for the clarification process of the absorption of the suspended solids by both the medium itself (curves 2 and 4) and the newly formed deposit (curves 1 and 2). The relatively small amount of the deposit retained after washing only minimally reduces the suspended solids content at the filter outlet during the non-detachment filtration.

The opposite filtration flow pattern is observed in **Figure 2**, where curves of the increase in the total head losses in the bed are shown also over time. Calculations of the quantity $\Delta\bar{h}$ were carried out using formula (4.21). They were again preceded by finding at each calculation step a set of two values of $\bar{S}_0$, $\bar{S}_e$. Naturally, the serious improvement of the absorption properties of the bed (example 2) caused the accelerated increase in its hydraulic resistance (curves 1 and 2). The passivity of the deposit with respect to a free impurity (example 3) contributed to the reduction of the head losses to the same extent (curves 2 and 4). The role of the primary deposit in reducing the bed flow capacity is not as important, although it is not advisable to neglect it due to its participation in the autocatalysis. The head losses due to this deposit increase by 0.161 (in its absence it equals 1) at the first time, and the excess reaches 0.574 at the end of the calculation period ($\bar{t} = 120$), although the contribution of the initial contamination to the clogging of the bed is markedly reduced in percentage terms.

Figure 1. Increase in the relative volumetric concentration of the suspended solids in the filtrate over time

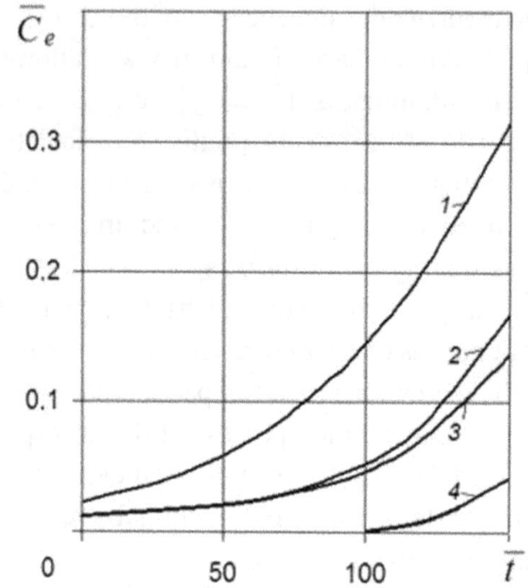

Note. 1 – example 3 (θ =0), 2 – example 1, 3 – example 4 ($\overline{S}^0$ = 0), 4 – example 2 ($\overline{\lambda}_0$ = 0.05)

Figure 2. Increase in the relative head losses in the bed over time

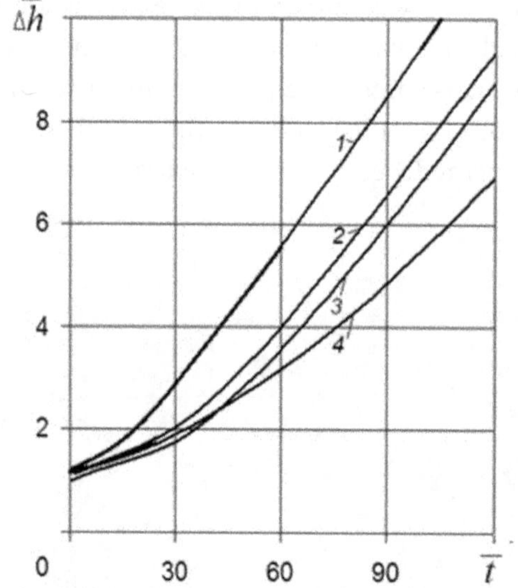

Note. 1 – example 2, 2 – example 1, 3 – example 4, 4 – example 3

The profiles $\overline{S}(\overline{z})$ shown in **Figure 3** have a complex shape, which also transforms significantly over time. Calculations of the concentration $\overline{S}$ were performed according to dependence (4.17) and equation (4.18) when the coordinate $\overline{z}$ changes from 0 to 1 for two fixed points in time corresponding to the initial (curve 1) and final (curves 2-4) stages of the clarification process. It is obvious that the nature of the deposit distribution by height changes sharply over time (curves 1 and 3) due to the depletion of the absorbing resource of a filtering material in the upper part of the bed (downward filtration) and gradual involvement of its lower part ($1 \geq \bar{z} \geq 0.3$) in the clarification process. The unevenness in the deposit distribution is much more pronounced in the second example (curve 2). Here, practically the entire upper half of the bed volume has exhausted its mentioned resource by the calculation time. In contrast, the lower part contains a minimum amount of the fresh deposit, since almost all impurities entering the bed are trapped at the top. The deposited matter is distributed more evenly along the height due to the autocatalysis.

The theoretical substantiation of the technological times is of particular importance in the operation of rapid filters, and its ultimate goal is to determine the duration of the filter run. Just the relative times $\bar{t}_p$, $\bar{t}_h$ became the subject of consistent calculations in the second series of examples. The value of $\bar{t}_f$ was specified on their basis. These times were calculated by a special method, that is based, firstly, on equation (4.28), which is used to determine $\bar{t}_p$; secondly, on the system of equations (4.29), (4.30), which, together with formula (4.31), is used to calculate $\bar{t}_h$. The curves of the change in $\bar{t}_f$ depending on four model parameters were calculated at the found set of two values of $\bar{t}_p$, $\bar{t}_h$ according to (1.41).

Figure 3. Profiles of the relative volumetric concentration of the suspended solids:

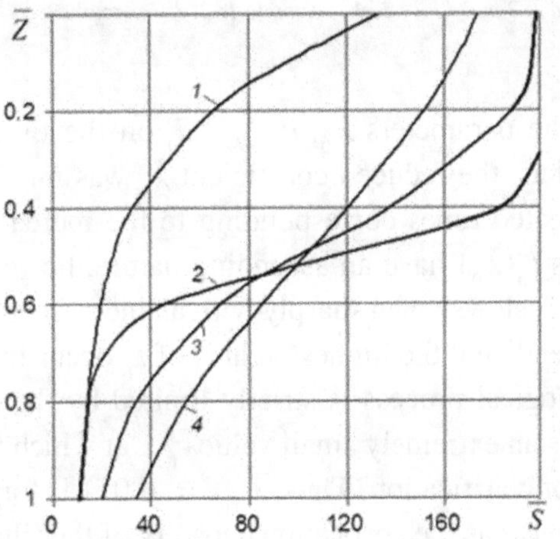

Note. 1, 3, 4 – $\bar{\lambda}_0$ = 0.02; 2 – $\bar{\lambda}_0$ = 0.05; 1 – $\bar{t}$ = 30; 2-4 – $\bar{t}$ = 100; 1-3 – θ =0.01; 4 – θ =0

Figure 4. Dependencies $\bar{t}_p(\bar{\lambda}_0)$, $\bar{t}_h(\bar{\lambda}_0)$, $\bar{t}_f(\bar{\lambda}_0)$

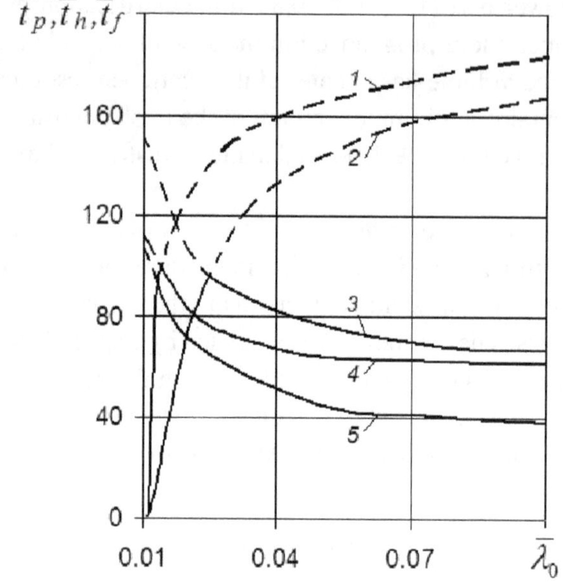

Note. 1, 2 – $\bar{t}_p$; 3-5 – $\bar{t}_h$; 1 and 4, 1 and 5, 2 and 3 – $\bar{t}_f$; 1, 4 – example 1; 2, 3 – example 3; 5 – example 5

To estimate the effect of the parameters $\bar{\lambda}_0$, θ, $\bar{S}_m$, C_0 on the times $\bar{t}_{p,h,f}$, the latter were varied continuously in wide range. Thus, the reduced coefficient $\bar{\lambda}_0$ was maximally increased by a factor of 9, and the changes in the indicated times corresponding to the increasing values of $\bar{\lambda}_0$ are shown in **Figure 4**. Naturally, the curves $\bar{t}_p(\bar{\lambda}_0)$ have an ascending nature, however, in a well-absorbing media $(\bar{\lambda}_0 = 0.05)$, the increment of $\bar{t}_p$ slows down sharply with an increase in $\bar{\lambda}_0$. On the contrary, the graphs of the dependence $\bar{t}_p(\bar{\lambda}_0)$ descend, and the highest values of $\bar{t}_h$ occur for moderately absorbing media. As a consequence, the technological process is strictly limited by the filtrate quality at lower values of $\bar{\lambda}_0$. At the same time, there is an extremely small value$\bar{\lambda}_{min}$, at which the bed is initially incapable of providing the required degree of clarification. Thus, $\bar{\lambda}_{min} = 0.011$ is for the basic example. The curves $\bar{t}_p(\bar{\lambda}_0)$ and $\bar{t}_h(\bar{\lambda}_0)$ initially converge as the absorption capacity of the filtering material increases. There is a value $\bar{\lambda}_{0m}$, at which these curves intersect. Further, the time of continuous operation of the filter is already limited on the section $\bar{\lambda}_0 > \bar{\lambda}_{0m}$ due to excessive head losses, so that $\bar{t}_p > \bar{t}_h$. Thus, in accordance with (1.41), the curve of the dependence $\bar{t}_f(\bar{\lambda}_0)$ should, firstly, be continuous, and secondly, have a fracture at $\bar{\lambda}_0 = \bar{\lambda}_{0m}$ with a clearly expressed maximum, which is confirmed by Figure 4. From the practical point of view, this fact indicates the importance of the correct selection of the filtering material and filtration mode. It is obvious that unsuccessfully selected material with far from optimal absorption properties ($\lambda_0 >> \lambda_{0m}$ or $\lambda_0 << \lambda_{0m}$) will cause inefficient operation of the filter. On the other hand, the relationship between the coefficient λ and rate V (Ison & Ives, 1968; Senyavin et al., 1977; Williams et al., 2007) established experimentally and theoretically may provide a basis for selecting the latter to approximate λ_0 to the value λ_{0m}. To get an idea of the possible implications for the predictions and substantiations of the commonly practiced averaging of the deposit composition characteristic $\gamma(S)$, a fifth example is calculated in addition to the basic (curves 1, 4) and third examples. Since the quantity $\bar{S}$ can in principle vary from 0 to$\bar{S}_m$, the average value γ_a is approximately 30 according to (4.14). Hence, its dimensionless analog $\bar{\gamma}_a = \gamma_a C_0$ is 0.003. Just curve 5 in Figure 4 is determined at $\bar{\gamma} = \bar{\gamma}_a = const$ and differs significantly from the reference curve 4, and they diverge even more with increasing $\bar{\lambda}_0$. Obviously, formal averaging of the function $\gamma(S)$ due to its nonlinearity and uncertainty of the limits of change can lead to significant errors in determining $\bar{t}_h$ and $\bar{t}_f$, which is consistent with the conclusions in Section 3.6 for the detachment filtration.

The significance of the autocatalytic θ and capacitive $\bar{S}_m$ parameters for the technological process was also estimated. The relative solids take-up capacity was varied from 150 to 400, and the parameter θ from 0 to 0.02, which made it possible to cover both normal and extreme conditions. A series of the calculated dependencies $\theta(\bar{t}_{p,h,t})$ and $\bar{S}_m(\bar{t}_{p,h,t})$ are described by the graphs in **Figure 5**. It follows that a noticeable (in this case up to 30%) prolongation of filter operation is possible due to the participation of a deposit in the removal of dispersed impurities. At the same time the autocatalysis does not allow to start filter washing later at higher values of θ. Formal increase in $\bar{S}_m$ from 150 to 175 causes prolongation of the filter operation by almost 40%. However, further increase in $\bar{S}_m$ gives the opposite result and the value of $\bar{t}_f$ decreases by 2 times and more. In general, according to the adopted model, the effect of the latter parameters on the suspension clarification and its hydraulic aspects is manifested by analogy with $\bar{\lambda}_0$ through the absorption rate functional coefficient $\bar{\lambda}$, which directly controls the removal of the suspended matter. Therefore, as in the case of variation of $\bar{\lambda}_0$, their hypothesized increase leads to the enhanced deposition of dispersed impurity which has twofold consequences. It has long been known that impurities in the form of the deposit with a high bound water content are localized at the top of the bed. Therefore, the calculated curves $\bar{t}_f(\theta)$, $\bar{t}_f(\bar{S}_m)$ also being continuous, have a distinct maximum in the form of a peak. Its position corresponds to very specific values θ_m and $\bar{S}_{mm}$. In practice, however, it is realistic to change only $\bar{S}_m$, even when the filter is in operation due to feeding it with a suspension with such a suspended solids content that $\bar{S}_m$ is equal to $\bar{S}_{mm}$.

Figure 5. Dependencies $\theta(\bar{t}_{p,h,f})$, $\bar{S}_m(\bar{t}_{p,h,f})$

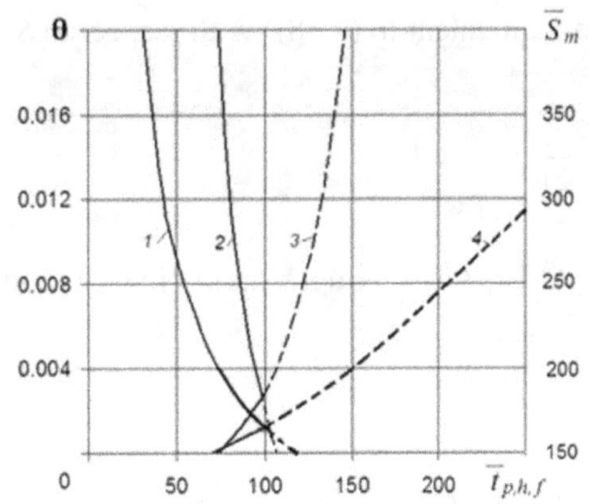

Note. 1 – $\bar{S}_m(\bar{t}_h)$, 2 – $\theta(\bar{t}_h)$, 3 – $\theta(\bar{t}_p)$, 4 – $\bar{S}_m(\bar{t}_p)$, 1, 4 – $\bar{S}_m(\bar{t}_f)$, 2, 3 – $\theta(\bar{t}_f)$

Figure 6. Dependencies $\bar{t}_{p,h,f}(\overline{C}_0)$

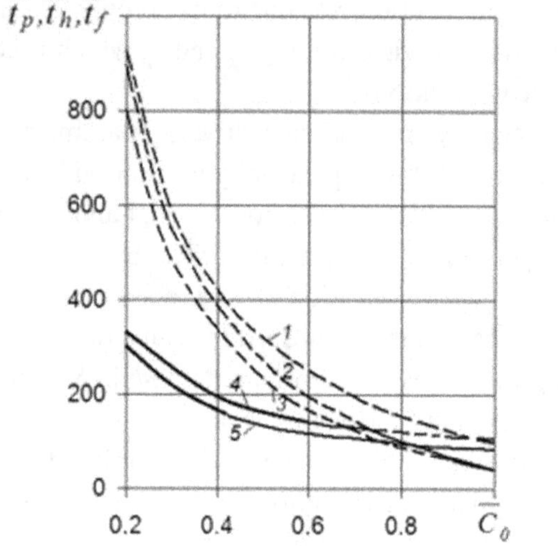

Note. 1-3 – $\bar{t}_p$; 4, 5 – $\bar{t}_h$; 2, 3 and 4 – $\bar{t}_f$ (solid lines); 1, 5 – θ =0.01, 2-4 – θ =0; 1, 2, 4, 5 – $\overline{C}_* = const$; 3 – $\overline{C}_* = const$

The nature of the dependence of the technological times on the initial concentration of impurity at both low and high water contamination is clearly shown in **Figure 6**. Here the results of the calculations of the times $\bar{t}_p$, $\bar{t}_h$ depending on C_0 are given in a graphical form. But since the quantity C_0 serving earlier as a scale for the required concentrations is now a variable, the calculation formulas and equations had to be transformed with the introduction of the new relative variable $\overline{C}_0 = C_0/C_{00}$ with the scale C_{00}. The exception was formula (4.17), the form of which did not change after the transition from $\bar{S}$ to $\overline{\overline{S}} = S/(n_0 C_{00})$. It is enough to replace the dimensionless initial θ, $\bar{S}_m$, $\bar{S}_0^0$ and complex ψ parameters by their constant analogs $\tilde{\theta}$ (equal to $\theta/\overline{C}_0$), $\overline{S}_m$, $\overline{S}^0$ and $\tilde{\psi}$, containing C_{00} instead of C_0. The variable $\overline{C}_0$ appears explicitly, firstly, in the equation (equivalent to (4.18)) with respect to $\overline{S}$

$$\bar{z} = \frac{\overline{C}_0}{\bar{\lambda}_0} \int_{\overline{S}}^{\overline{S}_0(\bar{t})} \frac{d\varsigma}{(\varsigma - \overline{S}^0)(\tilde{\theta}\varsigma + 1)(\overline{S}_m - \varsigma)}; \tag{4.32}$$

secondly, in the formulas for $\overline{C}_e = C_e/C_{00}$ and $\Delta\bar{h}$ taking into account (4.7), (4.15)

$$\overline{C}_e(\bar{t};\overline{C}_0) = \overline{C}_0 \frac{\overline{S}_e(\bar{t};\overline{C}_0) - \overline{S}^0}{\overline{S}_0(\bar{t}) - \overline{S}^0}, \tag{4.33}$$

$$\Delta\bar{h}(\bar{t},\overline{C}_0) = \frac{\overline{C}_0}{\bar{\lambda}_0} \int_{\overline{S}_e(\bar{t};\overline{C}_0)}^{\overline{S}_0(\bar{t})} \frac{(1 + b_s C_{00}\varsigma)^3 d\varsigma}{(\varsigma - \overline{S}^0)(\tilde{\theta}\varsigma + 1)(\overline{S}_m - \varsigma)\left[1 + (b_s - a_s) C_{00}\varsigma\right]^3}, \tag{4.34}$$

where $\overline{S}_e(\bar{t};\overline{C}_0)$ is found from equation (4.32) at $\bar{z} = 1$.

Thus, the time $\bar{t}_p$ should be determined by fitting from the equation

$$\overline{C}_0 \int_{\overline{S}_e(\overline{t}_p;\overline{C}_0)}^{\overline{S}_0(\overline{t}_p)} \frac{d\varsigma}{(\varsigma - \overline{S}^0)(\theta\varsigma + 1)(\overline{S}_m - \varsigma)} = \overline{\lambda}_0. \tag{4.35}$$

The way of expressing $\overline{S}_e$ through $\overline{S}_0$ depends on the requirements to the degree of the suspension clarification. At a fixed value C_* the ratio $\overline{C}_* = C_*/C_{00}$ will also be constant and according to (4.33)

$$\overline{S}_e(\overline{t}_p, \overline{C}_0) = \frac{\overline{C}_*}{\overline{C}_0}\overline{S}_0(\overline{t}_p) + \left(1 - \frac{\overline{C}_*}{\overline{C}_0}\right)\overline{S}^0. \tag{4.36}$$

If it is necessary to reduce any initial contamination by a certain number of times, for example, by an order of magnitude, then it is true that $\overline{C}_* = const$

$$\overline{S}_e(\overline{t}_p) = \overline{C}_*\overline{S}_0(\overline{t}_p) + \overline{S}^0(1 - \overline{C}_*). \tag{4.37}$$

To find the time $\overline{t}_h$, first of all, the system of the following equations with respect to $\overline{S}_0$, $\overline{S}_e$ should be solved using the expression for $f_k(\overline{S})$ (4.16)

$$\overline{C}_0 \int_{\overline{S}_e}^{\overline{S}_0} \frac{d\varsigma}{(\varsigma - \overline{S}^0)(\theta\varsigma + 1)(\overline{S}_m - \varsigma)} = \overline{\lambda}_0$$

$$\overline{C}_0 \int_{\overline{S}_e}^{\overline{S}_0} \frac{(1 + 30C_{00}\varsigma)^3 d\varsigma}{(\varsigma - \overline{S}^0)(\theta\varsigma + 1)(\overline{S}_m - \varsigma)(1 - 20C_{00}\varsigma)^3} = \overline{\lambda}_0$$

and the time $\overline{t}_h$ corresponding to them should calculated by formula (4.31), having previously replaced in it $\overline{\theta}$, $\overline{S}_m$, $\overline{S}_0$, $\overline{S}^0$ by $\tilde{\theta}$, $\overline{S}_m$, $\overline{S}_0$, $\overline{S}^0$.

Similarly, the dependences $\overline{t}_{p,h}(\overline{C}_0)$ were calculated at a change in $\overline{C}_0$ of 5 times for the data of the basic and third examples and first $\overline{C}_* = const$. Naturally, both technological times increase with decreasing initial impurity content, but to different degrees (Figure 6). In the case $\theta = 0$ adopted for the argument, the range of values can be conditionally divided into two intervals: $\overline{C}_0 \leq \overline{C}_{0m}$ and $\overline{C}_0 > \overline{C}_{0m}$, where $\overline{C}_{0m}$ is the relative characteristic value of $\overline{C}_0$, at which $\overline{t}_p$ and $\overline{t}_h$ are equalized. It is obvious that at the value of $\overline{C}_0$, belonging to the first interval, the filter has to be shut down because of an excessive expenditure of mechanical energy for the filtration flow. If it belongs to the second section, the technological process is limited directly by the quality of filtrate. For the basic example, the period of the suspension clarification is limited due to the rapidly increasing hydraulic resistance of the bed at any values of $\overline{C}_0$ from the given range. Thus, the corresponding curve $\overline{t}_h(\overline{C}_0)$ can also be interpreted as the curve of the dependence $\overline{t}_f(\overline{C}_0)$. In addition, $\overline{t}_p$ was calculated at $\theta = 0$ based on the requirement to reduce the impurity content in suspension after its clarification by an order of magnitude independently of the initial amount (curve 3). Therefore, expression (4.37) was used for $\overline{S}_e$ instead of (4.36). In the case of the linear kinetic equation ($\overline{\lambda} = const$), the time $\overline{t}_p$ is invariant with respect to C_0 due to the synchronism of the change in C_0 and C_*. However, due to the nonlinearity of the above equation ($\overline{\lambda} = \overline{\lambda}(\overline{S})$), and specifically the enhancement of the suspended solids absorption with decreasing C_0 due to increasing the difference S_m–S, the time $\overline{t}_p$ increases, although not to the same extent as for a fixed value of C_*. Such analysis helps to choose the method of intensification of filtration at its constant rate and a known value of C_0. Thus, if it is necessary to start the filter washing early due to severe clogging of the uniform bed ($C_0 < C_{0m}$), a special layered porous structure should probably be arranged instead.

4.2. FILTRATION WITH FLOW DIRECTION CHANGE

The main reason for incomplete implementation of the clarification resource of rapid filters is usually a significant unevenness of the deposit distribution by the height of a well-absorbing packed bed and, as a consequence, an accelerated increase in the head losses in it. In this section, the consequences of changing the flow direction in the bed, e.g. from downward to upward or vice versa, during the filter run have been subjected to the comprehensive theoretical analysis using exact analytical methods (Poliakov, 2019b; 2020). A more even accumulation of the deposit in the operating volume and, consequently, a more economical use of mechanical energy can be achieved in this way. Technically, such an operation (rapid filter switching) is not difficult to perform. However, it only makes practical sense in cases where the deposit is found to be stable with respect to the hydrodynamic effects. Then it is justified to use modern mathematical models of the non-detachment filtration for the theoretical studies and subsequent technological calculations.

Therefore, the main goal here was, based on an exact solution of the mathematical problem of the non-detachment filtration through the initially contaminated bed, to analyze its two stages with suspension flows of different orientation. The second stage starts after changing the direction of the filtration flow. The increase in filter efficiency due to its switching was evaluated by comparing with the results of traditional single-way filtration.

The filtration of aqueous suspension at the first stage is described by the mathematical model, which is given in Sections 2.1, 2.2 in a dimensionless form, namely (2.9) - (2.13), (2.28), and also includes (4.2). Since the emphasis here is on the effect due to the change in the direction of a suspension flow, for specificity, additional assumptions about S^0 and $\lambda(S)$ are made, which have little or no impact on the estimates of the above effect. Thus, in contrast to (4.13), it is assumed that

$$\bar{\lambda}(\bar{S}_j) = \bar{\lambda}_0 (\bar{S}_m - \bar{S}_j)^{m_\lambda} \tag{4.38}$$

where m_λ is the empirical constant, which can take any values, but the value $m_\lambda = 1$ is emphasized due to the greater possibilities in the analytical analysis; the index j here and hereinafter indicates the belonging of this characteristic to the first or second stages. Using the calculation equation (4.5) for the relative inlet concentration of the deposited particles, for example, it is easy to obtain the following formula

$$\bar{S}_{10}(\bar{t}) = \bar{S}_m - \left[\bar{S}_m^{1-m_\lambda} + \bar{\lambda}_0 (1 - m_\lambda)\bar{t}\right]^{\frac{1}{1-m_\lambda}}. \tag{4.39}$$

Further, along with the general approach using a function $\lambda(S)$ of an arbitrary form, a particular approach with $m_\lambda = 1$, is considered in parallel, which allows to obtain the expressions containing exclusively elementary functions. Then

$$\bar{S}_{10}(\bar{t}) = \bar{S}_m (1 - e^{-\bar{\lambda}_0 \bar{t}}) \tag{4.40}$$

$$\bar{S}_1(\bar{z},\bar{t}) = \frac{\bar{S}_m e^{-\bar{\lambda}_0 \bar{S}_m \bar{z}}(1 - e^{-\bar{\lambda}_0 \bar{t}})}{e^{-\bar{\lambda}_0 \bar{t}} + e^{-\bar{\lambda}_0 \bar{S}_m \bar{z}} - e^{-\bar{\lambda}_0(\bar{t}+\bar{S}_m \bar{z})}} \tag{4.41}$$

In general, the concentration $\bar{C}_1$ at already known quantities $\bar{S}_1(\bar{z},\bar{t})$, $\bar{S}_{10}(\bar{t})$ should be calculated by the following formula

$$\bar{C}_1(\bar{z},\bar{t}) = \frac{\bar{S}_1(\bar{z},\bar{t})}{\bar{S}_{10}(\bar{t})}. \tag{4.42}$$

In the limit at $\bar{t} \to 0$ follows from (4.42)

$$\bar{C}_1(\bar{z},0) = e^{-\bar{\lambda}(0)\bar{z}}.$$

Then, the relative outlet suspended solids concentration will be

$$\bar{C}_{1e}(\bar{t}) = \frac{e^{-\bar{\lambda}_0 \bar{S}_m}}{e^{-\bar{\lambda}_0 \bar{t}} + 1 - e^{-\bar{\lambda}_0(\bar{S}_m+\bar{t})}}.$$

The relative head losses $\Delta\bar{h}_1$ are obtained by integrating equation (2.11) along the bed height and should be calculated at f_k according to (2.25) by the formula

$$\Delta\bar{h}_1(\bar{t}) = \int_0^1 \frac{d\bar{z}}{f_k(\bar{S}_1(\bar{z},\bar{t}))} = \int_{\bar{S}_{10}(\bar{t})}^{\bar{S}_{1e}(\bar{t})} \frac{d\zeta}{\zeta\bar{\lambda}(\zeta)(1-\bar{\gamma}(\zeta)\zeta)^3} \tag{4.43}$$

Here $\bar{S}_{1e}$ is the relative volumetric concentration of the deposited suspension particles in the outlet section of the bed at the first stage, which should be found as a function of $\bar{t}$ from the equation

$$\int_{\bar{S}_{1e}(\bar{t})}^{\bar{S}_{10}(\bar{t})} \frac{d\varsigma}{\varsigma\bar{\lambda}(\varsigma)} = 1. \tag{4.44}$$

In principle, it is possible to calculate the integral in (4.43) and express $\Delta\bar{h}(\bar{t})$ only through elementary functions. But the corresponding representation is cumbersome and it is easier to use directly (4.43) in calculations.

The second stage begins at the moment of the reorientation of a suspension flow ($t=t_*$), which requires a change of signs in equations (2.11), (2.28), namely,

$$\frac{\partial\bar{C}_2}{\partial\bar{z}} - \frac{\partial\bar{S}_2}{\partial\bar{t}} = 0, \tag{4.45}$$

$$\bar{k}(\bar{S}_2)\frac{\partial\bar{h}_2}{\partial\bar{z}} = 1. \tag{4.46}$$

In addition, it is now necessary to take into account the initial bed contamination for the second stage, which is generally described by the equation

$$\bar{z} = \int_{\bar{S}_2^0}^{\bar{S}_{0*}} \frac{d\varsigma}{\varsigma\bar{\lambda}(\varsigma)}, \tag{4.47}$$

and at $m_\lambda = 1$

$$\bar{S}_2^0(\bar{z}) = \frac{\bar{S}_m \bar{S}_{0*} e^{-\bar{\lambda}_0 \bar{S}_m \bar{z}}}{\bar{S}_m - \bar{S}_{0*} + S_{0*} e^{-\bar{\lambda}_0 \bar{S}_m \bar{z}}}, \tag{4.48}$$

where $\bar{S}_{0*} = \bar{S}_{10}(\bar{t}_*)$, $\bar{S}_2^0(\bar{z})$ is the initial $(\bar{t} = \bar{t}_*)$ distribution function of the deposited suspension particles. Finally, the condition of a constancy of the impurity content at the inlet to the bed is transferred to its lower boundary, so that

$$\bar{z} = 1, \quad \bar{C}_2 = 1. \tag{4.49}$$

At the same time, the homogeneous condition of the first kind for $\bar{h}_2$ is assumed at the new bed outlet

$$\bar{z} = 0, \quad \bar{h}_2 = 0. \tag{4.50}$$

Thus, the desired functions at the second stage are the concentrations $\bar{C}_2(\bar{z},\bar{t})$, $\bar{S}_2(\bar{z},\bar{t})$ and head $\bar{h}_2$ $(\bar{z},\bar{t})$. The system of the first-order equations (4.2), (4.45) is reduced to the second-order equation

$$\frac{\partial}{\partial \bar{z}}\left(\frac{\partial F(\bar{S}_2)}{\partial \bar{t}}\right) - \frac{\partial \bar{S}_2}{\partial \bar{t}} = 0, \tag{4.51}$$

where $F(\bar{S}_2) = \int \frac{d\bar{S}_2}{\bar{\lambda}(\bar{S}_2)}$. Integration of (4.51) over the variable $\bar{t}$ starting from the time point $\bar{t}_*$ gives

$$\frac{\partial F(\bar{S}_2)}{\partial \bar{z}} - \bar{S}_2 = \frac{dF(\bar{S}_2^0)}{\partial \bar{z}} - \bar{S}_2^0. \tag{4.52}$$

Taking into account (4.47), an ordinary nonlinear equation with respect to $\bar{S}_2$ of the first order follows from (4.52)

$$\frac{1}{\bar{\lambda}(\bar{S}_2)}\frac{\partial \bar{S}_2}{\partial \bar{z}} - \bar{S}_2 = -2\bar{S}_2^0(\bar{z}). \tag{4.53}$$

If $\bar{\lambda}(\bar{S})$ has an arbitrary form, it is not possible to solve exactly the mathematical problem posed for the second stage. Therefore, it is reasonable to consider equation (4.53) and the related equations (4.46) and

$$\frac{\partial \bar{C}_2}{\partial \bar{z}} - \bar{\lambda}(\bar{S}_2)\,\bar{C}_2 = 0 \tag{4.54}$$

as a system of three ordinary equations of the first order and attract standard software of mathematical analysis for its solution. Preliminarily, the above system is represented in the canonical form

$$\frac{\partial \overline{C}_2}{\partial \overline{z}} = \overline{\lambda}(\overline{S}_2)\,\overline{C}_2,\ \frac{\partial \overline{S}_2}{\partial \overline{z}} = \overline{\lambda}(\overline{S}_2)\,\overline{S}_2 - 2\overline{S}_2^0(\overline{z})\overline{\lambda}(\overline{S}_2),\ \frac{\partial \overline{h}_2}{\partial \overline{z}} = \frac{1}{\overline{k}(\overline{S}_2)}, \tag{4.55}$$

where $\overline{S}_2^0(\overline{z})$ follows from (4.47). The boundary operator includes three conditions of the first kind at $\overline{z} = 1$ and first of all (4.49), the condition for $\overline{S}_2$

$$\overline{z} = 1, \quad \overline{S}_2 = \overline{S}_{2e}(\overline{t}). \tag{4.56}$$

The concentration function of the deposited particles in the new inlet cross-section $\overline{S}_{2e}(\overline{t}) = \overline{S}_2(1,\overline{t})$ is obtained from the equation

$$\int_{\overline{S}_{e*}}^{\overline{S}_{2e}} \frac{d\varsigma}{\overline{\lambda}(\varsigma)} = \overline{t} - \overline{t}_*, \tag{4.57}$$

where $\overline{S}_{e*} = \overline{S}_1(1,\overline{t}_*)$. Now, having set the time $\overline{t}$, we should carry out calculations by searching the values of $\overline{h}_2$ on the same boundary until condition (4.50) is satisfied.

The analytic expression for $\overline{S}_2$, containing only elementary functions, is derived in the distinguished case $\overline{\lambda} = \overline{\lambda}_0(\overline{S}_m - \overline{S}_2)$, namely,

$$\overline{S}_2(\overline{z},\overline{t}) = \overline{S}_m - \\ - \frac{e^{-\overline{\lambda}_0\overline{S}_m\overline{z}}\left[\overline{S}_m - \overline{S}_{2e}(\overline{t})\right]\left[e^{-\overline{\lambda}_0\overline{t}_*} + (1 - e^{-\overline{\lambda}_0\overline{t}_*})e^{-\overline{\lambda}_0\overline{S}_m}\right]^2}{\left[e^{-\overline{\lambda}_0\overline{t}_*}(1 - e^{-\overline{\lambda}_0\overline{t}_*})e^{-\overline{\lambda}_0\overline{S}_m\overline{z}}\right]^2\left\{e^{-\overline{\lambda}_0\overline{S}_m} - \overline{\lambda}_0 I(\overline{z})\left[\overline{S}_m - \overline{S}_{2e}(\overline{t})\right]\left[e^{-\overline{\lambda}_0\overline{t}_*} + (1 - e^{-\overline{\lambda}_0\overline{t}_*})e^{-\overline{\lambda}_0\overline{S}_m}\right]^2\right\}}, \tag{4.58}$$

where

$$I(\overline{z}) = \int_1^{\overline{z}} \frac{e^{-\overline{\lambda}_0\overline{S}_m\eta}\,d\eta}{\left[e^{-\overline{\lambda}_0\overline{t}_*} + (1 - e^{-\overline{\lambda}_0\overline{t}_*})e^{-\overline{\lambda}_0\overline{S}_m\eta}\right]^2} = \frac{1}{\overline{\lambda}_0\overline{S}_m(1 - e^{-\overline{\lambda}_0\overline{t}_*})} \times \\ \times \left[\frac{1}{(1 - e^{-\overline{\lambda}_0\overline{t}_*})e^{-\overline{\lambda}_0\overline{S}_m\overline{z}} - e^{-\overline{\lambda}_0\overline{t}_*}} - \frac{1}{(1 - e^{-\overline{\lambda}_0\overline{t}_*})e^{-\overline{\lambda}_0\overline{S}_m} - e^{-\overline{\lambda}_0\overline{t}_*}}\right],$$

$$\overline{S}_{2e}(\overline{t}) = \overline{S}_m - \left(\overline{S}_m - \overline{S}_{2e}^0\right)e^{-\overline{\lambda}_0(\overline{t} - \overline{t}_*)}, \tag{4.59}$$

and taking into account (4.48) dependence (4.58) reduces to the form

$$\overline{S}_{2e}(\overline{t}) = \overline{S}_m \frac{e^{-\overline{\lambda}_0\overline{t}_*} - e^{-\overline{\lambda}_0\overline{t}} + e^{-\overline{\lambda}_0\overline{S}_m}(1 - e^{-\overline{\lambda}_0\overline{t}_*})}{e^{-\overline{\lambda}_0\overline{t}_*} + e^{-\overline{\lambda}_0\overline{S}_m} - e^{-\overline{\lambda}_0(\overline{t}_* + \overline{S}_m)}}.$$

The concentration $\overline{C}_2$ is found from equation (4.54) under condition (4.49). The cumbersome calculations for the linear form of $\overline{\lambda}(\overline{S}_2)$ are also omitted and finally obtained

$$\overline{C}_2(\overline{z},\overline{t}) = e^{-\overline{\lambda}_0\overline{S}_m(1-\overline{z})}\left[\frac{e^{-\overline{\lambda}_0\overline{t}_*} + (1 - e^{-\overline{\lambda}_0\overline{t}_*})e^{-\overline{\lambda}_0\overline{S}_m}}{e^{-\overline{\lambda}_0\overline{t}_*} + (1 - e^{-\overline{\lambda}_0\overline{t}_*})e^{-\overline{\lambda}_0\overline{S}_m\overline{z}}}\right]^2 \frac{\overline{S}_m - \overline{S}_2(\overline{z},\overline{t})}{\overline{S}_m - \overline{S}_{2e}(\overline{t})}. \tag{4.60}$$

Thus, the increase in dispersed impurity content in the filtrate is described by the dependence

$$\overline{C}_{20}(\bar{t}) = \overline{C}_2(0,\bar{t}) = e^{-\bar{\lambda}_0 \overline{S}_m}\left[e^{-\bar{\lambda}_0 \bar{t}_*} + (1 - e^{-\bar{\lambda}_0 \bar{t}_*})e^{-\bar{\lambda}_0 \overline{S}_m}\right]^2 \frac{\overline{S}_m - \overline{S}_{20}(\bar{t})}{\overline{S}_m - \overline{S}_{2e}(\bar{t})}. \tag{4.61}$$

Finally, it is recommended to calculate the relative head losses at any time $\bar{t} > \bar{t}_*$ after the determination of $\overline{S}_2(\bar{z},\bar{t})$ similarly to (4.43) by the formula

$$\Delta \bar{h}_2(\bar{t}) = \int_0^1 \frac{d\bar{z}}{f_k(\overline{S}_2(\bar{z},\bar{t}))} = \int_0^1 \frac{d\bar{z}}{[1 - \bar{\gamma}(\overline{S}_2(\bar{z},\bar{t}))\overline{S}_2(\bar{z},\bar{t})]^3}. \tag{4.62}$$

It is convenient to evaluate the efficiency of the proposed method of the intensification of the non-detachment filtration from the technological point of view. The subject of calculations in this case are the technological times, namely, the time of the bed protective effect t_p and the time of reaching the maximum head losses in the bed t_h. They are generalized by the filter run duration t_f, which is identified according to (1.41).

The quantitative analysis of the prolonged non-detachment filtration of a suspension on the rapid filter was performed under the conditions when the downward filtration during one operating period was replaced by the upward filtration. The linear absorption function (4.38) (χ=1) and the values of the coefficient $\bar{\lambda}_0$ were taken, namely, or individual values corresponding to moderate and high absorption capacity of a filtering material, or its values from the range from 0.01 to 0.1. The values of other model parameters were fixed, specifically, $\overline{S}_m = 200$, $\bar{\gamma}_a = 0.003$. The time for changing the direction of suspension flow was varied discretely ($\bar{t}_* = 50$ and 100) or continuously from 0 to 100. The subject of many calculations were relative quantities – the head losses in the bed, the volumetric concentrations of the suspended and deposited particles, the technological times.

Figure 7. Increase in the relative head losses in the bed over time ($\bar{\lambda}_0 = 0.025$)

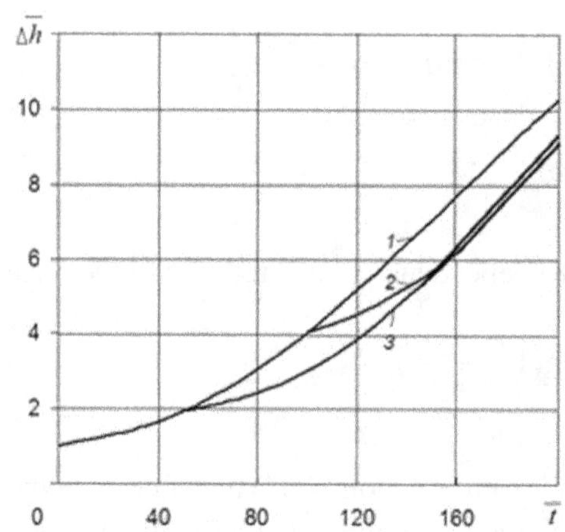

Note. 1 – $\bar{t}_* = 0$ or ∞, 2 – $\bar{t}_* = 100$, 3 – $\bar{t}_* = 50$

Figure 8. The relative volumetric concentration profiles of the deposited particles $\left(\bar{\lambda}_0 = 0.025\right)$

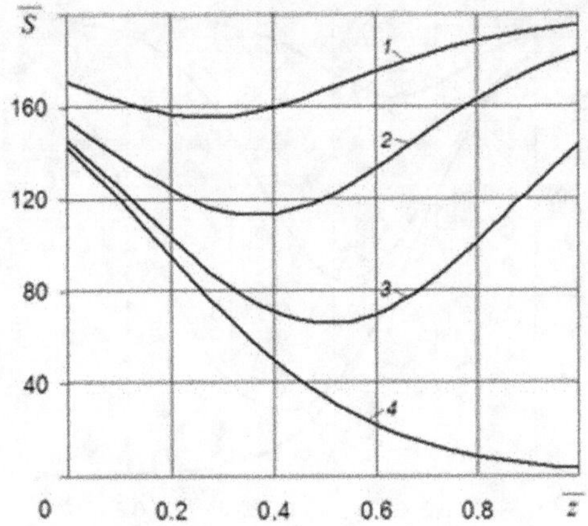

Note. 1 – $\bar{t}$ = 200, 2 – $\bar{t}$ = 150, 3 – $\bar{t}$ = 100, 4 – $\bar{t}$ = 50

First of all, the outlet concentration $\bar{C}_e$ and the time of the protective effect of the bed were calculated. The data obtained in this way showed that the change in the suspension flow direction does not lead to the deterioration of its separation quality in case of the non-detachment filtration in the absence of the autocatalysis. Therefore, the emphasis was made in the calculation part of this Section on the head losses and the time of their reaching the maximum allowable value (in all examples the value $\Delta\bar{h}_*$ was taken as 6). **Figure 7** shows the curves depicting the increase in these losses with progressive clogging of the bed at different filter switching times, including for comparison also $\bar{t}_* = \infty$ (or 0). Preliminary calculations were performed using equations (4.5), (4.6) by formula (4.43). It is evident that the rate of an increase in the hydraulic resistance of the porous layer slows down considerably due to the active involving the previously remote (lower) part of the layer in the clarification process (curves 2, 3). The reason for such a slowdown is a fundamentally different nature of a deposit distribution by the bed height after a suspension has been fed through its opposite surface. The results of calculating the profiles $\bar{S}_2(\bar{z})$ for different time points, starting from the point $\bar{t}_*$, are shown in **Figures 8 and 9**. They are obtained by fitting at two values of $\bar{\lambda}_0$ using (4.6) and (4.55) or (4.58). Due to an enhanced deposition near the new bed inlet, these profiles quickly acquire a shape opposite to the bell-shaped one with a clearly distinct minimum. At the same time, their curvature in the case of a heavy absorption $\left(\bar{\lambda}_0 = 0.05\right)$ turns out to be much larger, and the clarification resource is used obviously more fully.

Figure 9. The relative volumetric concentration profiles of the deposited particles $(\bar{\lambda}_0 = 0.05)$

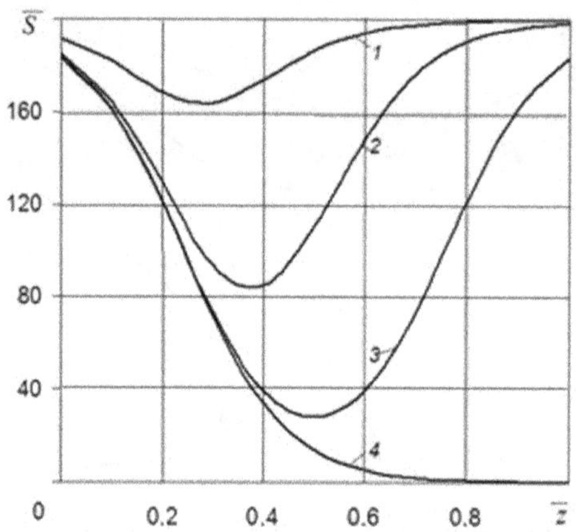

Note. $1 - \bar{t} = 200$, $2 - \bar{t} = 150$, $3 - \bar{t} = 100$, $4 - \bar{t} = 50$

Figure 10. Dependence $\bar{t}_h(\bar{\lambda}_0)$

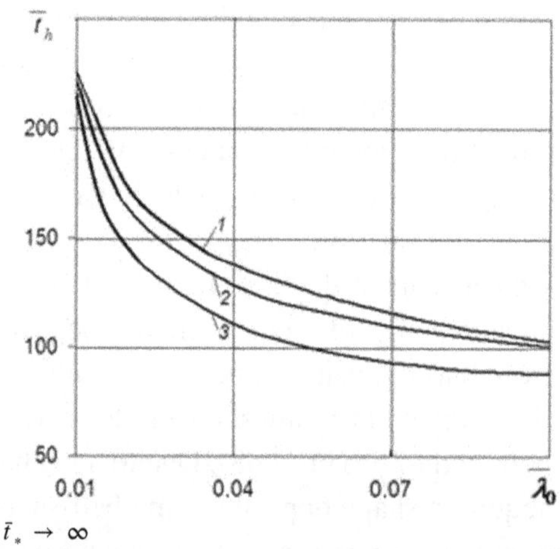

Note. $1 - \bar{t}_* = 50$, $2 - \bar{t}_* = 25$, $3 - \bar{t}_* \to \infty$

The efficiency of the discussed technological technique depends on the absorption properties of a filter media and the time of its implementation. It is preferable to estimate it quantitatively on the basis of the results of calculating the time $\bar{t}_h$, which gives a generalized idea of the degree and nature of a bed contamination. First of all, the value of $\bar{t}_h$ was calculated depending on the coefficient $\bar{\lambda}_0$, characterizing the potential absorption capacity of porous material along with $\bar{S}_m$. Formula (4.62) and criterion (1.36) were used, and the calculation data obtained at continuous change in $\bar{\lambda}_0$ by an order of magnitude are shown in **Figure 10** in the form of a series of graphs. Each of them corresponds to a certain value of $\bar{t}_*$. The effect achieved by redirecting the flow is easy to estimate visually by comparing curves 2, 3 with curve 1. It is most significant at the values of $\bar{t}_*$ between 0.04 and 0.07. Thus, it is advisable to use this technique if a filtering material has a moderate or better significant capability to absorb the suspended solids. In this case, the filter operation is limited due to the pronounced unevenness of the vertical deposit distribution and, as a consequence, significant head losses. At low values of $\bar{\lambda}_0$ up to the critical value $\lambda_{0C} = \ln \bar{C}_* / \bar{S}_m$, if the suspended solids concentration already exceeds the normative value C_* in the first filtrate portion, the quality of the filtrate becomes decisive when deciding whether to stop or continue filtration. The hydraulic aspects of the technological process are relegated to the background. And naturally, all the ways of increasing the time t_h lose their meaning from the technological point of view. However, the filtration process becomes an excessively expensive stage in modern technological schemes of water treatment in such situations due to frequent filter washing, and hence high costs of the treated water.

Figure 11 demonstrates the effect of redirecting the flow in the opposite direction in visual form and in percentage terms. Here its consequences are described by curves of the dependence of the relative increment $\Delta \bar{t}_h = (\bar{t}_{h0} - \bar{t}_{h\infty})/\bar{t}_{h\infty}$ ($\bar{t}_{ho}$, $\bar{t}_{h\infty}$) are the relative values of $\bar{t}_h$ at switching of the filter during and after the end of the calculation period) on the time $\bar{t}_*$ at different values of $\bar{\lambda}_0$. Naturally, the domain of the function $\Delta \bar{t}_h(\bar{t}_*)$ or, in other words, the range of the values of $\bar{t}_*$, at which there is an increase in $\bar{t}_h$, is limited to the values 0 and $\bar{t}_{h\infty}$. In particular, $\Delta \bar{t}_h$ turns to 0 at $\bar{\lambda}_0 = 0.06$, $\bar{t}_{h\infty} = 96.92$ (curve 2). In according to Figure 11, it is realistic to achieve the prolongation of the continuous operation of the filter by about 25% with the chosen initial data due to the timely change of the filtration direction ($\bar{t}_* \approx 50 - 85$ depending on the value of $\bar{\lambda}_0$) (curves 1, 2). In case of a poorly absorbing filtering material the effect is noticeably reduced (curve 3). Finally, Figure 11 shows the importance of the correct choice of the point in time t_*.

The consequences of changing the place of suspension supply are considered for comparison after the primary contamination of the bed, but with an intensive deposit mobilization (β>0). The dynamics of both relative concentrations ($\bar{C}_2$, $\bar{S}_2$) and head losses ($\Delta\bar{h}$), starting from the time point of changing the filtration flow direction ($t=t_*$), was analyzed on the basis of the exact solution of the corrected basic problem of the detachment filtration through the initially contaminated filter media from Section 2.1. In this case, the mass transfer equation (2.7) and the boundary condition (2.9) were replaced by (4.45) and (4.49) respectively. In addition, the initial vertical distribution of the deposited particles $\bar{S}_2^0(\bar{z}) = \bar{S}_1(\bar{z}, \bar{t}_*)$ for the second stage of filtration ($t \geq t_*$) was specified based on the solution presented in Section 2.1. Then, after introducing the new variable Z instead of $\bar{z}$, so that $Z = 1 - \bar{z}$, the specified distribution is described by the function

$$\bar{S}_2{}^0(Z) = \bar{\alpha}\, e^{-\bar{\alpha}(1-Z)} \int_0^{\bar{t}_*} e^{-\bar{\beta}\xi} I_0\left(2\sqrt{\bar{\alpha}\bar{\beta}(1-Z)\xi}\right) d\xi, \quad 1 \geq Z \geq 0. \tag{4.63}$$

Figure 11. Dependence $\Delta\bar{t}_h(\bar{t}_*)$

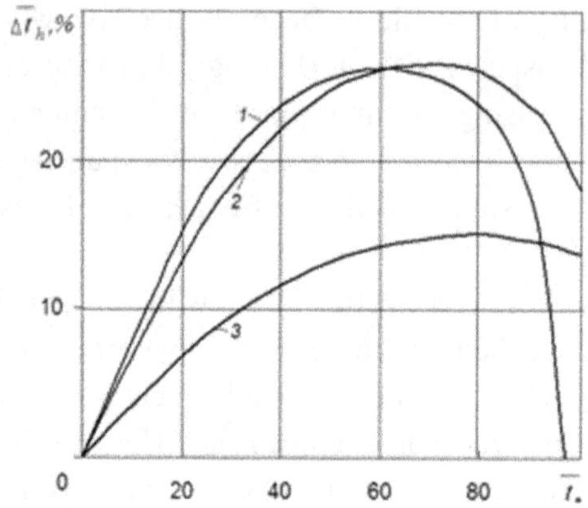

Note. $1-\bar{\lambda}_0 = 0.06$, $2-\bar{\lambda}_0 = 0.04$, $3-\bar{\lambda}_0 = 0.02$

Figure 12. Change in the relative outlet suspended solids concentration and head losses in the bed over time

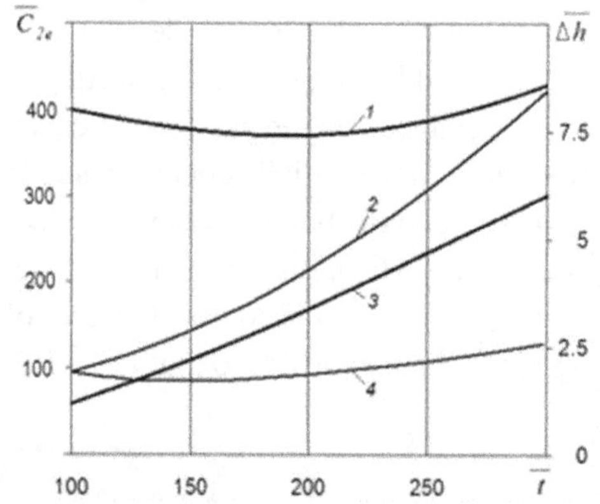

Note. 1, 3 - $\overline{C}_{2e}$; 2, 4 - $\Delta\bar{h}$; 1, 4 - with change in direction; 2, 3 - without change in direction

The solution of the new problem can be expressed, first of all, by the following calculation dependences for the desired concentrations

$$\overline{C}_{2e}(\bar{t}) = e^{-\bar{\alpha}}\left[e^{-\bar{\beta}(\bar{t}-\bar{t}_*)}I_0\left(2\sqrt{\bar{\alpha}\bar{\beta}\bar{z}(\bar{t}-\bar{t}_*)}\right)+\bar{\beta}\int_{\bar{t}_*}^{\bar{t}} e^{-\bar{\beta}\xi}I_0\left(2\sqrt{\bar{\alpha}\bar{\beta}\xi}\right)d\xi\right]+ \\ +\bar{\beta}\,e^{-\bar{\beta}(\bar{t}-\bar{t}_*)}\int_0^1 e^{-\bar{\alpha}(1-\eta)}\,\overline{S}^0(\eta)I_0\left(2\sqrt{\bar{\alpha}\bar{\beta}(1-\eta)(\bar{t}-\bar{t}_*)}\right)d\eta. \tag{4.64}$$

$$\overline{S}_2(Z,\bar{t}) = \overline{S}_2{}^0(Z)e^{-\bar{\beta}(\bar{t}-\bar{t}_*)}+\bar{\alpha}\,e^{-\bar{\alpha}Z}\int_{\bar{t}_*}^{\bar{t}} e^{-\bar{\beta}\xi}I_0\left(2\sqrt{\bar{\alpha}\bar{\beta}Z\xi}\right)d\xi+ \\ +\bar{\alpha}\bar{\beta}\,e^{-\bar{\beta}(\bar{t}-\bar{t}_*)}\int_0^Z e^{-\bar{\alpha}(Z-\eta)}\,\overline{S}^0(\eta)\int_{\bar{t}_*}^{\bar{t}} I_0\left(2\sqrt{\bar{\alpha}\bar{\beta}(Z-\eta)\varsigma}\right)d\varsigma d\eta. \tag{4.65}$$

It follows from (4.65) that the concentration $\overline{S}_2$ at the new inlet surface ($Z = 0,\ \overline{z} = 1$) varies non-monotonically according to the formula

$$\overline{S}_{20}(\overline{t}) = \overline{S}_2^0(0)e^{-\overline{\beta}(\overline{t}-\overline{t}_*)} + \frac{\overline{\alpha}}{\overline{\beta}}(e^{-\overline{\beta}\overline{t}_*} - e^{-\overline{\beta}\overline{t}}).$$

Finally, the complex nature of the head changes at the second stage of filtration follows from the formula

$$\Delta\overline{h}(\overline{t}) = \int_0^1 \frac{dZ}{f_k(\overline{S}_2(Z,\overline{t}))} \tag{4.66}$$

A typical example is calculated for clarity and its results are given in a graphical form (**Figure 12**). The initial information included the standard values $\overline{\alpha} = 5, \overline{\beta} = 0.01, \overline{\gamma}_a = 0.0015$ and also $\overline{t}_* = 100$. All the calculations were carried out for the conventional filtration and according to formulas (4.64) - (4.66) for filtration with the reorientation of a filtration flow. The subject of calculations were the relative concentrations $\overline{C}_{1e}$, $\overline{C}_{2e}$ and losses $\Delta\overline{h}(\overline{S}_1)$, $\Delta\overline{h}(\overline{S}_2)$. The set of curves 1, 2 in Figure 12 shows a jump-like increase of the suspended solids content in the filtrate at the time point of filter switching. The value of the jump $\Delta\overline{C}_e = \overline{C}_{2e} - \overline{C}_{1e}$, as follows from (4.64), will be

$$\Delta\overline{C}_e = \overline{\beta}\int_0^1 e^{-\overline{\alpha}(1-\eta)}\,\overline{S}_2^0(\eta)d\eta$$

Due to heavy contamination of the outlet section of the bed by the point in time $\overline{t}_*$, the quantity $\overline{C}_{2e}$ immediately and noticeably exceeds the normative value $\overline{C}_*$. Thereafter, $\overline{C}_{2e}$ decreases, but not sufficient to obtain the required filtrate quality. Due to the significant deposit removal at $\overline{t} > \overline{t}_*$, as well as the predominant deposition of the suspended solids in the still relatively clean inlet section of the bed, the head losses, at first, even slightly decrease and then increase much more slowly (curves 3, 4). Nevertheless, the fact that the suspended solids content in the filtrate increases excessively in this case is decisive for evaluating the expediency of changing the flow direction in the detachment filtration.

Summarizing the results of the theoretical studies and calculations carried out here, it should be emphasized that the reorientation of an aqueous suspension flow during the non-detachment filtration has little to no effect on the degree of its clarification. However, due to active involvement of filtration in the suspension separation of the removed and still weakly contaminated part of a bed at the second stage, the increase in the head losses in it can be significantly slowed down and, as a consequence, appropriately increase the time to reach their maximum allowable value. This effect is especially clear if the material of the bed has moderate or high absorption capacity. According to our data it is possible to prolong the continuous operation of the filter by 25% and more. Here, the switching time of the filter is of great importance. Specific examples show and it seems logical that it is reasonable to change the direction of a suspension flow after sufficiently heavy clogging of the part of the bed close to the inlet surface (in the examples $\overline{t}_* \approx 50 - 85$). In the case of the detachment filtration, an abrupt change in the flow direction leads to a sudden deterioration of the clarification quality, and the impurity content in the filtrate may immediately exceed the normative value and then remain unacceptably high for a long time.

4.3. CLARIFICATION OF HETEROGENEOUS SUSPENSION IN DYNAMIC LAYER (SURFACE FILTRATION)

The surface filtration is successfully used in the practice of water treatment along with the traditional deep bed filtration. The impurity is retained in the entire volume of the original filter media in the first method of the suspension clarification by filtration. At the same time, this material is used for a long time due to a regular washing and thus, restoring its former absorption capacity. If the second method of filtration is implemented, the surface (dynamic) layer is gradually formed either on a special retention surface with fine perforations or on the surface of the packed bed and is maintained for one filter run. In the first case the surface filtration is used independently, and in the second case as a supplement to the depth filtration

Technically, the surface filtration is implemented mainly due to the sieve effect, namely, depending on the ratio between the effective sizes of both the filter media pores and the larger particles of a suspension impurity. Here, two characteristic situations are possible. If the specified particles are larger than the pores (actually their necks) of a clean bed or perforation holes, the dynamic layer will begin to form immediately after the first portion of water with the dispersed impurity is fed to the filter. In the second situation, the size of large particles of a suspension may be somewhat smaller than the pore size of the bed. However, as a deposit accumulates in the inlet section of the bed, there comes a point in time when the cross-section porosity is so reduced that larger suspension particles are no longer able to penetrate the porous media. Naturally, a layer of the fine-grained material starts to form on the surface of the bed from this point onwards.

Sometimes the surface layer is formed unintentionally, for example, due to inefficient operation of sedimentation tanks and the presence of a coarse dispersed fraction in water that has not been properly prepared. It is obvious that with the simultaneous depth and surface filtration, the mechanical energy expenditure will increase at an accelerated rate due to the development of this layer.

The dynamic layer is often created purposefully in the practice of filtration (precoat filter), in membrane technologies (secondary membrane) to intensify deep bed filtration or to take full advantage of surface filtration (Kuberkar & Davis, 2000; Leichkis,1975; Leichkis & Wentzel, 1987). The advantages of the latter include, first of all, the slow increase in the head losses and the absence of the need for systematic washing. However, one should not fail to mention its essential disadvantage – low protective capability at the initial stage of filtration and formation of the operating layer. It is also important to note that the generation of a dynamic layer with specified properties can be really controlled – it is enough to select the appropriate dispersed matter (sorbent), to dose its addition to the initial contaminated water and to choose a hydraulic load.

Simulation of the surface filtration is difficult due to the expansion of the filtration flow region over time. If filtration is carried out at a given head (pressure) differential at the boundaries of the dynamic layer, the pattern of movement of the upper boundary of the specified area is not known in advance and must be found in the course of solving the corresponding mathematical problem. The application of the mathematical modeling methods is especially complicated in cases of the polydisperse impurities containing three or more particle fractions (Dharmappa et al., 1992; Veerapaneni & Wiesner, 1993), as well as significant compressibility of the surface layer at high operating pressures (Abboud & Corapcioglu, 1993; Ho & Zydnev, 2000).

This Section deals with the filtration of an aqueous suspension through a newly formed and expanding layer of a filter media over time. Its particles are initially contained in the suspension together with finer particles. Thus, the dispersed impurity essentially consists of particles of two varieties, with the volumetric concentration of the coarser (first) fraction being C_l, and the finer (second) fraction being C_0. All coarse particles, unlike fine particles, are trapped on a special surface with a large number of small holes, forming a porous layer. This dynamic layer ensures that the removal of the fine particles becomes more and more complete over time due to the absorption process.

Thus, analytical methods are used to study the non-detachment surface filtration sequentially in two modes: at a constant filtration rate (*V=const*) and a constant head differential at the boundaries of the dynamic layer (*Δh=const*) (Poliakov, 2019; 2023). Particles of both varieties do not interact with each other in the suspended state. The deposit is formed in the pores of the dynamic layer exclusively from small particles and a relatively large amount of the water bound by them.

The initial mathematical model of the surface filtration consists of the interrelated clarification and filtration flow compartments. The first compartment contains first of all the equations of the kinetics of fine particles removal (4.1), as well as their balance, which is justified to take at the vertical filtration and variable filtration rate in the following form

$$V(t)\frac{\partial C}{\partial z} - \frac{\partial S}{\partial t} = 0, \tag{4.67}$$

where *C, S* are the volumetric concentrations of the suspended and deposited fine particles of a suspension. The filtration coefficient in (4.1) is assumed to be a function of arbitrary form from the concentration *S*. Equations (4.1), (4.67) are supplemented by a set of two boundary conditions on the moving upper boundary of the filtration flow region

$$z = l(t), \quad C = C_0; \quad S = 0. \tag{4.68}$$

In this case, the lower boundary of the filtration flow region (*z*=0) is immobile, and the change in the concentrations *C, S* on it over time is determined in the course of solving the general problem.

The second compartment includes the motion equation with respect to linear filtration flow

$$V(t) = k(S_S)\frac{\partial h}{\partial z} \tag{4.69}$$

as well as the equations characterizing in general the hydraulic resistance of the dynamic layer (1.9) and the deposit composition (1.29). They are supplemented by the condition at the bottom boundary of the layer

$$z = 0, \quad h = H_d = const. \tag{4.70}$$

To make problem (1.9), (1.29), (4.67) - (4.70) dimensionless, we introduce a scale system including the fixed values of C_0, n_l, V_0, l_0, k_0. Then

$$\overline{C} = C/C_0,\ \overline{S} = S/(n_l C_0),\ \overline{z} = z/l_0,\ \overline{t} = V_0 t/(n_l l_0),$$
$$\overline{\lambda} = l_0\lambda/V_0,\ \overline{k} = k/k_0 = f_k(\overline{S}_S),\ \overline{S}_S = S_S/(n_l C_0),$$
$$\overline{V} = V/V_0,\ \overline{h} = k_0(h - H_d)/(l_0 V_0),\ \overline{l} = l/l_0,$$

n_l is the porosity of the clean layer and the initial problem is transformed as follows

$$\bar{V}(\bar{t})\frac{\partial \bar{C}}{\partial \bar{z}} - \frac{\partial \bar{S}}{\partial \bar{t}} = 0 \;; \tag{4.71}$$

$$\frac{\partial \bar{S}}{\partial \bar{t}} = \bar{\lambda}(\bar{S})\bar{C} \;; \tag{4.72}$$

$$\bar{V}(\bar{t}) = \bar{k}(\bar{S}_S)\frac{\partial \bar{h}}{\partial \bar{z}} \;; \tag{4.73}$$

$$\bar{k}(\bar{S}_S) = f_k(\bar{S}_S) \;; \tag{4.74}$$

$$\bar{S}_S = \gamma(\bar{S})\bar{S} \;; \tag{4.75}$$

$$\bar{z} = \bar{l}(\bar{t}), \quad \bar{C} = 1; \quad \bar{S} = 0 \;; \tag{4.76}$$

$$\bar{z} = 0, \quad \bar{h} = 0. \tag{4.77}$$

The surface filtration is investigated separately for the two main operating (rate) modes due to the fundamental differences in the course and results of solving the corresponding mathematical problems.

4.3.1. Surface Filtration with Constant Rate

Let the filtration rate be kept constant, so that $V=V_0=const$. Then $\bar{V} = 1$ and it is simple to find the pattern of the rise of the upper boundary of the dynamic layer at its known porosity n_l, which has in this case the linear form

$$l(t) = a_l t, \qquad \bar{l}(\bar{t}) = \bar{a}_l \bar{t}, \tag{4.78}$$

where $a_l = V_0 C_l/(1 - n_l)$, $\bar{a}_l = n_l a_l/V_0$.

To solve problem (4.71) - (4.77), the concentration $\bar{C}$ is expressed according to (4.72) through $\bar{S}$, so that

$$\bar{C}(\bar{S}) = \frac{1}{\bar{\lambda}(\bar{S})}\frac{\partial \bar{S}}{\partial \bar{t}}. \tag{4.79}$$

Then equation (4.71) reduces to the equation with respect to $\bar{S}$

$$\frac{\partial}{\partial \bar{t}}\left[\frac{1}{\bar{\lambda}(\bar{S})}\frac{\partial \bar{S}}{\partial \bar{z}} - \bar{S}\right] = 0 \tag{4.80}$$

As a result of integration of (4.80) by $\bar{t}$ we obtained

$$\frac{1}{\bar{\lambda}(\bar{S})}\frac{\partial \bar{S}}{\partial \bar{z}} = \bar{S} + A, \tag{4.81}$$

where A is the unknown integration constant. Now integrating (4.81) by $\bar{z}$ taking into account the second condition (4.76) gives

$$\int_0^{\bar{S}} \frac{d\varsigma}{\bar{\lambda}(\varsigma)(\varsigma + A)} = \bar{z} - \bar{a}_l\bar{t}. \tag{4.82}$$

By differentiation (4.82) on $\bar{t}$ is found

$$\frac{\partial \bar{S}}{\partial \bar{t}} = -\bar{a}_l\bar{\lambda}(\bar{S})(\bar{S} + A) \tag{4.83}$$

Then, taking into account (4.83), from (4.79) follows the expression for $\bar{C}$

$$\bar{C}(\bar{S}) = -\bar{a}_l(\bar{S})(\bar{S} + A). \tag{4.84}$$

The unknown constant A is determined from (4.84) using the first condition (4.76) and is equal to $-1/\bar{a}_l$. Thus, the equation with respect to (4.82) takes the final form

$$\bar{a}_l \int_0^{\bar{S}} \frac{d\varsigma}{\bar{\lambda}(\varsigma)\,(1 - \bar{a}_l\varsigma)} = \bar{a}_l\bar{t} - \bar{z}. \tag{4.85}$$

Correspondingly, for the concentration $\bar{C}$ the following is obtained

$$\bar{C}(\bar{S}) = 1 - \bar{a}_l\bar{S}. \tag{4.86}$$

Therefore, the outlet relative concentrations $\bar{C}_e$ and $\bar{S}_e$ will change over time as follows

$$\bar{C}_e(\bar{t}) = \bar{C}(0,\bar{t}) = 1 - \bar{a}_l\bar{S}_e(\bar{t}). \tag{4.87}$$

$$\int_0^{\bar{S}_e} \frac{d\bar{S}}{\bar{\lambda}(\bar{S})\,(1 - \bar{a}_l\bar{S})} = \bar{t}. \tag{4.88}$$

In the general case, the function $\bar{\lambda}(\bar{S})$ is represented as a series of the nonlinear multipliers (4.13). Then the integral in (4.88) can be calculated simply, but only numerically. If $\bar{\lambda}(\bar{S})$ is expressed as a series of the linear multipliers according to (4.14), then the specified integral is equal to

$$\frac{1}{\bar{\lambda}_0}\left[\frac{\bar{a}_l\theta}{(1 - \bar{a}_l\bar{S}_m)\,(1 + \theta\bar{S}_m)}\ln\frac{\bar{S}_m}{\bar{S}_m - \bar{S}_e} + \right.$$
$$\left. + \frac{\bar{a}_l^2\theta}{(1 - \bar{a}_l\bar{S}_m)\,(\bar{a}_l + \theta)}\ln(1 - \bar{a}_l\bar{S}_e) + \frac{\bar{a}_l\theta^2}{(1 + \theta\bar{S}_m)\,(\bar{a}_l + \theta)}\ln(1 + \theta\bar{S}_e)\right]$$

and taking into account (4.88) the relation between $\bar{S}_e$ and $\bar{t}$ is in this form

$$(\bar{S}_m - \bar{S}_e)(1 - \bar{a}_l\bar{S}_e)^{æ_1}(1 + \theta\bar{S}_e)^{æ_2} = \bar{S}_m \exp\left[-\frac{(1 - \bar{a}_l\bar{S}_m)(1 + \theta\bar{S}_m)}{\bar{a}_l\theta}\bar{\lambda}_0\bar{t}\right], \tag{4.89}$$

where

$$æ_1 = -\bar{a}_l(1 + \theta\bar{S}_m)/(\bar{a}_l + \theta),\ æ_2 = -\theta(1 - \bar{a}_l\bar{S}_m)/(\bar{a}_l + \theta).$$

Neglecting the effect of the autocatalysis (θ=0), it is possible in general to derive the representation for $\bar{S}_e$ in explicit form, viz,

$$\bar{S}_e(\bar{t}) = \bar{S}_m \frac{1 - e^{-(1-\bar{a}_l\bar{S}_m)\bar{\lambda}_0\bar{t}}}{1 - \bar{a}_l\bar{S}_m e^{-(1-\bar{a}_l\bar{S}_m)\bar{\lambda}_0\bar{t}}}. \tag{4.90}$$

Incidentally, in the simplest case $\bar{\lambda}(\bar{S}) = \bar{\lambda}_0$

$$\bar{C}(\bar{z},\bar{t}) = e^{-\bar{\lambda}_0(\bar{a}_l\bar{t}-\bar{z})},$$

$$\bar{S}(\bar{z},\bar{t}) = \frac{1}{\bar{a}_l}[1 - e^{-\bar{\lambda}_0(\bar{a}_l\bar{t}-\bar{z})}].$$

It is also succeeded to obtain an expression similar to (4.89) if $\bar{\lambda}(\bar{S})$ includes four linear multipliers. However, then the calculations become too cumbersome, and there is no practical sense in them, because $\bar{S}_e$ is necessary to find by fitting, as well as from (4.88).

The reduced head $\bar{h}$ is determined by the integration of equation (4.73) under condition (4.77). Then

$$\bar{h}(\bar{z},\bar{t}) = \int_0^{\bar{z}} \frac{d\eta}{\bar{k}(\bar{S}(\eta,\bar{t}))} \tag{4.91}$$

But since the required quantity $\bar{S}$ is obtained above in implicit form, the integral in (4.91) is transformed taking into account that

$$\frac{\partial\bar{S}}{\partial\bar{z}} = \left(\bar{S} - \frac{1}{\bar{a}_l}\right)\bar{\lambda}(\bar{S}). \tag{4.92}$$

Then

$$\bar{h}(\bar{S}) = \bar{a}_l \int_0^{\bar{S}} \frac{d\varsigma}{\bar{\lambda}(\varsigma)(1 - \bar{a}_l\varsigma)\bar{k}(\varsigma)}. \tag{4.93}$$

Obviously, the relative head losses $\Delta\bar{h}$ in the dynamic layer according to (4.93) are as follows

$$\Delta\bar{h}(\bar{S}_e(\bar{t})) = \bar{a}_l \int_0^{\bar{S}_e(\bar{t})} \frac{d\varsigma}{\bar{\lambda}(\varsigma)(1 - \bar{a}_l\varsigma)\bar{k}(\varsigma)}, \tag{4.94}$$

where the dependence $\bar{S}_e(\bar{t})$ should be found from (4.88).

The obtained calculation dependences can be useful for a special analysis to optimize the suspension filtration through the dynamic layer. Thus, it is easy to calculate from (4.87) the relative time $\bar{t}_*$, for which the relative outlet concentration $\bar{C}_e$ will decrease to the maximum allowable value $\bar{C}_*$. Taking into account (4.86), (4.87), this time is found by the formula

$$\bar{t}_* = \int_0^{(1-\bar{C}_*)/\bar{a}_l} \frac{d\bar{S}}{\bar{\lambda}(\bar{S})(1-\bar{a}_f\bar{S})}. \tag{4.95}$$

Obviously, the time $\bar{t}_*$ can be adjusted by changing appropriately $\bar{a}_l$. It should be noted that the parameter t_* coincides in meaning with the time of the protective capability of a bed t_p, widely used in the technological analysis of the depth filtration. However, here, unlike the previously comprehensively studied case, the situation looks directly opposite with respect to the filtrate quality. Indeed, the content of the suspended solids at the outlet of the dynamic layer steadily decreases over time from the initial maximum C_0. And its outlet concentration will reach the normative value C_* only after a very specific time t_*. But in the future the required quality of filtrate is practically guaranteed up to a new filter run. Thereby, necessary to bear in mind that at the initial stage of the surface filtration the suspension is clarified to an insufficient degree and it will have to be re-treated.

The behavior of the head losses function $\Delta h(t)$ at the depth and surface filtration has a similar nature. Indeed, the quantity Δh monotonically increases in both cases, but the increase rate is noticeably different. When evaluating the operability of the dynamic layer is of interest to determine the second technological time, namely, the time t_h to achieve the head losses of the maximum allowable value Δh_* and also to study the possibility of increasing t_h, varying a_l. It is proposed to calculate the required $\bar{t}_h$ at a given $\Delta \bar{h}_*$ depending on $\bar{a}_l$ in two stages. The value of $\bar{S}_{eh}$ is found at the first stage by fitting from the equation

$$\int_0^{\bar{S}_{eh}(\bar{a}_l)} \frac{d\bar{S}}{\bar{\lambda}(\bar{S})(1-\bar{a}_f\bar{S})\bar{k}(\bar{S})} = \frac{\Delta \bar{h}_*}{\bar{a}_l}, \tag{4.96}$$

and then $\bar{t}_h$ directly set according to the formula

$$\bar{t}_h(\bar{a}_l) = \int_0^{\bar{S}_{eh}(\bar{a}_l)} \frac{d\bar{S}}{\bar{\lambda}(\bar{S})(1-\bar{a}_f\bar{S})}. \tag{4.97}$$

Thus, the time interval Δt_f, during which the filter will provide the required quality of the filtrate at permissible mechanical energy expenditure, is

$$\Delta \bar{t}_f = \bar{t}_h - \bar{t}_* \tag{4.98}$$

In specific conditions of the surface filtration, it is necessary to strive to maximize the quantity Δt_f. In practice, it is possible to influence purposefully Δt_f through the concentration C_l, which is usually realistically change within very wide limits. And since C_l is linearly related to the coefficient a_l according to (4.78), it is convenient to operate a_l as an argument of the required dimensionless functional quantity $\Delta \bar{t}_f$, when choosing a rational control algorithm for the surface filtration. And then it is of practical interest to determine the value a_{lm}, at which the function $\Delta \bar{t}_f(a_l)$ reaches its maximum. The extremum of this function is simply found by the calculation of $\bar{t}_*$ directly using formula (4.95) and $\bar{t}_h$ by solving system (4.96), (4.97).

It should be noted that long extension of the dynamic layer is inexpedient for technological reasons. Indeed, in this way it is possible to minimize the content of the suspended solids in the filtrate, but at the same time the head losses in the entire layer will steadily increase, including due to the increase in the height of a porous layer. It is precisely these losses that mainly limit the surface filtration.

4.3.2. Surface Filtration with Constant Pressure Differential

Mathematical modeling of the surface filtration is significantly complicated in the second rate mode, when different constant pressures (heads) are maintained at both boundaries of the dynamic layer. In this case, the filtration rate decreases over time, and the corresponding pattern should be established in the course of solving the initial mathematical problem with variable coefficients. By analogy with the detachment filtration, when often and justifiably both mass-exchange coefficients are assumed to be directly proportional to the filtration rate, it is also logical to assume that the filtration coefficient is linearly related to the variable V. Then it is true

$$\bar{\lambda}(\bar{S},\bar{V}) = \bar{V}\bar{\bar{\lambda}}(\bar{S}) \tag{4.99}$$

and it is reasonable to introduce a new relative independent variable

$$\bar{\tau} = \int_0^{\bar{t}} \bar{V}(\xi)\,d\xi. \tag{4.100}$$

From the balance of a coarse impurity in the dynamic layer follows

$$\bar{l}(\bar{t}) = \bar{a}_l\bar{\tau}. \tag{4.101}$$

The clarification compartment was solved at the second rate mode in the same way as at the first mode. As a result, the dependence for the concentration $\overline{C}$ duplicating (4.86), as well as the equation similar in form to (4.85) were derived

$$\bar{a}_l \int_0^{\bar{S}} \frac{d\zeta}{\bar{\bar{\lambda}}(\zeta)(1-\bar{a}_l\zeta)} = \bar{a}_l\bar{\tau} - \bar{z}. \tag{4.102}$$

The solution of the filtration flow compartment (4.73) - (4.77) is first of all described by the reduced head distribution function

$$\tilde{h}(\bar{z},\bar{\tau}) = \bar{V}(\bar{\tau}) \int_0^{\bar{z}} \frac{d\eta}{f_k(\bar{S}(\eta,\bar{\tau}))} \tag{4.103}$$

Expression (4.103) is obtained by integrating equation (4.73) under condition (4.77) and contains the unknown variable $\bar{\tau}$. Therefore, it is necessary to find additionally for specific calculations the relationship between $\bar{\tau}$ and $\bar{t}$. Since, according to the specified conditions of the problem under consideration, a certain constant pressure p_0 is maintained at the upward moving boundary of the dynamic layer, the condition for the required head must be satisfied

$$\bar{z} = \bar{a}_l\bar{\tau}, \quad \bar{h} = \Delta\bar{h}_0 + \bar{a}_l\bar{\tau} \tag{4.104}$$

Due to the absence of a characteristic constant filtration rate in the formulation of this problem, it is convenient to use the coefficient k_0 as the appropriate scale. Then

$\bar{t} = k_0 t/(n_l l_0)$, $\bar{\bar{\lambda}} = l_0\bar{\lambda}$, $\bar{V} = V/k_0$, $\bar{h} = (h - H_d)/l_0$, ρ_w

is the suspension density, $\Delta\bar{h}_0 = \Delta h_0/l_0 = (p_0 - \rho g H_d)/(\rho g l_0)$, g is the gravity acceleration. Taking into account (4.104), the following ordinary nonlinear differential equation of the first order follows from (4.103) with respect to $\bar{\tau}$

$$\int_0^{\bar{a}_l\bar{\tau}} \frac{d\eta}{\bar{k}(\bar{S}(\eta,\bar{\tau}))}\frac{d\bar{\tau}}{d\bar{t}} = \Delta\bar{h}_0 + \bar{a}_l\bar{\tau}. \tag{4.105}$$

Solution of (4.105) under the homogeneous condition

$\bar{t} = 0, \quad \bar{\tau} = 0$

has the form of the following inverse function

$$\bar{t}(\bar{\tau}) = \int_0^{\bar{\tau}} (\Delta\bar{h}_0 + \bar{a}_l\sigma)\left[\int_0^{\bar{a}_l\sigma} \frac{d\eta}{\bar{k}(\bar{S}(\eta,\bar{\tau}))}\right]^{-1} d\sigma. \tag{4.106}$$

However, due to the implicit representation of the function $\bar{S}(\bar{z}, \bar{\tau})$ (4.102) the integral in (4.106) has to be transformed taking into account

$$\frac{\partial\bar{z}}{\partial\bar{S}} = -\frac{\bar{a}_l}{\bar{\lambda}(\bar{S})(1-\bar{a}_l\bar{S})}, \quad \frac{d\bar{\tau}}{d\bar{S}_e} = -\frac{1}{\bar{\lambda}(\bar{S}_e)(1-\bar{a}_l\bar{S}_e)}.$$

As a result, we derived the dependence of time $\bar{t}$ on the newly introduced parameter $\bar{S}_e$ in the following form

$$\bar{t}(\bar{S}_e) = \bar{a}_l\int_0^{\bar{S}_e} \frac{1}{[\Delta\bar{h}_0 + \bar{a}_l\bar{\tau}(\bar{S}_e)]\cdot\bar{\lambda}(\varsigma)(1-\bar{a}_l\varsigma)}\int_0^{\varsigma}\frac{d\bar{S}}{\bar{k}(\bar{S})\bar{\bar{\lambda}}(\bar{S})(1-\bar{a}_l\bar{S})}d\varsigma. \tag{4.107}$$

In this case, the relative filtration rate as a function of $\bar{S}_e$ should be calculated by the formula

$$\bar{V}(\bar{S}_e) = \left[\frac{\Delta\bar{h}_0}{\bar{a}_l} + \bar{\tau}(\bar{S}_e)\right]\left[\int_0^{\bar{S}_e}\frac{d\bar{S}}{\bar{k}(\bar{S})\bar{\bar{\lambda}}(\bar{S})(1-\bar{a}_l\bar{S})}\right]^{-1}. \tag{4.108}$$

Finally, the dynamics of the upper boundary of the dynamic layer must also be calculated essentially in two steps. At the given value of $\bar{S}_e$, the relative moment in time is successively specified at which the indicated boundary will rise at a height according to the formula (second stage)

$$\bar{l}(\bar{S}_e) = \int_0^{\bar{S}_e} \frac{1}{\bar{\lambda}(\bar{S})(1 - \bar{a}_l\bar{S})}. \tag{4.109}$$

Thus, the exact solution of the second problem of the surface filtration has been obtained in dimensionless parametric form. Some inconvenience in its practical use is due to the presence in the dimensionless variables and parameters of the length scale l_0, which is introduced purely formally. Therefore, in this case it is justified to get rid of l_0 and return to dimensional quantities. Then, for example, dependence (4.107) is transformed to the following form

$$t(S_e) = a_l \int_0^{S_e} \frac{1}{[\Delta h_0 + a_l \tau(\varsigma)]\,\tilde{\lambda}(\varsigma)\,(C_0 - a_l\varsigma)} \int_0^{\varsigma} \frac{dS}{k(S)\tilde{\lambda}(S)\,(C_0 - a_l S)} d\varsigma \tag{4.110}$$

Also, the expression for the dimensional filtration rate formally coincides with (4.108), in which all dimensionless quantities are replaced by similar dimensional quantities.

For the rational control of the surface filtration in the second rate mode it is necessary to calculate two technological times beforehand. The first time (t_*), as in the case of the first mode, is based on the maximum allowable concentration of impurities at the outlet of the dynamic layer C_* by the following dimensional formula

$$t_* = a_l \int_0^{(C_0 - C_*)/a_l} \frac{1}{[\Delta h_0 + a_l \tau(\varsigma)]\,\tilde{\lambda}(\varsigma)\,(C_0 - a_l\varsigma)} \int_0^{\varsigma} \frac{dS}{k(S)\tilde{\lambda}(S)\,(C_0 - a_l S)} d\varsigma. \tag{4.111}$$

The second technological time, namely, the time t_V of a reduction of the filtration rate to the minimum allowable value V_*, is determined on the basis of formula (4.108), as well as t_h, in two stages. First, the corresponding value S_{e*} is determined by fitting at a given value V_* from the equation

$$a_l V_* \int_0^{S_{e*}} \frac{dS}{k(S)\tilde{\lambda}(S)(1 - a_l S)} = \Delta h_0 + a_l \tau(S_{e*}), \tag{4.112}$$

and then the desired t_V also by formula (4.111), but at $S_e = S_{e*}$. In this case, the final goal of the technological calculations is the dimensional value of Δt_f, which is found by analogy with (4.98) as follows

$$\Delta t_f = t_V - t_*. \tag{4.113}$$

4.3.3. Quantitative Analysis

The quantitative analysis performed on a number of test examples allowed to illustrate clearly the obtained above exact solutions of two mathematical problems of the surface filtration, to demonstrate their wide possibilities in predicting the development of physical and chemical conditions in the newly formed dynamic layer, also in substantiating technological parameters. Initial data for them were selected and varied in order to estimate the significance of key model parameters for the clarification of the suspension with coarse and fine fractions of impurity. Previously repeatedly used expressions (4.14) and (1.10) were adopted for the relative functional coefficients of filtration $\bar{\lambda}(\bar{S})$ and flow $f_k(\bar{S})$ at m_{k1}=1, m_{k2}=3 and the average value of $\gamma(S)$. The subject of the predictive calculations for both rate modes were the relative values: volumetric concentrations of the suspended $\bar{C}$ and deposited $\bar{S}$ fine particles of the suspension, head losses in the entire dynamic layer$\Delta\bar{h}$, filtration flow rate $\bar{V}$, position of the upper boundary of the specified layer $\bar{l}$. Their changes over time, as well as the concentration profiles $\bar{S}(\bar{z})$ at given time points, were calculated in detail on the basis of the derived dependences in dimensionless form. The numerical values of the integrals contained in them were calculated using the Mathcad software package. The initial parameters from both problems can be conditionally divided into two groups. The values of the parameters from the first group were fixed: $\bar{S}_m = 200, \bar{\gamma}_a = 0.0035, \Delta\bar{h}_0 = 1$ (the choice of the value $\Delta\bar{h}_0$, and in fact l_0 is actually not fundamental, since it does not affect the dimensional characteristics of filtration). The parameters from the second group were alternately changed and finally four examples were compiled. For the basic example, the values $\bar{\lambda}_0 = 0.02$, θ=0, $\bar{a}_l = 0.01$ were given. The values of θ and $\bar{a}_l$ remained unchanged in the second example, and in the third example only θ changed (0.01), in the fourth example – $\bar{a}_l$ (0.02).

In the first series of calculations, the outlet concentration $\bar{C}_e$ was found using formula (4.87), and the parameter contained in it $\bar{S}_e$ – from (4.88). The series of four curves corresponding to the four examples is presented in **Figure 13**. Of particular note is the sharp decrease in the content of a fine impurity in the filtrate immediately after the start of filtration. Thus, the initial stage of the surface filtration with unsatisfactory filtrate quality should be terminated relatively quickly. It is natural that the autocatalysis (θ =0.01), accelerated formation of the surface layer $(\bar{a}_l = 0.02)$ considerably (in one and a half-two times and more) reduce the duration of the first stage.

Further, the profiles $\bar{S}(\bar{z})$ for the basic example were calculated for three moments in time, for the second - fourth examples – only when $\bar{t} = 100$ proceeding from equation (4.85). The corresponding curves are shown in **Figure 14**. It should be noted here that there is a significant nonlinearity in the distribution of the deposited fine particles by the height of the dynamic layer near its upper boundary, where their concentration increases sharply from 0. The calculated profiles become more filled with increasing absorption of fine particles (curves 4, 5). Curve 1 is a piece of the full profile corresponding to example 4 and determined on the interval [0, 2], since the dynamic layer boundary moves twice as fast in this case $(\bar{a}_l = 0.02)$. In general, it becomes evident from Figures 13 and 14, that there are significant differences in the nature of changes in $\bar{C}_e$ and $\bar{S}$ at the surface and depth filtration.

Figure 13. Decrease in the relative volumetric concentration of the suspended solids in the filtrate

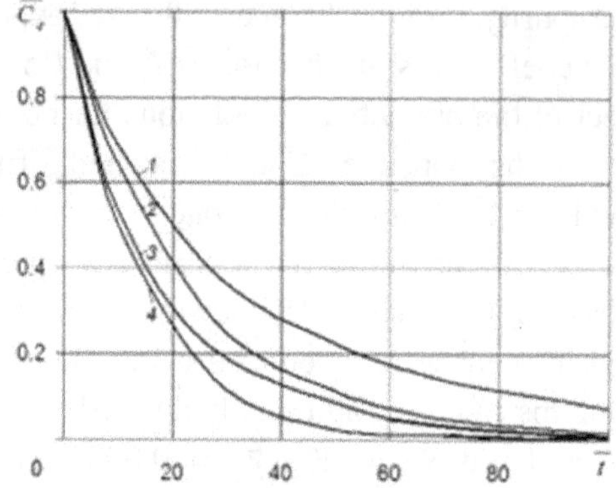

Note. 1 – basic example, 2– example 3, 3 – example 2, 4– example 4

Figure 14. Profiles of the relative volumetric concentration of the deposited fine particles

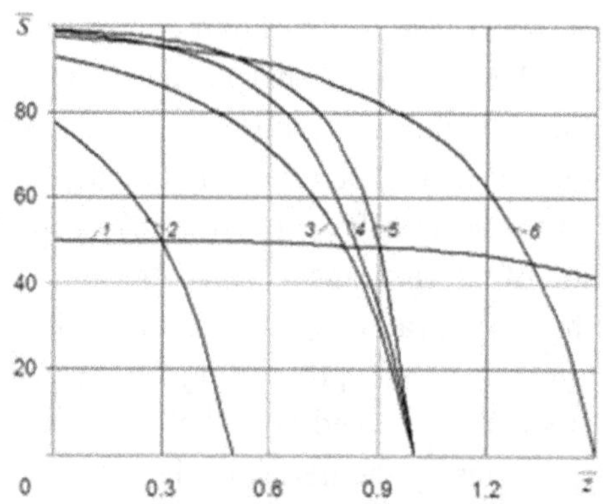

Note. 1, 3-5 – $\bar{t}$ = 100; 2 – $\bar{t}$ = 50; 6 –$\bar{t}$ = 150; 1 – example 4;2, 3, 6 – basic example; 4 – example 3; 5 – example 2

The behavior of the function $\Delta\overline{h}(\overline{t})$ with both methods of filtration (depth and surface), as follows, in particular, from **Figure 15**, is similar in general. However, the rate of an increase in the head losses at the surface filtration, calculated using (4.88), (4.94), is noticeably smaller due, first of all, to the initial past of a very large number of small particles through the still thin dynamic layer.

The theoretical substantiation of the technological parameters, in particular, the content of coarse particles in the suspension can be a useful application of the above solutions. **Figure 16** presents in a graphical form the relationship between the relative time of achieving the minimum required protective capability of the dynamic layer $\overline{t}_*$ and the coefficient of its formation rate $\overline{a}_l$. Three curves calculated by formula (4.95) and corresponding to different suspension clarification standards are shown here. It is clearly seen from here that, firstly, the dynamic layer will be ready for full-fledged operation in a short time already at the minimum amount of the structure-forming particles, and secondly, the calculation time $\overline{t}_*$ is clearly more sensitive to $\overline{a}_l$, and hence to C_e at their small values.

Figure 15. Increase in the relative head losses in the dynamic layer over time

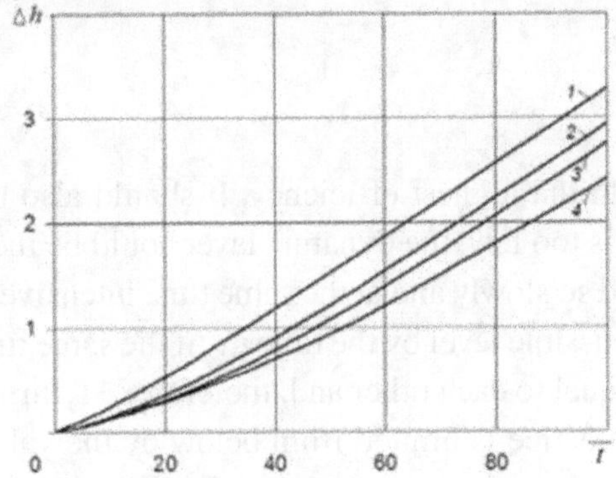

Note. 1 – example 4, 2 – example 2, 3 – example 3, 4 – basic example

Figure 16. Dependence $\bar{t}_*(\bar{a}_l)$

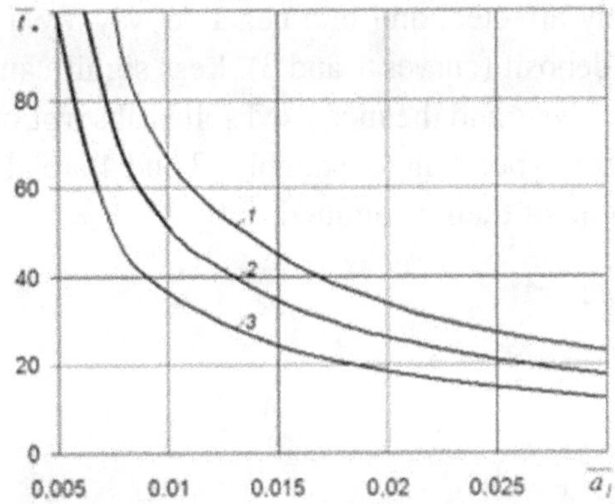

Note. 1 – $\bar{C}_*$ = 0.05, 2 – $\bar{C}_*$ = 0.1, 3 – $\bar{C}_*$ = 0.2

The above proposed approach to the technological optimization of the surface filtration is tested, and at the same time the possibilities of prolongation of continuous productive operation of the filter with a dynamic layer due to the rational choice of the content of a coarse-dispersed fraction in the heterogeneous suspension are studied also on four representative examples. The reference values 200 $(\bar{S}_m)$, 0.1 $(\bar{C}_*)$, 6 $(\Delta \bar{h}_*)$ are given, and, in addition, expressions are involved for $\bar{\lambda}(\bar{S})$ (4.14), for $f_k(\bar{S})$ (1.9), (1.10) with averaging $\bar{\gamma}(\bar{S})$. The input information for the baseline example also included:$\bar{\lambda}_0 = 0.02$, $\theta = 0$, $\bar{\gamma}_a = 0.0015$.In the second example, the coefficient $\bar{\gamma}_a$ was doubled (0.003), in the third example – also doubled $\bar{\lambda}_0$ (0.04), and in the fourth example – the autocatalysis was additionally taken into account (θ =0.01). Due to such selection of the examples it was possible to get an idea about sensitivity of the key technological parameter $(\Delta \bar{t}_f)$ in relation to the most important model coefficients $(\bar{\lambda}_0, \theta, \bar{\gamma}_a)$. The results of calculating the dependence of $\Delta \bar{t}_f$ on $\bar{a}_l$, carried out for the constant rate filtration mode according to (4.95) - (4.98), are given in **Figure 17**. First of all, we should note the pronounced nonlinear nature of the calculated curves and the presence of a maximum in them. From the practical point of view, the value of the argument $\bar{a}_{lm}$, at which it is reached, deserves special attention. It is obvious that at the value

$$C_l = \frac{1-n_l}{n_l}\bar{a}_{lm}$$

such a filter will operate with the highest efficiency. It should also be noted that if the content of coarse particles in a suspension is too low, the dynamic layer could be incapacitated altogether. Indeed, the specified layer will be formed so slowly and at the same time intensively clogged that the head losses in it will already exceed the permissible level by the time t_*. At the same time, there exists such a positive value $\bar{a}_{l0}$, at which t_h and t_* are equal to each other and, therefore, $\Delta\bar{t}_f$ turns to 0. Obviously, the range of the acceptable values of $\bar{a}_l$ for practice is limited from below by the value $\bar{a}_{l0}$ and, of course, includes the best value for controlling $\bar{a}_{lm}$. It is appropriate to emphasize a fundamental difference in the technological analysis of the surface and depth filtration. In the first case, the equality of the technological times t_h, t_* serves only as a guide for the subsequent search for the optimal clarification conditions and the values of the model parameters that ensure them. In the second case, the equality of t_h and t_p indicates just the achievement of the maximum duration of continuous operation of a rapid filter in accordance with both efficiency criteria. Varying the coefficients $\bar{\lambda}_0$, θ, $\bar{\gamma}_a$ affected the quantity $\Delta\bar{t}_f$ to different degrees. It was most significantly affected, and in a negative way, by a decrease in the proportion of fine suspension particles in the deposit (curves 1 and 3). Less significant were the changes in $\Delta\bar{t}_f$ due to taking into account the autocatalysis and the increased solids absorption (Examples 1 and 3, curves 2 and 1). Incidentally, the curves corresponding to examples 3 and 4 are almost merged and, in order not to overcrowd Figure 17, the second of them is omitted.

Figure 17. Dependence $\Delta\bar{t}_f(\bar{a}_l)$

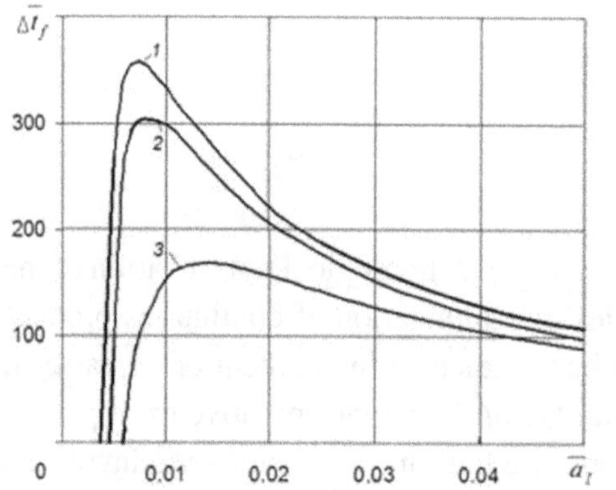

Note. 1 – example 3, 2 – basic, 3 – example 2

Figure 18. Reduction of the relative concentration of the suspended fine particles in the filtrate

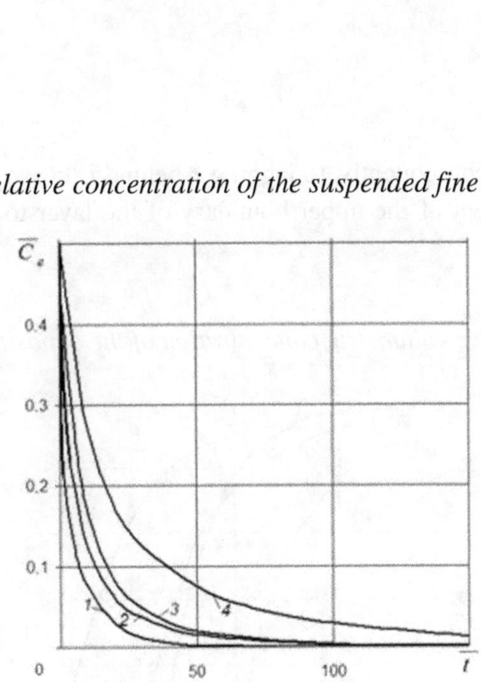

Note. 1 – example 4, 2 – example 2, 3 – example 3, 4 – basic example

The second series of calculations was again carried out for the second rate mode using the first four examples with a fixed value of $\overline{a}_l$. In this case, the exact dependences from Subsection 4.3.2 were used, which allowed to establish the regular changes in the concentrations $\overline{C}_e$, $\overline{S}$ under new conditions, and, in addition, now the initially unknown variable filtration rate and the position of the moving upper boundary of the surface layer. It is appropriate to notice that the form of the expressions intended for the direct calculation of $\overline{C}_e$ and $\overline{S}$ at a variable rate remained the same except for the formal replacement of $\overline{t}$ by $\overline{\tau}$, To find the values of the desired characteristics now in real time, it is necessary to establish the relationship between $\overline{t}$ and $\overline{\tau}$ in each particular case, relying on the corresponding dependence in dimensionless (4.107) or dimensional (4.110) forms. Instead of the relative coefficient $\overline{\lambda}$, its reduced equivalent analog $\overline{\overline{\lambda}}$ is adopted, which does not depend on the indicated rate

The filtrate quality characteristic $\overline{C}_e$ was determined sequentially in three stages. First of all, the value of $\overline{\tau}$ corresponding to the given time $\overline{t}$ was determined using formula (4.107); then the value of $\overline{S}_e$ corresponding to the obtained value of $\overline{\tau}$ was also found by fitting and, finally, the required value of $\overline{C}_e$ at the moment in time $\overline{t}$ was calculated using formula (4.87). The results of calculating the content of the suspended fine solids at the surface layer outlet in dynamics are given in **Figure 18**. It is obvious that the quality of filtrate improves obviously faster with an increase in the absorption capacity of the filtering material forming the layer, including due to the autocatalysis (curves 2, 3), and especially with increasing its formation rate (curve 1). It follows from the comparison of Figure 13 and Figure 18 that a sharper decrease in the amount of impurities in the filtrate is provided at the second rate mode.

Significant nonlinearity of the profiles $\overline{S}(\overline{z})$ is also preserved at a constant effective pressure differential (**Figure 19**). However, the domains in comparison with analogs at the first mode increase noticeably and are not known in advance. The intermediate calculation of the new variable $\overline{\tau}$ as a function of $\overline{t}$ is primarily necessary for the final determination of the concentrations on real time scale. But at the same time, the quantity $\overline{\tau}$ is uniquely related to the height of the growing dynamic layer according to (4.101). The data of calculations of the specified height proper in the time interval from 0 to 150 are presented in **Figure 20**. It is seen from here that a double increase in the concentration of coarse particles in the suspension leads to a commensurate increase in the specified layer, although not twice due to a reduction in the filtration flow rate and, consequently, to lagging $\bar{\tau}$ behind $\bar{t}$ (curve 1). The absorption properties are reflected in the displacement of the upper boundary of the layer to a much lesser extent than $\bar{a}_l$ (curves 2-4).

Figure 19. Profiles of the relative volumetric concentration of the deposited fine particles

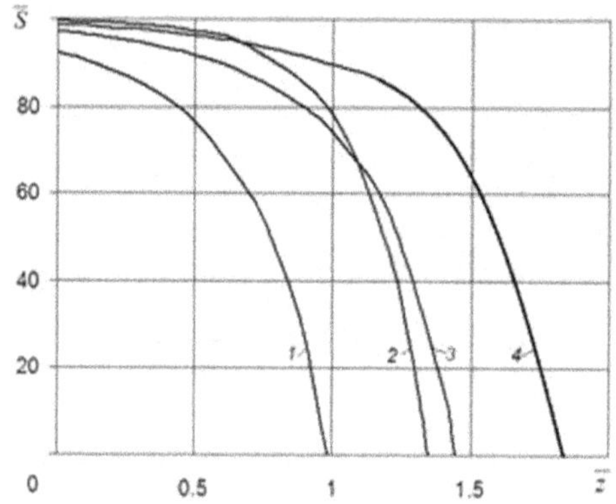

Note. 1 – $\bar{t}$ = 50; 2, 3 – $\bar{t}$ = 100; 4 – $\bar{t}$ = 150; 1, 2, 4 – basic example; 3 – example 3

Figure 20. Rise of the upper boundary of the dynamic layer over time

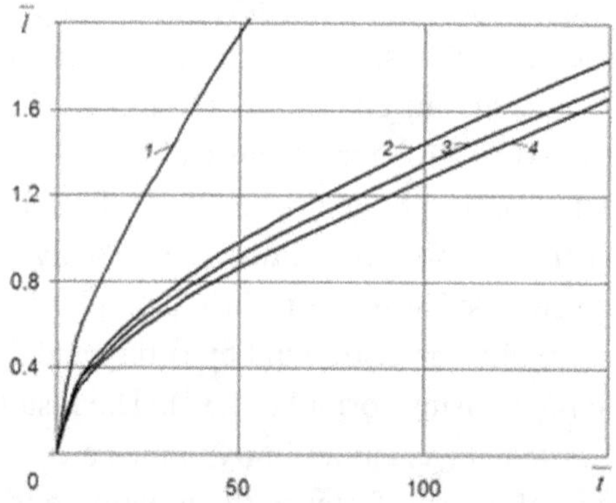

Note. 1 – example 4, 2 – base example, 3 – example 3, 4 – example 2

The determination of the filtration rate is of fundamental importance for evaluating the operability of the dynamic layer at the second rate mode. Its changes over time were directly calculated using formula (4.108) with the dependence $\bar{S}_e(\bar{t})$ found from (4.107). **Figure 21** shows plots describing the monotonic decrease in $\bar{V}$ for the data of the same four examples. As the coefficient $\bar{a}_l$ increases, on the one hand, the number of small particles passing through the dynamic layer decreases. At the same time, the filtration flow path of the liquid lengthens and, consequently, the resistance to its movement increases. On the other hand, a deposit is distributed much more evenly along the layer height. And if at first the first two factors prevail, then already the third factor dominates after a relatively short time, so that the desired rate at the developed surface filtration is noticeably higher at $\overline{a}_l = 0.02$ (curve 1), but not at $\overline{a}_l = 0.01$ (curve 2). The intensification of the absorption process naturally causes an accelerated decrease in the rate $\overline{V}$.

In conclusion of the calculation part of this Section, the time t_V of the filtration rate decrease to the minimum allowable value V_* was set at a continuous variation of the coefficient $\overline{a}_l$ as an example of the technological calculation for the filtration conditions at a constant pressure differential. The curves of the dependence $\overline{t}_V(\overline{a}_l)$ calculated using equation (4.108) and formula (4.107) at the different values of $\overline{V}_*$ are shown in **Figure 22**. It is obvious clearly a more violent reaction of the calculated time to the change in $\overline{a}_l$ at the large values of $\overline{V}_*$. Thus, if a serious decrease in the performance of the filter facility with a dynamic layer is allowed, it is realistic to achieve an appreciable prolongation of its effective operation by a slight increase in C_l and $\overline{a}_l$.

Even a fragmented technological analysis confirms that an increase in $\overline{a}_l$ easily implemented in practice leads, firstly, to a possible sharp improvement of the filtrate quality, and secondly, to a drop in the filtration rate in the second rate mode or to a rapid increase in the mechanical energy expenditure in the first mode. In general, it can be stated that the concentration of structure-forming impurity particles in essentially heterogeneous suspension is an important resource for its optimization at the surface filtration.

Figure 21. Decrease in the relative filtration rate over time

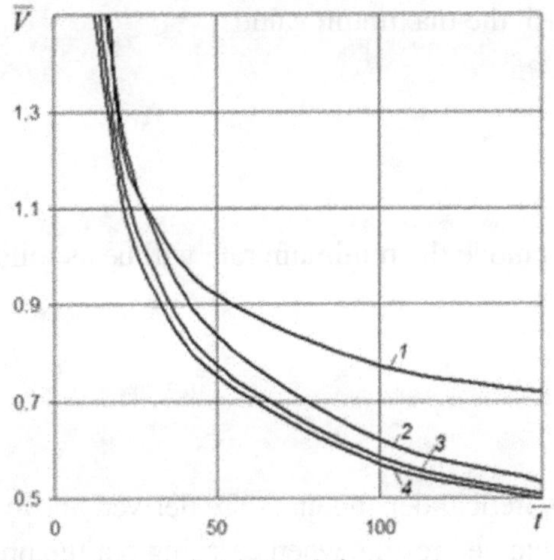

Note. 1 – example 4, 2 – baseline example, 3 – example 3, 4 – example 2

Figure 22. Dependence $\bar{t}_V(\bar{a}_l)$

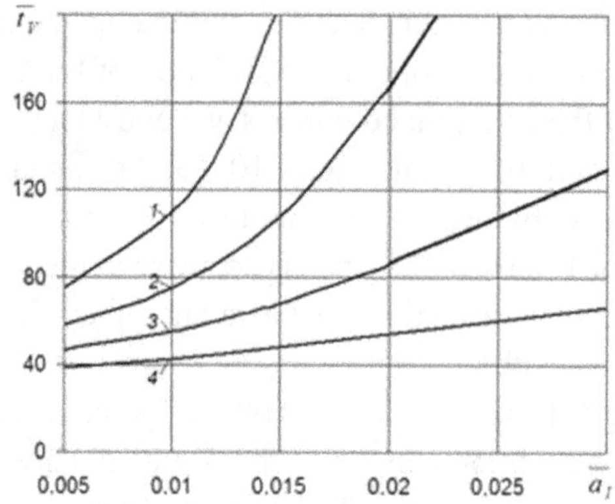

Note. 1 – $\bar{V}_*$ = 0.6, 2 – $\bar{V}_*$ = 0.7, 3 – $\bar{V}_*$ = 0.8, 4 – $\bar{V}_*$ = 0.9

To conclude this Section, it is appropriate to note that the behavior of the filtration characteristics at $t \to \infty$ can be asymptotic. If the values of $\bar{S}$, at which the function $\bar{\bar{\lambda}}(\bar{S})$ turns to 0, are greater than $1/\bar{a}_l$, then $\bar{S}_e$ will eventually stabilize at the value $1/\bar{a}_l$. Herewith, the total head losses in the case of the first filtration rate mode will reach the maximum value

$$\Delta\bar{h}_{\max} = \bar{a}_l \int_0^{1/\bar{a}_l} \frac{d\bar{S}}{\bar{k}(\bar{S})\bar{\bar{\lambda}}(\bar{S})\,(1-\bar{a}_l\bar{S})}.$$

At the filtration in the second mode the minimum rate will be as follows

$$\bar{V}_{\min} = \left[\bar{a}_l \int_0^{1/\bar{a}_l} \frac{d\bar{S}}{\bar{k}(\bar{S})\bar{\bar{\lambda}}(\bar{S})\,(1-\bar{a}_l\bar{S})}\right]^{-1}.$$

Thus, since the numerous theoretical dependences are derived above using the exact methods, they guarantee the reliability of the obtained results when carrying out the predictive and technological calculations in relation to the specific various conditions of the surface filtration (precoat filter, membrane module with cross-flow filtration). It is shown on typical examples that the purposeful change in the content of large particles in the polydisperse impurity enables efficient control of the heterogeneous suspension clarification in the surface layer of a newly formed deposit and thus, to optimize the technological process.

4.4. ESTIMATION OF FILTRATION RATE EFFECT ON TECHNOLOGICAL TIMES

Continuing the study of the possibilities of direct intensification of the technological process on ultra-rapid filters, i.e. due to increasing a hydraulic load, which was started in Section 3.4 in relation to the detachment filtration, below we consider in detail the behavior of the technological times at a purposeful change in the rate now of the non-detachment filtration. The filtration mode *V=const* does not change in this case. Thus, the quantitative analysis of the effect of the rate $\overline{V}$ on the times $\overline{t}_{p,h,f}$ at the firm fixation of suspension particles in the pore space of the packed bed is performed below on the basis of the exact solution of the problem for nonlinear kinetics of the general form obtained in Section 4.1, but in the absence of the second component of the interfacial mass exchange. By analogy with Section 3.4, the calculation equations containing the desired relative times $\overline{t}_{p,h}$ and criterion parameters $\overline{C}_*, \Delta\overline{h}_*$ are transformed in order to distinguish explicitly now the variable $\overline{V}$. For the relative reduced filtration coefficient $\overline{\lambda}_0$, previously constant, following the power form of the dependence of the coefficient $\overline{\alpha}$ (similar in physical meaning to $\overline{\lambda}_0$) on $\overline{V}$ (3.55), it is also taken as follows

$$\lambda_0(V) = \tilde{\lambda}_0 V^l, \ \overline{\lambda}_0(\overline{V}) = \overline{\tilde{\lambda}}_0 \overline{V}^{l-1} \tag{4.114}$$

where $\overline{\tilde{\lambda}}_0 = L V_0^{l-1} \tilde{\lambda}_0$, the range from 0 to 1 is chosen for l. The calculations are performed as in Section 3.4 with respect to the given technological times $\overline{t}_p, \overline{t}_h$. Then the time $\overline{t}_p$ is also calculated in two steps. First, the value $\overline{S}_{0p}$ is found by simple fitting at $\overline{S}^0 = const$ from the equation.

$$\int\limits_{\overline{S}^0+\overline{C}_*(\overline{S}_{0p}-\overline{S}^0)}^{\overline{S}_{0p}} \frac{d\zeta}{(\zeta - \overline{S}^0)\cdot\overline{\lambda}(\zeta)} = \overline{V}^{l-1} \tag{4.115}$$

and then the desired $\overline{t}_p$ immediately determined by the formula

$$\overline{t}_p(\overline{V}) = \frac{1}{\overline{V}^l}\int\limits_{\overline{S}^0}^{\overline{S}_{0p}(\overline{V})} \frac{d\zeta}{\overline{\lambda}(\zeta)} \tag{4.116}$$

In order to establish the second technological time, it is necessary, first of all, to find a set of two characteristic values of $\overline{S}$, corresponding to this moment in time, namely $\overline{S}_{0h}, \overline{S}_{eh}$. The system of equations, in which they are present as unknown quantities, has the following form

$$\int\limits_{\overline{S}_{eh}}^{\overline{S}_{0h}} \frac{d\zeta}{(\zeta - \overline{S}^0)\overline{\lambda}(\zeta) f_k(\zeta)} = \overline{V}^{l-2}\Delta\overline{h}_* \tag{4.117}$$

$$\int\limits_{\overline{S}_{eh}}^{\overline{S}_{0h}} \frac{d\zeta}{(\zeta - \overline{S}^0)\,\overline{\lambda}(\zeta)} = \overline{V}^{l-1} \tag{4.118}$$

Finally, the time $\overline{t}_h$ is simply calculated using the formula

$$\overline{t}_h(\overline{V}) = \frac{1}{\overline{V}^l}\int\limits_{\overline{S}^0}^{\overline{S}_{0h}(\overline{V})} \frac{d\zeta}{\overline{\lambda}(\zeta)} \tag{4.119}$$

It is useful to note that the procedure of finding the times $\overline{t}_p$ and $\overline{t}_h$ according to (4.115) - (4.119) using available software packages is carried out simply in two or three steps and requires a minimum of time. It is especially easy to calculate the time $\overline{t}_p$. First, the value $\overline{S}_{0p}$ corresponding to the given rate $\overline{V}$

is found by fitting from (4.117), and then the desired value of $\bar{\bar{t}}_p$ is calculated using formula (4.116). It is somewhat more difficult to determine the time $\bar{\bar{t}}_h$. First of all, it is necessary to determine a suitable set of two values of $\bar{S}_{0h}$, $\bar{S}_{eh}$ from the system of equations (4.117), (4.118). This can be done by implementing the iterative procedure, namely, it is reasonable to set the initial value $\bar{S}_{eh}$ and find the corresponding value $\bar{S}_{0h}$ from (4.118). Then it is necessary successively to substitute this set into equation (4.117) and calculate the corresponding discrepancy; in accordance with its sign to select a new value of $\bar{S}_{eh}$ and, repeating the calculations, determine the new discrepancy; referring to all the already available values of this discrepancy, to correct the quantity $\bar{S}_{eh}$ until the last discrepancy is less than the predetermined minimum value. Note that in this way a set of the values $\bar{S}_{0h}$, $\bar{S}_{eh}$, which satisfies the above system, is usually found in a few calculation steps. Finally, it remains only to substitute the value $\bar{S}_{0h}$ from this set into formula (4.119).

As a rule, the filtration rate limitation imposed on the filter operation is caused by the excessive reduction of the protective properties of the bed due to the clarification process intensification. Thus, there is a critical rate V_m at which the filter is unable to provide a proper degree of the clarification even of the first suspension portion. In the case of linear kinetics of the detachment filtration, the formula for the determination of $\bar{V}_m$ is given in Section 3.4. In the case of nonlinear kinetics of the general form and the non-detachment filtration, the similar formula is easily derived from equation (4.115). The analogous formula is easily derived for the nonlinear kinetics of general form and continuous filtration. Due to the smallness of the filtration time, it can be assumed that

$$\bar{\bar{\lambda}}(\bar{S}) = \bar{\bar{\lambda}}(0) = \bar{\bar{\lambda}}_0 \bar{S}_m$$

Then the simple formula follows from (4.115) after not complicated transformations

$$\bar{V}_m = \left[-\frac{\bar{\bar{\lambda}}(0)}{\ln \bar{C}_*} \right]^{\frac{1}{1-l}} \tag{4.120}$$

Consequently, $\bar{V}_m$ turns out to be infinitely large in the special case l=1 so that its finite value does not exist. Incidentally, involving formula (4.120), it is easy to find from (4.116) an approximate value $\bar{S}_{0p}$, corresponding to $\bar{V} = \bar{V}_m$. If we put $\bar{t}_p \approx 1$, then the expression follows from (4.116)

$$\bar{S}_{0p} \approx \bar{S}^0 + \bar{\bar{\lambda}}_0 \bar{S}_m V_m^l$$

The technological analysis of the filtration rate in the case of the non-detachment filtration is indicative due to the interpretation of the interfacial mass exchange as a single process, and hence to the reduction of the number of the mass-exchange coefficients, and was carried out at representation of the functional hydraulic conductivity according to (4.13). At that, only the reduced filtration coefficient depends on the rate $\bar{V}$ according to expression (4.114). Therefore, for completeness of the analysis it is expedient to estimate sensitivity of the calculation quantities $\bar{t}_p(\bar{V})$, $\bar{t}_h(\bar{V})$, $\bar{t}_f(\bar{V})$ with respect to the parameters $\bar{\bar{\lambda}}_0$ and l. And it is logical that with a continuous change in $\bar{V}$three to five times, the characteristic values of $\bar{\bar{\lambda}}_0$(0.02, 0.03) and l(0, 1) were also taken first. The first value of $\bar{\bar{\lambda}}_0$ corresponds to a moderate absorption capacity of the filtering material, and the second value corresponds to a high capacity. The value l=0 means that the absorption properties of the bed are independent of the filtration flow process, but l=1, on the contrary, indicates such an essential dependence. In addition, the values $\overline{S}_m(200)$, $\theta(0.01)$ and also $\overline{\gamma}_a(0.003)$ are fixed, since the corresponding factors are of secondary importance in these studies.

It is obvious that the increase in a hydraulic load causes, on the one hand, the increase in the volume of the clarified suspension and, on the other hand, serious negative consequences for the technological process. Thus, firstly, the suspended solids content in the filtrate, and secondly, the head losses in the bed increase faster. **Figure 23** shows the data of the calculations of the relative technological times $\overline{\overline{t}}_{p,h,f}$ for both values of $\overline{\overline{\lambda}}_0$ and l=0. If $\overline{\overline{\lambda}}_0 = 0.02$, then the filter operation is entirely limited by the filtrate quality and, therefore, curve 4 here describes the dependences of $\overline{\overline{t}}_p$ and $\overline{\overline{t}}_f$ on $\overline{V}$ simultaneously. In the case of a well-absorbing media, the technological process has to be interrupted due to the excessive mechanical energy expenditure at $1 \leq \overline{V} \leq 2.45$. The filter operation should be limited in time again in the range $2.45 < \overline{V} \leq 2.61$ due to the high outlet suspended solids content. Special attention should be paid to the fact that the filter efficiency drops to 0 when the rate $\overline{V}$ approaches the critical values of 1.7405 at $\overline{\overline{\lambda}}_0 = 0.02$ and 2.61 at $\overline{\overline{\lambda}}_0 = 0.03$. Thus, an increase in $\overline{V}$, for example, by a factor of two will entail an obligatory reduction of the duration $\overline{\overline{t}}_f$ by a factor of about 2.6 at $\overline{\overline{\lambda}}_0 = 0.03$. In general, the filter is inoperable if $\overline{\overline{\lambda}}_0 = 0.02$.

Figure 23. Dependencies $\overline{\overline{t}}_p(\overline{V})$, $\overline{\overline{t}}_h(\overline{V})$, $\overline{\overline{t}}_f(\overline{V})$

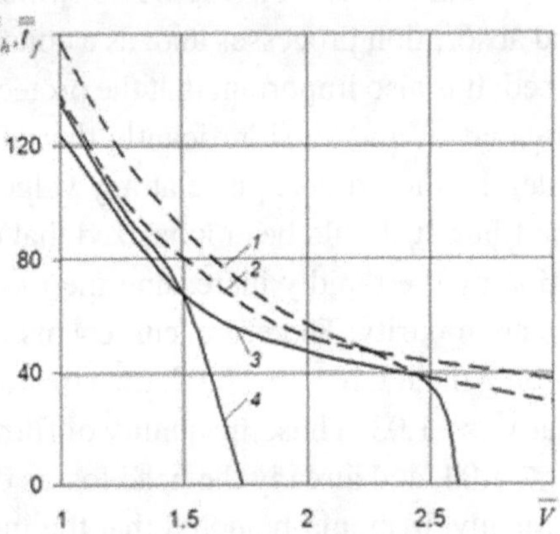

Note. 1, 3 – $\overline{\overline{\lambda}}_0 = 0.03$; 2, 4 – $\overline{\overline{\lambda}}_0 = 0.02$; 1, 4 – $\overline{\overline{t}}_p$; 2, 3 – $\overline{\overline{t}}_h$; the solid lines – $\overline{\overline{t}}_f$

Figure 24. Dependencies $\bar{t}_p(\bar{V})$, $\bar{t}_h(\bar{V})$, $\bar{t}_f(\bar{V})$

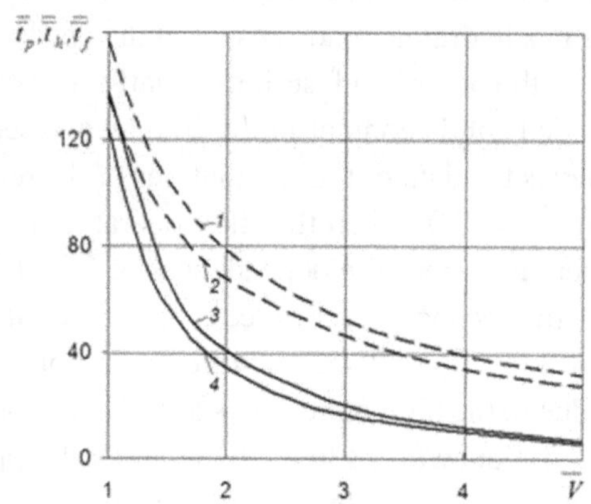

Note. 1, 4 – $\bar{\bar{\lambda}}_0$ = 0.03; 2, 3 – $\bar{\bar{\lambda}}_0$ = 0.02; 1, 2 – $\bar{t}_p$; 3, 4 – $\bar{t}_h$; solid lines – $\bar{t}_f$

A fundamentally different calculated filtration pattern is obtained with the same initial data, relatively large values of the hydraulic load and l=1 (**Figure 24**). Then both curves of the dependence $\bar{t}_h(\bar{V})$ are clearly lower in comparison with similar curves from Figure 23, which is explained by a commensurate increase in the filtration flow and absorption processes and, as a consequence, a higher deposit concentration at the inlet section of the bed. It is also important that the protective properties of the bed improve due to the formal increase in the rate $\bar{V}$, and so significantly that at least the short-term clarification of a suspension to the required degree should take place at any values of $\bar{V}$, and hence the amounts of the suspended solids entering the filter. It should be emphasized that the suspension filtration is almost completely limited in this situation by the rapidly increasing the mechanical energy expenditure on it due to the intensive deposition of an impurity. Therefore, curve 4 also characterizes the reduction of the filter run duration at a well-absorbing media. If $\bar{\bar{\lambda}}_0 = 0.02$, then the resulting curve $\bar{t}_f(\bar{V})$ is a continuous piecewise line with the fracture at $\bar{V} \approx 1.03$. Thus, the quality of filtrate does not meet the clarification standard in the interval $1 \leq \bar{V} \leq 1.03$, and already the head losses exceed the permissible value for $\bar{V}$ in the range from 1.03 up to 5. Finally, it should be noted that the increase in the rate $\bar{V}$, for example, in 5 times causes the reduction of the time of filter effective operation in 19.2 times at $\bar{\bar{\lambda}}_0 = 0.02$ and in 21.1 times at $\bar{\bar{\lambda}}_0 = 0.03$.

Figure 25. Dependencies $\bar{t}_p(\bar{\lambda}_0)$, $\bar{t}_h(\bar{\lambda}_0)$, $\bar{t}_f(\bar{\lambda}_0)$

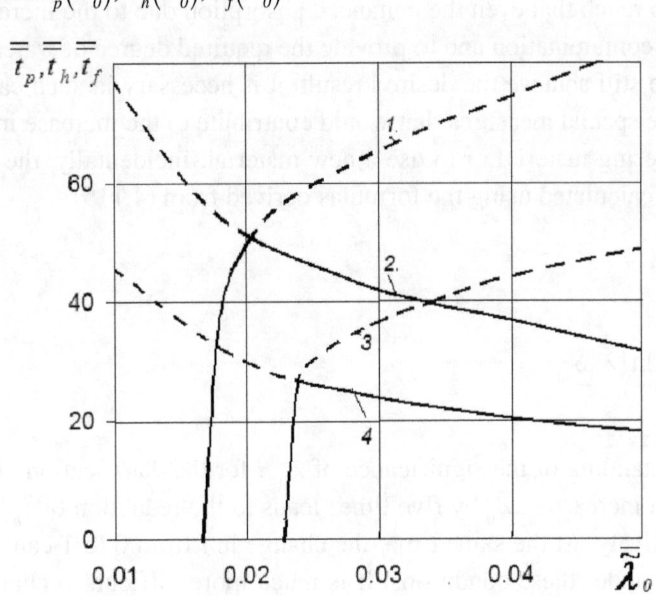

Note. 1, 3 – $\bar{t}_p$; 2, 4 – $\bar{t}_h$; 1, 2 – $\bar{V}$ = 2; 3, 4 – $\bar{V}$ = 3; solid lines – $\bar{t}_f$

Figure 26. Dependencies $\bar{t}_p(l)$, $\bar{t}_h(l)$, $\bar{t}_f(l)$

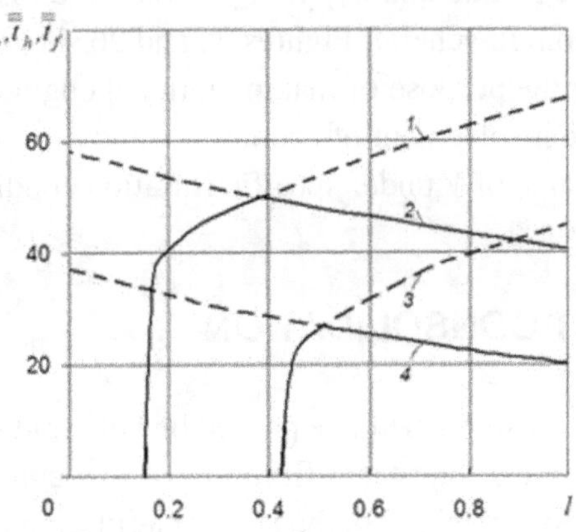

Note. 1, 3 – $\bar{t}_p$; 2, 4 – $\bar{t}_h$; 1, 2 – $\bar{V}$ = 2; 3, 4 – $\bar{V}$ = 3; solid lines – $\bar{t}_f$

In order to more fully estimate the sensitivity of the technological times with respect to the other two parameters ($\bar{\lambda}_0$, l), contained in the expression for the coefficient $\bar{\lambda}_0$ (4.13), the dependencies of $\bar{t}_{p,h}$ on them were calculated. At the same time, the values of $\bar{\lambda}_0$, l were alternately or continuously varied within wide ranges, or their single values (0.02; 1/3) were taken. In addition, two characteristic values (2,3) were chosen for $\bar{V}$. The results of the calculations of the functions $\bar{t}_{p,h}(\bar{\lambda}_0)$, $\bar{t}_{p,h}(l)$ according to the procedure described above are shown in **Figures 25 and 26**. We can see from here that there are intervals of the values $\bar{\lambda}_0$, l, for which actually $\bar{t}_p = 0$, i.e. the technological process is impossible in principle to implement under the restrictions imposed on it. Thus, the supply of dispersed contaminants to the filter increases so much that even the enhanced absorption due to the increase in $\bar{\lambda}$ is not able to cope with such a heavy contamination and to provide the required degree of water treatment at least for a short time. In order to still achieve the desired result, it is necessary in such cases either to reduce a hydraulic load or to take special measures that would contribute to the increase in $\bar{\bar{\lambda}}_0$, l, for example, to modify the existing filtering material or to use a new material. Incidentally, the approximate limiting values of $\bar{\bar{\lambda}}_{0m}$, l_m can be calculated using the formulas derived from (4.115)

$$\bar{\bar{\lambda}}_{0m} = -\frac{\ln \overline{C}_*}{\bar{\bar{S}}_m \bar{\bar{V}}^{l-1}}$$

$$l_m = 1 + \frac{\ln(-\ln \overline{C}_*) - \ln\left(\bar{\bar{\lambda}}_0 \bar{S}_m\right)}{\ln \bar{\bar{V}}}$$

For a general understanding of the significance of $\bar{\bar{\lambda}}_0$, l for the clarification effect of the filter, it is pertinent to note that an increase in $\bar{\bar{\lambda}}_0$ by five times leads to the reduction of $\bar{\bar{t}}_h$ by 2.43 and 2.5 times at $\bar{\bar{V}} = 2$ and 3 respectively. At the same time, the change in l from 0 to 1 causes the reduction of $\bar{\bar{t}}_h$ by 1.43 and 1.85 times under these conditions. It is much more difficult to characterize the response of the time $\bar{\bar{t}}_p$ to varying $\bar{\bar{\lambda}}_0$, l. The functions $\bar{\bar{t}}_p\left(\bar{\bar{\lambda}}_0\right)$, $\bar{\bar{t}}_p(l)$ formally describing its response are defined only in a part of the calculation intervals accepted for $\bar{\bar{\lambda}}_0$, l, namely, at $\bar{\bar{\lambda}}_0 \geq \bar{\bar{\lambda}}_{0m}$ and $l \geq l_{m1}$, where two characteristic sections are distinguished. At the small first section, these functions increase rapidly from 0 to relatively large values, the deviation from which will be insignificant at further increase in $\bar{\bar{V}}$. Naturally, the duration of the filter run makes sense to discuss only at these two sections, and the filter operation is limited partly by the filtrate quality, partly by the head losses at the first section, and only by the losses at the second section. In general, Figures 25 and 26 show the expediency, firstly, of an in-depth technological analysis for the purpose of making rational engineering decisions and selection of the rate mode, and secondly, the involvement of economic indicators for the final establishment of the optimal (in the applied sense) value of V under specific filtration conditions.

4.5. EFFECT OF DEPOSIT CONSOLIDATION

Usually, the properties of a deposit formed in a packed bed of rapid filters and, above all, its strength do not practically change during the long-term filtration of low-concentration aqueous suspensions. Therefore, the amount of the deposit retained in the layer of a filtering material is stable after its regular washing. This gives the right to perform the predictive and technological calculations during one arbitrary (representative) filter run, and extend the results obtained to all other filter runs. The exception may be the filter runs at the end of the service life of the bed, when the size of its elements noticeably decreases due to abrasion and crushing.

At the same time, there are also special cases with the formation of a deposit in the filtration practice, that has abnormal properties due to specific physical and chemical processes (Finno & Rechenmacher, 2003; Guo et al., 2015; Orlov, 2008; Sutzkover-Gutman et al., 2010; Sanda et al., 2009; Tiller et al., 1996). Importantly, their characteristic time appears to be commensurate with the characteristic time of the clarification process and the duration of the filter run. Thus, the newly formed deposit has enough time to transform noticeably during the current filter run and, in particular, its ability to withstand the significant hydrodynamic forces, that develop during high-rate wash flow, is progressing. As a consequence, not only the current deposit content within each filter run, but also the residual amount of the deposit in the sequence of filter runs steadily increases. Naturally, the filter operability inevitably decreases and its further operation becomes unprofitable for economic reasons in the end. It is reasonable to achieve at least a partial restoration of the former properties of the filtering material in such situations. However, this often requires too large additional expenses, so that it is easier and cheaper to replace the used bed with a new one. And here the question about the time of implementation of radical measures, and actually about the service life of the packed bed T_f becomes actual. For its theoretical solution, it is necessary first of all to preliminarily develop the procedure of the rational choice of the filter control algorithm at stable tendency of the filtering material to clogging. The first step is to determine the volume of the non-washable suspension particles, which were accumulated in the bed during the representative filter run ($t_{fi} \geq t \geq 0$). If to involve the empirical function of the hydrodynamic resistance, introduced according to (1.20), the concentration Δs_{ei} of firmly fixed particles of an impurity during an elementary time interval $[t, t+\Delta t]$ in a unit volume of the filtering material at depth z will be as follows

$$\Delta s_{ei}(z,t,t_{fi}) = \left[S_i(z,t+\Delta t) - S_i(z,t)\right] f_s(t_{fi} - t)$$

where S_i is the (volumetric) concentration of the suspension particles deposited over the course of *i-th* filter run. Then the concentration Δs_{ti} of the dispersed impurity fixed here for the entire filter run will be

$$\Delta s_{ti}(z,t_{fi}) = \int_0^{t_{fi}} \frac{\partial}{\partial t} S_i(z,t) f_s(t_{fi} - t)\, dt \tag{4.121}$$

Thus, the specific (per unit area of the inlet surface of a bed) volume W_{pi} of the fresh non-washable particles in the entire layer of a filtering material with the height L is equal to

$$W_{pi}(t_{fi}) = \int_0^L \Delta S_{ti}(z,t_{fi})\, dz = \int_0^L \int_0^{t_{fi}} \frac{\partial}{\partial t} S_i(z,t) f_s(t_{fi} - t)\, dt dz \tag{4.122}$$

From (4.122) follows

$$W_{pi}(t_{fi}) = L \int_0^{t_{fi}} \frac{\partial S_{ia}}{\partial t} f_s(t_{fi} - t)\, dt$$

where $S_{ia}(t) = \frac{1}{L}\int_0^L S_i(z,t)\, dz$.

To calculate W_{pi}^0 and, consequently, the concentration S_{i+1}^0 of the deposited particles in a bed before the beginning of the i+1-th filter run, first of all, it is necessary to find the quantities $S_i(z, t_{fi})$, $S_{ia}(t_{fi})$. Preliminarily, by double integration over height from 0 to L and over time from 0 to t_{fi} (the duration of the filter run t_{fi} is determined according to the procedure described in Section 1.3, using criteria (1.35) and (1.36) or (1.37)) both parts of the mass transfer equation (2.28), the equation of the balance of the suspended particles that have entered the bed during the period $t_{fi} \geq t \geq 0$ is derived

$$\int_0^{t_{fi}} V(t)\left[C_0(t) - C_{ei}(t)\right] dt = \int_0^{L} \left[S_i(z,\ t_{fi}) - S_i^0\right] dz \qquad (4.123)$$

Here C_{et} is the outlet (volumetric) concentration of the suspended solids, S_i^0 is the initial (volumetric) concentration of the deposited particles for a given filter run. Actually, in the case of a consolidated deposit S_i^0 means the total amount of the non-washable particles accumulated in the bed for all previous filter runs.

In general, the calculation of the suspension filtration with the consolidated deposit (as opposed to stable deposit) is seriously complicated, primarily due to a significant increase in the body of calculations. Indeed, it is now necessary to calculate the desired characteristics (concentrations, head losses, technological times) many times during the entire service life of the bed. The new and main task of the technological analysis becomes the substantiation of the specified time. And to a large extent, the reliability of the results of such analysis depends on the reliability of the hydrodynamic resistance function $f_s(t)$ and the criterion parameter f_{s*}. This function allows to calculate the initial concentration of the deposited particles (and deposit) for the successive filter run with a variable rate V and the suspended solids content in the initial suspension C_0 in a relatively simple way and is therefore used in the calculations of each filter run. The critical value of this function f_{s*} should also be specified for economic reasons. If this value is systematically exceeded, the optimum time T_f can be significantly shortened and the cost of the water treatment can increase to the same extent. The corresponding consolidation criterion is formulated in the form of inequality (1.39).

Thus, the relative volume $\overline{W}_{pi}$ of the suspension particles that have entered a bed over the period of time $\left[0,\ \bar{t}_{fi}\right]$ and are firmly fixed in it is calculated as follows

$$\overline{W}_{pi}(\bar{t}_{fi}) = \frac{W_{pi}}{n_0 C_0 L} = \overline{S}_{ia}(\bar{t}_{fi}) = \int_0^{\bar{t}_{fi}} \overline{V}(\bar{t})\left[\overline{C}_0(\bar{t}) - \overline{C}_{ei}(\bar{t})\right] f_s\left(\bar{t}_{fi} - \bar{t}\right) d\bar{t} \qquad (4.124)$$

In practice, aqueous suspensions are most often clarified at the first rate mode $V=V_0=const$ and $C_0 = const$ ($\overline{C}_0 = 1$). Then

$$\overline{W}_{pi}(\bar{t}_{fi}) = \int_0^{\bar{t}_{fi}} \left[1 - \overline{C}_{ei}(\bar{t})\right] f_s\left(\bar{t}_{fi} - \bar{t}\right) d\bar{t} = \int_0^{\bar{t}_{fi}} \left[1 - \overline{C}_{ei}\left(\bar{t}_{fi} - \bar{t}\right)\right] f_s(\bar{t})\, d\bar{t} \qquad (4.125)$$

where $\overline{C}_{ei} = C_{ei}/C_0$, $\Delta\overline{S}_i = \Delta S_i/(n_0 C_0)$, $\bar{t}_i = V_0 t_i/(n_0 L)$.

Then two approaches to the filtration implementation during the entire specified period (technological schemes) are analyzed sequentially. In case of the first scheme the duration of filter runs is gradually reduced. At the beginning of the i+1-th filter run the relative concentration of the non-washable suspension particles constant over the bed height will be as follows

$$\overline{S}_{i+1}^0 = \overline{S}_i^0 + \overline{W}_{pi}(\bar{t}_{fi}) = \overline{S}_i^0 + \int_0^{\bar{t}_{fi}} \left[1 - \overline{C}_{ei}(\bar{t}_{fi} - \bar{t})\right] f_s(\bar{t})\, d\bar{t}. \qquad (4.126)$$

Before the final (N_f-th) filter run, the initial concentration $\overline{S}_N^0$ reaches, in fact, the maximum permissible value $\overline{S}_{\max}^0$ under the conditions considered, which is determined depending on the relationship between the relative technological times $\bar{t}_{pN}$ and $\bar{t}_{hN}$. The required value is found at $\bar{t}_{hN}<\bar{t}_{pN}$ by fitting from the equation

$$\int_0^1 \frac{d\bar{z}}{f_k(\bar{S}(\bar{z},\bar{t}_{hN};\bar{S}^0_{\max}))} = \Delta\bar{h}_* \tag{4.127}$$

If $\bar{t}_{hN} > \bar{t}_{pN}$, then the equation is already used

$$\bar{C}_{eN}(\bar{t}_{pN};\bar{S}^0_{\max}) = \bar{C}_* \tag{4.128}$$

The sequence of calculations is limited due to the acceptance of the value $\bar{t}_{\min}$, which is minimum possible for $\bar{t}_{fN}$, involving the consolidation criterion (1.39). Obviously, the number of filter runs in this technological scheme is equal to the number of the calculation stages. The sum of all $\bar{W}_{pi}$ should be approximately equal to $\bar{S}^0_{\max}$ in this case, so that

$$\sum_{i=1}^{N_f} \bar{W}_{pi} = \sum_{i=1}^{N_f} \int_0^{\bar{t}_{fi}} [1 - \bar{C}_{ei}(\bar{t}_{fi} - \bar{t})] f_s(\bar{t})\, d\bar{t} \approx \bar{S}^0_{\max} \tag{4.129}$$

It should be noted that there must generally exist a discrepancy between the values of the parameter $\bar{S}_{\max}$ found by the two ways due to the conditions $\bar{t}_{fN} > \bar{t}_{\min}$ and $\bar{t}_{f,N+1} < \bar{t}_{\min}$. However, it is usually negligible. The formally described procedure of the step-by-step calculations (the first technological scheme) allows to extend the filter operation to the maximum after each washing, but at the same time the accumulation of the residual deposit in a bed is significantly accelerated. As a result, the clarification resource of the bed is quickly exhausted and it has to be replaced prematurely. Naturally, the relative service life $\bar{T}_f$ is equal to the sum of the relative durations of all N_f filter runs, so that

$$\bar{T}_f = \sum_{i=1}^{N_f} \bar{t}_{fi} \tag{4.130}$$

At practical implementation of the second technological scheme its theoretical substantiation begins with the determination of the most important technological parameter, namely, optimal duration of the operating periods t_{opt}. The relative value of $\bar{t}_{opt}$, as well as $\bar{t}_{\min}$, is determined using the consolidation criterion (1.39) from equality (1.44)

$$f_s(\bar{t}_{opt}) = f_{s*} \tag{4.131}$$

In the next step, two values $\left(\bar{S}^0_h, \bar{S}^0_p\right)$ of the relative concentration of suspension particles firmly attached to the packed bed are calculated before the respective operating periods (specified values are the initial values for it). One of them becomes the last period for a given bed change. In this case, two basic criteria for the constant rate depth filtration are involved. Using the formal representations for them (1.35) and (1.36), the equations with respect to the quantities $\bar{S}^0_h$ and $\bar{S}^0_h$ by analogy with (4.127), (4.128) take the following form

$$\int_0^1 \frac{d\bar{z}}{f_k(\bar{S}(\bar{t}_{opt}, \bar{z}; \bar{S}_h^0))} = \Delta \bar{h}_* \tag{4.132}$$

$$\bar{C}_e(\bar{t}_{opt}; \bar{S}_p^0) = \bar{C}_* \tag{4.133}$$

Obviously, the smaller of the two quantities, i.e.

$$\bar{S}^0_{\max} = \min(\bar{S}_p^0, \bar{S}_h^0)$$

like (4.129), must be approximately equal to the sum of all $\bar{W}_{pi}$ $(i=1,2,\ldots,N_f)$, so that

$$\sum_{i=1}^{N_f} \bar{W}_{pi} = \bar{W}_{tot} \approx \bar{S}^0_{\max} \tag{4.134}$$

and the relative bed service life will be

$$\sum_{i=1}^{N} \bar{t}_i = T_f = N_f \bar{t}_{opt} \tag{4.135}$$

Considering (4.125) and (4.129), the following is obtained

$$\int_0^{\bar{t}_{opt}} [N_f - \sum_{i=1}^{N} \bar{C}_{ei}(\bar{t}_{opt} - \bar{t})] f_s(\bar{t})\, d\bar{t} \approx \bar{S}^0_{\max} \tag{4.136}$$

Then two situations are consecutively considered, differing in the relation between $\bar{S}_h^0$ and $\bar{S}_p^0$. In practice, there is the relation $\bar{S}_h^0 \le \bar{S}_p^0$ more often due to a high absorption capacity of the filtering material. Then, taking into account regular early washings and the condition $\bar{C}_e << 1$ in accordance with the requirements of a high quality of the water treatment, it is justified to assume $\bar{C}_e \approx 0$ in the first approximation. As a consequence, the difference between $\bar{W}_{pi}$ is leveled out and the following equation results from (4.125), which is now valid for any operating periods

$$\bar{W}_{pi} \approx \int_0^{\bar{t}_{opt}} f_s(\bar{t})\, d\bar{t} = F_s(\bar{t}_{opt}) = \bar{t}_{opt} f_{sa} \tag{4.137}$$

where the average value f_{sa} will be

$$f_{sa} = \frac{1}{\bar{t}_{opt}} \int_0^{\bar{t}_{opt}} f_s(\bar{t})\, d\bar{t}$$

Thus,

$$\bar{W}_{tot} \approx \bar{S}_h^0 = N_f F_s(\bar{t}_{opt}) \tag{4.138}$$

Often the function $f_s(t)$ is almost linear in nature, i.e.

$$f_s(t) \approx \chi_a \bar{t}$$

Then it is approximately found

$$F_s(\bar{t}_{opt}) = \frac{\chi_a \bar{t}^2_{opt}}{2} = \bar{t}_{opt} f_s(0.5\bar{t}_{opt}) \tag{4.139}$$

It follows from (4.138) that the total number of operating periods, taking into account (4.139), will be as follows

$$N_f(\bar{t}_{opt}) = \frac{\bar{S}^0_h(\bar{t}_{opt})}{F_s(\bar{t}_{opt})} \approx \frac{\bar{S}^0_h(\bar{t}_{opt})}{\bar{t}_{opt} f_s(0.5\bar{t}_{opt})} \tag{4.140}$$

When N_f is known, the required total operating time of one bed change is calculated by formula (4.135).

In the second situation (a low absorption capacity of the filtering material) $\bar{S}^0_h > \bar{S}^0_p$ and the deposit removal from a filter in relation to its total inflow into the bed can already be noticeable. Then it is also reasonable to take into account in the integrand the quantities $\bar{C}_{ei}(\bar{t}_{opt} - \bar{t}; \bar{S}^0_i)$ from equality (4.136). Since here the values $\bar{C}_{ei}$ remain small and increase almost linearly over time, their generalized estimation can be considerably simplified with minimal additional errors. Thus, the central operating period is chosen from the sequence of N_f periods as a representative period. If N_f is odd, the considered period has the number $(N_f+1)/2$. If N_f is even, we can stop at the $0.5\,N_f$ period. N_f is assumed to be odd for certainty. Then it is approximately true

$$\sum_{i=1}^{N_f} \bar{C}_{ei}(\bar{t}_{opt} - \bar{t}; \bar{S}^0_i) \approx N_f \bar{C}_{e,(N_f+1)/2}(\bar{t}_{opt} - \bar{t}; \bar{S}^0_{(N_f+1)/2})$$

where the value $\bar{S}^0_{(N_f+1)/2}$ can be considered to be approximately equal to $0.5\bar{S}^0_p$. Thus, the specified number of the operating periods in the term $\bar{T}_f$ is

$$N_f = \frac{\bar{S}^0_h}{F_s(\bar{t}_{opt}) - \int_0^{\bar{t}_{opt}} \bar{C}_{e,(N_f+1)/2}(\bar{t}_{opt} - \bar{t}; \bar{S}^0_p) f_s(\bar{t})\, d\bar{t}} \tag{4.141}$$

Respectively, the longer term is determined taking into account (4.141) according to (4.135).

The exponential form of the hydrodynamic resistance function $f_s(t)$ (Martynov & Poliakov, 2022) was assumed in the examples calculated below, namely,

$$f_s(t) = 1 - e^{-a_h t} \tag{4.142}$$

Two limiting cases of the function $f_s(t)$ are indicative. In the first case $\bar{\alpha}_h \to \infty$, so that respectively $f_s \to 1$. This means that the entire deposit formed in the bed is non-washable. Then $\bar{T}_f$ at negligible breakthrough of the suspended solids through the filter media according to (4.135), (4.140) will be

$$\overline{T}_f\left(\overline{t}_{opt}\right) = \overline{S}^0_{max}\left(\overline{t}_{opt}\right)$$

where $\overline{t}_{opt}$ cannot be interpreted as the optimal time due to the impossibility to specify it on the basis of the consolidation criterion. It is only possible, using the criterion equations (1.35), (1.36), to determine the dependence $\overline{T}_f\left(\overline{t}_{opt}\right)$ corresponding to this case, and the required quantity will increase rapidly with decreasing $\overline{t}_{opt}$. Here we also find $\overline{S}^0_{max}$ from equations (4.132) or (4.133) at a given $\overline{t}_{opt}$. It is obvious that at equal duration of all the operating periods the maximum possible value of $\overline{t}_{opt}$ is equal to $\overline{t}_{f1}$, which the value $\overline{S}^0_2 = \overline{t}_{f1}$ is corresponded at $\overline{C}_e \approx 0$ after the first washing (and the last washing applied to the second technological scheme). In principle, it is possible to continue the suspension clarification due to the transformation of the essentially uneven deposit distribution along height at the end of the first filter run into the even distribution with the same total amount of the deposit due to the inefficiency of washing. However, the duration of the second filter run will necessarily be shorter than $\overline{t}_{f1}$. Thus, the second scheme actually degenerates into the simplest scheme.

In the second limiting case, all of the deposit is washed out regardless of its age $\left(\overline{\alpha}_h = 0\right)$. Then, the initial value $\overline{S}^0$ remains constant over the service life of the bed due to f_s=0 and $N_f \rightarrow \infty$. Consequently, the bed could be used indefinitely if there were no other limiting factors, such as changes in the size and shape of the structural elements. In reality, in case of a consolidated deposit there is $0<f_s<<1$, which predetermines the long productive life of a single bed change.

Thus, the consolidation criterion allows us to reasonably limit the sequence of single-type calculations in the first technological scheme. But it serves to select a single duration for all operating periods in the second scheme. If the function of the hydrodynamic resistance is unknown, the reliable information about the minimum permissible duration of the specified period in the conditions under consideration can be its equivalent substitute. Such information can be realistically obtained on the basis of the accumulated extensive experience in the operation of similar filters. The above-mentioned duration can be interpreted at the implementation of the first scheme as t_{min}, and of the second scheme – as t_{opt}. It is possible only formally to set an unlimited increase in $\overline{T}_f$ with a decrease in t_{opt} down to 0 in the absence of the reliable data on the economics of the suspensions filtration with consolidated deposit and, above all, the criterion parameter f_{s*}. Again, the question arises about the rational choice of the duration of operating periods, which cannot be solved without the technical and economic calculations. Relevant recommendations will be presented in the next monograph when describing the results of studies on iron removal from deep groundwater.

Realistic values of the model parameters were chosen for the quantitative analysis of the long-term operation of a rapid filter with consolidated deposit. However, together they meet the extreme filtration conditions, since the bed quickly exhausts its absorbing resource. Thus, the basic set of the specified fixed values includes:

$$\overline{\lambda}_0 = 0.05,\ \overline{S}_m = 300,\ \overline{\gamma}_a = 0.002,\ m_{k1} = 1,\ m_{k2} = 3;\ \Delta\overline{h}_* = 10,\ \overline{C}_* = 0.1\ ;$$

f_k according to (1.10). The parameters $\overline{\alpha}_h$, $\overline{t}_{min}$ were varying. Thus, the service life of the bed and the number of filter runs in it turn out to be relatively small. Then, on the one hand, the representativeness and completeness of the above-mentioned analysis are ensured, and on the other hand, the body of calculations is significantly reduced. Two technological schemes described above are considered on the basis of a rigorous solution of one problem of the non-detachment suspension filtration, which is expressed by the dependences for the relative concentrations $\overline{C}$(4.42) and $\overline{S}$(4.41). In this case, the second scheme is a sequence of filter runs of the same duration.

Calculations for the first scheme were carried out step by step based on the total number of filter runs. It was analyzed on two examples. In the first example, the required duration was determined strictly in accordance with the traditional procedure of the technological calculations for a constant rate filtration described in Section 1.3. In the second example, the time of the first five washings was intentionally and uniformly reduced (by 10 conventional units), and the durations of all subsequent filter runs were determined similarly to the first example. At full-scale modeling of the filter operation ($T_f \geq t \geq 0$) in accordance with the second technological scheme, the basic technological parameter t_{opt} was varied starting from the value, which is noticeably less than the actual minimum value $t_{\min}$.

The results of a set of calculations with respect to the first technological scheme are summarized in **Table 1**. The following generalized data are obtained according to the results of the calculation of the first example. For the total calculation period (service life of the bed) the suspended matter with the total volume in relative units 2191.5 has entered the filter. At the same time, the relative volume of the matter passed through the operating layer was 157.36. Thus, 92.82% was retained in the bed, and 7.18% of the initial impurity got into the filtrate. Accordingly, the error in the simplified calculations of the second technological scheme due to neglect of the quantity $\overline{C}_{eic}$ can be a few percent. In the second example similar indicators were: the total relative volume was equal to 2172.2; the volume of the suspended solids passed the bed was 147.6, which amounted to 6.8% of the total volume of impurity that entered the filter for the entire calculation period. Therefore, single washing of the filter earlier than the time found by the traditional technological calculations, despite the slower accumulation of the non-washable deposit in the bed, do not allow to increase its service life under these conditions. It is worth noting that the values of the technological times $\bar{t}_p$, $\bar{t}_h$ are getting closer and closer with each new filter run. Finally, $\bar{t}_p$ is even less than $\bar{t}_h$ for the last two filter runs. In such a situation, $\bar{t}_f$ is necessary to identify already with $\bar{t}_p$ and calculate the value $\overline{S}^0_{15}$ taking into account $\bar{t}_p = \bar{t}_f$. However, the difference between $\bar{t}_h$ and $\bar{t}_p$ is so small in the examples under consideration that it is acceptable to finish the calculations successively of $\overline{S}^0_{\max}$, N_f and eventually T_f also on the basis of formulas (4.127), (4.130).

Table 1. Technological parameters during the service life of one bed change

Filter run number	Example number	$\overline{S}^0_i$	$\bar{t}_{hi}$	$\bar{t}_{pi}$	$\bar{t}_{hi}\overline{C}_{eic} \cdot 10^{-3}$	$\overline{S}^0_{i+1}$ *exact*	$\overline{S}^0_{i+1}$ *appr.*
1	1	0	219	256	1.58	22.68	22.71
	2	0	209	256	1	20.72	20.74
2	1	22.68	202.5	233.4	2.32	19.45	19.5
	2	20.72	194	235.4	1.44	17.91	17.93
3	1	42.13	188.2	214	3.2	16.85	16.9
	2	38.59	180.7	217.5	1.04	15.6	15.61
4	1	59.03	175.6	197	4.25	14.7	14.76
	2	54.17	169.2	201.8	2.53	13.69	13.73
5	1	73.72	164.57	182.4	5.43	12.93	13
	2	67.86	159	188.3	3.2	12.11	12.15

continued on following page

Table 1. Continued

Filter run number	Example number	$\overline{S}^0_i$	$\bar{t}_{hi}$	$\bar{t}_{pi}$	$\bar{t}_{hi}\overline{C}_{eic}\cdot 10^{-3}$	$\overline{S}^0_{i+1}$ *exact*	$\overline{S}^0_{i+1}$ *appr.*
6	1	86.65	154.8	169.5	6.74	11.45	11.53
	2	79.79	159.9	176.2	6.04	12.21	12.29
7	1	98.1	146.1	158	8.16	10.21	10.29
	2	92.18	150.7	163.8	7.4	12.15	10.94
8	1	108.31	138.3	147.7	9.68	9.15	9.24
	2	104.335	141.3	151.5	11	8.25	8.34
9	1	117.46	131.3	138.5	11	8.25	8.34
	2	116.38	132.1	139.7	11	8.35	8.44
10	1	125.71	124.9	130.3	13	7.46	7.56
	2	124.73	125.7	131.4	13	7.56	7.66
11	1	133.175	119.1	122.8	15	6.78	6.885
	2	132.29	119.8	123.7	15	6.86	6.965
12	1	139.96	113.83	116.2	17	6.44	6.56
	2	139.16	114.45	116.8	16	6.26	6.37
13	1	146.36	108.78	109.7	18	5.65	5.76
	2	145.42	109.5	110.7	18	5.73	5.83
14	1	152.01	104.33	104	20	5.195	5.3
	2	151.15	105	105	20	5.26	5.37
15	1	157.21	100.2	98.8	22	4.79	4.9
	2	156.41	100.85	99.7	22	4.85	4.96

The calculations of the second series of examples in relation to the second technological scheme turned out to be significantly simpler than the first series due to the above presented estimates of the impurity removal from the bed for examples 1, 2 and recommendations for the theoretical substantiation of the rational technological parameters. First of all, it was found out that the relative term $\overline{T}_f$ demonstrates a high sensitivity with respect to the single relative duration t_{opt} (coincides with t_h due to the smallness of the impurity breakthrough). Calculations were carried out using formulas (4.140), (4.135) (at $\overline{C}_{ei} = 0$) and three values of the coefficient $\overline{\alpha}_h$, and their results are shown in **Figure 27**. The increase rate of T_f rises with decreasing t_h and at large values of $\overline{\alpha}_h$. Just **Figure 28** reflects the essentially nonlinear behavior of the dependence $\overline{T}_f(\overline{\alpha}_h)$ when the argument changes from 0 to 10^{-3}. Hence, the importance of studying the strength properties of a deposit by experimental methods and accurately determining the hydrodynamic resistance function. Finally, Figure 29 shows a series of curves describing the relationship between the number of the operating periods N_f, that fall within the total calculation period $\left[0, T_f\right]$, and their duration t_{opt}. The results of such calculations are particularly useful for the technical and economic analysis, since they give a clear pattern of the total non-productive losses of the treated water. It is logical that such losses increase sharply with decreasing $\overline{\alpha}_h$ and $\bar{t}_h$.

Figure 27. Dependence $\overline{T}_f(\bar{t}_h)$

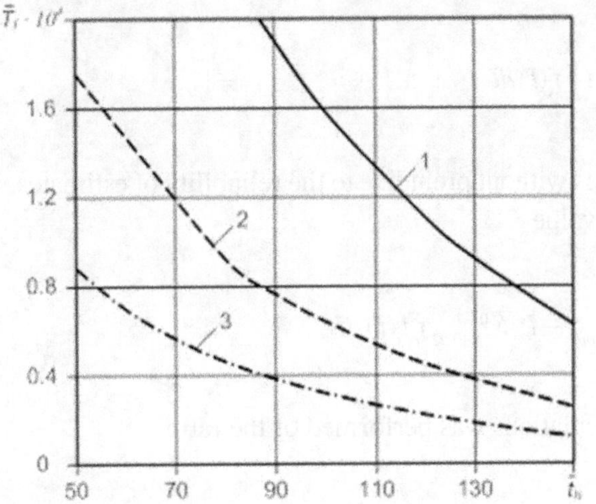

Note. 1 – $\bar{\alpha}_h$ = 0.0002, 2 – $\bar{\alpha}_h$ = 0.0005, 3 – $\bar{\alpha}_h$ = 0.001.

Figure 28. Dependence $\overline{T}_f(\bar{\alpha}_h)$

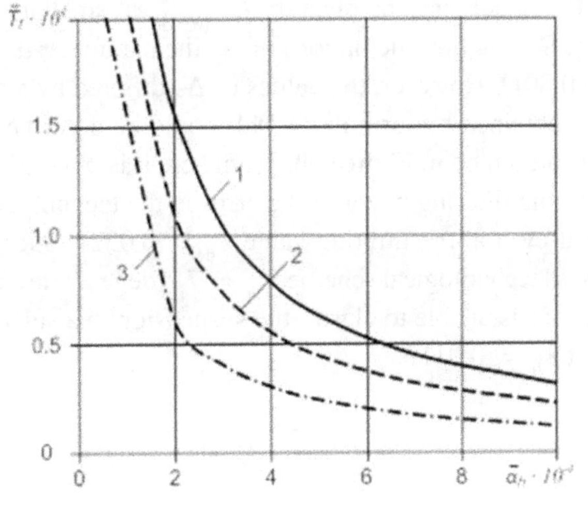

Note. 1 – $\bar{t}_h = \bar{t}_f$ = 100, 2 – $\bar{t}_h$ = 120, 3 – $\bar{t}_h$ = 150.

It is obvious that ignoring the removal of the suspended matter from a bed leads in the technological calculations of the filtration with a consolidatable deposit to overestimation of the calculated values of the concentrations S_S, S and closely related relative quantities $\overline{S}^0_i$, increments $\Delta\overline{S}_i$, maximum values $\overline{S}^0_{max}$, $\overline{S}^0_m$. However, the computational errors arising in this case cannot be large in principle due to strict limitations on the filtrate quality. It is easy to verify the validity of the assumption $\overline{C}_{ei} \approx 0$, and at the same time to identify the conditions, under which these errors become noticeable, by analyzing the different components of integral (4.136). It is proposed to compare actually two integrals, the difference between which should be equal to $\bar{S}^0_m$, namely, the basic integral $N_f F_s\left(\bar{t}_{opt}\right)$ and the evaluated integral itself

$$I_s\left(\bar{t}_{opt}\right) = \int_0^{\bar{t}_{opt}} \sum_{i=1}^{N_f} \bar{C}_{ei}\left(\bar{t}_{opt} - \bar{t}\right) f_s(\bar{t})\, d\bar{t} \tag{4.142}$$

To simplify such analysis without prejudice to the reliability of estimates, integral (4.142) is replaced by the expression close in value

$$I_s\left(\bar{t}_{opt}\right) \approx N_f \int_0^{\bar{t}_{opt}} \bar{C}_{e,(N_f+1)/2}\left(\bar{t}_{opt} - \bar{t};\, \bar{S}^0_{(N+1)/2}\right) f_s(\bar{t})\, d\bar{t} \tag{4.143}$$

and then a series of calculations was performed of the ratio

$$\Delta_C\left(\bar{t}_{opt}\right) = \frac{I_s\left(\bar{t}_{opt}\right)}{N_f F_s\left(\bar{t}_{opt}\right)} \tag{4.144}$$

The initial data here remained the same and, in addition, it was assumed $\bar{t}_{opt} = 100$, $\bar{\alpha}_h$ and $\bar{\lambda}_0$ were varied. Beforehand, the corresponding set of two values of $\bar{S}^0_h$, $\bar{S}^0_p$ was found for each accepted value of $\bar{\lambda}_0$. At $\bar{\lambda}_0 \leq 0.051$ $\bar{S}^0_p \leq \bar{S}^0_h$ was true, and at larger values of $\bar{\lambda}_0$, on the contrary, $\bar{S}^0_h \leq \bar{S}^0_p$. Therefore, the value $\bar{S}^0_{(N_f+1)/2}$ required for calculating the quantity $\bar{C}_{e,(N_f+1)/2}$ at smaller values of $\bar{\lambda}_0$ was specified based on $\bar{S}^0_p$, and then on $\bar{S}^0_h$. During the calculations of Δ_C the parameter $\bar{\alpha}_h$ was varied by an order of magnitude (from 0.0001 to 0.001). However, the values of Δ_C differed by a maximum of 1.5%. Therefore, the totality of the data obtained by formula (4.144) could be described by a single curve of the dependence $\Delta_C(\bar{\lambda}_0)$, which is presented in **Figure 30**. From here it is obvious that with the deterioration of the absorption capacity of the filtering material the error in the technological calculations increases rapidly, remaining still small even at the limiting value $\bar{\lambda}_0 = 0.0221$. Such value corresponds to the critical situation for the second technological scheme $\bar{t}_{opt} = \bar{t}_{f1}$ (degenerate case). Incidentally, the filter under the considered conditions is unable to clarify the suspension to a sufficient degree (to reduce C_0 by an order of magnitude) at $\bar{\lambda}_0 = 0.0112$.

Figure 29. Dependence $N_f(\bar{t}_h)$

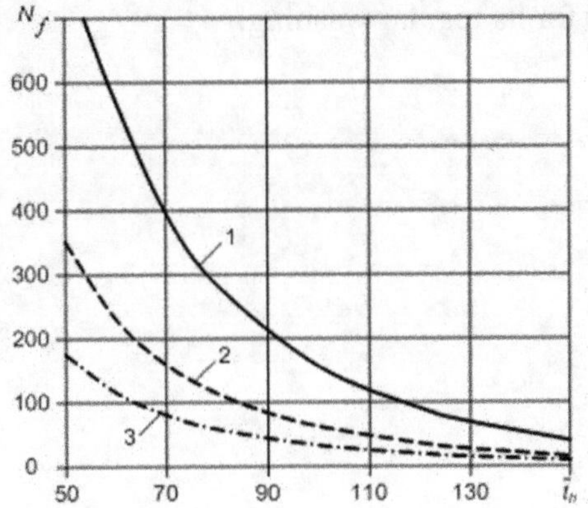

Note. 1 – $\bar{\alpha}_h$ = 0.0002, 2 – $\bar{\alpha}_h$ = 0.0005, 3 – $\bar{\alpha}_h$ = 0.001.

Figure 30. Dependence $\Delta_C(\bar{\lambda}_0)$

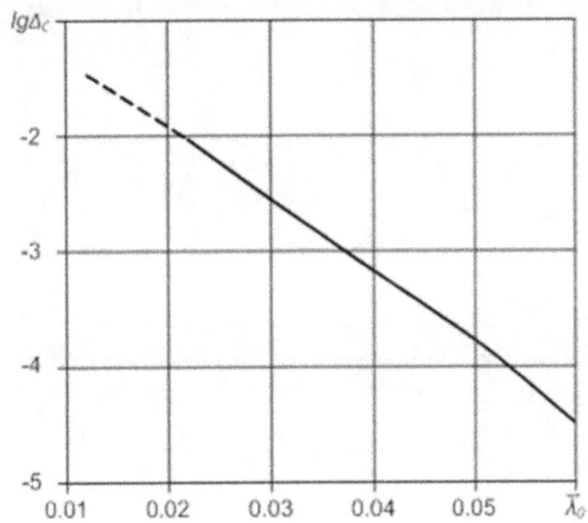

As a result, it has been established that

· reduction and unification of the filter run durations can contribute to a significant increase in the service life of one bed change;
· early washings do not allow prolonging the bed operation in the first technological scheme;
· it is possible in principle to significantly reduce the suspended solids removal from the bed for the entire service life due to the implementation of the second technological scheme, and at the same time there are good reasons for neglecting it in the technological calculations;

· rational choice of the duration of the operating period in the second technological scheme is of fundamental importance, since it determines not only the service life of the filter bed, but also the losses of the treated water for its regular washing.

REFERENCES

Abboud, N. M., & Corapcioglu, M. Y. (1993). Modeling of compressible cake filtration. *J. Colloid. Int. Sci.*, *160*, 304–316.

Alekseev, V. S., & Kommunar, G. M. (1974). Clogging of filters and prefilter zones of water intake wells. [in Russian]. *Water Resources*, (5), 170–177.

Boller, M. A., & Kavanauch, M. C. (1995). Particle characteristics and head loss increase in granular media filtration. *Water Research*, *29*(4), 1139–1149. DOI: 10.1016/0043-1354(94)00256-7

Chang, Y.-I., Cheng, W.-Y., & Chan, H.-C. (2009). A proposed correlation equation for predicting filter coefficient under unfavorable deposition conditions. Short communication. *Separation and Purification Technology*, *65*(3), 248–250. DOI: 10.1016/j.seppur.2008.10.031

Dharmappa, H. B., Verink, J., Fujiwara, O., & Vigneswaran, S. (1992). Optimization of granular bed filtration treating polydispersed suspension. *Water Research*, *26*(10), 1307–1318. DOI: 10.1016/0043-1354(92)90125-N

Finno, R. J., & Rechenmacher, A. L. (2003). Effects of consolidation history on critical state of sand. *Journal of Geotechnical and Geoenvironmental Engineering*, *129*(4), 350–360. DOI: 10.1061/(ASCE)1090-0241(2003)129:4(350)

Guo, S., Zhang, F., Song, X., & Wang, B. (2015). Deposited sediment settlment and consolidation mechanisms. *Water Science and Engineering*, *8*(4), 335–344. DOI: 10.1016/j.wse.2015.10.002

Ho, C.-C., & Zydnev, A. L. (2000). A combined pore blockage and cake filtration model for protein fouling during microfiltration. *Journal of Colloid and Interface Science*, *232*(2), 389–399. DOI: 10.1006/jcis.2000.7231 PMID: 11097775

Ison, C. R., & Ives, K. J. (1969). Removal mechanism in deep bed filtration. *Chemical Engineering Science*, *24*(4), 717–729. DOI: 10.1016/0009-2509(69)80064-3

Iwasaki, I. (1937). Some notes on sand filtration. *Journal - American Water Works Association*, *29*(10), 1591–1602. DOI: 10.1002/j.1551-8833.1937.tb14014.x

Khilar, K. C., & Fogler, H. S. (1998). *Migration of bines in porous media*. Kluwer Academic Publisher. DOI: 10.1007/978-94-015-9074-7

Kuberkar, V. T., & Davis, R. H. (2000). Modeling of fouling reduction by secondary membranes. *Journal of Membrane Science*, *168*(1-2), 243–258. DOI: 10.1016/S0376-7388(99)00324-5

Leichkis, I. M. (1975). *Filtering with the precoating of auxiliary substances*. Tehnika. (in Russian)

Leichkis, I. M., & Ventzel, A. D. (1987). Filtering with flush of a thin layer of auxiliary filtering agent. *Theoretical Foundations of Chemical Engineering*, *25*(6), 832.

Martynov, S. Yu., & Poliakov, V. L. (2022). Experimental studies on the hydrodynamic properties of a deposit in rapid filters during physicochemical removal of iron from groundwater. *Water Science and Technology: Water Supply*, *22*(11), 7603–7617. DOI: 10.2166/ws.2022.305

Mints, D. M., & Meltser, V. Z. (1970). Hydraulic resistance of granular porous medium in the process of clogging. *Rep.* [in Russian]. *USSR Academy of Sciences*, *192*(2), 304–306.

Ojha, C. S., & Graham, N. J. D. (1993). Theoretical estimates of bulk specific deposit in deep-bed filters. *Water Research*, *27*(3), 377–387. DOI: 10.1016/0043-1354(93)90037-I

Ojha, C. S. P., & Graham, N. J. D. (1992). Appropriate use of deep-bed filtration models. *Journal of Environmental Engineering*, *118*(6), 964–980. DOI: 10.1061/(ASCE)0733-9372(1992)118:6(964)

Orlov, V. O. (2008). *Iron removal from groundwater by simplified aeration and filtration*. NUWEE Press. (in Ukrainian)

Poliakov, V., Kravchuk, A., Kochetov, G., & Kravchuk, O. (2019). Clarification of aqueous suspensions with a high content of suspended solids in rapid sand filters. *Eureka: Physics and Engineering*, *1*(1), 28–45. DOI: 10.21303/2461-4262.2019.00827

Poliakov, V. L. (2019). Mathematical modeling of suspension clarification in dynamic layer. [in Russian]. *Rep. NAS of Ukraine*, (12), 33–40.

Poliakov, V. L. (2019). Theoretical substantiation of filtration intensification by changing the direction of the suspension flow. In: *Proceedings of* III International Scientific and Technical Conference "Water supply and drainage, design, construction, operation, monitoring", 194–196.

Poliakov, V. L. (2020). Mathematical modeling of the non-detachment filtration of water suspension with change in direction. [in Russian]. *Rep. NAS of Ukraine*, (5), 31–41.

Poliakov, V.L. (2023). Modeling of the surface filtration of an aqueous suspension. In: *Proceedings of 9 th International Scientific Conference* "Topical Problems of Fluid Mechanics", 42–47.

Sanda, A., Nopolaa, T., Hjelt, T., & Toivakka, M. (2009). A particle motion model for the study of consolidation phenomena. *Computers & Chemical Engineering*, *33*(7), 1227–1239. DOI: 10.1016/j.compchemeng.2009.03.001

Senyavin, M. M., Venitsianov, E. V., & Ayukaev, R. I. (1997). About mathematical models and engineering methods for calculating the process of natural water purification by filtration. [in Russian]. *Water Resources*, (2), 157–170.

Skouras, E. D., Burganos, V. N., Paraskeva, C. A., & Payatakes, A. C. (2011). Phenomenological description and network simulation of horizontal filtration dynamics. *Separation and Purification Technology*, *80*(1), 105–118. DOI: 10.1016/j.seppur.2011.04.016

Sutzkover-Gutman, I., Hasson, D., & Semiat, R. (2010). Humic substances fouling in ultrafiltration processes. *Desalination*, *261*(3), 218–231. DOI: 10.1016/j.desal.2010.05.008

Tan, H. K. S. (1994). Analysis of cyclic fixed bed sorption processes for second-order irreversible kinetics. *Chemical Engineering Science*, *49*(8), 1277–1285. DOI: 10.1016/0009-2509(94)85096-8

Tiller, F. M., Kirby, J. M., & Nguyen, H. L. (1996). Approximate theory for radial filtration / consolidation. *Journal of Geotechnical Engineering*, *122*(10), 797–805. DOI: 10.1061/(ASCE)0733-9410(1996)122:10(797)

Tufenkji, N., & Elimelech, M. (2004). Correlation equation for predicting single-collector efficiency in physicochemical filtration in saturated porous media. *Environmental Science & Technology*, *38*(2), 529–536. DOI: 10.1021/es034049r PMID: 14750730

Veerapaneni, S., & Wiesner, M. R. (1993). Role of suspension polydispersivity in granular media filtration. *Journal of Environmental Engineering*, *119*(1), 172–190. DOI: 10.1061/(ASCE)0733-9372(1993)119:1(172)

Williams, G. J., Sheikh, B., Holden, R. B., Kouretas, T. J., & Nelson, K. L. (2007). The impact of increased loading rate on granular media, rapid depth filtration of wastewater. *Water Research*, *41*(19), 4535–4545. DOI: 10.1016/j.watres.2007.06.018 PMID: 17632202

Zhurba, M.G. (2011). *Water treatment filters with floating filtering media. Scientific edition.* RNO VoSTU. (in Russian)

APPENDIX 4.1. SOLUTION BY SUCCESSIVE APPROXIMATIONS

If a rigorous solution of the suspension clarification compartment (II according Figure1 from Chapter 1) including equations (2.28), (4.2) and conditions (2.9), (2.10) cannot be obtained due to the complexity of the function $\bar{\lambda}$ ($\bar{S}$), then it is possible to obtain an approximate solution close to the exact solution in some cases by the method of successive approximations. The corresponding procedure is as follows. In the first approximation, the filter bed clogging is not taken into account, i.e., $\bar{S}^{(1)} = 0$. Then from (2.9), (2.28), (4.2) it follows that

$$\bar{C}^{(1)} = e^{-\bar{\lambda}(0)\bar{z}}$$

The second approximation for $\bar{S}$ is determined from the solution of the simple problem

$$\frac{\partial \bar{S}^{(2)}}{\partial \bar{t}} = \bar{\lambda}(\bar{S}^{(2)})\bar{C}^{(1)}\bar{z}$$

$$\bar{t} = 0,\ \bar{S}^{(2)} = 0$$

and then it is true

$$\int_0^{\bar{S}^{(2)}} \frac{d\varsigma}{\bar{\lambda}(\varsigma)} = \bar{t}\,\bar{C}^{(1)}\bar{z}. \tag{A.4.1}$$

When the explicit expression for $\bar{S}^{(2)}$ is known, the quantity $\bar{C}^{(2)}(\bar{z},\bar{t})$ is found by solving also the simple problem

$$\frac{\partial \bar{C}^{(2)}}{\partial \bar{z}} + \bar{\lambda}(\bar{S}^{(2)})\bar{C}^{(2)} = 0$$

$$\bar{z} = 0,\ \bar{C}^{(2)} = 1.$$

Hence

$$\bar{C}^{(2)}(\bar{z},\bar{t}) = \exp\left[-\int_0^{\bar{z}} \bar{\lambda}(\bar{S}^{(2)}(\eta,\bar{t}))d\eta\right]$$

The equation with respect to $\bar{S}^{(i)}(\bar{z},\bar{t})$ in an arbitrary i-th approximation (i>2)

$$\int_0^{\bar{S}^{(i)}} \frac{d\varsigma}{\bar{\lambda}(\varsigma)} = \int_0^{\bar{t}} \bar{C}^{(i-1)}(\bar{z},\xi)\,d\xi \tag{A.4.2}$$

and the expression for $\bar{C}^{(i)}$

$$\bar{C}^{(i)}(\bar{z},\bar{t}) = \exp\left[-\int_0^{\bar{z}} \bar{\lambda}(\bar{S}^{(i)}(\eta,\bar{t}))d\eta\right]$$

are successively derived in a similar way. It is not difficult to practically implement the described procedure if it is possible to calculate the integral in equations (A.4.1), (A.4.2) and to obtain an explicit representation for it. For example, if (4.38) is true for $\bar{\lambda}$, then at $m_\lambda \neq 1$

$$\bar{S}^{(i)}(\bar{z},\bar{t}) = \bar{S}_m\left\{1 - \left[1 - \bar{\lambda}_0(1 - m_\lambda)\bar{S}_m^{\,m_\lambda - 1}\int_0^{\bar{t}} \bar{C}^{(i-1)}(\bar{z},\xi)\,d\xi\right]^{1-m_\lambda}\right\}$$

and at m_{λ}=1

$$\overline{S}^{(i)}(\overline{z},\overline{t}) = \overline{S}_m \left\{ 1 - \exp\left[-\overline{\lambda}_0 \int_0^{\overline{t}} \overline{C}^{(i-1)}(\overline{z},\xi)\, d\xi \right] \right\}$$

The formula for $\overline{C}^{(i)}(\overline{z},\overline{t})$ remains the same.

Chapter 5
Detachment Filtration With Unregulated Rate

ABSTRACT

The depth detachment filtration with linear kinetics and an unregulated rate is considered in detail. A significant filtration rate is initially provided due to the suspension accumulation in the storage tank. The early and preparation stages are specially studied. The main interest is the final stage, during which the suspension is processed with a declining filtration flow rate. The technological analysis is carried out considering the limitation on the decrease in the filter performance. The hydraulic properties of the filter communications and variable head losses in them are taken into account. Mathematical modeling is significantly complicated due to the close relationship between the kinetic parameters and the filtration rate. An approximate solution is obtained with minimal errors for the general form of such relationship. An exact solution was obtained and a reliable calculation method developed based on it. The limiting storage tank height was evaluated, calculations of the filter run duration and applied optimization of the bed absorption properties were performed.

Along with the clarification of aqueous suspension on rapid filters, which is implemented at a constant filtration flow rate and thoroughly studied by analytical methods in the previous chapters, recently filtration with a variable flow rate is of considerable interest. The corresponding filtration rate mode is more and more widely used in the practice of water treatment due to a number of advantages and, first of all, simplicity of filter operation, greater reliability (Chaundry, 1987; Cleasby, 1993; Grabovsky & Gurinchyk, 2008; Poliakov, 2022; Saatci, 1990). In the process of the clogging of a filter media, its hydraulic resistance steadily increases, so that the filtration rate naturally decreases in the absence of special head losses compensators (rate regulators). At the same time, a short-term increase in the filtration rate is also possible, if a packed bed is unable to pass the entire amount of suspension fed to the filter in time. In this case, the initial suspension accumulates at the inlet to the medium, providing an additional reserve of mechanical energy. In order to maintain the high performance of the filter on a regular basis, it must be systematically washed at any rate mode. However, the time of the successive washing in the case of decreasing filtration rate should be set on the basis of the requirements for the filtrate quality and filtrate flow rate.

Below the mathematical modeling of the suspension separation in the second main rate mode (filtration), namely, at an unregulated filtration rate is performed. As a consequence, the performance of a filtration facility gradually decreases, and the hydraulic characteristics of the transportation system,

DOI: 10.4018/979-8-3693-8211-0.ch005

storage tank acquire a non-stationary character. Consequences of jointly occurring clarification and hydrodynamic processes consistently studied with the use of exact and approximate methods at two varieties of the above-mentioned mode. The operation of a rapid filter in the second rate mode is schematically illustrated in Figure 2 from Chapter 1.

In the first variety of the unregulated rate filtration, the stability of the energy resource in a packing bed or in the entire facility is ensured by maintaining a constant differential head (pressure) at the boundaries of the respective physical region of motion. In the second variety, a suspension is fed to the filter at a significant constant flow rate. In both cases, part of the suspension fed to the facility is retained in the storage tank due to the insufficient flow capacity of the clean and even more so of the clogged bed. However, in the first variety of the second rate mode, the rise of the initial suspension level in the storage tank is limited physically (head in the raw water collector), and in the second variety – structurally (overflow chute mark). In the first instance, the separation of a suspension was purposefully studied and, therefore, a constant differential head was set at the boundaries of the operating medium in Section 5.1. Numerous dependencies are derived and widely illustrated here, which allow to carry out the predictive and technological calculations with a high accuracy on the basis of two (approximate and special reference) solutions, in fact, the basic mathematical problem of the suspension filtration with a monotonically decreasing rate. Taking into account the proportionality of the head losses in the active porous bed and the transportation system of the filtration facility, and at the same time the significance of changes in the flow characteristics, Section 5.2 additionally considered the hydraulic aspects of the operation of the supply and offtake communications at constant head losses throughout the facility (the first type). The technological process is implemented at fixed heads at the collectors of a suspension (raw water) and filtrate (disposable or effective head). Finally, the theoretical studies of the filtration facility operation in the second variety of the second filtration rate mode (Section 4.3) were carried out in full (predictions, technological and design analysis). At the beginning of each operating period (filter run) due to a relatively high hydraulic resistance of the layer of a filtering material the surface layer of an initial suspension is formed above it. The characteristic time of its formation is significantly less than the similar time of the suspension clarification, which justifies the isolation of special (early) stages from the general technological process. At these stages, due to their short duration and, as a consequence, small amounts of the suspension and suspended matter entering the bed, it is reasonable to limit modeling only the hydrodynamics of the water flow in a clean porous medium and pipelines serving it, with emphasis on the dynamics of the suspension level in the storage tank. The porous material of the bed is saturated with the initial suspension in the shortest first stage. Nevertheless, this stage deserves a separate consideration, because quite a large amount of fluid can accumulate above the bed surface by its end. At the second stage, the main volume of the untreated suspension accumulates at the inlet of the bed and the offtake communications are activated. Therefore, the entire filter run at a variable filtration rate was conventionally divided into three stages, succeeding one another. The general approximate and particular exact solutions of the problems describing the fluid motion not only in the layer of a filtering material, but also in supply and offtake communications were found. The accuracy of all obtained approximate solutions was estimated.

The detailed technological analysis was based on two (first variety) or three (second variety) criteria of the clarification efficiency by filtration, the meeting of which guarantees the high quality of filtrate and economical operation of the facility. The hydraulic criterion (1.36) was no longer necessary. Instead of it the criterion was used, which provides control over the filter performance and prevents its excessive reduction. The new criterion can be written in two equivalent dimensionless forms, which are based on relation (1.37),

$$\bar{V} \geq \bar{V}_{*}, \bar{V}_{a} \geq \bar{V}_{a*} \tag{5.1}$$

where $\bar{V}_{a*}$ should be interpreted as a relative minimum allowable average hydraulic load for the filter run; it is recommended to find the limiting values $\bar{V}_{*}$, $\bar{V}_{a*}$ in advance on the basis of the technical and economic analysis of the filter operation, and the time point $\bar{t}_V$ corresponding to them by solving the equations $\bar{V}(\bar{t}_V) = \bar{V}_*$ or $\bar{V}_a(\bar{t}_V) = \bar{V}_{a*}$ Based on the corresponding exact solution, the suspension clarification on the rapid filter with the storage tank was analyzed in detail at the close relationship between the filtration rate and the absorption capacity of the bed. The expediency of involvement of the third, technical criterion (1.38) and continuation of the filter operation with the filled storage tank was discussed.

5.1. FILTRATION WITH CONSTANT HEAD DIFFERENTIAL AT FILTER BED BOUNDARIES

The detachment filtration of a suspension through the uniform layer of a filtering material is considered under linear kinetics of mass exchange between its solid and liquid phases, as well as a declining rate over time (Grabovsky, 2016; Poliakov, 2010). The progressive decrease in the filter performance is caused by a constant head differential at the boundaries of the specified layer and its increasing clogging. The creation of such conditions of filtration is not provided in a modern filtration facility by design, but they can be considered in theoretical studies as a first approximation to the actual situation, namely, a constant head difference at its inlet and outlet (at the corresponding collectors). And only in case of the filter operation continuation after the storage tank overflow, the boundary conditions of the first kind will allow to calculate correctly the subsequent development of physical and chemical situation in the operating layer. Hence, if all design elements of a given facility, providing the stable functioning of a filter bed, do not significantly affect the hydraulics of the whole structure, then changes in the heads at its both boundaries will be insignificant in comparison with the head losses inside it. Then the assumption of the fixed head losses in the bed is justified. At the same time, it is not difficult to maintain practically constant heads at the inlet and outlet of such structures (at the respective collectors). The corresponding filtration mode will be the subject of a detailed study in the following Sections.

It should be emphasized that, the models of the detachment suspension filtration with a variable rate are significantly more complicated from the mathematical point of view than similar models with a constant rate. As shown by numerous experimental studies, both mass exchange coefficients depend significantly on the indicated rate. For the dependences of the attachment α and the detachment β rate coefficients on V, the power form (3.54) is usually assumed (Logan et al., 1995; Senyavin et al., 1977; Venitsianov & Rubinstein, 1983). Thus, due to the lack of the control over the filtration flow rate, these coefficients appear to be not only variables, but also unknown values. They can be specified only by determining the filtration rate as a function of time, which requires solving a complex mathematical problem.

Thus, the mathematical model, which serves as a basis for the theoretical studies in this Chapter, now includes closely related the clarification (removal of dispersed contaminants) and filtration flow compartments (Figure 1) from Chapter 1. The first compartment contains the system of equations

$$V(t)\frac{\partial C}{\partial z}+\frac{\partial S}{\partial t} = 0 \quad (5.2)$$

$$\frac{\partial S}{\partial t} = \alpha_V V^l(t)C - \beta_V V^q(t)S \quad (5.3)$$

the standard boundary condition (2.3) and homogeneous initial (2.48) condition in the absence of an initial bed contamination.

The second compartment for a linear filtration flow includes the equation of motion

$$V(t) = -k(S_s)\frac{\partial h}{\partial z} \quad (5.4)$$

as well as the equations characterizing a regular decrease in the flow capacity of the bed being clogged (1.10) and a change in the deposit composition (1.29). The above system of equations is supplemented by the conditions at the outlet of the bed (2.6) and at its inlet

$$z=0,\ h=H_u; \quad (5.5)$$

where H_u is the constant head at the upper (downward filtration) boundary of the bed. Then dimensionless variables and parameters are introduced by analogy with Section 2.1:

$$\overline{C} = C/C_0,\ \overline{S} = S/(n_0 C_0),\ \overline{z} = z/L,\ f_k = k/k_0,\ \overline{\gamma} = \gamma C_0,$$

and

$$\overline{t} = V^0 t/(n_0 L),\ \overline{V} = V/V^0,\ \overline{h} = (h - H_d)/(H_u - H_d),$$

$$\overline{\alpha}_V = \alpha_V L V^{0^{l-1}}, \overline{V}_a = V_a/V^0, \overline{\beta}_V = \beta_V n_0 L V^{0^{q-1}},$$

V^0 is the initial filtration rate, which is expressed through other parameters of the problem as follows

$$V^0 = k_0 \frac{H_u - H_d}{L}$$

Then the relative independent variable $\overline{t}$ is replaced by

$$\overline{\tau} = \int_0^{\overline{t}} \overline{V}(\xi)d\xi$$

and the concentration S_s is excluded. Then equations and condition (5.2) - (5.5), taking into account (3.55), are transformed to the form

$$\frac{\partial \overline{C}}{\partial \overline{z}}+\frac{\partial \overline{S}}{\partial \overline{\tau}} = 0, \quad (5.6)$$

$$\frac{\partial \bar{S}}{\partial \bar{\tau}} = \bar{\alpha}_V \bar{V}^{l-1}(\bar{\tau})\bar{C} - \bar{\beta}_V \bar{V}^{q-1}(\bar{\tau})\bar{S}, \tag{5.7}$$

$$\bar{V}(\bar{\tau}) = -f_k(\bar{S})\frac{\partial \bar{h}}{\partial \bar{z}}, \tag{5.8}$$

$$\bar{z} = 0, \bar{h} = 1; \tag{5.9}$$

$$\bar{\tau} = 0, \quad \bar{S} = 0 \tag{5.10}$$

In contrast to the theoretical developments on the filtration with a constant rate presented in the first chapters, in this case it is not possible to solve the clarification and filtration flow compartments sequentially. Therefore, only the general approximate and partial exact solutions were found to the above formulated filtration problem with a declining rate due to a constant head differential at the boundaries of the clogged bed. Both solutions together with the results of the comparative analysis are given below and are of value, first of all, as an effective tool for the engineering calculations

The approximate solution of problem (1.9), (2.9), (2.12), (2.13), (5.6) - (5.10) is determined by averaging the values $\bar{V}, \bar{V}^{l-1}, \bar{V}^{q-1}$ over the calculation period $[0, \bar{T}]$ (dynamic averaging), so that

$$\bar{V}_a = \frac{1}{T}\int_0^T \bar{V}(\bar{t})d\bar{t}. \tag{5.11}$$

The quantities $\bar{V}^{l-1}, \bar{V}^{q-1}$ in equation (5.7) are replaced by their average values, and it is assumed that

$$[\bar{V}^{l-1}]_a = \bar{V}_a^{\ l-1}, [\bar{V}^{q-1}]_a = \bar{V}_a^{\ q-1} \tag{5.12}$$

The substantiation of equalities (5.12) is carried out taking into account the data on the behavior of the unregulated rate function $\bar{V}(\bar{t})$ under actual conditions. Considering the almost linear character of decreasing $\bar{V}$ over time and the comparatively small change during the filter run, the function $\bar{V}(\bar{t})$ is expanded into the Taylor series in the vicinity of the point $\bar{t} = 0$ and the first two terms are retained. Then

$$\bar{V}(\bar{t}) \approx 1 - a_V\bar{t},$$

where $a_V = -\frac{d\bar{V}}{d\bar{t}}$ at $\bar{t} = 0$ and $a_V > 0$. According to (5.11), the value of $\bar{V}_a$ in the time interval from 0 to T approximates to

$$\bar{V}_a(T) \approx 1 - \frac{a_V}{2}T \tag{5.13}$$

The Taylor series expansion of the function $\bar{V}_a^{\ r}(T)$ on T taking into account (5.13) gives

$$\bar{V}_a^{\ r}(T) \approx 1 - \frac{ra_V}{2}T \tag{5.14}$$

As a result of similar decomposition of the function $\bar{V}^r(T)$ by $\bar{t}$ we obtained

$$\bar{V}^r(\bar{t}) \approx 1 - ra_V\bar{t} \tag{5.15}$$

Averaging (5.15) over the same period leads to the approximate expression

$$[\overline{V}^r(\overline{t})]_a \approx 1 - \frac{ra_V}{2}\overline{t} \tag{5.16}$$

The equality of the expressions for $[\overline{V}_a(\overline{t})]^r$ and $[\overline{V}^r(\overline{t})]_a$ allows to present the kinetic equation (5.7) in a simplified form

$$\frac{\partial \overline{S}}{\partial \overline{\tau}} = \overline{\alpha}_V \overline{V}_a^{\,l-1}\overline{C} - \overline{\beta}_V \overline{V}_a^{\,q-1}\overline{S} \tag{5.17}$$

As before (Section 2.4), the quantity T introduced in the course of solving the problem plays the role of a parameter and it is identified with time $\overline{t}$ only in the calculation formulas.

The exact solution of the system of equations (5.6), (5.17) under conditions (2.9), (5.10) and constant coefficients in real time is presented in Section 2.1, and is expressed in this case under the homogeneous initial condition by the dependencies

$$\begin{aligned}\overline{C}(\overline{z},\overline{\tau};\overline{V}_a) &= e^{-\overline{\alpha}_V \overline{V}_a^{l-1}\overline{z}}\Big[e^{-\overline{\beta}_V \overline{V}_a^{q-1}\overline{\tau}} I_0\Big(2\sqrt{\overline{\alpha}_V\overline{\beta}_V\overline{V}_a^{l+q-2}\overline{z}\overline{\tau}}\Big)+ \\ &+\overline{\beta}_V \overline{V}_a^{q-1}\int_0^{\overline{\tau}} e^{-\overline{\beta}_V \overline{V}_a^{q-1}\eta} I_0\Big(2\sqrt{\overline{\alpha}_V\overline{\beta}_V\overline{V}_a^{l+q-2}\overline{z}\eta}\Big)d\eta\Big],\end{aligned} \tag{5.18}$$

$$\overline{S}(\overline{z},\overline{\tau};\overline{V}_a) = \overline{\alpha}_V \overline{V}_a^{l-1} e^{-\overline{\alpha}_V \overline{V}_a^{l-1}\overline{z}}\int_0^{\overline{\tau}} e^{-\overline{\beta}_V \overline{V}_a^{q-1}\eta} I_0\Big(2\sqrt{\overline{\alpha}_V\overline{\beta}_V\overline{V}_a^{l+q-2}\overline{z}\eta}\Big)d\eta \tag{5.19}$$

From formulas (5.18), (5.19) important consequences follow, namely, the increase of the suspended solids content in the filtrate and the deposit accumulation in the inlet cross-section of a bed are described at $\overline{t} \gg 0$ as follows

$$\begin{aligned}\overline{C}_e(\overline{t};\overline{V}_a) &= \overline{C}(0,\overline{t};\overline{V}_a) = e^{-\overline{\alpha}_V \overline{V}_a^{l-1}}\Big[e^{-\overline{\beta}_V \overline{V}_a^{q}\overline{t}} I_0\Big(2\sqrt{\overline{\alpha}_V\overline{\beta}_V\overline{V}_a^{l+q-1}\overline{t}}\Big)+ \\ &+\overline{\beta}_V \overline{V}_a^{q-1}\int_0^{\overline{V}_a\overline{t}} e^{-\overline{\beta}_B \overline{V}_a^{q-1}\eta} I_0\Big(2\sqrt{\overline{\alpha}_V\overline{\beta}_V\overline{V}_a^{l+q-2}\eta}\Big)d\eta\Big],\end{aligned} \tag{5.20}$$

$$\overline{S}_0(\overline{t};\overline{V}_a) = \overline{S}(0,\overline{t};\overline{V}_a) = \frac{\overline{\alpha}_V}{\overline{\beta}_V}\overline{V}_a^{l-q}(1 - e^{-\overline{\beta}_V \overline{V}_a^{q}\overline{t}}) \tag{5.21}$$

In order to use formulas (5.18) - (5.21) in engineering calculations, it is necessary to determine the unknown parameter $\overline{V}_a$. To establish the relationship between $\overline{V}_a$ and $\overline{t}$, first of all, equation (5.8) is integrated within the range from $\overline{z}$ to 1 and then, taking into account the boundary condition (2.13), the reduced head will be as follows

$$\tilde{h}(\overline{z},\overline{t};\overline{V}_a) = \overline{V}\int_{\overline{z}}^{1} \frac{d\eta}{f_k(\overline{S}(\eta,\overline{t};\overline{V}_a))}. \tag{5.22}$$

The expression for the filtration rate was found by involving the boundary condition (5.9)

$$\bar{V}(\bar{t};\bar{V}_a) = \left(\int_0^1 \frac{d\bar{z}}{f_k(\bar{S}(\bar{z},\bar{t};\bar{V}_a))}\right)^{-1}. \tag{5.23}$$

By averaging both parts of equality (5.23) in accordance with (5.11) the missing equation is derived relative to $\bar{V}_a$ in the implicit form

$$\int_0^{\bar{t}}\left(\int_0^1 \frac{d\bar{z}}{f_k(\bar{S}(\bar{z},\zeta;\bar{V}_a))}\right)^{-1} d\zeta - \bar{V}_a\bar{t} = 0. \tag{5.24}$$

The values of the function $\bar{V}_a(\bar{t})$ can be easily found by fitting from (5.24) using standard mathematical analysis software.

The high accuracy of the obtained approximate solution is evidenced by the results of its comparison with the exact partial solution (to be discussed below). According to the reference solution (2.55), (2.56), which corresponds to the only values of power exponents in (3.54), namely l=q=1, the required concentrations at $\bar{t} \gg 0$ have the following form

$$\bar{C}(\bar{z},\bar{\tau}) = e^{-\bar{\alpha}\bar{z}}\left[e^{-\bar{\beta}\bar{\tau}}I_0\left(2\sqrt{\bar{\alpha}\bar{\beta}\bar{z}\bar{\tau}}\right) + \bar{\beta}\int_0^{\bar{\tau}} e^{-\bar{\beta}\xi}I_0\left(2\sqrt{\bar{\alpha}\bar{\beta}\bar{z}\xi}\right)d\xi\right], \tag{5.25}$$

$$\bar{S}(\bar{z},\bar{\tau}) = \bar{\alpha}\, e^{-\bar{\alpha}\bar{z}}\int_0^{\bar{\tau}} e^{-\bar{\beta}\xi}I_0\left(2\sqrt{\bar{\alpha}\bar{\beta}\bar{z}\xi}\right)d\xi. \tag{5.26}$$

The time of the supply of the relative suspension volume $\bar{\tau}$ on a unit surface area of a bed is as follows

$$\bar{t} = \int_0^{\bar{\tau}}\int_0^1 \frac{d\bar{z}}{f_k(\bar{S}(\bar{z},\zeta))}d\zeta \tag{5.27}$$

The formula for the calculation of $\bar{V}(\bar{t})$ in this situation is the same as (5.23).

The approximate solution presented above allows us to reliably substantiate the technological, design parameters and, most importantly, the relative duration of the filter run $\bar{t}_f$ at a declining filtration rate. First of all, it is essential, relying on criteria (1.35), (5.1), to calculate the technological times $\bar{t}_p$, $\bar{t}_V$. For determination of $\bar{t}_V$ it is more convenient to set, for example, the minimum allowable average filtration rate $\bar{V}_{a*}$ and then it is necessary to solve the equation

$$\int_0^{\bar{t}_V}\left(\int_0^1 \frac{d\bar{z}}{f_k(\bar{S}(\bar{z},\zeta;\bar{V}_{a*}))}\right)^{-1} d\zeta - \bar{V}_{a*}\bar{t}_V = 0. \tag{5.28}$$

It is more difficult to calculate the time $\bar{t}_p$, since the rate $\bar{V}_{ap}$, denoting $\bar{V}_a(\bar{t}_p)$, is unknown and we have to find in pairs $\bar{t}_p$, $\bar{V}_{ap}$, by jointly solving the equations

$$\begin{aligned}\bar{C}_* e^{\bar{\alpha}_V\bar{V}_{ap}^{l-1}} &= e^{-\bar{\beta}_V\bar{V}_{ap}^{q-1}\bar{t}_p}I_0\left(2\sqrt{\bar{\alpha}_V\bar{\beta}_V\bar{V}_{ap}^{l+q-2}\bar{t}_p}\right)+ \\ &+\bar{\beta}_V\bar{V}_{ap}^{q-1}\int_0^{\bar{V}_{ap}\bar{t}_p} e^{-\bar{\beta}_V\bar{V}_{ap}^{q-1}\xi}I_0\left(2\sqrt{\bar{\alpha}_V\bar{\beta}_V\bar{V}_{ap}^{l+q-2}\xi}\right)d\xi,\end{aligned} \tag{5.29}$$

$$\int_0^{\bar{t}_p} \left(\int_0^1 \frac{d\bar{z}}{f_k\left(\bar{S}\left(\bar{z},\zeta;\bar{V}_{ap}\right)\right)} \right)^{-1} d\zeta - \bar{V}_{ap}\bar{t}_p = 0. \tag{5.30}$$

To find a suitable set of two values of $\bar{t}_p$, $\bar{V}_{ap}$, we can, first, graphically plot the curves of two relationships between the quantities $\bar{t}_p$ and $\bar{V}_{ap}$. The coordinates of the point of their intersection will give the desired set. Secondly, it is also convenient to use an iterative approach, when at the first step, given an actual value of $\bar{V}_{ap}$, we find from (5.30) the value of $\bar{t}_p$, corresponding to it. Both values are substituted into equation (5.29) and a discrepancy is calculated. Then, guided by it, the following values of $\bar{V}_{ap}$, $\bar{t}_p$, which should reduce this discrepancy after each next computational step up to its predetermined minimum value, are successively fitted. The latter is accepted on the basis of its comparison with the summands of equations (5.29), (5.30). It is recommended to determine the relative duration of the filter run at a declining filtration rate based on (1.45) by analogy with (1.46) as the lesser of the times $\bar{t}_p$, $\bar{t}_V$, i.e.

$$\bar{t}_f = \min\left(\bar{t}_p, \bar{t}_V\right). \tag{5.31}$$

Since the general solution of the initial problem in a dimensionless form (1.9), (2.9), (2.12), (2.13), (5.6)-(5.10), corresponding to arbitrary values of l and q, is approximate, and numerous calculations of the physical and chemical characteristics, technological parameters were carried out on its basis, then, first of all, we verified the validity of the application of the dynamic averaging technique of the filtration rate (5.11), due to which the basic calculation formulas and equations were derived. For this purpose, a number of examples were selected, the initial data for which are characteristic for the treatment of natural water and the secondary treatment of wastewater of different origin. The values of the model coefficients were accepted, which are typical for the operating conditions of modern filtration facilities, in particular, 4, 6, 8 for $\bar{\alpha}$; 0.005 for $\bar{\beta}$; 20 for γ_a, as well as the high inlet concentration (C_0=5•10^{-5}). It should be noted that to estimate the accuracy of the approximate solution, l=q=1 was accepted in the first series of examples, so that $\bar{\alpha}_V = \bar{\alpha}$ and $\bar{\beta}_V = \bar{\beta}$. The subject of the calculations using the exact and approximate formulas was the change over time of three key relative characteristics – the outlet concentration $\bar{C}_e$, rate $\bar{V}$ and specific volume of the clarified suspension τ. Figure 1 shows the curves of the dependence $\bar{C}_e(\bar{t})$, calculated by formulas (5.20), (5.25). A small discrepancy between the corresponding curves (1 and 2; 3 and 4; 5 and 6) is observed only at the end of the calculation period, the duration of which is noticeably longer than the duration of the filter run. In general, these curves are so close to each other that they almost merge. A relatively large value of C_0 was taken to achieve a sharp decrease in the quantity $\bar{V}$ during the filter operation. And here noticeable differences between the results of the exact calculations according to formulas (5.26), (5.23) and the approximate calculations according to (5.19) and also (5.23) are noted only after a reduction of the filtration flow rate by 60%. In practice, such reduction of the filter performance is unjustified for economic reasons. Therefore, it can be stated already on the basis of Figures 1 and 2 that the formulas recommended for applications provide the prediction of changes in filtration characteristics with a high accuracy. Finally, this conclusion is also confirmed by Figure 3, which shows the curves of increasing volumes of filtered suspension over time for two values of $\bar{\alpha}$. The exact and approximate values of $\bar{\tau}$ differ slightly in the case of a well-absorbing filter media. At the same time, at a moderate absorption capacity of a filtering material both curves practically coincide up to the moment of time, which corresponds to the relative value $\bar{t} = 250$. For example, this time will be 20 hours at n_0=0.4, L=1m, V^0=5m/h.

Figure 1. Change in the relative outlet volumetric concentration of the suspended solids over time

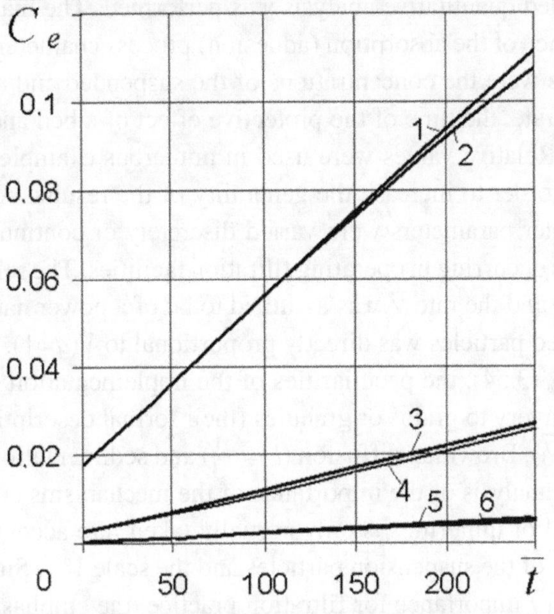

Note. 1, 3, 5 – approximate; 2, 4, 6 – exact; 1, 2 – $\overline{\alpha}$ = 4; 3, 4 – $\overline{\alpha}$ = 6; 5, 6 – $\overline{\alpha}$ = 8

Figure 2. Change in the relative filtration rate over time

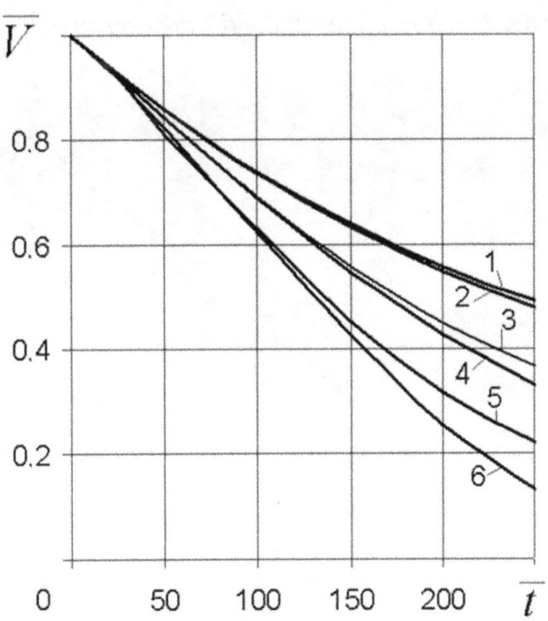

Note. 1, 3, 5 – exact; 2, 4, 6 – approximate; 1, 2 – $\overline{\alpha}$ = 4; 3, 4 – $\overline{\alpha}$ = 6; 5, 6 – $\overline{\alpha}$ = 8

In order to illustrate the serious possibilities of the approximate solution in predicting the development of physico-mechanical conditions in the bed, the substantiation of the rational technological and design parameters, a detailed quantitative analysis was performed. The main attention was paid to the estimation of the significance of the absorption (adhesion) process characterized by the parameters α_V, l. The subject of calculations were the concentrations of the suspended and deposited particles, the current and average filtration rate, the time of the protective effect of a bed and the permissible reduction of the filter performance. Relative values were used in numerous examples with the exception of the initial concentration C_0 in order to increase the generality of the results. A single value of $\overline{\beta}_V$ (0.005) was also accepted. The other parameters were varied discretely or continuously over a wide range to cover a variety of situations occurring in operating filtration facilities. The relationship between the mass exchange coefficients $\overline{\alpha}, \overline{\beta}$ and the rate $\overline{V}$ was assumed to be of a power nature. In this case, the mobilization rate of the deposited particles was directly proportional to $V(q{=}1)$. When accepting the values of l in the expression for $\overline{\alpha}$ (3.54), the peculiarities of the implementation of different mechanisms of the suspended particles delivery to grains or granules (their formal description is given in Section 2.6), namely, interception ($l = \frac{7}{8}$), Brownian diffusion ($l = \frac{1}{3}$) and sedimentation ($l = -\frac{1}{5}$) were taken into account. Nevertheless, the analysis of the importance of the mechanisms is rather tentative, since their contribution to the removal of impurities is only partially taken into account. In reality $\overline{\alpha}_V$, $\overline{\alpha}$ are also related to l through the size of the suspension particles and the scale V^{0l-1}. Since the time of the effective filter operation is of primary importance for filtration practice, the emphasis was placed on the calculation of the relative technological times $\overline{t}_p$, $\overline{t}_V$ and finally on the calculation of the relative duration of the filter run $\overline{t}_f$. Criteria (1.35), (5.1) were used for their determination and it was required, firstly, to provide reduction of impurity content by at least an order of magnitude ($\overline{C}_* = 0.1$), secondly, to prevent excessive reduction of the (average) filter performance $\left(\overline{V}_a = 0.9, 0.8, 0.7\right)$.

Figure 3. Change in the relative specific volume of the filtrate over time

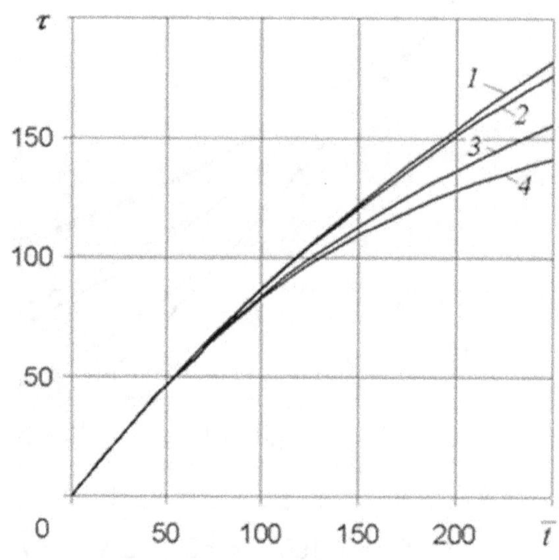

Note. 1, 3 – approximate; 2, 4 – exact; 1, 2 – $\overline{\alpha} = 4$; 3, 4 – $\overline{\alpha} = 8$

The initial concentration of the suspended solids was unchanged (C_0=2.5•10^{-5}) in the second series of examples. The effect of the above-mentioned mechanisms (l=7/8, 1/3, -1/5), the absorption properties of a filtering material ($\overline{\alpha}_V$=4, 6, 8) on the quality of the suspension clarification was studied. The curves of an increase in the outlet concentration $\overline{C}_e$ over time were calculated by formula (5.20), equation (5.24) and presented in Figure 4. The concentration $\overline{C}_e$ shows the greatest sensitivity in relation to the coefficient $\overline{\alpha}_V$. Thus, there is a significant breakthrough of the suspended solids through the bed at $\overline{\alpha}_V$=4 (curve 1) and $\overline{C}_e$ reaches 0.134 at $\overline{t} = 400$ (32 hours of the continuous filter operation at the values n_0=0.4, L=1m, V^0=5m/h), so that it should be washed out earlier at $\overline{C}_* < 0.134$ (at $\overline{C}_* = 0.1, \overline{t}_p \approx 265$). With the increase in $\overline{\alpha}_V$ up to 6 the number of solids in the filtrate sharply decreases. Now the maximum value of $\overline{C}_e$ is only 0.026 at $l = \frac{1}{3}$ and, if there are not too strict requirements to the quality of water treatment, it is reasonable to continue the filter operation, provided, certainly, that its performance has decreased insignificantly. The decision to turn off a filter should be made on the basis of the data of calculating the variables $\overline{C}_e(\overline{t})$ and $\overline{V}_a(\overline{t})$ (or $\overline{V}(\overline{t})$). Finally, the suspension particles at the filter outlet are practically absent during the entire calculation period with $\overline{\alpha}_V = 8$ (curve 6). The ways of the delivery of the mobile particles to fixed particles-collectors (including deposit) are less significant for the clarification process, as indicated by the small divergence of curves 2-4, which are obtained at $\overline{\alpha}_V = 6$ and different l. A decrease in the value of l, signifying a hypothesized weakening of the effect of the filtration flow process on the adhesion process, contributes to an increase in the removal of impurities from the filter over time. On the other hand, the reduction of a fluid velocity in the pores of the bed leads to the prolongation of the suspended particles stay here and, consequently, to the deposition of more of them. If the second factor prevails, the filtrate quality may even improve over time, which is confirmed by curves 5, 6.

The deposition rate of the suspended solids and the quantity S_0 primarily depend on the absorption capacity of a filtering material, which follows from the results of the calculations according to expressions (5.21, (5.24) shown in Figure 5. Naturally, the highest concentration of the immobilized particles in the inlet cross-section is observed at $\overline{\alpha}_V = 8$ (curve 1), and the lowest concentration at $\overline{\alpha}_V = 4$ (curve 5). The scatter of the curves corresponding to different values of l is noticeably smaller. The maximum value of $\overline{S}_0$ in the calculated examples was 1542, which is much inferior to the limiting value 4•10^4 (calculated at C_0=2.5•10^{-5}, S_0=n_0 as $\frac{n_0}{(n_0 C_0)}$). Thus, the dimensionless volumetric concentration of a deposit at the inlet to the bed is already $\gamma \overline{S}^0_0 \cong 3.084 \cdot 10^4$ at $\gamma(S) \approx \gamma_a$=20. However, the actual solids take-up capacity (in relation to deposit), which is not explicitly present in the adopted model but actually limits the deposit accumulation, is usually less than n_0 by 30-50%. Therefore, there is a high probability that the upper part of the porous column will become almost saturated with a deposit over time, and then the suspension particles will pass through it virtually in transit, being trapped in the lower part (downward filtration). This situation is considered separately under a constant hydraulic load in Section 2.5.

Figure 4. Change in the relative outlet concentration of the suspended solids over time

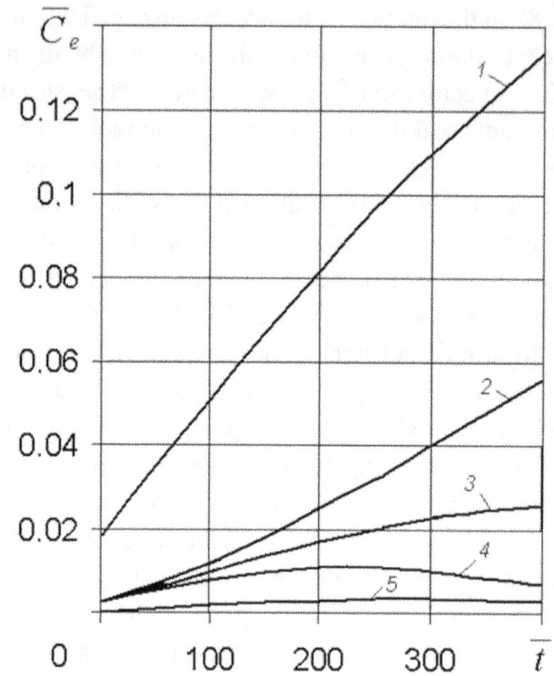

Note. 1, 3, 5 – $l = \frac{1}{3}$, 2 – $l = \frac{7}{8}$, 4 – $l = -\frac{1}{5}$; 1– $\overline{\alpha}_V = 4$, 2-4 – $\overline{\alpha}_V = 6$, 5 –$\overline{\alpha}_V = 8$

Figure 5. Change in the relative volumetric concentration of the deposit in the bed inlet cross-section over time

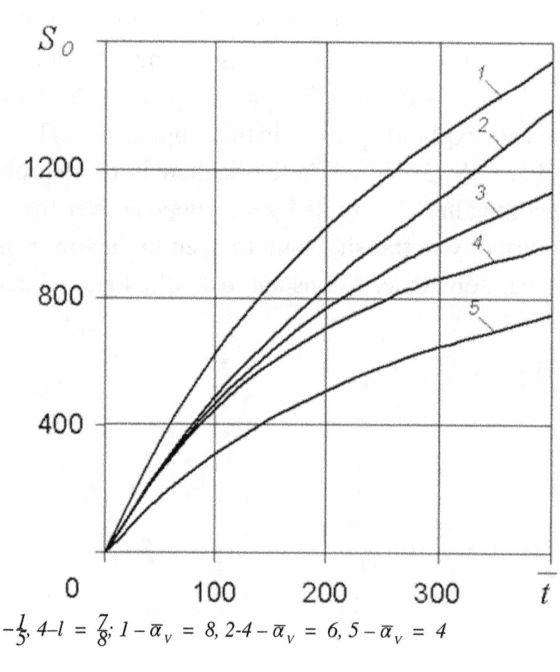

Note. 1, 3, 5 – $l = \frac{1}{3}$, 2 – $l = -\frac{1}{5}$, 4–$l = \frac{7}{8}$; 1 – $\overline{\alpha}_V = 8$, 2-4 – $\overline{\alpha}_V = 6$, 5 – $\overline{\alpha}_V = 4$

The determination of the dynamics of the average and current filtration rates acquires great importance due to the attraction for the estimation of the filter operation at a declining rate of a special criterion, introducing restrictions on its performance (1.32). If the corresponding curves of the dependences $\overline{V}_a(\overline{t})$, $\overline{V}(\overline{t})$ are known, it is relatively easy to find the relative times ($\overline{t}_{aV}$,$\overline{t}_V$) for which these rates will decrease to the economically justified minimum permissible values $\overline{V}_{a*}$, $\overline{V}_*$. Figure 6 demonstrates the decrease in the functional parameter $\overline{V}_a$ over time. The calculation of $\overline{V}_a$ was performed at the same values of l, $\overline{\alpha}_V$, as and $\overline{C}_e$. After a long time, it is advisable to decide whether to continue (or stop) the operation of the filter based on the results of the analysis of the curves of the dependences $\overline{C}_e(\overline{t})$ and $\overline{V}_a(\overline{t})$ (or $\overline{V}(\overline{t})$) with the involvement of their maximum permissible values $\overline{C}_*$ and $\overline{V}_{a*}$ (or $\overline{V}_*$). If we use Figures 4 and 5 (or Figure 7 with data of calculations of current filtration rate) for the illustration and choose the example with $\overline{\alpha}_V = 4$, curves 1 correspond to it. Due to a more even distribution of less amount of a deposit along the height, the decrease in the average and current filter performances in this situation is noticeably slower than in the other examples. Therefore, quite large rates $\overline{V}_a$ (0.833) and $\overline{V}$ (0.672) will remain even at $\overline{C}_* = 0.1$ and the point of time $\overline{t}_p = 265$. If the demands on the filtrate quality become stricter ($\overline{C}_* = 0.05$), the filter will have to be washed out much earlier ($\overline{t}_p = 100$), although the above-mentioned performances will only decrease by 7% and 14% respectively. A different calculated pattern is observed at the value $\overline{\alpha}_V = 8$, to which corresponds curves 5 in Figures 4 and 6, as well as curve 6 in Figure 7. Here the water treatment quality stops limiting the filter operation at all. At the same time, there is a sharp decrease in its performance and the time of washing should be determined only on the basis of the curves of the dependencies $\overline{V}_a(\overline{t})$ or $\overline{V}(\overline{t})$ and values $\overline{V}_{a*}$ or $\overline{V}_*$. Thus, the decisive value for obtaining the maximum output from the filter at a declining rate V becomes the time $\overline{t}_V$, as in other typical cases. Therefore, below the attention is focused on studying the peculiarities of its change when varying the absorption properties of a bed and the initial suspended solids content.

Figure 6. Change in the relative average filtration rate over time

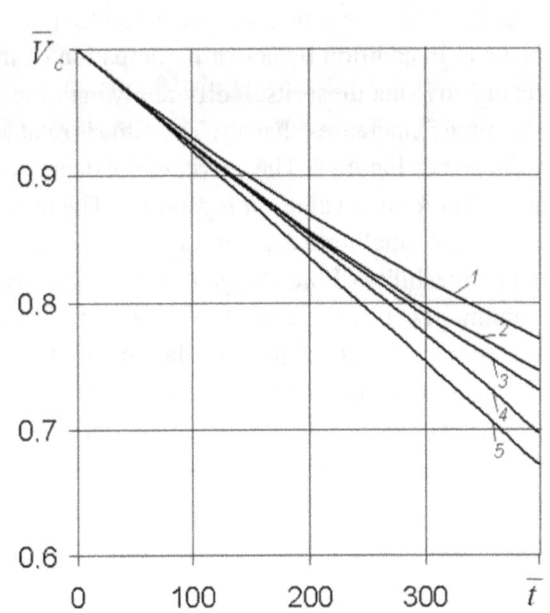

Note. 1, 3, 5 – $l = \frac{1}{3}$, 2 – $l = \frac{7}{8}$, 4 – $l = -\frac{1}{5}$; 1 – $\overline{\alpha}_V = 4$, 2-4 – $\overline{\alpha}_V = 6$, 5 – $\overline{\alpha}_V = 8$

Figure 7. Change in the relative filtration rate over time

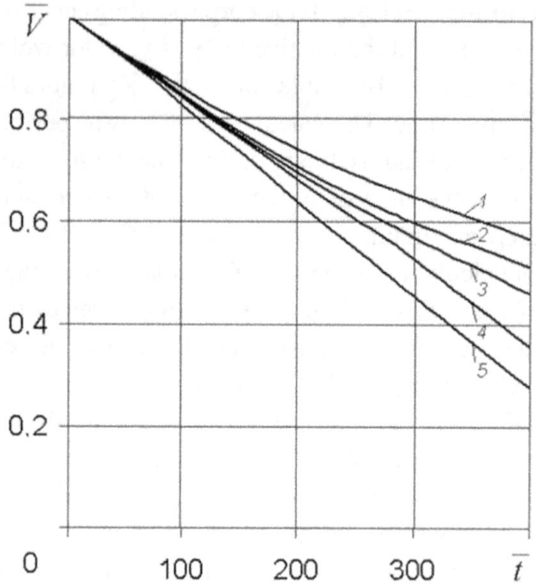

Note. $1, 3, 5 - l = \frac{1}{3}$, $2 - l = \frac{7}{8}$, $4 - l = -\frac{1}{5}$; $1 - \overline{\alpha}_V = 4$, $2\text{-}4 - \overline{\alpha}_V = 6$, $5 - \overline{\alpha}_V = 8$

First of all, Figure 8 illustrates the possible nature of the time $\bar{t}_V$ reduction at an continuous increase in the absorption capability of a filter media from weak ($\overline{\alpha}_V = 2$) to heavy $(\overline{\alpha}_V = 10)$ capability and also different mechanisms of the delivery of mobile particles to grains (l= 7/8, 1/3, -1/5). The time $\bar{t}_V$ was calculated at C_0=2.5•10^{-5} and three limiting values of $\overline{V}_{a*}$ (0.9, 0.8, 0.7) on the basis of equation (5.24) taking into account (5.19). Comparatively large values of $\bar{t}_V$ at $\overline{\alpha}_V = 2...2.5$ are explained by a significant breakthrough of solids, starting from the time point when the first portion of suspension reaches the outlet cross-section of a bed and in addition by active participation of its peripheral section in the clarification process. The effect of l on $\bar{t}_V$ manifests itself differently in the accepted wide range of the $\overline{\alpha}_V$ values. When decreasing $\overline{V}_{a*}$, the time $\bar{t}_V$ increases sharply. This time is related to the concentration C_0 in a similarly complex way, as evidenced by Figure 9. The curves of the dependence $\bar{t}_V(C_0)$ were calculated at $l = \frac{1}{3}$ using the same formulae, but for two values of $\overline{\alpha}_V$ and $\overline{V}_{a*}$.The decrease in C_0 to 2×10^{-5} or less leads to an accelerated increase in $\bar{t}_V$. A small rate deposition of the suspended solids at C_0<10^{-5} causes a very slow deterioration of the permeability of the clogged medium and, as a consequence, a long decrease in performance to the minimum permissible level. And here it is appropriate to emphasize that in contrast to the rate mode *V=const* the results of the calculations of all the characteristics, including both concentrations, at the filtration with a declining rate largely depend on the nature of the hydraulic conductivity reduction as the clogging of the bed. One special case of dependence (1.10) (m_{k1}=1, m_{k2}=3) is used in the monograph, the validity of which has been repeatedly confirmed experimentally and is more often implemented in the hydraulic studies of filtration.

Figure 8. Dependence $\bar{t}_V(\bar{\alpha}_V)$

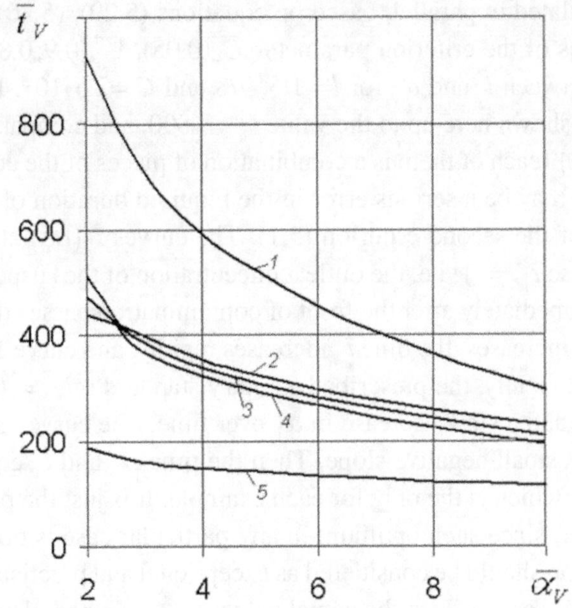

Note. 1, 4, 5 – $l = \frac{1}{3}$, 2 – $l = \frac{7}{8}$, 3 – $l = -\frac{1}{5}$; 1 – $\bar{V}_{a*} = 0.7$, 2-4 – $\bar{V}_{a*} = 0.8$, 5 – $\bar{V}_{a*} = 0.9$

Figure 9. Dependence $\bar{t}_V(C_0)$

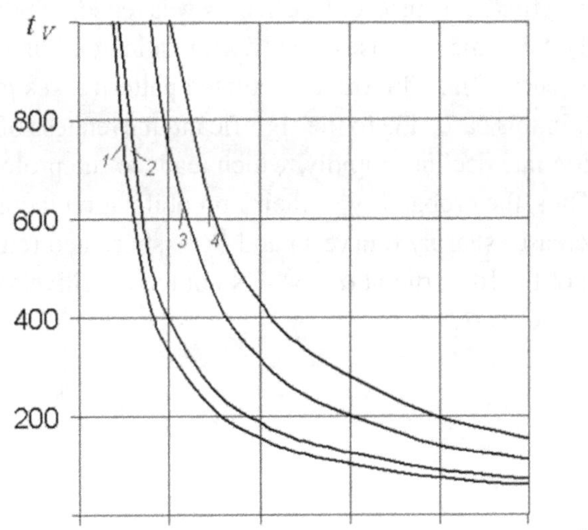

Note. 1, 3 – $\bar{\alpha}_V = 8$, 2, 4 – $\bar{\alpha}_V = 4$; 1, 2 – $\bar{V}_{a*} = 0.9$; 3, 4 – $\bar{V}_{a*} = 0.8$

To determine the filter run duration t_f according to (5.31), it is necessary to know the times t_p, t_V, the smaller of which is identified with t_f. The relation between the relative quantities t_p, t_V depends on the model parameters, especially on $\overline{\alpha}_V$, C_0. Therefore, two series of examples were specially calculated in order to reveal the specificity of the effect of the mentioned parameters on the relative duration $\overline{t}_f$. The times $\overline{t}_p$, $\overline{t}_V$ were calculated in parallel, based on equations (5.29), (5.30) and (5.28), respectively, using typical relative values of the criterion parameters$\overline{C}_*(0.05)$, $\overline{V}_{a*}(0.9, 0.8, 0.7)$. Figure 10 presents plots of the relationship between $\overline{t}_f$ and $\overline{\alpha}_V$ for l= 1/3, 7/8 and C_0=2.5•10^{-5}. In addition, the curves of the dependence $\overline{t}_p(\overline{\alpha}_V)$ are shown here up to the value $\overline{t}_p = 800$, and comparing them (1, 2) with their corresponding curves $\overline{t}_f(\overline{\alpha}_V)$ (each of them is a combination of pieces of the curves $\overline{t}_p(\overline{\alpha}_V)$ and $\overline{t}_V(\overline{\alpha}_V)$), we can conclude that there may be a serious error in the required duration of the operating period due to the unjustified neglect of the second criterion (5.1). The curves $\overline{t}_p(\overline{\alpha}_V)$ start at a single point with abscissa $\overline{\alpha}_V = 3$. In this case $\overline{t}_p = 1$, i.e. the outlet concentration of the suspended solids has the maximum permissible value immediately after the front of contamination passes the bed. As the absorption capacity of the filter media increases, the time $\overline{t}_p$ increases rapidly, and curve 1 has a vertical asymptote ($\overline{\alpha} \approx 5.3$) at $l = \frac{1}{3}$. Consequently, the prescribed sanitary standard ($\overline{C}_* = 0.05$) is not reached at all due to declining the rate $\overline{V}$ and a slow decrease in $\overline{\alpha}_V$ over time. The curves of the dependence $\overline{t}_V(\overline{\alpha}_V)$ in contrast to $\overline{t}_p(\overline{\alpha}_V)$ have a small negative slope. Then the times $\overline{t}_p$ and $\overline{t}_V$ equalize at the point of the intersection of these curves which is the only for each example. It is just the point where the value of $\overline{t}_f$ will be maximum. However, since such optimum in any particular case is observed at strictly explicit values of $\overline{\alpha}_V$, C_0, this situation should be considered as exceptional and practically unattainable. Usually, it will be true either or $\overline{t}_p > \overline{t}_V$, or $\overline{t}_p < \overline{t}_V$ at the actual values of $\overline{\alpha}_V$, C_0 and, therefore, $\overline{t}_f$ will be less than the maximum (at the optimal combination of $\overline{\alpha}_V$and C_0). At the same time, the decrease in $\overline{t}_f$ is insignificant at $\overline{\alpha}_V > 5$ and the choice of $\overline{\alpha}_V$ within such limits does not significantly affect $\overline{t}_f$. The graphs of the dependence of $\overline{t}_f$ on C_0, calculated at $l = \frac{7}{8}$; $\overline{\alpha}_V = 4, 6$; $\overline{V}_{a*} = 0.8, 0.9$ and shown in Figure 11 have a similar form, but essentially a different nature. Here the behavior of the curves of the dependence $\overline{t}_p(C_0)$ already is significantly different for both values of $\overline{\alpha}_V$. At moderate capability of porous medium to retain the suspended matter, firstly, its protective effect is weakened, which is expressed in a relative smallness of $\overline{t}_p$, and secondly, this time increases slowly with an increase in C_0 due to a significant particle removal from the filter (curve 2). A different calculated pattern takes place in the case of a well-absorbing filtering material. In this case, due to the significant unevenness of the deposit accumulation along the height, the filtration rate declines rapidly, which leads to the prolongation of the suspension particles stay in the pores. Thus, the probability of their immobilization increases significantly. For the same reason, the time $\overline{t}_p$ increases sharply (curve 1) and $\overline{S}_2$ is shortened (curves 3, 4) with increasing C_0. As a result, the duration of the filter run at $\overline{\alpha}_V = 4$ is not too sensitive to the concentration C_0, but reacts sharply to the change in C_0 at $\overline{\alpha}_V = 6$.

Figure 10. Dependencies $\bar t_f(\bar\alpha_V), \bar t_p(\bar\alpha_V)$

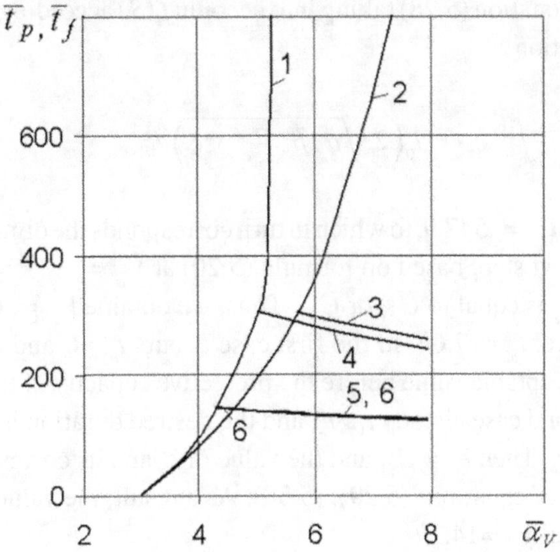

Note. 1, 2 – $\bar t_p$, 1 and 4, 1 and 6, 2 and 3, 2 and 5 – $\bar t_f$; 1, 3, 5 – l=3, 2, 4, 6 – l = $\frac{7}{8}$; 3, 4 – $\bar V_{a*}$ = 0.8, 5, 6 – $\bar V_{a*}$ = 0.9

Figure 11. Dependencies $\bar t_f(C_0)$, $\bar t_p(C_0)$

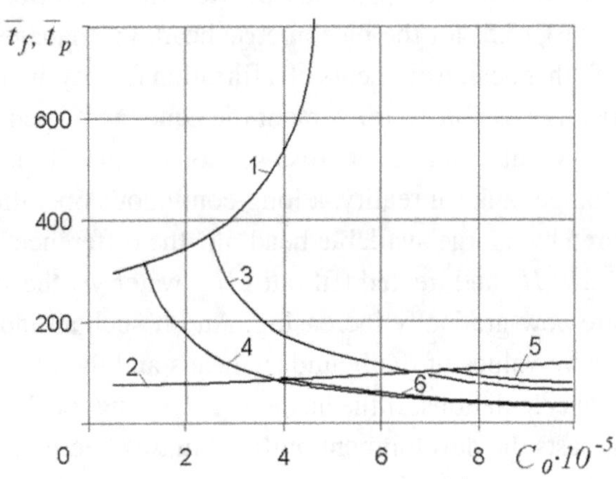

Note. 1, 2 – $\bar t_p$, 1 and 3; 1 and 4; 2 u 5; 2 and 6 – $\bar t_f$; 2, 5, 6 – $\bar\alpha_V$ = 4, 1, 3, 4 – $\bar\alpha_V$ = 6; 3, 5 – $\bar V_{a*}$ = 0.8, 4, 6 – $\bar V_{a*}$ = 0.9

In conclusion, the example of calculating the filter run duration at a fixed head differential at the bed boundaries is given for the specific conditions of the detachment filtration characterized by the following set of the values of the model and technological parameters: α_V=17.5h^{-1}, β_V=0.01h^{-1}, $l = \frac{1}{3}$ (fine solids with diameter of the order $10^{-6}m$), q=1, V^0=5mh^{-1}, L=1m, n_0=0.4m, γ_a=15, m_{k1}=1, m_{k2}=3, C_0=2.5•10^{-5}; V_{a*}=3.755mh^{-1} (filtration rate can be reduced approximately twice), C_{*1}=2.5•10^{-6}, C_{*2}=10^{-6} (satisfactory and high quality of water treatment).

The relative characteristic times $\bar{t}_V, \bar{t}_p$ were determined using the Mathcad software package. The dimensionless parameters were calculated beforehand: $\bar{\alpha}_V = 6, \bar{\beta}_V = 0.004, \bar{V}_{a*} = 0.75, \bar{C}_{*1} = 0.04, \bar{C}_{*2} = 0.1$. The time $\bar{t}_V$ was found from equation (5.28) taking into account $f_k(\bar{S})$ according to (1.10) and the following expression for the concentration $\bar{S}$

$$\bar{S}(\bar{z}, \bar{t}; \bar{V}_a) = \bar{\alpha}_V \bar{V}_a^{-\frac{2}{3}} e^{-\bar{\alpha}_V \bar{V}_a^{-\frac{2}{3}} \bar{z}} \int_0^{\bar{V}_a \bar{t}} e^{-\bar{\beta}_V \lambda \xi} I_0\left(2\sqrt{\bar{\alpha}_V \bar{\beta}_V \bar{V}_a^{-\frac{2}{3}} \bar{z} \xi}\right) d\xi$$

It was found that the value $\bar{t}_V = 547.9$, to which in turn corresponds the dimensional t_V=21.9h, satisfies the above equation. In the next step, based on formula (5.20) at $\bar{V}_a = \bar{V}_{a*} = 0.75$, we calculated such values of $\bar{t}_{a1}, \bar{t}_{a2}$, at which $\bar{C}_e$ is equal to $\bar{C}_{*1}$ or $\bar{C}_{*2}$. Thus, we obtained $\bar{t}_{a1} = 620$, so that t_{a1}=24.8h, and also $\bar{t}_{a2} = 440$ and, therefore, t_{a2}=17.6h. In the first case occurs $t_{a1} > t_V$ and the filter performance will decrease to the minimum acceptable value before the protective capacity of the bed is exhausted. Therefore, $t_f = t_V$=21.9h. In the second case already $t_V > t_{a2}$ and the desired duration is controlled by the criterion of the water treatment quality. Then $\bar{t}_f = \bar{t}_p$, and the value of $\bar{t}_p$ and its corresponding value of $\bar{V}_{ap}$ were found by solving the system of equations (5.29), (5.30). As a result, the values $\bar{t}_p = 352$, $\bar{V}_{ap} = 0.823$ were determined, so that finally t_f=14.1h.

5.2. FILTRATION WITH CONSTANT HEAD DIFFERENTIAL AT COLLECTORS OF INITIAL SUSPENSION AND FILTRATE

Due to setting in the problem from the previous Section the conditions of the first kind on the boundaries of a packed bed (2.6), (5.5) for the piezometric head, we managed to abstract in theoretical studies from the hydraulics of other design elements of a filtration facility, in particular, its transportation system. However, it is practically possible to implement the aqueous suspension filtration at a constant differential head (pressure) in the inlet and outlet cross-sections of the filtration flow only by maintaining a constant level in the storage tank. In reality, a long continuous operation of the filtration facility with a declining rate is ensured by a large available head, i.e. the difference between the fixed heads at the collectors of the initial (raw) H_{in} and treated (filtrate) H_{out} water. At the same time, the heads at the inlet and outlet of the bed are now gradually increasing. And if such an increase is insignificant, it is acceptable to introduce effective values of the boundary heads and then to use the dependencies from Section 5.1 for the engineering calculations. If the heads H_u, H_d change within a few tens of centimeters, it is preferable to use for forecasts the development outlined in this Section.

It should be emphasized that condition (5.5) is in principle unsuitable for describing the essentially nonlinear dynamics of head in the inlet cross-section of the bed and the suspension level in the storage tank. The role of the tank in an unregulated rate filtration (Section 1.2) is directly reduced to the accumulation of an excessive (due to insufficient flow capacity of the bed) amount of suspension. It is important that the filtration conditions in the layer of a granular material can be seriously improved by creating an increased inlet head. The specified head (and thus, also the delayed volume of the initial suspension) at a given constant head H_{in} is structurally limited to the specified value from above, so that it is always true $H_{in} > H_u$ due to the head losses in the supply communications.

Thus, if the operating height of the storage tank H_* (overflow chute mark) is greater, equal or insignificantly less than H_{in}, the overflow of the storage tank is impossible in principle. Then criterion (1.38) will not appear in the system of the filtration efficiency criteria. This is the essential difference between the technological analysis performed below on the basis of (5.31) and the similar analysis in the following Sections of this Chapter, where an additional restriction is imposed on the technological process, related specifically to the height of the storage tank and the dynamics of the suspension level in it.

5.2.1. First Stage

The first stage (index "1" is introduced) is characterized by the highest intensity of suspension inflow to a filter, formation and movement down a bed the front of media saturation. Here and hereinafter the more common downward direction is accepted, although the flow direction is not essentially important in the considered range of the filtration rates V. Then the mathematical model, jointly describing the hydrodynamics of a suspension in the supply communications, above and inside the bed at this stage, has the following form

$$H_{in} = H_{W1}(t) + R_{f1} Q_{in}(t)^2, \tag{5.32}$$

$$\omega \frac{dH_{W1}}{dt} = Q_{in}(t) - \omega V_1(t) \ ; \tag{5.33}$$

$$\frac{\partial^2 h_1}{\partial z^2} = 0 \ ; \tag{5.34}$$

$$z = Z_W, \quad h_1 = H_{W1}; z = Z_f, \quad h_1 = Z_f \ ; \tag{5.35}$$

$$t = 0, \quad H_{W1} = Z_W. \tag{5.36}$$

Here H_{W1}, Z_W, Z_f are the piezometric marks of the suspension level above the bed, the bed surface and the saturation front; Q_{in} is the flow rate of a suspension in the supply communications; V_1, h_1 are the filtration rate and piezometric head at the first stage. The capillary forces acting at the saturation front are not taken into account in the model (5.32) - (5.36) (Gusev, 1981; Poliakov, 2014), which is justified due to the short duration of this stage and high permeability of the porous medium. The depth of the wetted zone of the bed increases rapidly, and the mark Z_f is an unknown function of time.

The solution of equation (5.34) under conditions (5.35) gives the following expressions for the head h and rate V_1

$$h(z,t) = H_{W1}(t) + \frac{H_{W1}(t) - Z_f(t)}{Z_W - Z_f(t)}\left(z - Z_W\right), \tag{5.37}$$

$$V_1 = k_0 \frac{H_{W1}(t) - Z_f(t)}{Z_W - Z_f(t)}, \tag{5.38}$$

The kinematic condition at the saturation front is valid, which taking into account (5.38) will be

$$-\frac{dZ_f}{dt} = \frac{V_1}{n_0} = \frac{k_0}{n_0} \frac{H_{W1} - Z_f}{Z_W - Z_f}, \tag{5.39}$$

The transformed equation (5.33) is joined to condition (5.39), viz,

$$\omega \frac{dH_{W1}}{dt} = \sqrt{\frac{H_{in} - H_{W1}}{R_{f1}}} - k_0 \omega \frac{H_{W1} - Z_f}{Z_W - Z_f}, \tag{5.40}$$

as well as the initial condition (5.36) and

$$t = 0, \qquad Z_f = Z_W. \tag{5.41}$$

Problem (5.36), (5.39) - (5.41) can be easily reduced to a simpler problem with only one unknown quantity,

$$\omega \frac{dH_{W1}}{dZ_f} = \frac{n_0}{k_0 \sqrt{R_{f1}}} \left(\sqrt{H_{in} - H_{W1}} \frac{Z_W - Z_f}{H_{W1} - Z_f} - k_0 \omega \sqrt{R_{f1}} \right), \tag{5.42}$$

$$Z_f = Z_W, \qquad H_{W1} = Z_W. \tag{5.43}$$

Since analytical methods are ineffective in this case, and there are no high requirements for calculating the mark H_{W1} at the first stage, it can be assumed due to the transience of the bed wetting that

$$Q_{in}(t) \cong Q_{in}^0 = \sqrt{\frac{H_{in} - Z_W}{R_{f1}}} = const. \tag{5.44}$$

Then the form of system (5.39), (5.40) somewhat simplifies, so that

$$n_0 \frac{dZ_f}{dt} + k_0 \frac{H_{W1} - Z_f}{Z_W - Z_f} = 0, \tag{5.45}$$

$$\frac{dH_{W1}}{dt} = \frac{Q_{in}^0}{\omega} - k_0 \frac{H_{W1} - Z_f}{Z_W - Z_f}. \tag{5.46}$$

First of all, the relationship between H_{W1} and Z_f follows from system (5.45), (5.46)

$$H_{W1} = \frac{Q_{in}^0}{\omega} t + n_0 Z_f. \tag{5.47}$$

After substituting (5.47) into equation (5.45), the following is obtained

$$\frac{dZ_f}{dt} - \frac{k_0}{n_0 \omega} \frac{\omega (1 - n_0)(Z_W - Z_f) + Q_{in}^0 t}{Z_W - Z_f} = 0. \tag{5.48}$$

Problem (5.41), (5.48) is satisfied by a simple expression

$$Z_W - Z_f = \lambda_w t. \tag{5.49}$$

Then the quadratic equation with respect to λ_w follows from (5.48)

$$\lambda_w^{\,2} - \frac{k_0(1-n_0)}{n_0}\lambda_w - \frac{k_0 Q_{in}^0}{n_0 \omega} = 0. \tag{5.50}$$

The physical sense has such a root of equation (5.50)

$$\lambda_w = \frac{k_0(1-n_0)}{2n_0}\left(\sqrt{1+\frac{4n_0 Q_{in}^0}{k_0\omega(1-n_0)^2}}+1\right). \tag{5.51}$$

Taking into account expression (5.51) it is found from (5.36), (5.46)

$$H_{W1}(t) = Z_W + \left(\frac{Q_{in}^0}{\omega} - n_0\lambda_w\right)t. \tag{5.52}$$

The rate V_1, as follows from (5.38), (5.49), (5.52), will be constant

$$V_1 = \frac{k_0}{\lambda_w}\left(\frac{Q_{in}^0}{\omega} + \lambda_w - n_0\lambda_w\right). \tag{5.53}$$

Obviously, the bed will be completely saturated with liquid by the time

$$t_1 = \frac{Z_W - Z_d}{\lambda_w}, \tag{5.54}$$

where Z_d is the mark of the bed base. In this case the suspension level above the bed will be set at the following level

$$H_{W1}(t_1) = Z_W + \left(\frac{Q_{in}^0}{\omega} - n_0\lambda_w\right)\frac{Z_W - Z_d}{\lambda_w}. \tag{5.55}$$

The calculation dependencies and equations are given in a more general form due to the introduction of dimensionless variables and parameters:

$$\bar{h}_1 = \frac{h_1 - H_{out}}{\Delta H},\quad \bar{H}_{W1} = \frac{H_{W1} - H_{out}}{\Delta H},\quad \bar{z} = \frac{z}{L},\quad \bar{t} = \frac{k_0 t}{L},$$

$$\bar{Q}_{in}^0 = LQ_{in}^0/k_0\omega\Delta H,\quad \bar{V}_1 = LV_1/(k_0\Delta H),\quad \bar{R}_{fl} = k_0^2\omega^2\Delta H R_{fl}/L^2\ ; \tag{5.56}$$

$\Delta H=H_{in}-H_{out}$, $L=Z_W-Z_d$ is the bed height. Then the relative time of the end of the first stage will be

$$\bar{t}_1 = \frac{k_0 t_1}{L} = \frac{1}{\bar{\lambda}_w}, \tag{5.57}$$

where

$$\bar{\lambda}_w = \frac{\lambda_w}{k_0} = \frac{1-n_0}{2n_0}\left(\sqrt{1+\frac{4n_0\bar{Q}^0}{(1-n_0)^2(\bar{Z}_W-\bar{Z}_d)}}+1\right), \quad \bar{Z}_{W,d} = \frac{Z_{W,d}-H_{out}}{\Delta H}.$$

Thus, the reduced mark $\bar{H}_{W1}$ will reach by the time $\bar{t}_1$ the following value

$$\bar{H}_{W1}(\bar{t}_1) = \bar{Z}_W + \left[\bar{Q}^0 - n_0\bar{\lambda}_w(\bar{Z}_W-\bar{Z}_d)\right]\bar{t}_1. \tag{5.58}$$

5.2.2. Second Stage

The second stage is characterized by the rapid accumulation of the initial suspension above the bed surface and, as a consequence, a rise in its level due also to the insufficient permeability of the porous medium and the still relatively large difference between the head H_{in} and the mark H_{W2}. Below the index "2" means that the corresponding characteristic belongs to the second stage. Again, due to the low deposition rate, the filtration flow properties of the bed do not deteriorate significantly during the short period of the intensive formation of the surface liquid layer. Therefore, we can neglect the process of the suspension clarification and its effect on the dynamics and balance of liquid in the filter at the second stage. Then for the basis of the new hydraulic model it is possible to take the model from Section 5.1 and additionally to consider the effect of the offtake communications, consequences of the first stage. As a result, it is reasonable to present the mentioned model in the following form

$$\frac{dH_{W2}}{dt} = \frac{\sqrt{H_{in}-H_{W2}}}{\omega\sqrt{R_{f1}}} - V_2(t), \tag{5.59}$$

$$\frac{\partial^2 h_2}{\partial z^2} = 0, \tag{5.60}$$

$$V_2 = k_0\frac{\partial h_2}{\partial z}, \tag{5.61}$$

$$H_{d2}(t) = H_{out} + R_{f2}\omega^2 V_2(t)^2, \tag{5.62}$$

$$z = Z_W, \quad h_2 = H_{W2}; z = Z_d, \quad h_2 = H_{d2}; \tag{5.63}$$

$$t = 0, \quad H_{W2} = H_{W1}(t_1) = H^0_{W2}, \tag{5.64}$$

where H_{d2} is the head at the bed base; $V_2 h_2$ are the filtration rate and piezometric head at the second stage.

First of all, the head distribution in the bed is found

$$h_2(z,t) = H_{W2}(t) + \frac{H_{W2}(t)-H_{d2}(t)}{L}(z-Z_W). \tag{5.65}$$

Taking into account (5.62), the quadratic equation is derived from (5.63) for V_2

$$R_{f2}\omega^2 V_2^2 + \frac{L}{k_0}V_2 - H_{W2} + H_{out} = 0. \tag{5.66}$$

Thus,

$$V_2 = \frac{L}{2k_0 R_{f2}\omega^2}\left(\sqrt{1 + \frac{4k_0^2\omega^2 R_{f2}(H_{W2} - H_{out})}{L^2}} - 1\right). \tag{5.67}$$

Then the problem concerning the water level mark above the bed will include the equation

$$\frac{dH_{W2}}{dt} = \frac{\sqrt{H_{in} - H_{W2}}}{\omega\sqrt{R_{f1}}} - \frac{L}{2k_0 R_{f2}\omega^2}\left(\sqrt{1 + \frac{4k_0^2\omega^2 R_{f2}(H_{W2} - H_{out})}{L^2}} - 1\right) \tag{5.68}$$

and condition (5.64). In general case the solution of this problem has the following integral form

$$t = 2k_0\omega^2 R_{f2}\sqrt{R_{f1}}\, Y_1(H_{W2}), \tag{5.69}$$

where

$$Y_1(H_{W2}) = \int_{H_{W2}^0}^{H_{W2}} \frac{d\xi}{2k_0 R_{f2}\omega\sqrt{H_{in} - \xi} - \sqrt{R_{f1}}\sqrt{L^2 + 4k_0^2\omega^2 R_{f2}(\xi - H_{out})} + L\sqrt{R_{f1}}} \tag{5.70}$$

After transition to the relative quantities using (5.56), the form of the basic calculation formula (5.69) is considerably simplified and will be as follows

$$\bar{t} = 2\bar{R}_{f2}\sqrt{\bar{R}_{f1}}\, Y_1(\bar{H}_{W2}), \tag{5.71}$$

where

$$Y_1(\bar{H}_{W2}) = \int_{\bar{H}_{W2}^0}^{\bar{H}_{W2}} \frac{d\xi}{2\bar{R}_{f2}\sqrt{1 - \xi} - \sqrt{\bar{R}_{f1}(1 + 4\bar{R}_{f2}\xi)} + \sqrt{\bar{R}_{f1}}} \tag{5.72}$$

Analytical expressions follow for a number of special cases from formula (5.71). First of all, there will be $\bar{t} = 0$ at R_{f1}=0 (there are no head losses in the supply communications), i.e. the maximum suspension level (at the mark H_{in}) is instantly set above the bed.

If the head losses in the offtake communications are negligibly small (R_{f2}=0), then revealing the uncertainty of 0/0 type in (5.72), it is easy to obtain

$$\bar{t} = \ln\frac{\sqrt{1 - \bar{H}_{W2}^0} - \sqrt{\bar{R}_{f1}\bar{H}_{W2}^0}}{\sqrt{1 - \bar{H}_{W2}} - \sqrt{\bar{R}_{f1}\bar{H}_{W2}}} + \frac{1}{\sqrt{\bar{R}_{f1}(\bar{R}_{f1} + 4)}} \times$$
$$\times \ln\left(\frac{2\sqrt{\bar{R}_{f1}(1 - \bar{H}_{W2}^0)} + \sqrt{\bar{R}_{f1}(\bar{R}_{f1} + 4)} + 1}{2\sqrt{\bar{R}_{f1}(1 - \bar{H}_{W2}^0)} - \sqrt{\bar{R}_{f1}(\bar{R}_{f1} + 4)} + 1} \cdot \frac{2\sqrt{\bar{R}_{f1}(1 - \bar{H}_{W2})} - \sqrt{\bar{R}_{f1}(\bar{R}_{f1} + 4)} + 1}{2\sqrt{\bar{R}_{f1}(1 - \bar{H}_{W2})} + \sqrt{\bar{R}_{f1}(\bar{R}_{f1} + 4)} + 1}\right) \tag{5.73}$$

Finally, the infiltration practically stops at a very large hydraulic resistance of the mentioned communications ($R_{f2}\to\infty$) and then from (5.71), (5.72) follows the dependence of $\tilde{t}$ on $\bar{H}_W$ in such two forms

$$\tilde{t} = 2\sqrt{\bar{R}_{f1}}\left(\sqrt{1-\bar{H}^0_{W2}} - \sqrt{1-\bar{H}_{W2}}\right),$$

$$\bar{H}_W = 1 - \left(\sqrt{1-\bar{H}^0_{W2}} - \frac{\tilde{t}}{2\sqrt{\bar{R}_{f1}}}\right)^2. \tag{5.74}$$

Consequently, the reduced liquid level above the bed will rise to the maximum relative level 1 by the reduced point of time

$$\tilde{t}_m = 2\sqrt{\bar{R}_{f1}(1-\bar{H}^0_{W2})}. \tag{5.75}$$

Also, the expression for the filtration rate will be considerably simplified

$$\bar{V}_2 = \frac{1}{2\bar{R}_{f2}}\left(\sqrt{1+4\bar{R}_{f2}\bar{H}_{W2}} - 1\right). \tag{5.76}$$

In general, the rise of the suspension level in the filter is clearly uneven – very fast at the first stage and at the beginning of the second stage, and then slows down sharply, asymptotically tending to the maximum mark H_m in the general case. Taking into account the specific nature of a change in the quantity H_{W2}, it is legitimate to choose its maximum value at the second stage H_m as the initial mark for the third stage H^0_W. Formally, it is reached here at $t\to\infty$, and the formula for calculating the reduced quantity depending on the relative coefficients $\bar{R}_{f1}$, $\bar{R}_{f2}$ results from (5.71), (5.72) and is as follows

$$H_m = \frac{H_m - H_{out}}{\Delta H} = \frac{2\bar{R}^2_{f2} + \bar{R}_{f1}\bar{R}_{f2} - \bar{R}_{f1}}{2(\bar{R}_{f1}+\bar{R}_{f2})^2} + \\ +\sqrt{\frac{\left(2\bar{R}^2_{f2} + 2\bar{R}_{f1}\bar{R}_{f2} - \bar{R}_{f1}\right)^2}{4\left(\bar{R}_{f1}+\bar{R}_{f2}\right)^4} - \frac{\bar{R}^2_{f2} - \bar{R}_{f1}}{\left(\bar{R}_{f1}+\bar{R}_{f2}\right)^2}}. \tag{5.77}$$

The above obtained solutions of the filtration flow compartment of the general model at a variable filtration rate for the conditions occurring at its initial stages are illustrated by the calculations of the components of the filter liquid balance according to the formulas given in Subsection 5.2.2. Since the first stage actually plays an auxiliary role due to its transience, the results of the investigations from Subsection 5.2.1 were used only to determine the relative moment of the beginning of the second stage $\tilde{t}_1$ and the suspension level elevation above the bed $\bar{H}_{W1}(\tilde{t}_1)$ by this moment. Found by formulas (5.57), (5.58), they were then used as initial values in the calculations of the mark $\bar{H}_{W2}$, as well as the rates of flow $\bar{Q}_{in}$, filtration $\bar{V}_2$ and level rise $\frac{d\bar{H}_{W2}}{d\tilde{t}}$ as functions of time. The reduced marks of the bed boundaries were fixed ($\bar{Z}_W = 0.4,\ \ \bar{Z}_d = 0.1$) beforehand and hence $\bar{L} = \bar{Z}_W - \bar{Z}_d = 0.3$. The initial porosity was assumed to be 0.35 in all examples.

In the first series of examples, the dependence $\bar{H}_{W2}(\tilde{t})$ was calculated at discretely varying and equal characteristic values of $\bar{R}_{f1}$, $\bar{R}_{f2}$. The results of the calculations according to formula (5.71) taking into account (5.72) are presented in Figure 12 and indicate a significant unevenness of the suspension level rise in the filter. Although it formally continues indefinitely, the main increase in $\overline{H}_{W2}$ is observed up to the values $\bar{t} = 1 - 2$. This fact gives a reason to use the steady-state value of $\overline{H}_{W2}$ in the further consideration directly of the suspension clarification.

The effect of the resistances $\overline{R}_{f1,f2}$ on the level dynamics is more significant at their lower values and weakens as they increase. Naturally, the suspension accumulation above the bed leads to a change in the ratio between the various components of its balance, which is clearly illustrated in Figure 13. It shows curves of flow rate decline $\overline{Q}_{in}(1)$ and rise of the rate $\overline{V}_2(2)$ over time corresponding $\overline{R}_{f1} = \overline{R}_{f2} = 3$. These quantities practically equalize already at $\bar{t} = 5$, although a slight difference between them takes place even at lower $\bar{t}$. Increase in the flow capacity due to the head gain at its upper boundary causes the same rapid decrease in the suspension accumulation rate in the surface layer (curve 3).

Figure 12. Rise of the water level above the bed over time (first and second stages of filtration)

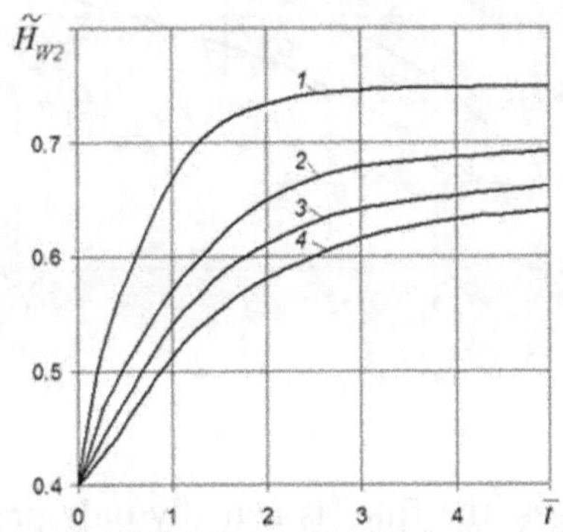

Note. 1 – $\overline{R}_{f1,f2} = 1$; 2 – $\overline{R}_{f1,f2} = 2$; 3 – $\overline{R}_{f1,f2} = 3$; 4 – $\overline{R}_{f1,f2} = 4$

Figure 13. Dynamics of the water balance components above the bed

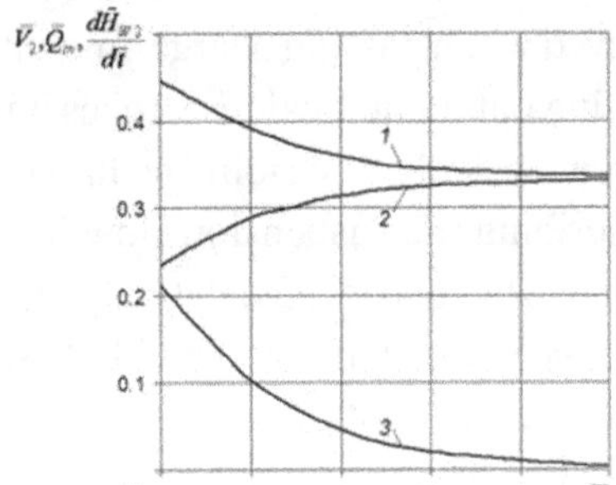

Note. 1 – $\overline{Q}_{in}$; 2 – $\overline{V}_2$; 3 – $\frac{d\overline{H}_{W2}}{d\bar{t}}$

The main result of the analysis of a filter facility operation at the first stages is the prediction of the suspension level position in the storage tank at the end of the second stage. The corresponding marks $\overline{H}_m$ were calculated by formula (5.77) at varying discretely the coefficient $\overline{R}_{f1}$ and continuously from 0 to 5 the coefficient $\overline{R}_{f2}$. The corresponding curves of the dependence $\overline{H}_m(\overline{R}_{f2})$ are shown in Figure 14. According to it, the effect of both kinds of communications on the quantity $\overline{H}_m$, on which the performance of the facility depends to a large extent, turns out to be very serious and intensifies with an increase in their resistances. However, since the third stage is determinant for the suspension separation, the main attention was paid to the analysis of the dispersed impurity behavior and, in this connection, the deterioration of filtration conditions in the bed.

Figure 14. Dependence $\overline{H}_m(\overline{R}_{f2})$

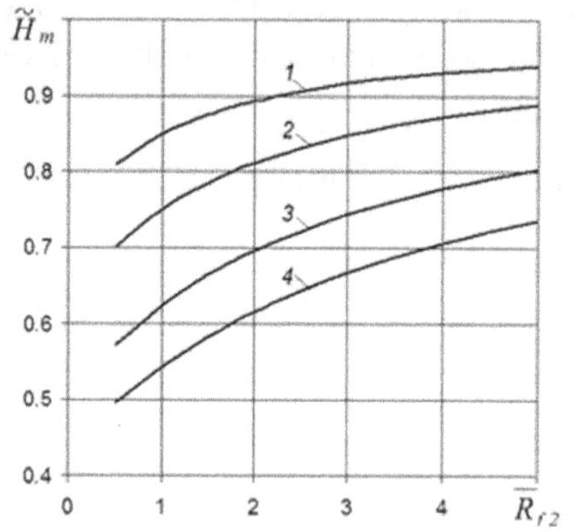

Note. 1 – $\overline{R}_{f1}$ = 0.5; 2 – $\overline{R}_{f1}$ = 1; 3 – $\overline{R}_{f1}$ = 2; 4 – $\overline{R}_{f1}$ = 3

During the first and second stages, the filter is actually only preparing for a long and productive operation in the third stage of filtration, which is extensively studied in the next Subsection.

5.2.3. Third Stage

During the preceding stages, a reserve of mechanical energy is generated in the storage tank of a filter, which further contributes to its long-term stable operation. It is during the third stage that the necessary clarification effect is achieved when passing a large amount of suspension (raw water) through the packed bed. And here when carrying out theoretical researches it is necessary to rely on the complete filtration mathematical model, comprehensively describing technological process. For the third stage, it primarily includes the equations describing the suspension flow in the supply communications (5.32); the suspension balance in the filter above the bed (5.33); the suspended solids transport in the bed (5.2); the suspension particles exchange between the liquid and solid phases of the bed (5.3); a change in the hydraulic resistance of a porous material during its clogging (1.30); the filtration flow of the clarified water in the porous medium (5.4); relationship between the volumetric concentrations of the deposit S_S and the deposited particles (1.29); and the filtrate flow in the offtake communications (5.62). In addition, the model contains the following initial conditions

$$t = 0, H_W = H_m; S = 0 ; \tag{5.78}$$

where H_m is the dimensional mark of the suspension level, calculated by formula (5.77), as well as conditions (2.9) and (5.63). The absence of indices "1", "2" in this model at the calculation characteristics (H_W, V, h, H_d) means that they refer to the third stage. Due to the short duration of the conditionally isolated initial stages of filtration, a minimum amount of the suspended solids will be retained in the bed to their end, which gives the right to assume a homogeneous initial condition for the desired quantity $\bar{S}$ at the main stage. The time is counted from the beginning of the third stage. The model presented in this Section includes actually seven equations with the same number of unknown quantities. To generalize the calculation formulas and equations, in addition to (5.56), dimensionless variables and parameters are introduced, containing a new scale for the filtration rate. There are different ways of its selection. The optimum way to normalize the rate is to use as a scale the expression

$$\frac{\sqrt{L^2 + 4k_0{}^2\omega^2\Delta H\left(R_{f1} + R_{f2}\right)} - L}{2k_0\omega^2\left(R_{f1} + R_{f2}\right)}$$

Then the relative rate is equal to 1 at the initial time point and declines monotonically over time. Nevertheless, here the expression $\frac{k_0\Delta H}{L}$ is preferred by analogy with V^0 from Section 5.1 due to its simplicity. In this case $\bar{t} = k_0\Delta Ht/n_0L^2, \bar{V} = LV/k_0\Delta H, \bar{\alpha}_V = (k_0\Delta H)^{l-1}\alpha_V/L^{l-2}, q{=}1, \bar{\gamma}_a = 0.0015, \bar{R}_{f1,f2} = k_0^2\omega^2\Delta HR_{f1,f2}/L^2$, and the initial relative rate will necessarily be less than unity now at $R_{f1}+R_{f2}>0$, and the deviation from it is the larger, the more is $R_{f1}+R_{f2}$.Thus, the initial model after the introduction of dimensionless quantities has the following form

$$\bar{H}_W(\bar{t}) + \bar{R}_{f1}\bar{Q}^2_{in} = 1, \tag{5.79}$$

$$\frac{d\bar{H}_W}{d\bar{t}} = \bar{Q}_{in} - \bar{V}(\bar{t}), \tag{5.80}$$

$$\bar{V}(\bar{t})\frac{\partial\bar{C}}{\partial\bar{z}} + \frac{\partial\bar{S}}{\partial\bar{t}} = 0, \tag{5.81}$$

$$\frac{\partial\bar{S}}{\partial\bar{t}} = \bar{\alpha}_V\bar{V}^l(\bar{t})\bar{C} - \bar{\beta}_V\bar{V}^q(\bar{t})\bar{S}\ ; \tag{5.82}$$

$$\bar{V}(\bar{t}) = -f_k(\bar{S})\frac{\partial\bar{h}}{\partial\bar{z}}\ ; \tag{5.83}$$

$$\bar{H}_d(\bar{t}) = \bar{R}_{f2}\bar{V}^2(\bar{t})\ ; \tag{5.84}$$

$$\bar{t} = 0, \bar{H}_W = \bar{H}_m; \bar{S} = 0\ ; \tag{5.85}$$

$$\bar{z} = 0, \bar{h} = \bar{H}_W; \bar{z} = 1, \bar{h} = \bar{H}_d. \tag{5.86}$$

The approximate solution of the above formulated complex nonlinear mathematical problem in partial derivatives with variable coefficients was also determined by averaging the rate $\bar{V}$ with the introduction of the parameter $\bar{V}_a$ according to (5.11). Problem (5.79) - (5.85) was solved in the general case in three stages. At the first stage, the solution of the clarification compartment (the solution procedure is described in Section 5.1) allowed us to obtain the expressions for the relative concentrations of the suspended and deposited particles, which, using the modified scale system, have the following form in real time

$$\begin{aligned} \bar{C}(\bar{z},\bar{t};\bar{V}_a) &= e^{-\bar{\alpha}_V \bar{V}_a^{l-1}\bar{z}}\Big[e^{-\bar{\beta}_V \bar{V}_a^{q}\bar{t}} I_0\Big(2\sqrt{\bar{\alpha}_V \bar{\beta}_V \bar{V}_a^{l+q-1}\bar{z}\bar{t}}\Big)+ \\ &+\bar{\beta}_V \bar{V}_a^{q}\int_0^{\bar{t}} e^{-\bar{\beta}_V \bar{V}_a^{q}\lambda} I_0\Big(2\sqrt{\bar{\alpha}_V \bar{\beta}_V \bar{V}_a^{l+q-1}\bar{z}\lambda}\Big)d\lambda\Big], \end{aligned} \tag{5.87}$$

$$\bar{S}(\bar{z},\bar{t};\bar{V}_a) = \bar{\alpha}_V \bar{V}_a^{l} e^{-\bar{\alpha}_V \bar{V}_a^{l-1}\bar{z}}\int_0^{\bar{t}} e^{-\bar{\beta}_V \bar{V}_a^{1}\lambda} I_0\Big(2\sqrt{\bar{\alpha}_V \bar{\beta}_V \bar{V}_a^{l+q-1}\bar{z}\lambda}\Big)d\lambda. \tag{5.88}$$

In general, all the dependences obtained below have a parametric form. At the second stage, the expressions for the head $\tilde{h}$ and rate $\bar{V}$ were derived from equation (5.83), conditions (5.86). Thus,

$$\tilde{h}(\bar{z},\bar{t};\bar{V}_a) = \bar{H}_W - \bar{V}(\bar{t};\bar{V}_a)\int_0^{\bar{z}} \frac{d\eta}{f_k(\bar{S}(\eta,\bar{t};\bar{V}_a))}, \tag{5.89}$$

$$\bar{V}(\bar{t};\bar{V}_a) = (\bar{H}_W - \bar{H}_d)\left(\int_0^1 \frac{d\eta}{f_k(\bar{S}(\eta,\bar{t};\bar{V}_a))}\right)^{-1}. \tag{5.90}$$

The quadratic equation with respect to the unknown rate $\bar{V}$ follows from equation (5.90), so that taking into account equality (5.83), it will be

$$\bar{R}_{f2}\bar{V}^2 + \bar{V}\int_0^1 \frac{d\eta}{\bar{k}(\bar{S}(\eta,\bar{t};\bar{V}_a))} - \bar{H}_W = 0, \tag{5.91}$$

and the expression for $\bar{V}$ becomes

$$\bar{V}(\bar{t};\bar{V}_a) = \frac{1}{2\bar{R}_{f2}}\left[\sqrt{\left(\int_0^1 \frac{d\eta}{\bar{k}(\bar{S}(\eta,\bar{t};\bar{V}_a))}\right)^2 + 4\bar{R}_{f2}\bar{H}_W(\bar{t};\bar{V}_a)} - \int_0^1 \frac{d\eta}{\bar{k}(\bar{S}(\eta,\bar{t};\bar{V}_a))}\right]. \tag{5.92}$$

The relationship between $\bar{H}_W$, $\bar{t}$ and $\bar{V}_a$ is determined at the final stage of the solution of the set mathematical problem. First of all, system (5.79), (5.80) is reduced to a single equation concerning $\bar{H}_W$

$$\frac{d\bar{H}_W}{d\bar{t}} = \frac{\sqrt{1-\bar{H}_W}}{\sqrt{\bar{R}_{f1}}} - \bar{V}(\bar{t}) \tag{5.93}$$

which is solved under condition (5.84). As a result of the first substitution

$$u = \sqrt{1-\bar{H}_W} \tag{5.94}$$

this problem is transformed in the following way

$$\frac{du^2}{d\bar{t}} + \frac{u}{\sqrt{\bar{R}_{f1}}} = \bar{V}(\bar{t}) \tag{5.95}$$

$$\bar{t} = 0, u = u^0 = \sqrt{1 - \bar{H}_m} \tag{5.96}$$

The integration of equation (5.95) under condition (5.96) gives

$$u^2 - u^{0^2} + \frac{1}{\sqrt{R_{fl}}} \int_0^{\bar{t}} u(\xi) d\xi = \bar{V}_a \bar{t} \tag{5.97}$$

The new dependent variable is introduced

$$y(\bar{t}) = \int_0^{\bar{t}} u(\xi) d\xi, \frac{dy}{d\bar{t}} = u(\bar{t}) \tag{5.98}$$

Then equality (5.97) is convenient to write in the following form

$$\bar{t} = G(y, u) = \frac{1}{\bar{V}_a} \left(\frac{y}{\sqrt{R_{fl}}} + u^2 - u^{0^2} \right) \tag{5.99}$$

According to Kamke (1976), the equivalent differential equation with respect to y as a function of u will be

$$\frac{dy}{du} = \frac{u G_u(y, u)}{1 - u G_y(y, u)} = 2\sqrt{R_{fl}} \frac{u^2}{\bar{V}_a \sqrt{R_{fl}} - u} \tag{5.100}$$

Here G_u, G_y denote the derivatives of the function $G(y,u)$, respectively, by u and y. Integrates (5.100) under the condition

$$u = u^0, y = 0 \tag{5.101}$$

and the result is

$$y = F(u) = 2\sqrt{R_{fl}} \left[\bar{V}_a R_{fl} \ln \frac{\bar{V}_a \sqrt{R_{fl}} - u^0}{\bar{V}_a \sqrt{R_{fl}} - u} - \frac{u^2 - u^{0^2}}{2} - \bar{V}_a \sqrt{R_{fl}} (u - u^0) \right] \tag{5.102}$$

Since it is true

$$\frac{dy}{d\bar{t}} = u = \frac{dF}{du} \frac{du}{d\bar{t}} \tag{5.103}$$

then the expression for the relative calculation time takes this general form

$$\bar{t}(u) = \int_{u^0}^{u} \frac{dF}{d\xi} \cdot \frac{d\xi}{\xi} \tag{5.104}$$

Considering (5.104) and the following from (5.102)

$$\frac{dF}{d\xi} = 2\sqrt{\overline{R}_{f1}}\frac{u^2}{\overline{V}_a\sqrt{\overline{R}_{f1}} - u} \tag{5.105}$$

the dependence is easily found

$$\bar{t}(\widetilde{H}_W;\overline{V}_a) = 2\sqrt{\overline{R}_{f1}}\left(\sqrt{1-\widetilde{H}_m} - \sqrt{1-\widetilde{H}_W} - \overline{V}_a\sqrt{\overline{R}_{f1}}\ln\frac{\overline{V}_a\sqrt{\overline{R}_{f1}} - \sqrt{1-\widetilde{H}_W}}{\overline{V}_a\sqrt{\overline{R}_{f1}} - \sqrt{1-\widetilde{H}_m}}\right) \tag{5.106}$$

Equality (5.106) is used to derive an equation that allows us to establish a single-valued relationship between $\overline{V}_a$ and $\widetilde{H}_W$ by simple fitting, and then to calculate the values of $\bar{t}$, $\overline{C}$, $\overline{S}$, $\overline{V}$, corresponding to the given $\overline{V}_a$ and found $\widetilde{H}_W$ values. The above equation is derived by integrating both parts of (5.92). Then, taking into account

$$\frac{d\bar{t}}{d\widetilde{H}_W} = \frac{\sqrt{\overline{R}_{f1}}}{\sqrt{1-\widetilde{H}_W} - \overline{V}_a\sqrt{\overline{R}_{f1}}} \tag{5.107}$$

obtained

$$2\overline{R}_{f2}V_a\bar{t}(\widetilde{H}_W;\overline{V}_a) = \\ = \sqrt{\overline{R}_{f1}}\int_{\widetilde{H}_m}^{\widetilde{H}_W}\left\{\sqrt{\left[\int_0^1\frac{d\eta}{f_k(\overline{S}(\eta,\zeta;\overline{V}_a))}\right]^2 + 4\overline{R}_{f2}\zeta} - \int_0^1\frac{d\eta}{f_k(\overline{S}(\eta,\zeta;\overline{V}_a))}\right\}\frac{d\zeta}{\sqrt{1-\zeta} - \overline{V}_a\sqrt{\overline{R}_{f1}}} \tag{5.108}$$

Here the expression for $\bar{t}$ transformed taking into account (5.107) will be

$$\overline{S}(\bar{z},\widetilde{H}_w;\overline{V}_a) = \bar{\alpha}_V\sqrt{\overline{R}_{f1}}\,\overline{V}_a^{\,l}e^{-\bar{\alpha}_V\overline{V}_a^{\,l-1}\bar{z}}\times \\ \times\int_{\widetilde{H}_m}^{\widetilde{H}_w} e^{-\bar{\beta}_V\overline{V}_a^{\,q}\bar{t}(\zeta;\overline{V}_a)}I_0\left(2\sqrt{\bar{\alpha}_V\bar{\beta}_V\overline{V}_a^{\,l+q-1}\bar{z}\bar{t}(\zeta;\overline{V}_a)}\right)\frac{d\zeta}{\sqrt{1-\zeta} - \overline{V}_a\sqrt{\overline{R}_{f1}}} \tag{5.109}$$

Thus, the sequence of the predictive calculations of the deep bed filtration is finally as follows. First of all, it is necessary to set the value of $\overline{V}_a$ (or $\widetilde{H}_W$) and find the corresponding value of $\widetilde{H}_W$ (or $\overline{V}_a$) from (5.108). Then, substituting the known set of two values ($\overline{V}_a$, $\widetilde{H}_W$) into (5.106), determine the calculated moment in time by which the level in the storage tank has risen to the already known mark. At that, the rate $\overline{V}$ will decline to the value following from formula (5.92). At last, only the initial value of $\overline{V}_a$ is enough to determine the distributions $\overline{C}(\bar{z})$, $\overline{S}(\bar{z})$ by formulas (5.87), (5.88) at the calculation time. And further it is possible to carry out the technological analysis on the basis of the mentioned formulas and equations.

The subject of the test calculations of the first series of examples at the third stage of the depth filtration were the relative outlet suspended solids concentration $(\overline{C}_e)$, concentration profiles of the deposited suspension particles $(\overline{S})$, filtration rate $(\overline{V})$ and suspension inflow from the first collector $(\overline{Q}_{in})$, hydraulic conductivity at the inlet to a bed $(\bar{k})$, as well as the reduced suspension level above the bed $(\widetilde{H}_W)$. The basis of all the calculations here was equation (5.108), the implementation of which by simple fitting using a standard mathematical analysis software allowed to determine a strict correspondence between $\overline{V}_a$

and $\overline{H}_W$ at any initial data. The relevant calculated time was determined using formula (5.106). Common for all examples were $\overline{R}_{f1} = \overline{R}_{f2} = 3$, C_0=5•10^{-5}, $\overline{\gamma}_a = 0.005$, q=1, $\overline{\beta}_V = 0.0075$. With the accepted values of $\overline{R}_{f1,f2}$, the value of $\overline{H}_m$ was 0.667 according to (5.77). Attention was focused in the examples on the behavior of different fractions of the suspended matter. The size of the suspension particles, usually characterized by the (equivalent) diameter d_p, largely determines the way (mechanism) of their delivery to the skeleton particles (grains). Such particles, and the deposit already formed over time, fulfill the role of collectors. Characterization of the main mechanisms (Brownian diffusion, sedimentation, interception) was given in Section 2.6. They correspond to the values of l, respectively, 1/3, -1/5, 7/8, which were used in the examples alternately, or the only value 1/3 was fixed. In the framework of the adopted model, the decrease in the filtration facility performance is controlled, first of all, by the rate coefficient of the suspension particles attachment, and hence by its components α_V, V^l. The acceptance of the relatively large values of α_V contributes to the intensive clogging of a bed porous material, which will be even more intensified (l>0) or, on the contrary, weakened (l<0) depending on the sign of l.

The importance of the constants α_V and l for the suspensions clarification in the second filtration mode is confirmed already in the first series of examples, the purpose of which was to determine the main elements of the water balance of a filter $(\overline{Q}_{in})$ and $(\overline{V})$ for a long time. Thus, there is an opportunity to analyze the operation of the facility both under favorable and unfavorable conditions and on this basis to make the right engineering decisions. Thereby, the initial value of the relative filtration rate $\overline{V}^0$ was accurately calculated as follows

$$\overline{V}^0 = \frac{\sqrt{4\left(\overline{R}_{f1} + \overline{R}_{f2}\right) + 1} - 1}{2\left(\overline{R}_{f1} + \overline{R}_{f2}\right)}$$

and amounted to 1/3. The dynamics of the flow rates $\overline{V}, \overline{Q}_{in}$ was calculated using equation (5.92) and the formula

$$\overline{Q}_{in}\left(\overline{t};\overline{V}_a\right) = \sqrt{\frac{1 - \overline{H}_W\left(\overline{t};\overline{V}_a\right)}{\overline{R}_{f1}}}, \tag{5.110}$$

where, in turn, the level $\overline{H}_W$ was systematically calculated at given values of $\overline{V}_a$ by fitting from equation (5.108). At the same time, the appropriate time was determined by formula (5.106). First, the index l was fixed (1/3) and the coefficient $\overline{\alpha}_V$ was varied (2, 3, 4). The plots of the dependencies $\overline{V}(\overline{t})$ and $\overline{Q}_{in}(\overline{t})$ are shown in Figure 15. A doubling $\overline{\alpha}_V$ resulted in a 14-fold reduction in the hydraulic load on the filter at the end of the calculation period. The suspension inflow to the filter also decreased significantly, though not so much (more than 3 times). Thus, we can tentatively conclude already based on Figure 15 about a very significant effect of the absorption properties of the filtering material on the facility performance. The effect of the physical and mechanical properties of the suspension on the absorption process can be seen from Figure 16 along with Figure 15. The graphs of similar dependences are shown here, but with a single value of $\overline{\alpha}_V$(3) and different of l (7/8, -1/5). Both figures confirm the fundamental importance for particles sticking and, in general, for the performance of a filtration facility, not just the rate V, but its degree V^l. Indeed, the decrease in V causes an appreciable decrease in $\overline{\alpha}$ over time at l>0 (curves 3 in Figure 15 and 2 in Figure 16) and, as a consequence, the clogging of a bed is inhibited and, consequently, the filtration flow rate decreases more slowly. The opposite situation takes place at l<0 (curve

4 in Figure 16). Now the mechanism determining the mechanical state and behavior of the suspension particles increases the absorption capability of a bed, which leads to an accelerated decrease in the filtration rate and thus, to even greater deposition of the suspended solids. As a result, the resistance of the bed (essentially its upper part) progressively increased and thus, the facility performance fell from about the time point $\bar{t} = 50$, so that the filtration flow of contaminated water at all practically ceased already at $\bar{t} = 85$. Similarly, though not to the same extent, the flow rate $\overline{Q}_{in}$ changed. It is obvious that in such cases the filter operation should be interrupted precisely because of too rapid clogging of a filter media.

Figure 15. Decrease in the relative filtration rate and water flow rate in the supply communications over time

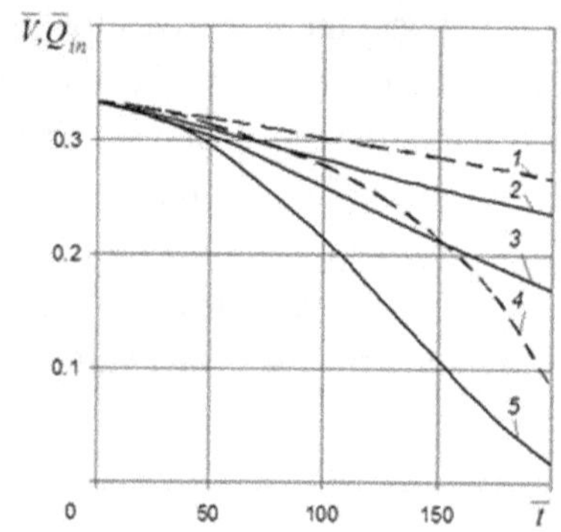

Note. 1, 4 – $\overline{Q}_{in}$; 2, 3, 5 – $\overline{V}$; 1, 2 – $\overline{\alpha}_V = 2$; 3 – $\overline{\alpha}_V = 3$; 4, 5 – $\overline{\alpha}_V = 4$

Figure 16. Decrease in the relative filtration rate and water flow rate in the supply communications over time

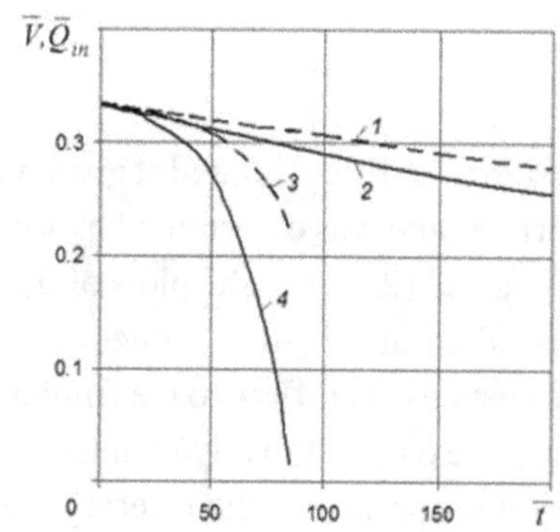

Note. 1, 3 – $\overline{Q}_{in}$; 2, 4 – $\overline{V}$; 1, 2 – $l = \frac{7}{8}$; 3, 4 – $l = -1/5t$.

The intermediate result of calculating the quantity $\overline{Q}_{in}$ was the reduced level $\overline{H}_W$, obtained by formula (5.97). The set of its values describing the dynamics of the suspension level above the bed at different values of $\overline{\alpha}_V$ and l is presented in Figure 17. Naturally, this level is closely related to the permeability of the bed. The best filtration conditions in the examples take place at its reduced absorption capability due to the smallness of $\overline{\alpha}_V(2)$ and rapidly decreasing $\overline{V}^l$ $\left(l = \frac{7}{8}\right)$. Then clogging of the bed occurs more slowly and, more importantly, more evenly along the height, which predetermines the lowest rate of the formation of the surface suspension layer (curves 4, 5). Higher values of $\overline{\alpha}_V(3,4)$ and lower value of $l\,(-\frac{1}{3})$ contribute to the accelerated deposit formation in the upper part of the bed and, as a consequence, the suspension level rises noticeably faster above it (curves 2, 3). The critical situation occurs at $l = -\frac{1}{5}$, when the flow capacity of the bed drops almost to 0 due to the very rapid clogging of the pores near the inlet cross-section. It is clear that now almost all the suspension entering the filter will be accumulated in the storage tank, which caused a sharp rise of curve 1. In this case, the quantity $\overline{H}_W$ reached the maximum value (1) in a finite time. The moment $\overline{t}_e$ corresponding to this value can be easily calculated on the basis of a simplified version of model (5.32) - (5.36). By neglecting the infiltration in it, the following formula is simply derived

$$\overline{t}_e = \overline{t}_{e0} + 2\sqrt{\overline{R}_{fl}(1 - \overline{H}_{W0})},$$

where $\overline{t}_{e0}$ is the relative time of the filtration process termination, $\overline{H}_{W0}$ is the reduced mark of the suspension level at this time point. The values $\overline{t}_{e0} = 85, \overline{H}_{W0} = 0.856$ were assumed in the last example and then $\overline{t}_e$ was 86.4.

One of the reasons for the premature and excessive increase in the bed hydraulic resistance is the uneven distribution of a deposit by its height. Its peculiarities at different stages of filtration and under the typical conditions can be seen in Figure 18. The profiles of the relative volumetric concentration of the deposited suspension particles $\overline{S}(\overline{z})$ at three characteristic time points are given here. During the initial stage of filtration ($\overline{t} \leq 50$), the deposit concentrates in the immediate vicinity of the bed inlet (curve 3). As the clarification process progresses, it involves the downstream sections (curve 2) down to the section adjacent to the base of the bed (curve 1). Thus, the contribution of its lower part to the removal of the suspended solids becomes significant, although the filtration flow conditions at the top of the bed remain significantly worse than at the bottom. In this case, a decrease in the permeability of the filter media in the inlet cross-section is especially indicative, which seriously affects the suspension infiltration and, at the same time, the entire technological process of its separation. Figure 19 demonstrates a possible decrease in the relative inlet hydraulic conductivity $f_k(0,\overline{t})$ at prolonged filtration under actual conditions. Its value decreases comparatively slowly in the example with the worst absorption properties of the bed ($\overline{\alpha}_V = 3,\ l = \frac{7}{8}$; curve 1). The other extreme situation is observed in the example where the reduction of the rate $\overline{V}$ favors the sticking of an impurity to the solid phase ($l = -\frac{1}{5}$). Therefore, this coefficient decreases almost to 0 in a short time (curve 5), which is unacceptable from a practical point of view.

Figure 17. Increase in the reduced mark of the suspension level above the bed over time

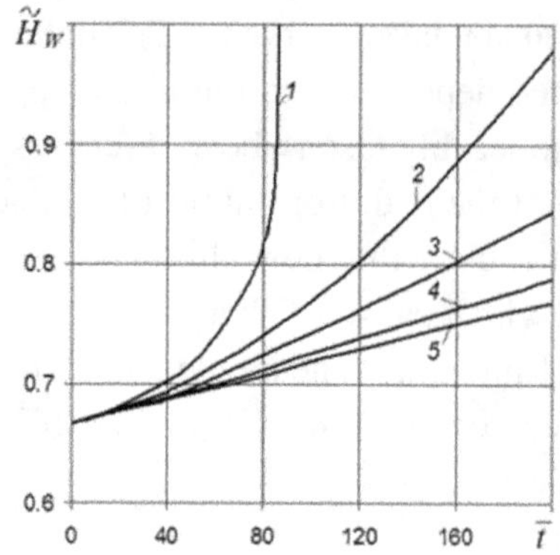

Note.1, 3, 5 – $\bar{\alpha}_v = 3$; 2 – $\bar{\alpha}_v = 4$; 4 – $\bar{\alpha}_v = 2$; 2-4 – $l = \frac{1}{3}$; 1 – $l = -\frac{1}{5}$; 5 – $l = \frac{7}{8}$

Figure 18. Profiles of the relative volumetric concentration of the deposited particles

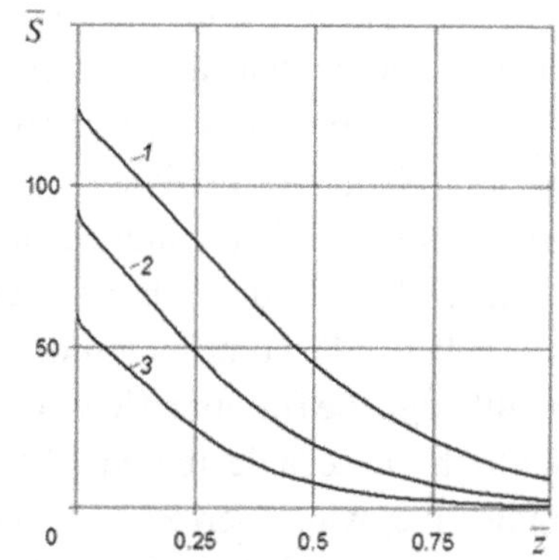

Note. 1 – $\bar{t} = 200$; 2 – $\bar{t} = 100$; 3 – $\bar{t} = 50$

Figure 19. Change in the relative hydraulic conductivity at the inlet cross-section

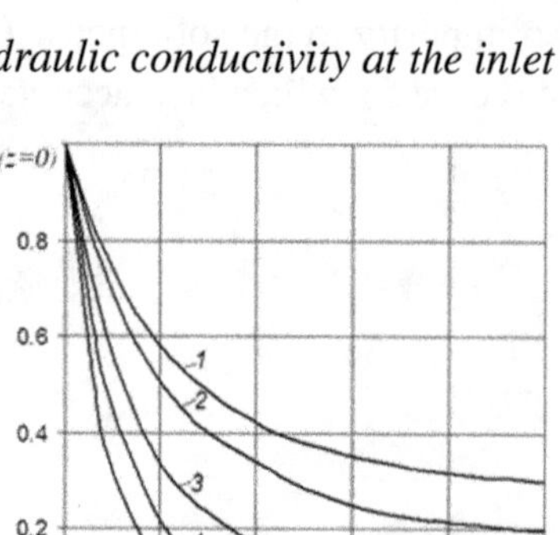

Note. 1, 3, 5 – $\bar{\alpha}_v = 3$; 2 – $\bar{\alpha}_v = 2$; 4 – $\bar{\alpha}_v = 4$; 1 – $l = \frac{7}{8}$; 2-4 – $l = \frac{1}{3}$; 5 – $l = -\frac{1}{5}$

Figure 20. Increase in the relative outlet volumetric concentration of the suspended over time

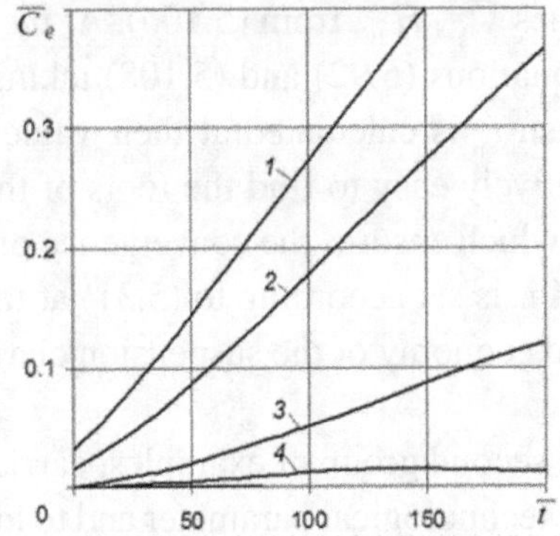

Note. $1, 3, 5 - \overline{\alpha}_V = 3;\ 2 - \overline{\alpha}_V = 2;\ 4 - \overline{\alpha}_V = 4;\ 2\text{-}4 - l = \frac{1}{3};\ 1 - l = \frac{7}{8};\ 5 - l = -\frac{1}{5}$

The reliable prediction of the suspended solids content in the filtrate is of great importance for the optimization of a filter facility operation. The solution obtained above can provide it at unstable flow characteristics. The outlet suspended solids concentration was directly calculated at a standard set of the values of the coefficients $\overline{\alpha}_V$, l by formula (5.87). At the same time, the data on the relationship between $\overline{V}_a$ and $\overline{t}$ according to (5.98) were used, as well as earlier in the analysis of the dependences $\overline{V}(\overline{t})$, $\overline{Q}_{in}(\overline{t})$, $\overline{S}(\overline{z},\overline{t})$, $\overline{k}(0,\overline{t})$ (Figures 15 - 18). The curves obtained in this way clearly illustrate in Figure 20 the nature of a decrease in the protective capability of the bed and its high sensitivity to the absorption parameters. Thus, at $\overline{\alpha}_V = 3$ and $l = -\frac{1}{5}$ the concentration $\overline{C}_e$ is lower than the maximum permissible concentration by several orders of magnitude and does not change much throughout the entire calculation period. In fact, the corresponding curve merges with the horizontal coordinate axis and, therefore, it cannot be highlighted in the figure. The increase in the exponent l up to 1/3 turns out to be so significant for the quality of the suspension clarification that even at a noticeably higher value of $\overline{\alpha}_V$ (4) the filtrate already contains a significant amount of the suspended solids (curve 4). Formal decrease in $\overline{\alpha}_V$ (3, 2) and increase in l $\left(\frac{7}{8}\right)$ contribute to a sharp deterioration of water treatment quality (curves 1, 2) and in such cases it is the protective properties of the bed that can become the main factor limiting the operation of the filtration facility.

The theoretical substantiation of the filter run duration also on the basis of (5.31) allows to achieve the maximum efficiency from the facility under the considered conditions. Since an excessive reduction of the hydraulic load on a filter is unprofitable even at a very good quality of water treatment, it is necessary along with the time t_p to find the time point t_V when the (average) filtration rate reaches the minimum permissible value. The analytical solution obtained in this Section provides just such an opportunity. The procedure of the technological calculations differs significantly when setting the limit values V_* and V_{a*}. The relative technological times $\overline{t}_p$, $\overline{t}_V$ are easier obtained based on V_{a*}. Then, at the first step, when V_{a*} and $\overline{C}_*$ are set, the time $\overline{t}_p$ is obtained from expression (5.87), which is transformed as follows

$$e^{-\overline{\beta}_V \overline{V}_a^q \overline{t}_p} I_0\left(2\sqrt{\overline{\alpha}_V \overline{\beta}_V \overline{V}_a^{l+q-1} \overline{t}_p}\right) + \\ +\overline{\beta}_V \overline{V}_a^{\,q} \int_0^{\overline{t}_p} e^{-\overline{\beta}_V \overline{V}_a^q \lambda} I_0\left(2\sqrt{\overline{\alpha}_V \overline{\beta}_V \overline{V}_a^{l+q-1} \lambda}\right) d\lambda = \overline{C}_* e^{\overline{\alpha}_V \overline{V}_a^{l-1}}; \tag{5.111}$$

at the second step, the corresponding value of $\bar{H}_{W*}$ is determined from equation (5.108) taking into account the dependence of $\bar{t}$ on $\bar{V}_a$, $\bar{H}_W$ (5.106) ; at the third step, the required value of $\bar{t}_V$ is determined with the now known set of two values $\bar{V}_{a*}$, $\bar{H}_{W*}$ from (5.106). At the initially given $\bar{V}_*$, first of all, it is necessary to solve the system of equations (5.92) and (5.108) taking into account (5.106) concerning the unknowns values $\bar{V}_{a*}$, $\bar{H}_{W*}$. Then $\bar{t}_p$ is calculated at their values found also from (5.111), and $\bar{t}_V$ from (5.106). Incidentally, it is relatively easy to find the roots of the above system due to a specially implemented search of the options, which ensures the convergence of the calculation data to the desired result. Finally, the required value of $\bar{t}_f$ is set according to (5.31) at the known values of $\bar{t}_p$, $\bar{t}_V$. Thereby both a high quality of the filtrate and economy of the suspension clarification by the depth filtration are guaranteed.

Naturally, the calculations of the second group of examples, carried out in order to test the proposed methodology for calculating the key technological parameter and to identify some patterns in its change, proved to be more labor-intensive, since it was necessary to determine $\bar{t}_p$, $\bar{t}_V$ at each calculation step. At the same time, the coefficient $\bar{\alpha}_V$ was varied continuously in a wide range (from 1 to 6), the values of l remained the same. In addition, the values of the criterion parameters were taken: $\bar{C}_* = 0.1$, $\bar{V}_{a*} = 0.9\bar{V}_{2m}$, $0.8\bar{V}_{2m}$, where $\bar{V}_{2m}$ was calculated by formula (5.76) at $\bar{H}_{W2} = \bar{H}_m$. First of all, it was assumed that $l = \frac{1}{3}$, and also the operation of the facility was allowed with the performance reduced by 10 and 20% on average. The corresponding graphs of the dependences $\bar{t}_p(\bar{\alpha}_V)$, $\bar{t}_V(\bar{\alpha}_V)$, calculated from equations (5.105), (5.106), are shown in Figure 21. Their upper fragments, which are not pieces of the graphs, are shown by dashed lines, while the lower fragments are shown by solid lines. The first fragment is a piece of the graph $\bar{t}_p(\bar{\alpha}_V)$ and describes the behavior of the quantity $\bar{t}_f$ on the interval $[1, \bar{\alpha}_{V*}]$. The second piece of the above-mentioned curve is identical to the corresponding fragment of the curve $\bar{t}_V(\bar{\alpha}_V)$ and is defined on the interval $[\bar{\alpha}_{V*}, 6]$. Thus, the main calculated curves have a clearly pronounced maximum at $\bar{\alpha}_V = \bar{\alpha}_{V*}$ and, therefore, at constant values of other model parameters there is a single value of $\bar{\alpha}_V$, at which the duration of the filter run will be the longest $\left(\bar{t}_{fm}\right)$. When decreasing $\bar{V}_{a*}$ from 0.3 to 0.266, the value of $\bar{t}_{fm}$ shifts slightly toward larger values of $\bar{\alpha}_V$ and at the same time increases one and a half times.

Then similar calculations were performed at $\bar{V}_{a*} = 0.9\bar{V}_{2m}$ and two other values of l (7/8, -1/5). The curves of the required dependence $\bar{t}_f\ (\bar{\alpha}_V)$, presented in Figure 22 with solid lines, also have maximums within the calculation interval at $\bar{\alpha}_{V*} = 5.67$ and 1.55 respectively. This fact is explained by the fundamentally different effect of the filtration rate on the absorption process at $l>0$ and $l<0$ from the physical point of view, when a declining filtration flow rate weakens or enhances the sticking of the suspended matter to particles-collectors. At that, the maximum durations $\bar{t}_{fm}$ in both cases are practically the same. Thus, the technologically best values of $\bar{\alpha}_{V*}$ and $\bar{t}_{fm}$ at an unregulated rate essentially depend on the suspension particles size.

Figure 21. Dependencies $\bar{t}_p(\bar{\alpha}_V)$, $\bar{t}_v(\bar{\alpha}_V)$, $\bar{t}_f(\bar{\alpha}_V)$

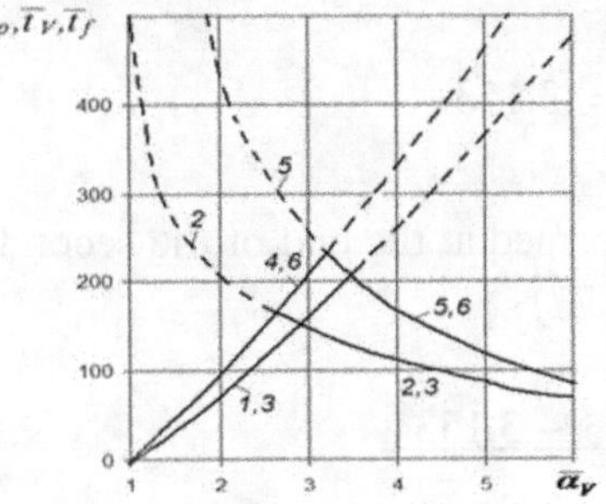

Note. 1, 4 – $\bar{t}_p$, 2, 5 – $\bar{t}_v$, 3, 6 – $\bar{t}_f$; 1-3 – $\bar{V}_{a*} = 0.9\bar{V}_{2m}$; 4-6 – $\bar{V}_{a*} = 0.8\bar{V}_{2m}$

Figure 22. Dependencies $\bar{t}_p(\bar{\alpha}_V)$, $\bar{t}_v(\bar{\alpha}_V)$, $\bar{t}_f(\bar{\alpha}_V)$

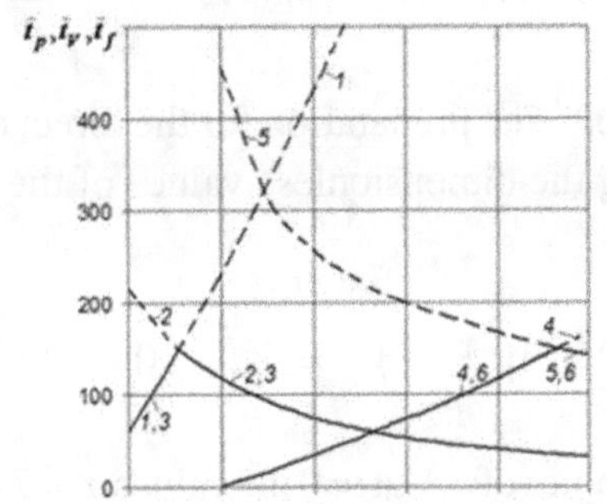

Note. 1, 4 – $\bar{t}_p$, 2, 5 – $\bar{t}_v$, 3, 6 – $\bar{t}_f$; 1-3 – $l = -\frac{1}{5}$; 4, 6 – $l = \frac{7}{8}$

In conclusion, the filter run duration is calculated in detail for two examples with the following initial information: H_{in}=3.5m, H_{out}=1m, L=1m, ω=1m^2, n_0=0.35, equivalent grain diameter d_g=0.8mm, $R_{f1}=R_{f2}=1.36{\bullet}10^{-2}h^2/m^2$, q=1, $\beta_V=0.005m^{-1}$, γ_a=20, $C_0=2.5{\bullet}10^{-5}$; the medium temperature is 10 . In both examples, a single initial value of the rate coefficient α is formally set ($30h^{-1}$). However, the absorption capacity of the bed significantly differs in the filtration process due to the acceptance of two limiting values of the exponent, namely, 0.7 in the first example and – 0.3 in the second example. The hydraulic conductivity of a clean bed was pre-calculated using the Kozeny-Karman formula

$$k_0 = \chi_t \cdot 0.0055\frac{g d_g^2 n_0^3}{\nu(1-n_0)^2} = 3600 \cdot 0.0055\frac{9.81(8\cdot 10^{-4})^2 0.35^3}{1.31\cdot 10^{-6}(1-0.35)^2} = 9.63m/h$$

where χ_t is the coefficient of the conversion of hours into seconds.

Filtration facilities should provide, firstly, the reduction of the suspended solids content in the contaminated water by an order of magnitude or more $(C_* = 0.1C_0,\ \bar{C}_* = 0.1)$, and secondly, the hydraulic load close to the maximum V_{max} on average for the filter run, namely. The value V_{max} is reached at the end of the second stage, and its relative value was determined by formula (5.76). But before we found the relative values of the total resistances of both types of communications in the facility according to (5.56)

$$\bar{R}_{f1,f2} = \frac{9.63^2 \cdot 1^2 \cdot 2 \cdot 5 \cdot 0.0136}{1^2} = 3.153.$$

Then the calculations were performed at the end of the second and beginning of the third periods, first, in accordance with formula (5.77)

$$\bar{H}_m = \frac{2 \cdot 3.153^2 + 2 \cdot 3.153 \cdot 3.153 - 3.153}{2(3.153 + 3.153)^2} +$$

$$+\sqrt{\frac{(2 \cdot 3.153^2 + 2 \cdot 3.153 \cdot 3.153 - 3.153)^2}{4(3.153 + 3.153)^4} - \frac{3.153^2 - 3.153}{(3.153 + 3.153)^2}} = 0.663,$$

secondly, by formula (5.76) at $\bar{H}_{W2} = \bar{H}_m = 0.663$. Then

$$\bar{V}_{max} = \frac{1}{2 \cdot 3.153}\left(\sqrt{1 + 4 \cdot 3.153 \cdot 0.663} - 1\right) = 0.327, \quad V_{max} = 7.8725 m/h$$

Thus, $\bar{V}_{a*} = 0.8 \cdot 0.327 = 0.262$. The preparation for the direct determination of the relative criterion times was completed by setting the dimensionless values of the following model coefficients and complexes:

$$\bar{\beta}_V = \Delta H \cdot \beta_V = 2.5 \cdot 0.005 = 1.25 \cdot 10^{-2}, \quad \bar{\gamma}_a = 2.5 \cdot 10^{-5} \cdot 20 = 5 \cdot 10^{-4}$$

Since $\alpha = \alpha_V V^l$, then based on the values of α, V at $t=0$ the values $\alpha_V = 7.077 m^{-0.7} h^{-0.3}$ and $\alpha_V = 55.713 m^{0.3} h^{1.3}$ are obtained at $l=0.7$ and $\theta=0$ respectively. Then at

$$\bar{\alpha}_V = \frac{(9.63 \cdot 2.5)^{l-1}}{1^{l-2}} \alpha_V = \begin{cases} 2.725 & at \quad l = 0.7, \\ 0.891 & at \quad l = -0.3, \end{cases}$$

the required $\bar{t}_p$, $\bar{t}_V$, which should satisfied equations (5.111), (5.108), respectively, were found by searching their possible values. Thus, the set of the values $\bar{t}_p = 233$, $\bar{t}_V = 385.7$ was determined in the first example, so that the relative duration of the filter run was $\bar{t}_f = \bar{t}_p = 233$. Similarly, $\bar{t}_p = 395$, $\bar{t}_V = 302.9$ in the second example and now $\bar{t}_f = \bar{t}_V = 302.9$. Consequently, it is necessary to start washing after 24.195 and 31.254 hours.

5.3. FILTRATION WITH CONSTANT RATE OF SUSPENSION SUPPLY ON FILTER FACILITY (GENERAL CASE)

In practice, the clarification of an aqueous suspension at a time-varying filtration rate is often carried out by feeding the filter with a stable flow rate Q_{in}. Due to the low flow capacity of the filter media in relation to the incoming flow rate, a surface layer of the untreated suspension (raw water) is formed and intensively develops above it. It is important that unlike the first variety of the second filtration rate mode, studied in detail in the previous Sections of this Chapter, the rise of the suspension level in the storage tank is strictly limited at the implementation of its second variety (split filter flow) due to the constancy of the quantity Q_{in}. Therefore, the value of Q_{in} at the operative separation of a large suspension volume and, as a consequence, the depth of its layer above the bed can be significant. At the same time, the indicated level rises also clearly unevenly, which justifies the isolation of early stages of filtration by analogy with the second problem (Section 5.2). The water-physical and physicochemical processes occurring in the bed and the offtake communications, in both cases of a variable filtration rate ($Q_{in}=Q_{in}(t)$ and $Q_{in}=const$) are identical and, thereby, are described by mathematical models close in form. Formal differences in the statement of the second and third problems of this Chapter are explained by different ways of the suspension supply to the filter and the role of the supply communications in it. Therefore, the theoretical analysis of the flooded filter operation carried out in this Section at a variable filtration rate and $Q_{in}=const$ is based on the approximate solution of the mathematical problem of the aqueous suspension clarification, in which also the mass exchange coefficients are variable and initially unknown. Its formulation standardly includes the clarification and filtration flow compartments, and at the same time the hydraulic compartment, which deserves a special attention in this situation due to the great importance for the filter performance of the level mode in the storage tank. Indeed, here a considerable volume of the initial suspension can be accumulated and a commensurate amount of mechanical energy can be stored, which is then spent for a long time to overcome the increasing hydraulic resistance of the clogging bed.

5.3.1. First and Second Stages

When considering the short first stage, the hydraulic compartment (5.32) - (5.36) is used, simplified by omitting equation (5.32) and setting a constant flow rate Q_{in}. Thus, the additional assumption (5.44) is automatically satisfied based on the conditions of the problem. As a consequence, the calculation formulas (5.53), (5.57) and (5.58) are retained, but they are now rigorous. Nevertheless, due to the change in some scales, their dimensionless form in this case is somewhat different. In principle, when introducing dimensionless variables and parameters in the mathematical model of the deep bed filtration with a stable suspension supply to the filter, which is analyzed in detail in this and the next Section, it is possible to use for the flow characteristics and for the current or constant heads and marks of different scales present in the statement of the problem to be solved. The flow characteristics are correlated to the reference value V_s, as which it is now reasonable to use either the reduced flow rate (average rate) $\frac{Q_{in}}{\omega}$, or the hydraulic conductivity of a clean bed k_0. To formally maximize the simplification of the original model, in the second case it is justified to take $Z_w-Z_d=L$ as a scale for the heads (Δh_s). Below, when modeling the operation of rapid filters and solving the corresponding general (first) and particular (second) mathematical problems, different systems of scales are deliberately used, differing just by the expressions for V_s. In this case, the representations for $\overline{C},\overline{S},\overline{z},\overline{\gamma}$ remain the same, but differ the representations for

$$\overline{t} = (V_s t)/(n_0 L), \overline{\alpha}_V = L V_s^{l-1}\alpha_V, \overline{\beta}_V = L V_s^{q-1}\beta_V, \overline{R}_{f2} = \left(\omega^2 V_s^{\,2} R_{f2}\right)/L.$$

Obviously, different ways of introducing dimensionless quantities can eventually lead to the calculation dependencies that may differ markedly in form, but their dimensional prototypes will coincide. Specifically, when adopting expressions for the functional coefficients α,β according to (3.49) in general form (arbitrary l,q), $V_s = \frac{Q_{in}}{\omega}$ is accepted. In the second problem (l=q=1) V_s=k_0. Equality of V_s to the value $\frac{Q_{in}}{\omega}$ means that the calculation quantity $\overline{V} = \frac{V}{V_s}$ will always be less than 1, and it can be reached only in the limit (at $t\to\infty$). If V_s is equal to k_0, then the initial value of $\overline{V}$, for example, at $\overline{Q}_{in} = \frac{Q_{in}}{(\omega k_0)} > 1$ will be 1, and will eventually increase to the maximum value $\overline{Q}_{in}$.

The introduction of a number of the scale parameters into the original model causes the appearance of three complex parameters in its dimensionless analog, namely, $n_0L/\Delta h_s$, $Q_{in}/(\omega V_s)$, $k_0\Delta h_s/(LV_s)$. Some of them can be eliminated by a suitable choice of the scales. Thus, if V_s= Q_{in}/ω, Δh_s= $LQ_{in}/(LV_s)$, then these complexes are equal to $\frac{n_0}{\overline{Q}_{in}}$, 1 and 1 respectively. However, the single scale $\overline{a}_l$ is below adopted for all the marks and heads involved in the conditions of the two problems, i.e. they are correlated to the height of the operating layer. In this case, the indicated complexes are n_0, 1, $\overline{Q}_{in}^{-1}$.

Thus, the problem describing the intensive wetting of the dewatered bed was formulated and solved approximately (due to an instability of the flow rate Q_{in}) in Section 5.2. Below, the previously obtained calculation dependencies are presented in a new form due to the use of a different scale system. In doing so, they become rigorous due to the constancy of Q_{in}. At this stage, the hydraulic characteristics of a practical interest are proposed to be determined as follows:

the reduced level in the storage tank and the relative constant filtration rate at an arbitrary point of time t ($t\leq t_1$) by the formulas

$$\overline{H}_{W1}(\bar{t},Q_{in}) = \overline{Z}_w + \left(1 - n_0\overline{\lambda}_w(\overline{Q}_{in})\right)n_0\bar{t}, \tag{5.112}$$

$$\overline{V}_1(\overline{Q}_{in}) = \frac{1}{\overline{Q}_{in}}\left(1 - n_0 + \frac{1}{\overline{\lambda}_w(\overline{Q}_{in})}\right), \tag{5.113}$$

where $\overline{\lambda}_w(\overline{Q}_{in}) = \frac{1 - n_0}{2n_0\overline{Q}_{in}}\sqrt{1 + \frac{4n_0\overline{Q}_{in}}{(1 - n_0)^2}}$; the relative time of the first stage end and the level position at the corresponding moment in time according to the formulas

$$\bar{t}_1(\overline{Q}_{in}) = \frac{1}{n_0\overline{\lambda}_w(Q_{in})} \tag{5.114}$$

$$\overline{H}_{W1}(\bar{t},Q_{in}) = \overline{Z}_w + \frac{1}{\overline{\lambda}_w(\overline{Q}_{in})} - n_0 \tag{5.115}$$

A similar problem with respect to the second stage of the filtration process, including (5.59) - (5.64) and reflecting the consequences of the first stage, as well as the connection of the offtake communications, was solved again by analytical methods in Section 5.2. The exact dependences derived in this case now, taking into account other scales, take the following form:

the inverse function describing the dynamics of the suspension level in the storage tank,

$$\bar{t} = \frac{2R_{f2}}{n_0}Y_2(\overline{H}_{W2}), \tag{5.116}$$

where

$$Y_2(\bar{H}_{W2}) = \bar{R}_{f2}\ln\frac{\bar{H}^0_{W2} - \bar{R}_{f2} - \bar{Q}_{in}}{\bar{H}_{W2} - \bar{R}_{f2} - \bar{Q}_{in}} + 2\sqrt{\bar{Q}^2_{in} + 4\bar{R}_{f2}\bar{H}^0_{W2}} - 2\sqrt{\bar{Q}^2_{in} + 4\bar{R}_{f2}\bar{H}_{W2}} +$$

$$+\bar{R}_{f2}\ln\frac{\left(\sqrt{\bar{Q}^2_{in} + 4\bar{R}_{f2}\bar{H}_{W2}} + 2\bar{R}_{f2}\bar{Q}_{in} + \bar{Q}^2_{in}\right)\left(\sqrt{\bar{Q}^2_{in} + 4\bar{R}_{f2}\bar{H}^0_{W2}} - 2\bar{R}_{f2}\bar{Q}_{in} - \bar{Q}^2_{in}\right)}{\left(\sqrt{\bar{Q}^2_{in} + 4\bar{R}_{f2}\bar{H}^0_{W2}} + 2\bar{R}_{f2}\bar{Q}_{in} + \bar{Q}^2_{in}\right)\left(\sqrt{\bar{Q}^2_{in} + 4\bar{R}_{f2}\bar{H}_{in}} - 2\bar{R}_{f2}\bar{Q}_{in} - \bar{Q}^2_{in}\right)}$$

the variable filtration rate

$$\bar{V}_2(\bar{t},\bar{Q}_{in}) = \frac{\sqrt{\bar{Q}_{in}{}^2 + 4\bar{R}_{f2}\bar{H}_{w2}(\bar{t},\bar{Q}_{in})} - \bar{Q}_{in}}{2\bar{R}_{f2}} \tag{5.117}$$

For practice it is of interest to calculate not so much the dynamics of the level on the surface of the bed, but its stabilized position, at which the flow rates of the contaminated water initially entering the filter and directly into the layer of a filtering material are equalized. Obviously, the subsequent rise of the level will be entirely due to the formation of a deposit and the deterioration in this regard the bed permeability. Then at unlimited supply of pure water in time its level in the storage tank, as follows from formula (5.116), should be settled at the mark

$$H_m = H_{W2}(\infty) = H_{out} + \frac{LQ_{in}}{k_0\omega} + R_{f2}Q^2_{in}$$

or in a dimensionless form

$$\bar{H}_m = \bar{Q}_{in} + \bar{R}_{f2}\bar{Q}^2_{in} \tag{5.118}$$

Just the limit value $\bar{H}_m$ according to (5.118) will be used at the main stage of filtration as an initial value for the desired function $\bar{H}_{w2}(\bar{t})$.

5.3.2. Third Stage

The filtration of an aqueous suspension at the main (third) stage is described by the system of equations in a dimensionless form, including

$$\frac{d\bar{H}_W}{d\bar{t}} = n_0[\bar{Q}_{in} - \bar{V}(\bar{t})] \tag{5.119}$$

(5.81) - (5.84), as well as the initial and boundary conditions (5.85), (5.86), and the parameter $\bar{H}_m$ in the first condition (5.85) should be calculated by formula (5.118). Since the clarification compartment for both ways of water supply to the filter is formulated in the same way, it is proposed to calculate the changes in the relative concentrations $\bar{C}, \bar{S}$ along the bed height and over time at Q_{in}=*const* using dependences (5.87), (5.88). When carrying out specific calculations, the above system should be supplemented with a representation for f_k, for example,

$$f_k(\bar{S}) = \{1 - [\bar{\gamma}(\bar{S})\bar{S}]^{m_{k1}}\}^{m_{k2}} \approx [1 - (\bar{\gamma}_a\bar{S})^{m_{k1}}]^{m_{k2}} \tag{5.120}$$

The key value in the calculations of the technological and design parameters is the equation relating the parameter $\bar{V}_a$ with time $\bar{t}$, which in this case differs from (5.108) only by additional multipliers, so that

$$\int_0^{\bar{t}} \left[\sqrt{ \left(\int_0^1 \frac{d\eta}{f_k(\bar{S}(\eta,\xi;\bar{V}_a))} \right)^2 + \frac{4\bar{R}_{f2}}{\bar{Q}_{in}^2} \bar{H}_W(\xi;\bar{V}_a) } - \int_0^1 \frac{d\eta}{f_k(\bar{S}(\eta,\xi;\bar{V}_a))} \right] d\xi = \frac{2\bar{R}_{f2}}{\bar{Q}_{in}} \bar{V}_a \bar{t}. \tag{5.121}$$

Now the dependence of the reduced level (mark) $\bar{H}_W$ on the variable $\bar{t}$ and parameter $\bar{V}_a$ has a linear nature

$$\bar{H}_W(\bar{t};\bar{V}_a) = \bar{H}_m + n_0(1 - \bar{V}_a)\bar{t}. \tag{5.122}$$

Then the relative filtration rate is the function of $\bar{t}$ and $\bar{V}_a$ of the following form

$$\bar{V}(\bar{t};\bar{V}_a) = \frac{\bar{Q}_{in}}{2\bar{R}_{f2}} \left(\sqrt{ \left(\int_0^1 \frac{d\eta}{f_k(\bar{S}(\eta,\bar{t};\bar{V}_a))} \right)^2 + \frac{4\bar{R}_{f2}}{\bar{Q}_{in}^2} \bar{H}_W(\bar{t};\bar{V}_a) } - \int_0^1 \frac{d\eta}{f_k(\bar{S}(\eta,\bar{t};\bar{V}_a))} \right) \tag{5.123}$$

The calculation dependences and equations derived in this Section allow to determine the rational values of the technological times $\bar{t}_p, \bar{t}_V, \bar{t}_H; \bar{t}_f$, at which the economical and qualitative operation of a filter would be provided in the second rate mode at stable suspension supply (Q_{in}=*const*). Due to the essential change in the filtration rate during the filter run, the same efficiency criteria were used, first of all, as in the case of the mode with $Q_{in}(t)$, i.e. (1.30) and (1.32). Since often in the case Q_{in}=*const* the reason for the filter successive turning off is an easily observable overflow of the storage tank, it is necessary to use for the operative control of its operability also the third criterion expressed by relation (1.33).

Equation (5.121) plays a key role in establishing here all the technological times. Thus, it is used to determine the time t_p and the uniquely related to it parameter $\bar{V}_{ap}$ together with the equation that follows from (5.87) and is analogous to (5.111),

$$\begin{aligned} \bar{C}_* e^{\bar{\alpha}_V \bar{V}_{ap}^{l-1}} &= e^{-\bar{\beta}_V \bar{V}_{ap}^q \bar{t}_p} I_0\left(2\sqrt{\bar{\alpha}_V \bar{\beta}_V \bar{V}_{ap}^{l+q-1} \bar{t}_p}\right) + \\ &+ \bar{\beta}_V \bar{V}_{ap}^q \int_0^{\bar{t}_p} e^{-\bar{\beta}_V \bar{V}_{ap}^q \xi} I_0\left(2\sqrt{\bar{\alpha}_V \bar{\beta}_V \bar{V}_{ap}^{l+q-1} \xi}\right) d\xi. \end{aligned} \tag{5.124}$$

Also, equation (5.116) and the following equation

$$\begin{aligned} &\frac{2\bar{R}_{f2}}{\bar{Q}_{in}} \bar{V}_* = \\ &= \sqrt{ \left(\int_0^1 \frac{d\eta}{f_k(\bar{S}(\eta,\bar{t}_V;\bar{V}_{aV}))} \right)^2 + \frac{4\bar{R}_{f2}}{\bar{Q}_{in}^2} \bar{H}_W(\bar{t}_V;\bar{V}_{aV}) } - \int_0^1 \frac{d\eta}{f_k(\bar{S}(\eta,\bar{t}_V;\bar{V}_{aV}))} \end{aligned} \tag{5.125}$$

form the system in relation to the interrelated values of $\bar{t}_V$ and $\bar{V}_{aV}$. The time $\bar{t}_H$ is more easily calculated due to the equality following from (5.122)

$$\overline{V}_a(\bar{t}_H) = 1 - \frac{\widetilde{H}_0 - \overline{R}_{f2} - \overline{Q}_{in}}{n_0 \bar{t}_H}. \tag{5.126}$$

Then the desired $\bar{t}_H$ is simply found by fitting from the single equation

$$\frac{2\overline{R}_{f2}}{\overline{Q}_{in}} \overline{V}_a(\bar{t}_H)\bar{t}_H = \\ = \int_0^{\bar{t}_H} \left[\sqrt{\left(\int_0^1 \frac{d\eta}{f_k(\overline{S}(\eta,\xi;\overline{V}_a(\bar{t}_H)))} \right)^2 + \frac{4\overline{R}_{f2}\widetilde{H}_*}{\overline{Q}_{in}^2}} - \int_0^1 \frac{d\eta}{f_k(\overline{S}(\eta,t;\overline{V}_a(\bar{t}_H)))} \right] d\xi \tag{5.127}$$

The relative duration of the filter run is identified with the smallest of the values $\bar{t}_p, \bar{t}_V, \bar{t}_H$ according to (1.31), i.e.

$$\bar{t}_f = \min(\bar{t}_p,\ \bar{t}_V,\ \bar{t}_H) \tag{5.128}$$

A quantitative analysis was carried out to demonstrate the actual possibilities of the approximate solution obtained in this Section with respect to, firstly, the prediction of changes in the main filtration characteristics $(\overline{C},\ \overline{V},\ \widetilde{H}_W)$, and secondly, the substantiation of rational values of the technological parameters $(\bar{t}_p, \bar{t}_V, \bar{t}_H)$ in the case of the implementation of the second variety of the filtration mode with a variable rate. Besides, an example of the practical application of the derived dependences and equations is given, when the initial information is specified in the dimensional form. Excluding it, only relative quantities with wide domains were calculated.

Figure 23. Dependence of $\widetilde{H}_{W1}(\bar{t}_1) - \widetilde{Z}_W$ on $\overline{Q}_{in}$:

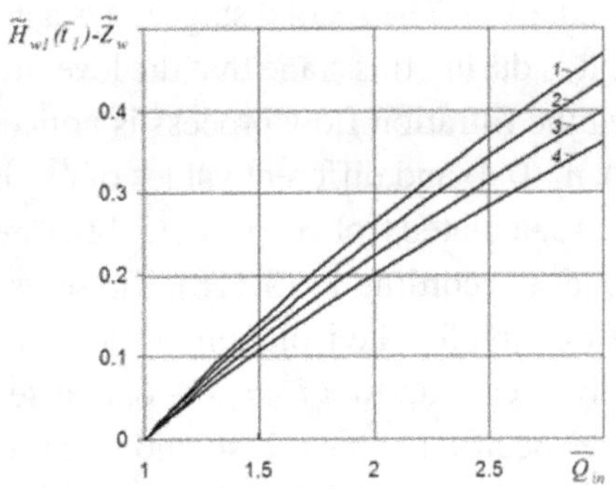

Note. 1 – n_0=0.45, 2 – n_0=0.4, 3 – n_0=0.35, 4 – n_0=0.3

Adhering to the conditional division of the technological process into three stages, first of all, the dynamics of the suspension level in the storage tank and the flow characteristic in the bed during the first and second stages were analyzed. And above all, the possible rise of the level above the initially dry bed by the end of the first stage ($t=t_1$) was estimated when $\overline{Q}_{in}$ continuously increased in maximum three times and n_0 discretely increased in one and a half times. Curves of the dependence $\widetilde{H}_{W2}(\overline{Q}_{in})$ are depicted for the four values of n_0 in Figure 23. From here we can see that the depth of the surface layer immediately after the complete saturation of the bed with suspension being clarified depends significantly on $\overline{Q}_{in}$ and to a lesser extent on n_0. This layer starts to form on the surface of the clean porous medium under the condition that $Q_{in}>\omega k_0$. If the filtering material initially (after the last washing) contained the unwashed deposit with the volumetric concentration S_S^0, then it is necessary to replace k_0 in this condition by the actual initial hydraulic conductivity k_{0S}. It is easy to determine the latter by using, for example, formula (1.10) or the dimensionless analog (5.120). Then the suspension will accumulate above the bed in case $\overline{Q}_{in} > \overline{k}_{0S}$, where $\overline{k}_{0S} = \frac{k_{0S}}{k_0} < 1$.

At lower flow rates $Q_{in}(\overline{Q}_{in} < 1)$, the infiltration into a dry bed occurs in two stages. First, the zone of an incomplete saturation is formed above the wetting front, where the moisture content θ_W is less than the total moisture capacity θ_{Wt}. The first stage ends at the moment when the front merges with the initial liquid level located below in the bed. At the second stage, the suspension continuing to enter the bed in the same volume partly passes through it in transit, partly is spent to replenish moisture reserves in the newly formed unsaturated zone up to its complete saturation. At that, the saturation front will move upward. The second stage and the first stage as a whole will end at the moment when this front reaches the inlet surface of the bed. Naturally, the storage tank will be included in the technological process in such a situation only at the second stage. Formal analysis of the suspension flow in a porous medium with an incomplete saturation is performed in Appendix 5.1. From a technological point of view, the trickling filtration is justified, in essence, when there is an acute need for an oxidizer, for example, to remove the concentrated dissolved organic contaminants, and will be detailed studied in future. An extensive bibliography is dedicated to the development of this applied problem and for example, we can mention the fundamental and modern works (Barbusiński et al., 2024; Gąszczak et al., 2018; Hekmat & Vortmeyer, 1994; Iliuta & Larachi, 2004; Liao et al., 2007; Liu et al., 2023; Oleynik & Kolpakova, 2012). However, when large volumes of a raw water or pre-treated wastewater need to be clarified on a regular basis, it is clearly preferable to use high-loaded filtration facilities.

Thus, the first stage is usually so short, that we can ignore it at all and assume $\overline{H}_{W2} = 0$ even at insignificant exceeding of Q_{in} over the value $k_0\omega$. The second stage plays a more important role in preparing a filter for the productive operation. It is during this stage that the level in the storage tank manages to rise to such a height, due to which then the filtration flow process is noticeably intensified. Specific nature of the level rise at the given R_{f2} (1), n_0 (0.4) and different values of $\overline{Q}_{in}$ is illustrated by Figure 24, where the dependences $\overline{H}_{Wi}(n_0\overline{t})(i = 1,2)$ calculated from (5.115), (5.116) are shown. The limiting values $\overline{H}_m$, reached at $t\rightarrow\infty$, here are 2, 2.5 and 3 according to (5.118). These values were taken as initial values for $\overline{H}_W$ at the third stage. The justification for the isolation of the preparatory stages is that, firstly, the level rise in the storage tank mainly occurs up to $n_0\overline{t} = 4 - 6$, and secondly, the duration of the filter run is always much longer than the specified values. It should be noted that in fact a small amount of suspension is still clarified by the beginning of the third stage and the bed is minimally contaminated. In part, the modeling still allows to account for the primary clarification due to some overestimation of the initial level in the storage tank and filtration rate. Logically, the values of $\overline{H}_m$ depend significantly on $\overline{Q}_{in}$. However, it is only feasible to rationally select a storage tank height by taking into account the maximum level rise above the clogged bed. The corresponding analysis will conclude this Chapter.

The calculation of the filtration rate at the first stages is also indicative (Figure 25). Thus, at the first stage ($t\leq t_1$) according to (5.108) the infiltration rate into the dewatered (unsaturated) bed is constant (short horizontal pieces of discontinuous curves 1-3). At the moment when the saturation front is lowered to the final mark (H_{out}), the nature of filtration changes dramatically - flow at a unit head gradient is replaced by pressure-controlled filtration flow. As a consequence, the filtration rate declines in jump-like manner to the value determined by the head difference at the boundaries of the bed by the end of the first stage. Further the level in the storage tank rises rapidly due to a low intensity of the filtration flow process and as a result, the calculated rate at first rapidly and gradually slowing down increases, asymptotically tending to the value $\overline{Q}_{in}$. The closeness of the calculated curves to each other is explained by the choice of the scale $\frac{Q_{in}}{\omega}$, which levels out the large difference in the values of the dimensional rates $V_{1,2}$.

Figure 24. Dependence of $\overline{H}_{Wi}(n_0\bar{t}) - \overline{Z}_W$ on $n_0\bar{t}$

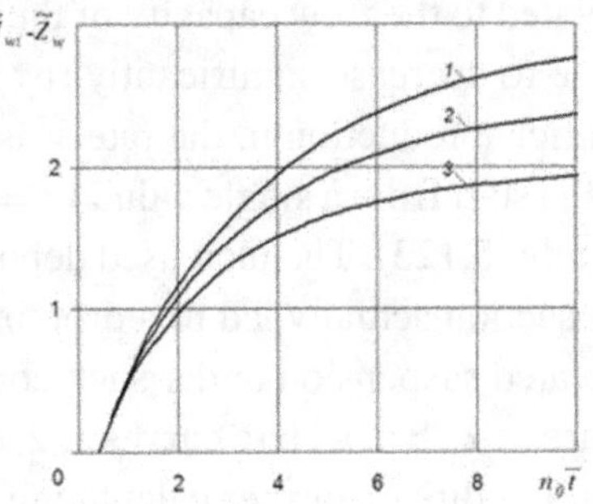

Note. 1 – $\overline{Q}_{in}$ = 2, 2 – $\overline{Q}_{in}$ = 1.5, 3 – $\overline{Q}_{in}$ = 1

Figure 25. Change in the relative filtration rate over time

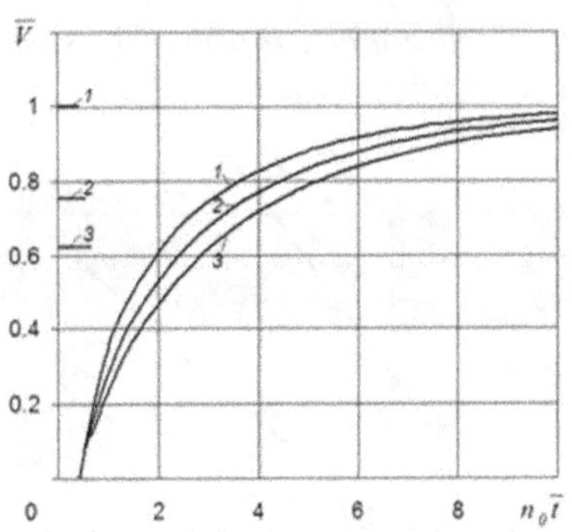

Note. 1 – $\overline{Q}_{in}$ = 1, 2 – $\overline{Q}_{in}$ = 1.5, 3 – $\overline{Q}_{in}$ = 2

Further, the values of a number of dimensionless model parameters were fixed, namely, $\overline{\beta} = 0.01$, q=1, $\overline{\gamma}_a = 0.0015$, $\overline{R}_{f2} = 1$. Taking into account the crucial importance of the absorption process (the sticking rate of the suspended matter) for the clarification operation of rapid filters, the coefficient α was varied in two ways. First, the reduced coefficient α_V was varied discretely or continuously within significant limits, and secondly, two values of the exponent l, reflecting the peculiarities of the absorption of small (reagent-free filtration) and large (agregated, reagent) particles of dispersed impurity, were taken. In the second series of examples, the increase in the suspended solids content in the filtrate, the decrease in the filter performance over time, and the increase in the suspension volume on the bed surface were predicted. Absorption capacity of the filtering material varied from weak $(\overline{\alpha}_V = 3)$ to severe $(\overline{\alpha}_V = 9)$. The curves of the dependence $\overline{C}_e(\overline{t})$, calculated by formula (5.87) and depicted in Figure 26, show the nature of the deterioration of the filtrate quality. The perceptible breakthrough of the suspension particles through a packed bed in the initial phase of the third stage (in the calculations of the filtration during the final stage and at t=0) takes place only at $\overline{\alpha}_V = 3{,}5$ $(\overline{C}_e(0) = 0.0498, 0.0068)$. Less deposit is formed in such cases and it is more evenly distributed by the height. This ensures a slower decline in the filtration rate (curves 1, 2 in Figure 27). It should be emphasized that in this Section, in contrast to Section 5.4, an approximate analytical solution is obtained and discussed on the test examples. As follows from the analysis of its accuracy performed in Section 5.1 (Figure 2), the calculations of the required characteristics, which are closely related to the flow capacity of the porous layer (rate and level modes), are trustworthy if it does not manage to decrease significantly (by a factor of 3, 4 or more) during the calculation period. Just such a significant reduction in the rate $\overline{V}$ is observed at $\overline{\alpha}_V = 7$ and 9. All the calculated curves according to (5.117) start from a single value $\overline{V} = 1$. The graphs of the dependence $\overline{V}$ $(\overline{t})$ in this figure were found by formula (5.123). The increased deposition of the suspended solids in the pores of the bed near its inlet cross-section actually led to a drop in $\overline{V}$ (curve 3). It is logical that there was a sharp rise in the level of untreated suspension under such conditions (curve 1 in Figure 28). The rate of the rise decreased several times at $\overline{\alpha}_V = 5$. Here curves 1, 2, corresponding to the values $\overline{\alpha}_V = 7$ and 9, are broken off due to the unreliability of further calculating $\overline{H}_W$. It is obvious that the intensive formation of a deposit can really cause rapid overfilling of the storage tank. And it is quite possible that the clarification resource of the filter will not be exhausted completely in this case. Further modeling of the technological process can be carried out only after an appropriate correction of the mathematical model, which will be the subject of a theoretical analysis in Section 5.4.

Figure 26. Increase in the relative volumetric concentration of the suspended matter in the filtrate over time

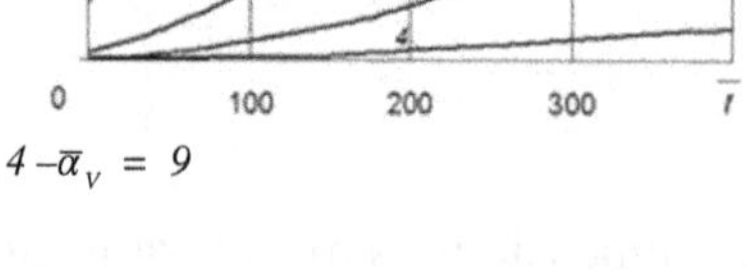

Note. 1 – $\overline{\alpha}_V = 3$, 2 – $\overline{\alpha}_V = 5$, 3 $\overline{\alpha}_V = 7$, 4 – $\overline{\alpha}_V = 9$

Figure 27. Decline of the relative filtration rate over time

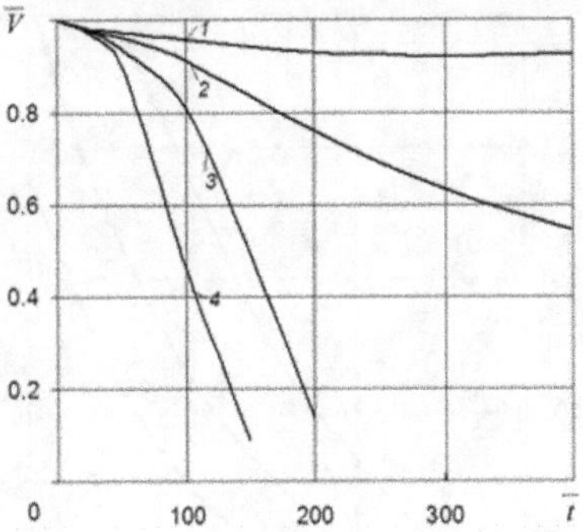

Note. $1 - \bar{\alpha}_V = 3$, $2 - \bar{\alpha}_V = 5$, $3 - \bar{\alpha}_V = 7$, $4 - \bar{\alpha}_V = 9$

In general, if to hypothetically consider the filter operation unlimited in time at $Q_{in}=const$, it is reasonable to distinguish four characteristic stages according to the results of analysis of the filtration rate. At the short first stage, the level mode in the storage tank and the rate mode in the bed are intensively formed with minimal contamination of the latter. At the second stage the rate of the initial suspension level rise sharply decreases, the filtration process stabilizes and a small amount of the suspended matter is carried away. At the third stage, due to the progressive clogging of the filter media and the suspension accumulation above it, the filtration rate declines slowly at first, and then faster and faster. The suspended solids content in the filtrate increases at an accelerated rate. At the final stage, the rate and especially the depth of the surface liquid layer increase, asymptotically approaching the limiting values, and the clarification of the suspension actually stops. The calculation of these values is given in Appendix 5.2. Thus, the behavior of the function $V(t)$ is determined mainly by the position of the suspension level in the storage tank and the degree of the filter media contamination. The first factor prevails at the first and fourth stages, the second factor – at the third stage. At the second stage the effects of these factors can even be compensated for each other at first. In reality, however, only the first two, and mainly the first step of the third stage, are of interest for practice, since the quality of filtration thereafter is too low. Therefore, it is them that will be the subject of all the subsequent tests.

Figure 28. Rise of the reduced suspension level in the storage tank over time

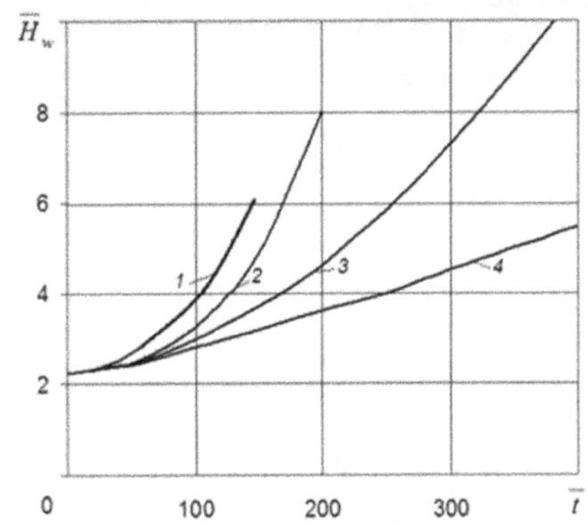

Note. $1 - \bar{\alpha}_V = 9$, $2 - \bar{\alpha}_V = 7$, $3 - \bar{\alpha}_V = 5$, $4 - \bar{\alpha}_V = 3$

Figure 29. Dependence $\bar{t}_p(\bar{\alpha}_V)$

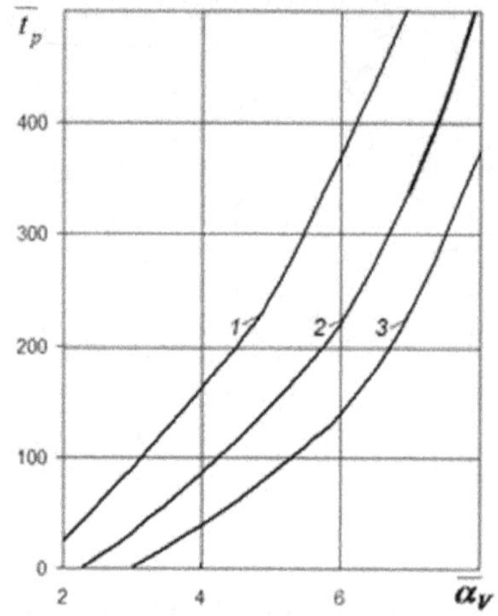

Note. 1 – $\bar{C}_*$ = 0.2, 2 – $\bar{C}_*$ = 0.1, 3 – $\bar{C}_*$ = 0.05

The calculation of the times is considerably more labor-intensive, since we have to solve two systems of equations, which relate implicitly $\bar{t}_p$ and $\bar{t}_V$ to $\bar{V}_a$ due to the given criterion parameters $\bar{C}_*$ and $\bar{V}_*$. Here, characteristic values were assumed in the third series of examples: 0.05, 0.1, and 0.2 for $\bar{C}_*$; 0.5, 0.7, and 0.9 for $\bar{V}_*$; 3, 4, 5, and 6 for $\bar{H}_*$.The key mass exchange coefficient $(\bar{\alpha}_V)$, however, was varied continuously by a factor of four. Plots of the dependence of the time $\bar{t}_p$ on $\bar{\alpha}_V$, calculated on the basis of the system of equations (5.121), (5.123), are shown in Figure 29. Conversion of $\bar{t}_p$ to 0 means that the required clarification of even the first suspension portion is not provided at the corresponding value of $\bar{\alpha}_V$, namely, the minimum values of $\bar{\alpha}_V$ here were 1.62 at $\bar{C}_*$ = 0.2, 2.3 at $\bar{C}_*$ = 0.1 and 3 at $\bar{C}_*$ = 0.05.

The situation is different when analyzing the hydraulic resistance and filter performance. The series of curves calculated on the basis of system (5.121), (5.123), which describe the relationship between $\bar{t}_V$ and $\bar{\alpha}_V$, is shown in Figure 30. Significant is the fact of a sharp decrease in the time $\bar{t}_V$ at the intensification of the absorption process. This technological parameter is much more sensitive to changes in $\bar{\alpha}_V$, than the time $\bar{t}_p$. All the calculated curves have vertical asymptotes. The values of $\bar{\alpha}_V$ corresponding to them are limiting in the sense that there is a strictly explicit value $\bar{V}_*$, which will be reached only asymptotically (at $t \to \infty$) in this case. This means that the filter performance ceases to be a limiting factor if the values of $\bar{\alpha}_V$ are smaller than the limiting value.

Figure 30. Dependence $\bar{t}_V(\bar{\alpha}_V)$

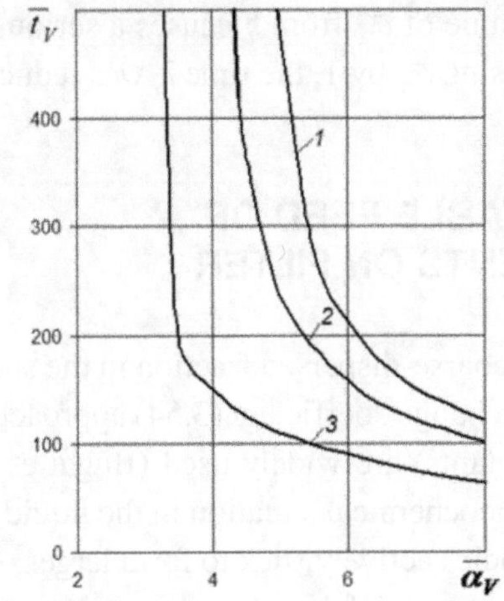

Note. 1 – $\bar{V}_* = 0.5$, 2 – $\bar{V}_* = 0.7$, 3 – $\bar{V}_* = 0.9$

Figure 31. Dependence $\bar{t}_H(\bar{\alpha}_V)$

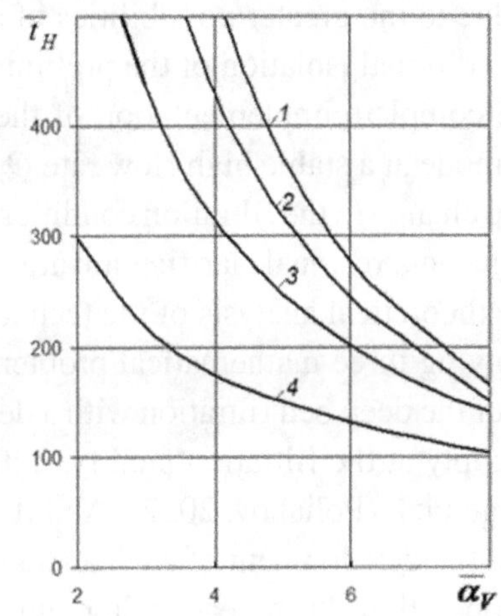

Note. 1 – $\bar{H}_* = 6$, 2 – $\bar{H}_* = 5$, 3 – $\bar{H}_* = 4$, 4 – $\bar{H}_* = 3$

If the level mode in a storage tank becomes important at the development of the algorithm of a filter control in case of an unregulated filtration rate and selection of filtering material, then the dependence $\bar{t}_H(\bar{\alpha}_V)$ becomes indicative. Figure 31 shows typical curves describing the specified dependence at different height of the storage tank (marks of its edge). They also have a distinct nonlinearity. With increasing absorption capacity of the material, the values of t_H sharply decrease, which means that the depth of the surface liquid layer can, in principle, limit the operation of the filter under the condition of obligatory turning off the filter with the overfilled storage tank. The following representative system of the criterion parameters is accepted: $\bar{C}_* = 0.1$, $\bar{V}_* = 0.5$, $\bar{H}_* = 5$. Then at $\bar{\alpha}_V \approx 6$ will be $\bar{t}_p = \bar{t}_V > \bar{t}_H$. Thus, under the given conditions it is reasonable to use a filtering material with such a value of $\overline{\alpha}_V$, which will allow to carry out the continuous suspension clarification as long as possible. It is appropriate to note that even a slight deviation of the value of $\overline{\alpha}_V$ from 5 causes a serious reduction in the $\overline{t}_f$ duration. For example, when increasing or decreasing $\overline{\alpha}_V$ by 1, the time $\overline{t}_f$ was reduced by almost one and a half time.

5.4. FILTRATION UNDER STABLE FEED OF SUSPENSION WITH REAGENTS ON FILTER

At the predominant content of a coarse-dispersed fraction in the suspension impurity, the exponent l in the expression for the key mass-exchange coefficient (3.54) approaches 1. To intensify rapid filtration, special reagents (coagulants, flocculants) are widely used (Bugai et al., 2007; Tobiason et al., 1993; Zapolsky & Baran, 1987). The physicochemical situation in the liquid phase of a packed bed is thus destabilized, the small suspended particles actively stick to form larger ones. The aggregated particles are not only better absorbed by a filtering material, but at the same time they show the increased sensitivity to the hydrodynamic factor during the impurity deposition. As a consequence, the dependences of both mass exchange coefficients α, β on the rate V acquire almost linear nature at linear filtration. Therefore, in fact, a special case of general representations for them, namely, l=q=1 (3.28), in relation to the filtration with the use of reagents at an unregulated rate deserves a particular study. In-depth theoretical studies of the operation of rapid filters with a storage tank under such conditions are justified not only because of their wide distribution, but also due to the greater possibilities of analytical methods. In this way, it is possible, in particular, to avoid conditional isolation of the preliminary stages.

It should be noted that for more complete implementation of the clarification resource of a rapid filter functioning in the second rate mode at a stable high flow rate Q_{in}, we must often take into account such technological features as a sharp change in the filtration conditions when continuing the suspension separation after overfilling the storage tank, residual clarification after the cessation of the contaminated water supply. As a consequence, the theoretical analysis of the technological process should be carried out step by step, formulating and solving three mathematical problems in general. Just below we consider three corresponding problems of the deep bed filtration with a declining rate, which fundamentally differ in the nature of suspension supply at the filtration facility, initial contamination of the bed, and the primary liquid stock in the storage tank (Poliakov, 2022). And it is the solutions of these problems in a complex can serve as a basis for the development of such an algorithm of the filter control, which would provide its most productive operation during each filter run, namely, from the moment of the successive turning on until the clarification of the last suspension portion from the storage tank. Thus, these problems describe the continuous operation of the filter as a single technological cycle.

The specificity of the filtration process at its different stages is formally reflected in the balance equation of the suspension being treated, the initial condition for a deposit, the boundary condition for the head at the inlet to the bed (z=0). Other elements of the mathematical problems statement are the same. Thus, the equations of mass transfer (5.2) and mass exchange kinetics (5.3) in the clarification compartment are common for all the problems under consideration. The second equation is specified as follows

$$\frac{\partial S}{\partial t} = \alpha_V V(t)C - \beta_V V(t)S, \tag{5.129}$$

as well as the condition of a constancy of the inlet suspended solids concentration (2.3), which is a traditional condition for the modeling of the suspension clarification on rapid filters. The filtration flow compartments are formed by unified equations of the suspension motion in the filter bed (5.4), it's hydraulic resistance (1.9), and the deposit state (1.29). Finally, the dynamics of the suspension level in the storage tank is described in the hydraulic compartments by the balance equation (5.33). The initial position of the indicated level and mechanical energy losses in the offtake communications characterize the initial condition in a general form

$$t = 0, \quad H_W = H_W^0 \; ; \tag{5.130}$$

the condition at the lower boundary of the bed

$$z = L, \quad h = H_{out} + R_{f2}\omega^2 V^2(t). \tag{5.131}$$

From a technological point of view serious differences in the filtration conditions under consideration, which lead to different calculation expressions at its three stages, are reflected in the missing boundary and initial conditions. Thus, it is assumed in the first problem corresponding to the first filtration stage that the bed was initially clean (5.78), and the head at its inlet is controlled by the increasing height of the surface liquid layer

$$z = 0, \quad h = H_W(t). \tag{5.132}$$

In the second problem, it is assumed that the suspension level in the storage tank is stably maintained at the maximum height (mark of its edge) H_*. In this case, during the time of the level rise from the initial mark H_W^0 to the specified maximum mark the bed has time to noticeably clogged up, so that the initial condition for the concentration S is taken in the general form (2.4). It is obvious that the suspension due to a constancy of H_* will actually enter the filter with the flow rate $Q_{in}(t)$ being declined, which is formally found from (5.33) at $dH_W/dt = 0$.

The third problem describes the clarification of the last suspension portion of the volume ωH_W^0 (or ωH_*), which was in the storage tank after the contaminated water supply from outside has been stopped (Q_{in}=0). Therefore, the basic system of equations and conditions is supplemented by the inhomogeneous initial condition (2.4) and the boundary condition (5.132), which, unlike the first problem, now reflects the peculiarities of declining the level above the bed.

The following dimensionless variables and parameters are used to summarize the results of the theoretical studies:

$$\bar{t} = \frac{k_0 t}{n_0 L}, \; \bar{V} = \frac{V}{k_0}, \; \bar{h} = \frac{h - H_{out}}{L}, \; \bar{H}_W = \frac{H_W - H_{out}}{L}, \; \bar{\alpha}_V = L\alpha_V,$$

$$\bar{\beta}_V = n_0 L\beta_V, \; \bar{\gamma} = \gamma C_0, \; \bar{R}_{f2} = \frac{\omega^2 k_0^2 R_{f2}}{L}, \; \bar{Q}_{in} = \frac{Q_{in}}{\omega k_0}$$

Then the basic system of the dimensionless equations after the introduction of the new independent variable $\bar{\tau}$ (Section 5.1) takes the final form

$$\frac{\partial \overline{C}}{\partial \overline{z}} + \frac{\partial \overline{S}}{\partial \overline{\tau}} = 0 \; ; \tag{5.133}$$

$$\frac{\partial \overline{S}}{\partial \overline{\tau}} = \overline{\alpha}_V \overline{C} - \overline{\beta}_V \overline{S} \; ; \tag{5.134}$$

$$\overline{V}(\overline{t}) = -\overline{k}(\overline{S}) \frac{\partial \widetilde{h}}{\partial \overline{z}} \; ; \tag{5.135}$$

$$\frac{d\overline{H}_W}{d\overline{t}} = n_0 \left[\overline{Q}_{in} - \overline{V}(\overline{t}) \right] \; ; \tag{5.136}$$

The system of equations (5.133) - (5.136) is supplemented by the unified boundary conditions

$$\overline{z} = 0, \quad \overline{C} = 1 \; ; \tag{5.137}$$

$$\overline{z} = 1, \quad \widetilde{h} = \overline{R}_{f2} \overline{V}^2(\overline{t}). \tag{5.138}$$

The following boundary and initial conditions in a dimensionless form are added to conditions (5.137), (5.138) for the above three problems: at $\overline{z} = 0$, for the first and third problems

$$\widetilde{h} = \overline{H}_W(\overline{t}) \; ; \tag{5.139}$$

for the second problem

$$\widetilde{h} = (H_* - H_{out})/L = \overline{H}_* \overline{t}_p \tag{5.140}$$

at $\overline{t} = 0$, for the first problem

$$\overline{S} = 0; \overline{H}_W = \overline{H}^0 \tag{5.141}$$

for the second and third problems

$$\overline{S} = \overline{S}^0(\overline{z}) \tag{5.142}$$

and also, for the third problem in addition to (5.142) the condition

$$\overline{H}_W = \overline{H}_* \tag{5.143}$$

The time countdown starts at each step from 0.

The solution of the clarification compartment of the first problem is expressed by the known dependencies

$$\overline{S}(\overline{z}, \overline{\tau}) = \overline{\alpha}_V e^{-\overline{\alpha}_V \overline{z}} \int_0^{\overline{\tau}} e^{-\overline{\beta}_V \varsigma} I_0 \left(2 \sqrt{\overline{\alpha}_V \overline{\beta}_V \overline{z} \varsigma} \right) d\varsigma, \tag{5.144}$$

$$\overline{C}(\overline{z},\overline{\tau}) = e^{-\overline{\alpha}_V \overline{z}}\left[e^{-\overline{\beta}_V \overline{\tau}} I_0\left(2\sqrt{\overline{\alpha}_V \overline{\beta}_V \overline{z}\overline{\tau}}\right) + \overline{\beta}_V \int_0^{\overline{\tau}} e^{-\overline{\beta}_V \varsigma} I_0\left(2\sqrt{\overline{\alpha}_V \overline{\beta}_V \overline{z}\varsigma}\right) d\varsigma\right]. \quad (5.145)$$

As a result of solving the filtration flow compartment of the first problem, the quadratic equation was derived with respect to the rate $\overline{V}$, which formally depends on $\overline{t}$ and $\overline{\tau}$

$$\overline{R}_{f2}\overline{V}^2 + \Psi(\overline{\tau})\overline{V} = \overline{H}_W(\overline{t},\overline{\tau}) \quad (5.146)$$

where $\Psi(\overline{\tau}) = \int_0^1 \frac{d\overline{z}}{f_k(\overline{S}(\overline{z},\overline{\tau}))}$. It follows that

$$\overline{V}(\overline{t},\overline{\tau}) = \frac{1}{2\overline{R}_{f2}}\left\{\sqrt{\Psi^2(\overline{\tau}) + 4\overline{R}_{f2}\overline{H}_W(\overline{t},\overline{\tau})} - \Psi(\overline{\tau})\right\}. \quad (5.147)$$

The integration of equation (5.136) under the second condition (5.141) gives

$$\overline{H}_W(\overline{t},\overline{\tau}) = \overline{H}^0_W + n_0(\overline{Q}_{in}\overline{t} - \overline{\tau}). \quad (5.148)$$

The Cauchy problem with respect to the function $\overline{\tau}(\overline{t})$ follows from (5.147) taking into account (5.148)

$$\frac{d\overline{\tau}}{d\overline{t}} = \frac{1}{2\overline{R}_{f2}}\left\{\sqrt{\Psi^2(\overline{\tau}) + 4\overline{R}_{f2}\left[\overline{H}^0_W + n_0(\overline{Q}_{in}\overline{t} - \overline{\tau})\right]} - \Psi(\overline{\tau})\right\}; \quad (5.149)$$

$$\overline{t} = 0, \quad \overline{\tau} = 0. \quad (5.150)$$

It is easy to solve problem (5.149), (5.150) numerically using the available software for mathematical analysis (Mathcad, Matlab, etc.). At the same time, the problem can be solved accurately in the special case of the hydraulically perfect offtake communications $\left(\overline{R}_{f2} = 0\right)$. Then equation (5.149) is simplified in the limit at $\overline{R}_{f2} \to 0$ as follows

$$\frac{d\overline{\tau}}{d\overline{t}} = \frac{\overline{H}^0_W + n_0\overline{Q}_{in}\overline{t} - n_0\overline{\tau}}{\Psi(\overline{\tau})}. \quad (5.151)$$

If we denote $\frac{d\overline{\tau}}{d\overline{t}}$ as y and solve (5.151) with respect to $\overline{t}$, then

$$\overline{t}(y,\overline{\tau}) = \frac{1}{n_0\overline{Q}_{in}}\left[\Psi(\overline{\tau})y - \overline{H}^0_W + n_0\overline{\tau}\right]. \quad (5.152)$$

The differential equation with respect to y that is equivalent to (5.151) has the form (Kamke, 1976)

$$\frac{dy}{d\overline{\tau}} = \frac{\overline{\tau}}{n_0\overline{Q}_{in} - \overline{\tau}\Psi(\overline{\tau})}\left(\frac{d\Psi}{d\overline{\tau}}y + n_0\right)$$

and is solved under the condition

$$\overline{\tau} = 0, \quad y = \overline{H}^0_W$$

After a series of transformations we find

$$y(\overline{\tau}) = e^{-\phi(\overline{\tau})}\left[\overline{H}_W^0 + n_0\int_0^{\overline{\tau}} \frac{\varsigma e^{\phi(\varsigma)} d\varsigma}{n_0\overline{Q}_{in} - \varsigma\Psi(\varsigma)}\right], \quad (5.153)$$

where $\phi(\overline{\tau}) = -2\int_0^{\overline{\tau}} \frac{\varsigma}{n_0\overline{Q}_{in} - \varsigma\Psi(\varsigma)}\frac{d\Psi}{d\varsigma}d\varsigma.$ Finally, the required relationship between $\overline{t}$ and $\overline{\tau}$ is obtained after returning from $\overline{y}$ to $\overline{t}$ in the following form

$$\overline{t}(\overline{\tau}) = \int_0^{\overline{\tau}} e^{-\phi(\eta)}\left[\overline{H}_W^0 + n_0\int_0^{\eta} \frac{\xi e^{\phi(\xi)} d\xi}{n_0\overline{Q}_{in} - \xi\psi(\xi)}\right]d\eta. \quad (5.154)$$

Once an unique relationship between $\overline{\tau}$ and $\overline{t}$ is established, changes over time are easily calculated: in $\overline{S}$ by (5.144), in $\overline{C}$ by (5.145), in $\overline{V}$ by (5.147), and in $\overline{H}_W$ by (5.148).

When the suspension level in the storage tank reaches its edge ($\overline{z} = \overline{H}_*$), the suspension filtration is usually stopped and the filter washing begins. However, it often does not exhaust its clarification resource by this moment in time and can still operate productively for some time. A solution of the second problem allows to theoretically substantiate the prolongation of the filter run. In such a situation, firstly, it is necessary to take into account the degree of the bed contamination by the time point of filling the storage tank (in the following calculations t=0) according to condition (5.142), secondly, equation (5.136) degenerates and now only reflects the fact of a synchronous change in Q_{in} and V over time.

Thus, the following expressions for both concentrations are found from the solution of the corresponding clarification compartment including condition (5.142),

$$\begin{aligned}\overline{S}(\overline{z},\overline{\tau}) &= \overline{S}^0(\overline{z})e^{-\overline{\beta}_V\overline{\tau}} + \overline{\alpha}_V e^{-\overline{\alpha}_V\overline{z}}\int_0^{\overline{\tau}} e^{-\overline{\beta}_V\varsigma} I_0\left(2\sqrt{\overline{\alpha}_V\overline{\beta}_V\overline{z}\varsigma}\right)d\varsigma + \\ &+ \overline{\alpha}_V\overline{\beta}_V e^{-\overline{\beta}_V\overline{\tau}}\int_0^{\overline{z}} \overline{S}^0(\eta)e^{-\overline{\alpha}_V(\overline{z}-\eta)}\int_0^{\overline{\tau}} I_0\left(2\sqrt{\overline{\alpha}_V\overline{\beta}_V(\overline{z}-\eta)\varsigma}\right)d\varsigma d\eta,\end{aligned} \quad (5.155)$$

$$\begin{aligned}\overline{C}(\overline{z},\overline{\tau}) &= e^{-\overline{\alpha}_V\overline{z}}\left[e^{-\overline{\beta}_V\overline{\tau}} I_0\left(2\sqrt{\overline{\alpha}_V\overline{\beta}_V\overline{z}\overline{\tau}}\right) + \overline{\beta}_V\int_0^{\overline{\tau}} e^{-\overline{\beta}_V\varsigma} I_0\left(2\sqrt{\overline{\alpha}_V\overline{\beta}_V\overline{z}\varsigma}\right)d\varsigma\right] + \\ &+ \overline{\beta}_V e^{-\overline{\beta}_V\overline{\tau}}\int_0^{\overline{z}} \overline{S}^0(\eta)e^{-\overline{\alpha}_V(\overline{z}-\eta)} I_0\left(2\sqrt{\overline{\alpha}_V\overline{\beta}_V\overline{\tau}(\overline{z}-\eta)}\right)d\eta.\end{aligned} \quad (5.156)$$

The analytical solution of the filtration flow compartment is represented in the following form

$$\overline{h}(\overline{z},\overline{\tau}) = \overline{H}_* - \overline{V}(\overline{\tau})\int_0^{\overline{z}} \frac{d\eta}{f_k(\overline{S}(\eta,\overline{\tau}))}, \quad (5.157)$$

where

$$\overline{V}(\overline{\tau}) = \frac{1}{2\overline{R}_{f2}}\left\{\sqrt{\Psi^2(\overline{\tau}) + 4\overline{R}_{f2}\overline{H}_*} - \Psi(\overline{\tau})\right\}. \quad (5.158)$$

In this case, time $\overline{t}$ is related to $\overline{\tau}$ in the following way

$$\bar{t}(\bar{\tau}) = 2\bar{R}_{f2}\int_0^{\bar{\tau}} \frac{d\varsigma}{\sqrt{\Psi^2(\bar{\tau}) + 4\bar{R}_{f2}\bar{H}_* } - \Psi(\bar{\tau})}. \tag{5.159}$$

In the special case of a negligibly small resistance of the offtake communications $\left(\bar{R}_{f2} \approx 0\right)$

$$\bar{t}(\bar{\tau}) = \frac{1}{\bar{H}_*}\int_0^{\bar{\tau}}\int_0^1 \frac{d\bar{z}d\varsigma}{f_k(\bar{S}(\bar{z},\varsigma))}. \tag{5.160}$$

The third problem is formally united with the first problem by a variable nature of the level above a bed, hence, of the boundary condition relative to $\bar{h}$ on the bed surface, and with the second problem – by the inhomogeneous initial condition for $\bar{S}$. Common for the third and second problems is the inhomogeneous initial condition for $\bar{S}$. Therefore, the strict expressions for the concentrations $\bar{C}, \bar{S}$ coincide with (5.155), (5.156). In view of Q_{in}=0

$$\bar{H}_W(\bar{\tau}) = \bar{H}_W^0 - n_0\bar{\tau} \tag{5.161}$$

and now it is possible not to resort to numerical solution of problem (5.149), (5.150). Indeed, the relation between $\bar{\tau}$ and $\bar{t}$ takes the form of an inverse integral function

$$\bar{t}(\bar{\tau}) = 2\bar{R}_{f2}\int_0^{\bar{\tau}} \frac{d\varsigma}{\sqrt{\Psi^2(\bar{\tau}) + 4\bar{R}_{f2}(\bar{H}_W^0 - n_0\bar{\tau})} - \Psi(\bar{\tau})}. \tag{5.162}$$

The desired characteristics $\left(\bar{S}, \bar{C}, \bar{V}, \bar{H}_W\right)$ are also determined according to formulas (5.144), (5.145), (5.147), (5.161). In this case, the concentration $\bar{S}$ in the inlet cross-section of the bed will be

$$\bar{S}_0(\bar{\tau}) = \bar{S}(0,\bar{\tau}) = \bar{S}^0(0)e^{-\bar{\beta}_V\bar{\tau}} + \frac{\bar{\alpha}_V}{\bar{\beta}_V}(1 - e^{-\bar{\beta}_V\bar{\tau}}). \tag{5.163}$$

The above presented calculation dependences and equations are mainly strict, and, therefore, provide high reliability of the results of the suspension clarification modeling on rapid filters with a variable rate, provided that the initial mathematical model and its information support are adequate to the actual conditions of depth filtration. However, the practical application of expressions (5.144), (5.145), (5.147), (5.148) for the conditions of the first filtration problem is difficult due to the inefficiency of analytical methods in establishing here the relationship between $\bar{t}$ and τ. Therefore, we have to resort, except for the special case $\bar{R}_{f2} = 0$, to the numerical solution of an ordinary nonlinear equation of the first order with integral coefficients (5.149) under the homogeneous condition (5.150). It is easy to obtain it with any accuracy and discreteness by involving the above-mentioned software. The set of pairs of the $\bar{t}$, $\bar{\tau}$ values found in this way allows us to relate the characteristics of the clarification and flow filtration calculated as functions of $\bar{\tau}$ with real time.

In addition, an approximate solution of the first problem is obtained below using dynamic averaging of the filtration rate, which can compete with a similar exact solution in engineering calculations, but has a wider area of application. Thus, the average quantity $\bar{V}_a$ of the relative rate $\bar{V}$ is introduced for the calculation period [0,T] according to (5.11). In this case, the duration of the period T can be any and, therefore, the parameter V_a is actually functional, but is considered constant in the operations of the differentiation and integration. Thus, the following equation and dependencies are derived

$$\begin{aligned} F(\bar{t},\bar{V}_a) &= 2\bar{R}_{f2}\bar{V}_a\bar{t}- \\ &-\int_0^{\bar{t}}\left\{\sqrt{\Psi^2(\xi;\bar{V}_a)+4\bar{R}_{f2}\left[\bar{H}_W^0+(\bar{Q}_{in}-\bar{V}_a)\,n_0\bar{t}\right]}-\Psi(\xi;\bar{V}_a)\right\}d\xi = 0, \end{aligned} \tag{5.164}$$

$$\bar{S}(\bar{z},\bar{t};\bar{V}_a) = \bar{\alpha}_V\bar{V}_a e^{-\bar{\alpha}_V\bar{z}}\int_0^{\bar{t}} e^{-\bar{\beta}_V\bar{V}_a\varsigma}I_0\left(2\sqrt{\bar{\alpha}_V\bar{\beta}_V\bar{V}_a\bar{z}\varsigma}\right)d\varsigma, \tag{5.165}$$

$$\bar{V}(\bar{t},\bar{V}_a) = \frac{1}{2\bar{R}_{f2}}\left\{\sqrt{\Psi^2(\bar{t};\bar{V}_a)+4\bar{R}_{f2}\left[\bar{H}_W^0+(\bar{Q}_{in}-\bar{V}_a)\,n_0\bar{t}\right]}-\Psi(\bar{t};\bar{V}_a)\right\}, \tag{5.166}$$

$$\bar{H}_W(\bar{t};\bar{V}_a) = \bar{H}_W^0+(\bar{Q}_{in}-\bar{V}_a)n_0\bar{t}, \tag{5.167}$$

$$\begin{aligned} \bar{C}_e(\bar{t},\bar{V}_a) &= \\ &= e^{-\bar{\alpha}_V}\left[e^{-\bar{\beta}_V\bar{V}_a\bar{t}}I_0\left(2\sqrt{\bar{\alpha}_V\bar{\beta}_V\bar{V}_a\bar{t}}\right)+\bar{\beta}_V\bar{V}_a\int_0^{\bar{t}} e^{-\bar{\beta}_V\bar{V}_a\xi}I_0\left(2\sqrt{\bar{\alpha}_V\bar{\beta}_V\bar{V}_a\xi}\right)d\xi\right]. \end{aligned} \tag{5.168}$$

It should be noted here that when modeling the rise of the suspension level in the storage tank, as shown in Sections 5.2, 5.3, it is reasonable to distinguish and study in detail two stages. During the short first stage, the level rises rapidly to a point at which the rate V is nearly equalized with $\frac{Q_{in}}{\omega}$. It is justified to consider the corresponding height as the initial value $\bar{H}_W^0$ in modeling of the subsequent slowed clogging of the filter media and an increase in the height of the surface liquid layer (the second stage).

The filtration rate is significantly affected by two differently directed factors. Their significance differs significantly both in degree and in time. Therefore, it is of interest to study in detail and separately the role of the suspension level in the storage tank and the clogging of a bed in the filtration flow process and mass transfer. It is relatively easy to conduct such a study with respect to the first factor. Let pure water is continuously supplied to the filter at the flow rate Q_{in}. Taking into account the resistance of the offtake communications, the filtration flow rate will be as follows

$$V(t) = \frac{L}{2k_0\omega^2R_{f2}}\left\{\sqrt{1+4R_{f2}\frac{\omega^2k_0^2}{L^2}\left[H_W(t)-H_{out}\right]}-1\right\}$$

or in a dimensionless form

$$\bar{V}(\bar{t}) = \frac{1}{2\bar{R}_{f2}}\left[\sqrt{1+4\bar{R}_{f2}\bar{H}_W(\bar{t})}-1\right]. \tag{5.169}$$

Then the equation with respect to the desired level $\bar{H}_W$ takes the following form

$$\frac{d\widetilde{H}_W}{d\bar{t}} = \frac{n_0}{2\overline{R}_{f2}}\left[2\overline{R}_{f2}\overline{Q}_{in} + 1 - \sqrt{1 + 4\overline{R}_{f2}\widetilde{H}_W(\bar{t})}\right]. \tag{5.170}$$

Solution of equation (5.170) at the initial condition

$$\bar{t} = 0, \quad \widetilde{H} = \widetilde{H}_W^0 \tag{5.171}$$

is expressed by the inverse function

$$\begin{aligned}\bar{t} &= \frac{2\overline{R}_{f2}}{n_0}\int_{\widetilde{H}_W^0}^{\widetilde{H}_W} \frac{d\varsigma}{2\overline{R}_{f2}\overline{Q}_{in} + 1 - \sqrt{1 + 4\overline{R}_{f2}\varsigma}} = \frac{1}{2n_0}\Big\{2\sqrt{1 + 4\overline{R}_{f2}\widetilde{H}_W^0} - \\ &-2\sqrt{1 + 4\overline{R}_{f2}\widetilde{H}_W} + \left(2\overline{R}_{f2}\overline{Q}_{in} + 1\right)\left[\ln\frac{\left(\sqrt{1 + 4\overline{R}_{f2}\widetilde{H}_W} + 2\overline{R}_{f2}\overline{Q}_{in} + 1\right)}{\left(\sqrt{1 + 4\overline{R}_{f2}\widetilde{H}_W^0} - 2\overline{R}_{f2}\overline{Q}_{in} - 1\right)} + \right. \\ &\left. + \ln\frac{\left(\sqrt{1 + 4\overline{R}_{f2}\widetilde{H}_W} - 2\overline{R}_{f2}\overline{Q}_{in} - 1\right)\left(\widetilde{H}_W^0 - \overline{R}_{f2}\overline{Q}_{in}^2 - \overline{Q}_{in}\right)}{\left(\sqrt{1 + 4\overline{R}_{f2}\widetilde{H}_W^0} + 2\overline{R}_{f2}\overline{Q}_{in} + 1\right)\left(\widetilde{H}_W - \overline{R}_{f2}\overline{Q}_{in}^2 - \overline{Q}_{in}\right)}\right]\Big\}.\end{aligned} \tag{5.172}$$

As $\widetilde{H}_W$ increases, $\overline{V}$ tends indefinitely to its limit value $\overline{Q}_{in}$. The steady-state value $\widetilde{H}_m$ corresponding to it, as follows from (5.169), is equal to

$$\widetilde{H}_m = \overline{R}_{f2}\overline{Q}_{in}^2 + \overline{Q}_{in}. \tag{5.173}$$

It is obvious that in the case of the hydraulically perfect communications of a filter

$$\widetilde{H}_m = \overline{Q}_{in}.$$

The great possibilities of the above proposed set of the calculation expressions, which can be interpreted in complex as an engineering method of calculating rapid filters at the depth filtration with reagents and a variable (unregulated) rate, demonstrate the results of a versatile quantitative analysis of jointly occurring clarification, flow filtration and hydrodynamic processes, conducted on a set of examples with characteristic initial data. The subject of such analysis was, first of all, the behavior of the most important relative filtration characteristics, namely, the volumetric concentration of the suspended solids at the filter outlet $\overline{C}_e$, the depth of the suspension layer on the bed surface $\widetilde{H}_W$, the filtration rate $\overline{V}$, the (specific) volume of the clarified suspension $\bar{\tau}$. Secondly, the technological times were determined, which control the filter operation based on the current filtrate contamination (time $\bar{t}_p$ of the bed protective effect), the filtration facility performance (time $\bar{t}_V$ of the filtration rate reduction to the minimum allowable value), the suspension stock in the storage tank (time of its filling $\bar{t}_H$). Finally, the questions are raised about the importance of the bed height, as the main design parameter, for the suspension separation; about its rational choice in order to intensify the filter operation at a given technological algorithm.

Due to the large number of parameters involved in the adopted two mathematical models, only some of them were varied continuously or discretely. The values of other parameters were fixed. In the first series of the test examples, changes in the above-mentioned characteristics were predicted over time for at least 1000 conventional units (approximately corresponding to 80 hours) at the given $\overline{\beta}_V = 0.01, \overline{\gamma}_a = 0.0005,$

n_0=0.47, $\overline{H}^0_W = 0$, $\overline{R}_{f2} = 1$ ($\overline{R}_{f2} = 0$ was assumed only in one example). Herewith, there were changes in $\overline{\alpha}_V$(5, 7,9) and $\overline{Q}_{in}$(0.2, 1). Here the value $\overline{\alpha}_V = 5$ means that the filter media is moderately absorbing; $\overline{\alpha}_V = 7$ – well-absorbing; $\overline{\alpha}_V = 9$ – severe absorbing. The selected and repeatedly used set of two $\overline{Q}_{in}$ values corresponds both to the low intensity of water supply and actual conditions of a low-productive filtration on rapid filters with a slow rise of the level above the bed $(\overline{Q}_{in} = 0.2)$, and to the high intensity $(\overline{Q}_{in} = 1)$ and extreme conditions, when it is possible to treat significant volumes of water. But at the same time, there are already initially significant head losses in the filter, and there is a serious risk of the accelerated clogging of the bed (upper section in downward filtration) and filling of the storage tank.

Of primary importance for controlling the filter operation is operational information on the quality of the suspension clarification, which is relatively easy to obtain theoretically. The content of the suspended matter in the filtrate was determined reliably by formula (5.145) and involving (5.149), (5.150), as well as in parallel by approximate formula (5.168) taking into account (5.164). It should be emphasized that the analytic-numerical solution of the first problem obtained above can provide any accuracy of the calculations of the required characteristics. Therefore, it can be considered as a reference and can be used for the estimation of the errors due to the application of approximate techniques and, in particular, the dynamic averaging of the filtration rate, due to which it was possible to derive the calculation expressions (5.164) - (5.168).

Two series of curves describing the increase in $\overline{C}_e$ and corresponding to $\overline{Q}_{in} = 0.2$ and 1 are shown in Figures 32 and 33. The discrepancy between the reference and approximate curves turned out to be so small that they merged in pairs in these figures, and up to the moment of the full depletion of the bed protective resource, when $\overline{C}_e$ approaches 1. Naturally, the increase in $\overline{Q}_{in}$ contributed to the passage of more suspension through the filter and, as a consequence, its accelerated clogging. Indeed, the impurity concentration at the filter outlet decreased by an order of magnitude at $\overline{\alpha}_V = 5, 7, 9$ and $\overline{Q}_{in} = 1$ after, respectively, times 142.75, 267.5, 402.9 and in the first case decreased at $\overline{Q}_{in} = 0.2$ by more than an order of magnitude after 700 units. It should be noted that the improvement of the filter communications and thus, minimization of their hydraulic resistance (ideally $\overline{R}_{f2} = 0$) practically did not affect the filtrate quality.

Figure 32. Increase in the relative volumetric concentration of the suspended matter in the filtrate over time $(\overline{Q}_{in} = 0.2)$

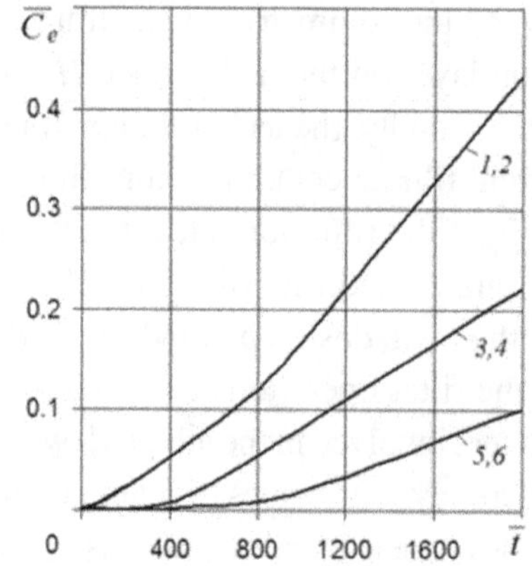

Note. 1, 3, 5 - approximate calculation; 2, 4, 6 - exact calculation; 1, 2 – $\overline{\alpha}_V = 5$; 3, 4 – $\overline{\alpha}_V = 7$; 5, 6 –

Figure 33. Increase in the relative volumetric concentration of the suspended matter in the filtrate over time ($\overline{Q}_{in} = 1$)

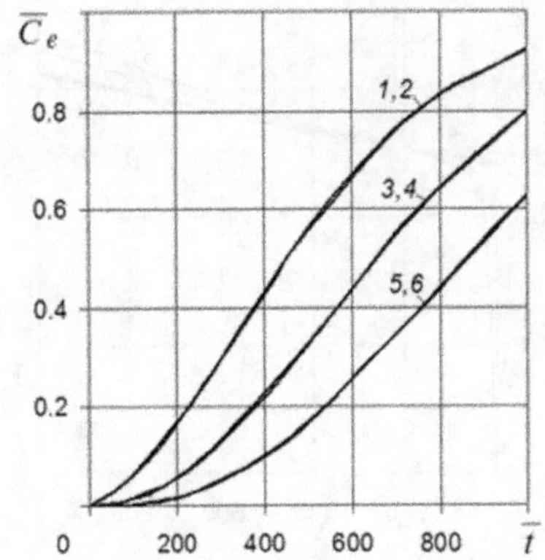

Note. 1, 3, 5 - approximate calculation; 2, 4, 6 - exact calculation; 1, 2 – $\overline{\alpha}_V = 5$; 3, 4 – $\overline{\alpha}_V = 7$; 5, 6 – $\overline{\alpha}_V = 9$

The applicability of the approximate solution for engineering calculations was also analyzed by comparing the theoretical data relative to the development of the surface suspension layer found using the reference analytic-numerical solution based on (5.148) -(5.150) and the approximate solution based on (5.164), (5.167). The obtained results are shown in the form of the dependence $\overline{H}_W(\overline{t})$ plots in Figure 34 for $\overline{Q}_{in} = 1$, and in Figure 35 for $\overline{Q}_{in} = 1.5$. Significant differences between the corresponding values of $\overline{H}_W$ occurred only at the preliminary stage of clarification ($\overline{t} < 10$), i.e. during the maximum first hour of the filter operation, and further did not exceed 4% at $\overline{\alpha}_V = 9$ and 8% at $\overline{\alpha}_V = 5$. Thus, it can be stated that the approximate solution is unsuitable at the specified stage, which is not of practical interest, but in general correctly reflects the consequences of the progressive clogging of the bed and intensive feeding of suspension to the filter. The quantity $\overline{H}_W$ was also determined at $\overline{R}_{f2} = 0$. Corresponding to the two values of $\overline{R}_{f2}$ (0 and 1) at $\overline{\alpha}_V = 9$, curves 1 and 3 differ considerably over the entire calculation period, so that, for example, a decrease in $\overline{R}_{f2}$ can appreciably slow down the rise of the suspension level and thus, prolong the filter operation. The importance of the bed absorption properties for the formation of the surface liquid layer is confirmed by Figure 35.

Figure 34. Rise of the relative suspension level above the clogged bed over time $(\overline{Q}_{in} = 1)$

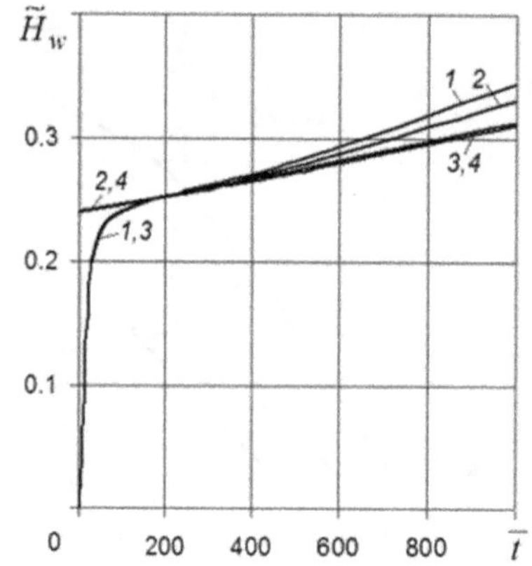

Note. 1, 3 - exact calculation; 2, 4 - approximate; 1, 2 – $\overline{\alpha}_V = 9$; *3, 4 –* $\overline{\alpha}_V = 5$

Figure 35. Rise of the relative suspension level above the clogged bed over time $(\overline{Q}_{in} = 1.5)$

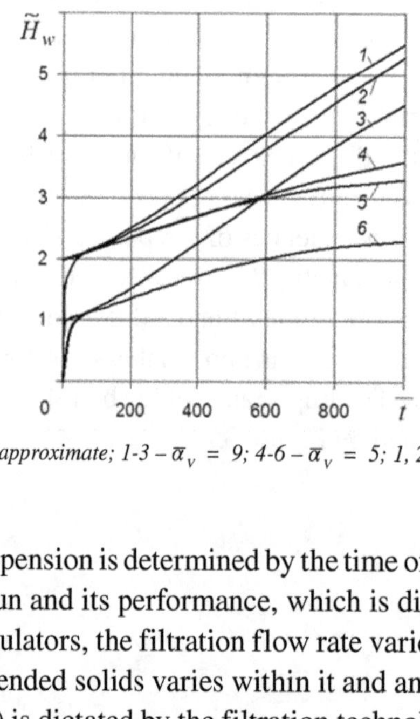

Note. 1, 3, 5, 6 - exact calculation; 2, 4 - approximate; 1-3 – $\overline{\alpha}_V = 9$; *4-6 –* $\overline{\alpha}_V = 5$; *1, 2, 4, 5 –* $\overline{R}_{f2} = 1$; *3, 6 –* $\overline{R}_{f2} = 0$

The volume of the clarified suspension is determined by the time of the filter operation with a constant hydraulic load during one filter run and its performance, which is directly proportional to the filtration rate. In the absence of the rate regulators, the filtration flow rate varies in porous medium over time due to both the deposition of the suspended solids varies within it and an increase in the inlet head. The behavior of the desired function *V*(*t*) is dictated by the filtration technology. Figure 36 presents the curves describing the changes in the relative rate $\overline{V}$ under the influence of these factors with respect to the conditions of the first (curves 4 - 9) and second (curves 1 - 3) problems. In the first case, the calculations were carried out at $\overline{Q}_{in} = 0.5$ and 1, as well as different values of $\overline{\alpha}_V$ using formulas (5.147), (5.148) and the solution of problem (5.149), (5.150). Their results indicate the balance of the factors throughout the calculation period, and the variation of $\overline{\alpha}_V$ had practically no effect on the quantity $\overline{V}$ at $\overline{Q}_{in} = 0.5$. As a consequence, the pieces of curves 7 - 9, corresponding to $\overline{\alpha}_V = 5, 7, 9$ and the time of the filter productive operation, diverge insignificantly. If, however, a constant level is maintained above the clean bed (for example, at the height of the storage bank edge), the dynamic outcomes pattern calculated by (5.155) - (5.158) changes significantly. Immediately the filtration flow process proceeds with a high intensity, which steadily decreases. And it is natural that the rate of a such decrease is controlled by the coefficient characterizing the absorption capacity of the filtering material. The acceptance of the relative initial value of the calculation rate $\overline{V}^0$ at the first stage of filtration is difficult due to the uncertainty of the initial filtration flow pattern. However, such a choice is not of principal importance due to the usually long continuous filtration time and rapid initial stabilization of the level in the storage bank near the mark $\overline{H}_m$ (5.173) followed by its slow rise. Nevertheless, at the calculations of the quantity $\overline{V}(\overline{t})$ it was supposed for concreteness in case of the suspension clarification at the first stage $\overline{H}_W^0 = 0$, and already $\overline{H}_W^0 = \overline{H}_* = 4$ at the second stage. Then according to the exact formula

$$\overline{V}^0 = \frac{\sqrt{\Psi_0^2 + 4\overline{R}_{f2}\overline{H}_W^0} - \Psi_0}{2\overline{R}_{f2}} \tag{5.174}$$

where $\Psi_0 = \int_0^1 \frac{d\overline{z}}{f_1(\overline{S}^0(\overline{z}))}$, the desired values of $\overline{V}^0$ at $\overline{S}^0 = 0$ and any values of $\overline{\alpha}_V$, respectively, were 0 and 1.562. In general, when performing the technological analysis at H_W=*const*, the flow rate characteristic $\overline{V}$ deserves special attention, because its rapid decrease can be a reason for early filter washing. The computational errors introduced by the above-mentioned technique were within a few percent at $\overline{Q}_{in} = 1$.

The volume of the clarified suspension from the beginning of the filter run minus its expenses for washing provides a clear indication of the filter operability. The reduced value of such volume is expressed in the reference solutions by the new variable τ, and in the approximate solution by $\overline{V}_a$. Figure 37 gives the curves of increasing $\overline{\tau}$ over time for three values of $\overline{Q}_{in}$(0.2, 1, 1.5) and two constant values of $\overline{H}_W$ (2, 4). It is obvious from here that the increase in $\overline{\tau}$ is more evenly when the effects of the bed clogging and level rise above it on the filtration flow process compensate each other $(\overline{Q}_{in} = const)$ At the same time, the quantities $\overline{Q}_{in}, \overline{H}_W$ significantly affect the rate and results of the suspension separation.

Figure 36. Change in the relative filtration rate over time

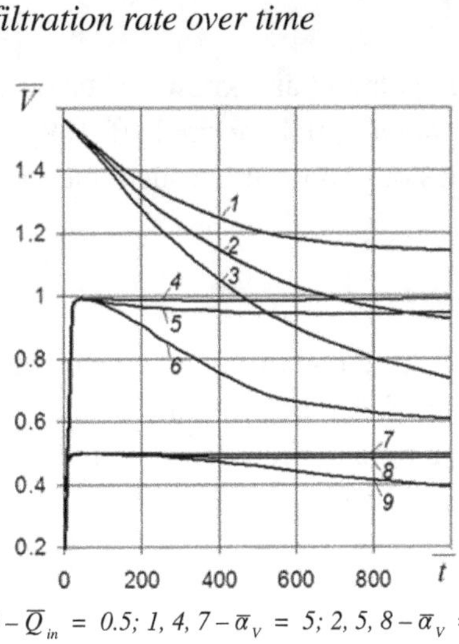

Note. 1-3 – H_W=const; 4-6 – $\overline{Q}_{in} = 1$; 7-9 – $\overline{Q}_{in} = 0.5$; 1, 4, 7 – $\overline{\alpha}_V = 5$; 2, 5, 8 – $\overline{\alpha}_V = 7$; 3, 6, 9 – $\overline{\alpha}_V = 9$

Figure 37. Increase in the relative volume of the clarified suspension over time

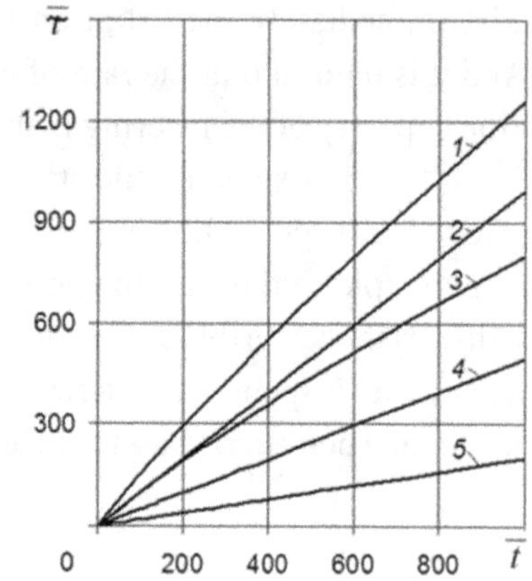

Note. $1 - \bar{H}_* = 4$, $2 - \bar{Q}_{in} = 1$, $3 - \bar{H}_* = 2$, $4 - \bar{Q}_{in} = 0.5$, $5 - \bar{Q}_{in} = 0.2$

When developing an algorithm for controlling the technological process of the suspension filtration here, as in Section 5.3, the decisive value is the consideration of the filter operation limitation due to an unsatisfactory quality and high cost of the clarification, as well as due to the design features. It is also necessary to simultaneously control the outlet suspended solids concentration, filtration flow rate, and the degree of the storage tank filling in order to determine the three technological times (t_p, t_V, t_H). The calculation of these times is complicated due to the necessity of the intermediate calculation of the values τ_p, τ_V, τ_H corresponding to them. Thus, $\bar{\tau}_p$ should be calculated in the general case $\bar{S}^0(z) > 0$ by fitting from the equation resulting from (5.156)

$$I_0\left(2\sqrt{\bar{\alpha}_V\bar{\beta}_V\bar{\tau}_p}\right) + \bar{\beta}_V\int_0^{\bar{\tau}_p} e^{-\bar{\beta}_V(\varsigma-\tau)} I_0\left(2\sqrt{\bar{\alpha}_V\bar{\beta}_V\varsigma}\right)d\varsigma + \\ +\bar{\beta}_V\int_0^{\bar{z}} \bar{S}^0(\eta)e^{-\bar{\alpha}_V(\bar{z}-1-\eta)} I_0\left(2\sqrt{\bar{\alpha}_V\bar{\beta}_V\bar{\tau}_p(\bar{z}-\eta)}\right)d\eta = \bar{C}_* e^{(\bar{\alpha}_V+\bar{\beta}_V\bar{\tau}_p)}. \tag{5.175}$$

Then the required relative time $\bar{t}_p$ is found at a known $\bar{\tau}_p$ or from the solution of problem (5.149), (5.150), or by formula (5.154), if Q_{in}=*const*≥0; by formula (5.159), if H_W=*cosnt*. Thus, criterion (1.35) is satisfied when the calculated values of $\bar{t}$ and $\bar{\tau}$ are in such intervals

$$0 \le \bar{t} \le \bar{t}_p, \qquad 0 \le \bar{\tau} \le \bar{\tau}_p.$$

To estimate the efficiency of the filtration with an unregulated rate the analysis of its performance is often of great importance. At their significant decrease due to the rapid clogging of the bed pore space, the technological process is limited in time in accordance with the relation for the rate V (1.32). In this case, the limits of a change in variables $\bar{t}$ and $\bar{\tau}$ will be

$$\bar{t}_V \ge \bar{t} \ge 0, \qquad \bar{\tau}_V \ge \bar{\tau} \ge 0.$$

To determine the relative time $\bar{t}_V$ for the rate $\bar{V}$ to reach the minimum allowable value $\bar{V}_*$ at the first filtration stage, first of all, $\bar{t}_V$ is expressed through $\bar{\tau}_V$ from (5.147) taking into account (5.148) in the following way

$$\bar{t}_V(\bar{\tau}_V) = \frac{1}{n_0 \bar{Q}_{in}}\left[n_0\bar{\tau}_V - \bar{H}_W^0 + \bar{R}_{f2}\bar{V}_*^2 + \Psi(\bar{\tau}_V)\bar{V}_*\right], \tag{5.176}$$

where $\Psi(\bar{\tau}_V) = \int_0^1 \frac{dz}{f_1(\bar{S}(z,\bar{\tau}_V))}$. The relationship between $\bar{\tau}_V$ and $\bar{t}_V$, found earlier numerically on the base of problem (5.149), (5.150), or analytically in the form (5.154), is used in (5.176). Thus, the representation for $\bar{\tau}_V$ is derived in the form of the implicit function $F(\bar{\tau}_V, \bar{t}(\bar{\tau}_V))$, whose root is the desired value of $\bar{\tau}_V$. The value of $\bar{t}_V$ corresponding to it is immediately calculated by (5.176). In practice, the rate V usually changes insignificantly up to the moment of the storage tank filling, remaining greater than V_*. At the second stage, however, it may decrease so rapidly that the second (economic) criterion (1.37) becomes decisive in the technological process control. Then the technological time $\bar{t}_V$ is easily determined in the following sequence – first the value of $\bar{\tau}_V$ is obtained by fitting from the equation

$$\Psi(\bar{\tau}_V) = \frac{\bar{H}_*}{\bar{V}_*} - \bar{R}_{f2}\bar{V}_*$$

and then $\bar{t}_V$ at the known $\bar{\tau}_V$ already from (5.159).

Since at present the filter, as a rule, starts to be washed immediately after the overfilling of the storage tank, then to check the validity of a such technological operation it is expedient to involve also the criterion allowing to control the level rise above the bed, namely (1.33). Thus, in accordance with (1.34) and the existing practice, it is acceptable to limit the filter operation time to the moment t_H provided that the two main criteria are not violated. The relative value of $\bar{t}_H$ is found by analogy with $\bar{t}_V$, namely, first of all, it is expressed through $\bar{\tau}_H$

$$\bar{t}_H(\bar{\tau}_H) = \frac{1}{n_0 \bar{Q}_{in}}\left(n_0\bar{\tau}_H + \bar{H}_* - \bar{H}_W^0\right), \tag{5.177}$$

and then a set of $\bar{\tau}_H$, $\bar{t}_H$values, which satisfies (5.177), is fitted by using the solution of problem (5.149), (5.150) or (5.154). Obviously, the third criterion turns out to be unclaimed at H_W=*const*. From the practical point of view, the third criterion is more convenient, since no measuring equipment is required to control its meeting, as in the case of the first and second criteria. But in principle the overflow of the storage tank cannot be a sufficient reason to turn off the filter. Indeed, it is not uncommon for a filter in this condition to continue to operate efficiently and with a high quality. On the other hand, it is not rarely the case that the concentration of impurities in the filtrate reaches a limit value when the storage tank is still unfilled. Nevertheless, all three criteria described below are assumed to be equally important and involved in the technological modeling of the suspension filtration. The decision to turn off the filter will be justified only after establishing the filter run duration, which is logically identified with the shortest of technological times (5.128).

The filter run duration can serve as an indicator of the filter efficiency during the filtration process in both rate modes (with constant or declining rate), since the value of $\bar{S}_{0h}$ takes into account both the sanitary and ecological aspects by means of the first criterion, and the economic aspects by means of the second criterion, and performance (without or taking into account washing losses) of the filter.

Therefore, the purpose of the following numerous technological calculations was to establish, first of all, the relative technological times $\bar{t}_p$, $\bar{t}_V$, $\bar{t}_H$, and, finally, the filter run duration $\bar{t}_f$ depending on the coefficients $\bar{\alpha}_V, \bar{\beta}_V$ directly controlling the interfacial mass exchange and the main design parameter L. Such calculations were performed for three examples differing by combinations of two values of $\bar{\gamma}_a$ (0.0005 and 0.001) and $\bar{R}_{f2}$ (1 and 0). Thus, thanks to the assumptions made for example 1 – $\bar{\gamma}_a = 0.0005$, $\bar{R}_{f2} = 1$; for example 2 – $\bar{\gamma}_a = 0.001$, $\bar{R}_{f2} = 1$; for example 3 – $\bar{\gamma}_a = 0.0005, \bar{R}_{f2} = 0$; it was possible to estimate the effect of two factors (deposit composition, resistance of the offtake communications) on the technological process, reducing the calculations to a minimum.

In the second series of examples, the absorption properties of the filter media were varied so that the coefficient $\bar{\alpha}_V$ changed continuously from 4 to 10. The only values of the model $(\bar{\beta}_V = 0.01, n_0 = 0.47, \bar{H}_W^0 = 0)$ and criterion $(\bar{C}_* = 0.1,\ \bar{V}_* = 0.75\,\bar{V}^0,\ \bar{H}_* = 4)$ parameters were involved. The calculations were performed sequentially for the conditions of the first (Q_{in}=*const*) and second (H_W=*const*) problems. Thus, two values (0.5 and 1) corresponding to the intensive and very intensive suspension feeding to the filter were taken for $\bar{Q}_{in}$. Then, due to the unique relationship between the initial rate $\bar{V}^0$ and $\bar{Q}_{in}$ the final values of $\bar{V}_*$ were 0.375 and 0.5 respectively. The calculation period was limited in all the considered cases by the value $\bar{t} = 1000$, which can be conventionally equated to about three days of real time. The series of curves of the dependences $\bar{t}_{p,H,V}(\bar{\alpha}_V)$ for $\bar{Q}_{in} = 0.5$ and 1 are given, respectively, in Figures 38 and 39. The behavior of the same-type curves for different $\bar{Q}_{in}$ is similar in many respects, but at the same time there are some important differences. The field of Figure 38 does not include curves of the dependence $\bar{t}_H(\bar{\alpha}_V)$ from examples 1, 3 at all. Consequently, if the suspension is fed to the filter with a small flow rate, the level rise in the storage tank will be slow, and hence the filter operation should be limited only by the content of a dispersed impurity in the filtrate. It is important to note that the filter performance was stable at $\bar{\gamma}_a = 0.0005$ (Examples 1 and 3) regardless of the bed absorption properties. But with more active binding of water by the deposited matter ($\bar{\gamma}_a = 0.001$, example 2) there was, firstly, a sharp decrease in the values of $\bar{t}_H$ (curves 4 in Figure 38 and low curve 7 in Figure 39), and secondly, the performance limitation due to second criterion is clearly shown in the form of the curves $\bar{t}_V(\bar{\alpha}_V)$ (5 in Figure 38 and 6 in Figure 39). The above limitation may become generally decisive and determine the duration $\bar{t}_f$ at even larger values of $\bar{\alpha}_V$. In turn, the parameter $\bar{R}_{f2}$ has almost no effect on the protective capability of the bed and, therefore, the curves actually merge. However, there is a noticeable increase in the parameter $\bar{t}_p$ with increasing the proportion of bound water in the deposit (example 2) and also at higher values of $\bar{\alpha}_V$. Thus, the filtrate quality in the substantiation of t_f becomes irrelevant and the second and especially the third criteria come to the forefront.

The effect of the reduced coefficient of the detachment rate of the deposited particles on the technological parameters is also manifested in a nonlinear form, as evidenced by Figure 40. Indeed, with a noticeable immediate increase in both stability of the deposit in relation to the hydrodynamic impacts $(\bar{\beta}_V < 0.0075)$, and the water binding ($\bar{\gamma}_a = 0.001$) an almost jump-like increase in time $\bar{t}_p$ is observed with a decrease in $\bar{\beta}_V$, even with a moderate absorption capability ($\bar{\alpha}_V = 7$) of a filtering material (curve 3). This fact is also explained by the intensive clogging of the inlet section of the bed and, consequently, by a serious increase in the hydraulic resistance of the filter facility. As a result, the rate $\bar{V}$ rapidly declines even though there is an accompanying slight reduction in the indicated resistance of its transportation system. In the case of example 3, the curve $\bar{t}_f(\bar{\beta}_V)$ is identified as the curve $\bar{t}_p(\bar{\beta}_V)$, because the latter is completely below the curve $\bar{t}_H(\bar{\beta}_V)$. In the other two examples, the sought $\bar{t}_f(\bar{\beta}_V)$ is represented by a continuous broken line consisting as before of two pieces. The filtration rate did not limit the filter operation in this series of calculations either.

Figure 38. Dependencies $\bar{t}_p(\bar{\alpha}_V)$, $\bar{t}_H(\bar{\alpha}_V)$, $\bar{t}_V(\bar{\alpha}_V)$, $\bar{t}_f(\bar{\alpha}_V)$ ($\bar{Q}_{in}$ = 0.5)

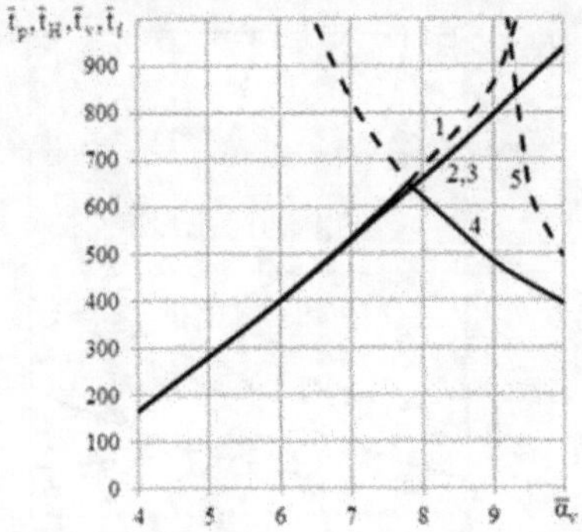

Note. 1-3 – $\bar{t}_p$; 4 – $\bar{t}_H$; 5 – $\bar{t}_V$; 1,4,5 – example 2; 2 – example 1; 3 – example 3; $\bar{t}_f(\bar{\alpha}_V)$ – solid lines

Figure 39. Dependencies $\bar{t}_p(\bar{\alpha}_V)$, $\bar{t}_H(\bar{\alpha}_V)$, $\bar{t}_V(\bar{\alpha}_V)$ ($\bar{Q}_{in}$ = 1)

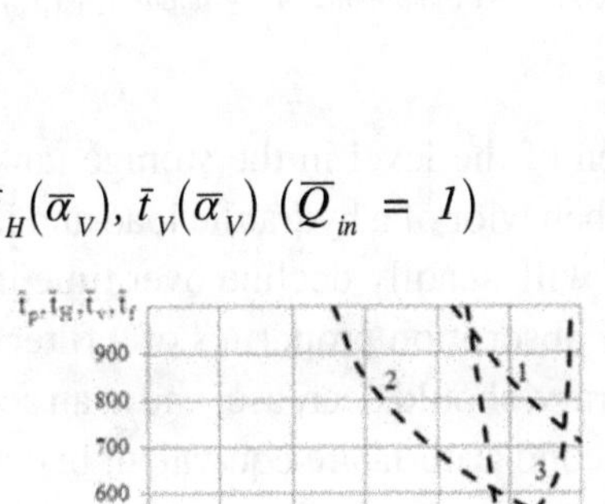

Note. 1,2,7 – $\bar{t}_H$; 3-5 – $\bar{t}_p$; 6 – $\bar{t}_V$; 1,5 – example 3; 2,4 – example 1; 3,6,7 – example 2; $\bar{t}_f(\bar{\alpha}_V)$ – solid lines

Figure 40. Dependencies $\bar{t}_p(\bar{\beta}_V)$, $\bar{t}_H(\bar{\beta}_V)$, $\bar{t}_f(\bar{\beta}_V)$

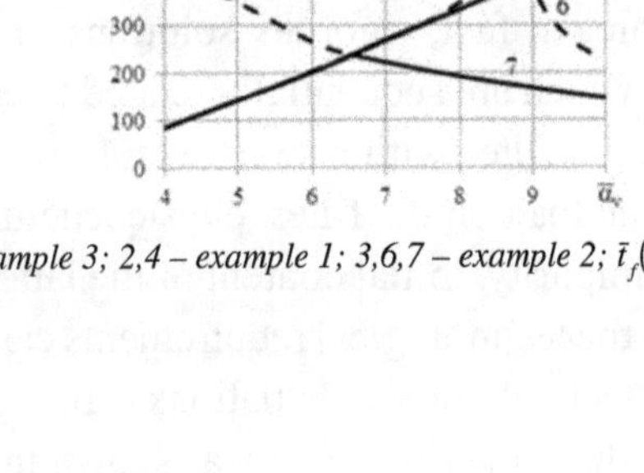

Note. 1,2,6 – $\bar{t}_H$; 3-5 – $\bar{t}_p$; 1,5 – example 3; 2,4 – example 1; 3,6 – example 2; $\bar{t}_f(\bar{\beta}_V)$ – solid lines

Figure 41. Dependencies $\bar{t}_p(\bar{\alpha}_V)$, $\bar{t}_V(\bar{\alpha}_V)$, $\bar{t}_f(\bar{\alpha}_V)$ $(\bar{H}_W = const)$

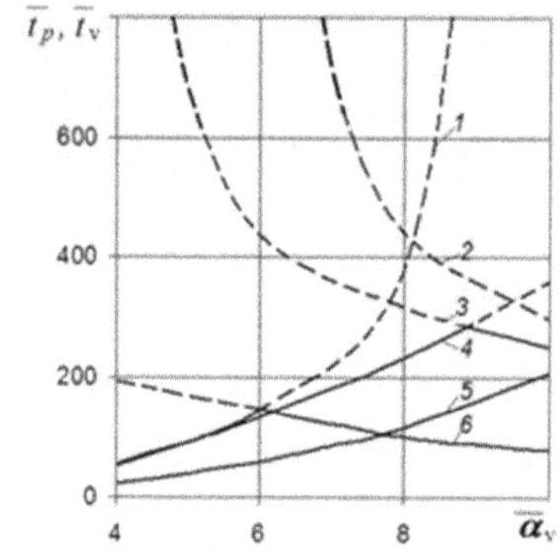

Note. 1, 4, 5 – $\bar{t}_p$; 2, 3, 6 – $\bar{t}_V$; 1, 6 –example 2; 2, 5 – example 3; 3, 4 – example 1; $\bar{t}_f(\bar{\alpha}_V)$ – solid lines

It is obvious that the stabilization of the level in the storage tank (usually at the mark of its edge) causes dramatic changes both in the behavior of a hydraulic load and in the approach to the technological calculations. Indeed, the quantity V will steadily decline over time in the absence of the level rise and inlet head increase. Naturally, as the absorption properties of a filter media improve, the time t_p should increase, and the time $\bar{t}_V$, on the contrary, should decrease, and at an accelerated rate. And since for all the chosen examples the first and the second criteria are equivalent in determining the time of the effective operation of the filter, the curves $\bar{t}_f(\bar{\alpha}_V)$ corresponding to them will again be broken lines, made up of suitable pieces of the curves $\bar{t}_p(\bar{\alpha}_V)$ and $\bar{t}_V(\bar{\alpha}_V)$, which are shown in Figure 41 by solid lines.

The solutions obtained above are, in fact, rigorous solutions of three mathematical problems of suspension separation by filtration, which are sequentially related to each other and form a single cycle. They are also useful when designing a filter with a storage tank, if the conditions of its operation are known in advance, namely: hydraulic load on the filter, physicochemical and filtration flow properties of the filter material and dispersed impurity, initial content of the latter. Then the developed calculation expressions after their adaptation to the technological requirements can contribute to the design analysis. In general, the main purpose of the technological calculations is the applied optimization of the control algorithm of a filtration facility. At the same time, as it was shown in Section 3.2, they can serve as an effective tool for a rational choice of the design parameters. And first of all, on the basis of the method of the technological modeling it is possible to establish such a height of a packed bed at the direct-flow filtration, at which the maximum duration of a filter run is achieved. Following the approach to the substantiation of the specified height described in Section 3.2, two characteristic situations with a filtering material are considered. From an economic point of view, they are limiting in the sense that the cost of such material is assumed to be either extremely low (production waste) or high (porous material with special properties). Then, analyzing the significance of the specified design parameter, it is enough to substantiate its value, that would allow to maximize the time of the continuous operation of a filter under given filtration conditions, and hence the volume of the clarified suspension.

Thus, the height of a bed actually does not affect the capital costs, when using a cheap filtering material, which besides is available in practically unlimited amount. Then it is justified to select it, relying only on the results of calculating the technological times. Here it is logical to assume that the optimal height from the economic and technological points of view will be such a height at which the time $\bar{t}_f$ will be the maximum. By analogy with Section 3.2, since the parameter L becomes an independent variable, and it was present in the previously introduced dimensionless variables and parameters as a length scale, the new scale L_0 and variable $\tilde{L} = \frac{L}{L_0}$ are introduced, so that $\tilde{z} = \frac{z}{L_0}$,

$$\tilde{t} = \frac{k_0 t}{(n_0 L_0)}, \tilde{h}, \tilde{H} = \frac{h, H}{L_0}, \tilde{\alpha}_V = L_0 \alpha_V, \tilde{\beta}_V = n_0 L_0 \beta_V, \tilde{R}_{f2} = \frac{\omega^2 k_0^2 R_{f2}}{L_0},$$

$$\bar{\tau} = \int_0^{\tilde{t}} \bar{V}(\xi) d\xi = \frac{1}{\tilde{L}} \int_0^{\tilde{t}} \bar{V}(\eta) d\eta = \frac{\tilde{\tau}}{\tilde{L}}.$$

Then dependencies (5.144), (5.145), (5.147), (5.148) and equation (5.149) are transformed after rescaling to the following form

$$\bar{C}_e(\tilde{\tau}; \tilde{L}) = e^{-\tilde{\alpha}_V \tilde{L}} \left[e^{-\tilde{\beta}_V \tilde{\tau}} I_0\left(2\sqrt{\tilde{\alpha}_V \tilde{\beta}_V \tilde{\tau} \tilde{L}}\right) + \tilde{\beta}_V \int_0^{\tilde{\tau}} e^{-\tilde{\beta}_V \varsigma} I_0\left(2\sqrt{\tilde{\alpha}_V \tilde{\beta}_V \tilde{L} \varsigma}\right) d\varsigma \right], \tag{5.178}$$

$$\bar{S}(\tilde{z}, \tilde{\tau}) = \tilde{\alpha}_V e^{-\tilde{\alpha}_V \tilde{z}} \int_0^{\tilde{\tau}} e^{-\tilde{\beta}_V \varsigma} I_0\left(2\sqrt{\tilde{\alpha}_V \tilde{\beta}_V \tilde{z} \varsigma}\right) d\varsigma, \tag{5.179}$$

$$\bar{V}(\tilde{t}, \tilde{\tau}; \tilde{L}) = \\ = \frac{1}{2\tilde{R}_{f2}} \left\{ \sqrt{\left[\int_0^{\tilde{L}} \frac{d\tilde{z}}{f_k(\bar{S}(\tilde{z}, \tilde{\tau}))}\right]^2 + 4\tilde{R}_{f2}\left[\tilde{H}^0 + n_0(\bar{Q}_{in}\tilde{t} - \tilde{\tau})\right]} - \int_0^{\tilde{L}} \frac{d\tilde{z}}{f_k(\bar{S}(\tilde{z}, \tilde{\tau}))} \right\}, \tag{5.180}$$

$$\tilde{H}_W(\tilde{t}, \tilde{\tau}) = \tilde{H}_W^0 + n_0(\bar{Q}_{in}\tilde{t} - \tilde{\tau}), \tag{5.181}$$

$$\frac{d\tilde{\tau}}{d\tilde{t}} = \frac{1}{2\tilde{R}_{f2}} \left\{ \sqrt{\left[\int_0^{\tilde{L}} \frac{d\tilde{z}}{f_k(\bar{S}(\tilde{z}, \tilde{\tau}))}\right]^2 + 4\tilde{R}_{f2}\left[\tilde{H}_W^0 + n_0(\bar{Q}_{in}\tilde{t} - \tilde{\tau})\right]} - \int_0^{\tilde{L}} \frac{d\tilde{z}}{f_k(\bar{S}(\tilde{z}, \tilde{\tau}))} \right\}. \tag{5.182}$$

If the filter media is expensive and scarce, it should be used as economically as possible. In this case it is logical to vary not only the height L, but also the cross-sectional dimensions of the bed (inlet surface area ω) so that the volume $W_f = \omega L$ remains unchanged. Naturally, a reduction of L means in this case a commensurate increase in area ω and a decrease in specific flow rate $\frac{Q_{in}}{\omega}$. Again, we have to transform the dimensionless expressions for $\bar{Q}_{in}, \tilde{R}_{f2}$ due to the variable nature of the quantity ω. Then the reduced quantities are introduced

$$\tilde{Q}_{in} = \frac{L_0 Q_{in}}{W_f k_0} = \frac{\bar{Q}_{in}}{\tilde{L}}, \quad \tilde{\tilde{R}}_{f2} = \frac{W_f^2 k_0^2}{L_0^3} R_{f2} = \tilde{L}^3 \tilde{R}_{f2} \tag{5.183}$$

and taking into account (5.149) the calculation expressions will be

$$\bar{V}(\tilde{t}, \tilde{\tau}; \tilde{L}) = \\ = \frac{\tilde{L}^2}{2\tilde{\tilde{R}}_{f2}} \left\{ \sqrt{\left[\int_0^{\tilde{L}} \frac{d\tilde{z}}{f_k(\bar{S}(\tilde{z}, \tilde{\tau}))}\right]^2 + 4\frac{\tilde{\tilde{R}}_{f2}}{\tilde{L}^2}\left[\tilde{H}_W^0 + n_0(\tilde{Q}_{in}\tilde{L}\tilde{t} - \tilde{\tau})\right]} - \int_0^{\tilde{L}} \frac{d\tilde{z}}{f_k(\bar{S}(\tilde{z}, \tilde{\tau}))} \right\}, \tag{5.184}$$

$$\bar{H}(\tilde{t},\tilde{\tau};\bar{L}) = \bar{H}_W^0 + n_0(\bar{Q}_{in}\bar{L}\tilde{t} - \tilde{\tau}), \tag{5.185}$$

$$\frac{d\tilde{\tau}}{d\tilde{t}} = \frac{\bar{L}^2}{2\bar{\bar{R}}_{f2}}\left\{\sqrt{\left[\int_0^{\bar{L}}\frac{d\bar{z}}{f_k(\bar{S}(\bar{z},\tilde{\tau}))}\right]^2 + 4\frac{\bar{\bar{R}}_{f2}}{\bar{L}^2}\left[\bar{H}_W^0 + n_0(\bar{Q}_{in}\bar{L}\tilde{t} - \tilde{\tau})\right]} - \int_0^{\bar{L}}\frac{d\bar{z}}{f_k(\bar{S}(\bar{z},\tilde{\tau}))}\right\}. \tag{5.186}$$

The expressions for the concentrations $\bar{C},\bar{S}$ (5.178), (5.179) do not change.

In the study of regular changes in the technological times due to an increase (or decrease) in the bed height, the first and third criteria were used if suspension was fed to a filter with a constant flow rate, and the first, second criteria if it was fed with a variable flow rate (at H_W=*cosnt*). In the case Q_{in}= *const* the rate $\bar{V}$ was pre-calculated for a long time (several thousand units) for these three basic examples, and also $\tilde{\alpha}_V = 3$, $\tilde{\beta}_V = 0.005$ and found that the deviations from $\bar{V}^0$ did not exceed a few percent. In the first situation (cheap filtering material), the values of 0.2, 0.5, and 1 were assumed for $\bar{Q}_{in}$, as before. The corresponding series of plots of the dependencies $\tilde{t}_p(\bar{L})$, $\tilde{t}_H(\bar{L})$ and the generalizing plot of $\tilde{t}_f(\bar{L})$ were calculated by (5.178), (5.181) taking into account (5.182) and are shown in Figures 42 and 43. At a low suspension feed $(\bar{Q}_{in} = 0.2)$, even a minimal increment of $\bar{L}$ (by a few tens of percent) resulted in a multiple increase in the time $\tilde{t}_p$. Much weaker $\tilde{t}_p$ responds to the height L change at a high feed $(\bar{Q}_{in} = 1)$. In this case, the parameters $\bar{\gamma}_a$, $\bar{R}_{f2}$ had almost no effect on the protective capability of the bed (curves 2-4 in Figure 43 almost merge), but significantly affected the position and slope of the curves $\tilde{t}_H(\bar{L})$ (1, 5, 6 in Figure 43). Naturally, the lengthening of the filtration flow path and a commensurate increase in the hydraulic resistance of the weakly clogged bed, as well as an increase in $\bar{Q}_{in}$, cause a decrease in $\tilde{t}_H$, which was particularly large at $\bar{Q}_{in} = 1$ and $\bar{L} > 3$. The parameter $\tilde{t}_H$ demonstrated a high sensitivity in relation to $\bar{L}$ at the conditions favorable for the filtration flow process (example 1, $\bar{Q}_{in} = 1$). The curves of the dependence $\tilde{t}_f(\bar{L})$, as well as similar to them curves of $\tilde{t}_f(\tilde{\alpha}_V)$, are often continuous broken lines with a single fracture. In less frequent cases, when only one or all three criteria are crucial for limiting the filter performance, the similar curves are, respectively, either smooth at all or have two fractures while remaining continuous. The abscissa of the fracture points $\bar{L}_m$ is of particular interest for design practice, since here the above curves reach their maximum. Consequently, it is reasonable to use in the new filter the layer of a cheap material with the height $\bar{L}_m$. It should be noted that the maximum values of $\tilde{t}_f$ in the basic examples, as well as the corresponding values of $\bar{L}_m$, differed significantly. This once again confirms the importance of the effect of the parameters $\bar{\gamma}_a$, $\bar{R}_{f2}$ on the clarification resource of a filter and the importance of a reasonable choice of L. A different pattern is observed at a constant suspension level above the bed (Figure 44). Here the broken curved lines made of pieces of curves 1, 6; 2, 5, and smooth curve 3 limit the area of effective application of such filtration mode (many suitable sets of the values of $\tilde{t}_f$ and $\bar{L}$). The highest value of $\tilde{t}_f$ turns out to be approximately the same in the examples, but again the values of $\bar{L}_m$ corresponding to the specified extrema are significantly different. An excessive increase in the bed thickness is practically dangerous because immediately after the first suspension portion has passed through it the rate will fall to the critical value $\bar{V}_*$ and the filter will be initially incapacitated as well as at too small thickness. The critical value of $\bar{L}(\tilde{t}_V = 0)$ was approximately the same and equal to 2.245 in all examples. The noted peculiarities of the behavior of the technological times when varying the height of an operating layer clearly indicate the importance of the theoretical substantiation of this design parameter.

Finally, the effects of simultaneously reducing the hydraulic load on a filter and the bed resistance are analyzed using the basic examples in the situation with a scarce filter media. Figure 45 shows the curves of a change in the times $\tilde{t}_p$ (1-3) and $\tilde{t}_H$ (4-6) under a hypothetical gradual and continuous thinning of the bed by a maximum of 2.5 times. These curves were calculated taking into account (5.186) and the above accepted values of $\overline{C}_*, \overline{H}_*$ by formulas (5.178), (5.185). With increased deposition (example 2) the technological process values was limited for the entire assumed range of the $\overline{L}$ by early overfilling of the storage tank and, as a result, the curve of the dependence $\tilde{t}_f(\overline{L})$ coincided with curve 6. In the other two cases, the desired curve has the usual form of a broken line with a single fracture. The hydraulics of the offtake communications has little effect on the maximum filter run duration and the corresponding value $\overline{L}_m$, which, however, decrease significantly with increasing $\overline{\gamma}_a$. Reducing the bed thickness with a simultaneous increase in its surface area will most often contribute to the prolongation of the filter operation. At the same time, the opposite situation is possible at a significant decrease in $\overline{L}$, when thinner filter beds will exhaust their clarification resource faster (curves 1, 3 at $\overline{L} < 0.55$).

Figure 42. Dependencies $\tilde{t}_p(\overline{L})$, $\tilde{t}_H(\overline{L})$ ($W_f \sim L$)

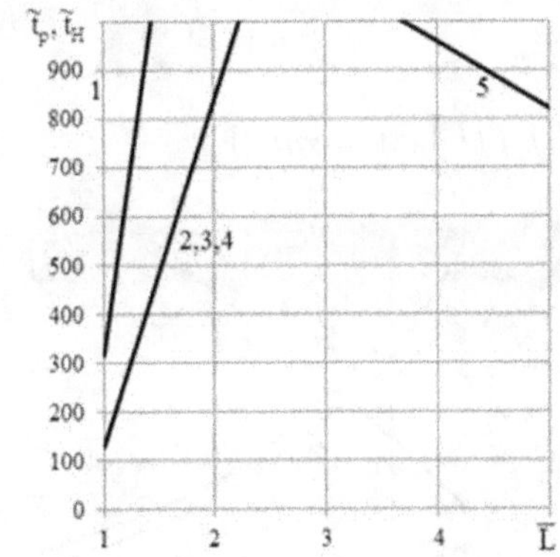

Note. 2-5 – $\overline{Q}_{in}$ = 0.5; 1– $\overline{Q}_{in}$ = 0.2; 5 – $\tilde{t}_H$; 1-4 – $\tilde{t}_p$; 1,3 – example 1; 2.5 – example 2; 4 – example 3

Figure 43. Dependencies $\tilde{t}_p(\overline{L})$, $\tilde{t}_H(\overline{L})$, $\tilde{t}_f(\overline{L})$ (Q_{in}=1, ($W_f \sim L$))

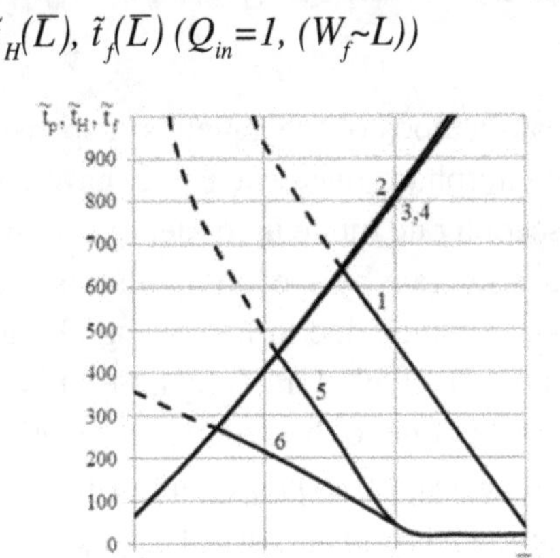

Note. 1,5,6 – $\tilde{t}_H(\overline{L})$; 2-4 – $\tilde{t}_p(\overline{L})$; 1, 4 – example 3; 2,6 – example 2; 3,5 – example 1; $\tilde{t}_f(\overline{\beta}_v)$ – solid lines

Figure 44. Dependencies $\tilde{t}_p(\bar{L})$, $\tilde{t}_V(\bar{L})$, $\tilde{t}_f(\bar{L})$ $\left(\bar{H}_W = const, W_f \sim L\right)$

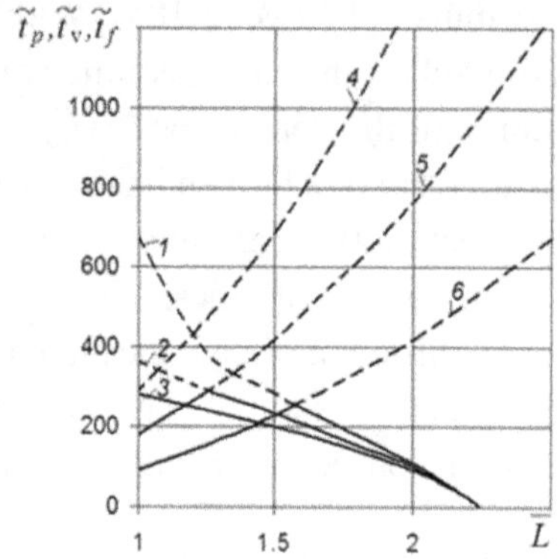

Note. 1-3 – $\tilde{t}_V$; 4-6 – $\tilde{t}_p$; 1, 6 – $\tilde{\alpha}_V = 5$; 2, 5 – $\tilde{\alpha}_V = 7$; 3, 4 – $\tilde{\alpha}_V = 9$; $\bar{t}_f(\bar{L})$ –solid lines

Figure 45. Dependencies $\tilde{t}_p(\bar{L})$, $\tilde{t}_H(\bar{L})$, $\tilde{t}_f(\bar{L})$ (W_f=const)

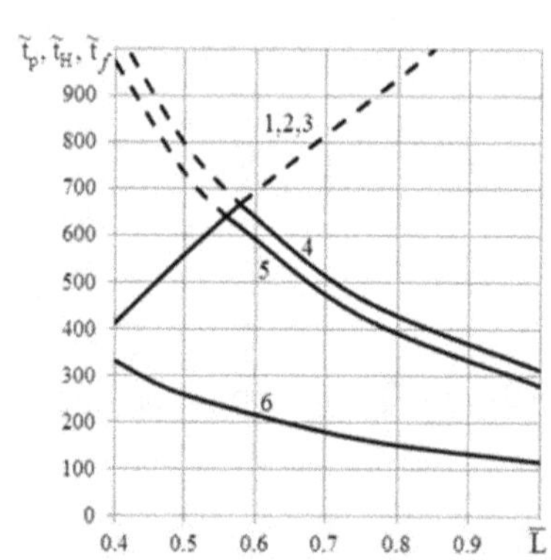

Note. 1-3 – $\tilde{t}_p(\bar{L})$; 4-6 – $\tilde{t}_H(\bar{L})$; 1, 5 – example 1; 2, 6 – example 2; 3, 4 – example 3; $\bar{t}_f(\bar{\beta}_V)$ – solid lines

The detailed studies by analytical methods of the aqueous suspension filtration with an unregulated rate, which is closely related to the interphase mass exchange, have convincingly demonstrated the expediency of the application of the second rate mode in modern water treatment practice. In this case the need for the strict control of the filtration rate by special regulators is eliminated. In spite of the steady decrease in the filtration facility performance due to increasing clogging of the bed, slow decrease in the hydraulic load allows it to operate effectively for a long time. It is possible to extend the continuous operation time of the facility even further due to the active participation in the technological process of the storage tank, which can accumulate a significant amount of initial suspension. The selection of the bed height and filtering material (absorption properties) based on a regular decrease in the filtration rate enables to fully implement the significant clarification resource of the facility, including the offtake communications and the storage tank in addition to the filter bed.

Here it should be emphasized that the choice of a filter media and bed height without proper justification can lead to a significant reduction in the efficiency of a rapid filter also at the declining filtration rate.

REFERENCES

Barbusiński, K., Szeląg, B., Parzentna-Gabor, A., Kasperczyk, D., & Rene, E. R. (2024). Application of a generalized hybrid machine learning model for the prediction of H2S and VOCs removal in a compact trickle bed bioreactor (CTBB). *Chemosphere*, *360*, 142181. DOI: 10.1016/j.chemosphere.2024.142181 PMID: 38685329

Bugay, N. G., Krivonog, A. I., & Krivonog, V. V. (2007). Contact reagent treatment of filtration water from fiber-porous material. [in Russian]. *Applied Hydromechanics*, *9*(1), 8–22.

Chaundry, F.H. (1987), Theory of declining rate filtration. I. Continuous operation. *J. Environ. Eng. Div., ASCE*, 113(4), 834–851.

Cleasby, J. L. (1993). Status of declining rate filtration design. *Water Science and Technology*, *27*(10), 151–164. DOI: 10.2166/wst.1993.0224

Gąszczak, A., Bartelmus, G., Burghardt, A., Rotkegel, A., & Sarzyński, R. (2018). Experiments and modelling of a biotrickling filter (BTF) for removal of styrene from airstreams. *Journal of Chemical Technology and Biotechnology*, *93*(9), 2659–2670. DOI: 10.1002/jctb.5620

Grabovsky, P. A. (2016). Water filtration through a granular layer with decreasing rate. [in Russian]. *Rep. NAS of Ukraine*, (8), 40–45.

Grabovsky, P. O., & Gurinchyk, N. A. (2008). Water filtration through a granular layer with decreasing rate in time. *Water and Ecology.* [in Russian]. *Problems and Solutions*, *37*(4), 3–11.

Gusev, E. M. (1981). Experimental studies of pressure seepage. In *Physics of soil waters* (pp. 195–205). Nauka. (in Russian)

Hekmat, D., & Vortmeyer, D. (1994). Modelling of biodegradation processes in trickle-bed bioreactors. *Chemical Engineering Science*, *49*(24), 4327–4345. DOI: 10.1016/S0009-2509(05)80025-9

Iliuta, I., & Larachi, F. (2004). Biomass accumulation and clogging in trickle-bed bioreactors. *AIChE Journal. American Institute of Chemical Engineers*, *50*(10), 2541–2551. DOI: 10.1002/aic.10201

Kamke, E. (1976). *Handbook on ordinary differential equations*. Nauka. (in Russian)

Kozeny, J. (1931). Grundwasserbewegung bei freiem Spiegel, Fluss- und Kanal – versickerung. *Wasserkraft und Wasserwirts chaft*, 4.3, 28–31.

Liao, Q., Tian, X., Chen, R., & Zhu, X. (2007). Mathematical model for gas-liquid two-phase flow and biodegradation of a low concentration volatile organic compound (VOC) in a trickling biofilter. Internal. J. Heat Mass Transfer, 1–13.

Liu, S., Gao, P. F., Li, S., Fu, H., Wang, L., Dai, Y., & Fu, M. (2023). A review of the recent progress in biotrickling filters: Packing materials, gases, micro-organisms, and CFD. *Environmental Science and Pollution Research International*, *30*(60), 125398–125416. DOI: 10.1007/s11356-023-31004-7 PMID: 38012483

Logan, B. E., Jewett, D. G., Arnold, D. G., Bouwer, E. J., & O'Melia, C. R. (1995). Clarification of clean-bed filtration models. *Journal of Environmental Engineering, 12*(12), 869–873. DOI: 10.1061/(ASCE)0733-9372(1995)121:12(869)

Oleynik, A. Ya., & Kolpakova, O. A. (2012). Theoretical substantiation of wastewater treatment on trickling filters. [in Russian]. *Rep. NAS of Ukraine*, (3), 120–126.

Poliakov, V. L. (2010). Clarification of suspension by filtration with decreasing flow rate. [in Russian]. *Rep. NAS of Ukraine*, (5), 57–63.

Poliakov, V. L. (2012). Filtration of suspensions with decreasing rate at linear kinetics of mass transfer. [in Russian]. *Journal of Water Chemistry and Technology, 34*(2), 107–130.

Poliakov, V. L. (2014). *Filtration flow deformations in drained soils: theory and applications*. Agrar Media Group. (in Russian)

Poliakov, V. L. (2022). Theoretical analysis of reagent filtration of aqueous suspensions with unregulated filtration rate. *Theoretical Foundations of Chemical Engineering, 56*(4), 477–496. DOI: 10.1134/S0040579522030101

Saatci, A.M. (1990). Application of the declining rate filtration theory. – Continuous operation. *J. Environ. Eng. Div., ASCE*, 116(1), 87–105.

Senyavin, M. M., Venitsianov, E. V., & Ayukaev, R. I. (1997). About mathematical models and engineering methods for calculating the process of natural water purification by filtration. [in Russian]. *Water Resources*, (2), 157–170.

Tobiason, J. E., Johnson, G. S., Westerhoff, P. K., & Vigneswaran, B. (1993). Particle size and chemical effects on contact filtration performance. *Journal of Environmental Engineering, 119*(3), 520–533. DOI: 10.1061/(ASCE)0733-9372(1993)119:3(520)

Venitsianov, E. V., & Rubinstein, R. N. (1983). *Dynamics of sorption from liquid media*. Nauka. (in Russian)

Zapolsky, A. K., & Baran, A. A. (1987). *Coagulants and flocculants in water purification processes*. Khimiya. (in Russian)

APPENDIX 5.1. SATURATION OF UNSATURATED PACKED BED (SECOND STAGE)

Saturation of the dewatered uniform layer of a filtering material, depending on the relationship between the flow rate Q_{in} and the parameter ωk_0 (characterizes its initial throughput capacity), can occur in one or two stages. The case $\overline{Q}_{in} > 1$ ($Q_{in} > \omega k_0$) is analyzed in detail in Section 5.2. If $\overline{Q}_{in} < 1$, then the pore space is filled with liquid actually in two stages. During the first stage ($t < t_{11}$), the wetting front is formed and lowered (downward filtration). The unsaturated zone with an increased (but less than the total moisture capacity) moisture content θ_* is formed and expands over time above it. Let initially the level of a residual liquid was within the bed at the mark H_{out}. After the front reaches this level, the externally supplied suspension is consumed partly for the bed saturation and partly for the transit flow. The end time of the first stage should be calculated by formula (5.57). Further the balance of liquid entering the bed is described by the equation

$$\omega(n_0 - \theta_*)\frac{dH_f}{dt} = Q_{in} - \omega V_{12}, \tag{A5.1.1}$$

where H_f is the mark of the saturation front (boundary between unsaturated and saturated zones), V_{12} is the filtration rate, which, taking into account head losses in offtake communications, is determined from the equation

$$V_{12} = k_0\frac{H_f - H_d}{H_f - Z_d} = k_0\frac{H_f - H_{out} - R_{f2}\omega^2 V_{12}^2}{H_f - Z_d}. \tag{A5.1.2}$$

Here Z_d is the piezometric mark of the lower bed boundary. Equation (A5.1.1) is supplemented by the initial condition

$$t = 0, \quad H_f = H_{out}. \tag{A5.1.3}$$

Further dimensionless variables and parameters are introduced by analogy with Section 5.3. Then problem (A5.1.1) - (A5.1.3) is transformed as follows

$$\frac{n_0 - \theta_*}{n_0}\frac{d\overline{H}_f}{d\overline{t}} = 1 - \overline{V}_{12}(\overline{H}_f), \tag{A5.1.4}$$

$$\overline{Q}_{in}\overline{V}_{12} = \frac{\overline{Q}_{in}^2\overline{H}_f - \overline{R}_{f2}\overline{V}_{12}^2}{\overline{Q}_{in}^2(\overline{H}_f - \overline{Z}_d)}, \tag{A5.1.5}$$

$$\overline{t} = 0, \quad \overline{H}_f = 0. \tag{A5.1.6}$$

From (A5.1.5) there is the expression for $\overline{V}_{12}$ as a function of $\overline{H}_f$

$$\overline{V}_{12}(\overline{H}_f) = \frac{\sqrt{\overline{Q}_{in}^6(\overline{H}_f - \overline{Z}_d)^2 + 4\overline{R}_{f2}\overline{Q}_{in}^2\overline{H}_f} - \overline{Q}_{in}^3(\overline{H}_f - \overline{Z}_d)}{2\overline{R}_{f2}} \tag{A5.1.7}$$

Then the relative time of the saturation front rise to the reduced mark $\overline{H}_f$ is proposed to be strictly determined by the formula

$$\overline{t} = \frac{n_0 - \theta_*}{2n_0\overline{R}_{f2}}\int_0^{\overline{H}_f}\frac{d\varsigma}{2\overline{R}_{f2} - \overline{V}_{12}(\varsigma)} \tag{A5.1.8}$$

Thus, the relative duration of the second stage will be as follows

$$\bar{t}_{12} = \frac{n_0 - \theta_*}{2n_0\bar{R}_{f2}}\int_0^{\bar{Z}_w}\frac{d\varsigma}{2\bar{R}_{f2} - \bar{V}_{12}(\varsigma)}. \quad \text{(A5.1.9)}$$

The integral in (A5.1.8), (A5.1.9) can be calculated in the special case $\bar{R}_{f2} = 0$. Then

$$\bar{V}_{12}\left(\bar{H}_f\right) = \frac{\bar{H}_f}{\bar{Q}_{in}\left(\bar{H}_f - \bar{Z}_d\right)}$$

and formula (A5.1.8) reduces to the following form

$$\bar{t} = \frac{\bar{Q}_{in}(n_0 - \theta_*)}{2n_0\bar{R}_{f2}\left(2\bar{R}_{f2}\bar{Q}_{in} - 1\right)}\left[\bar{H}_f + \frac{\bar{Z}_d}{2\bar{R}_{f2}\bar{Q}_{in} - 1}\ln\left(1 - \bar{H}_f\frac{2\bar{R}_{f2}\bar{Q}_{in} - 1}{2\bar{R}_{f2}\bar{Q}_{in}\bar{Z}_d}\right)\right]. \quad \text{(A5.1.10)}$$

APPENDIX 5.2. ESTIMATION OF LIMIT HEIGHT OF STORAGE TANK

When implementing the second variety of the second rate mode, the question of the rational substantiation of the storage tank height becomes relevant due to possible significant rise of the suspension level in it. It is obvious that the suspension level in it should reach the maximum level at full exhaustion of the absorbing resource of the operating layer under given conditions. Then, depending on the ratio between the dynamic S_d and true S_m (in relation to suspension particles) solids take-up capacities, two characteristic situations are possible.

In the first situation, which takes place under the condition

$$Q_{in} < \omega \left(\frac{\beta_V S_m}{\alpha_V C_0} \right)^{\frac{1}{l-q}}, \tag{A5.2.1}$$

the absorption capacity of a filtering material is finally determined by the ratio between the rates of the suspension absorption and the deposit detachment and is characterized by the quantity S_d, which is equal to

$$S_d = \frac{\alpha_V C_0}{\beta_V} V^{l-q}. \tag{A5.2.2}$$

Then the corresponding filtration rate V_∞ is related to the desired height of the liquid level in the storage tank $H_{W\infty}$ by the equation

$$V_\infty = k_{\min}\left(S_d(V_\infty)\right) \frac{H_{W\infty} + Z_w - R_{f2}\omega^2 V_\infty^2}{L}. \tag{A5.2.3}$$

Here and further the index "∞" indicates the implementation of the given and the second situations in the asymptotic sense, i.e. at $t \rightarrow \infty$, and also at the corresponding limit values of the calculation characteristics. Since the maximum value V_∞ is equal to $\frac{Q_{in}}{\omega}$, then the required mark is simply found from (A5.2.3), namely,

$$H_{W\infty} = \frac{LQ_{in}}{\omega k\left(Q_{in}\right)_{min} d_w f2^{2}_{in}} \tag{A5.2.4}$$

Here the coefficient k_{min} according to (A5.2.1), (1.6), (2.7) will be

$$k_{\min}\left(Q_{in}\right) = k_0 \left\{ 1 - \left[\frac{\gamma(S_{d\infty}) S_{d\infty}}{n_0} \right]^{m_{k1}} \right\}^{m_{k2}}, \quad S_{d\infty} = \frac{\alpha_V C_0}{\beta_V} \left(\frac{Q_{in}}{\omega} \right)^{l-q}. \tag{A5.2.5}$$

The second situation, which deserves separate consideration, occurs when the true solids take-up capacity is exhausted ahead of time. In other words, the concentrations S, S_S are actually limited to constant values S_m, S_{Sm}, which practically do not depend on the physical and chemical properties of the filter media, deposit and filtration flow mode. Then the constant k_{min} will be

$$k_{\min} = k_0 \left\{ 1 - \left[\frac{\gamma(S_m) S_m}{n_0} \right]^{m_{k1}} \right\}^{m_{k2}} \tag{A5.2.6}$$

and the height $H_{W\infty}$ is calculated taking into account (A5.2.6) also by formula (A5.2.4). It should be especially noted the determining importance here of the filtration flow properties of the bed being clogged, because even insignificant reduction of pore space can finally lead to an appreciable rise of the suspension level in the storage tank.

It is somewhat more difficult and less important from a practical point of view to estimate the value $H_{W\infty}$, when contaminant transport and phase transfer in the filtration facility is ensured by a constant head differential ΔH at both collectors. In this case, the suspension flow into the facility decreases steadily over time and the variable flow rate Q_{in} now reaches a minimum value Q_∞ at $t\rightarrow\infty$, which must be specifically calculated. In order for the deposit accumulation to be limited by the dynamic solids take-up capacity, condition (A5.2.1) must be met, where Q_{in} is replaced by the as yet unknown Q_∞. In this situation, the limiting volumetric concentrations of the deposited suspension particles $S_{d\infty}$ and deposit $S_{d\infty}$ are expressed through Q_∞ as follows

$$S_{d\,\infty}(Q_\infty) = \frac{\alpha_V C_0}{\beta_V}\left(\frac{Q_\infty}{\omega}\right)^{1-q}, \quad S_{dS\,\infty}(S_{d\,\infty}) = \gamma(S_{d\,\infty})S_{d\,\infty}. \tag{A5.2.7}$$

They are responded by the minimum value of the hydraulic conductivity

$$k_{\min}(Q_\infty) = k_0\left\{1 - \left[\frac{\gamma(S_{d\,\infty}(Q_\infty))\,S_{d\,\infty}(Q_\infty)}{n_0}\right]^{m_{k1}}\right\}^{m_{k2}}. \tag{A5.2.8}$$

In this case, the equation with respect to the flow rate Q_∞ takes the form

$$LQ_\infty = \omega\, k_{\min}(Q_\infty)\left[\Delta H - (R_{f1} + R_{f2})\,Q_\infty^2\right]. \tag{A5.2.9}$$

At the value Q_∞ established by fitting from (A5.2.9), or by solving the corresponding algebraic equation, the required $H_{W\infty}$ is calculated by the formula

$$H_{W\infty} = H_{in} - Z_w - R_{f1}\,Q_\infty^2. \tag{A5.2.10}$$

If Q_∞ is known, condition (A5.2.1) must be checked. If it is not satisfied, i.e. $S_d<S_m$, then k_{min} is determined first and again by formula (A5.2.6). Then equation (A5.2.9) becomes a quadratic, so that

$$Q_\infty = \frac{\sqrt{L^2 + 4\omega^2 k_{\min}^2\left(R_{f1} + R_{f2}\right)\Delta H} - L}{2\omega\, k_{\min}\left(R_{f1} + R_{f2}\right)}. \tag{A5.2.11}$$

The height $H_{W\infty}$ with the flow rate Q_∞ found according to (A5.2.1) is also calculated using formula (A5.2.10). In reality, the level position at the end of the filter run is often lower than the limit level. Therefore, it is better to substantiate the rational height of the storage tank on the basis of the technological calculations, e.g. following Appendix 5.3.

APPENDIX 5.3 EXAMPLE OF CALCULATING FILTER RUN DURATION

Initial data: $Q_{in}=12m/h$, $\alpha_V=18.5m^{-0.7}\bullet h^{-0.3}$, $l=0.7$, $\beta_V=0.03m^{-1}$, $\gamma_a=40$, $n_0=0.37$, $d_g=0.7mm$, $L=0.8m$, $\omega=1.2m^2$, $H_{out}=0.5m$, $R_{f2}=0.005m^2\bullet h^{-5}$, medium temperature 10°C, $C_0=2.5\bullet10^{-5}$. Required to calculate the dimensionless and dimensional values of the duration t_f.

According to (Kozeny, 1931), the hydraulic conductivity of a clean bed is as follows

$$k_0 = 3600 \cdot 0.0055 \frac{9.81(7 \cdot 10^{-4})^2 \cdot 0.37^3}{1.31 \cdot 10^{-6}(1 - 0.37)^2} = 9.27\frac{m}{h}$$

Let as a result of the clarification of raw water it is necessary to reduce the content of the suspended solids in it by an order of magnitude or more, so that $\overline{C}_* = 0.1$. It is also possible to reduce the performance of the filtration facility by 40%. At last, it is not allowed to operate with an overfilled storage tank, the edge of which is located at the reduced mark $\overline{H}_* = 6$.

The value $\overline{H}_m = 1.9787$ is preliminarily found. The relative values of the model parameters are then determined:

$$\overline{\alpha}_V = 18.5 \cdot 0.8\left(\frac{12}{1.2}\right)^{-0.3} = 7.416, \quad \overline{\beta}_V = 0.37 \cdot 0.8\left(\frac{12}{1.2}\right)^{-0.3} \cdot 0.03 = 8.88 \cdot 10^{-3},$$

$$\overline{\gamma}_a = 40 \cdot 2.5 \cdot 10^{-5} = 10^{-3}$$

The value of the parameter $\overline{V}_{ap}$ and its corresponding relative time of the bed protective effect were determined by solving the system of equations (5.116), (5.119) using fitting procedure and are equal to 0.963662; 344.22. Similarly, the value of the parameter $\overline{V}_{aV}$ and its corresponding relative time of declining the filtration rate to the set minimum value were also determined by fitting from system (5.116), (5.118) and are equal to 0.969347; 293. Finally, the suspension level in the storage tank will reach the edge by the time $\overline{t}_H = 438.79$. Thus, the sought relative duration $\overline{t}_f$ here will be

$$\overline{t}_f = \overline{t}_V = 293,$$

and in view of $t=0.0296\,\overline{t}$ finally $t_f=8.67$ hours.

APPENDIX 5.4 EXAMPLE OF APPLIED OPTIMIZATION OF FILTER BED ABSORPTION PROPERTIES

The possibility of prolonging the time of the continuous clarification of aqueous suspension due to the rational choice of a filtering material was analyzed on a specific example. Based on the solution of the first problem from Section 5.4, the change in the relative duration of the filter run $\bar{t}_f$ is calculated at a continuous variation of the key mass-exchange coefficient $\bar{\alpha}_V$ from 2 to 12. The initial information included, firstly, data on the criterion parameters, namely $\bar{C}_* = 0.1, \bar{V}_* = 0.8, \bar{H}_* = 4$; secondly, the fixed values of the model coefficients in the case of the filtration using reagents $\bar{\beta}_V = 0.01, \bar{\gamma}_a = 0.001, \bar{H}^0_W = 0, n_0$=0.3, .$\bar{R}_{f2} = 1, \bar{H}_{out} = 0.25, m_{k1}$=1, m_{k2}=3. The dependencies $\bar{t}_p(\bar{\alpha}_V), \bar{t}_V(\bar{\alpha}_V), \bar{t}_H(\bar{\alpha}_V)$ were determined sequentially. The first dependence was found on the basis of equations (5.141), (5.145); the second dependence - also from (5.145) and (5.143); the third dependence - from (5.144).

The results of the calculations in graphical form are shown in Figure 46. Their pieces, marked out according to (5.123) and depicted here by a solid line, together form the desired continuous broken line $\bar{t}_f(\bar{\alpha}_V)$. Thus, under the considered conditions the clarification process is limited by the protective factor (2.33 $< \bar{\alpha}_V <$ 6.6) at a low absorption capacity of the bed, by the design factor (6.6 $< \bar{\alpha}_V <$ 10.5) at a moderate to high absorption of the suspended solids by the filtering material. And only if $\bar{\alpha}_V >$ 10.5, the filter operation should be interrupted due to the excessive reduction of its performance. The maximum duration was 200 and reached at $\bar{\alpha}_V$ = 6.6. Increasing the height of the storage tank, i.e. in this case $\bar{H}_*$ can allow to significantly (almost 2 times) increase the time $\bar{t}_f$.

Figure 46. Dependencies $\bar{t}_p(\bar{\alpha}_V)$, $\bar{t}_V(\bar{\alpha}_V)$, $\bar{t}_H(\bar{\alpha}_V)$, $\bar{t}_f(\bar{\alpha}_V)$

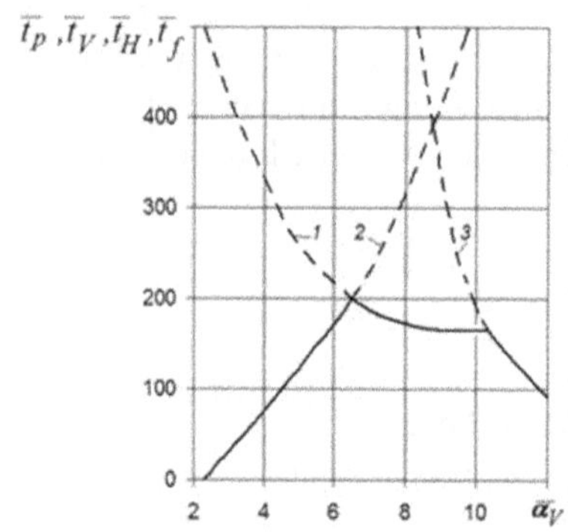

Note. 1 – $\bar{t}_H$, 2 – $\bar{t}_p$, 3 – $\bar{t}_V$; solid line – $\bar{t}_f$

Likewise, it is easy to analyze the influence of other initial parameters, which can be actually changed during the design or the operation of the rapid filters. There is a real opportunity to further intensify their running in this way.

Chapter 6
Nonlinear Detachment Filtration

ABSTRACT

Since the absorption rate of the suspended solids and the corresponding kinetic coefficient often decrease or increase as the deposit accumulates in the bed, Chapter 6 raises the issues of the detachment filtration with nonlinear kinetics. At the same time, the exact solutions of the suitable nonlinear problems were obtained in the presence of an initial deposit and a two-layer structure of the filter media, taking into consideration the autocatalysis phenomenon, progressive decrease in the absorptive capacity. Based on them the significance of the nonlinear effect was estimated, numerous technological calculations were performed. In view of the cumbersome calculation formalisms and the unambiguous choice of kinetics, an approximate solution of a similar problem with a kinetic equation of general form is obtained, which allows to calculate the characteristics of nonlinear filtration with minimal errors. The accuracy of the above solution is evaluated in case of linear filtration for the head losses due to severe clogging of the packed bed.

The deposit formed in filter facilities has often an active effect on the deposition of the suspended solids in the pore space of a packed bed. On the one hand, it can noticeably increase the number of vacancies for the suspended solids and thus, intensify the suspension clarification especially at the initial stage. A short-term enhancement of the absorption capability of the filtering material, the so-called "ripening", was regularly observed in experiments (Bugai et al. 2004; Calvin, 1992; Elimelech & O'Melia, 1990). On the other hand, the deposit formation means gradual filling of such places that existed in the filter bed initially and negatively affects the suspension separation. Thus, these effects have directly opposite consequences for the technological process of the aqueous suspension separation. In mathematical terms, however, both of them cause the nonlinearity of the mass exchange kinetics between liquid and solid phases in the bed. Naturally, linear kinetics is primarily suitable for the mathematical modeling of filtration if these effects compensate each other and then the deposit actually becomes inert with respect to the suspended matter. At the same time, if the granular and even more so the fibrous filtering material has a high absorption capability and the "ripening is negligible or stops quickly, the use of the linear mass exchange kinetics is also justified. Similar conditions are typical for the operation of many filtration facilities, which explains the wide use of linear kinetics equations in theoretical developments. If the above-mentioned effects are significant and mutually not compensated, then, first of all, they are reflected on the absorption process. And then the selection of the correct formal representation for the key mass exchange coefficient is very important. Being constant in linear kinetics, the coefficient α in nonlinear kinetics will already be a function of the concentration of the deposit S_S or deposited particles

DOI: 10.4018/979-8-3693-8211-0.ch006

S. Thus, in accordance with representation (1.14) and taking into account (1.15), (1.29), it is possible to consider that the absorption function of a dispersed contaminant will be as follows

$$f_a(d_g,\ d_p, V, T, S) = \alpha_0\left(d_g,\ d_p, V, T\right)\chi_S(S) \tag{6.1}$$

where the adhesion function $\chi_S(S)$ characterizes the effect of the deposited particles on the absorption rate of the suspended solids. If it is necessary to simultaneously take into account both of the above-mentioned effects, this function can be nonmonotonic for rapid filters in contrast to slow filters (Ojha & Graham, 1994), and then it is often justified to use polynomials of the second degree and higher, the coefficients of which are selected on the basis of experimental data. Below, however, its linear form, namely, the following two linear functions according to (1.17) and (4.14) (Alekseev & Kommunar, 1974; Jegatheesan & Vigneswaran, 2000; Kuhn & Briesen, 2016; Ornatsky et al., 1955; Poliakov & Martynov, 2021; Tan, 1994) will be involved in the theoretical studies of the nonlinear detachment filtration by exact analytical methods

$$\chi_S(S) = \chi_{Sm}(S) = S_m - S \tag{6.2}$$

$$\chi_S(S) = \chi_{Sa}(S) = 1 + \theta S \tag{6.3}$$

where S_m is actually the concentration of the limiting saturation by the deposited particles of a filtering material, which depends on the properties of a dispersed impurity, the method of the reagent treatment, and the filtration mode (Dzyubo & Alferova, 2007; Mintz, 1964; Senyavin et al., 1977). Just such a formalization of the absorption process is used in the theoretical analysis of mass exchange in the case of the contaminated water filtration in natural and artificial porous media for several decades, and its validity has received convincing experimental confirmation. For the nonlinear representations of $\chi_S(S)$, we have to use exclusively the approximate solution methods. According to (6.2), the formed deposit only complicates the subsequent removal of the suspended matter and, due to the progressive reduction of vacancies for the suspension particles, its accumulation rate will steadily and smoothly decrease. In accordance with (6.3), the suspended solids deposition is enhanced by the formation of the deposit and thus, the creation of new vacancies. Just such a formalization of the absorption process is used in the theoretical analysis of mass exchange in the case of the contaminated water filtration in natural and artificial porous media for several decades, and its validity has received convincing experimental confirmation. More general representations for the function $\chi_S(S)$ have been used both in this Chapter and in Chapter 4 in the study of the non-detachment filtration. For the nonlinear representations of $\chi_S(S)$ in case of the detachment filtration, we have to use exclusively the approximate solution methods.

The main content of this Chapter includes rigorous and approximate analytical solutions of a number of typical problems of the nonlinear direct-flow (plane) detachment filtration. Their practical expediency is demonstrated on numerous test examples, in which the factors essentially influencing the current physical and chemical state of the operating medium and considered in the formulation of the mentioned problems (instability of the contamination in natural water or pretreatment effluents, initial clogging of a filter bed, its non-uniform structure, limited absorbing resource of a filtering material, active participation of a deposit in the mass exchange) vary within wide limits. At the same time, the significance of the mentioned factors was assessed, which allows to outline the ways of their targeted change for the purpose of the technological process intensification. As usual, the forecast calculations are accompanied by the technological calculations. The results of the latter serve as a basis for the rational engineering decisions. In the statement of the first four mathematical problems, the linear form of the function $\chi_S(S)$ is assumed, which allowed us to find their exact solutions. In the last problem $\chi_S(S)$ is assumed to be nonlinear, and the type of nonlinearity is not initially specified. The approximation expression (6.2), also the combination of expressions (6.2), (6.3) are considered as important special cases.

Specifically, Section 6.1 deals with the nonlinear filtration through an initially clean bed in accordance with (6.2) at a known pattern of a change in the initial impurity content of a suspension over time. Section 6.2 assumes the same form of the function $\chi_S(S)$ and the presence of the initial deposit in a filter bed, which is unevenly distributed by its height after successive washing. In Section 6.3, it is assumed that the deposit has a catalytic effect on the suspended matter according to (6.3) and thus, contributes to the accelerated removal of contaminants. In Section 6.4 the suspension clarification is investigated within an arbitrary layer of an initially clean multilayer bed in the absence of the autocatalysis. The calculation dependences are given describing the spatial and temporal changes of both concentrations of a dispersed contamination. As a special case, the two-layer bed often encountered in the filtration practice is considered. Finally, an approximate solution is obtained in Section 6.5, which has a much wider area of application than the aforementioned exact solutions. The calculation dependences derived from it in parametric form make it possible to monitor the deterioration of the filtrate quality with a high accuracy over a long period of time, as well as the accumulation and distribution of a deposit, the head losses in a packed bed and, as a result, to perform the reliable technological and design analysis.

6.1. STRICT ANALYSIS FOR VARIABLE INITIAL CONCENTRATION OF SUSPENDED SOLIDS

By analogy with Section 2.2, we consider the filtration of aqueous suspension when the suspended solids content in it changes significantly during one filter run. The dependence $C_0(t)$, describing the known variation over time of the initial volumetric concentration of the suspended matter, is given in the boundary condition (2.26). Also, the clarification compartment includes the equation of mass transfer considering local changes in C (2.1) and the equation of the nonlinear mass-exchange kinetics under the linear adhesion function (6.2). Thus, the above equation in the presence of the significant deposit detachment takes the following form (Adin & Rebhun, 1987; Poliakov, 2011a; Saiers et al., 1994)

$$\frac{\partial S}{\partial t} = \alpha_0 \left(S_m - S\right) C - \beta S. \tag{6.4}$$

The filter media is assumed to be initially clean, thereby this compartment is supplemented by the homogeneous initial condition (2.29). Since the form of the filtration flow compartment remains the same, which is represented in a dimensionless form by equations (2.11), (2.12) and condition (2.13), then in the future, when carrying out the predictive and technological calculations, references will be made to the already derived formulas for the relative head losses and the equation with respect to the second technological time (t_h). To obtain generalized results, the initial problem considering the early stage of filtration is also represented in a dimensionless form

$$\frac{\partial \overline{S}}{\partial \overline{t}} = \overline{\alpha}_0\left(1-\overline{S}\right)\overline{C} - \overline{\beta}\overline{S}, \tag{6.5}$$

$$\frac{\partial \overline{C}}{\partial \overline{z}} + \overline{S}_{m0}\frac{\partial \overline{S}}{\partial \overline{t}} + \overline{n}_e\frac{\partial \overline{C}}{\partial \overline{t}} = 0, \tag{6.6}$$

$$\overline{z} = 0, \quad \overline{C} = \overline{C}_0(\overline{t}); \quad \overline{t} = \overline{n}_e\overline{z}, \quad \overline{C} = \overline{S} = 0. \tag{6.7}$$

Here the value S_m bounding the concentration S from above is taken as a new scale for it, so that $\overline{S} = S/S_m$, $\overline{C} = C/C_0^0$, C_0^0 is the initial volumetric concentration of the suspended solids;

$$\overline{\alpha}_0 = n_0 L C_0^0 \alpha_0/V,\ \overline{\beta} = n_0 L\beta/V,\ \overline{n}_e = n_e/n_0,\ \overline{z} = z/L,\ \overline{S}_{m0} = S_m/\left(n_0 C_0^0\right).$$

The solution of problem (6.5) - (6.7) is found, partly following the recommendations from (Alekseev et al., 1989; Verigin & Kommunar, 1978). A number of substitutions of dependent and independent variables are used, which can be reduced to the following sequence

$$\tau_t = \overline{t} - \overline{n}_e\overline{z}, \quad \overline{C} = \frac{\partial \Phi}{\partial \tau_t}, \quad \overline{S} = -\frac{1}{\overline{S}_{m0}}\frac{\partial \Phi}{\partial \overline{z}}, \tag{6.8}$$

$$U = \exp\left(\overline{\alpha}_0\Phi + \overline{\alpha}_0\overline{S}_{m0}\overline{z} + \overline{\beta}\tau_t\right). \tag{6.9}$$

Then the system of equations (6.5), (6.6) is transformed into the single linear equation with respect to U

$$\frac{\partial^2 U}{\partial \overline{z}\partial \tau_t} = \overline{\beta}\,\overline{\alpha}_0\overline{S}_{m0}U. \tag{6.10}$$

The boundary conditions for U follow from the solution of the equations.

$$\frac{\partial \ln U}{\partial \overline{z}} = \overline{\alpha}_0\overline{S}_{m0}\left(1-\overline{S}\right), \tag{6.11}$$

$$\frac{\partial \ln U}{\partial \tau_t} = \overline{\alpha}_0\overline{C} + \overline{\beta}. \tag{6.12}$$

Equation (6.11) is solved at τ_t=0, so that taking into account the second condition (6.7) will be

$$U(\overline{z},0) = U(0,0)\, e^{\overline{\alpha}_0\overline{S}_{m0}\overline{z}}. \tag{6.13}$$

The solution of equation (6.12) at $\overline{z} = 0$ taking into account the first condition (6.7) gives

$$U(0,\tau_t) = U(0,0)\exp\left[\overline{\beta}\tau_t + \overline{\alpha}_0\int_0^{\tau_t}\overline{C}_0(\varsigma)d\varsigma\right]. \tag{6.14}$$

By applying the Laplace transform on the variable $\overline{z}$, equation (6.10) and conditions (6.13), (6.14) are transformed into the problem with respect to the image function U_L

$$\frac{dU_L}{d\tau_t} - \frac{\bar{\beta}\bar{\alpha}_0\bar{S}_{m0}}{p}U_L = \frac{1}{p}\frac{d}{d\tau_t}U(0,\tau_t), \tag{6.15}$$

$$U_L(p,0) = \frac{U(0,0)}{p - \bar{\alpha}_0\bar{S}_{m0}} \tag{6.16}$$

The solution of (6.15), (6.16) has the form

$$\begin{aligned} U_L(p,\tau_t) = U(0,0)\Bigg[&\frac{\bar{\alpha}_0\bar{S}_{m0}}{p(p-\bar{\alpha}_0\bar{S}_{m0})}e^{\frac{\bar{\beta}\bar{\alpha}_0\bar{S}_{m0}\tau_t}{p}}+ \\ &+\frac{\widetilde{U}(\tau_t)}{p} + \frac{\bar{\beta}\bar{\alpha}_0\bar{S}_{m0}}{p^2}\int_0^{\tau_t}\widetilde{U}(\varsigma)\,e^{\frac{\bar{\beta}\bar{\alpha}_0\bar{S}_{m0}(\tau_t-\varsigma)}{p}}d\varsigma\Bigg], \end{aligned} \tag{6.17}$$

where $\widetilde{U}(\tau_t) = U(0,\tau_t)/U(0,0)$. When inverting (6.17), it is taken into account that

$$L^{-1}[e^{\frac{a}{p}\bar{z}}/p] = I_0(2\sqrt{a\bar{z}}), \qquad L^{-1}[e^{\frac{a}{p}\bar{z}}/p^2] = \sqrt{\frac{\bar{z}}{a}}\,I_1(2\sqrt{a\bar{z}}),$$

$$L^{-1}\left[\frac{e^{\frac{a}{p}\bar{z}}}{p(p-\bar{\alpha}_0\bar{S}_{m0})}\right] = \int_0^{\bar{z}} e^{\bar{\alpha}_0\bar{S}_{m0}(\bar{z}-\eta)}I_0(2\sqrt{a\eta})d\eta$$

Then the expression for the original function $U(\bar{z},\tau)$ takes the following form

$$\begin{aligned} U(\bar{z},\tau_t) = U(0,0)\Bigg[&\bar{\alpha}_0\bar{S}_{m0}\int_0^{\bar{z}} e^{\bar{\alpha}_0\bar{S}_{m0}(\bar{z}-\eta)}I_0\left(2\sqrt{\bar{\beta}\bar{\alpha}_0\bar{S}_{m0}\tau_t\eta}\right)d\eta+ \\ &+\frac{U(0,\tau_t)}{U(0,0)} + \frac{\sqrt{\bar{\beta}\bar{\alpha}_0\bar{S}_{m0}\bar{z}}}{U(0,0)}\int_0^{\tau_t}U(0,\varsigma)\frac{I_1\left(2\sqrt{\bar{\beta}\bar{\alpha}_0\bar{S}_{m0}\bar{z}(\tau_t-\varsigma)}\right)}{\sqrt{\tau_t-\varsigma}}d\varsigma\Bigg]. \end{aligned} \tag{6.18}$$

From (6.11), (6.12) the following representations for the desired concentrations $\bar{C}$, $\bar{\bar{S}}$ are derived

$$\bar{C}(\bar{z},\tau_t) = \frac{U_1(\bar{z},\tau_t)}{\bar{\alpha}_0 U(\bar{z},\tau_t)} - \frac{\bar{\beta}}{\bar{\alpha}_0}, \tag{6.19}$$

$$\bar{\bar{S}}(\bar{z},\tau_t) = 1 - \frac{U_2(\bar{z},\tau_t)}{\bar{\alpha}_0\bar{S}_{m0}U(\bar{z},\tau_t)}, \tag{6.20}$$

$$\begin{aligned} U_1(\bar{z},\tau_t) = \frac{\partial}{\partial\tau_t}U(\bar{z},\tau_t) = U(0,0)\Bigg\{&\left(\bar{\beta}\bar{\alpha}_0\bar{S}_{m0}\bar{z} + \bar{\alpha}_0\bar{C}_0(\tau_t) + \bar{\beta}\right)\frac{U(0,\tau_t)}{U(0,0)}+ \\ &+\sqrt{\frac{\bar{\beta}\bar{\alpha}_0^3\bar{S}_{m0}^3}{\tau_t}}\int_0^{\bar{z}}\sqrt{\varsigma}\,e^{\bar{\alpha}_0\bar{S}_{m0}(\bar{z}-\eta)}I_1\left(2\sqrt{\bar{\beta}\bar{\alpha}_0\bar{S}_{m0}\tau_t\eta}\right)d\eta + \frac{\sqrt{\bar{\beta}\bar{\alpha}_0\bar{S}_{m0}\bar{z}}}{U(0,0)}\times \\ &\times\int_0^{\tau_t}U(0,\varsigma)\left[\frac{\sqrt{\bar{\beta}\bar{\alpha}_0\bar{S}_{m0}\bar{z}}}{\tau_t-\varsigma}I_0\left(2\sqrt{\bar{\beta}\bar{\alpha}_0\bar{S}_{m0}\bar{z}(\tau_t-\varsigma)}\right) - \frac{I_1\left(2\sqrt{\bar{\beta}\bar{\alpha}_0\bar{S}_{m0}\bar{z}(\tau_t-\varsigma)}\right)}{(\tau_t-\varsigma)^{3/2}}\right]d\varsigma\Bigg\}, \end{aligned} \tag{6.21}$$

$$U_2(\bar{z},\tau_t) = \frac{\partial}{\partial \bar{z}} U(\bar{z},\tau_t) = U(0,0)\Big[\bar{\alpha}_0 \bar{S}_{m0} I_0\big(2\sqrt{\bar{\beta}\,\bar{\alpha}_0 \bar{S}_{m0}\bar{z}\,\tau_t}\big) + (\bar{\alpha}_0 \bar{S}_{m0})^2 \times$$

$$\times \int_0^{\bar{z}} e^{\bar{\alpha}_0 \bar{S}_{m0}(\bar{z}-\eta)} I_0\big(2\sqrt{\bar{\beta}\,\bar{\alpha}_0 \bar{S}_{m0}\tau_t \eta}\big)\, d\eta + \frac{\bar{\beta}\,\bar{\alpha}_0 \bar{S}_{m0}}{U(0,0)} \int_0^{\tau_t} U(0,\varsigma) I_0\big(2\sqrt{\bar{\beta}\,\bar{\alpha}_0 \bar{S}_{m0}\bar{z}(\tau_t-\varsigma)}\big)\, d\xi\Big]. \tag{6.22}$$

When dividing U_1 and U_2 in (6.19), (6.20) by U, the parameter $U(0,0)$ is reduced and hence the concentrations $\bar{C}$, $\bar{\bar{S}}$ are independent of it.

From formulas (6.19), (6.20) there follows a number of the important consequences for practice. First of all, the expressions for $\bar{C}$, $\bar{\bar{S}}$ essentially simplified at C_0=*const*, $\bar{C}_0 = 1$

$$\bar{C}(\bar{z},\tau_t) = \frac{I_0\big(2\sqrt{\bar{\beta}\,\bar{\alpha}_0 \bar{S}_{m0}\bar{z}\,\tau_t}\big) + (\bar{\alpha}_0+\bar{\beta})\, G_1(\bar{z},\tau_t)}{(\bar{\alpha}_0+\bar{\beta})\, G_1(\bar{z},\tau_t) - \bar{\beta}\, G_2(\bar{z},\tau_t) + e^{\bar{\alpha}_0 \bar{S}_{m0}\bar{z}+\bar{\beta}\tau_t}}, \tag{6.23}$$

$$\bar{\bar{S}}(\bar{z},\tau_t) = \frac{\bar{\alpha}\, G_1(\bar{z},\tau_t)}{(\bar{\alpha}_0+\bar{\beta})\, G_1(\bar{z},\tau_t) - \bar{\beta}\, G_2(\bar{z},\tau_t) + e^{\bar{\alpha}_0 \bar{S}_{m0}\bar{z}+\bar{\beta}\tau_t}}, \tag{6.24}$$

where

$$G_1(\bar{z},\tau_t) = \int_0^{\tau_t} e^{(\bar{\alpha}_0+\bar{\beta})\varsigma} I_0\big(2\sqrt{\bar{\beta}\,\bar{\alpha}_0 \bar{S}_{m0}\bar{z}(\tau_t-\varsigma)}\big)\, d\zeta,$$

$$G_2(\bar{z},\tau_t) = \int_0^{\tau_t} e^{\bar{\beta}\varsigma} I_0\big(2\sqrt{\bar{\beta}\,\bar{\alpha}_0 \bar{S}_{m0}\bar{z}(\tau_t-\varsigma)}\big)\, d\zeta. \tag{6.25}$$

The dependence describing the accumulation of the deposited suspension particles in the inlet cross-section of a bed in generalized terms has the following form

$$\bar{\bar{S}}_0(\bar{t}) = \bar{\alpha}_0 \int_0^{\bar{t}} \bar{C}_0(\xi) \exp\Big[\bar{\beta}(\xi-\bar{t}) - \bar{\alpha}_0 \int_0^{\bar{t}} \bar{C}_0(\varsigma)d\varsigma\Big] d\xi. \tag{6.26}$$

If formally the quantity $\bar{C}_0$ tends asymptotically (at $\bar{t} \to \infty$) to the value $\bar{C}_\infty$, then it is true that

$$\bar{\bar{S}}_0(\infty) = \frac{\bar{\alpha}_0 \bar{C}_\infty}{\bar{\alpha}_0 \bar{C}_\infty + \bar{\beta}}.$$

The solution of the filtration flow compartment is expressed by the formula for calculating the distribution of the reduced head (losses) $\tilde{h}$ within the layer $[\bar{z}, 1]$

$$\tilde{h}(\bar{z},\tau_t) = \int_{\bar{z}}^{1} \frac{d\eta}{f_k(\bar{\bar{S}}(\eta,\tau_t))}. \tag{6.27}$$

From (6.27) follows the dependence of the relative head losses on τ_t

$$\Delta\tilde{h}(\tau_t) = \int_0^{1} \frac{d\eta}{f_k(\bar{\bar{S}}(\eta,\tau_t))}. \tag{6.28}$$

When carrying out the technological calculations, it is reasonable to identify the variable τ_t with $\bar{t}$ due to $t >> n_e z$. Then the relative first technological time ($\bar{t}_p$) at a given value $\overline{C}_*$ should be determined according to (6.23) from the equation

$$I_0\left(2\sqrt{\beta\,\overline{\alpha}_0 \overline{S}_{m0} \bar{t}_p}\right) + (1 - \overline{C}_*)\,(\overline{\alpha}_0 + \beta)\, G_1\left(1, \bar{t}_p\right) + \beta\, \overline{C}_*\, G_2\left(1, \bar{t}_p\right) - \overline{C}_*\, e^{\overline{\alpha}_0 \overline{S}_{m0} + \beta \bar{t}_p} = 0. \tag{6.29}$$

Due to constancy of the filtration rate, the relative second technological time ($\bar{t}_h$) is proposed to be calculated from the equation

$$\int_0^1 \frac{d\eta}{f_k(\overline{S}(\eta, \bar{t}_h))} = \Delta \bar{h}_*, \tag{6.30}$$

involving the expression for $\overline{S}$ (6.24).

The possibilities that arise using the above exact solution to calculate some important filtration characteristics and parameters are demonstrated by a number of test examples. The results of numerous calculations are shown graphically. The subject of a quantitative analysis became the relative quantities of particular practical interest, namely, the outlet concentration $\overline{C}_e$, normalized hydraulic conductivity f_k, head losses in the filter bed $\Delta\bar{h}$ and finally the technological time $\bar{t}_p$. It was assumed that the filter operates under conditions of linearly increasing load of disperse impurity, i.e.

$$\overline{C}_0(\bar{t}) = 1 + a_c \bar{t},$$

where the coefficient a_c varied either discretely (0, 0.001, 0.002, 0.004) or continuously in the range from 0 to 0.004. The initial concentration of the dispersed impurity was quite high for the additional treatment of turbid (waste) water $\left(C_0^0 = 5 \cdot 10^{-5}\right)$. Then the relative parameter $\overline{S}_m$ was 500 at the typical porosity of a clean bed (n_0=0.4), the solids take-up capacity of a filter media and the deposit composition (S_m=0.01, $\gamma_a S_m$=0.5n_0, γ_a=20). The value of the coefficient β was fixed (0.005), and two values were accepted for the second mass exchange coefficient ($\overline{\alpha}_0$=0.008 and 0.016), which are characteristic of the filtering materials with a moderate and high absorption capacity. Finally, in order to estimate the effects of prolonged filtration on the hydraulic characteristics, the dependence of f_k on $\overline{S}$ of the standard form was taken

$$f_k(\overline{S}) = \left(1 - \bar{\gamma}_a \overline{S}_{m0} \overline{S}\right)^3,$$

where $\bar{\gamma}_a = \gamma_a C_0^0$.

Figure 1. Change in the relative volumetric concentration of the suspended solids over time

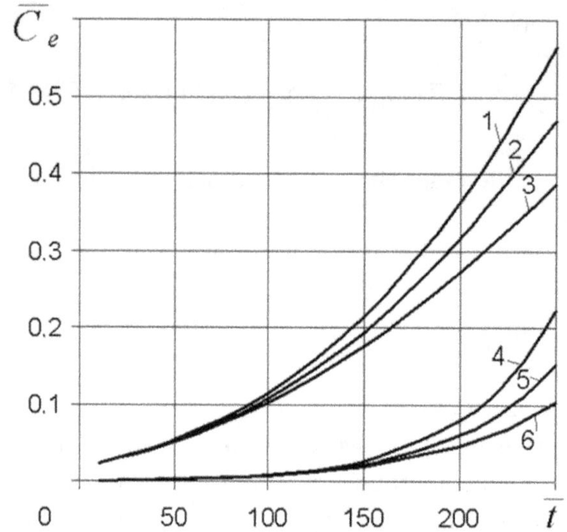

Note. 1-3 – $\overline{\alpha}_0 \overline{S}_{m0}$ = 4; 4-6 – $\overline{\alpha}_0 \overline{S}_{m0}$ = 8; 1, 4 – a_c=0.002; 2, 5 – a_c=0.001; 3, 6 – a_c=0

Figure 2. Profiles of the relative hydraulic conductivity

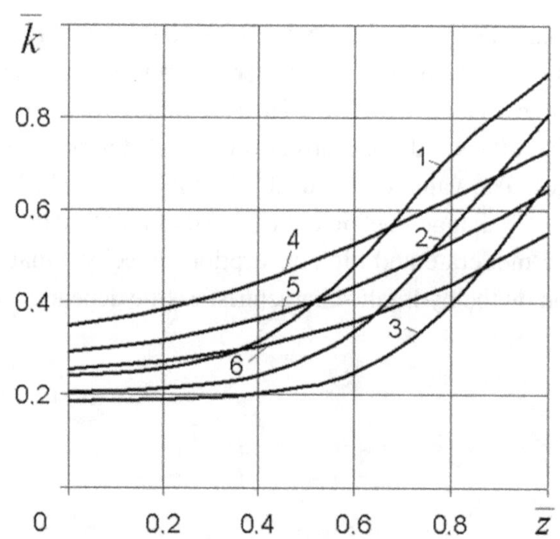

Note. 1-3 – $\overline{\alpha}_0 \overline{S}_{m0}$ = 8; 4-6 – $\overline{\alpha}_0 \overline{S}_{m0}$ = 4; 1, 4 – a_c=0; 2, 5 – a_c=0.002; 3, 6 – a_c=0.004

The progress of the separation of an aqueous suspension and a gradual increase in the amount of the suspended solids at the filter outlet is clearly illustrated by **Figure 1**, which shows the curves of increasing the quantity $\overline{C}_e$ over a long time. Since the calculations were performed from the beginning of the suspension entering the bed, it took a short time (approximately $\overline{n}_e$) to treat the first portion of the suspension. Therefore, the calculated curves are shifted right to the same extent of the ordinate axis. In this case, the maximum clarifying effect is achieved, which is characterized by the primary calculated value of the outlet concentration $e^{-\overline{\alpha}_0\overline{S}_{m0}}$ and $\overline{C}_e(\overline{n}_e)$ was 0.0183 at $\overline{\alpha}_0\overline{S}_{m0} = 4$ and 0.0003 at $\overline{\alpha}_0\overline{S}_{m0} = 8$ in the considered examples.

It is logical, that a significant improvement of the absorption properties of the filtering material leads to a sharp increase in the filtrate quality also in the case of nonlinear filtration, although the protective capability of the bed decreases faster than in the case of linear filtration (α=*const*). Also, the behavior of the function $\overline{C}_0(\overline{t})$ is significant for the suspension clarification. And it is natural that we should expect appreciable increase in the breakthrough of its particles with an increase in the initial content of impurity, namely, here by several tens of percent.

Figure 3. Change in the reduced head losses in the bed over time

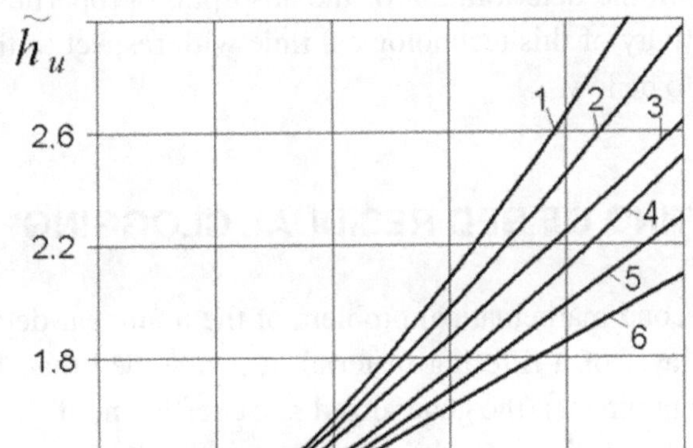

Note. 1-3 – $\overline{\alpha}_0\overline{S}_{m0} = 8$; 4-6 – $\overline{\alpha}_0\overline{S}_{m0} = 4$; 1, 4 – a_c=0.002; 2, 5 – a_c=0.001; 3, 6 – a_c=0

Figure 4. Dependence $\overline{t}_p(a_c)$

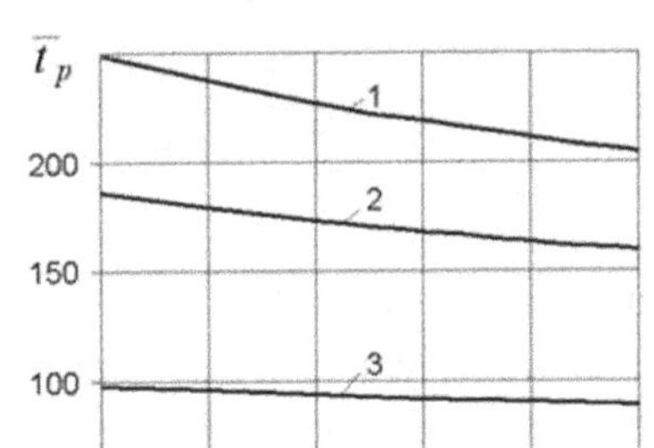

Note. 1 – $\overline{\alpha}_0\overline{S}_{m0} = 8$; 2 – $\overline{\alpha}_0\overline{S}_{m0} = 6$; 3 – $\overline{\alpha}_0\overline{S}_{m0} = 4$

Naturally, a decrease in the absorption rate of the suspended solids, as the deposit accumulates, has a significant and different effect not only on both concentrations, but also on the filtration flow characteristics. Attention on various consequences of the deterioration of the absorption capacity of the filter media will be emphasized in Sections 6.2, 6.4. For the time being it is only possible to state that instability of C_0 can seriously and comprehensively affects the technological process also at the nonlinear mass exchange kinetics. Firstly, a noticeable decrease in the permeability of a filter media is possible, both as it clogging and as a_c increases (**Figure 2**). The permeability of the porous medium decreases to the greatest extent at its inlet.

Obviously, such a decrease will be obviously larger at $f_\alpha=\alpha_0=const$. In general, the effect of a_c both on f_k, and on the relative head losses (**Figure 3**) is insignificant here, although it is still reasonable to take it into account in the predictive calculations of nonlinear filtration. Finally, the dependence $\bar{t}_p(a_c)$ was determined on the basis of (6.19). The corresponding curves in **Figure 4** are becoming more straight, and their slope decreases with the deterioration of the absorption properties of the above-mentioned media. In general, the sensitivity of this technological time with respect to the rate of a change in the inlet quantity C_0 turned out to be low.

6.2. STRICT ACCOUNTING OF BED RESIDUAL CLOGGING

When formulating the second mathematical problem of the nonlinear detachment filtration of suspension through a uniform layer of a filtering material, it is assumed that, firstly, for the absorption function of the dispersed contaminants (the generalized rate coefficient of suspension particles sticking or attachment α for the case of an active deposit) the representation is also valid in the form (6.1), (6.2); secondly, by the beginning of the successive run of the filter operation some amount of the deposit is preserved in the packed bed after the previous washing. The residual deposit is distributed more or less evenly along the bed height after its fluidization with intensive grains mixing. The initial concentration of the deposited particles S^0 is, therefore, assumed to be constant, although this assumption is not essential. Then the clarification compartment of the general suspension filtration model with nonlinear kinetics takes the following dimensionless form (Poliakov, 2014)

$$\frac{\partial \bar{C}}{\partial \bar{z}} + \bar{S}_m \frac{\partial \bar{S}}{\partial \bar{t}} = 0, \tag{6.31}$$

$$\frac{\partial \bar{S}}{\partial \bar{t}} = \bar{\alpha}_0 \bar{S}_m (1 - \bar{S}) \bar{C} - \bar{\beta} \bar{S} \; ; \tag{6.32}$$

$$\bar{t} = 0, \qquad \bar{S} = \bar{S}^0 \; ; \tag{6.33}$$

and also includes the standard boundary condition (2.9). Here $\bar{S}_m = S_m/(n_0 C_0)$, $\bar{S}^0 = S^0/S_m$. In order to avoid duplication of the solution procedure and deductions from Section 6.1, its course is given below with abbreviations and emphasis is placed on the features related to the primary clogging of the bed.

First of all, the following substitution of the dependent and independent variables is performed

$$Z = \bar{S}_m \bar{z}, \quad \bar{C} = \frac{\partial \Phi}{\partial \bar{t}}, \quad \bar{S} = -\frac{\partial \Phi}{\partial Z}. \tag{6.34}$$

Thus, equation (6.31) is automatically satisfied and we can proceed to the solution of a single equation, namely, the transformed kinetic equation. Then one more substitution of the already introduced function is made by analogy with (6.9)

$$U = \exp\left(\bar{\alpha}_0 \Phi + \bar{\alpha}_0 Z + \bar{\beta}\bar{t}\right). \tag{6.35}$$

As a result, the following equation with respect to the function $U(Z,\bar{t})$ is obtained after cumbersome transformations

$$\frac{\partial^2 U}{\partial Z \partial \bar{t}} = \bar{\alpha}_0 \bar{\beta} U. \tag{6.36}$$

The boundary and initial conditions follow from the system of equations

$$\frac{\partial \ln U}{\partial Z} = \bar{\alpha}_0\left(1 - \bar{S}\right), \tag{6.37}$$

$$\frac{\partial \ln U}{\partial \bar{t}} = \bar{\alpha}_0 \bar{C} + \bar{\beta}. \tag{6.38}$$

The values $\bar{t} = 0$ and $Z=0$ are successively accepted in equations (6.37), (6.38), and then they are integrated. As a result, the desired conditions are transformed to the form

$$U(Z,0) = U(0,0)\, e^{\bar{\alpha}_0(1-\bar{S}^0)Z}, \tag{6.39}$$

$$U(0,\bar{t}) = U(0,0)\, e^{(\bar{\alpha}_0+\bar{\beta})\bar{t}}. \tag{6.40}$$

We apply the integral Laplace transform on the variable $\bar{t}$ to equation (6.36). Then problem (6.36), (6.39), (6.40) reduces to the problem with respect to the image function $U_L(Z,p)$, namely,

$$\frac{dU_L}{dZ} - \frac{\bar{\alpha}_0\bar{\beta}}{p} U_L = \frac{\bar{\alpha}_0(1 - \bar{S}^0)\, U(0,0)}{p} e^{\bar{\alpha}_0(1-\bar{S}^0)Z}, \tag{6.41}$$

$$Z = 0, \quad U_L = \frac{U(0,0)}{p - \bar{\alpha}_0 - \bar{\beta}}. \tag{6.42}$$

The solution of problem (6.41), (6.42) is expressed by the dependence

$$U_L(Z,p) = U(0,0) e^{\frac{\bar{\alpha}_0\bar{\beta}Z}{p}} \left(\frac{1}{p - \bar{\alpha}_0 - \bar{\beta}} + \bar{\alpha}_0(1 - \bar{S}^0) \int_0^Z e^{\bar{\alpha}_0(1-\bar{S}^0)\xi} \frac{e^{-\frac{\bar{\alpha}_0\bar{\beta}\xi}{p}}}{p} d\xi \right). \tag{6.43}$$

As a result of the inversion of the integral function (6.43) we obtain

$$\frac{U(z,\bar{t})}{U(0,0)} = I_0\left(2\sqrt{\bar{\beta}\bar{\alpha}_0 \bar{S}_m z \bar{t}}\right) + \left(\bar{\alpha}_0 + \bar{\beta}\right) G_1(z,\bar{t}) + \bar{\alpha}_0 \bar{S}_m (1 - \bar{S}^0)\, G_3(z,\bar{t}). \tag{6.44}$$

Here

$$G_1(\bar{z},\bar{t}) = \int_0^{\bar{t}} e^{(\bar{\alpha}_0+\bar{\beta})\eta} I_0\left(2\sqrt{\bar{\beta}\bar{\alpha}_0\bar{S}_m\bar{z}(\bar{t}-\eta)}\right)d\eta,$$

$$G_3(\bar{z},\bar{t}) = \int_0^{\bar{z}} e^{\bar{\alpha}_0\bar{S}_m(1-\bar{S}^0)\xi} I_0\left(2\sqrt{\bar{\beta}\bar{\alpha}_0\bar{S}_m\bar{t}(\bar{z}-\xi)}\right)d\xi.$$

The sought concentrations $\bar{C}$, $\bar{S}$ are related to the function $U(\bar{z},\bar{t})$ and its two derivatives by analogy with (6.19), (6.20) as follows

$$\bar{C}(\bar{z},\bar{t}) = -\frac{\bar{\beta}}{\bar{\alpha}_0} + \frac{U_1(\bar{z},\bar{t})}{\bar{\alpha}_0 U(\bar{z},\bar{t})}, \tag{6.45}$$

$$\bar{S}(\bar{z},\bar{t}) = 1 - \frac{U_2(\bar{z},\bar{t})}{\bar{\alpha}_0\bar{S}_m U(\bar{z},\bar{t})}. \tag{6.46}$$

Here the well-known formula is used in the derivations

$$I_1(u)+u\frac{dI_1(u)}{du} = u\,I_0(u)$$

and the equality

$$\frac{\partial^2}{\partial z\,\partial t}I_0(2\sqrt{\beta\bar{\alpha}_0\bar{S}_m zt}\,)=\beta\bar{\alpha}_0\bar{S}_m\,I_0(2\sqrt{\beta\bar{\alpha}_0\bar{S}_m zt}\,).$$

Then after a number of transformations we obtain

$$\begin{aligned}U_1(\bar{z},\bar{t}) &= \frac{\partial}{\partial\bar{t}}U(z,t)= U(0,0)\left[(\bar{\alpha}_0+\bar{\beta})\,I_0\left(2\sqrt{\bar{\beta}\bar{\alpha}_0\bar{S}_m\bar{z}\bar{t}}\right)+\right.\\ &\left.+(\bar{\alpha}_0+\bar{\beta})^2G_1(\bar{z},\bar{t})+\bar{\beta}\bar{\alpha}_0\bar{S}_m G_3(\bar{z},\bar{t})\right],\end{aligned} \tag{6.47}$$

$$\begin{aligned}U_2(\bar{z},\bar{t}) &= \frac{\partial}{\partial\bar{z}}U(z,t)= U(0,0)\left[\bar{\alpha}_0\bar{S}_m I_0\left(2\sqrt{\bar{\beta}\bar{\alpha}_0\bar{S}_m\bar{z}\bar{t}}\right)+\right.\\ &\left.+\bar{\beta}\bar{\alpha}_0\bar{S}_m G_1(\bar{z},\bar{t}) + (\bar{\alpha}_0\bar{S}_m)^2G_3(\bar{z},\bar{t})\right].\end{aligned} \tag{6.48}$$

Substitution of (6.47), (6.48) into formulas (6.45), (6.46) gives such final expressions for the relative concentrations of the suspended and deposited solids as follows

$$\bar{C}(\bar{z},\bar{t}) = \frac{I_0\left(2\sqrt{\bar{\beta}\bar{\alpha}_0\bar{S}_m\bar{z}\bar{t}}\right) + (\bar{\alpha}_0+\bar{\beta})\,G_1(\bar{z},\bar{t}) + \bar{\beta}\bar{\alpha}_0\bar{S}_m\bar{S}^0G_3(\bar{z},\bar{t})}{I_0\left(2\sqrt{\bar{\beta}\bar{\alpha}_0\bar{S}_m\bar{z}\bar{t}}\right) + (\bar{\alpha}_0+\bar{\beta})\,G_1(\bar{z},\bar{t}) + \bar{\alpha}_0\bar{S}_m(1-\bar{S}^0)G_3(\bar{z},\bar{t})}, \tag{6.49}$$

$$\bar{S}(\bar{z},\bar{t}) = \frac{\bar{S}^0I_0\left(2\sqrt{\bar{\beta}\bar{\alpha}_0\bar{S}_m\bar{z}\bar{t}}\right) + \bar{\alpha}_0G_1(\bar{z},\bar{t}) + \bar{\alpha}_0\bar{S}_m(1-\bar{S}^0)\bar{S}^0G_3(\bar{z},\bar{t})}{I_0\left(2\sqrt{\bar{\beta}\bar{\alpha}_0\bar{S}_m\bar{z}\bar{t}}\right) + (\bar{\alpha}_0+\bar{\beta})\,G_1(\bar{z},\bar{t}) + \bar{\alpha}_0\bar{S}_m(1-\bar{S}^0)G_3(\bar{z},\bar{t})}. \tag{6.50}$$

Hence, in particular, a simple pattern of the deposited particles accumulation in the inlet cross-section of a bed $\bar{S}_0(\bar{t})$will be, namely,

$$\overline{S}_0(\bar{t}) = \frac{\overline{\alpha}_0}{\overline{\alpha}_0 + \overline{\beta}} + \left(\overline{S}{}^0_0 - \frac{\overline{\alpha}_0}{\overline{\alpha}_0 + \overline{\beta}}\right) e^{-(\overline{\alpha}_0 + \overline{\beta})\bar{t}}.$$

When the deposit detachment is negligible or it is possible to introduce an effective attachment rate coefficient that generalizes both sides of the mass exchange between the solid and liquid phases of a bed, it is justified to assume β=0 and then both concentrations are described by the following elementary functions

$$\overline{C}(\overline{z},\bar{t}) = \frac{e^{\overline{\alpha}_0\bar{t}} + \overline{S}{}^0 - 1}{e^{\overline{\alpha}_0\bar{t}} + e^{\overline{\alpha}_0\overline{S}_m(1-\overline{S}{}^0)\overline{z}} - 1}, \tag{6.51}$$

$$\overline{S}(\overline{z},\bar{t}) = \frac{e^{\overline{\alpha}_0\bar{t}} + \overline{S}{}^0 e^{\overline{\alpha}_0\overline{S}_m(1-\overline{S}{}^0)\overline{z}} - 1}{e^{\overline{\alpha}_0\bar{t}} + e^{\overline{\alpha}_0\overline{S}_m(1-\overline{S}{}^0)\overline{z}} - 1}. \tag{6.52}$$

If there was no initial deposit in a bed, the known simpler dependences follow from (6.49), (6.50)

$$\overline{C}(\overline{z},\bar{t}) = \frac{I_0\left(2\sqrt{\overline{\beta}\,\overline{\alpha}_0\overline{S}_m\overline{z}\bar{t}}\right) + (\overline{\alpha}_0 + \overline{\beta})\, G_1(\overline{z},\bar{t})}{I_0\left(2\sqrt{\overline{\beta}\,\overline{\alpha}_0\overline{S}_m\overline{z}\bar{t}}\right) + (\overline{\alpha}_0 + \overline{\beta})\, G_1(\overline{z},\bar{t}) + \overline{\alpha}_0\overline{S}_m G_3(\overline{z},\bar{t})}, \tag{6.53}$$

$$\overline{S}(\overline{z},\bar{t}) = \frac{\overline{\alpha}_0\, G_1(\overline{z},\bar{t})}{I_0\left(2\sqrt{\overline{\beta}\,\overline{\alpha}_0\overline{S}_m\,\overline{z}\bar{t}}\right) + (\overline{\alpha}_0 + \overline{\beta})\, G_1(\overline{z},\bar{t}) + \overline{\alpha}_0\overline{S}_m G_3(\overline{z},\bar{t})} \tag{6.54}$$

and when calculating the integral $G_3(\overline{z},\bar{t})$ we should assume $\overline{S}{}^0 = 0$.

The most important for the technological analysis of the filter operation is a special case of dependence (6.49), which allows to reliably predict changes in the filtrate quality and to establish a critical moment in time when the suspended solids content in it reaches the maximum permissible value. Then the formula for the calculation of the relative outlet concentration of the suspended solids takes the following form

$$\overline{C}_e(\bar{t}) = \frac{I_0\left(2\sqrt{\overline{\beta}\,\overline{\alpha}_0\overline{S}_m\,\bar{t}}\right) + (\overline{\alpha}_0 + \overline{\beta})\, G_1(1,\bar{t}) + \overline{\beta}\,\overline{\alpha}_0\,\overline{S}_m\overline{S}{}^0 G_3(1,\bar{t})}{I_0\left(2\sqrt{\overline{\beta}\,\overline{\alpha}_0\overline{S}_m\,\bar{t}}\right) + (\overline{\alpha}_0 + \overline{\beta})\, G_1(1,\bar{t}) + \overline{\alpha}_0\overline{S}_m(1 - \overline{S}{}^0)\, G_3(1,\bar{t})} \tag{6.55}$$

Filtration flow characteristics, as well as the second technological time (t_h) are determined identically for the conditions of the problems from the first Sections of this Chapter, and they differ only by the expressions for $\overline{S}$. The first technological time (t_p) is calculated in the standard way at the given $\overline{C}_*$ on the basis of (6.55).

The dependencies derived above were used for the calculations of the relative functions-concentrations $\overline{C}_e(\bar{t})$ and $\overline{S}(\overline{z};\bar{t})$, which were performed in order to estimate the significance of the initial clogging of a bed for the suspension clarification. Such information is useful for selecting the method of filter washing. Data on the model parameters primarily included their fixed values such as: $\overline{\alpha}_0$=1.6•10^{-3}, $\overline{\beta}$=5•10^{-3}, $\overline{S}_m$ = 5000. The degree of clogging of the washed bed was varied, namely, the values taken were $\overline{S}{}^0$ = 0, 0.02, 0.05. The series of curves of the dependence $\overline{C}_e(\bar{t})$ calculated by formula (6.55) is shown in **Figure 5**. If the bed was clean, then after the contamination front passing through it the concentration of the suspended solids at the outlet is equal only to 5.7•10^{-4}. At the same time 90% and more of contaminants should be removed from water up to the moment in time equal to 522 in relative units. The suspension separation significantly deteriorated already at $\overline{S}^0 = 0.02$. In this case, due to the intensive detachment of the deposit remaining after washing, the concentration of the suspended solids in the filtrate immediately amounted to 0.069, and reached the level of 0.1 by the relative moment in time 318. In general, if the preliminary contamination of the bed is sufficiently significant, the quality of the filtrate at the initial stage of filtration, determined by the deposit content and the filtration rate, may be too low. It is not difficult to find from the basic model the primary relative amount (specific volume) of the suspended solids in the clarified water $\overline{C}_e(0)$. This amount actually does not depend on the level of contamination of water entering the filter, can also be labeled $\overline{C}_e^0$ and is as follows

$$\overline{C}_e(0) = \overline{C}_e^0 = \frac{\overline{\beta}\,\overline{S}^0}{\overline{\alpha}_0\left(1-\overline{S}^0\right)} \tag{6.56}$$

It is obvious that the suspension clarification on a rapid filter has no sense at all, if the inequality is true

$$\overline{S}^0 \geq \frac{\overline{\alpha}_0\,\overline{C}_*}{\overline{\alpha}_0\,\overline{C}_* + \overline{\beta}}$$

where $\overline{C}_*$ is the relative maximum permissible volumetric concentration of the suspended matter in the filtrate. The filter is inoperable immediately after turning on at the accepted values of $\overline{\alpha}_0$, $\overline{\beta}$ and $\overline{C}_* = 0.1$ already when $\overline{S}^0 \geq 0.029$. If $\overline{S}^0 = 0.05$, then $\overline{C}_e(0)$ is equal to 0.176 according to (6.56).

Figure 5. Increase in the relative volumetric concentration of the suspended solids in filtrate over time

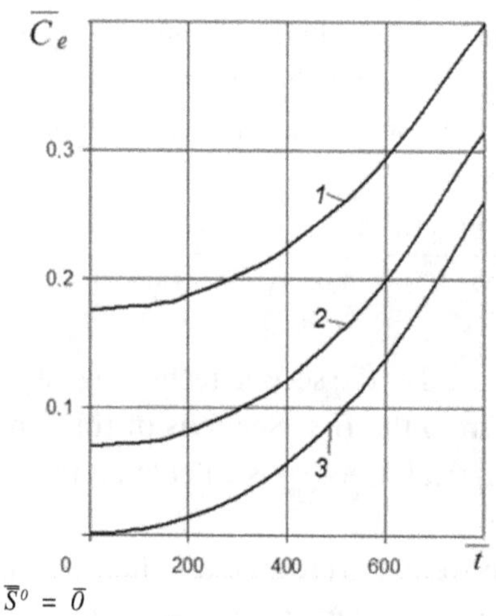

Note. 1 – $\overline{S}^0 = 0.05$, 2 – $\overline{S}^0 = 0.02$, 3 – $\overline{S}^0 = \overline{0}$

Figure 6. Profiles of the relative volumetric concentration of the deposited particles

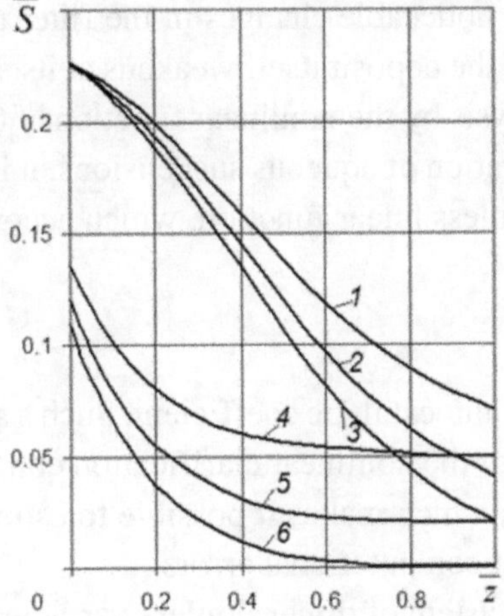

Note. 1-3 – $\overline{t}$ = 600, 4-6 – $\overline{t}$ = 100; 1,4 – $\overline{S}^o$ = 0.05, 2-5 – $\overline{S}^o$ = 0.02, 3-6 – $\overline{S}^o$ = 0

When analyzing the clogging of a bed, it is of practical interest to consider the changes in the concentration of the deposited particles (deposit) not only over time, but also along height. Indeed, it is the vertical distribution of the deposit that determines the total head losses in the filter, which can be the main limiting factor for the filtration process. Calculated using formula (6.50) for two time points, the concentration profiles $\overline{S}(\overline{z})$ corresponded to the start ($\overline{t}$ = 100) and end ($\overline{t}$ = 600) of the water clarification during one filter run (**Figure 6**). It is noteworthy that the initial deposit amount, despite its intensive detachment by the filtration flow over time, not only did not decrease, but increased, and even in the outlet cross-section of the bed.

This fact is explained by the constant presence of a quite large number of the suspended particles in the entire bed due to both their arrival to the filter with the untreated water and the mentioned detachment phenomenon, which ensures the mobilization of a significant amount of the deposited matter from the beginning of the filter operation. Just the absorption of suspended solids coming to the filtering material from different sources compensated excessively for the loss of deposit due to its mobilization.

A detailed comparative analysis of the main filtration characteristics and the technological times for a uniform and two-layer bed at linear and nonlinear mass exchange kinetics will be performed in Section 6.4.

6.3. STRICT ESTIMATION OF AUTOCATALYTIC EFFECT

When modeling the nonlinear non-detachment filtration the use of the linear function of absorption of dispersed contaminants $f_\alpha(S)$ in the kinetic equation allows in principle correctly to take into account not only gradual decrease in the absorption capacity of the layer of filtering material at its progressive clogging, but also the opposite effect, namely, reinforced fixation of the suspended solids by a deposit. On the one hand, the autocatalytic effect promotes their better deposition, and thus, increases the time of the protective effect of a bed (t_p). On the other hand, the deposit is distributed more unevenly by the bed height and, consequently, other technological time (t_h) are shortened.

In general, this effect can cause noticeable changes in the filter run time. Based on the few experimental data, the clarifying effect of the deposit itself weakens at its significant accumulation. Then it is preferable to describe the autocatalysis by the nonlinear functions $f_\alpha(S)$, $\chi_S(S)$. Nevertheless, when estimating its significance for the separation of aqueous suspensions, it is often sufficient to limit ourselves for $\chi_S(S)$ to the following dimensionless linear function, which according to (6.3) will be

$$\chi_S(\overline{S}) = 1 + \overline{\theta}\,\overline{S}, \tag{6.57}$$

where $\overline{\theta} = n_0 C_0 \theta$ is the relative autocatalytic coefficient. Such a simplified approach is also justified due to the following exact solution of the nonlinear clarification compartment in the case of the detachment filtration with a constant rate, which makes it possible to estimate the autocatalysis contribution to the technological process without computational errors.

Thus, returning to the previous system of dimensionless variables and parameters (Section 2.1) due to the lack of a suitable scale for S in the initial mathematical model, the above compartment contains the mass transfer equation (2.30), kinetic equation

$$\frac{\partial \overline{S}}{\partial \overline{t}} = \overline{\alpha}_0(1 + \overline{\theta}\,\overline{S})\overline{C} - \overline{\beta} S, \tag{6.58}$$

as well as the standard conditions at the inlet to a filter bed (2.9) and at the beginning of the first filter run (clean bed) (2.29). The filtration flow compartment remains unchanged and includes (2.11) - (2.13). The participation of the transportation system of the filtration facility in the implementation of the technological process is not analyzed due to the constancy of the filtration rate.

The solution of the mathematical problem of the suspension clarification in the presence of the autocatalysis largely repeats the solution of the nonlinear filtration problems from the previous Sections of this Chapter. Therefore, only new intermediate results are emphasized further. Thus, first of all, the replacement of variables similar to (6.34), (6.35) is performed

$$U = \exp(\overline{\alpha}_0\overline{z} + \overline{\beta}\,\overline{t} - \overline{\alpha}_0\Phi), \quad \Phi = \frac{1}{\overline{\alpha}_0}(\overline{\alpha}_0\overline{z} + \overline{\beta}\,\overline{t} - \ln U). \tag{6.59}$$

Then the initial system of equations is reduced to the equation with respect to the function $\Phi(\overline{z}, \vec{t})$

$$\frac{\partial^2 \Phi}{\partial \overline{z}\partial \overline{t}} + \overline{\alpha}_0\frac{\partial \Phi}{\partial \overline{t}} - \overline{\alpha}_0\frac{\partial \Phi}{\partial \overline{z}}\frac{\partial \Phi}{\partial \overline{t}} + \overline{\beta}\frac{\partial \Phi}{\partial \overline{z}} = 0. \tag{6.60}$$

In view of

$$\frac{\partial \Phi}{\partial \overline{t}} = \frac{1}{\overline{\alpha}_0}\left(\overline{\beta} - \frac{1}{U}\frac{\partial U}{\partial \overline{t}}\right), \qquad \frac{\partial \Phi}{\partial \overline{z}} = \frac{1}{\overline{\alpha}_0}\left(\overline{\alpha}_0 - \frac{1}{U}\frac{\partial U}{\partial \overline{z}}\right)$$

equation (6.60) is transformed to the analogous form (6.36) and is solved in this case under the conditions

$$\bar{z} = 0, \quad U = U(0,0)e^{-(\bar{\alpha}_0\bar{\theta}-\bar{\beta})\bar{t}}. \tag{6.61}$$

$$\bar{t} = 0, \quad U = U(0,0)e^{\bar{\alpha}_0\bar{z}}. \tag{6.62}$$

The procedure of finding the solution, using the Laplace transform, first in the intermediate form (via images) and then in the final form is omitted. As a result, first of all, obtained

$$U(\bar{z},\bar{t}) = U(0,0)\Big[I_0\big(2\sqrt{\bar{\alpha}_0\bar{\beta}\bar{z}\bar{t}}\big) - \\ -(\bar{\alpha}_0\bar{\theta}-\bar{\beta})\int_0^{\bar{t}} e^{-(\bar{\alpha}_0\bar{\theta}-\bar{\beta})(\bar{t}-\xi)} I_0\big(2\sqrt{\bar{\alpha}_0\bar{\beta}\bar{z}\xi}\big)d\xi + \bar{\alpha}_0\int_0^{\bar{z}} e^{\bar{\alpha}_0(\bar{z}-\eta)} I_0\big(2\sqrt{\bar{\alpha}_0\bar{\beta}\bar{t}\eta}\big)d\eta\Big]. \tag{6.63}$$

The expressions for the desired relative concentrations are derived from the following representations

$$\bar{C}(\bar{z},\bar{t}) = \frac{1}{\bar{\alpha}_0\bar{\theta}}\frac{U_1(\bar{z},\bar{t})}{U(\bar{z},\bar{t})}, \quad \bar{S}(\bar{z},\bar{t}) = \frac{1}{\bar{\alpha}_0\bar{\theta}}\frac{U_2(\bar{z},\bar{t})}{U(\bar{z},\bar{t})}. \tag{6.64}$$

Here

$$U_1(\bar{z},\bar{t}) = U(0,0)\Big[\bar{\beta}U(\bar{z},\bar{t}) - \frac{\partial}{\partial\bar{t}}U(\bar{z},\bar{t})\Big] \\ = \bar{\alpha}_0\bar{\theta}U(0,0)\Big[I_0\big(2\sqrt{\bar{\alpha}_0\bar{\beta}\bar{z}\bar{t}}\big) - (\bar{\alpha}_0\bar{\theta}-\bar{\beta})G_4(\bar{z},\bar{t})\Big], \\ U_2(\bar{z},\bar{t}) = U(0,0)\Big[\frac{\partial}{\partial\bar{z}}U(\bar{z},\bar{t}) - \bar{\alpha}_0U(\bar{z},\bar{t})\Big] = \bar{\alpha}_0^2\bar{\theta}U(0,0)G_4(\bar{z},\bar{t}),$$

where

$$G_4(\bar{z},\bar{t}) = \int_0^{\bar{t}} e^{-(\bar{\alpha}_0\bar{\theta}-\bar{\beta})(\bar{t}-\xi)} I_0\big(2\sqrt{\bar{\alpha}_0\bar{\beta}\bar{z}\xi}\big)d\xi.$$

Then finally

$$\bar{C}(\bar{z},\bar{t}) = \frac{I_0\big(2\sqrt{\bar{\alpha}_0\bar{\beta}\bar{z}\bar{t}}\big) - (\bar{\alpha}_0\bar{\theta}-\bar{\beta})G_4(\bar{z},\bar{t})}{I_0\big(2\sqrt{\bar{\alpha}_0\bar{\beta}\bar{z}\bar{t}}\big) - (\bar{\alpha}_0\bar{\theta}-\bar{\beta})G_4(\bar{z},\bar{t}) + \bar{\alpha}_0G_3(\bar{z},\bar{t})}, \tag{6.65}$$

$$\bar{S}(\bar{z},\bar{t}) = \frac{\bar{\alpha}_0G_4(\bar{z},\bar{t})}{I_0\big(2\sqrt{\bar{\alpha}_0\bar{\beta}\bar{z}\bar{t}}\big) - (\bar{\alpha}_0\bar{\theta}-\bar{\beta})G_4(\bar{z},\bar{t}) + \bar{\alpha}_0G_3(\bar{z},\bar{t})}. \tag{6.66}$$

In the absence of the deposit detachment ($\bar{\beta} = 0$) formulas (6.65), (6.66) will be simplified as follows

$$\bar{C}(\bar{z},\bar{t}) = \frac{e^{-\bar{\alpha}_0\bar{\theta}\bar{t}}}{e^{\bar{\alpha}_0\bar{z}} + e^{-\bar{\alpha}_0\bar{\theta}\bar{t}} - 1},$$

$$\bar{S}(\bar{z},\bar{t}) = \frac{1 - e^{-\bar{\alpha}_0\bar{\theta}\bar{t}}}{\bar{\theta}(e^{\bar{\alpha}_0\bar{z}} + e^{-\bar{\alpha}_0\bar{\theta}\bar{t}} - 1)}.$$

Also, from (6.66) follows the pattern of the increase of the concentration $\bar{S}$at the bed inlet over time

$$\bar{S}_0(\bar{t}) = \bar{S}(0,\bar{t}) = \frac{\bar{\alpha}_0}{\bar{\alpha}_0\bar{\theta} - \bar{\beta}}\left[e^{(\bar{\alpha}_0\bar{\theta}-\bar{\beta})\bar{t}} - 1\right]. \tag{6.67}$$

According to (6.66), the deposit accumulation in the inlet cross-section of a bed formally has an unlimited at $\bar{\alpha}_0\bar{\theta} - \bar{\beta} \geq 0$ or limited at $\bar{\alpha}_0\bar{\theta} - \bar{\beta} < 0$ increasing nature depending on the sign of the difference $\bar{\alpha}_0\bar{\theta} - \bar{\beta}$. In the first case the concept of the dynamic solids take-up capacity loses its meaning at all, and in the second case its relative value will be

$$\bar{S}_d = \frac{\bar{\alpha}_0}{\bar{\alpha}_0\bar{\theta} - \bar{\beta}}.$$

Also, in the first case it is obligatory, and in the second case it is possible that the increase in $\bar{S}$ (and $\bar{S}_S$) due to finiteness (and smallness) of the pore space volume is strictly limited. At least, the following inequalities are valid

$$\bar{\gamma}(\bar{S}_0)\,\bar{S}_0 < 1, \qquad \bar{\gamma}_a\bar{S}_0 < 1.$$

In practice, the quantities $\bar{\gamma}(\bar{S}_0)\bar{S}_0$, $\bar{\gamma}_a\bar{S}_0$ should be less than 1 by a few tenths due to the inaccessibility of a part of the specified space to the deposit. The above estimates are valid for any bed cross-section. The function $\bar{S}_0(\bar{t})$ degenerates from exponential into the linear function $\bar{S}_0(\bar{t}) = \bar{\alpha}_0\bar{t}$ at $\bar{\alpha}_0\bar{\theta} - \bar{\beta} = 0$.

The importance of the autocatalysis phenomenon for the separation of a low-concentration aqueous suspension is discussed on a number of test cases with the fixed values $\bar{\alpha}_0 = 5$, $\bar{\beta} = 0.01$, $\bar{\gamma}_a = 0.0015$. The relative autocatalysis coefficient was varied discretely or continuously in the range from 0 to 0.002, which made it possible to cover both situations characteristic of the detachment filtration with filter ripening, namely, either the sticking of the suspended solids to a deposit or the detachment of the deposited particles prevails in the mass exchange. The reliability of the conducted predictive and technological calculations is guaranteed by the rigor of the above obtained solution of the corresponding problem of nonlinear filtration, from which the exact calculation formulas and equations are derived. The subject of the predictions were two characteristics, which directly determine the efficiency of a filtration facility in the first rate mode (constant rate). **Figures 7 and 8** present the results of the calculations of the dynamics of the relative outlet concentration of the suspended solids carried out by formula (6.65) in a graphical form. Figure 7 clearly demonstrates the significance of the coefficient $\bar{\theta}$ at the prolonged clarification of a suspension. The calculated curves corresponding to $\bar{\theta} > 0$ deviate noticeably downward from the baseline curve 1 ($\bar{\theta} = 0$), starting at about moment $\bar{t} = 70$. Figure 8 shows the importance of accounting for the autocatalytic effect especially at the final stage of filtration, provided that it remains closely related to the amount of the deposit, for example, according to (6.57), when the filter media is heavily clogging.

The effect of the autocatalysis on the mechanical energy expenditure is similarly manifested, as indicated by **Figures 9 and 10**. Its relative value was determined by formula (2.24) at $f_k(\bar{S})$ according to (2.90) and m_{k1}=1. Here, there is even greater divergence of the calculated curves at $\bar{t} > 70$ in comparison with Figure 7. The divergence of the curves sharply increases over time, and becomes uncontrollable $\bar{\theta} > 0.001$ (Figure 9). A singularity in the behavior of $\Delta\bar{h}$ with respect to $\bar{\theta}$ is clearly visible at $\bar{t} = 150$ (curve 1). The nature of the relation between the head losses and the coefficient $\bar{\theta}$ became essentially nonlinear at a long duration of filtration. The maximum effect of the autocatalysis was naturally reached in the calculated examples at t=300, $\bar{\theta}$ = 0.0015 and was expressed in their almost tenfold increase (Figure 10).

Figure 7. Increase in the relative volumetric concentration of the suspended solids in the filtrate over time

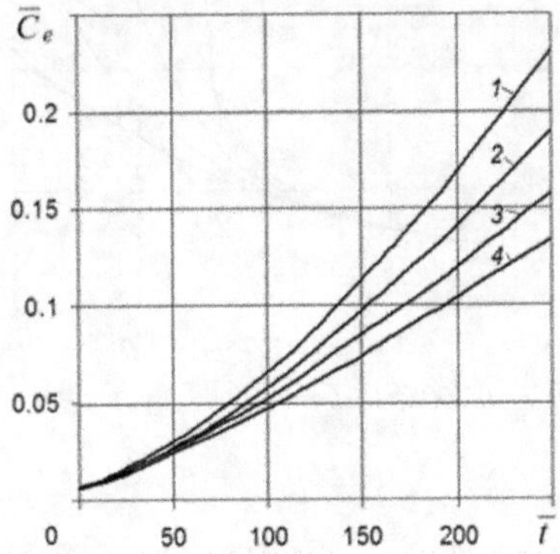

Note. 1 – $\bar{\theta}$ = 0, 2 – $\bar{\theta}$ = 0.0005, 3 – $\bar{\theta}$ = 0.001, 4 – $\bar{\theta}$ = 0.0015

Figure 8. Dependence $\bar{C}_e(\bar{\theta})$

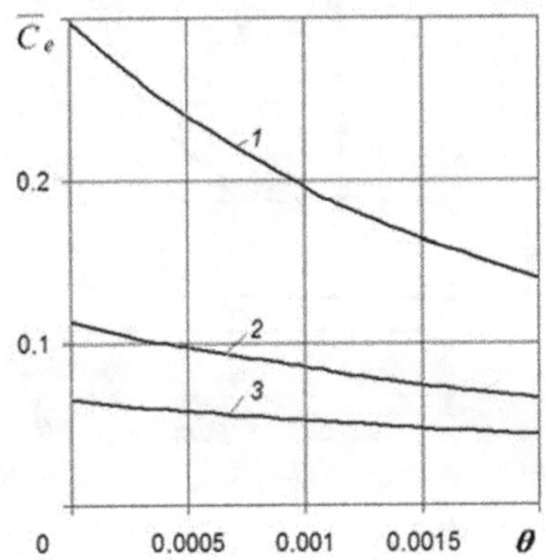

Note. 1 – $\bar{t}$ = 300, 2 – $\bar{t}$ = 200, 3 – $\bar{t}$ = 100

Figure 9. Increase in the relative head losses over time

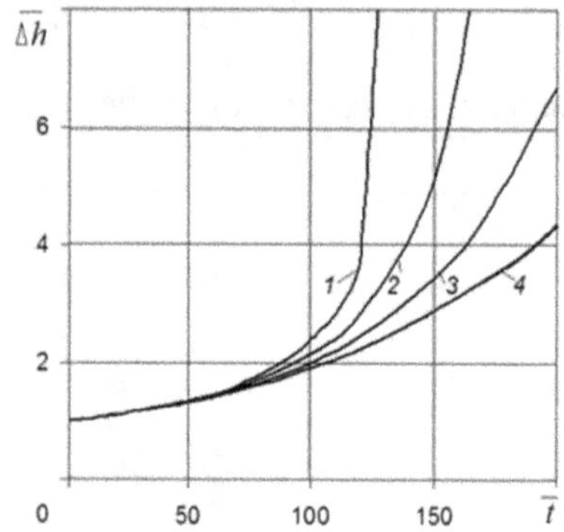

Note. 1 – $\bar{\theta}$ = 0.0015, 2 – $\bar{\theta}$ = 0.001, 3 – $\bar{\theta}$ = 0.0005, 4 – $\bar{\theta}$ = 0

Figure 10. Dependence $\Delta\bar{h}(\bar{\theta})$

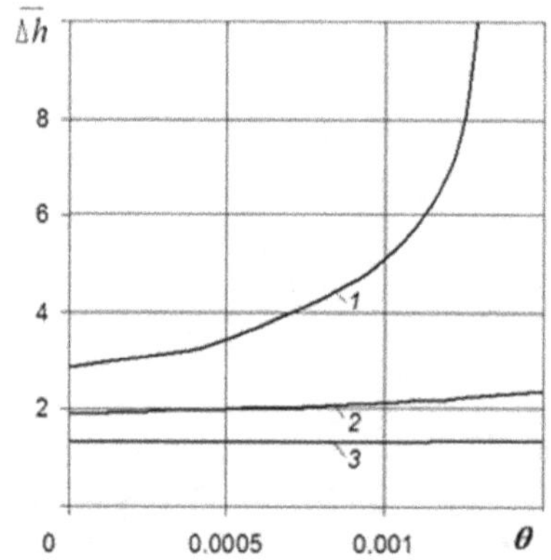

Note. 1 – $\bar{t}$ = 150, 2 – $\bar{t}$ = 100, 3 – $\bar{t}$ = 50

Finally, the relative technological times $\bar{t}_p$, $\bar{t}_h$ were calculated as functions of $\bar{\theta}$. The corresponding curves are depicted in **Figure 11** and obtained on the basis of (6.65), (6.66) at given single values of the relative criterion parameters $\bar{C}_* = 0.1$, $\Delta\bar{h}_* = 6$. Attention is drawn to the fact of the existence of the optimal (in technological sense) value of the coefficient $\bar{\theta}$. Indeed, the duration of the filter run appeared to be maximum at $\bar{\theta} = 8.5 \cdot 10^{-4}$ and noticeably decreased at a change in $\bar{\theta}$ in both directions. Thus, the expediency of in-depth study of the above-mentioned phenomenon and its careful consideration in modeling of the detachable or non-detachment filtration is once again confirmed.

It is appropriate to note that the accepted linear form for the adhesion function and the values of the coefficient $\bar{\theta}$ may lead to the overestimation of the autocatalytic effect especially at a long filtration. Therefore, in order to obtain a more plausible physicochemical pattern in the operating layer in such cases, the nonlinear adhesion functions should be involved in theoretical studies. Just such an approach to modeling of the nonlinear non-detachable filtration will be implemented by approximate analytical methods in Section 6.5.

Figure 11. Dependencies $\bar{t}_{p,h,f}(\bar{\theta})$

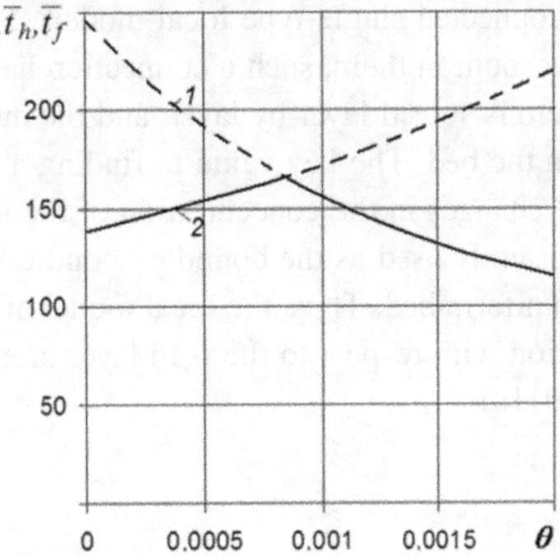

Note. 1– $\bar{t}_h$, 2 – $\bar{t}_p$; solid line – $\bar{t}_f$

6.4. STRICT ANALYSIS OF LAYERED BED OPERATION

The uneven distribution of a deposit along the bed height is reduced and thus, the mechanical energy expenditure is reduced by using porous structures made of two or more layers. Then it is often possible to carry out washing at a later date. The feasibility of implementing suspension filtration through multilayer (three-layer) and two-layer bed at linear mass exchange kinetics has been discussed in detail in Sections 2.3, 2.4. In the following, the essentially nonlinear filtration is discussed in the general case where the filter bed consists of an arbitrary number of layers. However, in this Section, unlike Section 2.3, a single layer distinguishes with number *j*, whose position is not specified, so that in principle it can be the top, bottom, or any intermediate layer in a multilayer column. And then, when modeling the technological process, the filtration characteristics related to the isolated layer are in the focus of attention. In this case, the input information includes, along with a set of the values of the model parameters related to the specified layer, also the function $C_{0j}(t)$, which describes the change in the concentration of the suspended solids at the inlet to it. This function is the main result of modeling of the nonlinear filtration of a suspension through the overlying stratum of j–1 layers, as well as the function of the deposit distribution here. The purpose of the investigations carried out in this Section using exact methods is to predict changes in the suspended solids concentration at the outlet of this layer (C_{ej}), head losses in it (Δh_j) and the transformation of the deposit concentration profiles ($S_j(z)$). To obtain a complete pattern in the bed, such studies should be continued, and in fact duplicated up to the bottom layer inclusive. And then the concentration C_{ej} provides a direct indication of the amount of the suspended matter that have passed through the bed or, in other words, about the filtrate quality. Finally, the scale system used in the introduction of dimensionless quantities is here selected with respect to a representative (*j*- th) layer, the technological process within which is detailed studied below.

The above model of the nonlinear filtration of a suspension in a considered arbitrary single layer of a multilayer bed is close to the model for a uniform bed from the first Sections. However, the formalization of the technological process as a whole and especially its theoretical study becomes much more complicated due to the layered structure of the bed and filtration flow. Now the initial mathematical model should consist of j interconnected single-type local models. Since the diffusion component of mass transfer is not taken into account in them, such a connection has a one-sided nature and the corresponding mathematical problem is solved layer by layer, and the number of stages in the solution is equal to the number of layers in the bed. The key value in finding a solution to the above problem is the establishment of patterns of changes in the concentration of the suspended solids at the interlayer boundaries, since they are consistently used as the boundary conditions instead of a constant value of the inlet concentration C_0 for a uniform bed. Thus, the local model of the clarification compartment at the nonlinear detachment filtration with respect to the j- th layer after all justified simplifications has the following form (Poliakov, 2011a)

$$V\frac{\partial C_j}{\partial z} + \frac{\partial S_j}{\partial t} = 0, \tag{6.68}$$

$$\frac{\partial S_j}{\partial t} = \alpha_{0j}\left(S_{mj} - S_j\right) C_j - \beta_j S_j \ ; \tag{6.69}$$

$$t = 0, \quad S_j = 0 \ ; \tag{6.70}$$

$$z = L_{j-1}, \quad C_j = C_{0j}(t). \tag{6.71}$$

The depth of the j - th layer is $L_{j-1} = \sum_{q=1}^{j-1} l_q$, l_q is the thickness of the q - th layer ($j-1 \geq q \geq 1$); C_{0j} is the variable concentration of the suspended matter at the upper boundary of the j-th layer, which should be determined at the previous stage of the solution of the clarification compartment, and, in fact, as a result of a sequential theoretical analysis of the clarifying effect of all overlying layers the parameters S_{mj}, n_{0j} (porosity of the clean j - th layer) are used as scales in addition to C_0, V, L at the transition to dimensionless variables and parameters, then $\overline{\overline{S}}_j = S_j/S_{mj}$, $\bar{t} = Vt/(n_{0j}L)$, $\bar{\alpha}_{0j} = n_{0j}LC_0\alpha_{0j}/V$, $\bar{\beta}_j = n_{0j}L\beta_j/V$, $\psi_j\overline{S}_{mj} = S_{mj}/\left(n_{0j}C_0\right)$, $\overline{L}_j = L_j/L$ and problem (6.68) - (6.71) is transformed in the following way

$$\frac{\partial \overline{C}_j}{\partial \bar{z}} + \overline{S}_{mj}\frac{\partial \overline{\overline{S}}_j}{\partial \bar{t}} = 0, \tag{6.72}$$

$$\frac{\partial \overline{\overline{S}}_j}{\partial \bar{t}} = \bar{\alpha}_{0j}\left(1 - \overline{\overline{S}}_j\right)\overline{C}_j - \bar{\beta}_j\overline{\overline{S}}_j \ ; \tag{6.73}$$

$$\bar{t} = 0, \quad \overline{\overline{S}}_j = 0 \ ; \tag{6.74}$$

$$\bar{z} = \overline{L}_{j-1}, \quad \overline{C}_j = \overline{C}_{0j}(\bar{t}). \tag{6.75}$$

Intermediate derivations are omitted. The exact solution of problem (6.72) - (6.75) can be expressed by the following dependences for the desired concentrations $\overline{C}_j$, $\overline{\overline{S}}_j$, which are defined on the interval $\overline{L}_{j-1} \leq \bar{z} \leq \overline{L}_j$,

$$\overline{C}_j(\bar{z},\bar{t}) = -\frac{\bar{\beta}_j}{\bar{\alpha}_{0j}} + \frac{U_{1j}(\bar{z},\bar{t})}{\bar{\alpha}_{0j}U_j(\bar{z},\bar{t})}, \tag{6.76}$$

$$\overline{S}_j(\overline{z},\overline{t}) = 1 - \frac{U_{2j}(\overline{z},\overline{t})}{\overline{\alpha}_{0j}\overline{S}_{mj}U_j(\overline{z},\overline{t})}. \tag{6.77}$$

$$\frac{U_j(\overline{z},\overline{t})}{U_j(0,0)} = \overline{U}_j(\overline{t}) + \overline{\alpha}_{0j}\overline{S}_{mj}G_6(\overline{z},\overline{t}) + \sqrt{\overline{\alpha}_{0j}\overline{\beta}_j\overline{S}_{mj}\left(\overline{z}-\overline{L}_{j-1}\right)} \times$$
$$\times \int_0^{\overline{t}} \overline{U}_j(\xi)\frac{I_1\left(2\sqrt{\overline{\alpha}_{0j}\overline{\beta}_j\overline{S}_{mj}(\overline{z}-\overline{L}_{j-1})(\overline{t}-\xi)}\right)}{\sqrt{\overline{t}-\xi}}d\xi, \tag{6.78}$$

$$\frac{U_{1j}(\overline{z},\overline{t})}{U_j(0,0)} = \left[\overline{\alpha}_{0j}\overline{C}_{0j}(\overline{t}) + \overline{\beta}_j + \overline{\alpha}_{0j}\overline{\beta}_j\overline{S}_{mj}\left(\overline{z}-\overline{L}_{j-1}\right)\right]\overline{U}_j(\overline{t})+$$
$$+\sqrt{\frac{\overline{\beta}_j\overline{\alpha}_{0j}^3\overline{S}_{mj}^3}{\overline{t}}}\int_0^{\overline{z}-\overline{L}_{j-1}}\sqrt{\eta}\,e^{\overline{\alpha}_{0j}\overline{S}_{mj}(\overline{z}-\overline{L}_{j-1}-\eta)}I_1\left(2\sqrt{\overline{\alpha}_{0j}\overline{\beta}_j\overline{S}_{mj}\overline{t}\eta}\right)d\eta+$$
$$+\sqrt{\overline{\alpha}_{0j}\overline{\beta}_j\overline{S}_{mj}\left(\overline{z}-\overline{L}_{j-1}\right)}\int_0^{\overline{t}}\overline{U}_j(\xi)\left[\frac{\sqrt{\overline{\alpha}_{0j}\overline{\beta}_j\overline{S}_{mj}\left(\overline{z}-\overline{L}_{j-1}\right)}}{\overline{t}-\xi}\times\right. \tag{6.79}$$
$$\left.\times I_0\left(2\sqrt{\overline{\alpha}_{0j}\overline{\beta}_j\overline{S}_{mj}(\overline{z}-\overline{L}_{j-1})(\overline{t}-\xi)}\right) - \frac{I_1\left(2\sqrt{\overline{\alpha}_{0j}\overline{\beta}_j\overline{S}_{mj}(\overline{z}-\overline{L}_{j-1})(\overline{t}-\xi)}\right)}{(\overline{t}-\xi)^{3/2}}\right]d\xi,$$

$$\frac{U_{2j}(\overline{z},\overline{t})}{U_j(0,0)} =$$
$$= \overline{\alpha}_{0j}\overline{S}_{mj}I_0\left(2\sqrt{\overline{\alpha}_{0j}\overline{\beta}_j\overline{S}_{mj}\overline{t}\left(\overline{z}-\overline{L}_{j-1}\right)}\right) + \overline{\alpha}_{0j}\overline{\beta}_j\overline{S}_{mj}G_7(\overline{z},\overline{t}) + \overline{\alpha}_{0j}^2\overline{S}_{mj}^2G_6(\overline{z},\overline{t}), \tag{6.80}$$

where

$$G_6(\overline{z},\overline{t}) = \int_{\overline{L}_{j-1}}^{\overline{z}} e^{\overline{\alpha}_{0j}\overline{S}_{mj}(\overline{z}-\eta)}I_0(2\sqrt{\overline{\alpha}_{0j}\overline{\beta}_j\overline{S}_{mj}\overline{t}(\eta-\overline{L}_{j-1})})\,d\eta,$$
$$G_7(\overline{z},\overline{t}) = \int_0^{\overline{t}}\overline{U}_j(\xi)\,I_0(2\sqrt{\overline{\alpha}_{0j}\overline{\beta}_j\overline{S}_{mj}\overline{t}(\overline{z}-\overline{L}_{j-1})\,(\overline{t}-\xi)})\,d\xi.$$

Naturally, the function $G_6(\overline{z},\overline{t})$ coincides with $G_3(\overline{z},\overline{t})$ at $j = 1$, $\overline{S}^0 = 0$. Using further the formula

$$\sqrt{\overline{\alpha}_{oj}\overline{\beta}_j\overline{S}_{mj}\overline{t}}\;\; I_0\left(2\sqrt{\overline{\alpha}_{oj}\overline{\beta}_j\overline{S}_{mj}\overline{t}\left(\overline{z}-\overline{L}_{j-1}\right)}\right) =$$
$$= \sqrt{\overline{z}-\overline{L}_{j-1}}\;\;\frac{d}{d\overline{z}}I_1\left(2\sqrt{\overline{\alpha}_{oj}\overline{\beta}_j\overline{S}_{mj}\overline{t}\left(\overline{z}-\overline{L}_{j-1}\right)}\right) + \frac{I_1\left(2\sqrt{\overline{\alpha}_{oj}\overline{\beta}_j\overline{S}_{mj}\overline{t}\left(\overline{z}-\overline{L}_{j-1}\right)}\right)}{2\sqrt{\overline{z}-\overline{L}_{j-1}}},$$

expressions (6.78), (6.79) can be represented in a more compact form, namely,

$$\frac{U_j(\overline{z},\overline{t})}{U_j(0,0)} = I_0\left(2\sqrt{\overline{\alpha}_{oj}\overline{\beta}_j\overline{S}_{mj}\overline{t}\left(\overline{z}-\overline{L}_{j-1}\right)}\right) + \overline{\alpha}_{oj}\overline{S}_{mj}G_6(\overline{z},\overline{t}) + G_8(\overline{z},\overline{t}), \tag{6.81}$$

$$\frac{U_{1j}(\bar z,\bar t)}{U_j(0,0)} =$$

$$= \left(\bar\alpha_{0j}\bar C_{0j}(0) + \bar\beta_j\right) I_0\left(2\sqrt{\bar\alpha_{0j}\bar\beta_j\bar S_{mj}\bar t\,(\bar z - \bar L_{j-1})}\right) + \bar\alpha_{0j}\bar\beta_j\bar S_{mj} G_6(\bar z,\bar t) + G_9(\bar z,\bar t), \tag{6.82}$$

where

$$G_8(\bar z,\bar t) = \int_0^{\bar t} \bar U_j(\xi)\left[\bar\alpha_{0j}\bar C_{0j}(\xi) + \bar\beta_j\right] I_0\left(2\sqrt{\bar\alpha_{0j}\bar\beta_j\bar S_{mj}(\bar z - \bar L_{j-1})(\bar t - \xi)}\right) d\xi,$$

$$G_9(\bar z,\bar t) = \int_0^{\bar t} \bar U_j(\xi)\left[\left(\bar\alpha_{0j}\bar C_{0j}(\xi) + \bar\beta_j\right)^2 + \bar\alpha_{0j}\frac{d}{d\xi}\bar C_{0j}(\xi)\right] I_0\left(2\sqrt{\bar\alpha_{0j}\bar\beta_j\bar S_{mj}(\bar z - \bar L_{j-1})(\bar t - \xi)}\right) d\xi,$$

the parameter $\bar C_{0j}(0)$ means the concentration $\bar C_j$ at the boundary between the j–1 -th and j-th layers at the moment in time to which the contamination front has time to pass through the upper j–1 layers. It is easy to calculate this parameter by the formula

$$\bar C_{0j}(0) = \exp\left(-\sum_{q=1}^{j-1}\bar\alpha_{0q}\bar l_q\right).$$

Then finally the required concentrations will be

$$\bar C_j(\bar z,\bar t) = \frac{\bar C_{0j}(0) I_0\left(2\sqrt{\bar\alpha_{0j}\bar\beta_j\bar S_{mj}\bar t\,(\bar z - \bar L_{j-1})}\right) + G_{10}(\bar z,\bar t)}{I_0\left(2\sqrt{\bar\alpha_{0j}\bar\beta_j\bar S_{mj}\bar t\,(\bar z - \bar L_{j-1})}\right) + \bar\alpha_{0j}\bar\beta_j\bar S_{mj} G_6(\bar z,\bar t) + G_8(\bar z,\bar t)}, \tag{6.83}$$

$$\bar S_j(\bar z,\bar t) = \frac{\bar\alpha_{0j} G_{11}(\bar z,\bar t)}{I_0\left(2\sqrt{\bar\alpha_{0j}\bar\beta_j\bar S_{mj}\bar t\,(\bar z - \bar L_{j-1})}\right) + \bar\alpha_{0j}\bar\beta_j\bar S_{mj} G_6(\bar z,\bar t) + G_8(\bar z,\bar t)}, \tag{6.84}$$

where

$$G_{10}(\bar z,\bar t) =$$

$$= \int_0^{\bar t} \bar U_j(\xi)\left[\bar\alpha_{0j}\bar C_{0j}^2(\xi) + \bar\beta\,\bar C_{0j}(\xi) + \frac{d}{d\xi}\bar C_{0j}(\xi)\right] I_0\left(2\sqrt{\bar\alpha_{0j}\bar\beta_j\bar S_{mj}(z - \bar L_{j-1})(\bar t - \xi)}\right) d\xi,$$

$$G_{11}(\bar z,\bar t) = \int_0^{\bar t} \bar C_{0j}(\xi)\,\bar U_j(\xi) I_0\left(2\sqrt{\bar\alpha_{0j}\bar\beta_j\bar S_{mj}(\bar z - \bar L_{j-1})(\bar t - \xi)}\right) d\xi.$$

In the case of the simplest and most common in the practice of filtration variety of a layered bed, namely, two-layer bed (j=2), first of all, the function of the suspended solids concentration at the lower boundary of the upper layer $\bar C_{1e}(\bar t)$ is calculated by formula (6.55) and the concentration profile of the deposited particles at an arbitrary point in time $\bar S_1(\bar z,\bar t)$ by formula (6.54). Here we can also use the dependences for $\bar C_{1e}$, $\bar S_1$ in the equivalent form

$$\overline{C}_{1e}(\overline{t}) = \frac{I_0(2\sqrt{\overline{\alpha}_{01}\overline{\beta}_1\overline{S}_{m1}\overline{l}_1\overline{t}}) + (\overline{\alpha}_{01}+\overline{\beta}_1)\, G_1(\overline{l}_1,\overline{t})}{(\overline{\alpha}_{01}+\overline{\beta}_1)\, G_1(\overline{l}_1,\overline{t}) - \overline{\beta}_1 G_2(\overline{l}_1,\overline{t}) + e^{\overline{\alpha}_{01}\overline{S}_{m1}\overline{l}_1+\overline{\beta}_1\overline{t}}}, \tag{6.85}$$

$$\overline{S}_1(\overline{z},\overline{t}) = \frac{\overline{\alpha}_{01}\, G_1(\overline{z},\overline{t})}{(\overline{\alpha}_{01}+\overline{\beta}_1)G_1(\overline{z},\overline{t}) - \overline{\beta}_1 G_2(\overline{z},\overline{t}) + e^{\overline{\alpha}_{01}\overline{S}_{m1}\overline{z}+\overline{\beta}_1\overline{t}}}, \tag{6.86}$$

where G_1, G_2 should be calculated from (6.25) after replacing in them $\overline{\alpha}_0, \overline{\beta}, \overline{S}_{mj}$ by $\overline{\alpha}_{01}, \overline{\beta}_1, \overline{S}_{m1}$.

Changes in the content of the suspended and deposited particles of a suspension within the second (lower at downward filtration) layer are described by dependences (6.76) - (6.80) or (6.83), (6.84), in which it is necessary to put j=2, $\overline{L}_{j-1} = \overline{l}_1$, $\overline{C}_{oj}(\overline{t}) = \overline{C}_{1e}(\overline{t})$,

$$\overline{U}_2(\overline{t}) = \exp\left[\overline{\beta}_2\overline{t} + \overline{\alpha}_{02}\int_0^{\overline{t}}\overline{C}_{1e}(\xi)d\xi\right].$$

Finally, if a suspension, the impurity concentration in which varies over time according to the known law $\overline{C}_0 = \overline{C}_0(\overline{t})$, is fed to the filter with a uniform bed then the calculation dependences (6.83), (6.84) remain, but it is necessary to remove the index j in them and put $\overline{C}_{oj}(0) = 1$, $\overline{L}_{j-1} = 0$. Now the integrals contained in them will take the following form

$$G_6(\overline{z},\overline{t}) = \int_0^{\overline{z}} e^{\overline{\alpha}_0\overline{S}_m(\overline{z}-\eta)} I_0\left(2\sqrt{\overline{\beta}\overline{\alpha}_0\overline{S}_m\overline{t}\eta}\right)d\eta,$$

$$G_8(\overline{z},\overline{t}) = \int_0^{\overline{t}}\overline{U}(\xi)[\overline{\alpha}_0\overline{C}_0(\xi)+\overline{\beta}]I_0\left(2\sqrt{\overline{\beta}\overline{\alpha}_0\overline{S}_m\overline{z}(\overline{t}-\xi)}\right)d\xi,$$

$$G_{10}(\overline{z},\overline{t}) = \int_0^{\overline{t}}\overline{U}(\xi)\left[\overline{\alpha}_0\overline{C}_0^2(\xi)+\overline{\beta}\,\overline{C}_0(\xi)+\tfrac{d}{d\xi}\overline{C}_0(\xi)\right]I_0\left(2\sqrt{\overline{\beta}\overline{\alpha}_0\overline{S}_m\overline{z}(\overline{t}-\xi)}\right)d\xi,$$

$$G_{11}(\overline{z},\overline{t}) = \int_0^{\overline{t}}\overline{C}_0(\xi)\overline{U}(\xi)I_0\left(2\sqrt{\overline{\beta}\overline{\alpha}_0\overline{S}_m\overline{z}(\overline{t}-\xi)}\right)d\xi$$

Incidentally, the introduced integrals are connected by the relations $G_6 = \overline{\alpha}_{0j}G_7 + \overline{\beta}_jG_5$, $G_5 = \overline{\alpha}_{0j}G_8 + \overline{\beta}_jG_4$. It should be noted that the solutions of the first problem and the fourth problem generalizing it are presented in this Chapter in two equivalent forms. The disadvantage of the calculation relationships derived from the solution of the problem of filtration through a layered bed is the presence in them of the derivatives $d\overline{C}_{0j}/d\overline{t}$, which can complicate the calculations. However, they have a more compact form.

The head losses in both one arbitrary layer and the entire multilayer bed should be calculated as recommended in Section 2.3. The expression for the function $\overline{S}_j(\overline{z},\overline{t})$ in the formula for the relative head losses in the j - th layer, viz.

$$\Delta\overline{h}_j = \frac{1}{\overline{k}_{0j}}\int_{\overline{L}_{j-1}}^{\overline{L}_j}\frac{d\eta}{f_{kj}(\overline{S}_j(\eta,\overline{t}))}$$

must then be of form (6.77), (6.78), (6.80) or (6.84).

The calculation dependences and equations derived above, in the presence of reliable initial information in full body, allow us to perform a detailed theoretical analysis of the detachment filtration of a suspension through a multilayer bed, taking into account (in contrast to Section 2.3) the gradual decrease in its absorption capacity (Poliakov, 2011b). Below a similar detailed analysis will be carried out for the two-layer bed, which is widespread in water treatment practice. For the designers and operators of filtration facilities it is especially important that these developments, provided that the basic mathematical models are adequate to the actual pattern of filtration, ensure a reliable choice of the design and technological parameters due to the rigor of the analytical solutions. Thus, in principle, it is realistic to achieve a significant increase in the filter efficiency by reducing the total head losses. In order to demonstrate the advantages of a layered media over a uniform media in the case of the nonlinear filtration, the most important characteristics and parameters were calculated in parallel using the formulas from Section 6.2 at $S^0=0$. In addition, in order to reveal changes in the concentrations, heads, technological times due to the negative effect of the formed deposit on the absorption process, the data of their calculations on the basis of the nonlinear and linear kinetics were compared. In the latter case, appropriate formulas and equations were used from Sections 2.3, 3.1.

All the parameters involved in the initial model can be conditionally divided into two groups. It is reasonable to include to the first group those parameters, concerning which there is too limited information in the specialized literature, or which do not play a determining role in the intensification of filtration due to the arrangement of nonuniform beds. The only characteristic values were accepted for such parameters. Parameters from the second group were changed either continuously or discretely in the course of numerous calculations, thus, reflecting the variety of the conditions of the aqueous suspension clarification on rapid filters.

As is known, the initial content of dispersed contaminants determines the intensity of the deposit accumulation and the head loss increase. In order to estimate more fully the effect of the concentration C_0 on the filtration duration, either specific and sharply different values ($5 \bullet 10^{-6}$, $5 \bullet 10^{-5}$) were taken for it, or it was varied within significant limits (up to 10^{-4}). Also, the single value of the coefficient γ_a (20) was assumed in the calculations.

To reveal regular differences in the process of the suspension separation at nonlinear and linear kinetics, the filtration characteristics were calculated by the corresponding formulas (6.76), (6.77) and (2.57), (2.58) in parallel. When choosing the coefficients α_{0j}, α_j, they were assumed to be connected by the relation $\alpha_j = \alpha_{0j} S_{mj}$.

Thus, the same rate of the absorption of the suspended solids at the beginning of the filter run was ensured for both types of mass exchange kinetics. In relative terms this ratio takes the form

$$\overline{\alpha}_j = \overline{\alpha}_{0j} \overline{S}_{mj}$$

When calculating the coefficients $\overline{S}_{mj}$, it was taken into account that porosities and true solids take-up capacities (with respect to a deposit) of both bed layers are the same and equal to $n_{0j}=0.4$, $S_{Smj}=0.2$ respectively.

First of all, efforts were made to clearly demonstrate the undeniable advantages of layered media over uniform media also in the nonlinear filtration (mass exchange kinetics). For this purpose, the above-mentioned characteristics calculated for the two-layer and uniform media were regularly compared. In addition, the thickness and the absorption capacity of the top layer changed smoothly in the examples, so that the two-layer structure eventually degenerated into a well- or moderately absorbing uniform porous medium. The mobilization of the deposited contaminants begins to play a responsible role in the filter performance only when significant amount of a deposit accumulates in the pore space. This important process was taken into account in the examples but not specifically studied. Therefore, a single value of the coefficient $\overline{\beta}$ was used (0.005).

We considered a two-layer bed with layers of the same thickness, i.e. $l_1=l_2=L/2$, and a uniform bed with the thickness L. For certainty, a downward filtration is assumed as usual. The capability of the upper and lower layers to absorb the suspended matter under nonlinear kinetics was characterized by the values of the deposition parameters $\alpha_{01}\overline{S}_{m1} = 4$ and $\alpha_{02}\overline{S}_{m2} = 8$ respectively. The similar parameter for uniform beds $\alpha_0\overline{S}_m$ was assumed to be equal either to 4 or 8. In the case of linear kinetics, the relative rate coefficients were for both layers $\overline{\alpha}_1 = 4, \overline{\alpha}_2 = 8$. The initial concentration C_0 was selected to be ($5{\bullet}10^{-6}$) or high ($5{\bullet}10^{-5}$).

A key role in the evaluation of the quality of water treatment and in general the efficiency of a filter facility, as well as in the determination of the technological and design parameters is played by the outlet volumetric concentration of the suspended solids C_e. And naturally, first of all, the change in the relative concentration $\overline{C}_e$ over time was found. The curves of the dependence $\overline{C}_e(\overline{t})$ calculated by formulas (6.83) and (2.59) are presented in **Figure 12**. The initial values of this function are calculated at $\overline{t} \approx 0$ (corresponds to the moment when the contamination front reaches the base of the bed in the absence of its ripening) by the formulas:

for nonlinear kinetics

$$\overline{C}_e(0) = \overline{C}_e^0 = exp\left[-\overline{\alpha}_{01}\overline{S}_{m1}\overline{l}_1 - \overline{\alpha}_{02}\overline{S}_{m2}\left(1 - \overline{l}_1\right)\right], \quad (6.87)$$

for linear kinetics

$$\overline{C}_e(0) = exp\left[-\overline{\alpha}_1\overline{l}_1 - \overline{\alpha}_2\left(1 - \overline{l}_1\right)\right]. \quad (6.88)$$

The best filtrate quality was naturally achieved under the clarification action of the uniform bed with a well-absorbing material and at lower concentration C_0. However, this situation cannot be considered optimal because, as will be shown below, the deposit was extremely unevenly distributed by the height. As a result, the head losses can be too high. Also, a prolonged removal of almost all the suspended solids from suspension was observed in the two-layer bed both with linear kinetics (curve 6) and nonlinear kinetics (curve 5), if $C_0=5{\bullet}10^{-6}$. It should be noted that in the case of linear mass exchange, the concentration C_0 does not affect the quantity $\overline{C}_e$ and, therefore, a single curve is obtained for its different values (different scales for C). Minimal differences between the mentioned curves are explained by the relatively small deposition during the calculation period and, as a consequence, also by a small decrease in the functional coefficient f_α according to (6.1), (6.2). It is clear that with a significant increase in the impurity content in suspension, the protective effect of a bed sharply weakens, which is expressed in the rapid rise of curves 1,2,4.

Figure 12. Change in the relative suspended solids concentration at the filter outlet over time

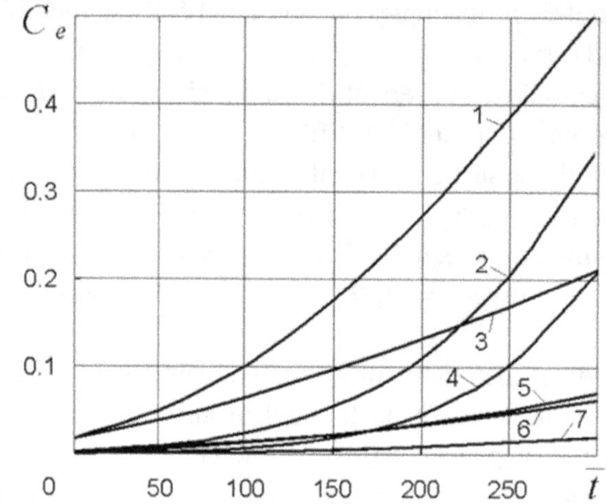

Note. 1, 5, 7 – nonlinear kinetics, 6 – linear; 2, 5, 6 – two-layer; 1, 3, 4, 7 – uniform; 1, 2, 4 – C_0=5·10^{-5}, 3, 5, 7 – C_0=5·10^{-6}; 1, 3 – $\overline{a}_0\overline{S}_m$ = 4, 4, 7 – $\overline{a}_0\overline{S}_m$ = 8

Figure 13. Relative concentration profiles of the deposited particles

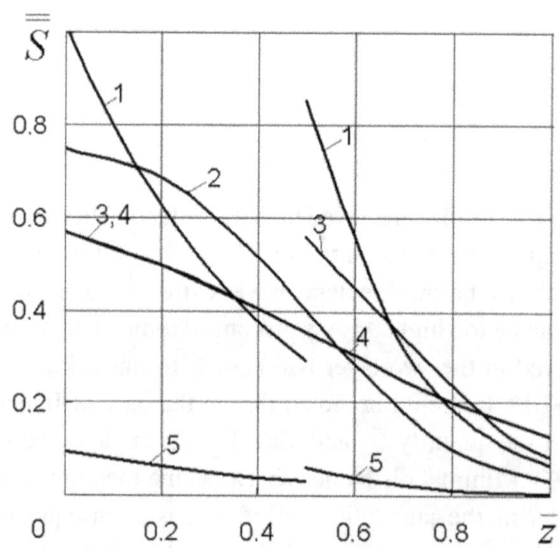

Note 1, 5–linear kinetics, 2 - 5–nonlinear; 1, 3, 5–two-layer; 2, 4 – uniform; 1 - 4 – C_0=5·10^{-5}, 5 – C_0=5·10^{-6}; 2 – $\overline{a}_0\overline{S}_m$ = 8, 4 $\overline{a}_0\overline{S}_m$ = 4

The vertical distribution of the deposited suspension particles is of great importance in modeling the flow capacity of the clogged bed. This distribution was also calculated in different ways. The profiles of the relative concentration of the deposited particles are shown in **Figure 13** and were calculated by formulas (6.84) and (2.56), (2.60) from Chapter 2 with the recalculation at the transition from $\overline{S}$ to $\overline{\overline{S}}$ at the moment corresponding to the final stage of the clarification process ($\overline{t}$ = 200). The bed was assumed to be two-layer or uniform, kinetics was accepted to be nonlinear or linear, and the initial suspended solids concentration was considered to be moderate or high. The function $\overline{S}(\overline{z})$ had a discontinuity of the first kind at the boundary between the layers of a two-layer bed (curves 1, 3, 5). With increasing the absorption capacity of the filter media, the deposit was more and more concentrated in the upper part of the bed. The deposited particles accumulated more intensively in the pore space of the two-layer load at linear kinetics and C_0=5•10^{-5} (discontinuous curve 1), and the bottom layer also took an active part in water treatment. The same situation was observed at nonlinear kinetics, C_0=5•10^{-6} (curve 5). The values of $\overline{\overline{S}}$ exceeded 1 near the upper bed boundary with an intensive deposition (curve 1). This means that here the true solids take-up capacity is completely exhausted and the implemented mathematical model of linear filtration, the calculation formulas developed on its basis actually become incorrect and need to be improved. At the given initial data, the error due to ignoring the limitation of the solids take-up capacity turned out to be minimal. However, with increasing C_0 the formation and development of the zone of maximum saturation with a deposit in the upper part of a bed must be taken into account. A similar situation is considered in Section 2.5 on the basis of linear kinetics, but in this case the corresponding correction makes no sense.

Figure 14. Dependence $\overline{t}_p(\overline{\alpha}_{01}\overline{S}_{ml})$

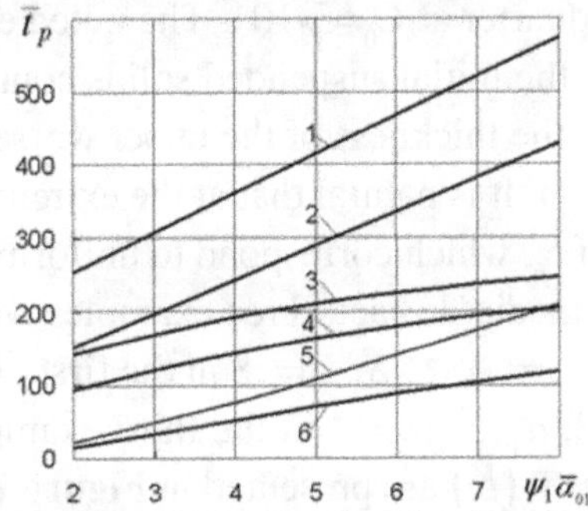

Note. 1, 2, 4 – C_0=5·10^{-5}, 3, 5, 6 – C_0=5·10^{-6}; 1, 3 – $\overline{C}_e$ = 0.1, 2, 4 – $\overline{C}_e$ = 0.05, 5, 6 – $\overline{C}_e$ = 0.01

Figure 15. Dependence $\overline{t}_p(\overline{l}_1)$

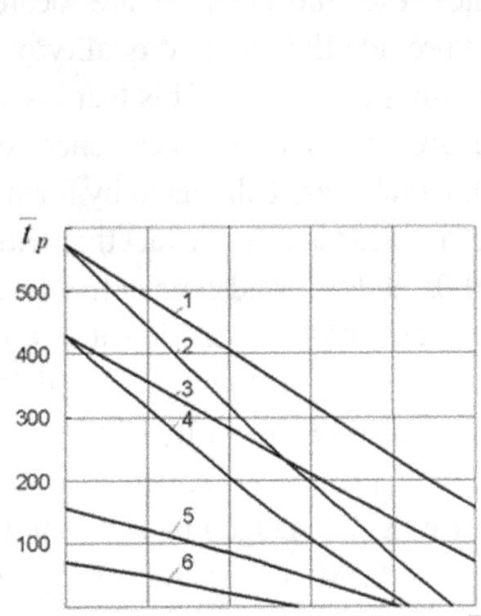

Note. 1, 2, 5 – $\overline{C}_e$ = 0.05, 3, 4, 6 – $\overline{C}_e$ = 0.01; 1, 3 – example 1, 2, 4 – example 2, 5, 6 – example 3

The concentration of the suspended particles in the filtrate C_e is closely related to the time $\bar{t}_p$, which is identical to the duration of the filter run at the insignificant head losses in the bed. Therefore, in the next series of examples we analyzed the influence on the relative technological time $\bar{t}_p$ of those parameters of the problem of filtration through the two-layer bed, in relation to which it is particularly sensitive, namely, the parameter $\bar{\alpha}_{01}\bar{S}_{m1}$, thickness $\bar{l}_1$ and concentration C_0. The value of $\bar{t}_p$ here and below was set on the basis of the given maximum permissible values for filtrate $\bar{C}_*$ (0.01, 0.05, 0.1) by fitting according to (6.76) from the equation

$$U_{1j}(1,\ \bar{t}_p) = (\bar{\alpha}_{0j}\bar{C}_* + \bar{\beta})U_j(1,\ \bar{t})$$

In the first series of the time $\bar{t}_p$ calculations the thicknesses of layers were assumed to be the same $\bar{l}_1 = \bar{l}_2 = 0.5$; the deposition parameter for the lower layer $\bar{\alpha}_{02}\bar{S}_{m2}$ was equal to 8, and for the upper layer $\bar{\alpha}_{01}\bar{S}_{m1}$ varied from 2 to 8; the values of C_0 were $5 \cdot 10^{-6}$ and $5 \cdot 10^{-5}$, respectively, $\bar{S}_{m1} = \bar{S}_{m2} = 5000$ or 500. The results of the calculations of the dependence $\bar{t}_p(\bar{\alpha}_{01}\bar{S}_{m1})$ by formulas (6.76), (6.78), (6.79) and (6.85) are shown in **Figure 14** in the case of a two-layer bed. The deterioration of the absorption properties of the upper layer, on the one hand, activates the suspension particles deposition in the lower layer, and, on the other hand, promotes the breakthrough of more of them due to a noticeable weakening of the protective effect of the bed as a whole. Thus, the decrease of $\bar{\alpha}_{01}\bar{S}_{m1}$ in four times causes the reduction of the time $\bar{t}_p$ by a factor of two and a quarter at $C_0 = 5 \cdot 10^{-6}$. The noted effect is more pronounced at lower values of $\bar{C}_*$. A tenfold reduction of the initial suspended solids content causes about a doubling of $\bar{t}_p$.

In the second series of examples, the thickness of the upper worse absorbing layer was varied in the maximum possible range (from 0 to 1). It is natural that at the extreme values of $\bar{l}_1$ (0 and 1) there were the largest and the smallest values of $\bar{t}_p$, which correspond to uniform beds with better ($\bar{\alpha}_0\bar{S}_m = 8$) and worse ($\bar{\alpha}_0\bar{S}_m = 4$) absorbing filter media. In fact, three examples were calculated, differing in the parameters $\bar{\alpha}_{0j}\bar{S}_{mj}$. We accepted $\bar{\alpha}_{01}\bar{S}_{m1} = 4$, $\bar{\alpha}_{02}\bar{S}_{m2} = 8$ in the first example, $\bar{\alpha}_{01}\bar{S}_{m1} = 2$, $\bar{\alpha}_{02}\bar{S}_{m2} = 8$ in the second example, $\bar{\alpha}_{01}\bar{S}_{m1} = 2$, $\bar{\alpha}_{02}\bar{S}_{m2} = 4$ in the third example. The data of $\bar{t}_p$ calculations in the form of graphs of the dependence $\bar{t}_p(\bar{l}_1)$ are presented in **Figure 15**. It is obvious that the calculated time sharply decreased, and practically linearly with an increase in the thickness $\bar{l}_1$ in all the considered cases. Curves 1,3 over the entire interval of the $\bar{l}_1$ values are located above the abscissa axis, which indicates the capability of the bed to provide the required quality of the filtrate at any ratio of $\bar{l}_1$ to $\bar{l}_2$, although the time $\bar{t}_p$ decreases many times at $\bar{l}_1 \rightarrow 1$. This fact is explained by a relatively large value of the parameter $\bar{\alpha}_{01}\bar{S}_{m1}$ in the first example and, as a consequence, small values of $\bar{C}_e$ after passing the contamination front through the bed, which were calculated by formula (6.87) at $\bar{t} \approx 0$. Reducing this coefficient by half in the remaining examples leads to the fact that with a thin bottom layer and due to the quite large normative values of $\bar{C}_*$ (0.05 and 0.1) and even at the time point when the above-mentioned front reaches the base of the filter bed, the outlet suspended solids concentration will be higher than the permissible concentration $\bar{C}_*$. Thus, according to Figures 14 and 15, changing the parameters of the top layer can significantly affect the protective capability of the bed. The initial concentration C_0 has an equally significant effect on the time $\bar{t}_p$, which is confirmed by **Figure 16**. Here are the data for $\bar{t}_p$, obtained at $\bar{l}_1 = \bar{l}_2$ and two forms of mass exchange kinetics with the fixed coefficients and parameters $\bar{\alpha}_1 = \bar{\alpha}_{01}\bar{S}_{m1} = 4$, $\bar{\alpha}_2 = \bar{\alpha}_{02}\bar{S}_{m2} = 8$. Also, the values 0.05, 0.1, 0.2 are selected for $\bar{C}_*$. The inlet concentration of the suspended solids varied continuously from the small value ($5 \cdot 10^{-7}$) to very high value (10^{-4}). The decline in the curves corresponding to nonlinear kinetics is due to a gradual decrease in the rate of the contaminants sticking to the granular material of the bed as a consequence of a reduction in the number of vacancies on the surface of the grains. In the case of linear kinetics, the time $\bar{t}_p$ is independent of C_0. Indeed, an increase in C_0 while keeping the value of $\bar{C}_*$ the same leads to a similar increase in C_*. The time $\bar{t}_p$ for the outlet suspended solids concentration to reach a higher value of C_* remains the same due to the enhanced particle deposition and mass transfer, namely, $\bar{t}_p$ are found to be 477, 656 and 907 at $\bar{C}_* = 0.05$, 0.1 and 0.2 respectively. Significant differences in the values of $\bar{t}_p$ at different forms of kinetics indicate the importance of the correct selecting the mass exchange equation and kinetic coefficients.

Figure 16. Dependence $\bar{t}_p(C_0)$

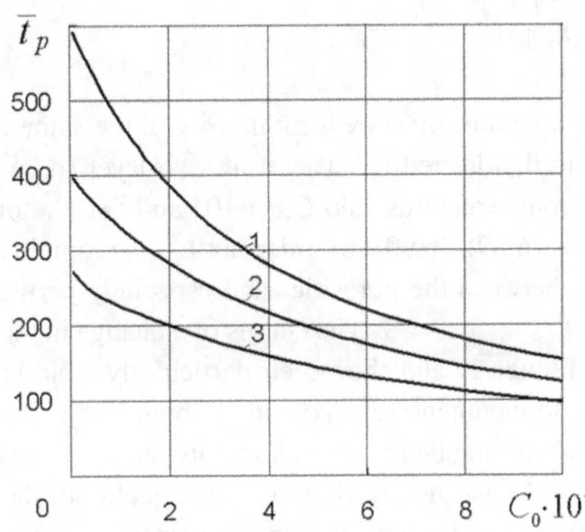

Note. 1 – $\bar{C}_e = 0.2$, 2 – $\bar{C}_e = 0.1$, 3 – $\bar{C}_e = 0.05$

The progressive deposition of the dispersed contaminants in the pores of the bed at long-term filtration can lead to the loss of the filter control and, as a consequence, a significant reduction in its performance. In this connection there is a necessity to establish the dynamics of the head losses and the point in time when further operation of the filtration facility becomes economically inexpedient. It is realistic to predict them reliably only as a result of the implementation of the filtration flow compartment of the general mathematical model of the suspension clarification, which was done in Chapter 2 with respect to the linear filtration mode with a constant rate and for an arbitrary form of the dependence of the hydraulic conductivity k on the concentration of the deposited particles S. Thus, the formulas for calculating the relative reduced head, head losses dynamics, and the second technological time (t_h) depend on the form of kinetics not directly, but indirectly - through the expressions for the concentrations S_j, obtained as a result of the solution of the clarification compartment. As before, to calculate the above characteristics, and, most importantly, the filter run duration, the relationship between the hydraulic conductivities k_j and S_j was specified. The dependence between them in a dimensionless form was taken in accordance with (1.10), (2.148) as follows:

for linear kinetics

$$f_{kj}(\bar{S}_j) = \left(1 - \frac{\bar{\gamma}_{aj}}{\bar{n}_{0j}}\bar{S}_j\right)^3, \tag{6.89}$$

for nonlinear kinetics

$$f_{kj}(\overline{S}_j) = \left(1 - \frac{\overline{S}_{mj}\overline{\gamma}_{aj}}{\overline{n}_{0j}}\overline{S}_j\right)^3, \tag{6.90}$$

where $\overline{n}_{0j} = n_{0j}/n_{01}$ and, therefore, $\overline{n}_{01} = 1, \overline{\gamma}_{aj} = \gamma_{aj}C_0$.

The reduced head at the upper boundary of the two-layer bed $\widetilde{h}_1(0,\overline{t})$ and the equivalent relative total head losses in the bed $\Delta\overline{h}$, as it follows from (2.72), increase in the following way

$$\Delta\overline{h} = \int_0^{l_1} \frac{d\eta}{f_{k1}(\overline{S}_1(\eta,\overline{t}))} + \frac{1}{\overline{k}_{02}}\int_{l_1}^{1} \frac{d\eta}{f_{k2}(\overline{S}_2(\eta,\overline{t}))}. \tag{6.91}$$

For the calculations of the function $\Delta\overline{h}(\overline{t})$ by formula (6.91), the same initial data are accepted as in the examples with respect to the desired $\overline{C}_e$. These are two-layer ($\overline{\alpha}_{01}\overline{S}_{m1} = 4$, $\overline{\alpha}_{02}\overline{S}_{m2} = 8$) and uniform ($\overline{\alpha}_0\overline{S}_m = 4$ and 8) porous structures; and C_0=5•10^{-6} and 5•10^{-5}; nonlinear, linear kinetics and their corresponding expressions (6.89), (6.90). In order not to overcomplicate a quantitative analysis, the usually existing difference between the porosities and especially permeabilities of the layers was not taken into account, so that $\overline{k}_{02} = \overline{n}_{02} = 1$. The curves of a change in the relative head losses in the bed over time are depicted in **Figure 17** and show their particularly rapid increase with linear kinetics and at C_0=5•10^{-5} (curve 1) due to the intensive deposition in both layers. Also, the head losses sharply increased in the uniform well-absorbing bed (curve 2), mainly due to the extremely uneven deposition of contaminants along its height. Consequently, despite a high quality of the filtrate (Figure 12, curves 4, 6), the operation of such a filter may become ineffective over time. At the same time, due to a much more even deposit distribution, as well as the breakthrough of more suspended solids amount, the losses $\Delta\overline{h}$ in the two-layer bed under nonlinear kinetics increased much slower over time (curve 4). The reduction of the impurity content in suspension by a factor of ten contributed to a multiple reduction of the mechanical energy expenditure for the filtration process (curve 5).

Figure 17. Increase in the relative head losses in the bed over time

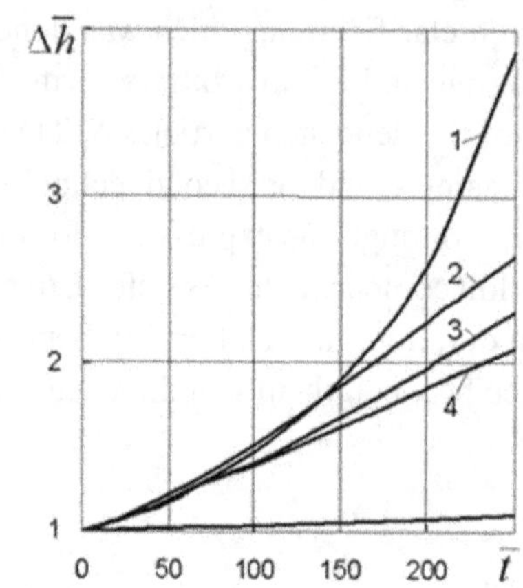

Note. 1 - 4 – C_0=5·10^{-5}, 5 – C_0=5·10^{-6}; 1, 4, 5 – two-layer; 2 – uniform, $\overline{\alpha}_0\overline{S}_m = 8$; 3 – uniform, $\overline{\alpha}_0\overline{S}_m = 4$; 1, 5 – linear kinetics; 2 - 4 – nonlinear kinetics

Figure 18. Dependence $\bar{t}_f(\bar{\alpha}_{01}\bar{S}_{m1})$

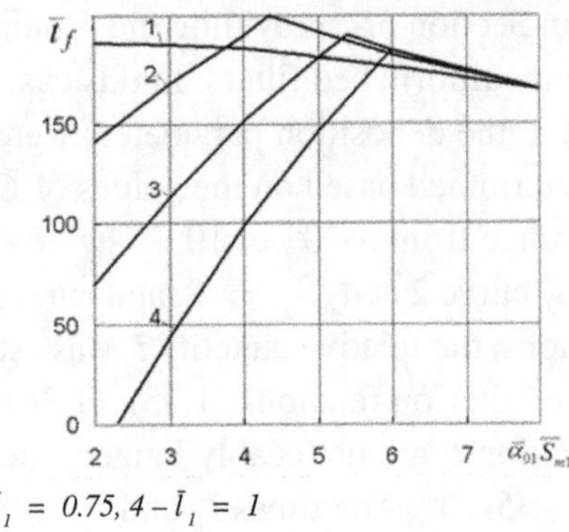

Note. $1-\bar{l}_1 = 0.25$, $2-\bar{l}_1 = 0.5$, $3-\bar{l}_1 = 0.75$, $4-\bar{l}_1 = 1$

Further, the significance of such model parameters as $\bar{\alpha}_{01}\bar{S}_{m1}$, $\bar{l}_1$ and C_0 for the most important technological parameter ($\bar{t}_f$) was analyzed with respect to a two-layer bed, its uniform analogs and nonlinear kinetics. In contrast to the characteristic times $\bar{t}_p$, $\bar{t}_h$, determined on the basis of one criterion of the filter performance estimation, namely, the filtrate quality ($\bar{C}_* = 0.1$) or the permissible head losses ($\Delta\bar{h}_* = 4$), the duration $\bar{t}_f$ was substantiated using both criteria. In the first series of examples, C_0 was fixed ($5\bullet10^{-5}$), and with given $\gamma_{aj}\bar{S}_{mj}=0.2$, $n_{01}=0.4$, $\bar{S}_{m1} = \bar{S}_{m2} = 500$ the relative thickness of the top layer was being changed ($\bar{l}_1 = 0.25, 0.5, 0.75, 1$), and the parameter $\bar{\alpha}_{01}\bar{S}_{m1}$ varied continuously from 2 to 8. In the case $\bar{l}_1 = 1$ there is a uniform bed also when varying $\bar{\alpha}_0\bar{S}_m$ from 2 to 8. The corresponding curve 4 in **Figure 18**, calculated, as well as other similar curves, on the basis of (2.24), (2.148), (6.76), shows that the maximum duration $\bar{t}_f$ in the uniform beds will be approximately at $\bar{\alpha}_0\bar{S}_m = 6$ (it should be remembered that this result is obtained based on the strictly specified $\bar{t}_f$ is reached if the thicknesses of the layers of a two-layer bed are the same. Then the time $\bar{t}_f$ is equal to approximately 200 at $\bar{\alpha}_{01}\bar{S}_{m1} = 4.3$. To a somewhat lesser extent, the effect of the deposit redistribution into the depth of the bed affects the duration of the effective filter operation at $\bar{l}_1 = 0.75$ (curve 3). The monotonic nature of curve 1 is explained by the predominant retention of the suspended solids in the thicker bottom layer, where the contaminants are deposited again very unevenly. The noticeable additional head losses in the bed in this case ultimately limit the duration of the filter operation over the entire considered range of $\bar{\alpha}_{01}\bar{S}_{m1}$ values. At small values of $\bar{\alpha}_{01}\bar{S}_{m1}$, the last option is preferable, because the breakthrough of suspension particles increases markedly with increasing the values of $\bar{l}_1$ and the time t_p, which will be relatively short in such a situation, becomes decisive for t_f.

The plots of the duration $\bar{t}_f$ vs. the thickness $\bar{l}_1$ dependence have a largely similar nature (**Figure 19**). Here we also assumed in the examples $\bar{\alpha}_{02}\bar{S}_{m2} = 8$, $C_0 = 5\cdot10^{-5}$, and the same values of $\bar{C}_*$, $\Delta\bar{h}_*$. At low absorption capacity of the upper layer ($\bar{\alpha}_{01}\bar{S}_{m1} = 2$ or 4) and small values of $\bar{l}_1$ almost all the suspended matter, which was not retained in it, was deposited in the lower layer. Thus, on the one hand, the protective effect of the bed due to the large capacity of the second well-absorbing layer will be maintained for a long time, and on the other hand, the head losses will not be too large. Therefore, the duration $\bar{t}_f$ will be the longest at $\bar{l}_1 < 0.2$ if $\bar{\alpha}_{01}\bar{S}_{m1} = 2$ (curve 1). With increasing $\bar{l}_1$ the underlying layer retains increasingly smaller amounts of impurity, so that the time $\bar{t}_p$, and the time $\bar{t}_f$ decreases sharply with it. The protective capability of the bed increases so much at $\bar{\alpha}_{01}\bar{S}_{m1} = 6$, that $\bar{t}_p$ is greater than $\bar{t}_h$ even for the values of $\bar{l}_1$ close to 1. Therefore, the corresponding curve 3 increases monotonically.

Finally, the results of the calculations of the last series of examples are presented in **Figure 20** and demonstrate a fundamental possibility to determine the range of the values of the initial volumetric concentration C_0 using formulas from Section 6.2, providing the prolongation of the operation of filters with two-layer bed in comparison with uniform bed filters. In this case, the thickness $\bar{l}_1$ was alternately taken equal to 0, 0.25, 0.5, 0.75 and 1, the deposition parameters were $\bar{\alpha}_{01}\bar{S}_{m1} = 4$, $\bar{\alpha}_{02}\bar{S}_{m2} = 8$, and the characteristic times $\bar{t}_f$, $\bar{t}_p$ were determined based on the values of $\bar{C}_*$, $\Delta\bar{h}_*$, respectively, 0.1, 4. The concentration C_0 was varied in the range from $5{\bullet}10^{-7}$ to 10^{-4}. The head losses in the bed exceeded the specified value $\Delta\bar{h}_*$ in four examples (curve 2 at $\bar{\alpha}_0\bar{S}_m = 8$, and curves 3-5 at $\bar{l}_1 = 0.25$, 0.5 and 0.75) at large values of C_0. For such situations, the relative quantity $\bar{t}_f$ was established by the joint calculation of the times $\bar{t}_p$, $\bar{t}_h$ on the basis of the criterion relations (1.35), (1.36) and formulas (6.76), (6.77). As can be seen from Figure 20, the duration $\bar{t}_f$ are noticeably longer when the bed consists of two layers of equal thickness ($\bar{l}_1 = 0.5$) and $C_0 > 5{\bullet}10^{-5}$. The times $\bar{t}_p$ and $\bar{t}_h$ become equal at $C_0 = 5{\bullet}10^{-5}$. Further increase in C_0 leads to an accelerated decrease in $\bar{t}_h$, as evidenced by the significantly higher location of curve 1 with respect to curve 2 at $C_0 > 2.5{\bullet}10^{-5}$. Naturally, the layered bed has a weaker effect on the filter operation time at $\bar{l}_1 = 0.25$ and 0.75. It is necessary to note that the data of the calculations of the heads, their losses, and, consequently, the time $\bar{t}_h$, moreover, $\bar{t}_f$ at relatively large values of C_0 have indicative nature because of assuming the initial permeabilities and porosities of the layers equal. When analyzing or predicting the operation of a particular filter with the layered bed in depth, inevitable differences between the hydraulic conductivities and porosities of different clean layers should be taken into account. Such consideration would lead in Figure 20, to some shift of the curves $\bar{t}_f(C_0)$ in their piece, which is determined on the basis of the calculation of $\bar{t}_h$.

Figure 19. Dependence $\bar{t}_f(\bar{l}_1)$

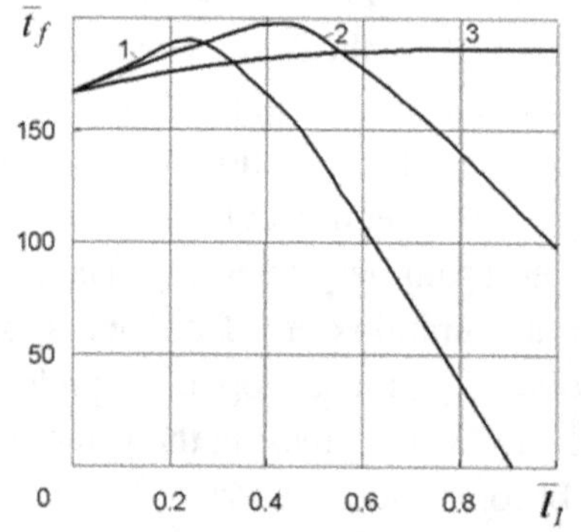

Note. 1 – $\bar{\alpha}_{01}\bar{S}_{m1} = 4$, 2 – $\bar{\alpha}_{01}\bar{S}_{m1} = 6$, 3 – $\bar{\alpha}_{01}\bar{S}_{m1} = 2$

Figure 20. Dependencies $\bar{t}_f(C_0)$, $\bar{t}_p(C_0)$

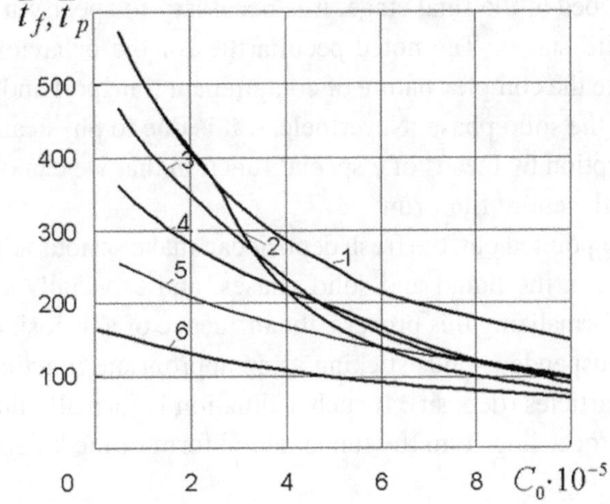

Note. 1, 2, 6 – uniform; 3 - 5 – two-layer; 1 – $\bar{t}_p$, 2 - 6 – $\bar{t}_f$; 1, 2 – $\bar{\alpha}_0\bar{S}_m$ = 8, 6 – $\bar{\alpha}_0\bar{S}_m$ = 4; 3 – $\bar{l}_1$ = 0.25, 4 – $\bar{l}_1$ = 0.5, 5 – $\bar{l}_1$ = 0.75

6.5. APPROXIMATE ANALYSIS

The rigorous solutions of the four problems presented in the previous Sections of this Chapter actually exhaust the possibilities of exact analytical methods for nonlinear filtration (in fact mass exchange kinetics). However, often the actual interfacial mass exchange during the clarification of aqueous suspensions in the layer of a filtering material cannot be reliably described by simple nonlinear kinetics when the absorption function of dispersed particles $f_\alpha(S)$ in (1.13), (6.1) is linear, much less by linear kinetics of the detachment filtration. Therefore, there is a demand for methods of the calculation of the technological process with kinetics of more complex form in the practice of the suspension separation on rapid filters. At present, only approximate analytical methods are able to satisfy such a need. Just the corresponding approximate solution, which has a very wide field of application, is given below. Its generalized nature is due to an arbitrary kind of the function $f_\alpha(S)$, which is initially allowed.

Since this Section focuses on different forms of nonlinearity in the formalization of the absorption process, which significantly complicate the application of analytical methods, equations (2.28), (6.92) are supplemented by the standard first-order boundary (2.9) and homogeneous initial (2.29) conditions. The filtration flow compartment is transformed only due to changes in the expression for the concentration $\bar{\bar{S}}$. Its general form remains the same and still includes system (2.11) - (2.13).

As evidenced by the numerous data of the experimental studies of aqueous suspensions filtration, the dynamics of the outlet concentration of the suspended solids has a complex nature, which is expressed both in the different direction and magnitude of its changes. Nevertheless, the similarity of such changes at the different stages of the technological process is clearly seen, which served as a basis for the conventional division of the operating period (filter run) into a number of characteristic stages, namely four stages according to (Jegatheesan & Vigneswaran, 2005). At the same time, a steady tendency to the deterioration of the filtrate quality is clearly manifested only in a certain time after the successive turning on of a filter (the fourth and main stage). When developing a filter control algorithm, it is this stage that becomes crucial. However, in order to reliably predict the development of the physicochemical situation in the packed bed at the final stage, it is necessary to specify it during the first, then the second and, finally, the third stages. The noted peculiarities of the behavior of the outlet suspended solids concentration indicate the complex nature of contaminant transport and phase transfer in the bed, primarily its absorption by the solid phase. Nevertheless, it is due to physically justified description of the suspended matter absorption by means of a special function that we can obtain a calculated pattern close to reality throughout the entire filter run.

As it has been repeatedly pointed out, the fresh deposit can make serious adjustments in the exchange of dispersed particles between the liquid and solid phases, and especially significantly it affects the absorption process. When formalizing this process, the influence of a deposit on it is allowed for taking the rate coefficient of the suspended solids sticking as an appropriate function of the concentration of the deposited suspension particles (deposit). If such a situation is formally interpreted as a generalization of linear filtration or proceeding from the fundamental form of the kinetic equation (1.13), then it should be assumed that

$$\alpha = \alpha(S), \; f_\alpha(S) = \alpha_0 \chi_S(S) \tag{6.92}$$

To more adequately describe the prolonged immobilization of suspended matter under the influence of deposit, we have to operate with the functions $\alpha(S)$, $\chi_S(S)$ of the complex form (Bai & Mackie, 1995; Bai & Tien, 2000; Boller & Kavanauch, 1995; Ison & Ives, 1969; Mackie & Zhao, 1999; Skouras et al., 2011). This significantly hinders, although not excludes, the application of analytical methods. If we adhere to the generally accepted assumption of a linear relationship between f_β and the concentration S, then the dimensionless equation of mass exchange kinetics is generalized as follows

$$\frac{\partial \bar{S}}{\partial \bar{t}} = \bar{\alpha}(\bar{S})\,\bar{C} - \bar{\beta}\bar{S} \tag{6.93}$$

When developing engineering methods for calculating nonlinear filtration, to avoid averaging of the above-mentioned functions one has to compromise. In this situation it consists in some simplification of the empirical absorption functions, which, nevertheless, allows to take into account all nonlinear effects in modeling of the suspension separation. From the physical point of view, it is sufficient to take into account in applications the two nonlinear effects distinguished and considered separately in the first Sections of this Chapter. Expressions (6.1) - (6.3) are suitable for their description and estimation. In the first approximation, it is justified to base on the kinetics of the detachment filtration without significant damage to the accuracy of calculations, and to describe the limited absorption resource and the autocatalysis by linear functions (6.2), (6.3), so that, for example, $\alpha(S)$ takes the form similar to (4.14)

$$\alpha(S) = \alpha_0 (S_m - S)(1 + \theta S). \tag{6.94}$$

In this case, selecting the values of the constants S_m and θ on the basis of the experimental data and the exact solution of the corresponding mathematical problem, it is really plausible to model the transition of contaminants from the liquid phase to the solid phase, and, as a result, to provide a reliable prediction of the changes in the characteristics of nonlinear filtration. What is fundamental here is the joint consideration of the aforementioned effects.

To solve the nonlinear problem (2.28), (6.93), (2.9), (2.10) (actually the clarification compartment), the parameter $\overline{C}_a$ is introduced as follows

$$\overline{C}_a(\overline{z},T)=\frac{1}{T}\int_0^T \overline{C}(\overline{z},\overline{t})\,d\overline{t} \tag{6.95}$$

Then we have the problem for the required quantity $\overline{S}$ which after the replacement of $\overline{C}$ by $\overline{C}_a$ in (6.94) includes the equation

$$\frac{\partial\overline{S}}{\partial\overline{t}} = \overline{\alpha}(\overline{S})\,\overline{C}_a-\beta\overline{S} \tag{6.96}$$

and condition (2.10). Its solution is represented as an inverse integral function

$$\overline{t} = F(\overline{S},\overline{C}_a)=\int_{\overline{S}^0}^{\overline{S}}\frac{d\zeta}{\overline{\alpha}(\zeta)\overline{C}_a-\beta\zeta} \tag{6.97}$$

It is important to note that this expression allows us to accurately calculate the increase in the relative concentration $\overline{S}$ in the inlet cross-section ($\overline{S}_0$), viz,

$$\overline{t} = \int_{\overline{S}^0}^{\overline{S}_0}\frac{d\zeta}{\overline{\alpha}(\zeta)-\beta\zeta} \tag{6.98}$$

Below the method of the nonlinear filtration calculation is developed for those cases when it is possible to invert function (6.97) and thus, to derive the relation

$$\overline{S}(\overline{z},\overline{t})=F^{-1}(\overline{C}_a(\overline{z},\overline{t}),\overline{t}). \tag{6.99}$$

Such inversions are possible, in particular, if the image function $F(\overline{S})$ is a polynomial of degree one, two or three. The inversion procedure and the final results obtained using it are given below for the first two cases.

At the next stage of solving the problem, the mass transfer equation is integrated over $\overline{t}$ in the range from 0 to T and as a result the equation is obtained

$$T\frac{d\overline{C}_a}{d\overline{t}}+S(\overline{z},T)-\overline{S}^0 = 0 \tag{6.100}$$

which is solved under the condition

$$\overline{z} = 0,\;\; \overline{C}_a = 1 \tag{6.101}$$

The sought relation between $\overline{C}_a$, $\overline{z}$, $\overline{t}$ also has the form of an inverse integral function

$$\bar{z} = T\int_{\bar{C}_a}^{1} \frac{d\zeta}{F^{-1}(\zeta,\bar{t}) - \bar{S}^0} \tag{6.102}$$

In order to establish the pattern of the decrease in the content of the dispersed contaminants in the liquid phase in the direction of its movement and, most importantly, to predict the deterioration of the filtrate quality, it is necessary to solve the equation with respect to $\bar{C}$

$$\frac{\partial \bar{C}}{\partial \bar{z}} + \bar{\alpha}(F^{-1}(\bar{C}_a,\bar{t}))\,\bar{C} = \bar{\beta}\,F^{-1}(\bar{C}_a,\bar{t}) \tag{6.103}$$

under condition (2.9). Preliminarily, taking into account the expression

$$\frac{\partial \bar{z}}{\partial \bar{C}_a} = \frac{T}{\bar{S} - F^{-1}(\zeta,\bar{t})} \tag{6.104}$$

the auxiliary quantity μ is specified

$$\mu(\bar{C}_a,\bar{t},T) = \int_0^{\bar{z}} \bar{\alpha}(F^{-1}(\bar{C}_a(\eta,\bar{t})))d\eta = T\int_{\bar{C}_a}^{1} \frac{\bar{\alpha}(F^{-1}(\zeta,\bar{t}))}{F^{-1}(\zeta,\bar{t}) - \bar{S}^0} \tag{6.105}$$

Then the solution of (6.103), (2.9) in a general form allowing for (6.104) is as follows

$$\begin{aligned} \bar{C}(\bar{C}_a,\bar{t},T) &= e^{-\mu(\bar{C}_a,\bar{t},T)}\left[1 + \bar{\beta}\int_0^{\bar{z}} F^{-1}(\bar{C}_a(\eta,\bar{t}),\bar{t})\,e^{\mu(\bar{C}_a(\eta,\bar{t}),\bar{t},T)}\,d\eta\right] = \\ &= e^{-\mu(\bar{C}_a,\bar{t},T)}\left[1 + \bar{\beta}T\int_{\bar{C}_a}^{1} \frac{F^{-1}(\zeta,\bar{t})}{\bar{S}^0 - F^{-1}(\zeta,\bar{t})}\,e^{\mu(\zeta,\bar{t},T)}\,d\zeta\right]. \end{aligned} \tag{6.106}$$

For the specific calculations of the quantity $\bar{C}(\bar{z},\bar{t})$, it is necessary to find in parallel the distribution of $\bar{C}_a$ along height, using (6.102). And since in determining the dynamics of both concentrations an arbitrary calculated time actually coincides with the duration of the calculation period, it is necessary to assume $T = \bar{t}$ in the final expressions.

To calculate the characteristics of nonlinear filtration, it is necessary to first establish the expression for the concentration $\bar{S}$ in explicit form. It should be emphasized that it is possible to find the function $F^{-1}(\bar{C}_a,\bar{t})$ only by adopting for $\bar{\alpha}(\bar{S})$ of strictly explicit representations. Two options of the expression $\bar{\alpha}$ $(\bar{S})$, namely, according to (6.2) and (6.94) are of greatest interest for theoretical analysis and practical application.

If the function $\bar{\alpha}(\bar{S})$ is linear according to (6.2), then in such a situation there is an alternative strict solution, which is given and discussed in detail in Section 6.2 for the case S^0=*const*. Consequently, there is a unique opportunity to exhaustively estimate the errors in the calculations of all the filtration characteristics due to the simplification of the kinetic equation. It is appropriate to note that the corresponding simplification technique, which consists in partial averaging of one of the multipliers in the term directly describing the removal of contaminants, allows us to preserve the nonlinearity of the model (kinetics) of suspension clarification.

Thus, in the first option ($\overline{\alpha}(\overline{S})$ according to (6.2)) from (6.97) follows

$$\bar{t} = \frac{1}{\overline{\alpha}_0 \overline{C}_a + \overline{\beta}} \ln \frac{\overline{S}^0 - \sigma_1(\overline{C}_a)}{\overline{S} - \sigma_1(\overline{C}_a)} \tag{6.107}$$

where $\sigma_1(\overline{C}_a) = \overline{\alpha}_0 \overline{S}_m \overline{C}_a / (\overline{\alpha}_0 \overline{C}_a + \overline{\beta})$. As a result of the inversion of (6.107) we obtain

$$F^{-1}(\overline{C}_a, \bar{t}) = \overline{S}(\overline{C}_a, \bar{t}) = \sigma_1(\overline{C}_a) + \left[\overline{S}^0 - \sigma_1(\overline{C}_a)\right] e^{-(\overline{\alpha}_0 \overline{C}_a + \overline{\beta})\bar{t}} \tag{6.108}$$

Then the change in the parameter $\overline{C}_a$ along the bed height at the moment in time T is described by the following expression in integral form

$$\bar{z} = T \int_1^{\overline{C}_a} \frac{d\zeta}{[\overline{S}^0 - \sigma_1(\zeta)](1 - e^{-(\overline{\alpha}_0 \zeta + \overline{\beta})T})} \tag{6.109}$$

Thus,

$$\frac{d\bar{z}}{d\overline{C}_a} = \frac{T}{[\overline{S}^0 - \sigma_1(\overline{C}_a)](1 - e^{-(\overline{\alpha}_0 \overline{C}_a + \overline{\beta})T})} \tag{6.110}$$

so that

$$\mu(\overline{C}_a, \bar{t}, T) = \overline{\alpha}_0 T \int_{\overline{C}_a}^{1} \frac{\overline{S}_m - \overline{S}(\zeta, \bar{t})}{\overline{S}^0 - \sigma_1(\zeta)} \frac{d\zeta}{1 - e^{-(\overline{\alpha}_0 \zeta + \overline{\beta})T}} \tag{6.111}$$

Consequently, it is proposed to calculate the spatial and temporal changes in the relative concentration of the suspended solids in the bed volume using the formula

$$\overline{C}(\overline{C}_a, \bar{t}) = e^{-\mu(\overline{C}_a, \bar{t})} \left[1 + \overline{\beta}\bar{t} \int_1^{\overline{C}_a} \frac{\overline{S}(\zeta, \bar{t})}{[\overline{S}^0 - \sigma_1(\zeta)](1 - e^{-(\overline{\alpha}_0 \zeta + \overline{\beta})\bar{t}})} e^{\mu(\zeta, \bar{t})} d\zeta \right] \tag{6.112}$$

and also, equality (6.102), assuming in it $T = \bar{t}$.

At the second option ($\overline{\alpha}(\overline{S})$ according to (6.94)) the calculation dependencies are derived in a similar sequence and, first of all, from (6.113) follows

$$\bar{t} = \frac{1}{\overline{\alpha}_0 \overline{\theta}\, \overline{C}_a [\sigma_3(\overline{C}_a) - \sigma_2(\overline{C}_a)]} \ln \frac{[\overline{S}^0 - \sigma_3(\overline{C}_a)][\overline{S} - \sigma_2(\overline{C}_a)]}{[\overline{S}^0 - \sigma_2(\overline{C}_a)][\overline{S} - \sigma_3(\overline{C}_a)]} \tag{6.113}$$

Then after inversing (6.113) is obtained

$$\overline{S}(\overline{C}_a, \bar{t}) = \frac{\sigma_2(\overline{C}_a) - \sigma_3(\overline{C}_a)\phi_1(\overline{C}_a) e^{\phi_2(\overline{C}_a)\bar{t}}}{1 - \phi_1(\overline{C}_a) e^{\phi_2(\overline{C}_a)\bar{t}}}$$

Here

$$\sigma_{2,3}(\overline{C}_a) = -\frac{1}{2}\left(\frac{\overline{\alpha}_0\overline{C}_a + \beta}{\overline{\alpha}_0\theta\,\overline{C}_a} - \overline{S}_m\right) \pm \sqrt{\frac{1}{4}\left(\frac{\overline{\alpha}_0\overline{C}_a + \beta}{\overline{\alpha}_0\theta\,\overline{C}_a} - \overline{S}_m\right)^2 + \frac{\overline{S}_m}{\theta}}$$

$$\sigma_2(\overline{C}_a) - \sigma_3(\overline{C}_a) = \sqrt{\left(\frac{\overline{\alpha}_0\overline{C}_a + \beta}{\overline{\alpha}_0\theta\,\overline{C}_a} - \overline{S}_m\right)^2 + \frac{4\overline{S}_m}{\theta}}$$

$$\phi_1(\overline{C}_a) = \frac{\overline{S}^0 - \sigma_2(\overline{C}_a)}{\overline{S}^0 - \sigma_3(\overline{C}_a)},\; \phi_2(\overline{C}_a) = \overline{\alpha}_0\theta\,\overline{C}_a\sqrt{\left(\frac{\overline{\alpha}_0\overline{C}_a + \beta}{\overline{\alpha}_0\theta\,\overline{C}_a} - \overline{S}_m\right)^2 + \frac{4\overline{S}_m}{\theta}}.$$

Further,

$$\overline{z} = T\int_1^{\overline{C}_a} \frac{\left(1 - \phi_1(\zeta)\, e^{-\phi_2(\zeta)T}\right) d\zeta}{\overline{S}^0 - \sigma_2(\zeta)\,[\overline{S}^0 - \sigma_3(\zeta)]\,\phi_1(\zeta)\, e^{-\phi_2(\zeta)T}} \tag{6.114}$$

therefore,

$$\frac{d\overline{z}}{d\overline{C}_a} = \frac{T\left(1 - \phi_1(\overline{C}_a)\, e^{-\phi_2(\overline{C}_a)T}\right)}{\overline{S}^0 - \sigma_2(\overline{C}_a)\,[\overline{S}^0 - \sigma_3(\overline{C}_a)]\,\phi_1(\overline{C}_a)\, e^{-\phi_2(\overline{C}_a)T}} \tag{6.115}$$

Then

$$\mu(\overline{C}_a, \overline{t}) = \overline{\alpha}_0\overline{t}\int_1^{\overline{C}_a} \frac{[\overline{S}_m - \overline{S}(\zeta,\overline{t})]\left(1 - \phi_1(\zeta)\, e^{-\phi_2(\zeta)\overline{t}}\right) d\zeta}{\overline{S}^0 - \sigma_2(\zeta)\,[\overline{S}^0 - \sigma_3(\zeta)]\,\phi_1(\zeta)\, e^{-\phi_2(\zeta)\overline{t}}} \tag{6.116}$$

As a result, the following formula is derived

$$\overline{C}(\overline{C}_a, \overline{t}) = \\ = e^{-\mu(\overline{C}_a,\overline{t})}\left\{1 - \beta\overline{t}\int_{\overline{C}_a}^{1} \frac{\overline{S}(\zeta,\overline{t})\left(1 - \phi_1(\zeta)\, e^{-\phi_2(\zeta)\overline{t}}\right)}{\overline{S}^0 - \sigma_2(\zeta) - [\overline{S}^0 - \sigma_3(\zeta)]\,\phi_1(\zeta)\, e^{-\phi_2(\zeta)\overline{t}}}\, e^{\mu(\zeta,\overline{t})}\, d\zeta\right\}, \tag{6.117}$$

which is complemented in the calculations of $\overline{C}(\overline{z},\overline{t})$ by (6.114) and it is assumed that $T = \overline{t}$.

The relative head losses in a bed are calculated with the given permeability function $f_k(\overline{S})$ on the basis of formula (2.24). However, in order to calculate the integral contained in it, it is necessary to use $\overline{C}_a$ instead of $\overline{z}$. After the appropriate transformation, the indicated formula takes the following form

$$\Delta\overline{h}(\overline{t}) = \int_{\overline{C}_{ae}}^{1} \frac{d}{d\zeta}\overline{z}(\zeta,\overline{t})\, \frac{d\zeta}{f_k(\overline{S}(\zeta,\overline{t}))} \tag{6.118}$$

where $\overline{C}_{ae}$ is calculated from (6.114) at $\overline{z} = 1$. In the first option using (6.110) is obtained

$$\Delta\bar{h}(\bar{t}) = \int_{\bar{C}_{ae}}^{1} \frac{d\zeta}{f_k(\bar{S}(\zeta,\bar{t}))\left[\bar{S}^0 - \sigma_1(\zeta)\right]\left(1 - e^{-(\bar{\alpha}_0\zeta+\bar{\beta})\bar{t}}\right)} \tag{6.119}$$

Similarly, involving (6.115), the formula is derived for the second option

$$\Delta\bar{h}(\bar{t}) = \int_{\bar{C}_{ae}}^{1} \frac{\left(1 - \phi_1(\zeta)\, e^{-\phi_2(\zeta)\bar{t}}\right) d\zeta}{f_k(\bar{S}(\zeta,\bar{t}))\left\{\bar{S}^0 - \sigma_2(\zeta) - \left[\bar{S}^0 - \sigma_3(\zeta)\right] \phi_1(\zeta)\, e^{-\phi_2(\zeta)\bar{t}}\right\}} \tag{6.120}$$

In the quantitative analysis, the main attention was paid to the estimation of the errors due to the simplification of the kinetic equation and the introduction of a new independent variable. The rigorous solution of the mathematical problem of nonlinear filtration with a constant rate and the linear adhesion function (6.2), obtained in Section 6.2, served as a reference for the first series of test examples. The increase, firstly, in the relative content of the suspension particles in free and deposited states, and secondly, in the mechanical energy expenditure was predicted over time and along the bed height. Preliminary the values were fixed of $\bar{\beta}$ (0.005) and $\bar{S}_m$ (5000). The filtering material was assumed to be either moderately ($\bar{\alpha}_0 = 0.001$), or well-absorbing ($\bar{\alpha}_0 = 0.0015$).

Figure 21 demonstrates the results of the systematic calculations of the dispersed contaminants content in the filtrate over 500 relative time units, which corresponds to approximately 40 hours of real time. The values of $\bar{C}_e$ were determined exactly by formula (6.49), and approximated by (6.112) taking into account (6.111). The errors due to the simplified description of the absorption process were significantly larger at $\bar{\alpha}_0 = 0.0015$, reaching the maximum value 9.5%. However, they decreased markedly as the filtration continued ($\bar{t} > 500$). If the absorption capacity of the bed was one and a half times weaker, the maximum error did not exceed 2% and both calculated curves (1, 2) almost merged during the calculation period.

Figure 21. Increase in the relative concentration of the suspended solids in the filtrate over time

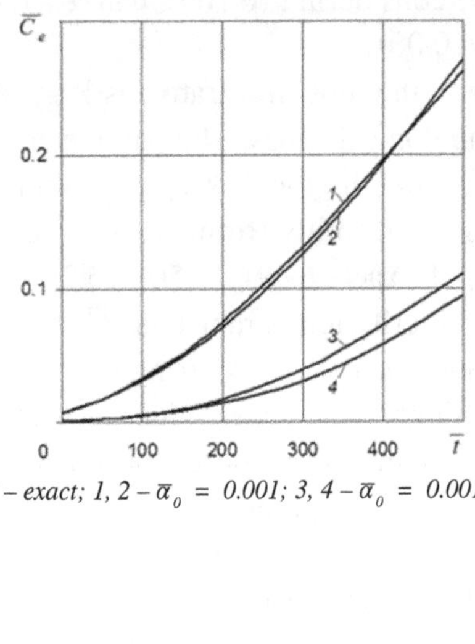

Note. 1, 3 – approximate calculation; 2, 4 – exact; 1, 2 – $\bar{\alpha}_0 = 0.001$; 3, 4 – $\bar{\alpha}_0 = 0.0015$

Figure 22. Profiles of the relative concentration of the deposited particles

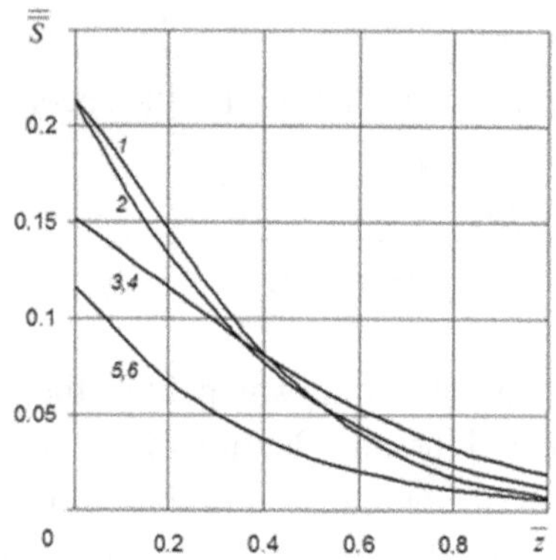

Note. 1, 3, 5 – exact calculation; 2, 4, 6 – approximate; 1, 2 – $\bar{\alpha}_0 = 0.0015$*; 3-6 –* $\bar{\alpha}_0 = 0.001$*; 1-4 –* $\bar{t} = 400$*; 5, 6 –* $\bar{t} = 200$

The concentration profiles of the deposited suspended solids $\bar{S}(\bar{z})$ were calculated for the time points corresponding to the intermediate ($\bar{t} = 200$) and final stages ($\bar{t} = 400$) of the productive filtration. The quantity $\bar{S}$ was exactly found for the same values of $\bar{\alpha}_0$ by formula (6.50), and approximated by formula (6.108) taking into account (6.109). The calculated profiles are shown in **Figure 22**. Small discrepancies of the corresponding approximate and reference curves occurred only at the actively absorbing contaminant bed and were leveled at a smaller value of $\bar{\alpha}_0$ so much that the pairs of such curves are depicted for both moments in time by single lines. It should be noted that in the approximate calculations of linear filtration (Section 2.4), the deposit distribution and then the head losses versus the impurity concentration were determined with clearly large errors, which also increased rapidly over time. In this situation, however, the approximate solution makes it possible to calculate these characteristics with obviously higher accuracy, which is also confirmed by **Figure 23**. Here the curves are presented of an increase in the relative head losses over time, calculated exactly by formula (6.28) taking into account (2.148), (6.50) and approximated by (6.119) considering (6.109). Curves 1,2, corresponding to $\bar{\alpha}_0 = 0.0015$ merge, and diverge slightly at $\bar{\alpha}_0 = 0.001$.

But from a practical point of view the most illustrative is **Figure 24**, which illustrates the data of the calculations of the relative technological times. The parameters $\bar{t}_p$, $\bar{t}_h$ and then $\bar{t}_f$ were calculated at continuous change of the coefficient $\bar{\alpha}_0$ in a wide range, which made it possible to cover materials with significantly different absorption capability (from weak to very high). The values of $\bar{t}_p$, $\bar{t}_h$ were also found in parallel on the basis of exact (6.49), (6.50), (6.28) and approximate (6.112), (6.108), (6.109), (6.119) formulas at the given criterion parameters ($\bar{C}_* = 0.1$, $\Delta\bar{h}_* = 6$). The curves of the dependences $\bar{t}_p(\bar{\alpha}_0)$, $\bar{t}_h(\bar{\alpha}_0)$ have monotonically decreasing or increasing nature. When calculating $\bar{t}_p$, $\bar{t}_h$ the errors increased markedly at higher and lower values of $\bar{\alpha}_0$. Nevertheless, a noticeable decrease in the accuracy of the technological calculations was observed only in situations that are not of interest for applications. The key technological time as a function of $\bar{\alpha}_0$ is depicted by solid broken lines, which have the pronounced maximum $\bar{t}_{fm}$ at the fracture point $\bar{\alpha}_0 = \bar{\alpha}_{opt}$. Of particular note is the closeness of the exact and approximate curves of the dependence $\bar{t}_f(\bar{\alpha}_0)$, in particular, the values $\bar{t}_{fm}$ and $\bar{\alpha}_{opt}$. Unsuccessful selection of filtering material (it has a reduced ($\bar{\alpha}_0 > \bar{\alpha}_{opt}$) or increased ($\bar{\alpha}_0 < \bar{\alpha}_{opt}$) capability to retain dispersed impurities) can cause a significant reduction in the duration of the filter run and a commensurate increase in the operating costs.

Figure 23. Increase in the relative head losses in the bed

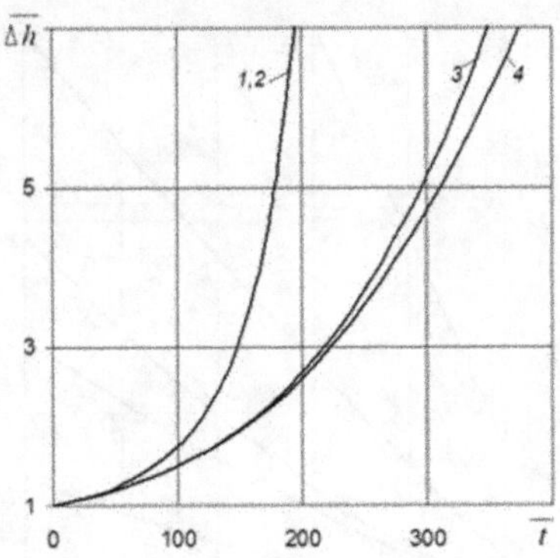

Note. 1, 3 – exact calculation; 2, 4 – approximate; 1, 2 – $\bar{\alpha}_0$ = 0.0015; 3, 6 – $\bar{\alpha}_0$ = 0.001

Figure 24. Dependencies $\bar{t}_p(\bar{\alpha}_0)$, $\bar{t}_h(\bar{\alpha}_0)$, $\bar{t}_f(\bar{\alpha}_0)$

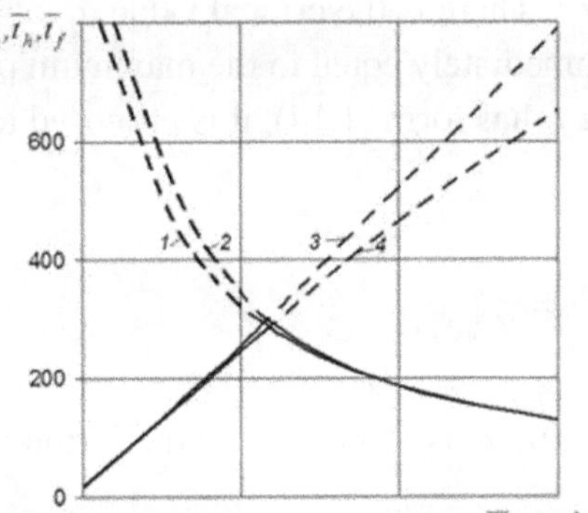

Note. 1, 3 – exact calculation; 2, 4 – approximate; 1, 2 – $\bar{t}_h$; 3, 4 – $\bar{t}_p$; the solid line – $\bar{t}_f(\bar{\alpha}_0)$

The second series of examples shows the application of the developed calculation methodology in the case of the nonlinear adhesion function $\chi_S(S)$, when it is unrealistic to obtain an exact solution of the initial mathematical problem of nonlinear filtration. Therefore, only approximate calculations were performed, first, of two relative filtration characteristics ($\bar{C}_e$, $\Delta\bar{h}$), second, of a standard set of the technological times ($\bar{t}_p$, $\bar{t}_h$, $\bar{t}_f$). The initial data used in the first series of examples were supplemented with two values of $\bar{\theta}$ (0.001, 0.002). Here the subject of analytical analysis was also, first of all, the filtrate quality. The curves describing the increase in the outlet suspended solids content with time up to $\bar{t}$ = 500 are presented in **Figure 25**. They correspond to four values of $\bar{\alpha}_0$ and two values of $\bar{\theta}$. The amount of the contaminants free passed through the bed was rapidly increasing at a weak absorption capability of the filtering material ($\bar{\alpha}_0$ = 0.0005), in spite of the significant autocatalytic effect. At the same time according to (6.87) the relative concentration of the suspended solids amounted to 0.082 already in the first portion of filtrate ($\bar{C}_e^0$). With increasing $\bar{\alpha}_0$ the value of $\bar{C}_e^0$ sharply decreased. Similarly obtained $\bar{C}_e^0$ = 0.0235 at $\bar{\alpha}_0$ = 7.5 · 10^{-4} and $\bar{C}_e^0$ = 0.0019 at $\bar{\alpha}_0$ = 1.25 · 10^{-3}.

Figure 25. Increase in the relative concentration of the suspended solids in the filtrate over time

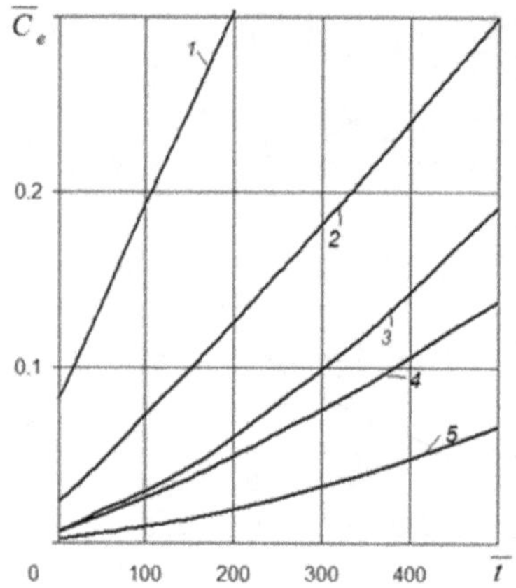

Note. 1, 2, 4, 5 – $\overline{\theta}$ = 0.002; 3 – $\overline{\theta}$ = 0.001; 1 – $\overline{\alpha}_0$ = $5 \cdot 10^{-4}$; 2 – $\overline{\alpha}_0$ = $7.5 \cdot 10^{-4}$; 3, 4 – $\overline{\alpha}_0$ = 10^{-3}; 5 – $\overline{\alpha}_0$ = $1.25 \cdot 10^{-3}$

It should be noted that the contamination level of the first portion of the filtered water increases with a hypothetical decrease in α (or α_0). There is the critical value α_{cr} of the coefficient α, at which the relative outlet concentration C_e is immediately equal to the maximum permissible C_*. In the general case, when $S^0>0$ and the expression for f_α has form (1.14), it is proposed to determine the specified value by fitting from the equation

$$\frac{\overline{\beta}\,\overline{S}^0}{\overline{\alpha}_{cr}\chi_S(\overline{S}^0)} + \left(1 - \frac{\overline{\beta}\,\overline{S}^0}{\overline{\alpha}_{cr}\chi_S(\overline{S}^0)}\right)e^{-\overline{\alpha}_{cr}\chi_S(\overline{S}^0)} = C_* \tag{6.121}$$

If the bed is initially clean ($S^0=0$), α_{cr} is calculated by the formula

$$\alpha_{cr} = -\frac{\ln \overline{C}_*}{\chi_S(0)} \tag{6.122}$$

At the given values of $\overline{C}_*$ and $\chi_S(0)$ α_{cr} is $4.6 \cdot 10^{-4}$. Of special note are curves 3 and 4, which were determined at the same value $\overline{t}$ = 400, but different values of $\overline{\theta}$. Therefore, we can clearly evaluate the significance of the autocatalysis for the suspension clarification by their position relative to each other. The discrepancy between the calculated values reached 30 or 40% at $\overline{\alpha}_0$ = 0.0015 (depending on the choice of the reference value)

Figure 26. Increase in the relative head losses in the bed

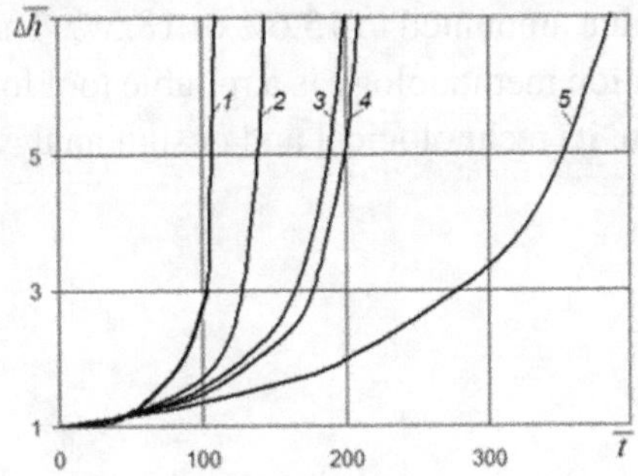

Note. 1, 2, 4, 5 – $\bar{\theta}$ = 0.002; 3 – $\bar{\theta}$ = 0.001; 1 – $\bar{\alpha}_0$ = 1.25 · 10^{-3}; 2, 3 – $\bar{\alpha}_0$ = 10^{-3}; 4 – $\bar{\alpha}_0$ = 7.5 · 10^{-4}; 5 – $\bar{\alpha}_0$ = 5 · 10^{-4}

Figure 27. Dependencies $\bar{t}_p(\bar{\alpha}_0)$, $\bar{t}_h(\bar{\alpha}_0)$, $\bar{t}_f(\bar{\alpha}_0)$

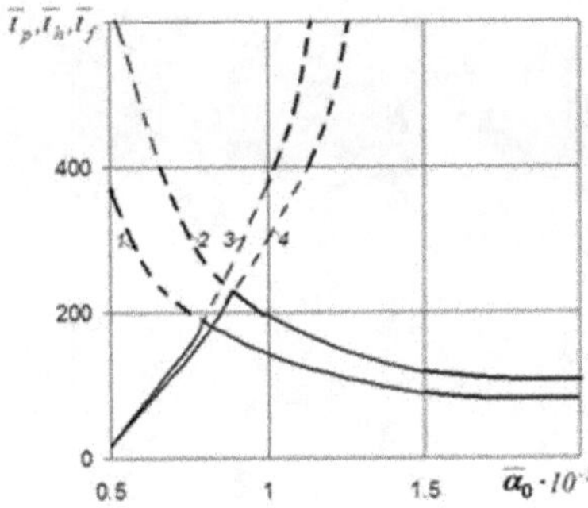

Note. 1, 2 – $\bar{t}_h$; 3, 4 – $\bar{t}_p$; 1, 3 – $\bar{\theta}$ = 0.002; 2, 4 – $\bar{\theta}$ = 0.001

The results of calculating the dynamics of the relative head losses in the bed are shown in **Figure 26**. It is logical that the lowest rate of the losses increment occurred in the case $\bar{\alpha}_0 = 0.001$ (curve 5). The effect of the autocatalysis was reflected in curves 2, 3, the discrepancy between which is explained by a twofold change in the value of $\bar{\theta}$. Significant nonlinearity of the calculated curves is caused by the existence of the vertical asymptotes in the function $\Delta\bar{h}(\bar{t})$ during the calculation period, which is associated with the choice of the approximation expression for $\bar{k}(\bar{S})$ and increased clogging of the inlet section of the bed. When the quantity $\Delta\bar{h}$ exceeds a certain value ($\Delta\bar{h}_*$), further operation of the filter becomes inexpedient for economic reasons (1.31). The array of the technological calculations data, based on the same values of the criterion parameters $\bar{C}_*$, $\Delta\bar{h}_*$, is described by the curves of the dependences $\bar{t}_{p,h}(\bar{\alpha}_0)$ in **Figure 27**. As a result, the graphs of the final dependence $\bar{t}_f(\bar{\alpha}_0)$ are plotted on the basis of (1.41). They are depicted in Figure 27 by the solid continuous broken lines. The fracture point ($\bar{\alpha}_0 = \bar{\alpha}_{opt}$) indicates the optimum absorption properties of the filter material. It is possible to achieve the maximum technological effect in the considered situation at $\bar{\theta} = 0.001$ and 0.002, if $\bar{\alpha}_{opt} \approx 7.9 \cdot 10^{-4}$ and $8.9 \cdot 10^{-4}$.

Then the highest values $\bar{t}_{fm}$ corresponding to them are approximately 225 and 190. Thus, the indicated effect is expressed in a shift of the value of $\bar{\alpha}_{opt}$ by 10^{-4}, and also in a change in $\bar{t}_{fm}$ by 35 units, so that the deviation from its reference value amounted to 15.6% or 18.5%. Therefore, the described and tested on characteristic examples calculation methodology is a reliable tool for forecasting the development of the nonlinear detachment filtration, its technological and design analysis.

REFERENCES

Adin, A., & Rebhun, M. (1987). Deep-bed filtration: Accumulation-detachment model parameters. *Chemical Engineering Science*, *42*(5), 1213–1219. DOI: 10.1016/0009-2509(87)80072-6

Alekseev, V. S., & Kommunar, G. M. (1974). Clogging of filters and prefilter zones of water intake wells. [in Russian]. *Water Resources*, (5), 170–177.

Alekseev, V.S., Kommunar, G.M., & Sherzhukov, B.S. (1989). *Mass transfer in water-saturated rocks. Results of Science and Technology. V.11.* Hydrogeology, Engineering Geology. (in Russian)

Bai, R., & Mackie, R. I. (1995). Modeling the transition between deposition modes in deep bed filtration. *Water Research*, *29*(1), 2601–2604. DOI: 10.1016/0043-1354(95)00078-Y

Bai, R., & Tien, C. (2000). Effect of deposition in deep-bed filtration determination and search of rate parameters. *Journal of Colloid and Interface Science*, *231*(2), 299–311. DOI: 10.1006/jcis.2000.7130 PMID: 11049680

Boller, M. A., & Kavanauch, M. C. (1995). Particle characteristics and head loss increase in granular media filtration. *Water Research*, *29*(4), 1139–1149. DOI: 10.1016/0043-1354(94)00256-7

Bugay, N.G., Krivonog, A.I., & Krivonog, V.V. (2004). Reagentless water purification by filters from fiber-porous polyethylene. *Applied Hydromechanics*, 6(78), 2, 27–43. (in Russian)

Calvin, R. M. (1992). Ripening of silica sand used for filtration. *Water Research*, *26*(5), 683–688. DOI: 10.1016/0043-1354(92)90245-Y

Dzyubo, V. V., & Alferova, L. I. (2007). Natural water filtration in the mode of non-uniform rates. [in Russian]. *Bulletin of TSASU*, (2), 180–190.

Elimelech, M., & O'Melia, C. R. (1990). Kinetics of deposition of colloidal particle in porous media. *Environmental Science & Technology*, *24*(10), 1528–1536. DOI: 10.1021/es00080a012

Ison, C. R., & Ives, K. J. (1969). Removal mechanism in deep bed filtration. *Chemical Engineering Science*, *24*(4), 717–729. DOI: 10.1016/0009-2509(69)80064-3

Jegatheesan, V., & Vigneswaran, S. (2000). Transient stage deposition of submicron particles in deep bed filtration under unfavorable conditions. *Water Research*, *34*(7), 2119–2131. DOI: 10.1016/S0043-1354(99)00381-4

Jegatheesan, V., & Vigneswaran, S. (2005). Deep bed filtration: Mathematical models and observations. *Critical Reviews in Environmental Science and Technology*, *35*(6), 515–569. DOI: 10.1080/10643380500326432

Kuhn, M., & Briesen, H. (2016). Dynamic modeling of filter-aid filtration including surface and filtration effects. *Chemical Engineering & Technology*, *39*(3), 425–434. DOI: 10.1002/ceat.201500347

Mackie, R. I., & Zhao, Q. (1999). A framework for modeling removal in the filtration of polydisperse suspensions. *Water Research*, *33*(3), 794–806. DOI: 10.1016/S0043-1354(98)00267-X

Mints, D. M. (1964). *Theoretical bases of water purification technology*. Publishing House on Construction. (in Russian)

Ojha, C. S., & Graham, N. J. D. (1994). Computer-aided simulation of slow sand filter performance. *Water Research*, *28*(5), 1025–1030. DOI: 10.1016/0043-1354(94)90187-2

Ojha, C. S. P., & Graham, N. J. D. (1992). Appropriate use of deep-bed filtration models. *Journal of Environmental Engineering*, *118*(6), 964–980. DOI: 10.1061/(ASCE)0733-9372(1992)118:6(964)

Ornatsky, N. S., Sergeev, E. M., & Shehtman, Y. M. (1955). *Investigation of the process of sand clogging*. Moscow State University Press. (in Russian)

Poliakov, V., & Martynov, S. (2021). Mathematical modeling of physicochemical iron removal from groundwater at rapid filters. *Chem. Eng. Sci.*, 231

Poliakov, V. L. (2011). Filtration of suspensions through multilayer packed bed at nonlinear kinetics of mass exchange. 1. Theory. [in Russian]. *Journal of Water Chemistry and Technology*, *33*(1), 3–14.

Poliakov, V. L. (2011). Filtration of suspensions through multilayer packed bed at nonlinear kinetics of mass exchange. 2. Quantitative analysis. [in Russian]. *Journal of Water Chemistry and Technology*, *33*(2), 111–127.

Poliakov, V.L. (2014). To the calculation of suspension filtration at nonlinear mass transfer kinetics. *Problems of Water Supply, Sewage and Hydraulics. Collected papers*, 23, 145–153. (in Russian)

Saiers, , J.EHornberger, , G.MLiang, , Z. (1994). First- and second-order kinetics approaches for modeling the transport of colloidal particles in porous media. *Water Resources Research*, *30*(9), 2499–2506.

Senyavin, M. M., Venitsianov, E. V., & Ayukaev, R. I. (1997). About mathematical models and engineering methods for calculating the process of natural water purification by filtration. [in Russian]. *Water Resources*, (2), 157–170.

Skouras, E. D., Burganos, V. N., Paraskeva, C. A., & Payatakes, A. C. (2011). Phenomenological description and network simulation of horizontal filtration dynamics. *Separation and Purification Technology*, *80*(1), 105–118. DOI: 10.1016/j.seppur.2011.04.016

Tan, H. K. S. (1994). Analysis of cyclic fixed bed sorption processes for second-order irreversible kinetics. *Chemical Engineering Science*, *49*(8), 1277–1285. DOI: 10.1016/0009-2509(94)85096-8

Verigin, N. N., & Kommunar, G. M. (1978). Filtration of suspensions through porous media. *Proceedings of VNII VODGEO*, 74, 120–124. (in Russian)

APPENDIX 6.1. APPROXIMATE CALCULATION OF LINEAR FILTRATION

In the following, the linear analog of nonlinear filtration is distinguished and considered separately as a special case due to its wide spreading in the theory and practice. The corresponding approximate solution is actually an alternative to the solution obtained in Section 2.4. The former solution provides a high accuracy in determining the outlet concentration of the dispersed contaminants for a long time (much longer than the filter run duration). However, the reliability of the prediction of the dynamics of the impurity deposition in the operating volume and, as a consequence, of the head losses in it turned out to be significantly less. The new solution and the resulting calculation dependencies can guarantee minimal errors in the determination of the spatial and temporal changes in the concentration of the deposited suspension particles and head losses, provided that the empirical data on the deposit composition (composition factor γ or its average value γ_a) are reliable.

Now, if $\alpha(S)=\alpha_0=const$, then according to the methodology of solving the nonlinear filtration problem described in Section 6.5, it is easy to obtain

$$\overline{S}(\bar z,\bar t)=\overline{S}^0 e^{-\bar\beta\bar t}+\frac{\bar\alpha_0}{\bar\beta}(1-e^{-\bar\beta\bar t})\,\overline{C}_a(\bar z,\bar t), \tag{A.6.1}$$

where $\overline{S}^0$ is the relative initial concentration of the deposited particles. The following expression is derived for the relative average concentration of the suspended solids

$$\overline{C}_a(\bar z,\bar t)=\frac{\bar\beta}{\bar\alpha_0}\overline{S}^0+\left(1-\frac{\bar\beta}{\bar\alpha_0}\overline{S}^0\right)e^{-\delta(\bar t)\bar z},\ \text{where } \delta(\bar t)=\bar\alpha_0(1-e^{-\bar\beta\bar t})/(\bar\beta\bar t). \tag{A.6.2}$$

Then the concentration $\overline{S}$ taking into account (A.6.2) takes the final form

$$\overline{S}(\bar z,\bar t)=\overline{S}^0+\left(\frac{\bar\alpha_0}{\bar\beta}-\overline{S}_0\right)(1-e^{-\bar\beta\bar t})e^{-\delta(\bar t)\bar z}. \tag{A.6.3}$$

Concentration $\overline{C}$ is expressed through $\overline{C}_a$ as follows

$$\overline{C}(\bar z,\bar t;\overline{C}_a)=e^{-\bar\alpha_0\bar z}\left\{1+\bar\beta\int_0^{\bar z}e^{\bar\alpha_0\eta}\left[\overline{S}^0e^{-\bar\beta\bar t}+\frac{\bar\alpha_0}{\bar\beta}(1-e^{\bar\beta\bar t})\,\overline{C}_a(\eta,\bar t)\right]d\eta\right\}. \tag{A.6.4}$$

After a number of transformations, it is obtained

$$\overline{C}(\bar z,t)=e^{-\bar\alpha_0\bar z}+\frac{\bar\beta}{\bar\alpha_0}\overline{S}^0(1-e^{-\bar\beta\bar t})+\frac{\bar\alpha_0-\bar\beta\,\overline{S}^0}{\bar\alpha_0-\delta(\bar t)}(1-e^{-\bar\beta\bar t})(e^{-\delta(\bar t)\bar z}-e^{-\bar\alpha_0\bar z}). \tag{A.6.5}$$

Finally, the relative head losses in the filter bed are calculated by formula (2.24). If the function $f_k(\overline{S})$ has form (2.148), the required losses will be as follows

$$\begin{aligned}\Delta\bar h(\bar t)=\frac{1}{(1-\bar\gamma_a\overline{S}^0)^3}\Bigg\{&1+\frac{1}{\delta(\bar t)}\ln\frac{a(\bar t)-e^{-\delta(\bar t)}}{a-1}+\\&+\frac{a(\bar t)}{\delta(\bar t)}\left[\frac{1}{a(\bar t)-1}-\frac{1}{a(\bar t)-e^{-\delta(\bar t)}}\right]+\frac{a^2(\bar t)}{2\delta(\bar t)}\left[\frac{1}{(a(\bar t)-1)^2}-\frac{1}{(a(\bar t)-e^{-\delta(\bar t)})^2}\right]\Bigg\},\end{aligned} \tag{A.6.6}$$

where $a(\bar t)=\dfrac{\bar\beta(1-\bar\gamma_a\overline{S}^0)}{\bar\gamma_a(\bar\alpha_0-\bar\beta\,\overline{S}^0)(1-e^{-\bar\beta\bar t})}$.

In the special case $\overline{S}^0=0$, the basic calculation formulas are simplified to the following form

$$\overline{C}(\bar z,t)=e^{-\bar\alpha_0\bar z}+\frac{\bar\alpha_0}{\bar\alpha_0-\delta(\bar t)}(1-e^{-\bar\beta\bar t})(e^{-\delta(\bar t)\bar z}-e^{-\bar\alpha_0\bar z}), \tag{A.6.7}$$

$$\Delta\bar{h}(\bar{t}) = 1 + \frac{1}{\delta(\bar{t})}\ln\frac{a_0(\bar{t}) - e^{-\delta(\bar{t})}}{a_0(\bar{t}) - 1} + \\ + \frac{a_0(\bar{t})}{\delta(\bar{t})}\left[\frac{1}{a_0(\bar{t}) - 1} - \frac{1}{a_0(\bar{t}) - e^{-\delta(\bar{t})}}\right] + \frac{a_0^2(\bar{t})}{2\delta(\bar{t})}\left[\frac{1}{(a_0(\bar{t}) - 1)^2} - \frac{1}{(a_0(\bar{t}) - e^{-\delta(\bar{t})})^2}\right], \tag{A.6.8}$$

where $a_0(\bar{t}) = \frac{\bar{\beta}}{\bar{\gamma}_a \bar{\alpha}_0 (1 - e^{-\bar{\beta}\bar{t}})}$.

The errors in the calculation of the key filtration characteristics due to the use of the alternative approximate solution were estimated based on the typical examples. The fixed values were assumed: $\bar{\beta}$ (0.005), γ_a(20), C_0(0.00005), $\bar{S}^0$ (0), so that $\bar{\gamma}_a$ is equal to 0.001. Three characteristic values of the coefficient $\bar{\alpha}_0$ were set (5,7,9). The subject of the calculations were relative quantities: the concentration $\bar{C}_e$ and head losses $\Delta\bar{h}$ as functions of relative time, as well as the concentration profile $\bar{S}(\bar{z})$ at the time moment $\bar{t}$ = 300. The time intervals were accepted so that the maximum values of $\bar{C}_e$ and $\Delta\bar{h}$ significantly exceeded the standard values $\bar{C}_*$, $\Delta\bar{h}_*$. The results obtained in this way are illustrated in three Figures.

Change in the outlet suspended solids concentration over time was calculated exactly by formula (2.57) and approximated by formulas (2.88) and (A.6.5). Two series of curves of the dependence $\bar{C}_e(\bar{t})$, corresponding to $\bar{\alpha}_0$ = 5, 7, are shown in **Figure 28**. In general, the errors of the approximate calculations using both approximate formulas were acceptable for engineering applications. However, in the case of the moderate absorption capacity of the filtering material, the alternative solution was preferable, namely, the maximum error at $\bar{t}$ = 600 (far beyond the standard filter run) was 5% (second solution) and 11.4% (first solution). The situation changed with an increase in the absorption capacity, and then the best results were obtained using the solution from Section 2.4. Accordingly, the maximum computational errors were 16.2 and 6.9%.

The prerequisite for a reliable calculation of mechanical energy expenditure for overcoming the resistance to the suspension flow from the clogged porous medium is the proximity of the approximate deposit distribution within the bed to the exact distribution (within the framework of the adopted mathematical model). **Figure 29** shows three profiles $\bar{S}(\bar{z})$ at the point in time corresponding to the final stage of the clarification and $\bar{\alpha}_0$ = 5. The outlet concentration of contaminants in the filtrate has increased by this moment to the value 0.112. In principle, at the most hydraulically critical inlet bed section, the reference (2) and approximate (3) profiles according to formulas (2.56) and (A.3), respectively, practically coincided. Their minimal divergence took place only at a distance from the inlet cross-section, which, as the subsequent head losses calculation showed, is not of principal importance. The first approximate solution (formula (2.86)) gave obviously worse results, and exactly at the inlet to the bed and in its vicinity. The consequences of such a fact for the filtration hydraulics can be easily estimated also by the next series of calculations.

The curves of the head losses increase for nine options (three values of $\bar{\alpha}_0$ and three calculation methods) were determined by formula (2.24) taking into account (2.148) at different expressions for $\bar{S}(\bar{z})$ and are shown in **Figure 30**. The losses increase rate sharply rises over time, which is caused by the accelerated contamination of the inlet section of the bed, especially at large values of $\bar{\alpha}_0$. The first approximate solution allowed to reliably calculate the dynamics of the losses only at their increase up to about 3-4 units, which is insufficient for the realistic beds due to the smallness of their initial hydraulic resistance. The second approximate solution provided substantially smaller errors in the determination of $\Delta\bar{h}$, which only in situations, uncharacteristic for the suspension clarification on rapid filters, reached 11%. With the use of the first solution in similar conditions the obtained values of $\Delta\bar{h}$ exceeded the reference values in 2.5 times. Under normal filtration conditions the errors in determining, in particular, the technological time $\bar{t}_h$ were within 3-5% (the second solution). With an improvement in the bed absorption properties the computational errors decreased so much that the approximate curves 3, 6 practically merged with the reference curves 2, 5 and only at the corresponding set of two curves (8, 9) minimally diverged.

Figure 28. Increase in the relative concentration of the suspended solids in the filtrate

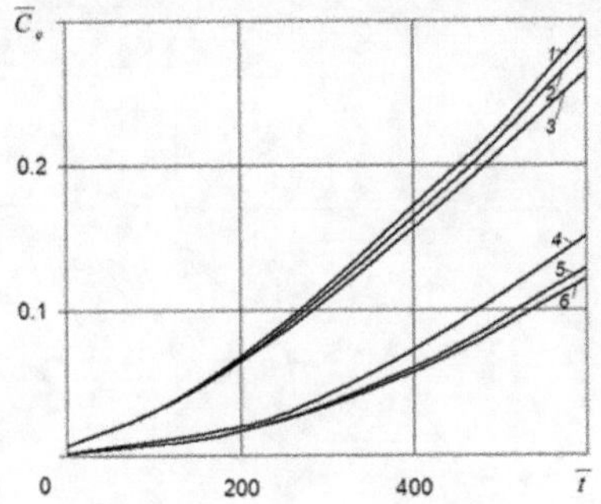

Note. 1, 5 – exact solution; 2, 4 – second approximate; 3, 6 – first approximate; 1-3 –$\bar{\alpha}$ = 5; 4-6 – $\bar{\alpha}$ = 7

Figure 29. Relative concentration profiles of the deposited impurity

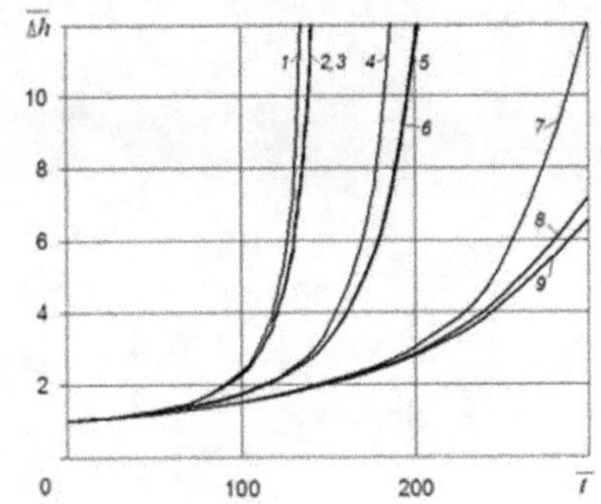

Note. 1 – first approximate solution, 2 – exact, 3 – second approximate

Figure 30. Increase in the relative head losses over time

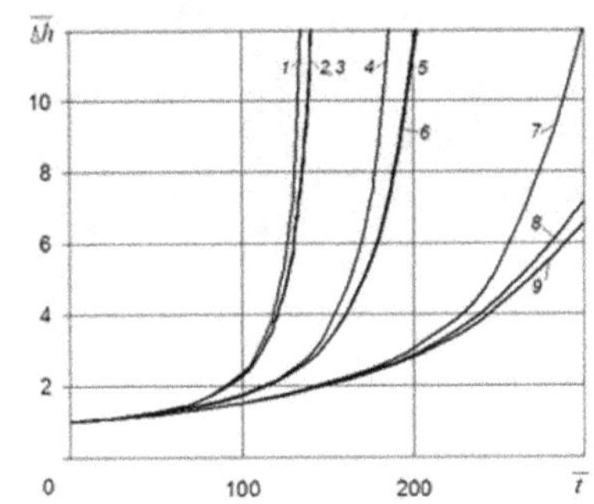

Note. 2, 5, 8 - exact solution; 1, 4, 7 – first approximate; 3, 6, 9 – second approximate; 1-3 –$\bar{\alpha}$ = 9; 4-6 – $\bar{\alpha}$ = 7; 7-9 – $\bar{\alpha}$ = 5

Chapter 7
Plane-Radial and Radial Detachment Filtration

ABSTRACT

The subject of the in-depth research by exact and approximate methods in Chapter 7 is an understudied variety of the deep bed filtration, namely filtration with an axisymmetric suspension flow, which is implemented in a filter media having the shape of a cylindrical layer. The stable supply of contaminated water to the filter bed through its external or internal surfaces is thoroughly studied for linear kinetics. According to the results of the technological analysis, it has been concluded that the operation of the filter can be noticeably prolonged due to replacing the plane shape of the filter medium and the suspension flow with an axisymmetric shape and thus, using filter materials with high absorption capacity. In turn, the design analysis allowed to substantiate the rational dimensions (radii) of the bed. Chapter ends with the research on the radial filtration with spherical symmetry of the suspension flow in the functional layer of a spherical or semi-spherical shape.

Fundamental importance for the theory and practice of the deep bed filtration has the long-established fact of the essential mutual influence of the filtration flow and mass exchange processes. Thus, there is a principal opportunity to increase the efficiency of water treatment filters by a purposeful change of filtration rate in time, which is confirmed by the results of the studies conducted in Chapter 5, and also within the filter media volume. Especially promising in technological terms seems to be the arrangement of the filtration flow in the filter bed, the rate of which decreases or increases with distance from the inlet cross-section. Then, taking into account the expressions for the mass exchange coefficients (3.54) and physical and chemical properties of suspension, filtering material, it is realistic to achieve a significant intensification or weakening of the absorption process, which plays a key role in the dispersed impurity removal. When materials with a high absorption capability are used, as a rule, the clarification resource of a packed bed is not fully utilized. The reason of the unsatisfactory operation of the filter, which is expressed in premature termination of filtration and overconsumption of wash water, is the excessive expenditure of mechanical energy due to the accelerated clogging of the section adjacent to the inlet surface. It is possible to significantly prolong the filter operation in such situations by creating unfavorable conditions for the sticking of suspension particles to the bed elements (grains) at the specified section and, on the contrary, favorable conditions at the peripheral section. Thus, if the thickness of the filter media is sufficient, on the one hand, the required amount of the suspended solids will be retained within it and, on the other hand, its hydraulic resistance will increase much more slowly. As a result,

DOI: 10.4018/979-8-3693-8211-0.ch007

the amount of the clarified suspension with the normative outlet suspended solids content will increase commensurately by extending the continuous operation of the filter. Stable filtration flow with its moderate spatial irregularity can be provided using filters, structurally equipped with the rate regulators and a uniform bed of small thickness and special shape, for example, in the form of a cylindrical or spherical layer (Dzyubo, 2006; Poliakov, 2014, 2015b).

In general, axisymmetric (plane-radial) filtration flow of liquids in natural porous media has been given extremely great attention for several decades due to the wide use of vertical drainage in engineering practice (Oleinik, 1978; Bear, 1979). Modeling of flows of such geometry is significantly complicated in the case of heterogeneous fluids in the presence of intensive interfacial mass transfer due to the change in filtration rate in the radial direction. Therefore, the relevant studies conducted by experimental (Voitov, 1974; Bolton et al., 2004) and theoretical (Dzyubo & Alferova, 2007; Liu, et al., 2016; Tiller, et al., 2004; Vigneswaran, et al., 1985) methods were scattered. Nevertheless, they have clearly demonstrated the advantages of the plane-radial filtration over the plane filtration. Such advantages are especially clear in the case of a high absorption capacity of the filtering material. The theory of the (plane-) radial filtration was thoroughly developed in a series of works by the author (Poliakov, 2013a; 2013b; 2015b; 2015c), which were used to write this Chapter. The results obtained using analytical methods provide a significant contribution to the above theory and can serve as a basis for the applied optimization of the corresponding technological and design parameters.

In five Sections of Chapter 7, two types of the uneven detachment filtration (plain-radial and radial) at linear kinetics of the interfacial mass exchange became the subject of research. The main attention was paid to the plain-radial filtration, which is dedicated to the first four Sections. An important design advantage of this method of the suspension clarification is the compactness of a filter facility. Indeed, it is possible to increase the performance of the cylindrical bed at a constant hydraulic load proportionally to increasing its height without complicating the filter design. The ultimate goal of this development was a reliable prediction of changes in the characteristics of the plane-radial filtration, as well as an engineering methodology for the substantiation of the technological and design parameters, which is based on approximate solutions of two similar axisymmetric non-stationary problems of the joint mass transfer, mass exchange, and filtration flow. This methodology is presented in fragments in Sections 7.2, 7.3 and 7.4. It should be noted that the efficiency of the (plane-)radial filtration depends significantly on the place of suspension feeding into the bed. In the first three Sections, plane-radial filtration is analyzed when the suspension enters through the outer cylindrical surface, and in Section 7.4 – through the inner surface. In order to assess the validity of both approximate solutions, and thus, of the engineering methodology based on them, particular exact solutions were obtained to serve as a reference for the comparative analysis. They are presented in the first and fourth Sections. The radial filtration of suspension through a bed in the form of either a full spherical layer or a spherical half-layer is studied in the last Section. Certainly, the use of such beds due to the complexity of their design, operation and especially washing can make sense especially at a very high cost of absorbing material. Nevertheless, the study of this type of the uneven detachment filtration is useful in that it provides a deeper understanding and evaluating the capabilities of water treatment filters with a spatial unevenness of the filtration flow.

7.1. PLANE-RADIAL FILTRATION WITH SUSPENSION FEEDING FROM OUTSIDE (EXACT SOLUTION).

The basic mathematical model of the plane-radial linear detachment filtration describes the axisymmetric processes of mass transfer, mass exchange and filtration flow in a complex. In this case, two components of the interfacial exchange of dispersion impurity (attachment and detachment of the suspension particles) are considered separately by analogy with the previous Chapters, with the exception of Chapter 4, and a hydraulic load is considered constant. The suspension is fed from outside and separated in a filter bed having the form of a cylindrical layer (**Figure 1**). Since special importance is given in this Chapter to non-standard forms of the suspension flow and its spatial unevenness caused by them, the particular approximation expression (1.10) is accepted for the dependence of k on S, which, nevertheless, due to the presence of two parameters (m_{k1}, m_{k2}) allows approximating the experimental data with respect to the flow capacity of the bed being clogged with high accuracy. Due to the scarcity of relevant information, the filtration flow compartment contains instead (S) its constant analog (average value) γ_a, although, if necessary, the theoretical results obtained below can be easily generalized to the case of a variable γ. Then the above-mentioned model after accepting a number of previously discussed assumptions is simplified to the following form (Poliakov, 2013a)

$$V(r)\frac{\partial C}{\partial r} - \frac{\partial S}{\partial t} = 0, \tag{7.1}$$

$$\frac{\partial S}{\partial t} = \alpha_V V^l(r)C - \beta_V V^q(r)S, \tag{7.2}$$

$$V(r) = k(S)\frac{\partial h}{\partial r}, \tag{7.3}$$

$$k(S) = k_0\left[1 - \left(\gamma_a \frac{S}{n_0}\right)^{m_{k1}}\right]^{m_{k2}}; \tag{7.4}$$

$$t = 0, \quad S = 0\ ; \tag{7.5}$$

$$r = r_0, \quad C = C_0\ ; \tag{7.6}$$

$$r = r_e, \quad h = H_d. \tag{7.7}$$

Here V is the filtration rate changing in the radial direction; α_V, β_V are the reduced rate coefficients of the attachment and detachment of the specified particles given in accordance with (3.54); l,q are the empirical coefficients varying most often in the range from 0 to 1; r_0, r_e are the radii of the outer and inner cylindrical surfaces of the bed; H_d is the head on the inner bed surface.

Figure 1. Scheme of the plane-radial suspension filtration (external supply)

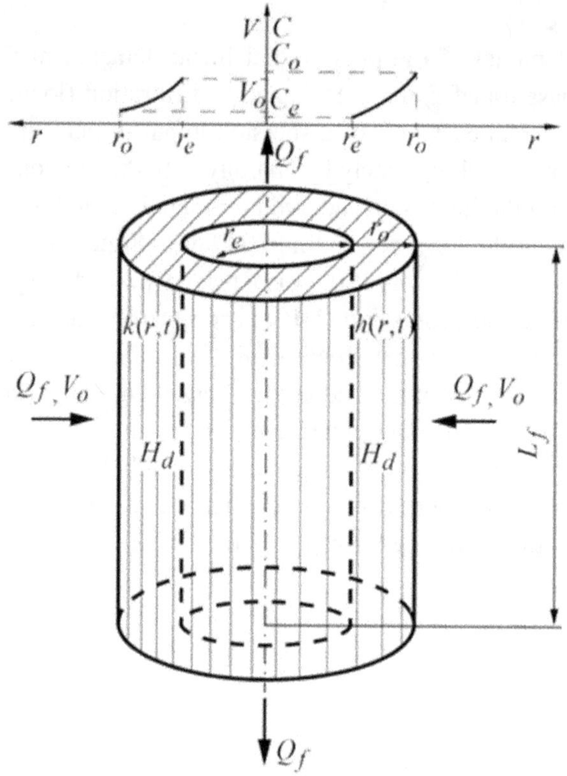

The final, generalized form of model (7.1) - (7.7) is given after the introduction of dimensionless variables and parameters, including as the scales of the length and rate the quantities r_0 and V_0, where the hydraulic load $V_0=Q_f/(2\pi r_0 L_f)$, Q_f is the filter performance, L_f is the height of a cylindrical bed. As a result, it is obtained

$$\bar{V}(\bar{r})\frac{\partial \bar{C}}{\partial \bar{r}} - \frac{\partial \bar{S}}{\partial \bar{t}} = 0, \tag{7.8}$$

$$\frac{\partial \bar{S}}{\partial \bar{t}} = \bar{\alpha}_V \bar{V}^l(\bar{r})\bar{C} - \bar{\beta}_V \bar{V}^q(\bar{r})\bar{S}, \tag{7.9}$$

$$\bar{V}(\bar{r}) = -\bar{k}(\bar{S})\ln \bar{r}_e \frac{\partial \bar{h}}{\partial \bar{r}}, \tag{7.10}$$

$$\bar{k}(\bar{S}) = f_k(\bar{S}) = \left(1 - \bar{\gamma}_a^{m_{k1}} \bar{S}^{m_{k1}}\right)^{m_{k2}} ; \tag{7.11}$$

$$\bar{t} = 0, \quad \bar{S} = 0 ; \tag{7.12}$$

$$\bar{r} = 1, \quad \bar{C} = 1 ; \tag{7.13}$$

$$\bar{r} = \bar{r}_e, \quad \bar{h} = 0. \tag{7.14}$$

Here

$$\bar{V} = V/V_0,\ \bar{C} = C/C_0,\ \bar{S} = S/(n_0 C_0),\ \bar{r} = r/r_0,\ \bar{r}_e = r_e/r_0,\ \bar{t} = V_0 t/(n_0 r_0),$$

$$\bar{\alpha}_V = r_0 V_0^{l-1}\alpha_V,\ \bar{\beta}_V = n_0 r_0 V_0^{q-1}\beta_V,\ \bar{k} = k/k_0,\ \bar{\gamma}_a = \gamma_a C_0,$$

$$\bar{h}(\bar{r},\bar{t}) = \frac{h(r,t) - H_d}{h(r_0,0) - H_d} = k_0\frac{h(r,t) - H_d}{r_0 V_0}.$$

Since the clarification compartment, including equations (7.8), (7.9), initial (7.12), boundary (7.13) conditions, due to the known quantity $\bar{V}(\bar{r})$ and neglecting of the local derivative in (7.8) does not depend on the filtration flow compartment, it is solved separately using the integral Laplace transform. From (7.9) follows the expression for the image $\bar{S}_L$

$$\bar{S}_L(\bar{r},p) = \frac{\bar{\alpha}_V \bar{V}^l(\bar{r})}{p + \bar{\beta}_V \bar{V}^q(\bar{r})}\bar{C}_L(\bar{r},p). \tag{7.15}$$

Then the transformed equation (7.8) with respect to the image function $\bar{C}_L(\bar{r},p)$ takes the form

$$\bar{V}(\bar{r})\frac{d\bar{C}_L}{d\bar{r}} - \frac{\bar{\alpha}_V \bar{V}^l(\bar{r})p}{p + \bar{\beta}_V \bar{V}^q(\bar{r})}\bar{C}_L = 0 \tag{7.16}$$

and is solved under the condition

$$\bar{r} = 1,\quad \bar{C}_L = 1/p. \tag{7.17}$$

As a result, the representation for the sought $\bar{C}_L$ is obtained

$$\bar{C}_L(\bar{r},p) = \frac{1}{p}\exp\left[-\bar{\alpha}_V p\int_{\bar{r}}^{1}\frac{\bar{V}^{l-1}(\eta)d\eta}{p + \bar{\beta}_V \bar{V}^q(\eta)}\right]. \tag{7.18}$$

It is possible to derive exact dependences convenient for calculations using software packages such as Mathcad, Matlab, Mathematica, Maple in the special case q=0. Then expression (7.18) is considerably simplified, so that

$$\bar{C}_L(\bar{r},p) = \frac{1}{p}\exp\left[\frac{\bar{\beta}_V U_1(\bar{r})}{p + \bar{\beta}_V} - U_1(\bar{r})\right], \tag{7.19}$$

where $U_1(\bar{r}) = \bar{\alpha}_V\int_{\bar{r}}^{1}\bar{V}^{l-1}(\eta)d\eta$. In the case of the plane-radial filtration

$$\bar{V} = 1/\bar{r}. \tag{7.20}$$

The inversion of (7.19) gives for the original $\bar{C}(\bar{r},\bar{t})$ this expression

$$\overline{C}(\overline{r},\overline{t}) = e^{-U_1(\overline{r})}\left[e^{-\overline{\beta}_V\overline{t}}I_0\left(2\sqrt{\overline{\beta}_V U_1(\overline{r})\overline{t}}\right) + \overline{\beta}_V\int_0^{\overline{t}} e^{-\overline{\beta}_V\xi}I_0\left(2\sqrt{\overline{\beta}_V U_1(\overline{r})\xi}\right)d\xi\right], \qquad (7.21)$$

and considering (7.20)

$$\overline{U}_1(\overline{r}) = \overline{\alpha}_V\frac{1-\overline{r}^{2-l}}{2-l}. \qquad (7.22)$$

Thus, at l=0 or 1 $\overline{U}_1(\overline{r})$ will be, respectively,

$$U_1(\overline{r}) = \overline{\alpha}_V\frac{1-\overline{r}^2}{2}, \qquad U_1(\overline{r}) = \overline{\alpha}_V(1-\overline{r}). \qquad (7.23)$$

Taking into account (7.18), we obtain the image function $\overline{S}_L(\overline{r},p)$, namely,

$$\overline{S}_L(\overline{r},p) = \frac{\overline{\alpha}_V\overline{V}^l(\overline{r})}{p\left[p+\overline{\beta}_V\overline{V}^q(\overline{r})\right]}\exp\left[-\overline{\alpha}_V p\int_{\overline{r}}^1\frac{\overline{V}^{l-1}(\eta)d\eta}{p+\overline{\beta}_V\overline{V}^q(\eta)}\right]. \qquad (7.24)$$

It is possible to simply invert expression (7.24) also at q=0. Then it is transformed to the form

$$S_L(\overline{r},p) = \frac{\overline{\alpha}_V\overline{V}^l(\overline{r})}{p(p+\overline{\beta}_V)}\exp\left[\frac{\overline{\beta}_V U_1(\overline{r})}{p+\overline{\beta}_V} - U_1(\overline{r})\right]. \qquad (7.25)$$

As a result of the inversion of (7.25) is found

$$\overline{S}(\overline{r},t) = \overline{\alpha}_V\overline{V}^l(\overline{r})e^{-U_1(\overline{r})}\int_0^{\overline{t}} e^{-\overline{\beta}_V\xi}I_0\left(2\sqrt{\overline{\beta}_V U_1(\overline{r})\xi}\right)d\xi. \qquad (7.26)$$

Thus, follows from (7.26), for example, at l=1 and the plane-radial filtration

$$\overline{S}(\overline{r},\overline{t}) = \overline{\alpha}_V\frac{e^{-\overline{\alpha}_V(1-\overline{r})}}{\overline{r}}\int_0^{\overline{t}} e^{-\overline{\beta}_V\xi}I_0\left(2\sqrt{\overline{\alpha}_V\overline{\beta}_V(1-\overline{r})\xi}\right)d\xi. \qquad (7.27)$$

Since $\overline{V} = 1$ at the inlet to the bed for any values of l and q, the exact expression for the relative inlet concentration $\overline{S}_0(\overline{t}) = \overline{S}(1,\overline{t})$ will formally be the same as for vertical filtration, viz,

$$\overline{S}_0(\overline{t}) = \frac{\overline{\alpha}_V}{\overline{\beta}_V}(1-e^{-\overline{\beta}_V\overline{t}}). \qquad (7.28)$$

Significant differences will appear here after the transition to dimensional quantities, since l and q are present in the complexes $\overline{\alpha}_V$, $\overline{\beta}_V$.

Finally, formula (7.21) allows to estimate accurately the primary protective capability of the bed. Immediately after the contamination front has passed through it (the first stage), the residual amount of the suspended solids, which is characterized by the quantity $\overline{C}(\overline{r}_e,0)$, according to (7.21) will be

$$\overline{C}_e^0 = \overline{C}(\overline{r}_e,0) = \exp\left(\alpha_V \frac{\overline{r}_e^{2-l} - 1}{2-l}\right). \tag{7.29}$$

If the quantity C_e^0 is predetermined, the required initial level of the suspension clarification is provided at the relative radius of the inner surface of the bed

$$\overline{r}_e = \left(1 + \frac{2-l}{\overline{\alpha}_V}\ln \overline{C}_e^0\right)^{\frac{1}{l-2}}. \tag{7.30}$$

In general, the function $\overline{C}^0(\overline{r})$, characterizing the amount and location of the suspended matter in the bed by the end of the short-term first stage, has the following form

$$\overline{C}^0(\overline{r}) = \exp\left[-\overline{\alpha}_V \int_{\overline{r}}^{1} \overline{V}^{l-1}(\eta)d\eta\right].$$

With the established dynamics of the distribution of the deposited particles in the bed, the filtration flow compartment is easily solved exactly. Then profiles of the reduced head $\widetilde{h}(\overline{r},\overline{t})$ taking into account (7.11) are described by the general integral

$$\widetilde{h}(\overline{r},\overline{t}) = -\frac{1}{\ln \overline{r}_e}\int_{\overline{r}_e}^{\overline{r}} \frac{\overline{V}(\eta)d\eta}{[1 - \overline{\gamma}_a^{\,m_{k1}}\overline{S}^{m_{k1}}(\eta,\overline{t})]^{m_{k2}}}, \tag{7.31}$$

and with the plane-radial filtration

$$\widetilde{h}(\overline{r},\overline{t}) = -\frac{1}{\ln \overline{r}_e}\int_{\overline{r}_e}^{\overline{r}} \frac{d\eta}{\eta\,[1 - \overline{\gamma}_a^{\,m_{k1}}\overline{S}^{m_{k1}}(\eta,\overline{t})]^{m_{k2}}}, \tag{7.32}$$

Thus, the total reduced head losses in the bed at the initial point in time is 1 according to (7.32), and they reach the quantity $\widetilde{h}(1,\overline{t})$ to an arbitrary point in time $\overline{t}$.

In the quantitative analysis carried out on the basis of the strict dependencies (7.21) and (7.32), the parameter l was in the center of attention, which, as well as the coefficient α_V, controls the intensity and nature of the process of the suspended solids sticking to the elements of the filtering material and the formed deposit. Therefore, in accordance with the available literature data, the widest possible range of the l values was accepted for it, which was used in the calculations of the times of achieving the outlet suspended solids concentration $\overline{C}_e$ and the total reduced head losses $\Delta\widetilde{h}$ predetermined values. The stated times are considered as functions of this indicator. As an exception was taken l=1, which allowed to carry out an additional comparison of the approximate solution with a special exact solution. Thus, the first calculation formula took the following form

$$\overline{C}_e(\overline{t};l) = \overline{C}(\overline{r}_e,\overline{t};l) =$$

$$= e^{-U_1(\overline{r}_e;l)}\left[e^{-\overline{\beta}_V\overline{t}}I_0\left(2\sqrt{\overline{\beta}_V U_1(\overline{r}_e;l)\overline{t}}\right) + \overline{\beta}_V\int_0^{\overline{t}} e^{-\overline{\beta}_V\xi}I_0\left(2\sqrt{\overline{\beta}_V U_1(\overline{r}_e;l)\xi}\right)d\xi\right], \tag{7.33}$$

where $U_1(\bar{r}_e;l) = \bar{\alpha}_V(1 - \bar{r}_e^{2-l})/(2 - l)$. Identical conditions for the absorption process are provided at the inlet to the bed in all examples. This is achieved by a special selection of $\alpha_V(l)$, so that the value of the coefficient $\alpha_0 = \alpha(r_0) = \alpha_V(l) \cdot V_0^l$ and hence $\bar{\alpha}_V$ are independent of l. At the same time, the suspended solids content in the filtrate will differ significantly at the extreme values of l. The subject of calculations was, first of all, the relative time $\bar{t}_p$, up to which the quality of the clarified water met the existing sanitary and hygienic norms. The standard assumption was that 90% of the initial amount of the dispersed contaminants should be removed. The set of $\bar{t}_p$ values was found by fitting from equation (7.33) at $\bar{\beta}_V = 0.005$, $\bar{r}_e = 0.25$ and a number of the values of the coefficient $\bar{\alpha}_V$ (8, 10, 12). The corresponding curves of the dependence $\bar{t}_p(l)$ are shown in **Figure 2**. According to them, the limiting weakening of the effect of the filtration flow process on the absorption process leads to a significant reduction of the time $\bar{t}_p$ (by 3 - 4 times) and can cause a premature washing of the filter.

Figure 2. Dependence $\bar{t}_p(l)$

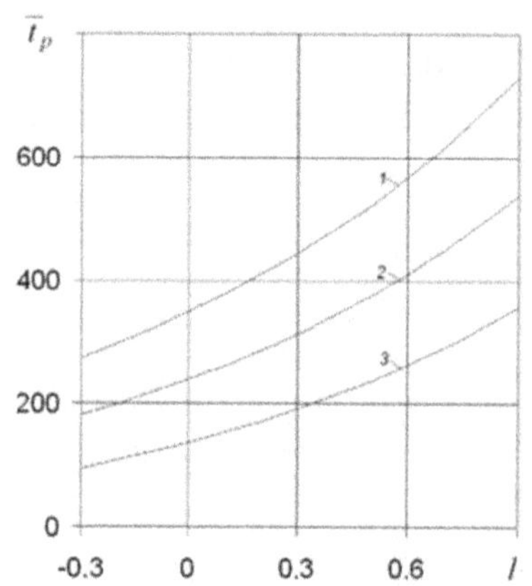

Note. $1 - \bar{\alpha}_V = 12$, $2 - \bar{\alpha}_V = 10$, $3 - \bar{\alpha}_V = 8$

Figure 3. Dependence $\bar{t}_h(l)$

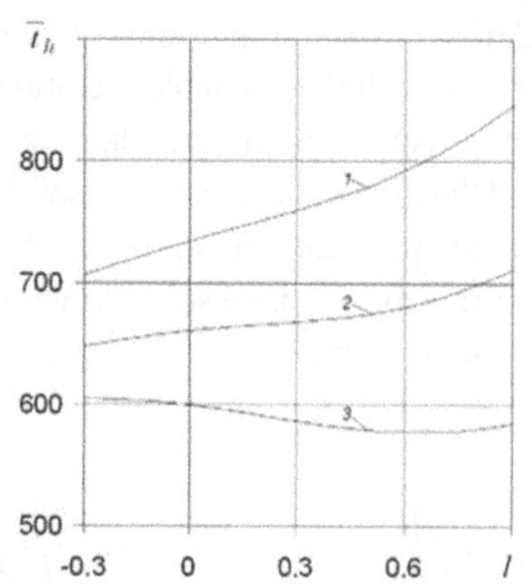

Note. $1 - \bar{\alpha}_V = 12$, $2 - \bar{\alpha}_V = 10$, $3 - \bar{\alpha}_V = 8$

In the second series of examples, the relative time $\bar{t}_h$ was calculated, over which the initial head losses (in a clean bed) increased by a factor of 8, so that$\Delta\tilde{h} = \tilde{h}(\bar{r}_e, \bar{t}_h; l) = 8$. Additionally, it was assumed that $\bar{\gamma}_a = 0.001$, m_{k1}=1, m_{k2}=3. Equation (7.32), taking into account the fixed value of $\tilde{h}$, was again solved by fitting with respect to $\bar{t}_h$, which allowed us to present a series of indicative curves in **Figure 3**. From here we can see that the relationship between the time $\bar{t}_h$ and the exponent l is expressed relatively weakly. Indeed, mechanical energy is expended to overcome the hydraulic resistance mainly near the outer surface of the bed. It is here that the suspension particles are deposited with the greatest intensity, which is practically independent of l due to the constancy of the coefficient α_0. With increasing l from -0.3 to 0.9 $\bar{t}_h$ increases only by 20% in the case of a well-absorbing bed $(\bar{\alpha}_V = 12)$, and even there is the minimal decrease in $\bar{t}_h$ (by 3.4%) if a filter media is moderately absorbing.

7.2. PLANE-RADIAL FILTRATION WITH SUSPENSION FEEDING FROM OUTSIDE (APPROXIMATE SOLUTION)

The field of application of the rigorous solution of one problem of the plane-radial filtration obtained in Section 7.1 is largely limited due to the simplified interpretation of the phenomenon of the detachment of the deposited suspension particles. We assumed q=0 and, therefore, the detachment turns out to be unrelated to the filtration flow process, although just the hydrodynamic force is the prime cause of the deposit mobilization. Therefore, this solution is of a little practical value. However, it is valuable as a reference for assessing the accuracy of approximate solutions. One such solution is given and discussed below (Poliakov, 2015b; d). The underlying mathematical model of the axisymmetric filtration is given in a generalized form earlier and contains equations (7.8) - (7.11), initial (7.12) and boundary (7.13), (7.14) conditions. However, now the exponent q in the mass exchange equation formally can take any actual values. The procedure of the approximate solution of the analogous problem of the conventional vertical filtration at constant hydraulic load, as well as the substantiation of its high accuracy are outlined in Section 2.4. Below attention is emphasized on the technological aspects of the suspension clarification in the case of the axisymmetric shape of the liquid flow with the suspended solids in order to show its serious advantages. Therefore, cumbersome derivations are mostly omitted.

First of all, the concentration $\bar{S}$ in equation (7.9) is averaged over the calculation period [0,T], viz,

$$\bar{S}_a(\bar{r};T) = \frac{1}{T}\int_0^T \bar{S}(\bar{r},\bar{t})d\bar{t}.$$

Then the suspended solids concentration $\bar{C}$ after solving equation (7.8) with consideration of (7.9) under condition (7.13) is expressed through the new parameter T and the averaged characteristic $\bar{S}_a$ as follows

$$\bar{C}(\bar{r};T) = e^{-U(\bar{r})}\left[1 + \bar{\beta}_V\int_{\bar{r}}^{1} e^{-U(\eta)}\,\bar{V}^{q-1}(\eta)\,\bar{S}_a(\eta;T)d\eta\right], \tag{7.34}$$

where $U(\bar{r}) = \bar{\alpha}_V \int_{\bar{r}}^{1} \bar{V}^{l-1}(\eta)d\eta$. By substituting (7.34) into equation (7.9), integrating the latter and averaging the expression for $\bar{S}(\bar{r},\bar{t})$ obtained in this way, we finally derive the integral equation with respect to $\bar{S}_a$

$$\left[2+\bar{\beta}_V T\bar{V}^q(\bar{r})\right]e^{U(\bar{r})}\bar{S}_a(\bar{r};T) = \\ = \bar{\alpha}_V\bar{\beta}_V T\bar{V}^l(\bar{r})\int_{\bar{r}}^{1} e^{U(\eta)}\bar{V}^{q-1}(\eta)\bar{S}_a(\eta;T)d\eta + \bar{\alpha}_V T\bar{V}^l(\bar{r}). \tag{7.35}$$

The replacement is being made

$$y = \int_{\bar{r}}^{1} e^{U(\eta)}\bar{V}^{q-1}(\eta)\bar{S}_a(\eta;T)d\eta \tag{7.36}$$

and then equation (7.35) reduces to the form

$$\frac{dy}{d\bar{r}} + \bar{\beta}_V E(\bar{r};T)y + E(\bar{r};T) = 0, \tag{7.37}$$

where $E(\bar{r};T) = \dfrac{\bar{\alpha}_V T\bar{V}^{l+q-1}}{2+\bar{\beta}_V T\bar{V}^q(\bar{r})}$. As a result of its solution under the condition

$$\bar{r} = 1, \qquad y = 0 \tag{7.38}$$

was found

$$y(\bar{r};T) = \frac{1}{\bar{\beta}_V}\left\{\exp\left[\bar{\beta}_V\int_{\bar{r}}^{1} E(\eta;T)d\eta\right] - 1\right\}. \tag{7.39}$$

Based on (7.36) and (7.39), the parameter $\bar{S}_a$ will be

$$\bar{S}_a(\bar{r};T) = \frac{\bar{\alpha}_V T\bar{V}^l(\bar{r})}{2+\bar{\beta}_V T\bar{V}^q(\bar{r})}\exp\left[-2\bar{\alpha}_V\int_{\bar{r}}^{1}\frac{\bar{V}^{l-1}(\eta)d\eta}{2+\bar{\beta}_V T\bar{V}^q(\eta)}\right]. \tag{7.40}$$

Integration of equation (7.9), in which the time-averaged summand describing the deposit detachment is averaged over time, under the initial condition (7.12) and taking into account (7.40) gives

$$\bar{S}(\bar{r},\bar{t}) = \frac{2\bar{\alpha}_V\bar{t}\bar{V}^l(\bar{r})}{2+\bar{\beta}_V\bar{t}\bar{V}^q(\bar{r})}\exp\left[-2\bar{\alpha}_V\int_{\bar{r}}^{1}\frac{\bar{V}^{l-1}(\eta)d\eta}{2+\bar{\beta}_V\bar{t}\bar{V}^q(\eta)}\right] \tag{7.41}$$

Finally, expression (7.41) is substituted into equation (7.34) and after a number of transformations is obtained

$$\bar{C}(\bar{r},\bar{t}) = 2\exp\left[-2\bar{\alpha}_V\int_{\bar{r}}^{1}\frac{\bar{V}^{l-1}(\eta)d\eta}{2+\bar{\beta}_V\bar{t}\bar{V}^q(\eta)}\right] - \exp\left[-\bar{\alpha}_V\int_{\bar{r}}^{1}\bar{V}^{l-1}(\eta)d\eta\right]. \tag{7.42}$$

At the plane-radial filtration the relative rate $\overline{V}$ changes according to (7.20) and then the formula for calculating the outlet concentration $\overline{C}_e$ is transformed to the following form

$$\overline{C}_e(\overline{t}) = 2\exp\left[-2\overline{\alpha}_V \int_{\overline{r}_e}^{1} \frac{\eta^{1+q-l} d\eta}{2\eta^q + \overline{\beta}_V \overline{t}}\right] - \exp\frac{\overline{\alpha}_V(\overline{r}_e^{2-l} - 1)}{2 - l}. \tag{7.43}$$

It is possible to express the integrals in formulas (7.41), (7.42) through elementary functions only at some values of the exponents l,q. Thus, if mass exchange does not depend on the filtration rate ($l=q=0$), then

$$\overline{S}(\overline{r},\overline{t}) = \frac{2\overline{\alpha}_V \overline{t}}{2 + \overline{\beta}_V \overline{t}} \exp\frac{\overline{\alpha}_V(\overline{r}^2 - 1)}{2 + \overline{\beta}_V \overline{t}}, \tag{7.44}$$

$$\overline{C}(\overline{r},\overline{t}) = 2\exp\frac{\overline{\alpha}_V(\overline{r}^2 - 1)}{2 + \overline{\beta}_V \overline{t}} - \exp\frac{\overline{\alpha}_V(\overline{r}^2 - 1)}{2}. \tag{7.45}$$

If, on the contrary, the sticking of impurity depends essentially on the specified rate, and the deposit mobilization does not depend on it ($l=1$, $q=0$), then

$$\overline{S}(\overline{r};\overline{t}) = \frac{2\overline{\alpha}_V \overline{t}}{\overline{r}(2 + \overline{\beta}_V \overline{t})} \exp\frac{2\overline{\alpha}_V(\overline{r} - 1)}{2 + \overline{\beta}_V \overline{t}}, \tag{7.46}$$

$$\overline{C}(\overline{r};\overline{t}) = 2\exp\frac{2\overline{\alpha}_V(\overline{r} - 1)}{2 + \overline{\beta}_V \overline{t}} - \exp\left(\overline{\alpha}_V \overline{r} - \overline{\alpha}_V\right). \tag{7.47}$$

Finally, if the filtration flow process has a strong effect on both sides of mass exchange ($l=q=1$)

$$\overline{S}(\overline{r},\overline{t}) = \frac{2\overline{\alpha}_V \overline{t}}{2\overline{r} + \overline{\beta}_V \overline{t}} \left(\frac{2 + \overline{\beta}_V \overline{t}}{2\overline{r} + \overline{\beta}_V \overline{t}}\right)^{\frac{\overline{\alpha}_V \overline{\beta}_V \overline{t}}{2}} e^{\overline{\alpha}_V(\overline{r} - 1)} \tag{7.48}$$

$$\overline{C}(\overline{r},\overline{t}) = \frac{2 + \overline{\beta}_V \overline{t}}{2\overline{r} + \overline{\beta}_V \overline{t}} e^{\overline{\alpha}_V(\overline{r} - 1)}. \tag{7.49}$$

The changes in the reduced head $\overline{h}$ within the bed volume and over time are calculated at the established distribution of the deposited particles in the radial direction by formulas (7.31), (7.32).

The duration of the continuous filter operation (time t_f) at a constant hydraulic load is determined in accordance with the recommendations from Section 3.1 based on two criteria. First of all, the concentration of the suspended solids at the outlet of the bed is limited from above by the value C_*, prescribed by the sanitary and hygienic standards. The relative time of the bed protective effect $\overline{t}_p$ meeting this requirement is calculated at arbitrary values of l,q by fitting from the equation

$$2\exp\left[-2\overline{\alpha}_V \int_{\overline{r}_e}^{1} \frac{\overline{V}^{l-1}(\eta) d\eta}{2 + \overline{\beta}_V \overline{t}_p \overline{V}^q(\eta)}\right] = \overline{C}_* + \exp\left[-\overline{\alpha}_V \int_{\overline{r}_e}^{1} \overline{V}^{l-1}(\eta) d\eta\right] \tag{7.50}$$

For the above considered special cases of the interrelation of the filtration flow and mass exchange processes, formulas are derived in the explicit form from (7.50) to calculate $\bar{t}_p$. Thus, at $l=q=0$

$$\bar{t}_p = \frac{\bar{\alpha}_V(\bar{r}_e^2 - 1)}{\bar{\beta}_V \ln\left(\frac{\bar{C}_*}{2} + \frac{1}{2}\exp\frac{\bar{\alpha}_V \bar{r}_e^2 - \bar{\alpha}_V}{2}\right)} - \frac{2}{\bar{\beta}_V}; \tag{7.51}$$

at $l=1$, $q=0$

$$\bar{t}_p = \frac{\bar{\alpha}_V(\bar{r}_e - 1)}{\bar{\beta}_V \ln\left(\frac{\bar{C}_*}{2} + \frac{1}{2}e^{\bar{\alpha}_V(\bar{r}_e - 1)}\right)} - \frac{2}{\bar{\beta}_V}; \tag{7.52}$$

at $l=q=1$

$$\bar{t}_p = \frac{2}{\bar{\beta}_V} \cdot \frac{(1 - \bar{C}_* + e^{\bar{\alpha}_V(\bar{r}_e - 1)})^{-\bar{\alpha}_V}}{(\bar{C}_* + e^{\bar{\alpha}_V(\bar{r}_e - 1)})^{-\bar{\alpha}_V} - 1}. \tag{7.53}$$

At the same time, it is undesirable that the head losses in the bed Δh exceeds the value Δh_*, which will lead to a decrease in the performance of the filter facility. It is suggested to calculate the moment in time t_h, when the head losses increasing due to the progressive clogging of the filtering material are equalized with the maximum allowable losses Δh_*, by fitting from the equation based on (7.32)

$$\int_{\bar{r}_e}^{1} \frac{d\eta}{\eta\left[1 - \bar{\gamma}_a^{\,m_{k1}} \bar{S}^{m_{k1}}(\eta, \bar{t}_h)\right]^{m_{k2}}} + \Delta\bar{h}_* \ln \bar{r}_e = 0, \tag{7.54}$$

where $\Delta\bar{h}_* = \Delta h_* / \left[h(r_0, 0) - H_d\right]$.

A thorough quantitative analysis of the plane-radial filtration was performed on numerous examples in order to show its high efficiency and serious advantages over the traditional vertical (downward or upward) filtration. Since the analysis was based on the above approximate analytical solution, special attention was paid to its accuracy. It was estimated by comparing the values of the outlet suspended solids concentration $\bar{C}_e$, calculated by the exact formula (7.33) and the approximate formula (7.43). In this case, the key mass exchange coefficient $\bar{\alpha}_V$ varied in the range from 4 to 9 and, in addition, values of $\bar{\beta}_V$, $\bar{r}_e$ were fixed (0.005, 0.333). The exact and approximate curves of the dependence $\bar{C}_e(\bar{t})$, corresponding to the special case of $l=q=0$, are shown in **Figure 4**. It should be emphasized that a very large calculation period was accepted for the completeness of the comparison, so that its duration far exceeds the time of the bed protective effect. Nevertheless, the difference between the approximate and reference values of $\bar{C}_e$ did not exceed 6.7% even at the end of the mentioned period. If the filtrate quality is predicted, the calculation time is much shorter and, therefore, the quantity $\bar{C}_e$ was calculated with an error of less than 1-1.5%. Within the same limits, as evidenced by **Figure 5**, the approximate values deviate from the exact values in another representative case ($l=1$, $q=0$), although the maximum error in the examples slightly decreases - up to 6.3%. And also, it turns out to be negligible in the time interval of interest for designers and technologists. And the dynamics of the suspension particles deposition at the inlet to the bed ($\bar{r} = 1$) was determined in conclusion of the substantiation of the approximate solution. At that, the corresponding inlet concentration $\overline{S}_0$ was calculated by the exact formula (7.28) and by the formula derived from (7.41) taking into account $\overline{V}(1) = 1$, so that

$$\overline{S}_0(\overline{t}) = \frac{2\overline{\alpha}_V\overline{t}}{2+\overline{\beta}_V\overline{t}}. \tag{7.55}$$

Figure 4. Increase in the relative outlet volumetric concentration of the suspended solids over time (l=0)

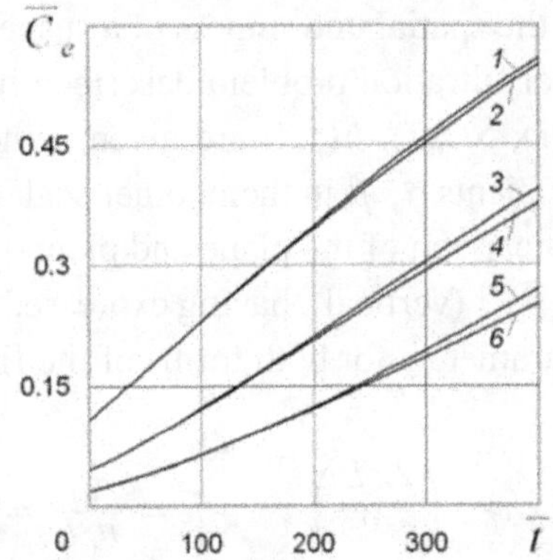

Note. 1, 3, 5 – exact calculation; 2, 4, 6 – approximate; 1, 2 – $\overline{\alpha}_V = 5$; *3, 4 –* $\overline{\alpha}_V = 7$; *5, 6, –* $\overline{\alpha}_V = 9$

Figure 5. Increase in the relative outlet volumetric concentration of the suspended solids over time (l=1)

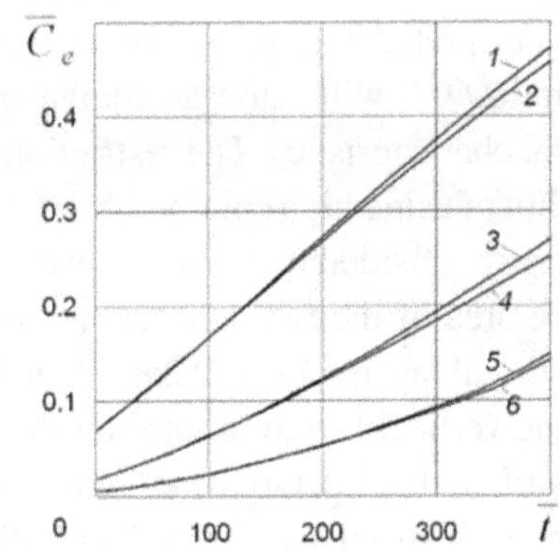

Note. 1, 3, 5 – exact calculation; 2, 4, 6 – approximate; 1, 2 – $\overline{\alpha}_V = 4$; *3, 4 –* $\overline{\alpha}_V = 6$; *5, 6, –* $\overline{\alpha}_V = 8$

It should be noted that the concentration $\overline{S}_0$ does not depend on l,q and in this sense the results of comparison of its exact and approximate values are generalized. Formula (7.55) coincides in terms of writing with the analogous formula for the plane (vertical) filtration (2.89). However, their identity takes place only due to the use of dimensionless coefficients and time, and they are already different at the transition to the dimensional coefficients and time. The plots of the dependence $\overline{S}_0(\overline{t})$ corresponding to the three values of $\overline{\alpha}_V$ are shown in **Figure 6**. The discrepancy between the curves calculated at the same $\bar{\alpha}_V$ using different formulas was also of the order of 1 - 2%. It is important to note that the considered examples actually cover the entire range of possible values of the exponent l due to the choice of its values 0, 1, whereas according to the literature data the minimum values of l are -0.3, -0.2, and the maximum values are 0.7, 0.875 (Rajagopalan & Tien, 1976; Senyavin et al., 1977). In general, the following various analyses will almost exclusively involve the experimentally determined values -0.3 and 0.7. The results presented in Figures 4 - 6 give a reason to consider the approximate solution as a reliable tool for in-depth theoretical studies of the plane-radial filtration, and the scientific substantiation of the technological and design parameters of the cylindrical filter beds prospectively.

Only approximate formulas and equations were used for further calculations. At the same time, in order to estimate the significance of the spatial unevenness of a suspension flow for filter operation, we used a similar solution of the vertical filtration problem described in Section 2.4, where the resulting dependencies for the quantities $\bar{C}_e(\bar{t})$, $\bar{S}(\bar{z},\bar{t})$, $\bar{h}(\bar{z},\bar{t})$ are given. When introducing the dimensionless independent variables $\bar{t}$, $\bar{z}$ and coefficients $\bar{\alpha}$, $\bar{\beta}$ in them, other scales of the length and rate were used. Therefore, for correctness of the comparison of the plane and plane-radial filtration it was necessary to recalculate the relative quantities $\bar{\alpha}$, $\bar{\beta}$, $\tilde{t}$ (vertical), having expressed them preliminary through $\bar{\alpha}_V$, $\bar{\beta}_V$, $\bar{t}$ (plane-radial) and geometrical parameters for both forms of the filter beds as follows

$$\bar{\alpha} = \bar{\alpha}_V L r_0^{l-2}\left(\frac{2\pi L_f}{\omega}\right)^{l-1}, \quad \bar{\beta} = \bar{\beta}_V L r_0^{q-2}\left(\frac{2\pi L_f}{\omega}\right)^{q-1}, \quad \tilde{t} = \frac{Vt}{n_0 L} = \frac{2\pi r_0^2 L_f}{\omega L}\bar{t},$$

where L_f is the height of the filter media capacity volume, ω is as before the area of its inlet surface.

The basic information intended for the comparison of two types of filtration flow (uneven and even detachment filtration) under similar conditions included, first of all, the same values of the operating volumes (W_f=1m^3) and heights (L, L_f=1m) of the packed beds of the conventional and axisymmetric forms, as well as the filter performance (Q_f=10m^3/h). In this situation, the suspension flow through the outer cylindrical surface of the bed can be interpreted as a specific (per unit of its height) flow rate. Obviously, the increase in L_f at the unchanged ratio Q_f/L_f will cause an equivalent increase in the performance Q_f, which will not affect the technological characteristics. The restrictions on the height are imposed on the basis of economic and design feasibility of using high columns-beds. Filtration flow properties of a clean filtering material of beds of both forms considered in parallel were characterized by the parameters: k_0=10m/h, n_0=0.35. The inlet surface area of the first bed (cubic, ω) was assumed to be 1m^2, and the inner radius of the second bed (cylindrical layer, r_e) was 0.2m. Then for a given volume of the layer W_f, the outer radius (r_0) is 0.6m. Thus, the vertical filtration rate was constant, namely, V=V_0=10m/h. The rate V_0 was several times less (2.653m/h) at the inlet to the cylindrical bed and monotonically increased with decreasing the radius up to the maximum value at the outlet (7.958m/h). The complex parameter $\bar{\gamma}_a$, which serves simultaneously as a measure of the suspension particles content in the deposit (γ_a) and at the inlet to the bed (C_0), was accepted or equal to 0.001, or varied within wide limits, which is characteristic not only for the secondary wastewater treatment, but also for the treatment of turbid water with a large amount of the dispersed impurity. Finally, the values 1 and 3 were assumed as usual for the coefficients m_{k1}, m_{k2} in the expression for the current hydraulic conductivity (7.11).

Thus, all the subsequent calculations were carried out in parallel using approximate formulas and equations for the traditional direct-flow and non-traditional plane-radial flow types given in this Section and in Section 2.4. Their subject were relative quantities – the outlet suspended solids concentration $\bar{C}_e$, the head losses in the bed $\Delta\bar{h}$ (are equivalent to $\bar{h}$ $(0,\bar{t})$ or $\bar{h}$ $(1,\bar{t})$), the technological times $\bar{t}_p$, $\bar{t}_h$ based on which the key technological parameter $\bar{t}_f$ was then specified.

Figure 6. Increase in the relative volumetric concentration of the deposited particles in the inlet cross-section of the bed over time

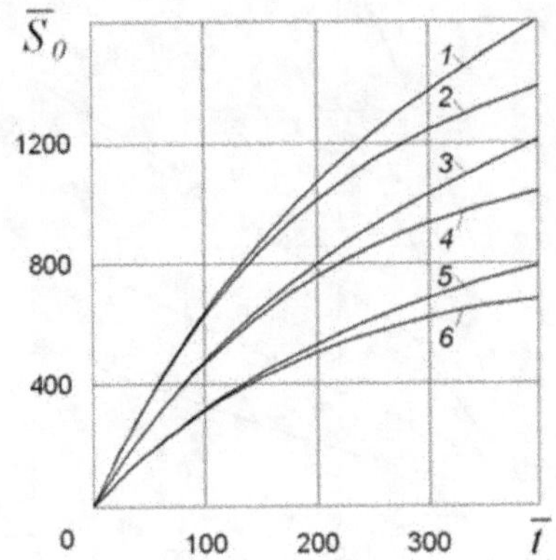

Note. 1, 3, 5 – approximate calculation; 2, 4, 6 – exact; 1, 2 – $\bar{\alpha}_V$ = 12; 3, 4 – $\bar{\alpha}_V$ = 10; 5, 6 – $\bar{\alpha}_V$ = 8

Figure 7. Change in the relative outlet volumetric concentration of the suspended solids over time (at l=0.7)

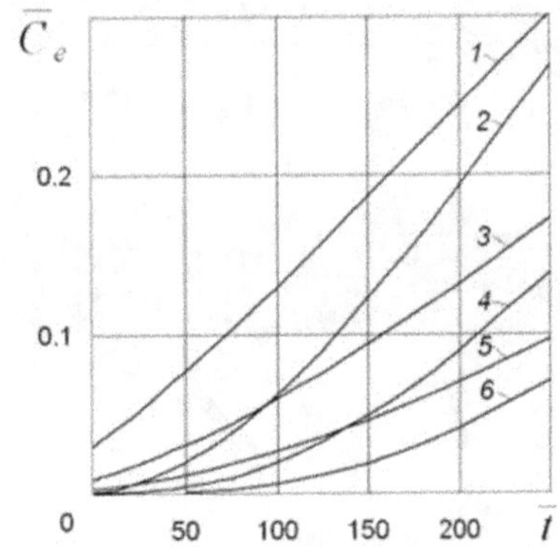

Note. 1, 2 – $\bar{\alpha}_V$ = 6; 3, 4 – $\bar{\alpha}_V$ = 8; 5, 6 – $\bar{\alpha}_V$ = 10; 1, 3, 5 – plane-radial; 2, 4, 6 – vertical

First of all, the increase of the suspended solids content in the filtrate was calculated. Graphs of the dependence $\bar{C}_e(\bar{t})$, corresponding to different types of filtration flow, values of $\bar{\alpha}_V$, are presented in **Figures 7** and **8**. Naturally, the breakthrough of the suspended particles is at first noticeably less in the case of the cubic bed than of the cylindrical bed at their increased absorption to the filtering material due to a high filtration rate and $l>0$. However, approximately the same amount of impurity was retained in both filtration media over time. If the filtration flow process inhibits the absorption process ($l<0$), the protective capability of the second bed was much greater. The suspended solids content in the filtrate is high already at the beginning of the suspension clarification at lower values of $\bar{\alpha}_V$ and l= -0.3. In this series of examples both beds could provide the required water treatment quality for the longest time at $\bar{\alpha}_V = 10$ and l=0.7.

Figure 8. Change in the relative outlet volumetric concentration of the suspended solids over time at l= -0.3

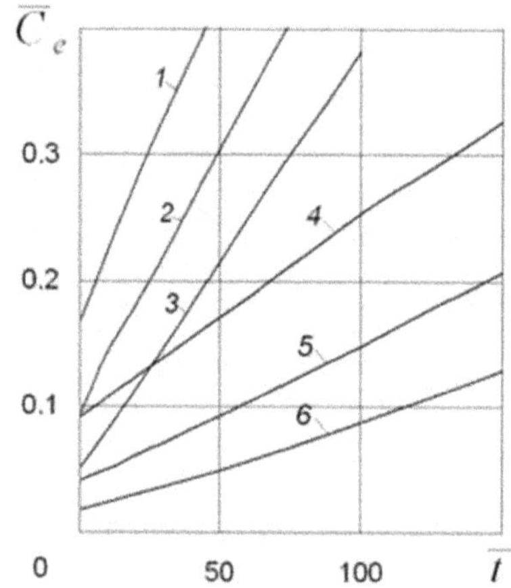

Note. 1,4 – $\bar{\alpha}_V = 6$; 2, 5 – $\bar{\alpha}_V = 8$;3, 6 6 – $\bar{\alpha}_V = 10$; 1-3 – vertical; 4-6 – plane-radial

Figure 9. Increase in the relative head losses in bed over time at l=0.7

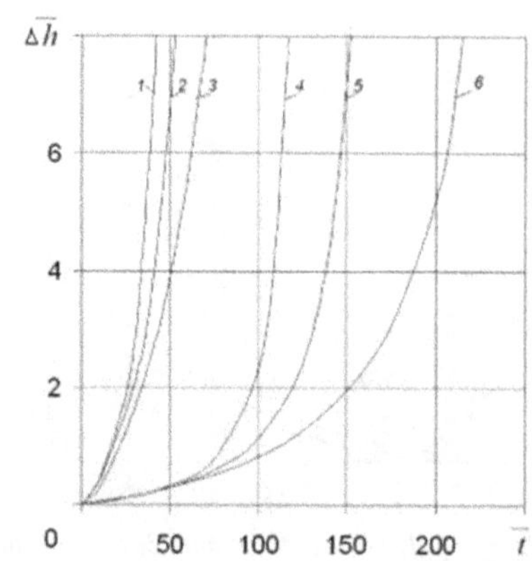

Note. 1, 4 – $\bar{\alpha}_V = 10$; 2, 5 – $\bar{\alpha}_V = 8$; 3, 6 – $\bar{\alpha}_V = 6$; 1-3 – vertical; 4-6 – plane-radial

To an even greater extent, the acceleration of the suspension flow and the resulting change in the deposition rate of the suspended solids is reflected in the head distribution and head losses inside both considered beds. **Figures 9** and **10** show the increase in the relative head losses in the bed over time, calculated using the approximate formulas for the plane-radial (7.32), (7.41) and vertical (2.93) filtration at the same values of $\bar{\alpha}_V$ and $\bar{\gamma}_a = 0.001$. Figure 9 corresponds to especially favorable conditions for the suspension particles absorption, while Figure 10, on the contrary, corresponds to unfavorable conditions. From the hydraulic point of view, the advantage of the unconventional type of filtration flow is obvious in the first situation. Due to more active participation in the clarification process of the peripheral section of the cylindrical bed, the sharp increase in the hydraulic resistance here occurred much later than in the cubic bed as a result of the relatively low filtration rates. The curves of the dependence $\Delta\bar{h}$ $(\bar{t})$ for different filtration flow types with weakened absorption are quite different (Figure 10). Thus, the plane-radial filtration is clearly preferable at first and for a very long time, but in the end the excessive accumulation of a deposit near the outer cylindrical surface caused a rapid increase in the head losses, which exceeded similar losses in the second bed. However, the value of the formulas used above for the prediction of the filtrate quality and mechanical energy consumption is mainly in the fact that it is possible to reliably establish on their basis the most important technological parameters characterized in Section 1.3, namely, the relative time of the bed protective effect $\left(\bar{t}_p\right)$ and the time of reaching the maximum permissible value of the head losses $\left(\bar{t}_h\right)$. Therefore, the results of calculating these times and thus, the duration of the filter run $\bar{t}_f$ are the most indicative in the comparative estimation of different filtration flow varieties.

Figure 10. Increase in the relative head losses in the bed over time at l= -0.3

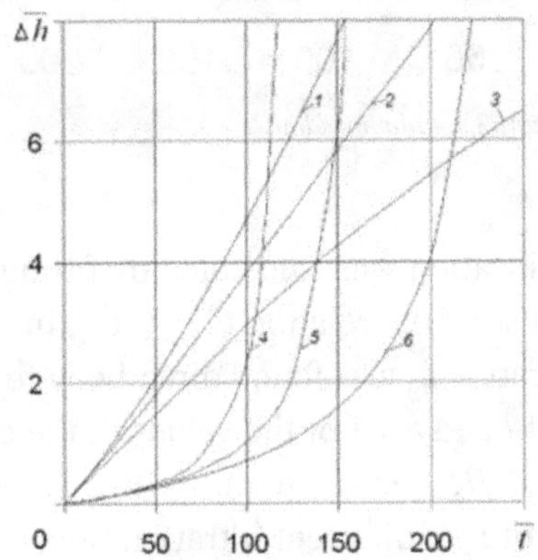

Note. 1, 4 – $\bar{\alpha}_v$ = 10; 2, 5 – $\bar{\alpha}_v$ = 8; 3, 6 – $\bar{\alpha}_v$ = 6; 1-3 – vertical; 4-6 – plane-radial

Figure 11. Dependence $\bar{t}_p(\bar{\alpha}_V)$

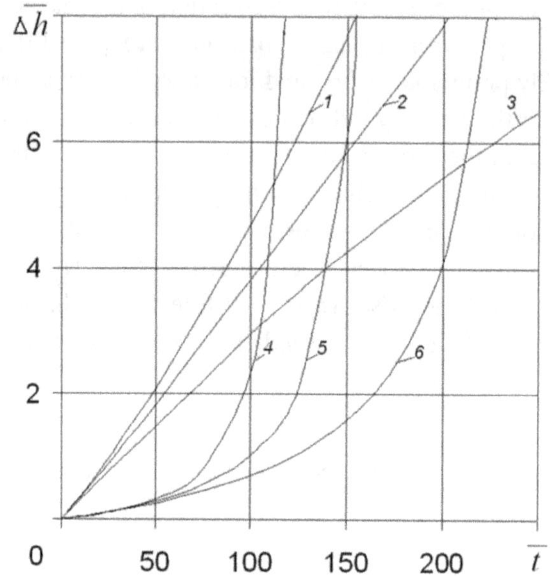

Note. 1, 2 – l=0.7; 3, 4 – l= -0.3; 1,4 – vertical; 2,3 – plane-radial

The time $\bar{t}_p$ for the plane-radial filtration was calculated by fitting from equation (7.33), and the criterion of the filtrate quality was, as usual, the relation $\bar{C}_e \leq \bar{C}_*$, in which $\bar{C}_*$ was assumed to be equal to 0.1. In the case of the vertical filtration $\bar{t}_p$ was found directly by formula (2.88). The data concerning the time $\bar{t}_p$ for two extreme values of l and a set of the values of the coefficient $\bar{\alpha}_V$ from a wide interval [2, 12] are shown in **Figure 11**. If l=0.7, then the impurity was somewhat better retained by the cubic bed. But the difference in the values of $\bar{t}_p$ at different filtration flow types leveled out (curves 1 and 2) as the absorption properties of the filter media improved. In addition, the cylindrical bed provided a longer filter operation (until the next washing) at very high values of $\bar{\alpha}_V$. In case of the complicated absorption of the suspended solids due to the hydrodynamic effect of the flow, the suspension separation in the cylindrical bed was significantly more effective than in the cubic bed. First of all, the boundary value of $\bar{\alpha}_V$, at which the limiting level of the suspension clarification is reached immediately after the passage of the contamination front through the bed, is 5.756 in the first case and 7.757 in the second case. The value of $\bar{t}_p$ is several times higher for the cylindrical bed at the same values of $\bar{\alpha}_V$.

The numerous calculations of the time $\bar{t}_h$ were performed on the basis of the criterion equations (7.54) and (2.93) with the same initial information as before, including $\bar{\gamma}_a = 0.001$. Because of the thinness of the cylindrical bed layer and the consequent smallness of the initial head losses, a large value of the relative maximum allowable losses $\Delta\bar{h}_*$ was assumed. Due to the differences in scaling of the plane-radial and vertical filtration, $\Delta\bar{h}_*$ was recalculated for the second case and its equivalent value turned out to be 2.341. As can be seen from **Figure 12**, a significant gain in the calculated time was achieved due to the unevenness of the filtration flow at l=0.7 (curves 3 and 4). On the contrary, if l= -0.3, then a large hydraulic load promoted the redistribution of a significant part of the deposit deep into the cubic bed, so that mechanical energy was even more economically used in this case, albeit insignificantly. It should be noted that the effect of the exponent l on the time $\bar{t}_h$, according to the location of curves 2, 3 relative to each other, is minimal, which agrees with the data of similar studies from Section 7.1.

But especially indicative for the estimation of the effect of the plane-radial filtration is the comparison of the filter run duration at different forms of a bed and filtration flow, because it is it, along with the hydraulic load and the area of the inlet surface, that determines the amount of the clarified suspension for one operating period and the consumption of the washing water. The results of the calculations of the duration $\bar{t}_f$, as a function of $\bar{\alpha}_V$, are shown in **Figure 13**. The broken curves here are obtained by combining suitable pieces of the plots of the dependences $\bar{t}_p(\bar{\alpha}_V)$ and $\bar{t}_h(\bar{\alpha}_V)$ in Figs. 11 and 12, relying on the representation for $\bar{t}_f$ (1.41). It is obvious from this that the vertical filtration can provide a longer filter operation only in the case of weakly absorbing filtering material $(\bar{\alpha}_V < 5,\ l = 0.7)$. However, the required duration at the unconventional filtration will be 2 - 3 times longer already at $\bar{\alpha}_V > 7$ and l=0.7. Even more convincingly in favor of this variety of the uneven filtration are curves 3, 4, which correspond to l= -0.3.

Figure 12. Dependence $\bar{t}_h(\bar{\alpha}_V)$

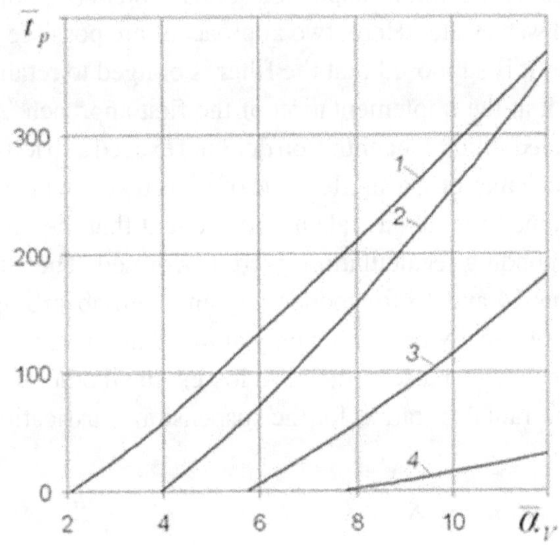

Note. 1, 2 – l=0.7; 3, 4 – l= -0.3; 1,4 – vertical; 2,3 – plane-radial

Figure 13. Dependence $\bar{t}_f(\bar{\alpha}_V)$

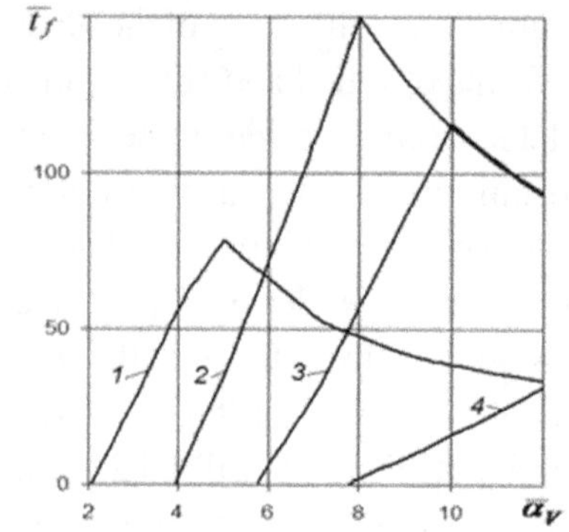

Note. 1, 2 – l=0.7; 3, 4 – l= -0.3; 1,4 – vertical; 2,3 – plane-radial

Finally, this Section concludes with a discussion of the relative time $\bar{t}_f$ data obtained from the already mentioned criterion equations for the initial suspended solids content (C_0), that varied over a large range for the natural and pretreated wastewater. Here, two approaches are possible when selecting the relative criterion value $\overline{C}_*$. As before, it is supposed that the filter is obliged to retain 90% of the volume of the dispersed impurity entering it at the implementation of the first approach. At the same time, it can be ensured that the outlet suspended solids concentration does not exceed a strictly explicit (normative) value in all cases. In the following calculations, a single value of $\overline{C}_*$ is taken, which facilitates calculations and is not fundamental. At the same time, it was taken into account that the quantities $\overline{S}$ and $\bar{\gamma}$ depend on C_0 and, therefore, the corresponding recalculations were performed. The curves of the dependences $\bar{t}_f$ (C_0) are presented in **Figures 14** and **15** for moderately and well-absorbing beds. It can be seen that the vertical filtration is technologically preferable only at *l*=0.7 and a very low initial suspended solids content and, consequently, at slow increase in the head losses. In all other situations it is certainly more advantageous to use the plane-radial filtration for the suspension clarification.

Figure 14. Dependence $\bar{t}_f(C_0)$ at $\bar{\alpha}_V = 8$

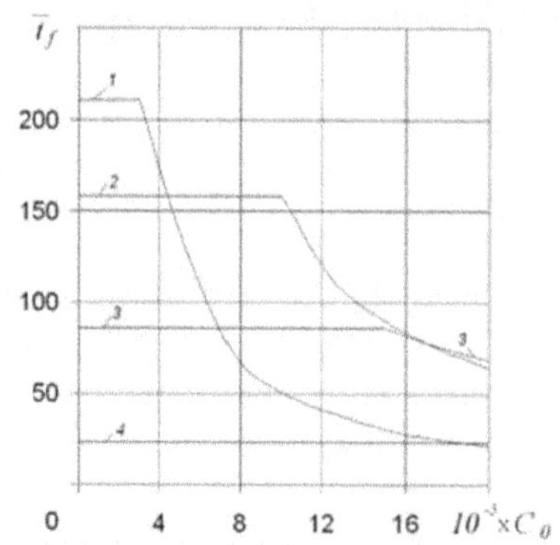

Note. 1, 2 – l=0.7; 3, 4 – l= -0.3; 1,4 – vertical; 2,3 – plane-radial

Figure 15. Dependence $\bar{t}_f(C_0)$ at $\bar{\alpha}_V = 10$

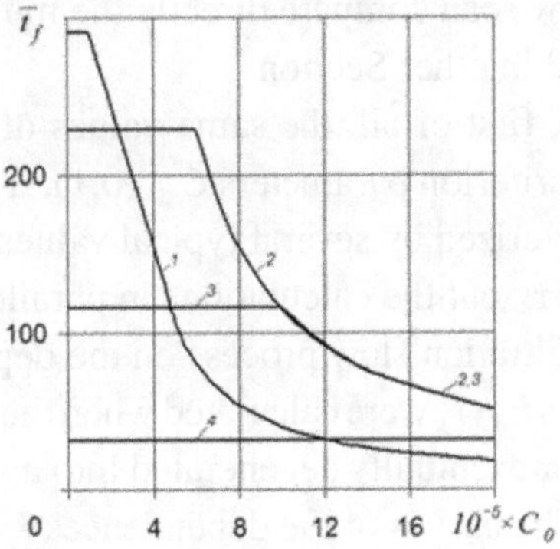

Note. 1, 2 – l=0.7; 3, 4 – l= -0.3; 1,4 – vertical; 2,3 – plane-radial

7.3. SUBSTANTIATION OF DESIGN PARAMETERS OF CYLINDRICAL BED

An important application of the solution obtained in Section 7.2 to the problem of the plane-radial linear detachment filtration is the rational choice on its basis of the design parameters of a filter bed in the form of the cylindrical layer (Poliakov, 2013b; 2015c). In the presence of a non-alternative filtering material, these parameters include, first of all, the height of the bed L_f, its inner (r_e) and outer (r_0) radii. The height L_f is selected on the basis of the technical possibilities of the operation of high (column) filters and does not affect the technological process of filtration. The increase in L_f allows to increase the performance of a water treatment filter while maintaining its compactness. The scientific substantiation of the above-mentioned radii, however, contributes to the filter operation with maximum efficiency. Therefore, it is their relation to the key technological parameter (t_f) that is analyzed in detail below. Equations (7.50), (7.54) became the basis in determining the relative values of the technological times $\left(\bar{t}_p, \bar{t}_h, \bar{t}_f\right)$. In the first series of examples, they were used in their original form, since the value of r_0 was fixed. The radius r_e was varied to the maximum extent – from 0 to r_0. In the other two series, the radius r_0 became a variable quantity and, therefore, in principle, could not serve as a scale. In the second series, the radius r_e was unchanged, which predetermined its use as a length scale. This required a minor transformation of the original criterion equations. Finally, in the third series of examples, the filter bed volume W_f (and height L_f) was initially set and independent of the radii r_e, r_0, which varied continuously and now in a consistent manner. Therefore, to obtain complete information on the effect of the bed geometry on the clarifying resources of the filter, it is sufficient to specify the relationship between the time $\bar{t}_f$ and, for example, the radius $\bar{r}_e$. If necessary, it is easy to calculate $\bar{r}_0$ using a simple formula with the known values of $\bar{r}_e$ and W_f, L_f. In such a case, it is logical to relate all quantities having the dimension of the length to the minimum value r_{min} of the variable r_0, which corresponds to the selected volume W_f and the smallest possible value r_e (0).

Due to the introduction of the new scale, the calculation equations have to be transformed accordingly and they will be presented below in their final form immediately before the presentation of the results obtained on their basis. In general, it should be noted that all the criterion equations involved in this

Section in the technological calculations are dimensionless. Thus, firstly, it is possible to significantly reduce the number of specified parameters (complexes) despite even its complication due to the replacement of the length scale. Secondly, we can compare directly the multiple data of the calculations of the filter run duration obtained here and in other Sections.

The initial information includes, first of all, the same values of the model parameters $\bar{\beta}_V$ (0.005), $q(1)$, $\bar{\gamma}_a$ (0.001), $m_{k1}(1)$, $m_{k2}(3)$ and criterion parameters $\bar{C}_*$ (0.1), $\Delta\bar{h}_*$ (8) in all examples. The absorption properties of the bed are characterized by several typical values of the coefficient $\bar{\alpha}_V$ and exponent l (0.7, -0.3), which allowed us to carry out the calculations in parallel for two physically fundamentally different forms of the effect of the filtration flow process on the deposition of the suspended particles.

First of all, the technological times $\bar{t}_p$, $\bar{t}_h$ were calculated when the formal increasing the radius $\bar{r}_e$ from 0 to 1. Thus, the cylindrical bed layer eventually degenerated into a surface with no hydraulic resistance and unable to trap fine impurities. The curves of the dependences $\bar{t}_p(\bar{r}_e)$, $\bar{t}_h(\bar{r}_e)$ and generalizing them $\bar{t}_f(\bar{r}_e)$, which correspond to $\bar{\alpha}_V = 10$ and two values of l, are shown in **Figure 16**. The graphs $\bar{t}_h(\bar{r}_e)$ for different l practically merged and, therefore, are depicted by one line (3). Reducing the thickness of the filter bed has a sharply negative effect on its protective properties and a weakly positive effect on the mechanical energy losses. The latter fact is explained by the pronounced unevenness of the deposit distribution, which mainly concentrates near the outer boundary of the bed ($\bar{r} = 1$) even at a low hydraulic load, as well as by the essential nonlinearity of the relationship between the hydraulic conductivity and deposit concentration. Fast and unlimited increase in the time $\bar{t}_h$ takes place only at the asymptotic tendency of $\bar{r}_e$ to 1. It is important to note the existence in each particular case of such a characteristic value $\bar{r}_{e*}$, that the bed will not be able to provide the required filtrate quality even immediately after passing through it the contamination front (in calculations at t=0) at $\bar{r}_e \geq \bar{r}_{e*}$. Obviously, the quantity $\bar{r}_e$ recommended in the design of the filter should be noticeably smaller than $\bar{r}_{e*}$. A filtration flow has a serious effect on the suspended solids retention inside the bed indirectly through the absorption process, and hence on the time $\bar{t}_p$. And at the same time, it does not affect the formation of the deposit near the inlet surface of the bed at all, so that the time $\bar{t}_h$ is practically independent of the exponent l.

Later on, the expansion of the cylindrical bed was modeled and thereby new length scales and corresponding modification of the criterion equations became necessary. At first, due to the stable position of the inner surface, its radius played just such a role, so that $\bar{r}_0 = r_0/r_e$. In this case, the calculation equations were transformed to the following form

$$2\exp\left[-2\bar{\alpha}_{V0}(\bar{r}_0)\int_1^{\bar{r}_0}\frac{\eta^{1+q-l}d\eta}{2\eta^q+\bar{\beta}_{V0}(\bar{r}_0)\bar{t}_{p0}}\right] = \bar{C}_* + \exp\frac{\bar{\alpha}_{V0}(\bar{r}_0)\left(1-\bar{r}_0^{2-l}\right)}{2-l}, \tag{7.56}$$

$$\int_1^{\bar{r}_0}\frac{d\eta}{\eta\left[1-\bar{\gamma}^{m_1}\bar{S}^{m_1}(\eta,\bar{t}_{h0};\ \bar{r}_0)\right]^{m_2}} = \Delta\bar{h}_* = 2\pi k_0\frac{L_f\Delta h_*}{Q_f}. \tag{7.57}$$

Here

$$\bar{S}(\bar{r},\bar{t}_0;\bar{r}_0) = \frac{\bar{\alpha}_{V0}(\bar{r}_0)\bar{t}_0\bar{r}^{q-l}}{2\bar{r}^q+\bar{\beta}_{V0}(\bar{r}_0)\bar{t}_0}\exp\left[-2\bar{\alpha}_{V0}(\bar{r}_0)\int_{\bar{r}}^{\bar{r}_0}\frac{\eta^{1+q-l}d\eta}{2\eta^q+\bar{\beta}_{V0}(\bar{r}_0)\bar{t}_0}\right];$$

$$\bar{\alpha}_{V0}(\bar{r}_0) = \bar{r}_0^{l-2}\bar{\alpha}_V,\quad \bar{\beta}_{V0}(\bar{r}_0) = \bar{r}_0^{q-2}\bar{\beta}_V,\quad \bar{t}_0 = \bar{r}_0^2\bar{t}.$$

In the course of calculating the technological parameters $\bar{t}_{p0}$, $\bar{t}_{h0}$, performed by fitting on the basis of equations (7.56), (7.57) also at $\bar{\alpha}_V = 10$, and then calculating $\bar{t}_p$, $\bar{t}_h$ the design parameter $\bar{r}_0$ was increased continuously from 1 (the layer of filtering material degenerated into a surface) in five times. The results of the calculations are demonstrated in **Figure 17** already in the final form of the dependence of the duration $\bar{t}_f$ on $\bar{r}_0$ at the absorption of the suspended matter either intensified or weakened by the suspension flow. Hence it is obvious that as the volume of the filter media increases and the filtration flow path lengthens, its protective capability improves very quickly at first and then more and more slowly, especially at the intense absorption (l=0.7, curve 1). At the same time, at the values of $\bar{r}_0$ close to unity and corresponding thicknesses of the cylindrical layer $(\bar{r}_0 - 1)$, the latter is no longer able to remove the required amount of the suspended solids from a suspension for a long time and thus, the efficiency of the filter is low. Besides, by analogy with the previous series of examples and the parameter $\bar{r}_{e*}$ introduced there, it is possible to specify some limiting value $\bar{r}_{0*}$ so that the filter for which the relation $\bar{r}_0 \leq \bar{r}_{0*}$ is true, in principle cannot cope with its task of reducing the impurity content in the suspension to the normative value. It should be emphasized that at higher values of r_0, e.g. $\bar{r}_0 > 3$ at l= -0.3, the time of the continuous suspension separation is limited only by the hydraulic factor.

Figure 16. Dependencies $\bar{t}_p(\bar{r}_e)$, $\bar{t}_h(\bar{r}_e)$, $\bar{t}_f(\bar{r}_e)$

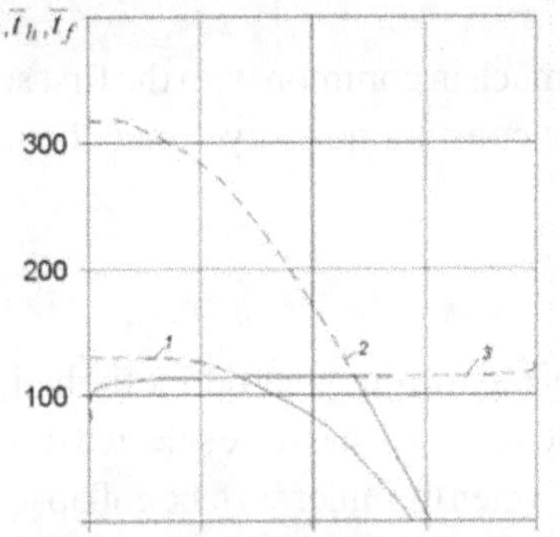

Note. solid line – $\bar{t}_f$; 1, 2 – $\bar{t}_p$; 3 – $\bar{t}_h$; 2, 3 –l=0.7; 1, 3 – l= -0.3

Figure 17. Dependencies $\bar{t}_p(\bar{r}_0)$, $\bar{t}_h(\bar{r}_0)$, $\bar{t}_f(\bar{r}_0)$

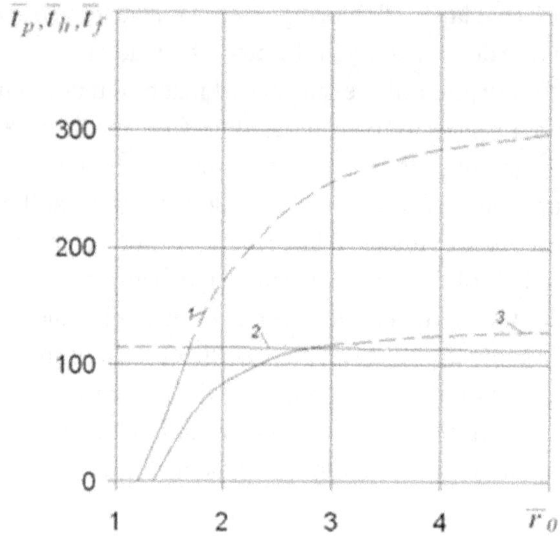

Note. solid line – $\bar{t}_f$; 1, 3 – $\bar{t}_p$; 2 – $\bar{t}_h$; 1, 2 –l=0.7; 2, 3 – l= -0.3

The final series of examples has much in common with the first series, but also an important difference, namely, the bed volume W_f is now a constant quantity. Since the design radii are related by the relation

$$\pi L_f(r_0^2 - r_e^2) = W_f, \tag{7.58}$$

then it is sufficient to specify one of them to uniquely define the bed geometry. Both radii vary widely in this series and, therefore, the appropriate length scale has to be selected from additional considerations. It is logical to proceed from the limiting situation when the inner surface collapses into a line (vertical axis) and its corresponding minimum value r_e becomes 0. Then the radius of the corresponding outer surface $r_{\min}$ is calculated by the following formula from (7.58)

$$r_{\min} = \sqrt{\frac{W_f}{\pi L_f}}.$$

Just such, in fact, initial value of the radius r_0 was then used to rescale the calculation equations, as a result of which they took the final form

$$2\exp\left[-2\bar{\alpha}_{Ve}\int_{\bar{r}_e}^{\bar{r}_0(\bar{r}_e)}\frac{\eta^{1+q-l}d\eta}{2\eta^q + \bar{\beta}_{Ve}\bar{t}_{pe}}\right] = \bar{C}_* + \exp\frac{\bar{\alpha}_{Ve}\bar{r}_e^{2-l} - \bar{\alpha}_{Ve}\bar{r}_0^{2-l}(\bar{r}_e)}{2-l}, \tag{7.59}$$

$$\int_{\bar{r}_e}^{\bar{r}_0(\bar{r}_e)}\frac{d\eta}{\eta\left[1 - \bar{\gamma}^{m_1}\bar{S}^{m_1}(\eta, \bar{t}_{he};\ \bar{r}_e)\right]^{m_2}} = \Delta\bar{h}_*. \tag{7.60}$$

Here

$$\bar{S}(\bar{r},\bar{t}_e;\bar{r}_e) = \frac{2\bar{\alpha}_{V2}\bar{t}_e\bar{r}^{q-l}}{2\bar{r}^q+\bar{\beta}_{Ve}\bar{t}_e}\exp\left[-2\bar{\alpha}_{Ve}\int_{\bar{r}}^{\bar{r}_0(\bar{r}_e)}\frac{\eta^{1+q-l}d\eta}{2\eta^q+\bar{\beta}_{Ve}\bar{t}_e}\right],\ \bar{r}_0 = \sqrt{1+\bar{r}_e^2},$$

the index "e" at the coefficients $\bar{\alpha}_V$, $\bar{\beta}_V$ and time $\bar{t}$, also the technological parameters t_p, t_h indicates the substitution of the earlier constant in the corresponding dimensionless complexes and now a variable quantity r_0 by a characteristic special value ($r_{\min}$), so that

$$\bar{\alpha}_{Ve} = r_{\min}^{2-l}\left(\frac{Q_f}{2\pi L_f}\right)^{l-1}\alpha_V,\quad \bar{\beta}_{Ve} = n_0 r_{\min}^{2-q}\left(\frac{Q_f}{2\pi L_f}\right)^{q-1}\beta_V,\quad \bar{t}_e = \frac{Q_f t}{2\pi n_0 L_f r_{\min}^2}.$$

Thus, the most important technological parameter $(\bar{t}_f)$ was calculated as a function of the relative radius-argument $\bar{r}_e$ according to equations (7.59), (7.60) presented in a convenient form. In addition to the same values of l the mass exchange coefficient $(\bar{\alpha}_V)$, which is key for describing the removal of the suspended solids from a suspension, was discretely varied, which allowed to partially reflect the diversity of the absorption properties of actual filtering materials. The consequences of the coordinated increase of the radii r_e and r_0 directly are the thinning of the filter media layer, the reduction of both the hydraulic load (at fixed filter performance Q_f) and the unevenness of the filtration flow. Due to the complex cross-effect of the listed factors and the identification of the duration $\bar{t}_f$ with the smallest of the times $\bar{t}_p$, $\bar{t}_h$, the dependence between it and the radius $\bar{r}_e$ acquires a specific nature with a pronounced maximum and significantly differs at the intensified (**Figure 18**) and weakened (**Figure 19**) attachment of the suspended solids to the elements (grains) of the bed. Particularly, at first, there is a significant prolongation of the continuous operation of the filter up to the maximum (185.8 at $\bar{\alpha}_{Ve} = 12$, $\bar{r}_e = 2.47$) at l=0.7 as $\bar{r}_e$ increases, and then its efficiency decreases due to the progressive reduction in the thickness of the cylindrical layer and the hydraulic load. Moreover, such a bed in general is unable to clarify the suspension to the desired level at $\bar{r}_e \geq \bar{r}_{e*}$, which once again confirms the need for the careful substantiation of the technological and design parameters of the plane-radial filtration. The use of a well-absorbing material with a small inner surface area of the cylindrical bed can even complicate the filter operation due to a significant hydraulic resistance of the relatively thick operating layer, but due to the rational selection of its both radii, it is possible to carry out washing noticeably later. The rate of a change in the duration $\bar{t}_f$ is considerably reduced if the filtration flow prevents the deposit formation (**Figure 19**). Nevertheless, in general, the quantity $\bar{t}_f$ does not differ too much in this (l= -0.3) and the first (l=0.7) cases, because the reduction of the hydraulic load and filtration flow path actually affects only the protective properties of the filter media, and to a greater extent the filter operation is limited with the selected initial data by the excessive head losses.

Figure 18. Dependence $\bar{t}_f(\bar{r}_e)$ at l=0.7

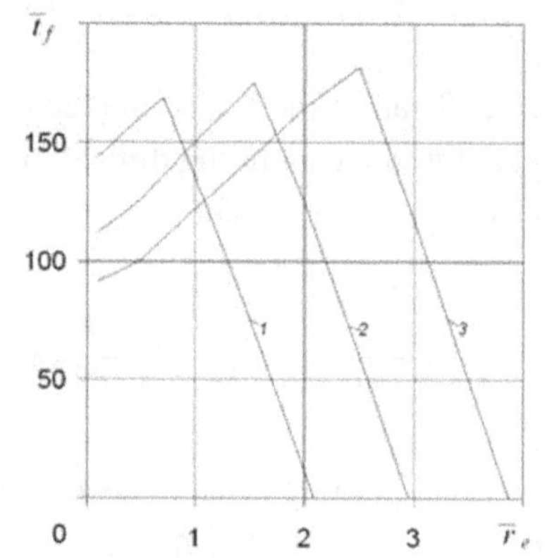

Note. 1 - $\bar{\alpha}_{Ve}$ = 8, 2 - $\bar{\alpha}_{Ve}$ = 10, 3 - $\bar{\alpha}_{Ve}$ = 12

Figure 19. Dependence $\bar{t}_f(\bar{r}_e)$ at l= -0.3

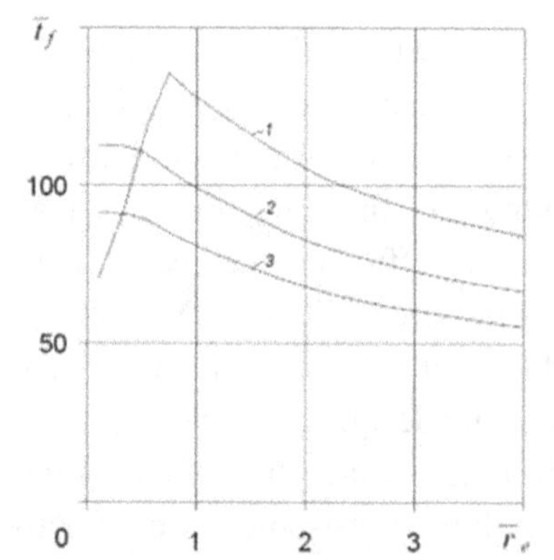

Note. 1 - $\bar{\alpha}_{Ve}$ = 8, 2 - $\bar{\alpha}_{Ve}$ = 10, 3 – $\bar{\alpha}_{Ve}$ = 12

In conclusion, it is important to note that the results of a targeted study of the relationship between the geometry and technology of the filtration process should first of all be regarded as an illustrative material. And, for example, sometimes it is enough to specify the parameter γ (or γ_a) characterizing the deposit composition as a consequence of the interaction of filter media, water and impurities, so that the relation between the times $\bar{t}_p$, $\bar{t}_h$ becomes different in principle and the duration $\bar{t}_f$ changes abruptly as a consequence. Nevertheless, the developments of this Section allow us to make a general and, most importantly, a realistic idea about the intensification of the plane-radial filtration by varying the size of the cylindrical bed, and the basic solution is a reliable tool for their practical implementation.

The algorithm of the theoretical substantiation of the best radii of a cylindrical bed from the technological point of view is presented on a specific example. The initial information contains the data on four groups of the model parameters. The first group (clarification parameters) includes such values: l=1/3, q=1, α_V=40$m^{-1/3}h^{-2/3}$, β_V=0.03m^{-1}, γ_a=28. The second group includes filtration flow parameters with the following values: k_0=10m/h, n_0=0.35, H_d=0, m_{k1}=1, m_{k2}=3. The technology of suspension separation is determined by the parameters: $Q_f(25m^3/h)$, $C_0(2.5 \cdot 10^{-5} m^3/m^3)$, $C_*(2.5 \cdot 10^{-6} m^3/m^3)$, Δh_*=1.5m. The design parameters are partially set, namely, there is available filtering material with the volume W_f=2.5m^3 at the limiting height of the filter column L_f=2.5m. It is necessary to set the radii of the inner (r_0) and outer (r_e) surfaces of the bed, at which the duration of the continuous operation of the filter will be the longest.

Since all the calculations were carried out using formulas and equations in dimensionless form and W_f=*const*, we previously determined

$$r_{\min} = \sqrt{\frac{2.5}{\pi \cdot 2.5}} = 0.564m,$$

$$\bar{\alpha}_{Ve} = 0.564^{5/3}\left(\frac{25}{2\pi \cdot 2.5}\right)^{-2/3} \cdot 40 = 11.297, \quad \bar{\beta}_{Ve} = 0.35 \cdot 0.564 \cdot 0.03 = 0.006,$$

$$\bar{\gamma}_a = 28 \cdot 2.5 \cdot 10^{-5} = 7 \cdot 10^{-4}, \quad \bar{C}_* = 0.1, \quad \Delta\bar{h}_* = 2\pi \cdot 10\frac{2.5 \cdot 1.5}{25} = 9.425.$$

Further, the dependences of the times $\bar{t}_p$, $\bar{t}_h$ on the radius $\bar{r}_e$ were calculated sequentially from equations (7.59), (7.60). The corresponding curves are shown in **Figure 20**. It can be seen that the relative filter run duration can reach the value 244.6 when $\bar{r}_e$ is approximately equal to 1.647. In dimensional quantities it means that the filter with the cylindrical bed, the inner and outer radii of which are r_e=0.929m, r_0=1.087m, can operate efficiently and efficiently for the maximum time, specifically, for 17.1 hours under the considered conditions.

Figure 20. Dependencies $\bar{t}_p(\bar{r}_e)$, $\bar{t}_h(\bar{r}_e)$, $\bar{t}_f(\bar{r}_e)$

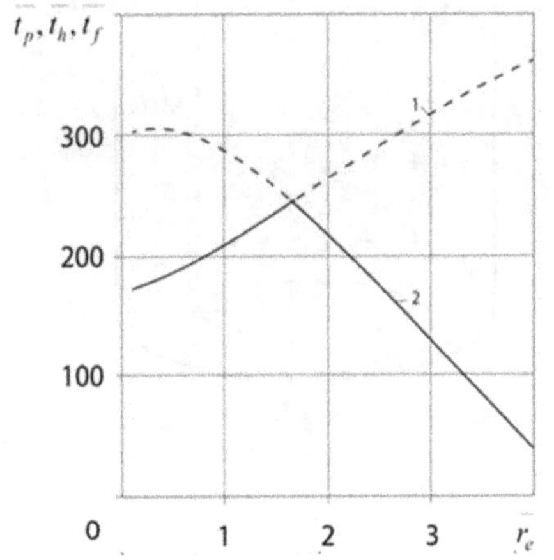

Note. solid line – $\bar{t}_f$, 1 – $\bar{t}_p$, 2 – $\bar{t}_h$

7.4. PLANE-RADIAL FILTRATION WITH SUSPENSION FEEDING FROM INSIDE

When implementing the plane-radial filtration the directionality of a suspension flow becomes important due to the spatial change in its rate. Incidentally, at the traditional vertical filtration the fundamental difference in modeling of the upward and downward flows of the suspension being clarified takes place only at a high hydraulic load, when in the first case the elements (grains) of a filter bed go into the suspended state under the impact of the significant hydrodynamic force. In order to avoid the fluidization of a cohesionless filtering material, it is necessary to limit the filtration rate. Changing the direction of the suspension flow in combination with its unevenness in the cylindrical bed can significantly affect the absorption process and indirectly through it the depth distribution of a deposit, and ultimately the duration of the continuous operation of the filter.

Figure 21. Scheme of plane-radial suspension filtration (feeding from inside)

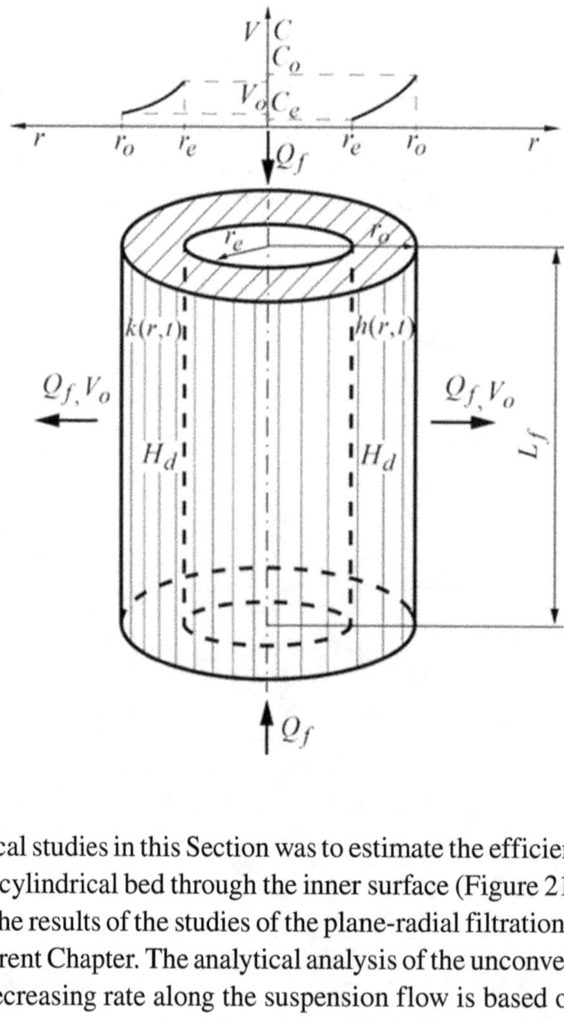

The aim of the theoretical studies in this Section was to estimate the efficiency of the suspension separation when it is fed into a cylindrical bed through the inner surface (Figure 21) (Poliakov, 2015a). It was performed by comparing the results of the studies of the plane-radial filtration obtained in the second and present Sections of the current Chapter. The analytical analysis of the unconventional uneven detachment filtration variety with a decreasing rate along the suspension flow is based on the mathematical model (7.8) - (7.14), which after proper correction of signs and boundary conditions takes the following form

$$V(r)\frac{\partial C}{\partial r} + \frac{\partial S}{\partial t} = 0, \tag{7.61}$$

$$\frac{\partial S}{\partial t} = \alpha_V V^l(r)C - \beta_V V^q(r)S, \tag{7.62}$$

$$V(r) = -k(S)\frac{\partial h}{\partial r}, \tag{7.63}$$

$$k(S) = k_0\left[1 - \left(\gamma_a \frac{S}{n_0}\right)^{m_{k1}}\right]^{m_{k2}}; \tag{7.64}$$

$$r = r_e,\ C = C_0;\ r = r_0,\ h = H_d\ ; \tag{7.65}$$

$$t = 0, \quad S = 0. \tag{7.66}$$

Therefore, now the inlet suspended solids concentration (C_0) is set on the inner surface of a bed, and the outlet head (H_d) is fixed on the outer surface. The new problem is formulated in dimensionless form as follows

$$\overline{V}(\overline{r})\frac{\partial \overline{C}}{\partial \overline{r}} - \frac{\partial \overline{S}}{\partial \overline{t}} = 0, \tag{7.67}$$

$$\frac{\partial \overline{S}}{\partial \overline{t}} = \overline{\alpha}_V \overline{V}^l(\overline{r})\overline{C} - \overline{\beta}_V \overline{V}^q(\overline{r})\overline{S}, \tag{7.68}$$

$$\overline{V}(\overline{r}) = -\overline{k}(\overline{S})\ln \overline{r}_e \frac{\partial \overline{\overline{h}}}{\partial \overline{r}}, \tag{7.69}$$

$$\overline{k}(\overline{S}) = \left(1 - \overline{\gamma}_a^{\,m_{k1}} \overline{S}^{m_{k1}}\right)^{m_{k2}}; \tag{7.70}$$

$$\overline{r} = \overline{r}_e,\ \overline{C} = 1;\ \overline{r} = 1,\ \overline{\overline{h}} = 0\ ; \tag{7.71}$$

$$\overline{t} = 0, \quad \overline{S} = 0. \tag{7.72}$$

Here $\overline{V} = V/V_{oe}$, $V_{oe} = Q_f/(2\pi r_e L_f)$, V_{oe} is the hydraulic load;

$$\overline{t} = V_{oe}t/(n_0 r_0),\ \overline{\alpha}_V = r_0 V_{oe}^{l-1}\alpha_V,\ \overline{\beta}_V = n_0 r_0 V_{oe}^{q-1}\beta_V\ ;$$

$$\overline{\overline{h}}(\overline{r},\overline{t}) = \frac{h(r,t) - H_d}{h(r_e,0) - H_d} = \frac{k_0}{r_0 V_{oe}\overline{r}_e \ln \overline{r}_e}\left[H_d - h(r,t)\right].$$

The procedure for solving problem (7.67) - (7.72) is omitted, since it is described with minimal differences in Section 7.2. As a result, a number of dependences and equations are recommended at once for calculating the characteristics and parameters of the considered option of the plane-radial filtration with feeding from inside. And first of all, the high accuracy of the filtrate quality prediction is provided by the following formula

$$\overline{C}_e(\overline{t}) = 2\exp\left[-2\overline{\alpha}_V\int_{\overline{r}_e}^{1}\frac{\overline{V}^{l-1}(\eta)d\eta}{2+\overline{\beta}_V\overline{t}\,\overline{V}^q(\eta)}\right] - \exp\left[-\overline{\alpha}_V\int_{\overline{r}_e}^{1}\overline{V}^{l-1}(\eta)d\eta\right]. \tag{7.73}$$

Since the filter performance does not change over time and the hydraulic load is again taken as the rate scale, then

$$\overline{V}(\overline{r}) = \frac{\overline{r}_e}{\overline{r}} \tag{7.74}$$

Thus, formula (4.73) is transformed to the final form

$$\overline{C}_e(\overline{t}) = 2\exp\left(-2\overline{\alpha}_V\overline{r}_e^{l-1}\int_{\overline{r}_e}^{1}\frac{\eta^{1+q-l}d\eta}{2\eta^q+\overline{\beta}_V\overline{r}_e^q\overline{t}}\right) - \exp\frac{\overline{\alpha}_V\overline{r}_e(1-\overline{r}_e^{l-2})}{2-l}. \tag{7.75}$$

The profile of the relative concentration of the deposited suspension particles in a cylindrical (spherical) bed at an arbitrary moment in time is described by the function

$$\overline{S}(\overline{r},\overline{t}) = \frac{2\overline{\alpha}_V\overline{t}\,\overline{V}^l(\overline{r})}{2+\overline{\beta}_V\overline{t}\,\overline{V}^q(\overline{r})}\exp\left[-2\overline{\alpha}_V\int_{\overline{r}_e}^{\overline{r}}\frac{\overline{V}^{l-1}(\eta)d\eta}{2+\overline{\beta}_V\overline{t}\,\overline{V}^q(\eta)}\right]. \tag{7.76}$$

Taking into account (7.74), it is easy to obtain

$$\overline{S}(\overline{r},\overline{t}) = \frac{2\overline{\alpha}_V\overline{r}_e^l\overline{t}\,\overline{r}^{q-l}}{2\overline{r}^q+\overline{\beta}_V\overline{r}_e^q\overline{t}}\exp\left(-2\overline{\alpha}_V\overline{r}_e^{l-1}\int_{\overline{r}_e}^{\overline{r}}\frac{\eta^{1+q-l}d\eta}{2\eta^q+\overline{\beta}_V\overline{r}_e^q\overline{t}}\right). \tag{7.77}$$

From (7.77) results that the concentration $\overline{S}$ will increase at the inlet to the bed over time as follows

$$\overline{S}_0(\overline{t}) = \overline{S}(\overline{r}_e,\overline{t}) = \frac{2\overline{\alpha}_V\overline{t}}{2+\overline{\beta}_V\overline{t}}. \tag{7.78}$$

Expression (7.78) has the same form as (7.55). However, the quantities S_0 are essentially different in both cases of the plane-radial filtration, in what it is possible to be convinced after transition in (7.78) and (7.55) to the dimensional time and mass exchange coefficients.

It is proposed to determine the reduced head, and in fact the relative head losses at the bed section from $\overline{r}$ to 1 by the formula

$$\tilde{\overline{h}}(\overline{r},\overline{t}) = \frac{1}{\overline{r}_e\ln\overline{r}_e}\int_{1}^{\overline{r}}\frac{\overline{V}(\eta)d\eta}{f_k(\overline{S}(\eta,\overline{t}))}. \tag{7.79}$$

In view of (7.70) and (7.79) it takes the following form

$$\tilde{\overline{h}}(\overline{r},\overline{t}) = \frac{1}{\ln\overline{r}_e}\int_{1}^{\overline{r}}\frac{d\eta}{\eta\left[1-\overline{\gamma}_c^{\,m_{k1}}\overline{S}^{m_{k1}}(\eta,\overline{t})\right]^{m_{k2}}}. \tag{7.80}$$

Consequently, the relative total head losses in the bed $\Delta\overline{h}$ are equal to $\overline{\overline{h}}(\overline{r}_e,\overline{t})$.

On the basis of dependences (7.73) and (7.80) the technological times are calculated simply and reliably. Thus, the relative time of the protective effect of a bed $\overline{t}_p$ should be found by fitting at a given maximum permissible value of the outlet concentration C_* based on the equation

$$2\exp\left(-2\overline{\overline{\alpha}}_V\overline{r}_e^{l-1}\int_{\overline{r}_e}^{1}\frac{\eta^{1+q-l}d\eta}{2\eta^q+\overline{\overline{\beta}}_V\overline{r}_e^q\overline{t}}\right) = \overline{C}_* + \exp\frac{\overline{\overline{\alpha}}_V\overline{r}_e(1-\overline{r}_e^{l-2})}{2-l}. \tag{7.81}$$

The relative time $\overline{t}_h$ for the head losses Δh to reach the limit value Δh_* is similarly calculated from the equation

$$\int_{\overline{r}_e}^{1}\frac{d\eta}{\eta\left[1-\overline{\gamma}^{m_{k1}}\overline{S}^{m_{k1}}(\eta,\overline{t}_h)\right]^{m_{k2}}} + \Delta\overline{\overline{h}}_*\ln\overline{r}_e = 0, \tag{7.82}$$

where $\Delta\overline{\overline{h}}_* = \Delta h_*/\left[h(r_e,0)-H_d\right] = -2\pi k_0 L_f\Delta h_*/\left(Q_f\ln\overline{r}_e\right) = \Delta\overline{h}_*$, since the scales for the heads are the same at different options of suspension entering the bed.

The accuracy of the second approximate solution of the plane-radial filtration problem can be accurately estimated by the exact particular (at q=0) solution of the same problem. Since finding an analogous solution to the first problem of the plane-radial filtration (suspension is supplied from outside) is described in detail in Section 7.1, the calculated characteristics for both relative concentrations and reduced head in the second case are given without derivations. Thus, the exact expressions for them will be

$$\overline{C}(\overline{r},\overline{t}) = e^{-U_e(\overline{r})}\left[e^{-\overline{\overline{\beta}}_V\overline{t}}I_0\left(2\sqrt{\overline{\overline{\beta}}_V U_e(\overline{r})\overline{t}}\right) + \overline{\overline{\beta}}_V\int_0^{\overline{t}}e^{-\overline{\overline{\beta}}_V\xi}I_0\left(2\sqrt{\overline{\overline{\beta}}_V U_e(\overline{r})\xi}\right)d\xi\right], \tag{7.83}$$

$$S(\overline{r},\overline{t}) = \overline{\overline{\alpha}}_V\left(\frac{\overline{r}_e}{\overline{r}}\right)^l e^{-U_e(\overline{r})}\int_0^{\overline{t}}e^{-\overline{\overline{\beta}}_V\xi}I_0\left(2\sqrt{\overline{\overline{\beta}}_V U_e(\overline{r})\xi}\right)d\xi. \tag{7.84}$$

Here

$$U_e(\overline{r}) = \overline{\overline{\alpha}}_V\overline{r}_e^{l-1}\frac{\overline{r}^{2-l}-\overline{r}_e^{2-l}}{2-l}, \tag{7.85}$$

so that at l=0 and 1 respectively

$$U_e(\overline{r}) = \frac{\overline{\overline{\alpha}}_V\left(\overline{r}^2-\overline{r}_e^2\right)}{2\overline{r}_e}, \qquad U_e(\overline{r}) = \overline{\overline{\alpha}}_V\left(\overline{r}-\overline{r}_e\right).$$

Formulas (7.83) and (7.84) allow us to accurately determine the concentration $\overline{C}_e$ and head $\overline{\overline{h}}$, if we put $\overline{r} = 1$ in (7.83), and use expression (7.84) in formula (7.80). Finally, equation (7.82), after substituting expression (7.84) into it, provides a reliable calculation of the parameter $\overline{t}_h$. The time $\overline{t}_p$ is reliably found from the equation

$$e^{-\overline{\overline{\beta}}_V \overline{\overline{t}}_p} I_0\left(2\sqrt{\overline{\overline{\beta}}_V U_e(1)\overline{\overline{t}}_p}\right) + \overline{\overline{\beta}}_V \int_0^{\overline{\overline{t}}_p} e^{-\overline{\overline{\beta}}_V \xi} I_0\left(2\sqrt{\overline{\overline{\beta}}_V U_e(1)\xi}\right) d\xi = \overline{C}_* e^{U_e(1)}. \qquad (7.86)$$

Quantitative analysis was carried out on a number of examples for typical and extreme conditions in order to estimate the computational errors due to the use of the approximate solution, to establish the area of its rational application. Since the directionality of a filtration flow is not reflected in the protective effect of a cylindrical bed, all the attention here was paid to the head losses and closely related the technological time $\overline{\overline{t}}_h$. Therefore, the subject of calculations were the relative quantities $\Delta\overline{\overline{h}}, \overline{\overline{t}}_h$ at the standard fixed values of the radius $\overline{r}_e$ (0.333) and coefficients $\overline{\overline{\beta}}_V$ (0.005), $\overline{\gamma}_a$ (0.001), m_{k1} (1), m_{k2} (3). At the same time, the most important mass exchange coefficient $(\overline{\overline{\alpha}}_V)$ was varied in wide ranges.

First of all, the exact and approximate values of the time $\overline{\overline{t}}_h$, obtained by fitting from equation (7.82) at $\overline{S}(\overline{r},\overline{t})$ according to (7.84) and (7.77), respectively, were compared. It was assumed either l=0 or 1 and q=0, and the coefficient $\overline{\overline{\alpha}}_V$ was continuously varied from 3 to 15 and thus, filtering materials with both the very low absorption capacity $(\overline{\overline{\alpha}}_V = 3-5)$, and the very high absorption capacity $(\overline{\overline{\alpha}}_V = 13-15)$ were covered. Reference and approximate curves of the dependence $\overline{\overline{t}}_h(\overline{\overline{\alpha}}_V)$ corresponding to the specified initial information are shown in **Figure 22** and together with the data of the similar time $\overline{\overline{t}}_p$ calculations from Section 7.2 indicate the legitimacy of involving formulas (7.75), (7.77) and criterion equations (7.81), (7.82) for the engineering calculations of the plane-radial filtration with the suspension feeding from inside and, in particular, for the substantiation of the technological parameters. Incidentally, the very large deviation of the approximate curves from the exact curves at $\overline{\overline{\alpha}}_V < 6$ in Figure 22 does not contradict the made conclusion from the practical point of view, as such filtering materials in general are inexpedient to apply in the water treatment technologies.

For the following series of examples, the approximate solution from the second Section was also used. Thus, the hydraulic characteristics $\Delta\overline{h},\ \overline{t}_h$ were determined in parallel for two varieties of the plane-radial filtration (suspension feeding into a bed from different sides) and q=1. Since different scales of the rate are involved in the representation of the above-mentioned solutions in dimensionless form, and the previous solution, containing the coefficients $\overline{\alpha}_V$, $\overline{\beta}_V$ and variable $\overline{t}$, is taken as a base to draw comparisons, then preliminary the coefficients $\overline{\overline{\alpha}}_V$, $\overline{\overline{\beta}}_V$ and time $\overline{\overline{t}}$ were expressed through them as follows

$$\overline{\overline{\alpha}}_V = \overline{\alpha}_V \overline{r}_e^{1-l}, \quad \overline{\overline{\beta}}_V = \overline{\beta}_V \overline{r}_e^{1-q}, \quad \overline{\overline{t}} = \overline{t}\,\overline{r}_e^{-1}.$$

The value $\overline{\alpha}_V = 10$ and also two fundamentally physically different values of the exponent l in (3.54) were accepted. As can be seen from **Figure 23**, which shows the curves of the head losses increase in the bed over time, the peculiarities of the effect of the filtration process on the absorption process almost do not affect the dynamics of the hydraulic resistance of the bed when the suspension enters through the outer lateral surface. On the contrary, they play an important role in the reverse direction of the filtration flow. Moreover, we can conclude that it is preferable to feed the suspension from the outside at l>0, and from the inside if l<0. The validity of this statement follows from **Figure 24**, which presents curves of the dependence $\overline{t}_h(\overline{\alpha}_V)$ for the same technological conditions. It should be noted that the time t_h at l= -0.3 is much longer if the suspension enters the bed through its inner surface.

Figure 22. Dependence $\bar{t}_h(\bar{\alpha}_V)$

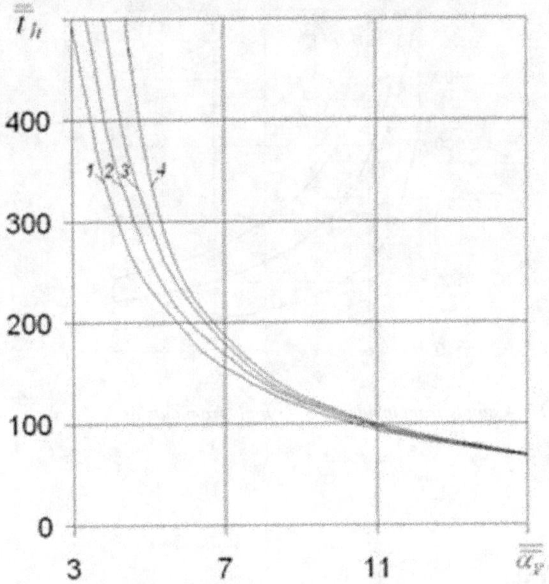

Note. 1, 3 – l=0; 2, 4 – l=1; 1, 2 – approximate; 3, 4 – exact

Figure 23. Increase in the relative head losses in the bed over time

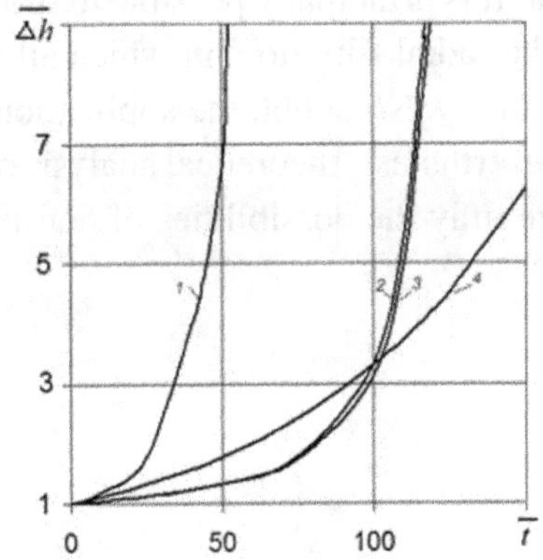

Note. 1, 2 –l=0.7; 3, 4–l= -0.31, 4 – supply from inside; 2, 3 – supply from outside

Figure 24. Dependence $\bar{t}_h(\bar{\alpha}_V)$

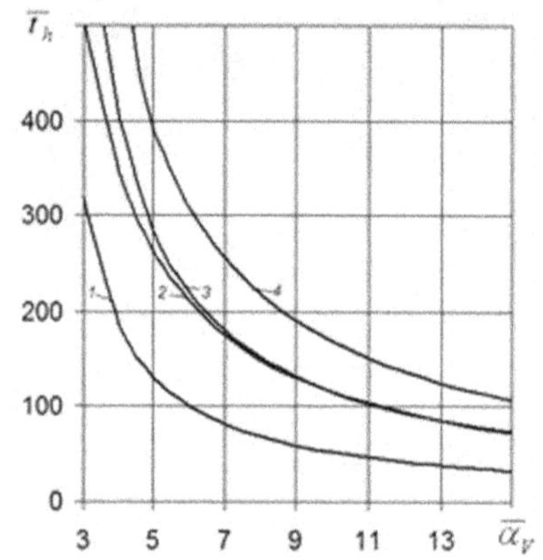

Note. 1, 2 – l=0.7; 3, 4 – l= -0.3; 1-4 – supply from inside; 2, 3 – supply from outside

7.5. RADIAL FILTRATION

As the previous theoretical studies of the plane-radial filtration have shown, less pronounced spatial unevenness of the axisymmetric filtration flow in the cylindrical bed of a water treatment filter, in comparison with the plane filtration flow, can contribute to a significant increase in the efficiency of its operation. It should be expected that the weakening of the noted unevenness will affect the operation of a rapid filter to an even greater extent. It is principally possible to further improve the filter performance due to the implementation of only the radial filtration, in which all the processes occurring in the bed are characterized by a spherical symmetry. No doubt, the application of beds in the form of a spherical layer in practice is hardly realistic. Nevertheless, theoretical analysis of the radial filtration is of scientific interest, as it allows to estimate more fully the possibilities of non-traditional uneven filtration and the ways of its intensification.

Figure 25. Scheme of the radial suspension filtration through a spherical half-layer

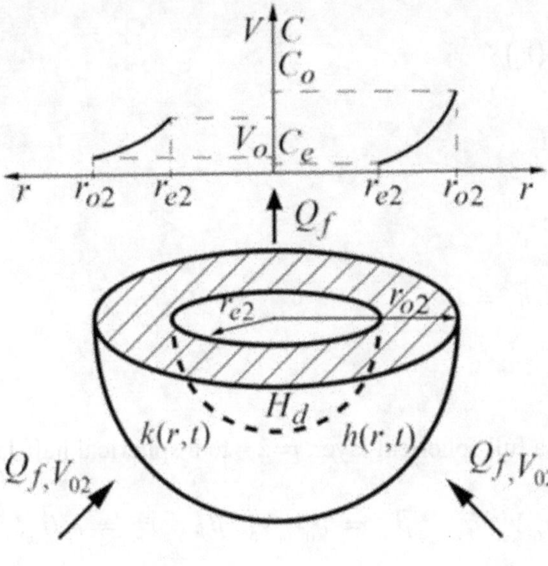

Below the operation is studied in parallel also by analytical methods of two varieties of a spherical bed, firstly, in the form of a full layer with the radii of the outer (r_{01}) and inner (r_{e1}) surfaces, and secondly, in the form of a half-layer with the radii of corresponding surfaces r_{02}, r_{e2} (**Figure 25**). From the practical point of view the second variety of a filter media is obviously preferable, as in this case it is much easier to wash an excessively clogged filtering material. Then, at a given constant filter performance, the variable filtration rate in both media (V_1, V_2) will depend on the radius r as follows

$$V_1(r) = \frac{Q_f}{4\pi\, r^2}, \qquad V_2(r) = \frac{Q_f}{2\pi\, r^2}. \tag{7.87}$$

If the corresponding hydraulic loads (suspension supply from outside) are chosen as the rate scales as before

$$V_{01} = \frac{Q_f}{4\pi r_{01}^2}, \qquad V_{02} = \frac{Q_f}{2\pi r_{02}^2}, \tag{7.88}$$

then the relative rates $\overline{V}_i$ $(i = 1,2)$ formally take a single form

$$\overline{V}_i = \frac{V_i}{V_{0i}} = \frac{1}{\overline{r}_i^2}, \tag{7.89}$$

where $\overline{r}_i = r/r_{0i}$.

The underlying radial filtration model can be simplified considerably after adopting some of the usual assumptions described in Chapter 1. As a result, the corresponding mathematical problem for two varieties of the spherical bed in dimensionless form is formulated as follows

$$\bar{V}_i(\bar{r}_i)\frac{\partial \bar{C}_i}{\partial \bar{r}_i} - \frac{\partial \bar{S}_i}{\partial \bar{t}_i} = 0, \tag{7.90}$$

$$\frac{\partial \bar{S}_i}{\partial \bar{t}_i} = \bar{\alpha}_{Vi}\bar{V}_i^l(\bar{r}_i)\bar{C}_i - \bar{\beta}_{Vi}\bar{V}_i^q(\bar{r}_i)\bar{S}_i, \tag{7.91}$$

$$\bar{V}_i(\bar{r}_i) = \bar{k}(\bar{S}_i)\left(\frac{1}{\bar{r}_{ei}} - 1\right)\frac{\partial \bar{h}_i}{\partial \bar{r}_i}, \tag{7.92}$$

$$\bar{k}(\bar{S}_i) = \left(1 - \bar{\gamma}_a^{\,m_{k1}}\bar{S}_i^{m_{k1}}\right)^{m_{k2}} ; \tag{7.93}$$

$$\bar{r}_i = 1,\ \bar{C}_i = 1;\ \bar{r}_i = \bar{r}_{ei},\ \bar{h}_i = 0 ; \tag{7.94}$$

$$\bar{t}_i = 0, \quad \bar{S}_i = 0. \tag{7.95}$$

Here i=1 corresponds to a full spherical layer, i=2 – to a spherical half-layer;

$$\bar{t}_i = V_{0i}t/(n_0 r_{0i}), \quad \bar{\alpha}_{Vi} = r_{0i}V_{0i}^{l-1}\alpha_V, \quad \bar{\beta}_{Vi} = n_0 r_{0i}V_{0i}^{q-1}\beta_V, \quad \bar{r}_{ei} = r_{ei}/r_{0i},$$

$$\bar{h}_i(\bar{r}_i,\bar{t}_i) = \frac{h_i(r,t) - H_d}{h_i(r_{0i}, 0) - H_d} = \frac{r_{0i}V_{0i}}{k_0}\left[h_i(r,t) - H_d\right].$$

The procedure of obtaining an approximate solution of problem (7.90) - (7.95) is omitted and the calculation formulas and equations are given at once. They allow to determine all required characteristics of the clarification and filtration flow for the radial filters with a high accuracy, reliably substantiate their design parameters.

Of primary importance for the radial filter control is the prediction of the filtrate quality, which is recommended based on the following relationship between the relative outlet concentration $\bar{C}_{ei}$ and the time $\bar{t}_i$

$$\bar{C}_{ei}(\bar{t}_i) = 2\exp\left(-2\bar{\alpha}_{Vi}\int_{\bar{r}_{ei}}^{1}\frac{\eta^{2+2q-2l}}{2\eta^{2q} + \bar{\beta}_{Vi}\bar{t}_i}d\eta\right) - \exp\frac{\bar{\alpha}_{Vi}\left(\bar{r}_{ei}^{3-2l} - 1\right)}{3 - 2l}. \tag{7.96}$$

To calculate the local and total head losses in spherical beds, the approximate expressions for the relative functions-concentrations of the deposited suspension particles were used

$$\bar{S}_i(\bar{r}_i,\bar{t}_i) = \frac{2\bar{\alpha}_{Vi}\bar{t}_i\bar{r}_i^{2q-2l}}{2\bar{r}_i^{2q} + \bar{\beta}_{Vi}\bar{t}_i}\exp\left(-2\bar{\alpha}_{Vi}\int_{\bar{r}_i}^{1}\frac{\eta^{2+2q-2l}}{2\eta^{2q} + \bar{\beta}_{Vi}\bar{t}_i}d\eta\right). \tag{7.97}$$

Then the change in the reduced head $\bar{h}_i$ within the i - th filter bed over time is described by the following formula

$$\bar{h}_i(\bar{r}_i,\bar{t}_i) = \left(\frac{1}{\bar{r}_{ei}} - \frac{1}{\bar{r}_i}\right)^{-1}\int_{\bar{r}_{ei}}^{\bar{r}_i}\frac{d\xi}{\xi^2\left[1 - \bar{\gamma}_a^{\,m_{k1}}\bar{S}_i^{m_{k1}}(\xi,\bar{t}_i)\right]^{m_{k2}}}. \tag{7.98}$$

To optimize the operation of the filter facility, first of all, it is necessary to find the relative time of the protective effect of the i - th filter media $\bar{t}_{pi}$ from the equation

$$2\exp\left(-2\bar{\alpha}_{Vi}\int_{\bar{r}_i}^{1}\frac{\eta^{2+2q-2l}}{2\eta^{2q}+\bar{\beta}_{Vi}\bar{t}_{pi}}d\eta\right) = \bar{C}_* + \exp\frac{\bar{\alpha}_{Vi}(\bar{r}_e^{3-2l}-1)}{3-2l}. \tag{7.99}$$

Then the second relative technological time ($\bar{t}_{hi}$) should be determined from the equation

$$\int_{\bar{r}_{ei}}^{1}\frac{d\eta}{\eta^2\left[1-\gamma^{m_1}\bar{S}_i^{m_1}(\eta,\bar{t}_{hi})\right]^{m_2}} - \left(\frac{1}{\bar{r}_{ei}}-1\right)\Delta\bar{h}_{*i} = 0, \tag{7.100}$$

where $\Delta\bar{h}_{*i} = \Delta h_* / \left[h_i(r_{0i}, 0) - H_d\right]$. Then $\bar{t}_{fi}$ is finally set as the smaller of the values $\bar{t}_{pi}$, $\bar{t}_{hi}$.

The exact solution of the corresponding mathematical problem described in Section 7.1 for the special case q=0 can be useful in the theoretical studies of the radial filtration. Thus, dependence (7.21) allows to accurately calculate the suspended solids content in the filtrate and dependence (7.31) with consideration of (7.26) – to calculate the head losses. In this case it is necessary to assume

$$U(\bar{r}_i) = \bar{\alpha}_{Vi}\frac{1-\bar{r}_i^{3-2l}}{3-2l}.$$

The efficiency of the radial filtration was analyzed by comparing it with the vertical and plane-radial filtration. The subject of the calculations of numerous examples with various initial data were the main relative characteristics of the complex process of the suspension clarification, such as the outlet concentration $\bar{C}_e$, the total head loss $\Delta\bar{h}$ and the filter run duration $\bar{t}_f$. The procedure of calculations was complicated due to a dimensionless form of the final dependencies and equations, while in four basic problems (filtration vertical, plane-radial and radial with two varieties of a spherical bed) the hydrodynamics and geometry of filtration flows and, as a consequence, the length and rate scales differed significantly. The radius r_0 and hydraulic load V_0, corresponding to the plane-radial filtration, were chosen as the fundamental scales. Therefore, in the final the determination of the desired characteristics of the radial filtration, and in view of their subsequent comparison with similar characteristics of other varieties of uneven filtration and filtration flow, the following formulas were used for the conversion of the relative mass exchange coefficients and time

$$\bar{\alpha}_{V1} = \left(\frac{2}{L_f}\right)^{1-l}\frac{r_{01}^{3-2l}}{r_0^{2-l}}\bar{\alpha}_V,\ \bar{\alpha}_{V2} = \frac{r_{02}^{3-2l}}{r_0^{2-l}L_f^{1-l}}\bar{\alpha}_V,$$

$$\bar{\beta}_{V1} = \left(\frac{2}{L_f}\right)^{1-q}\frac{r_{01}^{3-2q}}{r_0^{2-q}}\bar{\beta}_V,\ \bar{\beta}_{V2} = \frac{r_{02}^{3-2q}}{r_0^{2-q}L_f^{1-q}}\bar{\beta}_V,$$

$$\bar{t}_1 = \frac{r_0^2 L_f}{2r_{01}^3}\bar{t},\ \bar{t}_2 = \frac{r_0^2 L_f}{2r_{02}^3}\bar{t}.$$

When selecting the initial information, special attention was paid to studying the significance of the absorption process, which directly ensures the removal of impurity from a suspension. To reflect the diversity of the absorption (in the generalized sense) properties of actual filtering materials, the coefficient $\bar{\alpha}_V$ was varied and, in addition, two extreme values of l (-0.3 and 0.7) were taken. As before, the values of $\bar{\beta}_V$ and q were fixed, respectively, 0.005 and 1.

To ensure the correctness of the comparison of the operation of the four options of the filter beds having the form of a cylindrical layer (the first bed), a cube (the second), a spherical full layer (the third) and a spherical half-layer (the fourth), their volume (W_f=1m^3) and performance (Q_f=10m^3/h) were assumed to be the same and also L_f=L=1m. Moreover, equal values of the inner radius of both layers (r_e=r_{e1}=r_{e2}=0.2m) were taken, which is not fundamental. Then we found the dimensional values of the radii of outer surfaces r_0=0.6m, r_{01}=0.627m, r_{02}=0.786m and dimensionless values of the radii of the inner surfaces $\bar{r}_e$ = 0.333, $\bar{r}_{e1}$ = 0.319, $\bar{r}_{e2}$ = 0.2545 from the equality of the volumes of the cylindrical and spherical beds.

First of all, the increase of the suspended solids content in the filtrate for a considerable time was monitored, so that its quality did not meet the sanitary and hygienic standards at all at the end of the calculation period. Curves of the dependence $\bar{C}_e(\bar{t})$ corresponding to different types of filtration flow and two varieties of a spherical bed are calculated by formulas (7.43), (7.96) and presented in **Figure 26** for l=0.7, and in **Figure 27** for l= -0.3. In the case of intensifying the suspended solids deposition due to a filtration flow, the protective capability of the filter bed of the curvilinear forms deteriorates insignificantly (curves 4 and 1-3 in Figure 26). If the filtration flow process complicates the immobilization of the suspended particles, then the absorption process proceeds precisely in the third bed variety with the highest intensity due to the lowest filtration rate and thus, the impurity breakthrough is minimized.

Figure 26. Increase in the relative outlet suspended solids concentration over time at l=0.7

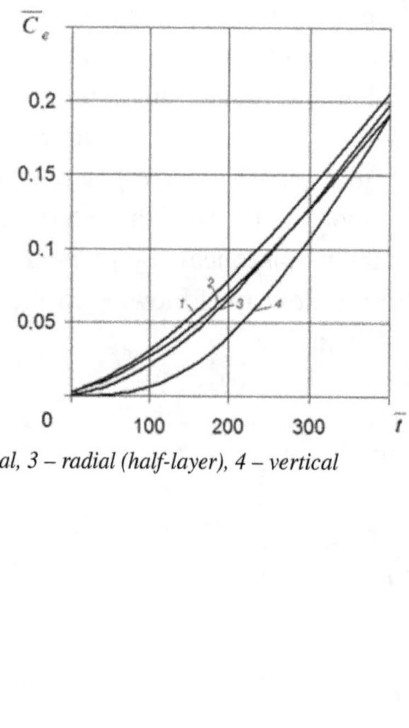

Note. 1 – radial (full layer), 2 – plane-radial, 3 – radial (half-layer), 4 – vertical

Figure 27. Increase in the relative outlet suspended solids concentration over time at l= -0.3

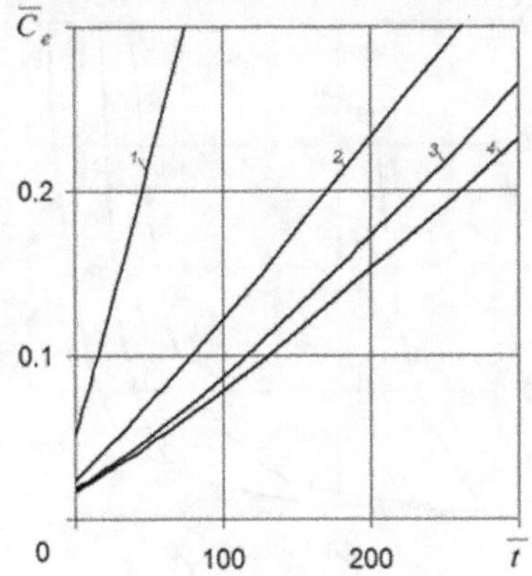

Note. 1 – vertical, 2 – radial (half=layer), 3 – plane-radial, 4 – radial (full layer)

It is obvious that the accelerated deposition of the suspended particles in the pores of the bed, accompanied by binding of a large amount of free water, causes its rapid clogging and a significant increase in the hydraulic resistance. **Figure 28** shows the dynamics of the relative head losses in four beds, according to formulas (2.93), (7.32), (7.98) at also $\overline{\alpha}_V = 10$ and additionally $\overline{\gamma}_a = 0.001$. It follows that the use of spherical beds along with cylindrical beds makes it possible to slow down the increase of the indicated losses in them, and especially significantly with l=0.7. Just the main advantage of the plane-radial and radial filtration is a more rational use of the energy resource of filters. From the hydraulic point of view, spherical beds provide a noticeably better result, and irrespective of the nature of the filtration flow effect on mass exchange. And only if l= -0.3 and $\overline{t} > 100$ (**Figure 29**), the head losses at the vertical filtration appear to be less than at the plane-radial and radial filtration. Nevertheless, such, in fact, exceptional situation, as it will be shown below at the determination of the main technological parameter, does not mean that it is more advantageous to arrange descending (or ascending) motion of a suspension under the considered conditions. Indeed, then the time t_f will be determined by the protective capability of a bed, which is better in the layers of the filter media of a curvilinear shape.

Figure 28. Increase in the relative head losses in the bed over time at l=0.7

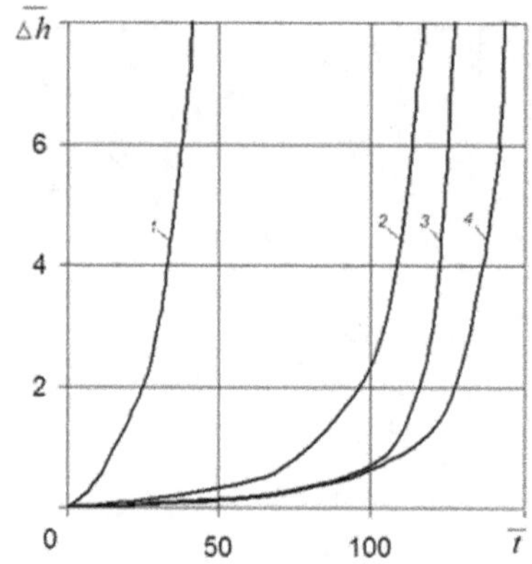

Note. 1 - vertical, 2 - plane-radial, 3 - radial (half-layer), 4 - radial (full layer)

Figure 29. Increase in the relative head losses in the bed over time at l= -0.3

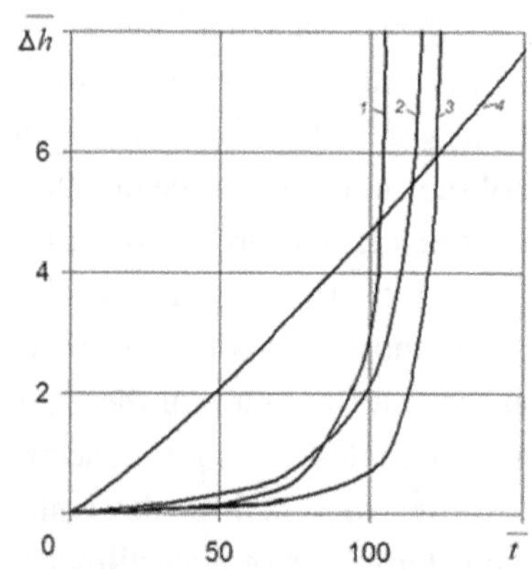

Note. 1 - radial (full layer), 2 - plane-radial, 3 - radial (half-layer), 4 - vertical

However, the most illustrative for the comparative estimations of different types of the detachment filtration are the data of the calculations of the relative duration of the filter run when the absorption properties of a filtering material vary within significant limits. First of all, we systematically found the values of the technological times $\bar{t}_p$, $\bar{t}_h$ by fitting procedure from the criterion equations (1.26), (1.27). Their generalizing graphs of the dependence $\bar{t}_f(\bar{\alpha}_V)$ for the radial form of the bed are shown at *l*=0.7 in **Figure 30** (curves 2, 4), and at *l*= -0.3 in **Figure 31** (curves 1, 3). Also, similar curves are given on these figures, reflecting the relationship between $\bar{t}_f$ and $\bar{\alpha}_V$ for previously studied in detail the direct-flow and plain-radial filtration. It is obvious that the continuous operation of a filter is considerably prolonged due to the application of the beds in the form of a curvilinear layer. At the same time the difference in the values of $\bar{t}_f$ calculated for the cylindrical and spherical beds is so small that the practically best of them is chosen from the considerations of washing processability. In this case, the first form of the bed is preferable, although, in principle, it is quite realistic to provide regular washing of the bed having the fourth form.

Figure 30. Dependence $\bar{t}_f(\bar{\alpha}_v)$ at l=0.7

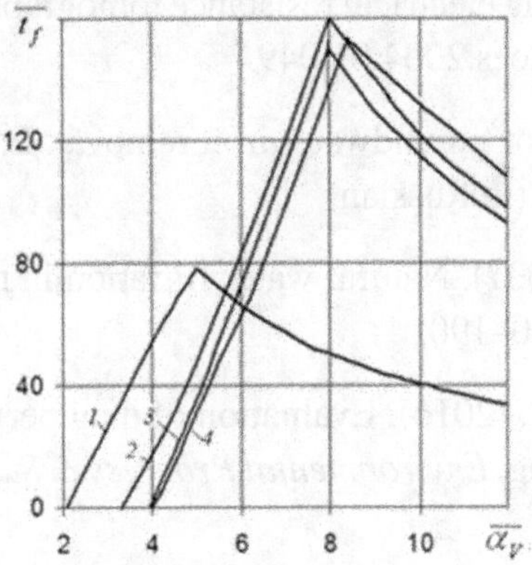

Note. vertical, 2 - radial (half-layer), 3 - plane-radial, 4 - radial (full layer)

Figure 31. Dependence $\bar{t}_f(\bar{\alpha}_v)$ at l= -0.3

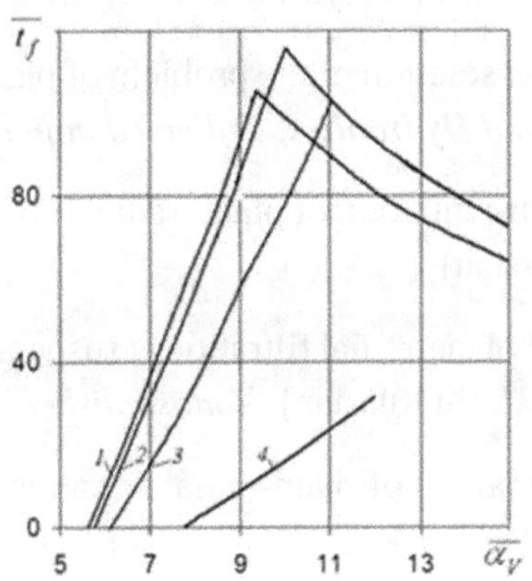

Note. 1 - radial (full layer), 2 - plane-radial, 3 - radial (half-layer), 4 – vertical

REFERENCES

Bear, J. (1988). *Dynamics of fluids in porous media*. Mineola.

Bolton, G. T., Hooper, C. W., Mann, R., & Stitt, T. H. (2004). Flow distribution and velocity measurement in a radial fixed bed reactor using electrical resistance tomography. *Chemical Engineering Science*, *59*(10), 1989–1997. DOI: 10.1016/j.ces.2004.01.049

Dzyubo, V.V. (2006). Radial filters of groundwater iron removal. Engineering and technological solutions. *Sanitary Engineering*, 16–19. (in Russian)

Dzyubo, V. V., & Alferova, L. I. (2007). Natural water filtration in the mode of non-uniform rates. [in Russian]. *Bulletin of TSASU*, (2), 180–190.

Liu, L., Xue, Q., Wan, Y., & Tian, Y. (2016). Evaluation of dual permeability of gas flow in municipal solid waste: Experiment and modeling. *Environmental Progress & Sustainable Energy*, *35*(1), 1–7. DOI: 10.1002/ep.12184

Oleynik, A. Ya. (1978). *Flow calculations of vertical drainage*. Naukova dumka. (in Russian)

Poliakov, V. L. (2013). Theoretical analysis of the plane-radial filtration of suspension. [in Russian]. *Rep. NAS of Ukraine*, (12), 52–59.

Poliakov, V.L. (2013). About rational choice of design parameters of cylindrical bed of water purification filter. *Problems of Water Supply, Sewage and Hydraulics. Collected papers*, 22, 113–123. (in Russian)

Poliakov V.L. (2013). About one strict solution of the problem of plane-radial filtration of a suspension. *Problems of Water Supply, Sewage and Hydraulics. Collected papers*, 22, 124–131. (in Russian)

Poliakov, V.L. (2014). Theoretical substantiation of plane-radial filtration of suspension. *Applied hydromechanics*, 16(88), 3, 52–62. (in Russian)

Poliakov, V. L. (2015). Calculation of plane-radial filtration at suspension feeding into the filter bed from inside. *Vodoochistka, Vodopodgotovka* [in Russian]. *Vodosnabzheniy*, *5*(89), 16–21.

Poliakov, V. L. (2015). Theoretical analysis of plain-radial filtration. *Water and Ecology.* [in Russian]. *Problems and Solutions*, *2*(62), 14–24.

Poliakov, V. L. (2015). About rational choice of the form and sizes of bed of a water treatment filter. *Vodoochistka, Vodopodgotovka* [in Russian]. *Vodosnabzheniya*, *9*(93), 8–16.

Poliakov V.L. (2015). On the theory of plane-radial filtration and its some applications. *Problems of Water Supply, Sewage and Hydraulics. Collected papers*, 25, 217–222. (in Russian)

Rajagopalan, R., & Tien, C. (1976). Trajectory analysis of deep bed filtration with sphere in cell porous media model. *AIChE Journal. American Institute of Chemical Engineers*, *22*(3), 523–533. DOI: 10.1002/aic.690220316

Senyavin, M. M., Venitsianov, E. V., & Ayukaev, R. I. (1997). About mathematical models and engineering methods for calculating the process of natural water purification by filtration. [in Russian]. *Water Resources*, (2), 157–170.

Tiller, F. M., Kirby, J. M., & Nguyen, H. L. (1996). Approximate theory for radial filtration / consolidation. *Journal of Geotechnical Engineering*, *122*(10), 797–805. DOI: 10.1061/(ASCE)0733-9410(1996)122:10(797)

Tiller, F. M., & Yeh, C. S. (1985). The role of porosity infiltration. Part X: Deposition of compressible cakes on external radial surfaces. *AIChE Journal. American Institute of Chemical Engineers*, *31*(8), 1241–1248. DOI: 10.1002/aic.690310803

Vigneswaran, S., Thiyagaram, M., & Ben Ami, R. (1985). Mathematical modeling of radial flow filtration and its application to groundwater recharge problems. *AIChE Journal. American Institute of Chemical Engineers*, *31*(11), 1929–1933. DOI: 10.1002/aic.690311121

Voytov, E. L. (1974). *Pretreatment of biologically treated urban sewage on radial filters.* Ph.D. dissertation abstract. Novosibirsk State University Press. (in Russian)

Chapter 8
Infiltration of Aqueous Suspension Into Uniform Ground Bed

ABSTRACT

The problems of environmental protection related to the need to be flexible and timely in regulating groundwater resources are addressed in Chapter 8. Analytical methods are used to solve the problems of saturation of hydrogeological complexes with water from surface sources. The consequences of significant volume of water with a considerable amount of dispersed impurities passing through the ground bed overlapping aquifers are analyzed. The joint absorption of an unsaturated medium and the transfer of the suspended matter are considered as a three-stage technological process (usually implemented in order to replenish the depleted groundwater reserves).

The unbalanced use of (deep) groundwater for the drinking and technical water supply needs leads to its depletion in the long term. This situation is especially dangerous in the areas where the surface water sources are absent or their resources are significantly limited and insufficient to supply the local economy with water in full. Thus, the unjustifiably active exploitation of the underground water sources threatens and may already in the near future acute shortage, first of all, of drinking water. Therefore, it is important to constantly take care of replenishing groundwater reserves, taking appropriate preventive measures (Plotnikov, 1983; Recommendations, 1976; Usenko, 1972). Such measures should include, first of all, the implementation of the downward flow of large volumes of water from the ground surface. Periodically, the excess surface water is generated by heavy precipitation events, creating a surface runoff that predominantly flows directly or through the local drainage systems into temporary or permanent functioning watercourses. As a result, such water is essentially useless for water management except, perhaps, for hydropower. Moreover, the mentioned runoff can silt up their channels and create emergency situations at barrier structures. The arrangement of even the partial transfer of the excess surface water into the aquifers contributes to solving two practical tasks at once, namely, mitigation of the negative consequences for the environment from the intensive surface runoff and the restoration of the groundwater resources.

Traditionally, special water management facilities, namely, infiltration fields, are used to solve the second problem and, in parallel, the first problem. The analogs of the above-mentioned fields are special sites (infiltration basins), which allow solving a smaller-scale problem of the rainfall runoff regulation in

DOI: 10.4018/979-8-3693-8211-0.ch008

order to ensure a stable operation of the local drainage system (Berdanov et al., 1978; Bochever, 1977; Burchak, 1978; Kommunar & Bogolyubov, 1977; Tkachuk & Shevchuk, 2016; Yakovenko, 2009). Due to the temporary accumulation of the excess water on the infiltration site, it is possible to significantly reduce, and in some cases even in general to remove the peak load on the above system.

It should be emphasized that many technological schemes for the infiltration process into a dewatered ground bed are implemented in water management practice, which is caused by a variety of natural factors (water-physical properties of ground bed and conditions in underlying and hydraulically connected pressure aquifers), technological parameters (volumes of supplied water, intensity of inflow), design aspects (drains for the intensification of the water flow process), etc. A frequently practiced scenario of the mentioned process development is accepted below for the in-depth theoretical analysis. In general, it is important to note that the purpose of this theoretical development was not universal recommendations for practice, but the creation of a basic calculation tool, which can be easily adapted to various natural and technogenic conditions. Such a tool has wide possibilities in terms of studying the patterns of the combine filtration flow and clarification processes and, at the same time, the estimation of the errors introduced into applied calculations by commonly used methods of the simplification of initial mathematical models (linearization, averaging of variable characteristics, introduction of effective coefficients, etc.). To ensure long-term normal operation, the filtration flow properties of the above-mentioned fields and the infiltrate quality are subject to higher requirements and, first of all, the filtering column should have a high flow capacity. Also, the water supplied to such fields may contain only insignificant amounts of impurities. In reality, even a short-term but violent surface flow can mobilize and carry away a lot of solid matter. As a consequence, the infiltrate from the surface runoff is often a suspension with a very high content of dispersed impurities. At the same time, it often contains other forms of contaminants such as organic compounds, heavy metals, biogenic elements and so on. However, the five Sections of this Chapter analyze only the patterns of the redistribution and removal of the dispersed form of contamination from the downward suspended filtration flow. Naturally, when a suspension is filtered through a porous medium, which can extract the suspended matter from the suspension flow and retain it for a long time, the clarification process also takes place. Since the thickness of a ground bed is usually large, this matter and other impurities are practically absent in the filtered water when it reaches a deep aquifer. They are mainly deposited near the surface of the ground. Incidentally, in the presence of the large suspended particles, the formation of a dynamic layer is also possible. In case of repeated use of the infiltration field and high content of dispersed impurities in the infiltrate, a significant amount of a deposit may accumulate here. Thus, a special near-surface layer is formed in the soil, the porosity of which is significantly lower than its initial value. As a consequence, the hydraulic resistance of the entire ground bed increases significantly and the subsequent application of such a field becomes economically unjustified. Therefore, it is necessary, on the one hand, to prepare the infiltrate by its preliminary clarification, and on the other hand, to develop a calculation method. Such a method should provide a reliable forecast of the interrelated changes in the physical and mechanical state of the filtering column and the development of the filtration process under water-physical conditions complicated by the prolonged separation of aqueous suspension.

Other characteristic examples of the clarification process in natural porous media are irrigation with polluted water (wastewater, drainage-discharge water, etc.) in reclamation systems, in particular, rice systems (Romashchenko & Balyuk, 2010; Shtykov et al., 1987), and water treatment in wetlands (Kadlec & Knight, 1996). Incidentally the relevant studies are set out in next monograph. In the first case, it is sometimes expedient to intensify the filtration flow process by providing uniform water supply over the field area with a sufficiently high flow rate. Intensification of the above process can seriously affect both components of the interfacial mass exchange. Coincidentally, the periodic supply of a clean water may mobilize and remove some part of the accumulated deposit or at least redistribute it down the ground profile. This Chapter focuses on establishing the theoretical basis for the engineering calculations of the infiltration of the surface water with a fine-dispersed impurity into a uniform ground bed. Previously, the infiltration process was studied exclusively by experimental and theoretical methods with respect to the normal and anomalous fluids in an extensive literature, e.g. (Dunin, 1980; Geiger & Durnford, 2000; Gusev, 1981; Guyonnet et al., 2000; Poliakov, 2008; Polubarinova-Kochina, 1977; Verigin & Vasiliev, 1977). Mathematical modeling of the downward flow of pure water under complex water-physical conditions (nonuniform structure and fractured strata, unsaturated flow, presence of an unstable surface water layer) was performed in (Kosterin & Skvortsov, 2002; Latifi et al., 1994; Oleinik & Tkach, 1972; Parlange et al., 1985; Watson & Awadalla, 1986). Fundamentally new in the results of the theoretical studies reported below is the presence of the suspended matter in water, as well as the clarification process in the porous medium being intensively wetted. As a consequence, its filtration flow properties gradually deteriorate, and the application of the mathematical modeling methods becomes much more difficult. Nevertheless, it was possible to obtain calculation expressions in an analytical form due to the partial averaging of the required characteristics (previously thoroughly substantiated for the nonlinear problems of aqueous suspensions filtration). These expressions just allow to easily and reliably predict spatial and temporal changes in the desired filtration characteristics using modern application software packages (Mathcad, Matlab, Maple, etc.) and thus, facilitate the search for rational technological and design decisions.

The efficiency of the clarification of raw waters by filtration under natural conditions and technogenic waters in filtration facilities is often significantly different, which is partly explained, first of all, by the lower permeability and worse absorption capacity of a soil strata, their great thickness and nonuniformity. As a consequence, the interfacial mass exchange takes place in them at a clearly slower rate with the negligible deposit detachment. Nevertheless, the processes of mass transfer, mass exchange, filtration flow in both cases are similar and, therefore, the corresponding mathematical models are similar. The existing differences between them are reflected mainly in the empirical functions for the model coefficients. At the same time, sometimes it is reasonable to make minor adjustments to the basic models, for example, due to the increasing importance of the diffusion mechanism of mass transfer. The theoretical study of the clarification process occurring against the background of the intensive infiltration in a soil column makes sense from the scientific point of view, as it allows us to expand our understanding of the patterns of the aqueous suspension filtration, which is often encountered in water management engineering and nature.

8.1. FEEDING SUSPENSION TO INFILTRATION FIELD WITH SPECIFIED FLOW RATE. FIRST STAGE

In general, when solving the practical issues on the regulation and renewal of the groundwater reserves, the control of the deterioration of the filtration flow properties of the soil column in the infiltration field over time is of particular importance. Indeed, such ground beds due to their usually large thickness, even despite the relatively weak absorption capacity of the soil, can reliably protect groundwater from contaminants contained in the infiltration water. However, at the same time, the flow capacity of the filtering ground bed in the process of its clogging, and, therefore, and the efficiency of the field can be significantly reduced in case of its repeated use. And since the washing of the clogged ground bed is problematic to implement in practice and inexpedient in principle, the forecast of the development of filtration conditions in it under the implementation of various technological algorithms becomes especially important. And then, it is reasonable to distinguish three characteristic stages in the technological cycle for completeness of the theoretical analysis of the groundwater recharge at the expense of surface water sources (**Figure 1**) (Poliakov, 2008; 2014). The surface water layer is formed and its depth increases at the first stage (Figure 1a). At the same time, in a porous uniform column, initially without gravitational moisture, a saturation zone is formed in which the entire pore space is filled with water (a small amount of the trapped air may be present). This zone is bounded from below by the saturation front ($z = -Z_i(t)$, Z_i is the variable mark of the front at the i- th stage, with a constant ground surface mark accepted as the zero mark in the absence of a dynamic layer). The extent (thickness) of this zone increases over time due to the front moving deep into the ground bed. The first stage can end as a result of the water supply shutdown to the infiltration field (at the point in time t_1) under the condition that $n_W \Omega M < W_{in}$, where n_W is the effective porosity, which takes into account the initial moisture reserves (actual porosity minus firmly bound moisture initially contained in the medium); Ω is the area of the infiltration field, M is the thickness of the ground bed; W_{in} is the total volume of aqueous suspension supplied to the field. Exactly such option of the filtration flow development is studied in detail below. It is also possible the mentioned front to reach the aquifer roof at the stable long-term inflow of liquid. Further the second stage takes place in both cases. Attention is emphasized on the situation when the surface layer is gradually depleted, recharging the filtration flow and thus, providing the progressive expansion of the saturation zone (Figure 1a). This stage ends at the moment of the infiltration shutdown from the surface of the ground bed, which should be calculated unlike the first characteristic time t_1. Also, the end of the second stage is possible at the convergence of the infiltration and ground waters. Finally, the third stage (Figure 1b) comes, during which the saturated zone, which has accumulated all the water supplied to the field except for the minor evaporation losses, moves downward from the ground surface to the aquifer. Incidentally, it may make sense to analyze the inflow of the infiltration water residuals on the basis of the spatial filtration flow models at its rapid spreading in the mentioned well-permeable horizon.

Often the physical properties of a soil column change significantly with depth. Then, to simplify the applied analysis, it is reasonable to hypothetically reduce such a ground bed to an equivalent (according to the relevant features) uniform column, characterizing the accumulation, throughput and retention capacities by the effective (reduced) porosity, hydraulic conductivity, and mass exchange coefficients.

Figure 1. Schemes for the calculation of the filtration flow and clarification processes at the first, second (a) and third (b) stages

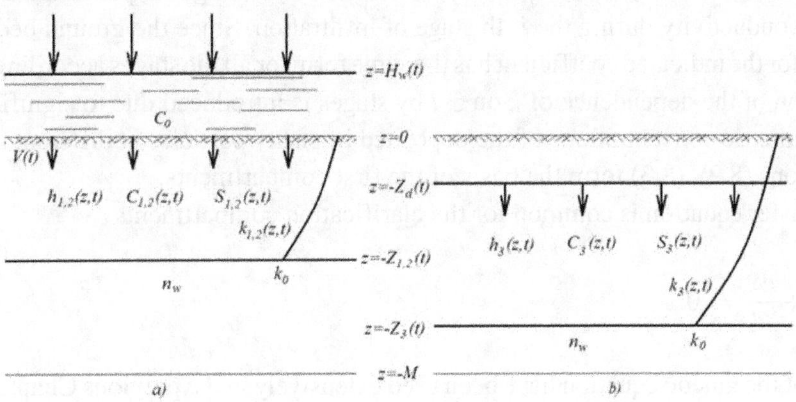

It should be noted that the described technological cycle may be incomplete in practice due to the actual absence of the first or third stages. However, all three stages are analyzed separately and sequentially in this Chapter using analytical methods. Previously, a similar detailed approach to the study of the infiltration process was implemented in the monograph (Poliakov, 2014), and the ground bed was assumed to be layered and subjected to the filtration flow deformations. Here the ground is uniform, but the infiltrate contains a dispersed impurity, the deposition of which leads to the progressive deterioration of the seepage properties of the porous medium. Thus, the subject of mathematical modeling now becomes the clarification process along with the filtration flow process. In contrast to the developments outlined in the previous Chapters, this Chapter prefers the dimensional form of the presentation of mathematical models and the results of their implementation due to the lack of a suitable length scale in the statement of the initial problems. And only when conducting the quantitative analysis of the combine infiltration and clarification processes, dimensionless coefficients and variables are used to generalize the calculated data. The mathematical models for different stages have both compartments, namely, the infiltration and clarification compartments. The common components for the first compartment are:

filtration flow equation in the saturation zone ($-Z_i \le z \le 0$)

$$\frac{\partial}{\partial z}\left[k_i(z,t)\frac{\partial h_i}{\partial z}\right] = 0 \tag{8.1}$$

kinematic condition at the saturation front

$$z = -Z_i(t),\ n_W\frac{dZ_i}{dt} = k_0\frac{\partial h_i}{\partial z} \tag{8.2}$$

and also, the form of the condition at the front for heads is conserved for all stages

$$z = -Z_i(t),\ h_i = -Z_i(t) - \psi_w \tag{8.3}$$

Here and hereinafter, the index "i" indicates that the corresponding characteristic belongs to the i-th stage (i=1,2,3); ψ_w is the head increment at the saturation front due to the capillary forces (Gusev, 1981; Mein & Farrell, 1974; Neuman, 1976); $k_i(z,t)$ is the function describing the spatial and temporal changes in the hydraulic conductivity during the i- th stage of infiltration. Since the ground bed is uniform, the basic expression for the indicated coefficient has the same form for all the stages according to (1.9), (1.10). The differentiation of the dependence of k on z, t by stages is introduced due to significant differences between the volumetric concentrations of the deposited impurity particles at different stages. Equation (8.1) and conditions (8.2), (8.3) form the basis of the first compartments.

The mass transfer equation is common for the clarification compartments

$$n_W \frac{\partial C_i}{\partial t} - V(t)\frac{\partial C_i}{\partial z} + \frac{\partial S_i}{\partial t} = 0 \tag{8.4}$$

Three forms of the kinetic equation have been used extensively in the previous Chapters. Two of them are given below in dimensional terms. The first equation corresponds to the non-detachment filtration (4.1)

$$\frac{\partial S_i}{\partial t} = \lambda(S_i)\, C_i \tag{8.5}$$

The second and third equations refer to the detachment filtration and have the form, respectively, at a constant rate (2.2), and also at a variable rate according to (5.3)

$$\frac{\partial S_i}{\partial t} = \alpha_V V^l C_i - \beta_V V^q S_i. \tag{8.6}$$

In the following, attention is focused on the form 1 and form 3 kinetics. The value of the second form mainly lies in the possibility of obtaining mathematically rigorous design expressions in such situations. For each stage of the filtration process, a detailed analysis of the filtration flow conditions in the intensively wetted ground bed is performed separately in order to develop the practical recommendations for the rational and long-term use of the infiltration field.

In order to generalize (by taking into account the formation and draining of the surface water layer) the filtration flow compartment (can now be interpreted as hydrodynamic compartment), namely, equation (8.1) and boundary conditions (8.2), (8.3), first of all, are supplemented by the equation describing the water balance on the infiltration field, according to which the water supplied to it is spent in parallel for the infiltration and surface accumulation. Thus,

$$\frac{dH_W}{dt} + n_W \frac{dZ_1}{dt} = \frac{dw}{dt}, \tag{8.7}$$

where H_W is the depth of the surface water layer; w is the volume of water supplied per unit area of the field during the calculation period. It should be noted that the volume w is related to the flow rate Q_{in} (Sections 5.3, 5.4) as follows

$$w(t) = \int_0^t Q_{in}(\xi)\, d\xi$$

and the nature of its distribution during this period is not decisive. However, when determining the required characteristics at the end of the period, it is really important to know the final value of w. Obviously, if the water inflow to the field is uniform over time, its rate dw/dt will be constant. If the inflow is non-uniform, then it is justified to consider it in engineering calculations as uniform with an obligatory conservation of the total volume delivered to the field. Finally, the above-mentioned equations and conditions are supplemented by the homogeneous initial conditions

$$t = 0, \quad H_W = Z_1 = 0 \; ; \tag{8.8}$$

ground condition

$$z = 0, \quad h_1 = H_W(t). \tag{8.9}$$

Under heavy either snow melting or precipitation, a volley water supply is possible, so that its entire volume W_{in} enters the field within a short time. Since the infiltration process in comparison with the free water movement has much longer characteristic time, the ground bed is only minimally wetted during this time. In such a situation, it is justified to assume that initially a water layer instantly formed on the field surface with the height

$$H_0 = \frac{W_{in}}{\Omega}$$

and only then did it start to filter down into the soil column.

The solution of the hydrodynamic compartment allows to strictly find the relationship between H_W and Z_1 under the condition $w(0)=0$ (first stage)

$$H_W(t) + n_W Z_1 = w(t), \tag{8.10}$$

and also, the head distribution in the expanding filtration flow region

$$h_1(z,t) = H_W(t) - \left[H_W(t) + Z_1(t) + \psi_w\right] \int_z^0 \frac{d\eta}{k_1(\eta,t)} \left[\int_{-Z_1(t)}^0 \frac{dz}{k_1(z,t)}\right]^{-1}. \tag{8.11}$$

From (8.11) follows the expression for the variable filtration rate

$$V(t) = \left[H_W(t) + Z_1(t) + \psi_w\right] \left[\int_{-Z_1(t)}^0 \frac{dz}{k_1(z,t)}\right]^{-1}. \tag{8.12}$$

The kinematic condition (8.2), allowing for the relation of k with S and (8.10), takes the following form

$$n_W \frac{dZ_1}{dt} = \left[(1 - n_W)\, Z_1 + w(t) + \psi_w\right] \left[\int_{-Z_1(t)}^0 \frac{dz}{k(S_1(z,t))}\right]^{-1}. \tag{8.13}$$

To derive the calculation expressions, it is necessary to solve equation (8.13) under condition (8.8). And here, first of all, it is required to calculate the integral contained in (8.13), which is realistic to do only by preliminarily establishing the function - concentration $S_i(z,t)$ by solving the clarification compartment. It should be emphasized that the application of analytical methods to solve the mathematical problems formulated above is extremely problematic because of the close relation existing between both compartments. This relation is especially significant at a variable filtration rate and necessitates the joint solution of the compartments. Only approximate solution of the set problems is possible in this situation. The only exception is the so-called essentially limiting case, which will be analyzed separately in Section 8.3 and for which an exact solution was obtained.

As a rule, the filtration rate in the infiltration field changes within small limits during one technological cycle, because, firstly, there are no conditions for the accumulation of a significant amount of water in the surface layer, secondly, the effect of the capillary forces is generally weak, thirdly, the absorption capacity of the ground bed is often insignificant. Sometimes sharp changes of the flow characteristics up to unlimitedly large values observed at the mathematical modeling are explained by the non-physicality of the mathematical problem formulation. For example, when pumping fluid from a reservoir at a constant head, the initial and boundary conditions usually do not agree at the initial moment (the influence of the wellbore is not discussed in this case). In the problem under consideration, the same consequences result from the assumption $\psi_w>0$, which causes an infinitely large head difference at $t=0$. Indeed, the initial rate V^0 will be finite according to (8.13) if the condition is satisfied

$$\lim_{t\to 0}\frac{w(t)+\psi_w}{t} < \infty .$$

Therefore, in order V^0 to have a reasonable value from the physical point of view, dw/dt must tend at $t\to 0$ to a constant positive value, and ψ_w must tend to 0. However, the flow rate stabilizes quickly in the noted cases and then changes smoothly. Both in the presence of a crisis phenomenon due to the incorrect formulation of the problem and in the absence of such a phenomenon, it is convenient to resort to its parameterization. As a parameter it is reasonable to use the average rate V_a for the calculation period. Since the duration of such a period T can in principle be any, V_a actually turns out to be a function of time. However, when performing differentiation and integration operations, the quantity V_a decreasing over time can be treated as a constant. Thus, we introduce the functional parameter V_a for the period $[0,T]$ according to (5.11). Then integration of condition (8.2) within the range from 0 to T gives

$$n_W Z_1(T) = V_a T, \tag{8.14}$$

so that from (8.10) follows

$$H(T) = w(T) - n_W Z_1(T) = w(T) - V_a T. \tag{8.15}$$

Taking into account (8.14), (8.15) the parameter V_a for an arbitrary point in time should be found by fitting from the equation.

$$V_a t = \int_0^t \left[w(\xi) + \psi_w + \frac{1-n_W}{n_W} V_a \xi \right] \left[\int_{-\frac{V_a}{n_W}\xi}^{0} \frac{dz}{k_1(z,\xi,V_a)} \right]^{-1} d\xi. \tag{8.16}$$

It is easier to find an approximate solution if $V(0)=V^0<\infty$. Then it is justified to use an iterative procedure, and the first approximation for the desired $V(t)$ is V^0. Thus, the first approximation for Z_1 will be

$$Z_1^{(1)}(t) = \frac{V^0}{n_W}t \tag{8.17}$$

and further V and Z_1 are refined by finding their second approximations. According to (8.12)

$$V^{(2)}(t) = \left[w(t) + \psi_w + \frac{1-n_W}{n_W}V^0 t\right]\left[\int_{-\frac{V^0}{n_W}t}^{0}\frac{dz}{k(S_1(z,t;V^0))}\right]^{-1}, \tag{8.18}$$

and equation (8.13) is transformed to the form with respect to $Z_1^{(2)}$

$$n_W\frac{dZ_1^{(2)}}{dt} = \left[w(t) + \psi_w + \frac{1-n_W}{n_W}V^0 t\right]\left[\int_{-\frac{V^0}{n_W}t}^{0}\frac{dz}{k(S_1(z,t;V^0))}\right]^{-1}. \tag{8.19}$$

At the parameterization of the considered problem, when the initially unknown functional parameter V_a is involved, the form of the equations relatively $V(t)$, $Z_1(t)$ turns out to be identical to (8.18), (8.19) after replacing in them V^0 by V_a. Thus, there is no need to find two approximations of the solution, but we have to calculate the corresponding value of V_a from (8.16) at each computational step. Incidentally, it is easy to formally write down the solution of equation (8.19) under condition (8.8) and, for example, it is expressed relatively Z_1 by the dependence

$$Z_1(t) = \frac{1}{n_W}\int_0^t\left[w(t) + \psi_w + \frac{1-n_W}{n_W}V_a\xi\right]\left[\int_{-\frac{V_a\xi}{n_W}}^{0}\frac{dz}{k(S_1(z,\xi;V_a))}\right]^{-1}d\xi. \tag{8.20}$$

In addition, it is easy to calculate, the amount of impurity that has entered the ground bed at a given rate $V(t)$ by the current point in time w_{S1}, viz,

$$w_{S1}(t;V_a) = C_0\int_0^t V(\xi;V_a)d\xi =$$

$$= C_0\int_0^t\left[w(\xi) + \psi_w + \frac{1-n_W}{n_W}V_a\xi\right]\left[\int_{-\frac{V_a\xi}{n_W}}^{0}\frac{dz}{k(S_1(z,\xi;V_a))}\right]^{-1}d\xi.$$

To perform specific calculations, it is necessary to solve the clarification compartment, setting the quantity $S_1(z,t;\ V_a)$. The specified compartment includes at any stage, firstly, equation (8.4), secondly, one of the kinetic equations (8.5), (8.6), finally, the boundary condition with respect to C_i and the initial condition with respect to S_i. In the presence of initial water on the ground surface (first and second stages, i=1,2), it is assumed that the suspended solids concentration in it, and hence at z=0 does not change over time, so that

$$z = 0, \quad C_i = C_0 = const \tag{8.21}$$

There should already be a certain amount of the deposit in the filter bed at the beginning of any technological cycle, except for the first one. It is natural that this amount will increase after each new cycle. Therefore, in the general case for arbitrary stages and cycles it is reasonable to assume the following initial condition

$$t = t_{i-1}, \quad S_i = S_i^0(z), \text{(8. 22)}$$

where t_{i-1} is the end time of the previous stage (the time is counted from the beginning of the cycle under consideration, to which corresponds t=0). When putting the infiltration field into operation, in fact $S_1^0 = 0$, and in other cases $S_1^0 > 0$.

Further the solutions of the above compartment are presented for the first and third forms of mass exchange kinetics described by the dimensional equations (8.5) and (8.6) (the second type is a special case of the third type). Thus, if equation (8.5) is adequate to the actual conditions of filtration, the deposited impurity particles at the first stage are distributed along the depth of the ground bed as follows

$$z = V_a \int_{S_0(t)}^{S_1} \frac{d\varsigma}{\varsigma \lambda(\varsigma)} \tag{8.23}$$

where the concentration of impurity at the upper boundary of the bed S_0 should be found from the equation

$$t = \frac{1}{C_0} \int_{S_0^0}^{S_0} \frac{dS}{\lambda(S)}, \tag{8.24}$$

S_0^0 is the inlet initial concentration (t=0). When deriving (8.23), it was assumed that the original system of equations is valid at the beginning of calculations. This assumption is strictly satisfied at $S_1{}^0 = 0$. At the saturation front, the concentrations of the suspended (C_1) and deposited (S_1) matter are related by the relationship

$$C_1(-Z_1, t; V_a) = C_0 \frac{S_1(-Z_1, t; V_a)}{S_0(t; V_a)}. \tag{8.25}$$

In a particular case

$$\lambda = \lambda_0(S_m - S_i), \tag{8.26}$$

expressions (8.23) - (8.26) are considerably simplified. Here λ_0 is the reduced filtration coefficient, S_m is solids take-up capacity in relation to the specified particles. First of all, it turns out that

$$S_0(t) = S_m(1 - e^{-\lambda_0 C_0 t}) \tag{8.27}$$

and then (8.23) becomes

$$z = \frac{V_a}{\lambda_0 S_m} \ln \frac{S_1[S_m - S_0(t)]}{S_0(t)(S_m - S_1)} \tag{8.28}$$

or explicitly

$$S_1(z, t; V_a) = \frac{S_m S_0(t)}{S_0(t) + [S_m - S_0(t)] e^{-\frac{\lambda_0 S_m}{V_a} z}}, \tag{8.29}$$

and also

$$C_1(-Z_1, t; V_a) = \frac{C_0 S_m}{S_0(t) + [S_m - S_0(t)] e^{\frac{\lambda_0 S_m}{V_a} Z_1}} \tag{8.30}$$

In the case of the third form of mass exchange kinetics and in the presence of the initial contamination of the ground bed, the change in the concentration of the deposited particles over time and depth ($-Z_1 \le z \le 0$) is described by the dependence

$$\begin{aligned} S_1(z, t; V_a) = {} & S_1^0(z)\, e^{-\beta_V V_a^q t} + \alpha_V V_a^{l-1} e^{\alpha_V V_a^{l-1} z} C_0 \\ & \times \int_0^{t + \frac{n_W}{V_a} z} e^{-\beta_V V_a^q \xi} I_0\left(2\sqrt{-\alpha_V \beta_V V_a^{l+q-1} z \xi}\right) d\xi + \alpha_V \beta_V V_a^{l+q-1} e^{-\beta_V V_a^q t} \\ & \times \int_z^0 S_1^0(\eta)\, e^{(\alpha_V V_a^{l-1} - n_W \beta_V V_a^{q-1})(z-\eta)} \int_{\frac{n_W}{V_a}(\eta - z)}^{t} I_0\left(2\sqrt{\alpha_V \beta_V V_a^{l+q-1}(\eta - z)\xi}\right) d\xi\, d\eta. \end{aligned} \tag{8.31}$$

Thus, at initially clean bed (first run)

$$S_1(z, t; V_a) = \alpha_V V_a^{l-1} e^{\alpha_V V_a^{l-1} z} C_0 \int_0^{t + \frac{n_W}{V_a} z} e^{-\beta_V V_a^q \xi} I_0\left(2\sqrt{-\alpha_V \beta_V V_a^{l+q-1} z \xi}\right) d\xi. \tag{8.32}$$

In this case the suspended solids concentration within the saturation zone will be as follows

$$\begin{aligned} C_1(z, t; V_a) = {} & C_0 e^{\alpha_V V_a^{l-1} z} \left[e^{-\beta_V V_a^q \left(t + \frac{n_W}{V_a} z\right)} I_0\left(2\sqrt{\alpha_V \beta_V V_a^{l+q-1} z\left(-\frac{n_W}{V_a} z - t\right)}\right) + \right. \\ & \left. + \beta_V V_a^q \int_0^{t + \frac{n_W}{V_a} z} e^{-\beta_V V_a^q \xi} I_0\left(2\sqrt{-\alpha_V \beta_V V_a^{l+q-1} z \xi}\right) d\xi \right] + \beta_V V_a^q \times \\ & \times \int_z^0 S_1^0(\eta) e^{\alpha_V V_a^{l-1}(z-\eta)} e^{-\beta_V V_a^q \left(t + \frac{n_W}{V_a} z - \frac{n_W}{V_a} \eta\right)} I_0\left(2\sqrt{\alpha_V \beta_V V_a^{l+q-1}(\eta - z)\left(t + \frac{n_W}{V_a} z - \frac{n_W}{V_a} \eta\right)}\right) d\eta. \end{aligned} \tag{8.33}$$

If the filtration rate during the calculation period is constant or changes slightly ($V \approx const = V_a$), which is often characteristic of the problems of the intensive wetting of soil columns with solution or suspension, then the mass exchange kinetics is justified to refer to the second type. Then the following strict dependence with the constant mass exchange coefficients α, β is valid for S_1

$$S_1(z,t;V_a) = S_1^0(z)\, e^{-\beta t} + \frac{\alpha C_0}{V_a} e^{\frac{\beta}{V_a}z} \int_0^{t+\frac{n_W}{V_a}z} e^{-\beta\xi} I_0\left(2\sqrt{-\frac{\alpha\beta}{V_a} z\xi}\right) d\xi +$$
$$+\frac{\alpha\beta}{V_a}\, e^{-\beta t} \int_z^0 S_1^0(\eta)\, e^{\frac{\alpha - n_W\beta}{V_a}(z-\eta)} \cdot \int_{\frac{n_W}{V_a}(\eta - z)}^{t} I_0\left(2\sqrt{\frac{\alpha\beta}{V_a}(\eta - z)\xi}\right) d\xi.$$

It should be noted that dependences (8.28) - (8.33), although allowing to calculate the clarification process with a high accuracy, are still approximate. However, it is possible to derive a rigorous expression similar to (8.27) for the concentration of the deposited particles on the surface of the ground bed ($z=0$) as a function of $V(t)$, which corresponds to both the first and the second stages

$$S_0(t) = \alpha_V C_0 \int_0^t V^l(\xi)\, \exp\left[-\beta_V \int_\xi^t V^q(\varsigma)\, d\varsigma\right] d\xi. \tag{8.34}$$

Thus, expression (8.33) can be useful in estimating errors due to the application of various approximate techniques in solving the problems of the joint filtration flow and clarification, in particular, dynamic averaging of the filtration rate.

The dependencies presented above for S_1 already allow to carry out all the necessary predictive and technological calculations with respect to the first stage of the filtration flow process. Now thanks to (8.29), (8.31) it is possible to reliably determine changes over time of the filtration rate, water level on the surface, head losses, saturation front propagation, and what is especially important for the estimation of the prospects in the infiltration field utilization, progressive clogging of the filtering ground bed. But the direct calculations of the hydraulic and concentration characteristics are preceded by the determination of the values of the parameter V_a, corresponding to the calculation moments of time, which are set by fitting from equation (8.16) after the substitution of a suitable representation $k(S_1)$ into it. And only then at a known set of the values of V_a it is easy to calculate Z_1 in dynamics according to (8.20), H_W at Z_1 obtained according to (8.10), V at Z_1, H_W obtained according to (8.12), head losses in the saturation zone Δh determined according to the formula

$$\Delta h = h(0,t) - h(-Z_1,t) = \tau(t) + \psi_w + \frac{1-n_W}{n_W} t. \tag{8.35}$$

Finally, it is possible to specify the deposit distribution along the depth of the indicated zone.

Since the contamination of the ground bed will continue in the future during the second and third stages, the fixation of the physical and mechanical state of the field at the end of the first stage is of great importance. It serves as a starting point for subsequent calculations. Therefore, it is necessary to calculate the maximum surface water level $H_W^0 = H_W(t_1)$, depth $Z^0 = Z_1(t_1)$ and the final profile for the first stage $S_1^0(z) = S_1(z,t_1)$ by a known moment in time (the end of the first stage). But preliminary the value of V_a is found corresponding to t_1 from (8.16). Incidentally, the change in H_W during the first stage can be non-monotonic, if water supply to the field is sharply reduced but not stopped.

8.2. SECOND AND THIRD STAGES

The second stage starts from the moment when water inflow to the infiltration field stops, i.e. at $t=t_1$. The mentioned stage ends at the moment in time t_2, when all the water accumulated on the surface of the ground infiltrates into it. Obviously, the time t_2 must be calculated in contrast to t_1. Thus, the formulation of the hydrodynamic (filtration flow) and clarification compartments for the second stage with respect to the sought quantities h_2, Z_2, V, S_2, first of all, includes general elements (8.1) - (8.7), as well as the equation of water balance on the field in the absence of its inflow from outside

$$\frac{dH_W}{dt} + n_W \frac{dZ_2}{dt} = 0 \tag{8.36}$$

and a series of initial conditions

$$t = t_1, \quad H_W = H_W^0; \quad Z_2 = Z^0; \quad S_2 = S_1(z, t_1) = S_2^0(z). \tag{8.37}$$

From (8.36), (8.37) follows a linear relationship between H_W and Z_2

$$H_W(t) = H_W^0 + n_W\left[Z^0 - Z_2(t)\right] \tag{8.38}$$

The solution of the hydrodynamic compartment is first of all expressed by the dependence for the piezometric head

$$h_2(z,t) = W_{in} - n_W Z_2(t) - \left[W_{in} + \psi_w + (1 - n_W)\, Z_2(t)\right] \int_z^0 \frac{d\eta}{k_2(\eta,t)} \left[\int_{-Z_2(t)}^0 \frac{dz}{k_2(z,t)}\right]^{-1} \tag{8.39}$$

Consequently, the filtration rate decrease during the second stage is described by the following function

$$V(t) = \left[W_{in} + \psi_w + (1 - n_W)\, Z_2(t)\right] \left[\int_{-Z_2(t)}^0 \frac{dz}{k_2(z,t)}\right]^{-1}. \tag{8.40}$$

Then taking into account (8.40), the equation similar to (8.13) follows from condition (8.2) with respect to Z_2

$$n_W \frac{dZ_2}{dt} = \left[W_{in} + \psi_w + (1 - n_W)\, Z_2(t)\right] \left[\int_{-Z_2(t)}^0 \frac{dz}{k_2(z,t)}\right]^{-1}. \tag{8.41}$$

As in the first stage, the filtration rate has to be averaged, but now over the period $[t_1,T]$, so that

$$V_a = \frac{1}{T - t_1} \int_{t_1}^{T} V(t) dt \tag{8.42}$$

and from condition (8.2) it follows

$$n_W \left[Z_2(T) - Z^0 \right] = V_a \left(T - t_1 \right), \tag{8.43}$$

and according to equation (8.38)

$$H_W(T) = H_W^0 - V_a \left(T - t_1 \right). \tag{8.44}$$

The equation relatively V_a is derived by integrating (8.40) from t_1 to t and taking into account (8.43), (8.44) will be

$$V_a \left(t - t_1 \right) = \int_{t_1}^{t} \left[W + \psi_w + \frac{1 - n_W}{n_W} \left(V_a \xi - V_a t_1 + n_W Z^0 \right) \right] \left[\int_{-Z_2 - \frac{V_a}{n_W}(\xi - t_1)}^{0} \frac{dz}{k_2(z, \xi)} \right]^{-1} d\xi. \tag{8.45}$$

Now to determine the specific values of S_2, k_2, it is necessary to solve the clarification compartment accepting a suitable form of mass exchange kinetics.

In the case of kinetics of the first form, the system of equations (8.4), (8.5) is reduced to a single equation with respect to S_2

$$\frac{V_c}{\lambda(S_2)} \frac{\partial S_2}{\partial z} - S_2 = \left[\frac{V_a}{\lambda(S_1)} \frac{\partial S_1}{\partial z} - S_1 \right]_{t = t_1} \tag{8.46}$$

As it is easy to see from (8.23), the right part of equation (8.46) is equal to 0. Therefore, the solution of the latter under the general boundary condition for both stages, viz.

$$z = 0, \quad S_i = S_0(t); \quad i = 1, 2 \tag{8.47}$$

by analogy with (8.23) will be expressed by such inverse function

$$z = \int_{S_0(t)}^{S_2} \frac{d\zeta}{\zeta \lambda(\zeta)} \tag{8.48}$$

where $S_0(t)$ is calculated in the interval $[t_1, t_2]$ by fitting from (8.24). The concentration of the suspended solids C_2 in the saturation zone and at its lower boundary (front) is determined in the same way as at the first stage. In particular, formula (8.25) remains valid for the second stage – it is enough just to replace C_1, S_1 by C_2, S_2. In this case, V_a should be found from equation (8.45).

In addition to (8.26) let

$$\lambda = \lambda_0 (S_m - S_i)^2 \tag{8.49}$$

where λ_0 is the constant reduced filtration coefficient. Then it follows from (8.48)

$$z = \frac{V_a}{\lambda_0 S_m^2}\left[\frac{S_2}{S_m - S_2} - \frac{S_0}{S_m - S_0} + \ln\frac{(S_m - S_0)\, S_2}{S_0\,(S_m - S_2)}\right]. \tag{8.50}$$

and to solve (8.50) relatively S_2 impossible. If

$$\lambda = \lambda_0 \sqrt{S_m - S_2}, \tag{8.51}$$

then $z(S_2)$ will be

$$z = \frac{V_a}{\lambda_0 \sqrt{S_m}} \ln\left(\frac{\sqrt{S_m - S_0} + \sqrt{S_m}}{\sqrt{S_m - S_0} - \sqrt{S_m}} \cdot \frac{\sqrt{S_m - S_2} - \sqrt{S_m}}{\sqrt{S_m - S_2} + \sqrt{S_m}}\right) \tag{8.52}$$

or in an equivalent form

$$S_2(z,t) = S_m\left\{1 - \left[\frac{1 + \phi_*(z,t)}{1 - \phi_*(z,t)}\right]^2\right\}, \tag{8.53}$$

where

$$\phi_*(z,t) = \frac{\sqrt{S_m - S_0(t)} - \sqrt{S_m}}{\sqrt{S_m - S_0(t)} + \sqrt{S_m}}\, e^{\frac{\lambda_0\sqrt{S_m}}{V_a}z}.$$

The quantity $S_0(t)$ at an arbitrary exponent δ in the power function $\lambda(S_i)$ will be

$$S_0(t) = S_m - \sqrt[1-\delta]{(S_m - S_0^0)^{1-\delta} + C_0(1-\delta)t}. \tag{8.54}$$

If the third form of kinetics takes place, the concentration of the deposited particles changes in the saturation zone ($-Z_2 \leq z \leq 0$) over time and depth according to the following expression

$$\begin{aligned} S_2(z,t;V_a) &= S_2^0(z)\, e^{-\beta_V V_a^q (t-t_1)} + \alpha_V V_a^{l-1} e^{\alpha_V V_a^{l-1} z} C_0 \\ &\times \int_{t_1}^{t} e^{-\beta_V V_a^q \xi} I_0\left(2\sqrt{\alpha_V \beta_V V_a^{l+q-1} z\xi}\right) d\xi + \alpha_V \beta_V V_a^{l+q-1} e^{-\beta_V V_a^q (t-t_1)} \\ &\times \int_{z}^{0} S_2^0(\eta)\, e^{\alpha_V V_a^{l-1}(z-\eta)} \int_{t_1}^{t} I_0\left(2\sqrt{\alpha_V \beta_V V_a^{l+q-1}(\eta - z)\xi}\right) d\xi\, d\eta. \end{aligned} \tag{8.55}$$

At that, the increase in the suspended matter concentration at the saturation front is described by the dependence

$$\begin{aligned} C_2(-Z_2, t; V_a) &= C_0 e^{\alpha_V V_a^{l-1} Z_2}\Big[e^{-\beta_V V_a^q (t-t_1)} I_0\big(2\sqrt{\alpha_V \beta_V V_a^{l+q-1}(t-t_1)}\big)+ \\ &+\beta_V V_a^q \int_{t_1}^{t} e^{-\beta_V V_a^q \xi} I_0\big(2\sqrt{\alpha_V \beta_V V_a^{l+q-1} Z_2 \xi}\big)d\xi\Big]+ \\ &+\beta_V V_a^{q-1} \int_{-Z_2}^{0} S_2^0(\eta)\, e^{-\alpha_V (Z_2+\eta)} I_0\big(2\sqrt{\alpha_V \beta_V V_a^{l+q-1}(Z_2+\eta)(t-t_1)}\big)d\eta. \end{aligned} \tag{8.56}$$

In the case of the second form kinetics

$$\begin{aligned} S_2(z,t;V_a) &= S_2^0(z)\, e^{-\beta_V (t-t_1)} + \frac{\alpha_V C_0}{V_a} e^{\frac{\alpha_V z}{V_a}} \int_{t_1}^{t} e^{-\beta_V \xi} I_0\left(2\sqrt{-\frac{\alpha_V \beta_V}{V_a} z\xi}\right)d\xi+ \\ &+\frac{\alpha_V \beta_V}{V_a} e^{-\beta_V (t-t_1)} \int_{z}^{0} S_2^0(\eta)\, e^{\frac{\alpha_V}{V_a}(z-\eta)} \int_{t_1}^{t} I_0\left(2\sqrt{\frac{\alpha_V \beta_V}{V_a}(\eta - z)\xi}\right)d\xi d\eta. \end{aligned} \tag{8.57}$$

Now we can state that the basis is fully prepared for the theoretical analysis of the above-mentioned processes at the second stage. It is obvious that the calculations of all the required characteristics here can be performed from the point in time t_1 and up to the point of the end of this stage t_2, which corresponds to the service life of the surface layer and should be determined in accordance with (8.39) by fitting from the equation

$$Z_2(t_2, V_a) = Z_2^0 + \frac{H_W^0}{n_W} \tag{8.58}$$

The final (third stage, Figure 1b) is of less interest in analyzing the ground bed clogging as a result of passing through it a relatively small amount of suspension, when all the water entering the field is in the bed (except for possible losses to surface runoff). Indeed, during this stage there is actually the secondary treatment of already the partially clarified water. Therefore, it is sufficient to adjust the amount of the deposit accumulated in the wetted part of the bed after the end of the second stage. In addition to the concentration profile of the deposited particles $S_2(z,t_2)$, it is also necessary to know the distribution of the suspended solids along the depth of the saturated zone at the same time, i.e. the function $C_2(z,t_2)$. From the solution obtained above it will be in the case of the kinetics of the third form

$$\begin{aligned} C_2(z,t_2) &= C_0 e^{\alpha_V V_a^{l-1} z}\Big[e^{-\beta_V V_a^q t_2} I_0\big(2\sqrt{-\alpha_V \beta_V V_a^{l+q-1} z t_2}\big)+ \\ &+\beta_V V_a^q \int_{t_1}^{t_2} e^{-\beta_V V_a^q \xi} I_0\big(2\sqrt{-\alpha_V \beta_V V_a^{l+q-1} z\xi}\big)d\xi\Big]+ \\ &+\beta_V V_a^{q-1} \int_{z}^{0} S_1^0(\eta)\, e^{\alpha_V (z-\eta)} I_0\big(2\sqrt{-\alpha_V \beta_V V_a^{l+q-1}(z-\eta)t}\big)d\eta \end{aligned} \tag{8.59}$$

Since the porosity of the clogged ground bed by the end of the second stage changes with depth and amounts to

$$n(z,t_2) = n_0 - \gamma_a S_2(z,t_2)$$

then, neglecting the insignificant transfer of the dispersed matter during the third stage and the time of its immobilization, it is easy to approximately calculate the corresponding addition ΔS to the quantity S_2 by the formula

$$\Delta S(z) = \left[n_0 - \gamma_a S_2(z,t_2)\right] C_2(z,t_2) \tag{8.60}$$

Then it is proposed to use the following formula for further estimations of the flow capacity of the ground bed with the newly formed deposit

$$S(z) = S_2(z,t_2) + \Delta S(z) . \tag{8.61}$$

Certainly, the assumptions made regarding the behavior of the suspended solids in the third stage introduce the additional errors in the subsequent estimates, which, however, should not be serious.

8.3. LIMITING CASE

Special attention in the theoretical studies of the suspension filtration with a variable rate through a dewatered column was paid, as in Section 5.4, to the special clarification case (l=q=1, the third form of kinetics), since it does not contradict actual conditions and it is possible to obtain rigorous solutions of the corresponding mathematical problems for both the first and the second stages. Thus, the above-mentioned solutions can serve as a reference for estimating the errors introduced into the calculations due to the rate averaging and other approximation techniques.

Preliminarily, the dimensional independent variable t is replaced by the cumulative variable τ by analogy with Section 5.1 as follows

$$\tau = \int_0^t V(\xi) d\xi. \tag{8.62}$$

Then the formulation of the problem of the intensive wetting of a ground bed by water with a high content of dispersed impurity at the first stage, first of all, includes the system of equations

$$\frac{dH}{d\tau} + n_W \frac{dZ_1}{d\tau} = \frac{dw}{d\tau}, \tag{8.63}$$

$$\frac{\partial C_1}{\partial z} - \frac{\partial S_1}{\partial \tau} = 0, \tag{8.64}$$

$$\frac{\partial S_1}{\partial \tau} = \alpha_V C_1 - \beta_V S_1, \tag{8.65}$$

operator of the boundary and initial conditions

$$z = 0, \quad h = H_W(\tau) \; ; \tag{8.66}$$

$$z = -Z_1(\tau), \quad h = -Z_1(\tau) - \psi_w; \quad n_W \frac{dZ_1}{d\tau} = 1 ; \tag{8.67}$$

$$\tau = 0, \quad H_W = Z_1 = 0 \tag{8.68}$$

$\partial C_1/\partial\tau$ is not taken into account in equation (8.64) due to its smallness.

The system (8.63) - (8.65) is joined by equations (8.1), (8.4) and conditions (8.21), (8.22). The type of the dependence between $H_W(\tau)$ and $Z_1(\tau)$ remains the same, i.e., formally (8.10) is valid. The solution of equation (8.1) under conditions (8.3), (8.9) now gives the following expression for the head h_1

$$h_1(z,\tau) = H_W(\tau) - \left[H_W(\tau) + Z_1(\tau) + \psi_w\right] \int_z^0 \frac{d\eta}{f_k(\eta,\tau)} \left[\int_{-Z(\tau)}^0 \frac{dz}{f_k(z,\tau)}\right]^{-1}. \tag{8.69}$$

From (8.67) follows that

$$n_W Z_1 = \tau. \tag{8.70}$$

As a result of solving system (8.64), (8.65) under conditions (8.21), (8.22), the following exact formula is derived for S_1

$$S_1(z,\tau) = \alpha_V C_0 e^{\alpha_V z} \int_0^\tau e^{-\beta_V\varsigma} I_0\left(2\sqrt{-\alpha_V\beta_V z\varsigma}\right) d\varsigma. \tag{8.71}$$

It follows from (8.69) that

$$V(\tau) = \frac{d\tau}{dt} = \left[H_W(\tau) + Z_1(\tau) + \psi_w\right]\left[\int_{-Z_1(\tau)}^0 \frac{dz}{k(S_1(z,\tau))}\right]^{-1}. \tag{8.72}$$

Integration of (8.72) in the range from 0 to τ and from 0 to t gives the desired relationship between t and τ

$$t = \frac{n_W}{k_0} \int_0^\tau \frac{\displaystyle\int_{-\varsigma/n_W}^0 \frac{dz}{f_k(S_1(z,\varsigma))}}{(1-n_W)\varsigma + \left[w(\varsigma) + \psi_w\right] n_W} d\varsigma. \tag{8.73}$$

Finally, the head losses within the saturation zone are as follows

$$\Delta h(\tau) = h(0,\tau) - h\left(-Z_1(\tau),\tau\right) = \frac{(1-n_W)}{n_W}\tau + w(\tau) + \psi_w. \tag{8.74}$$

Conclusively

$$V(\tau) = k_0\left[w(\tau) + \psi_w + \frac{1-n_W}{n_W}\tau\right]\left[\int_{-\tau/n_W}^0 \frac{dz}{f_k(z,\tau)}\right]^{-1}, \tag{8.75}$$

$$h_1(z,\tau) = w(\tau) - \tau + \left[w(\tau) + \psi_w + \frac{1-n_W}{n_W}\tau\right]\int_z^0 \frac{d\eta}{f_k(\eta,\tau)} \cdot \left[\int_{-\tau/n_w}^0 \frac{dz}{f_k(z,\tau)}\right]^{-1} \tag{8.76}$$

From (8.10), (8.75) it is obvious that the surface layer will increase if

$$\frac{dw}{d\tau} > \frac{1}{n_W}$$

The formation of the filtration flow and the specified layer will begin under the condition that $V^0 > k^0$. According to (8.75)

$$V^0 = \lim_{\tau\to 0} V(\tau) = k_0\left[n_W \lim\frac{w(\tau)+\psi_w}{\tau} + 1 - n_W\right]$$

It follows that the following condition must be satisfied

$$\lim_{\tau\to 0}\frac{w(\tau)+\psi_w}{\tau} > 1$$

Thus, V^0 can formally turn out to be infinitely large. Thus, accepting the constant $\psi_w>0$ leads to a result that contradicts the physics of the infiltration phenomenon. In the subsequent quantitative analysis, however, $\psi_w=0$ and the assumption of the linearity of $w(\tau)$ are assumed, which guarantees the finiteness of V during any calculation period.

To determine the empirical dependence $w(\tau)$, characterizing the infiltration process, it is necessary to have reliable data, at least N synchronous measurements of the specific volume w and height H_W for a number of time moments ($t^{(u)}$, $u=1,2,\ldots,N$). Then the value $\tau^{(u)}$ corresponding to $t^{(u)}$ is found in the following way: an arbitrary (j- th) time interval is selected and the average filtration rate within its limits is calculated at known values of $w^{(j-1)}$, $w^{(j)}$, $H_w^{(j-1)}$, $H_w^{(j)}$ by the formula

$$V^{(j)} = \frac{w(t^{(j)}) - w(t^{(j-1)}) - H_W(t^{(j)}) + H_W(t^{(j-1)})}{t^{(j)} - t^{(j-1)}}$$

Hence, in accordance with (8.62)

$$\tau^{(j)} = \sum_{u=1}^{j} V^{(u)}(t^{(u)} - t^{(u-1)})$$

and will be $t_0=0$ at $u=1$. The desired approximation function $w(\tau)$ is selected according to the set of N pairs of the values $w(t^{(u)})$, $\tau^{(u)}$ obtained in this way.

At the second stage of the filtration flow process, the behavior of the sought characteristics S_2, H_W, Z_2, h_2 is described by a similar model except for the equation describing the water balance on the infiltration field,

$$\frac{dH_{W2}}{d\tau} + n_W\frac{dZ_2}{d\tau} = 0, \tag{8.77}$$

and the initial conditions

$$\tau = \tau_1, \quad S_2 = S(z, \tau_1) = S_2^0(z) \ ; \tag{8.78}$$

$$H_{W2} = H_W(\tau_1) = H_W^0 \ ; \tag{8.79}$$

$$Z_2 = Z(\tau_1) = Z^0 \ ; \tag{8.80}$$

where $\tau_1 = \int_0^{t_1} V(\xi)d\xi$; the expressions for $S_2^0(z)$, H_W^0, Z^0 follow, respectively, from (8.71), (8.38), (8.70), so that

$$H_W^0 = w(\tau_1) - \tau_1, \quad Z^0 = \frac{\tau_1}{n_W}. \tag{8.81}$$

Integration of (8.77) taking into account (8.81) gives

$$H_{W2}(\tau) = H_W^0 - n_W[Z_2(\tau) - Z^0]. \tag{8.82}$$

It follows from (8.67) that

$$Z_2(\tau) = Z^0 + \frac{\tau - \tau_1}{n_W}, \tag{8.83}$$

and, therefore,

$$H_{W2}(\tau) = H_W^0 + \tau_1 - \tau. \tag{8.84}$$

As a result of solving (8.1) under conditions (8.66), (8.67) it is obtained

$$h_2(z, \tau) = H_W^0 + \tau_1 - \tau + \\ + \left[H_W^0 + Z^0 + \psi_w + \frac{1 - n_W}{n_W}(\tau - \tau_1)\right] \times \int_z^0 \frac{d\eta}{f_k(\eta, \tau)} \left[\int_{-Z_2(\tau)}^0 \frac{dz}{f_k(z, \tau)}\right]^{-1}. \tag{8.85}$$

System (8.65), (8.66) under inhomogeneous conditions and a constant filtration rate is strictly solved in Chapter 1. In the case of a variable rate the strict expression for the concentration S_2 will be

$$S_2(z, \tau) = S^0(z)e^{-\beta_V(\tau-\tau_1)} + \alpha_V e^{\alpha_V z} \int_{\tau_1}^{\tau} e^{-\beta_V \varsigma} I_0\left(2\sqrt{-\alpha_V\beta_V z\varsigma}\right) d\varsigma + \\ + \alpha_V\beta_V e^{-\beta_V(\tau-\tau_1)} \int_0^z e^{\alpha_V(z-\eta)} S^0(\eta) \int_{\tau_1}^{\tau} I_0\left(2\sqrt{\alpha_V\beta_V(\eta - z)\varsigma}\right) d\varsigma d\eta. \tag{8.86}$$

Incidentally, in this case the suspended matter in the saturation zone will be distributed according to the following dependence

$$C_2(z,\tau) = e^{\alpha_V z}\left[e^{-\beta_V(\tau-\tau_1)} I_0\left(2\sqrt{-\alpha_V\beta_V z(\tau-\tau_1)}\right) + \beta_V\int_{\tau_1}^{\tau} e^{-\beta_V\varsigma} I_0\left(2\sqrt{-\alpha_V\beta_V z\varsigma}\right)d\varsigma\right] + \\ + \beta_V e^{-\beta_V(\tau-\tau_1)}\int_0^z e^{\alpha_V(z-\eta)} S_0(\eta) I_0\left(2\sqrt{\alpha_V\beta_V(\tau-\tau_1)(\eta-z)}\right)d\eta. \tag{8.87}$$

The relationship between t and parameter τ is established from the problem

$$\frac{d\tau}{dt} = \left[H_W^0 + Z^0 + \psi_w + \frac{1-n_W}{n_W}(\tau-\tau_1)\right]\left[\int_{-Z_2(\tau)}^{0}\frac{dz}{f_k(S_2(z,\tau))}\right]^{-1}; \tag{8.88}$$

$$t = t_1, \quad \tau = \tau_1 \tag{8.89}$$

and is represented as the integral function

$$t = t_1 + \frac{n_W}{k_0}\int_{\tau_1}^{\tau}\frac{1}{n_W(H_W^0 + Z^0 + \psi_w) + (1-n_W)(\varsigma-\tau_1)}\int_{-Z^0-\frac{\varsigma-\tau_1}{n_W}}^{0}\frac{dz}{f_k(S_2(z,\varsigma))}d\varsigma. \tag{8.90}$$

The head losses in the saturation zone are as follows

$$\Delta h(\tau) = Z^0 + \psi_w + \frac{\tau-\tau_1}{n_W}. \tag{8.91}$$

It is proposed to calculate the change in the filtration rate over time at the second stage by the formula

$$V(\tau) = k_0\left[H_W^0 + Z^0 + \psi_w + \frac{1-n_W}{n_W}(\tau-\tau_1)\right]\left[\int_{-Z^0-\frac{\tau-\tau_1}{n_W}}^{0}\frac{dz}{f_k(S_2(z,\tau))}\right]^{-1}. \tag{8.92}$$

To determine the end time of the second stage t_2, first of all, we find the corresponding value τ_2 from the condition $H_{W2}(\tau_2)=0$, so that

$$\tau_2 = H_W^0 + \tau_1, \tag{8.93}$$

and then it calculates directly from (8.90) at $\tau=\tau_2$.

Finally, using the dependencies $S_1(z,\tau)$, $S_2(z,\tau)$ or $V(\tau)$ it is not difficult to calculate the total volume of the dispersed impurity W_S, which entered the ground bed on the field with area Ω to an arbitrary point in time t during the first or second stages. Indeed,

$$W_S(t) = \Omega C_0\int_0^t V(\xi)d\xi = W_S(\tau) = \Omega C_0\tau, \tag{8.94}$$

and, therefore,

$$w_S(\tau) = C_0\tau.$$

8.4. QUANTITATIVE ANALYSIS

The quantitative analysis of the processes of infiltration, clarification, jointly occurring under the conditions of the formation and depletion of the surface layer of water with dispersed impurities, was carried out on numerous examples. It was based on the materials of the theoretical studies, which are presented in the previous three Sections. At the same time, only the calculation expressions related to the kinetics of the third form were used. The subject of calculations were changes in the most important characteristics, first of all, over time actually during one technological cycle, namely, the filtration rate and also head H_W, hydraulic conductivity k_{in}, concentration S_0 on the ground surface (z=0). The distribution of the concentration S in the wetted part of the ground bed was also calculated up to the saturation front at fixed moments in time. Incidentally, the determination of the deposit accumulation in case of the repeated use of the infiltration field is significantly complicated due to the presence of high multiplicity integrals in the calculation relationships for the last cycle. To generalize the results of calculations, the following relative variables and parameters were introduced:

$$\overline{C}_i = C_i/C_0,\ \overline{S}_i = S_i/(n_W C_0),\ \overline{z} = z/M,\ \overline{t} = k_0 t/(n_W M),\ \overline{V} = V/k_0,$$

$$\overline{H}_W = H_W/M,\ \overline{Z}_i = Z_i/L,\ \overline{k} = k/k_0,\ \overline{w} = w/M,$$

$$\overline{\alpha}_V = M\alpha_V,\ \overline{\beta}_V = n_W M\beta_V,\ \overline{\psi}_w = \psi_w/M,\ \overline{\gamma}_a = n_W C_0\gamma_a/n_0.$$

Due to the large number of the model parameters and different significance for the above-mentioned processes, the single values were accepted for some of them, which are characteristic for the suspension filtration with quite high the suspended matter content through a thick, highly permeable mineral soil column: $\overline{\beta}_V = 0.1$, n_W=0.3, $\overline{\psi}_w = 0$, $\overline{S}^0 = 0$, m_1=1, m_2=3. The other parameters were alternately changed, and the new values differed from the reference ones by several tens of percent. Thus, the coefficients $\overline{\alpha}_V$, $\overline{\gamma}_a$ and the linear feed function $\overline{w}(\overline{t}) = \sigma\overline{t}$ (or $\overline{w} = \sigma\overline{\tau}$) were varied.

First of all, the limiting case of clarification (l=1) was thoroughly investigated on specific examples, since for it we managed to obtain the exact solutions of the mathematical problems for the first and second stages of the filtration flow process. The results of the calculations on their basis are presented in four Figures. The strict equations, dependencies from Section 8.3 and initial data of the basic example and three other examples with changed parameters were used.

In the first example, the above set of the parameter values was supplemented with the values $\overline{\alpha}_V = 100$, $\overline{\gamma}_a = 0.02$ and the linear function $\overline{w}(\overline{\tau}) = 1.25\overline{\tau}$. In the second example, the coefficient in the expression for $\overline{w}$, and hence the intensity of water supply to the field were increased by 20%; in the third example, the larger value of $\overline{\alpha}_V$ (150) was assumed; in the fourth example, $\overline{\gamma}_a$ increased in the same degree (0.03). The filtration flow process proceeded in two stages, with the second stage beginning at the moment $\overline{t}_1$ when the saturation front reached the middle of the wetted bed, i.e., at $\overline{Z}_1(\overline{t}_1) = 0.5$, and ending at the moment $\overline{t}_2$ (or $\overline{\tau}_2$), when the equality $\overline{Z}_2(\overline{t}_2) = 1$ is satisfied.

The graphs of the dependence $\overline{V}(t)$, which are shown in **Figure 2**, were calculated up to the complete depletion of the initial water layer on the ground surface ($\overline{t} \leq \overline{t}_2$, first and second stages) for all examples taking into account the relationship between $\overline{\tau}$ and $\overline{t}$ (8.73) at the first stage, by (8.92) at (8.90) at the second stage. Throughout the first stage, except for a short period of time at its beginning, the filtration rate decreased slowly due to the fact that the increase in the head losses in the saturation zone was largely compensated by the increase in the head at its upper boundary. Similarly, the rate V also changed at the suspension clarification on a rapid filter with the storage tank (Sections 5.3, 5.4). With the enhancement of the absorption capacity of the medium (the coefficient $\overline{\alpha}_V$ has been increased) and the activation of water binding by the deposited impurity ($\overline{\gamma}_a$ has been increased) the deposit formation accelerates, which caused a noticeably greater decrease in the rate by the end of the first stage (curves 3, 4). Obviously, more favorable conditions for the development of the filtration flow process should be formed with a higher water supply to the field (example 2, curve 1). In this case, the infiltration intensity increased by about 10%, and the duration of the first and second stages increased by 15%. Also, the initial water level on the ground surface and the depth of the saturated zone were calculated in dynamics for the basic and second examples. The corresponding graphs were calculated using dependences (8.10), (8.70) and (8.83), (8.84) for the first and second stages, respectively, and are shown in **Figure 3**. Naturally, the surface layer formed significantly faster under an enhanced water inflow to the field, with its height reaching the maximum mark of 0.075. In the basic example, this mark was half as low. At the same time, the rate of the ground bed wetting depended on the volume of the supplied water clearly weaker. Incidentally, the calculated curves $\overline{H}_W(\overline{t})$ can be useful for assessing the reliability of the function $\overline{w}(\overline{\tau})$, which has to be determined using the information on the actual change in the height of the surface water layer.

Figure 2. Decline in the relative filtration rate over time

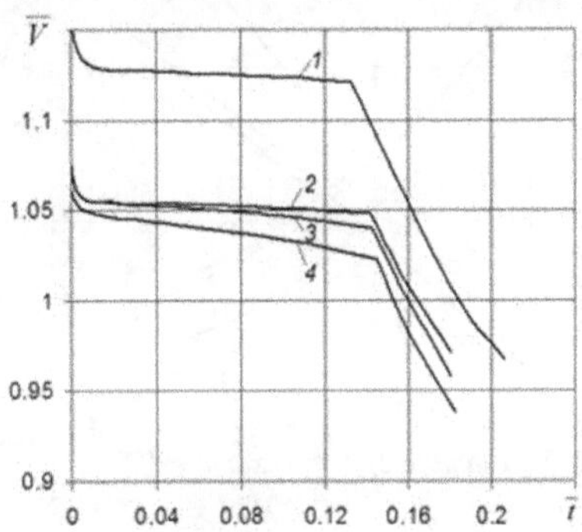

Note. 1 – example 2, 2 – example 1, 3 – example 3, 4 – example 4

Figure 3. Changes in the relative depths of the saturation front and surface layer over time

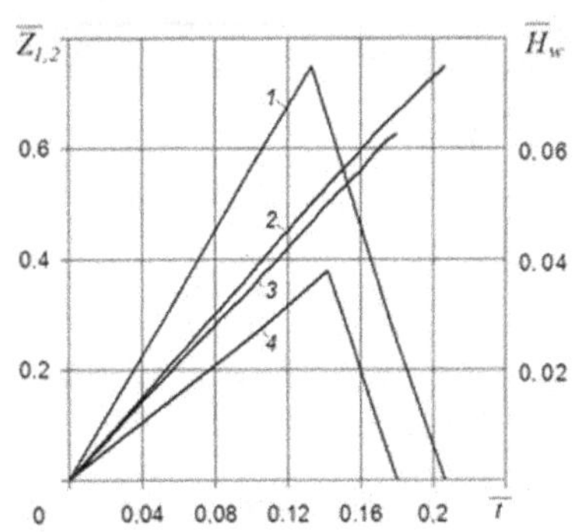

Note. 1, 4 – $\overline{H}_W(\overline{t})$; 2, 3 – $\overline{Z}_{1,2}(\overline{t})$; 1, 2 – example 2; 3, 4 – example 1

When studying the process of the suspension clarification, the inlet section of the filtering column (packed bed) is indicative, where the impurity is deposited especially quickly. Just **Figure 4** represents graphically the consequences of the natural intensification of the clogging of the medium and a decrease in its permeability here (the values of $\bar{k}_{in}$ were determined by formulas (8.71), (8.86) and then by (1.9), (1.10) at $\bar{z} = 0$) for three considered examples. Curves 1, 6 are distant from the other curves, which indicates the importance for the suspension separation of the absorption properties of the operating medium. It is relevant to note that when using linear filtration models, it is possible to essentially overestimate the degree of the clogging of the inlet section of a soil column, which would entail underestimation of the medium permeability here and, as a consequence, of the filtration rate. In such a case, as mentioned in Section 2.7, the dynamic solids take-up capacity will be greater than the similar true capacity and the deposit accumulation will be inhibited due to the inaccessibility of a significant part of the pore space. Then the noted limitation on the deposition, which is characteristic of natural porous structures, should be taken into account, for example, in accordance with the recommendations outlined in Section 2.5.

Figure 4. Changes in the relative concentration of the deposited particles ($\bar{S}_0$) and hydraulic conductivity ($\bar{k}_{in}$) over time at the ground surface

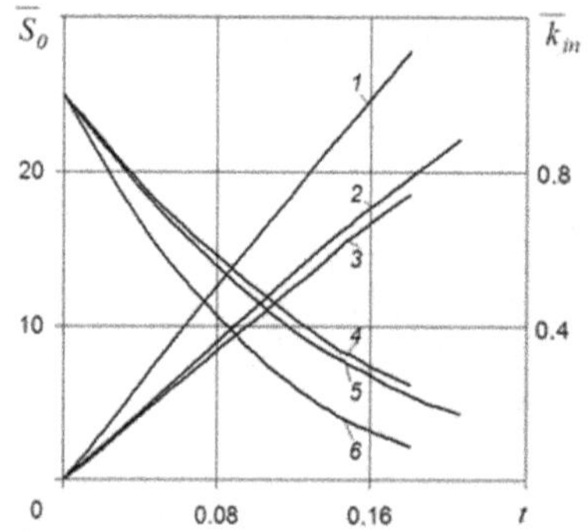

Note. 1-3 – $\bar{S}_0$; 4-6 – $\bar{k}_{in}$; 1, 6 – example 3; 2, 5 – example 2; 3, 4 – example 1

Figure 5. Profiles of the relative concentration of the deposited particles ($\bar{S}$) and hydraulic conductivity ($\bar{k}_{1,2}$)

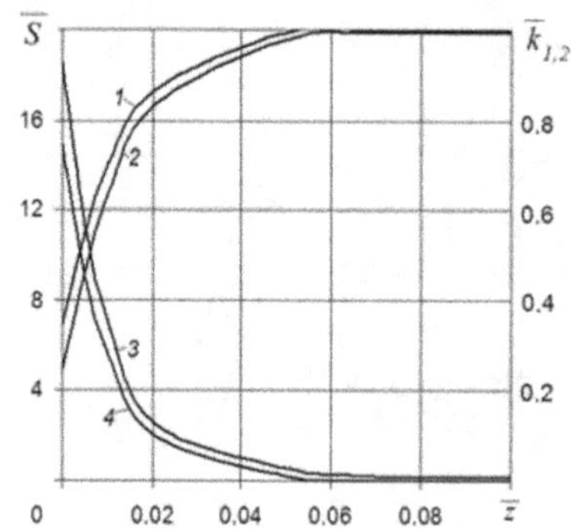

Note. 1 – $\bar{k}_1$; 2 – $\bar{k}_2$; 3, 4 – $\bar{S}$; 1, 4 – $\bar{\tau} = \bar{\tau}_2$; 2, 3 – $\bar{\tau} = \bar{\tau}_1$

If the ground bed still has a significant capacity with respect to the dispersed impurity, it can be used for the passage of the contaminated water for a long time. Nevertheless, its filtration flow properties significantly deteriorate after the implementation of a certain number of technological cycles, especially near and at the ground surface. The profiles of the concentration $\bar{S}(\bar{z})$ and hydraulic conductivities (permeabilities) $\bar{k}_{1,2}(\bar{z})$ at the end of the first $(\bar{\tau} = \bar{\tau}_1)$ and second $(\bar{\tau} = \bar{\tau}_2)$ stages of the first cycle are depicted in **Figure 5** and illustrate the actual situation in the bed with a highly uneven distribution of the deposit by the depth of its wetted part. Thus, even at an insignificant water contamination, but passing large volumes of water in the end, so much deposit will accumulate near the surface of the bed, that it will become a serious obstacle for the filtration flow in the future. In parallel, the calculations of the desired characteristics were carried out without taking into account the local derivative in the mass transfer equations. It turned out that the error due to such simplification of the initial models, for example, did not exceed 0.5% in determining $\bar{Z}_i$, and was not more than 2.2% for $\bar{H}_W$, which is consistent with the results from Section 2.1.

In reality, the exponent l is usually less than 1 and can even sometimes take negative values close to 0. In such cases, we have to use the results of the studies performed by the approximate analytical methods in Sections 8.1, 8.2. Some of them, which are the most illustrative and carried out by analogy with the limiting case (l=1, Figure 2 - 5), but already at l=1/3, are presented graphically below. Notably that here it is enough to know in advance the function $\bar{w}(\bar{t})$, which is not difficult to determine only on the basis of data on the volumes of the water suspension supplied to the field for several time intervals. In contrast to the limiting case, the water level on the ground surface does not have to be measured, but is reliably calculated using the formulas given in the above Sections. In the following series of examples, the values n_W(0.3), $\bar{\psi}_w$ (0), $\bar{S}^0$ (0), q(1), m_1(1), m_2(3) remained as before. The values of σ and $\bar{\alpha}_V$, $\bar{\beta}_V$ were varied. The intensive wetting of a dry ground bed (in the absence of the gravity moisture) when developing a water layer on its surface (the first stage, $\bar{w} = \sigma\bar{t}$), as well as an inflow into a partially wetted bed during the depletion of a preformed surface layer (the second stage, $\bar{w} = 0, \bar{Z}^0 = 0.2$) were calculated separately. The desired relative characteristics (the filtration rate $\bar{V}$, the depths of the saturation front $\bar{Z}$ and water layer on the ground surface $\bar{H}_W$, the inlet hydraulic conductivity $\bar{k}_{in}$ and concentration of the deposited impurity $\bar{S}_0$) were determined up to the moment $\bar{t}_*$ when the saturation front reaches the lower boundary of the bed (confluence of the infiltration and ground waters).

Five examples were calculated in parallel for the first stage. The following values of the model coefficient were taken: $\bar{\alpha}_V = 25$, $\bar{\beta}_V = 0.025$ in examples 1 to 4; $\bar{\alpha}_V = 50$, $\bar{\beta}_V = 0.05$ in example 5, as well as $\sigma = n_W = 0.3$ in example 1; σ=0.5 in examples 2, 5; σ=0.7 in example 3, σ=1 in example 4. In addition, to estimate the effect of the clogging of the bed, a control example 6 was calculated, corresponding to the cases of the pure water flow or no absorption of impurity, so that $\bar{k} = 1$. Data of the calculations of the $\bar{V}$ as a function of $\bar{t}$ by formula (8.12) at $\bar{k}(\bar{z},\bar{t})$ according to (1.9), (1.10); $\bar{S}_1(\bar{z},\bar{t})$ according to (8.31) are presented in **Figure 6**. Naturally, the limiting time $\bar{t}_*$ depended on the parameter σ and slightly increased with decreasing the latter, for example, amounting to 0.89 at σ=1. The filtration rate of suspension during the first stage decreased insignificantly, namely, maximally in the first example (by 11%), and minimally (by 7.2%) in the fourth example due to the relatively low absorption capacity of the filtering soil. The formal doubling of $\bar{\alpha}_V$, $\bar{\beta}_V$ led to the increased deposit accumulation near the upper boundary of the ground bed and eventually (at $\bar{t} > 0.5$) to a sharp decrease in the quantity $\bar{V}$, which will not correspond to reality at a low true solids take-up capacity. It should be kept in mind that such behavior of $\bar{V}$ is caused by the form of expression (1.10), which is often unable to approximate with the same quality the actual values of $\bar{k}$ in weakly and heavily clogged bed. At the same time, the accuracy of the approximate solution and the resulting calculation expressions should deteriorate, since the technique of averaging the filtration rate is less effective when it changes within wide limits. At supplying clean water to the field, its level rise on the ground surface provides constancy of $\overline{V}$ at $\overline{\psi}_w = 0$, which is equal to

$$\overline{V} = \frac{1 - n_W}{2} + \sqrt{\left(\frac{1 - n_W}{2}\right)^2 + \sigma}. \tag{8.95}$$

Such filtration flow situation in Figure 6 corresponds to the straight line 3 in Figure 6 calculated at σ=0.5. In the remaining examples, formula (8.95) was used to determine the initial value $\overline{V}^0$, since it was initially assumed that $\overline{S} = \overline{S}_0 = 0$. The filtration flow is formed under the condition $\overline{V} \geq 1$. If $\overline{V} < 1$, then the liquid moves with incomplete saturation. In such situations, filtration flow models appreciably coarsen the nature of the water-physical process and it is more correct to use more complex models of the moisture transfer containing two hydrophysical characteristics determined by experience. Curve 6 is indicative in this respect, which is located below the conditional horizontal line $\overline{V} = 1$ except for the initial point ($\overline{V} = \overline{V}^0 = 1$). This straight line corresponds to the minimum value of $\sigma(\sigma_{min}=0.3)$, at which the filtration flow is still formed in the clean ground bed, and in this sense it can be considered as a limiting line. Then, all the water supplied to the field is used for the infiltration. Therefore, the moisture content of the wetted section of the bed is less than the total moisture capacity at $\sigma<\sigma_{min}$. If $\sigma>\sigma_{min}$, then the depth of the surface layer should increase along with the expansion of the saturation zone, where the moisture of the medium is maximum. From physical considerations it is obvious that the quantity $\overline{V}$ should be less than 1 (curve 6) at the progressive contamination of the bed, $\sigma=n_W$ and $\overline{t} > 0$. For this reason, the filtration flow model turns out to be incorrect, and the indicated curve describes the corresponding process only qualitatively correctly.

Figure 6. Change in the relative filtration flow rate over time

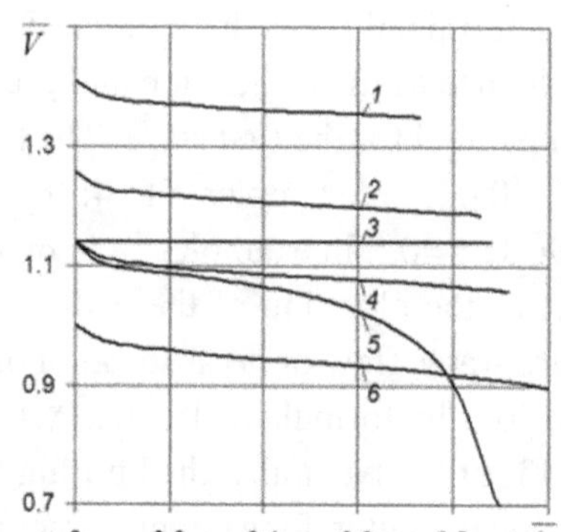

Note. 1 – example 4, 2 – example 3, 3 – example 6 (control), 4 – example 2, 5 – example 5, 6 – example 1

Figure 7. Expansion of the saturation zone over time

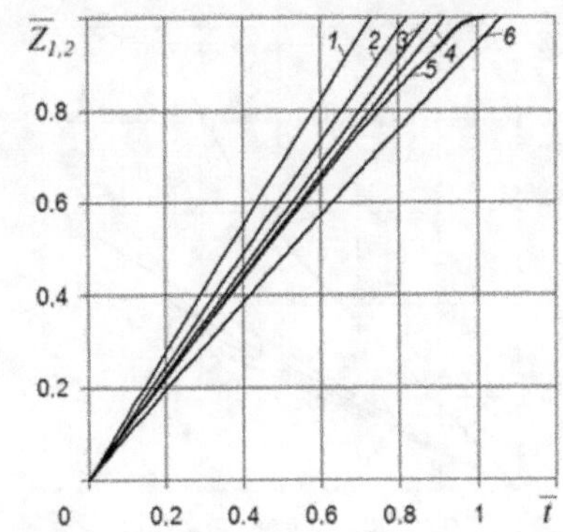

Note. 1 – example 4, 2 – example 3, 3 – example 6 (control), 4 – example 2, 5 – example 5, 6 – example 1

The nature of the saturation zone expansion in all examples, as evidenced by **Figure 7**, is similar. The deposit accumulation in the ground bed restrains the front movement, while the intensification of the suspension supply to the field, on the contrary, accelerates it. The second factor dominates and, therefore, the time $\bar{t}_*$ was the smallest (0.73) in the fourth example (σ=1). The effect of clogging can be conveniently estimated here by comparing curves 3 and 4, which were calculated at the same value of σ(0.5) for the clean and clogged bed respectively. The enhanced absorption (curve 5) caused not only a slower expansion of the saturation zone at $\bar{t} \geq 1$, but also a sharp inhibition of the above-mentioned front due to a virtually complete clogging of pores in the inlet cross-section of the column. However, this situation does not correspond to reality.

The water level at the surface of the ground bed is closely related to the position of the saturation front. The additional water volume entering the field due to a hypothetical increase in σ, is distributed between the wetted and surface layers and $\overline{H}_W$ increases faster (curves 1, 2, 4, 6 in **Figure 8**). Incidentally, the specified layer is absent only if $\sigma = \sigma_{\min}$ and the water is pure, but a slow rise of the specified level is observed already at $\overline{\alpha}_V = 25$, $\overline{\beta}_V = 0.025$. Naturally, when the absorption of impurity increases, the flow capacity of the wetted part of the bed decreases and the $\overline{H}_W$ rise rate increases (curves 3 and 4).

Figure 8. Increase in the depth of the surface suspension layer over time

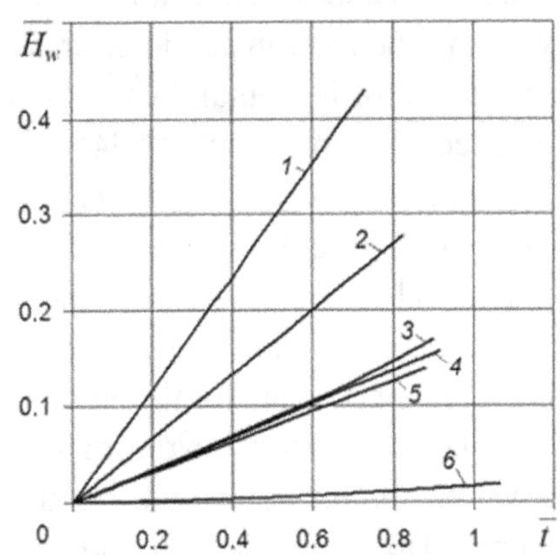

Note. 1 – example 4, 2 – example 3, 3 – example 5, 4 – example 2, 5 – example 6 (control), 6 – example 1

Figure 9. Increase in the relative concentration of the deposited solids at the upper boundary

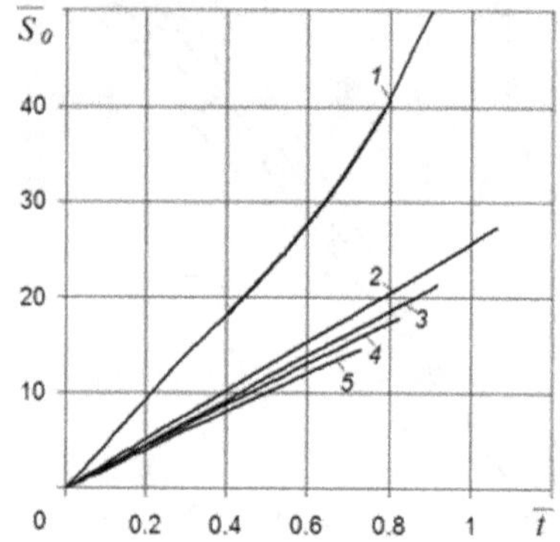

Note. 1 – example 5, 2 – example 1, 3 – example 2, 4 – example 3, 5 – example 4

Figure 10. Decrease in the relative permeability of the ground bed at the upper boundary

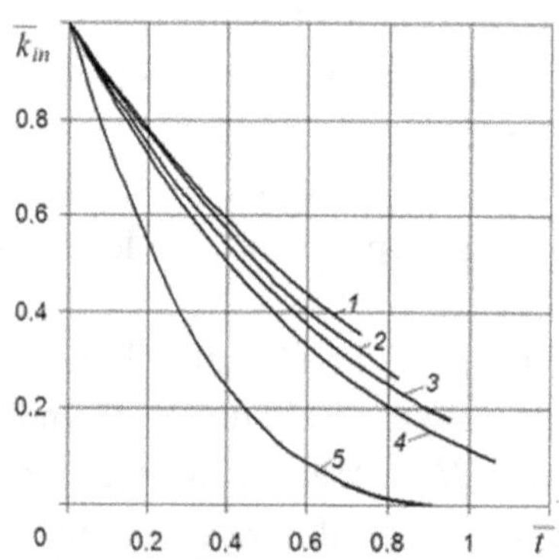

Note. 1 – example 4, 2– example 3, 3 – example 2, 4 – example 1, 5 – example 5

Since the suspended solids, irrespective of the value of the exponent l, are retained mainly near the upper boundary of the ground bed, an increase in the concentration of the deposited particles (deposit) and a decrease in the permeability of the medium in the inlet cross-section (z=0) deserve a special attention when analyzing the general pattern of its contamination. Corresponding data of the approximate calculations of the values of $\overline{S}_0$ according to formula (8.34) and $\overline{k}_{in}$ by formula (1.10) are shown in **Figures 9 and 10**. Here, we should treat critically the significant increase in $\overline{S}_0$ and decrease in $\overline{k}_{in}$, which take place for example 5 (curves 1 in Figure 9 and 5 in Figure 10). Nevertheless, if the limitation of the specified resource is actually dictated by the ratio between the mass exchange processes, then the data obtained on the basis of the calculation expressions given in the previous Sections of this Chapter are trustworthy during the entire calculation period and for the entire bed. In general, the reliable results of the mathematical modeling of the suspension separation during its infiltration into both dry (without gravity moisture) and saturated bed can be obtained only by using the values of mass exchange coefficients found on the basis of experimental data under similar conditions.

For better clarity, when comparing the results of calculating the infiltration process at the second stage, it is reasonable to abstract from the first stage and assume a single initial filtration flow pattern in the considered examples, in particular, the fixed values $\bar{H}_W^0$ (0.3) and Z^0 (0.2). The rational choice of Z^0, as well as the eligibility of applying the technique of averaging filtration rate to solve the problem set at the beginning of this Chapter and the expediency of introducing the parameter $\bar{V}_a$ in this way were justified in the case of the pure water supply to the field. Preliminarily, the errors were calculated

$$\Delta\bar{V} = \left|\frac{V_{ex} - V_{ap}}{V_{ex}}\right|, \quad \Delta\bar{Z} = \left|\frac{Z_{ex} - Z_{ap}}{Z_{ex}}\right|,$$

where the known exact solution of the corresponding problem and the resulting representations for quantities Z, V with index ex (index ap for approximations) served as a reference. Formally, the exact expression for $\bar{Z}(\bar{t})$ in the form of inverse function follows from (8.13) when $k = k_0$ $(\bar{k} = 1)$ and will be

$$\bar{t} = \int_{Z^0}^{Z_{exact}} \frac{\varsigma d\varsigma}{\bar{W}_{in} + \bar{\psi}_w + (1 - n_W)\varsigma} =$$

$$= \frac{1}{1 - n_W}\left[Z_{ex} - Z^0 - \frac{\bar{W}_{in} + \bar{\psi}_w}{1 - n_W}\ln\frac{\bar{W}_{in} + \bar{\psi}_w + (1 - n_W)\,Z_{ex}}{\bar{W}_{in} + \bar{\psi}_w + (1 - n_W)\,Z^0}\right]$$

where $\bar{W}_{in} = W_{in}/M$. When the relation between $\bar{Z}$ and $\bar{t}$ is known, the sought $\bar{V}_{ex}$ is exactly found according to (8.12), namely,

$$\bar{V}_{ex} = \frac{\bar{W}_{in} + \bar{\psi}_w + (1 - n_W)Z_{ex}}{Z_{ex}}.$$

Conducting such a study here is all the more useful because the filtration flow problem of the intensive wetting of a soil column from the surface liquid layer has a mathematical peculiarity, manifested at the initial moment of time, when the flow rate turns out to be unlimitedly large. The data of the calculations of $\Delta\bar{V}$ and $\Delta\bar{Z}$ at different values of Z^0 are shown in **Figures 11 and 12** and indicate the smallness of the errors due to rate averaging already at $Z^0 \geq 0.1$.

This fact gives a reason for accepting the value $Z^0 = 0.2$ in the two main examples with suspension and a well-absorbing ground bed ($\bar{\alpha}_V = 50$, $\bar{\beta}_V = 0.05$ in example 1 and $\bar{\alpha}_V = 100$, $\bar{\beta}_V = 0.1$ in example 2). Thus, the ratio $\bar{\alpha}_V/\bar{\beta}_V$, characterizing the potential absorptive capacity of the bed to retain impurities, is the same in both examples. The subject of calculations here were the relative rate $\bar{V}$ and the hydraulic conductivity $\bar{k}$, for which its the inlet $\bar{k}_{in}$ and effective $\bar{k}_e$ values were determined in parallel. It is $\bar{k}_e$ what gives a generalized idea of the flow capacity at first of the expanding and gradually clogging saturation zone, and the entire saturated bed after the confluence of the infiltration and ground waters. The function $\bar{k}_e(\bar{t})$ was established formally based on the equality of the reduced filtration flow rate, the expression for which is as follows

$$\bar{V}(\bar{t}) = \bar{k}_e(\bar{t})\frac{\bar{H}_W(\bar{t}) - \bar{h}(Z_i(\bar{t}))}{Z_i(\bar{t})},$$

as well as the strict expression in accordance with Darcy's law. As a result, for the second stage the desired $\bar{k}_e(\bar{t})$ was

$$\bar{k}_e(\bar{t}) = \frac{\bar{Z}_2(\bar{t})\,\bar{V}(\bar{t})}{\bar{W}_{in} + \bar{\psi}_w + (1 - n_W)\,\bar{Z}_2(\bar{t})}.$$

Figure 11. Dependence $\Delta\bar{V}(\bar{t})$

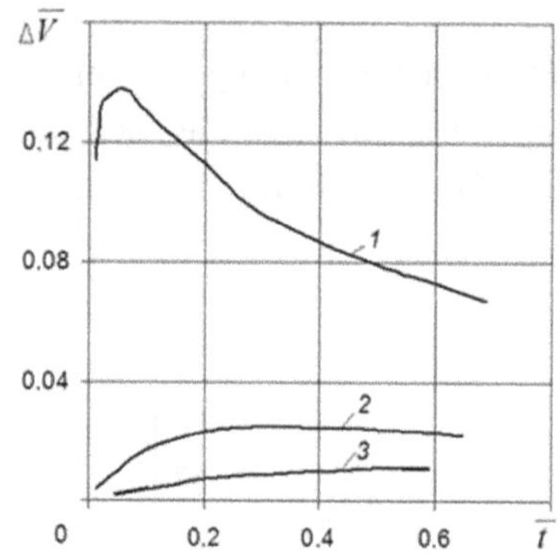

Note. $1 - Z^0 = 0.01$; $2 - Z^0 = 0.1$; $3 - Z^0 = 0.2$

Figure 12. Dependence $\Delta\bar{Z}_2(\bar{t})$

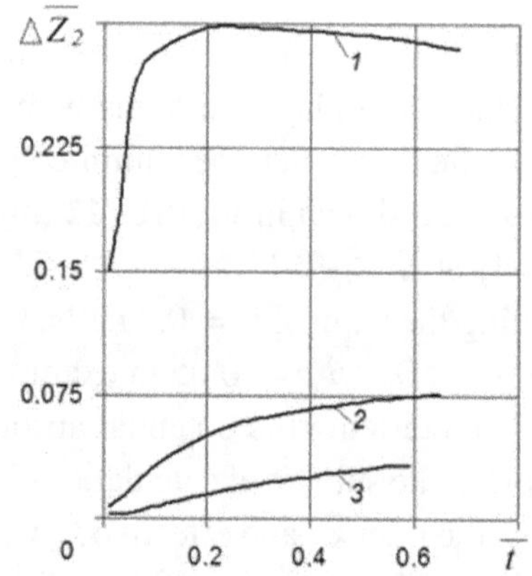

Note. $1 - Z^0 = 0.01$; $2 - Z^0 = 0.1$; $3 - Z^0 = 0.2$

Examples 1 and 2 were calculated directly using formulas (8.40) and (1.10), (8.55) with $\bar{S}_2^0 = \bar{t}_1 = 0$ and the same values of l,q. To estimate the effect of the clogging of a ground bed, the results of the approximate calculations of $\bar{V}$ in the case of water without impurities were also involved, using equation (8.45) and expressions (8.40), (8.43) to ensure the correctness of the comparison. The information relatively $\bar{V}$, $\bar{k}_{in}$, $\bar{k}_e$ obtained in this way is presented graphically in **Figures 13 and 14** respectively.

Due to the short duration of the stage under consideration and the technological cycle as a whole, the above-mentioned effect turns out to be insignificant with weaker absorption (example 1). However, this conclusion is valid if only one technological cycle is implemented. With heavy absorption (example 2), a sharp decrease in both $\overline{V}$ (curve 3 in Figure 13) and $\overline{k}$ (curves 2, 4 in Figure 14) is possible by the end of the stage.

Figure 13. Decrease in the relative filtration rate over time

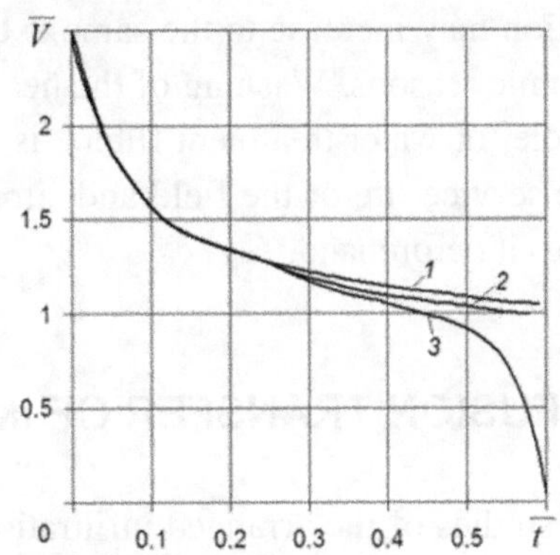

Note. 1 – example 3 (control), 2 – example 2, 3 – example 1

Figure 14. Decrease in the relative effective $(\overline{k}_e)$ and inlet $(\overline{k}_{in})$ hydraulic conductivity over time

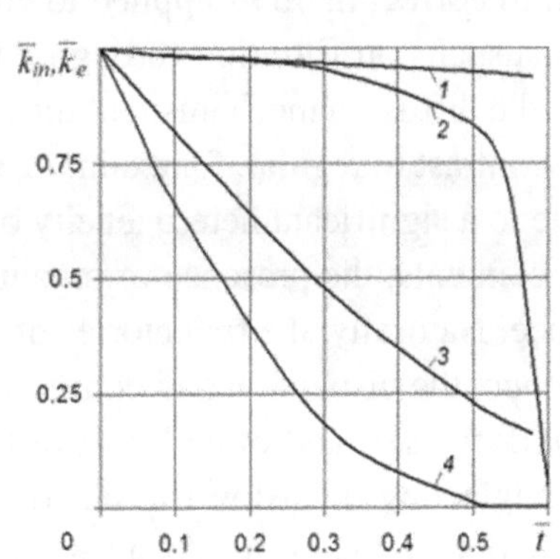

Note. 1, 3 – example, 2, 4 – example; 1, 2 – $\overline{k}_e$; 3, 4 – $\overline{k}_{in}$

In conclusion, it should be emphasized that the main result of the theoretical research conducted in this Chapter is an engineering method, which allows, using the software packages Mathcad, Matlab, Mathematica etc., to reliably predict the joint proceeding of the infiltration, clarification and hydrological processes when water with an increased content of dispersed impurities is supplied in a given way to a special area (infiltration field). Obviously, this method should be provided with the reliable initial information, the full composition of which it outlines. The large number of the examples calculated above serves mainly to illustrate the patterns of the accumulation of a non-washable deposit in the initially dewatered ground bed and, as a consequence, the gradual deterioration of its flow capacity during one technological cycle with a possible change of water flow conditions in the pore space (with full and partial saturation), and, at the same time, to substantiate the developed calculation tool. Successive performance of similar calculations for a number of such cycles makes it possible to predict the consequences of the repeated use, first of all, of the infiltration field, as well as other similar water management objects with water-physical conditions regulated by means of the infiltration process. And here it is appropriate to emphasize that the repeated implementation of the technological cycle causes a steady decrease in the flow capacity of the filtering column, so that the filtration rate, in the end, may significantly decrease, and its duration may increase to the same extent. Then further operation of the field does not make sense for economic reasons. Washing of the heavily clogged ground bed, which is an integral part of the operating cycle for water treatment filters, is practically impossible. Therefore, the only realistic way to extend the service life of the field and similar objects is to remove a part of impurities from the water supplied to it beforehand.

8.5. ACCOUNTING FOR DIFFUSION TRANSFER OF IMPURITY

The materials of our theoretical studies of the arranged infiltration of the surface water with a significant amount of the suspended matter, which are described in the previous Sections of this Chapter, give a generally correct idea of the interrelated filtration flow and clarification processes occurring in the ground bed, as well as their consequences. The similarity of these processes in natural and artificial (packed bed) porous media allows us to use the same type of mathematical models in both cases. Differences between them are not essential, so that it is justified to take a simpler option and its known solution as a basis, and then, in fact, to correct them as applied to complicated conditions. In the considered case of the natural aqueous suspension filtration through a thick uniform bed, the increasing importance should be emphasized of the diffusion mechanism of the dispersed impurity transfer. In general, in natural porous formations in contrast to regular, for example, granular structures, the suspended solids are usually dispersed more due to a significant heterogeneity of the mechanical composition and non-sphericity of structure-forming elements, the presence of organic compounds and, as a result, the complex configuration of a pore space. Naturally, the trajectories of moving particles are curved more in such situations and, as a consequence, the dispersion and convective diffusion coefficients increase, and the number *Pe*, on the contrary, decreases, and often by an order and more in comparison to the media from sorted particles. Then neglecting diffusion transfer of impurity can lead to the obvious discrepancy between the calculated and actual patterns. As a result, severe underestimation of the filtration flow rates and also transport of the suspended matter deep into the ground bed is possible. But since the convection mechanism in reality, as a rule, remains dominant, even with such a significant decrease in *Pe*, it is enough to refine the basic mathematical model and its solution ($Pe \to \infty$). And then the asymptotic methods (successive approximations, small parameter) are effective. And taking into account moderate requirements to the accuracy of engineering calculations, it is usually sufficient to limit oneself to the refinement of the first approximation, i.e., to determine the second approximation. Thus, Pe^{-1} plays the role of a small parameter. Then, for the linear detachment filtration and combined mass transfer, the iterative procedure is implemented as follows. The first iteration here are exact solutions of the corresponding filtration problems at a constant (Section 2.1) and variable (Section 5.1) filtration rates, which are expressed by the dependences for the relative quantities $\overline{C}_1$, $\overline{S}_1$, $\tilde{h}_1$. They are also found in the second approximation $(\overline{C}_2, \overline{S}_2)$ on the basis of the clarification compartment refined taking into account the impurity diffusion, which includes at $l=q=1$ in (5.3) the system of dimensionless equations

$$\frac{\partial \overline{C}_2}{\partial \overline{z}} + \frac{\partial \overline{S}_2}{\partial \overline{\tau}} = \frac{1}{Pe}\frac{\partial^2 \overline{C}_1}{\partial \overline{z}^2} \tag{8.96}$$

$$\frac{\partial \overline{S}_2}{\partial \overline{\tau}} = \overline{\alpha}_V \overline{C}_2 - \overline{\beta}_V \overline{S}_2 \tag{8.97}$$

and the boundary conditions

$$\overline{z} = 0,\ \overline{C}_2 = 1 \tag{8.98}$$

$$\overline{\tau} = 0,\ \overline{S}_2 = \overline{S}^0(\overline{z}) \tag{8.99}$$

Here $\overline{\tau}$ is a relative generalized characteristic of time, which is identical to real time at the filtration with a constant rate, and is expressed through it in the case of an unregulated rate in Section 5.1. Incidentally, the representation of the initial model in a dimensionless form in contrast to the developments in Sections 8.1, 8.2 is justified here due to the need to formally distinguish the Peclet number $Pe=V_aL/D$ in it (at a constant filtration rate $V=V_a$).

First of all, problem (8.96) - (8.99) is solved with respect to the image functions $\overline{C}_{2L}, \overline{S}_{2L}$. It is taken into account that

$$\frac{d^2\overline{C}_{1L}}{d\overline{z}^2} = \frac{\overline{\alpha}_V^2 p}{(p+\overline{\beta}_V)^2} e^{-\frac{\overline{\alpha}_V p \overline{z}}{p+\overline{\beta}_V}} + \frac{\overline{\beta}_V}{p+\overline{\beta}_V}\frac{d\overline{S}^0}{d\overline{z}} - \frac{\overline{\alpha}_V \overline{\beta}_V p}{(p+\overline{\beta}_V)^2}\overline{S}^0(\overline{z}) + \frac{\overline{\alpha}_V^2 \overline{\beta}_V p^2}{(p+\overline{\beta}_V)^3}\int_0^{\overline{z}} \overline{S}^0(\eta)\, e^{-\frac{\overline{\alpha}_V p}{p+\overline{\beta}_V}(\overline{z}-\eta)}\, d\eta. \tag{8.100}$$

Then the required quantity $\overline{C}_{2L}$ is determined by solving the following problem

$$\frac{d\overline{C}_{2L}}{d\overline{z}} + \frac{\overline{\alpha}_V p}{p+\overline{\beta}_V}\overline{C}_{2L} = \frac{1}{Pe}\left[\frac{\overline{\alpha}_V^2 p}{(p+\overline{\beta}_V)^2} e^{-\frac{\overline{\alpha}_V p \overline{z}}{p+\overline{\beta}_V}} + \frac{\overline{\beta}_V}{p+\overline{\beta}_V}\frac{d\overline{S}^0}{d\overline{z}} - \frac{\overline{\alpha}_V \overline{\beta}_V p}{(p+\overline{\beta}_V)^2}\overline{S}^0(\overline{z}) + \frac{\overline{\alpha}_V^2 \overline{\beta}_V p^2}{(p+\overline{\beta}_V)^3}\int_0^{\overline{z}} \overline{S}^0(\eta)\, e^{-\frac{\overline{\alpha}_V p}{p+\overline{\beta}_V}(\overline{z}-\eta)}\, d\eta\right]; \tag{8.101}$$

$$\overline{z} = 0,\ \overline{C}_{2L} = \frac{1}{p} \tag{8.102}$$

Thus, it is found

$$\overline{C}_{2L}(\overline{z}, p) = \overline{C}_{1L}(\overline{z}, p) + \Delta \overline{C}_{2L}(\overline{z}, p) \tag{8.103}$$

where

$$\overline{C}_{1L}(\overline{z}, p) = e^{-\overline{\alpha}_V \overline{z}}\frac{e^{\frac{\overline{\alpha}_V \overline{\beta}_V \overline{z}}{p+\overline{\beta}_V}}}{p} \tag{8.104}$$

$$
\Delta \overline{C}_{2L}(\overline{z}, p) = \frac{1}{Pe}\left\{ \overline{\alpha}_V^2 \overline{z}\, e^{-\overline{\alpha}_V \overline{z}} \frac{p}{p+\overline{\beta}_V} e^{\frac{\overline{\alpha}_V \overline{\beta}_V \overline{z}}{p+\overline{\beta}_V}} + \frac{\overline{\beta}_V}{p+\overline{\beta}_V}\left[\overline{S}^0(\overline{z}) - \overline{S}^0(0)\, e^{-\overline{\alpha}_V \overline{z}}\, e^{\frac{\overline{\alpha}_V \overline{\beta}_V \overline{z}}{p+\overline{\beta}_V}}\right] + \right.
$$

$$
\left. + \frac{\overline{\alpha}_V^2 \overline{\beta}_V p^2}{(p+\overline{\beta}_V)^3} \int_0^{\overline{z}} \int_0^{\eta} e^{-\overline{\alpha}_V(\overline{z}-\xi)}\, \overline{S}^0(\xi)\, e^{\frac{\overline{\alpha}_V \overline{\beta}_V (\overline{z}-\xi)}{p+\overline{\beta}_V}}\, d\xi d\eta \right\}. \tag{8.105}
$$

Here it was taken into consideration that

$$
\int_0^{\overline{z}} \frac{d}{d\eta} \overline{S}^0(\eta) e^{-\frac{\overline{\alpha}_V p}{p+\overline{\beta}_V}(\overline{z}-\eta)}\, d\eta = \overline{S}^0(\overline{z}) - \overline{S}^0(0) e^{-\frac{\overline{\alpha}_V p \overline{z}}{p+\overline{\beta}_V}} + \frac{\overline{\alpha} p}{p+\overline{\beta}_V} \int_0^{\overline{z}} \overline{S}^0(\eta) e^{-\frac{\overline{\alpha}_V p}{p+\overline{\beta}_V}(\overline{z}-\eta)}\, d\eta
$$

The transition from the image functions (8.103) - (8.105) to the original functions results in the previously presented expression for $\overline{C}_1(\overline{z}, \tau)$ (5.25) and also

$$
\Delta \overline{C}_2(\overline{z}, \overline{\tau}) = \frac{\overline{\alpha}_V^2 \overline{z}}{Pe} e^{-\overline{\alpha}_V \overline{z} - \overline{\beta}_V \overline{\tau}} \left[I_0\left(2\sqrt{\overline{\alpha}_V \overline{\beta}_V \overline{z}\overline{\tau}}\right) - \overline{\beta}_V \int_0^{\overline{\tau}} I_0\left(2\sqrt{\overline{\alpha}_V \overline{\beta}_V \overline{z}\xi}\right) \right] +
$$

$$
+ \frac{\overline{\beta}_V e^{-\overline{\beta}_V \overline{\tau}}}{Pe} \left[\overline{S}^0(\overline{z}) - \overline{S}^0(0)\, e^{-\overline{\alpha}_V \overline{z}} I_0\left(2\sqrt{\overline{\alpha}_V \overline{\beta}_V \overline{z}\overline{\tau}}\right) + \right. \tag{8.106}
$$

$$
\left. + \overline{\alpha}_V^2 G_1(\overline{z}, \overline{\tau}) - \overline{\alpha}_V^2 \overline{\beta}_V G_2(\overline{z}, \overline{\tau}) + \overline{\alpha}_V^2 \overline{\beta}_V^2 G_3(\overline{z}, \overline{\tau}) \right].
$$

Here

$$
G_1(\overline{z}, \overline{\tau}) = \int_0^{\overline{z}} \int_0^{\eta} \overline{S}^0(\overline{z} - \eta)\, e^{-\overline{\alpha}_V \zeta} I_0\left(2\sqrt{\overline{\alpha}_V \overline{\beta}_V \zeta \overline{\tau}}\right) d\zeta d\eta
$$

$$
G_2(\overline{z}, \overline{\tau}) = \int_0^{\overline{z}} \int_0^{\eta} \int_0^{\overline{\tau}} \overline{S}^0(\overline{z} - \eta)\, e^{-\overline{\alpha}_V \zeta} I_0\left(2\sqrt{\overline{\alpha}_V \overline{\beta}_V \zeta \xi}\right) d\xi d\zeta d\eta,
$$

$$
G_3(\overline{z}, \overline{\tau}) = \int_0^{\overline{z}} \int_0^{\eta} \int_0^{\overline{\tau}} \overline{S}^0(\overline{z} - \eta)\, e^{-\overline{\alpha}_V \zeta} \zeta\, I_0\left(2\sqrt{\overline{\alpha}_V \overline{\beta}_V \zeta \xi}\right) d\xi d\zeta d\eta.
$$

In the special case of the initially clean porous medium

$$
\Delta \overline{C}_2(\overline{z}, \overline{\tau}) = \frac{\overline{\alpha}_V^2 \overline{z}}{Pe}\, e^{-\overline{\alpha}_V \overline{z} - \overline{\beta}_V \overline{\tau}} \left[I_0(2\sqrt{\overline{\alpha}_V \overline{\beta}_V \overline{z}\overline{\tau}}) - \overline{\beta}_V \int_0^{\overline{\tau}} I_0\left(2\sqrt{\overline{\alpha}_V \overline{\beta}_V \overline{z}\xi}\right) d\xi \right]. \tag{8.107}
$$

Similarly, it is presented

$$
\overline{S}_2(\overline{z}, \overline{\tau}) = \overline{S}_1(\overline{z}, \overline{\tau}) + \Delta \overline{S}_2(\overline{z}, \overline{\tau}) \tag{8.108}
$$

where (5.26) is valid for $\overline{S}_1(\overline{z}, \overline{\tau})$, and (8.102) for $\Delta \overline{S}_2$

$$
\Delta \overline{S}_2(\overline{z}, \overline{\tau}) = \overline{\alpha}_V \int_0^{\overline{\tau}} e^{-\overline{\beta}_V \xi} \Delta \overline{C}_2(\overline{z}, \xi)\, d\xi \tag{8.109}
$$

Thus, due to the determination of the required concentrations in the second approximation all previously found solutions of the mathematical problems of linear filtration, which, in essence, are limiting ($Pe \to \infty$), and in practice are applicable approximately at $Pe>250$, are actually generalized to the case when the diffusion begins to play a significant role in mass transfer. Formally, it becomes possible to perform the calculations of a suspension filtration in a wider range of Pe values ($Pe>25$). Certainly, in each specific case at $250>Pe>25$ it is expedient to conduct special theoretical studies in order to refine the filtration characteristics. At the same time, in such a case, taking into account firstly, the corrective nature of the quantities $\Delta \overline{C}_2$, $\Delta \overline{S}_2$, secondly, the practicality of modeling and calculations of filtration, the difficulties with obtaining the initial information, it is enough to limit ourselves for the concentrations $\overline{C}$, $\overline{S}$ by introducing correction coefficients χ_c, χ_s on the basis of the above results as follows

$$\chi_c(\overline{z},\overline{t}) = \frac{\overline{C}_2(\overline{z},\overline{t})}{\overline{C}_1(\overline{z},\overline{t})} = 1 + \frac{\Delta \overline{C}_2(\overline{z},\overline{t})}{\overline{C}_1(\overline{z},\overline{t})} \tag{8.110}$$

$$\chi_s(\overline{z},\overline{t}) = \frac{\overline{S}_2(\overline{z},\overline{t})}{\overline{S}_1(\overline{z},\overline{t})} = 1 + \frac{\Delta \overline{S}_2(\overline{z},\overline{t})}{\overline{S}_1(\overline{z},\overline{t})} \tag{8.111}$$

Here $\Delta \overline{C}_2$, $\Delta \overline{S}_2$ are determined by (8.106), (8.109), and the dependences are taken for $\overline{C}_1$, $\overline{S}_1$, which were previously derived at $Pe \to \infty$ based on the considered conditions. Obviously, such coefficients allow to get only a general idea of the additional contribution of the diffusion transfer of dispersed impurity to the technological process in general and separately to its clarification and filtration flow components. An in-depth study of filtration in various conditions under the enhanced dispersion of the suspended solids has yet to be performed.

Obviously, the refinement of the quantity S makes it possible to correct the hydraulic calculations. And first of all, the distribution of the piezometric head along the height of the porous layer (ground bed) is more accurately determined. In the second turn the hydraulic characteristics that are of greatest practical interest more reliably set by taking into account diffusion transfer of impurity in addition to convection transfer. Such can be the head losses in a filtering medium (filtration facility), filtration rate, liquid level above its surface, wetting depth. In general, the composition of the hydraulic calculations is specified by the filtration flow conditions. The principles of such calculations for a constant or variable flow filtration rate and a variety of the physical and chemical conditions (uniform and layered structures, plane and axisymmetric flows, filtration flow region with fixed and moving boundaries, linear and nonlinear mass exchange kinetics, hydrodynamically stable or unstable deposit) are described in the previous and this Chapter. Following them, a special attention should be paid to the head losses in the saturated medium Δh at a constant filtration rate. Here, the dimensional formula similar to (2.24) is applicable for the refinement of the indicated losses

$$\Delta h_2 = \frac{V}{k_0} \int_0^{Z_i} \frac{dz}{f_k(S_2(z,t))} \tag{8.112}$$

where Z_i generalized means the depth of the final cross-section of the filtration flow, which is either an unknown variable when the ground bed is being saturated with liquid or a constant value and equals to the thickness of the fully saturated bed.

In case of an unsteady filtration flow, the hydraulic calculations become more complicated, because along with refinement of the filtration rate it is also necessary to make corrections in the relationship between the cumulative variable τ and real time corresponding $Pe \rightarrow \infty$, as well as to the dependence of the surface layer depth and the position of the wetting front on it. Just at the arrangement of the filtration flow process by feeding the infiltration field with an excessive amount of surface water, the change in the saturated flow rate is caused, firstly, by the instability of the liquid level position above the wetted column and the expansion of flow region, secondly, by progressive clogging of a pore space. Then the dimensional rate V is more exactly calculated, when a surface layer with the depth H_W forms, by the formula

$$V(\tau) = k_0[H_W(\tau) - H_{out}] \left[\int_{-Z_i}^{0} \frac{dz}{f_k(S_2(z,\tau))} \right]^{-1} \tag{8.113}$$

Accordingly, the dependence $t(\tau)$ is refined as follows

$$t = \frac{1}{k_0} \int_0^{\tau} \left[\int_{-Z_i(\xi)}^{0} \frac{dz}{f_k(S_2(z,\xi))} \right] \frac{d\xi}{H_W(\xi) - H_{out}} \tag{8.114}$$

Now for predicting changes in V over time and uniquely related quantities H_W, Z_i, it is necessary to use, for example, formulas (8.10), (8.20) for the first stage, preliminarily replacing S in them (actually S_1) by S_2. It is recommended to determine the quantity S_2 and together with it C_2 on the basis of the representations for S, C derived in Sections 8.1 -8.3. In this case, in view of the relative smallness of the corrections ΔS_2, ΔC_2, it is acceptable to use the coefficients χ_c, χ_s for the refinement of S_1, C_1 in accordance with (8.110) - (8.114). These coefficients change over time and along the height of the saturated zone ($0 \leq z \leq Z_i$) except for the inlet cross-section of the filtration flow, where $\Delta \overline{C}_2 = 0$, and hence $\chi_{c,s} = 1$ at any stage of the detachment filtration.

Incidentally, here the exact expression for $S_0(\tau) = S(0,\tau)$ according to (5.163) has the dimensional form

$$S_0(\tau) = S_0^0 e^{-\beta_V \tau} + \frac{\alpha_V}{\beta_V} C_0 (1 - e^{-\beta_V \tau})$$

and is useful for estimating the accuracy of the approximate solutions of problem (8.101) -(8.104). If l and q are arbitrary, then the general expression is valid for $S_0(\tau)$

$$S_0(\tau) = S_0^0 e^{-\mu(\tau)} + \alpha_V \int_0^{\tau} V^{l-1}(\xi) e^{-\mu(\xi)} d\xi, \quad \mu(\tau) = \beta_V \int_0^{\tau} V^{q-1}(\xi) d\xi$$

Figure 15. Dependence $\chi_c(\lg Pe)$

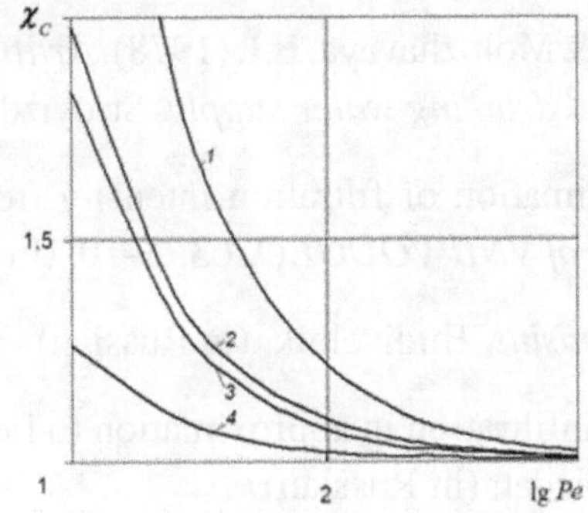

Note. 1, 3 – $\bar{\alpha}_V$ = 7; 2, 4 – $\bar{\alpha}_V$ = 5; 1, 2 – $\bar{\tau}$ = 100; 3, 4 – $\bar{\tau}$ = 300

Figure 16. Dependence $\chi_s(\lg Pe)$

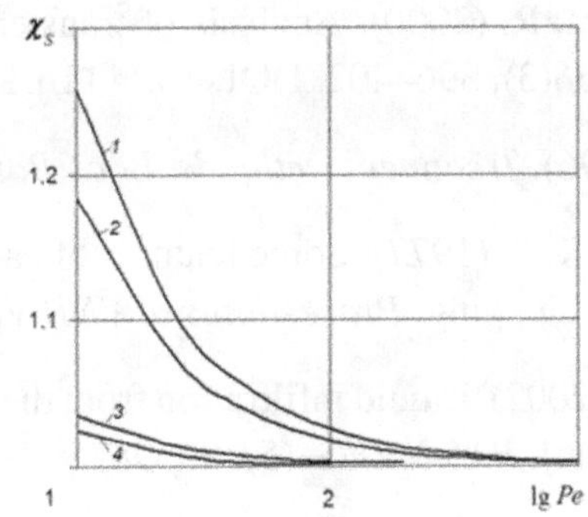

Note. 1, 3 – $\bar{\alpha}_V$ = 7; 2, 4 – $\bar{\alpha}_V$ = 5; 1, 2 – $\bar{\tau}$ = 100; 3, 4 – $\bar{\tau}$ = 300

To illustrate the functional coefficients $\chi_{c,s}(\bar{z},\bar{\tau})$, their calculations were performed depending on *Pe* by formulas (8.107), (8.110) and (8.109), (8.111). In this case, the remote cross-section of the filtration flow ($\bar{z}$ = 1), the relative values of $\bar{\tau}$ (100 and 300) and also $\bar{\beta}_V$ = 0.01, $\bar{\alpha}_V$ = 5 and 7 were accepted. The series of curves of the dependence $\chi_{c,s}(\lg Pe)$ are depicted in **Figures 15** and **16**. Figure 15 shows that the refinement of the base quantity $\bar{C}_2$ becomes significant in accordance with the above proposed procedure, namely, $\bar{C}_2$ exceeds $\bar{C}_1$ by 20% at $\bar{\tau}$ = 100 and $\bar{\alpha}_V$ = 5 in the case $Pe \approx 50$, and the similar *Pe* increases to 100 at the same $\bar{\tau}$ and $\bar{\alpha}_V$ = 7. For a larger value of $\bar{\tau}$, a similar increment of $\Delta \bar{C}_2$ is achieved at $\bar{\alpha}_V$ = 5 and 7, if $Pe \approx 15$ and 40 respectively. Similarly, the coefficient χ_S changes in the range of *Pe* values from 10 to 1000 (Figure 16). However, the values of χ_s are significantly smaller than the corresponding values of χ_c at the same values of *Pe*. As a consequence, the range of *Pe* values, at which a reliable calculation of the concentration *Pe* is possible taking into account the suspension diffusion, is even wider. In particular, a similar correction (ΔS_2=0.2S_1, χ_s=1.2) at $\bar{\tau}$ = 100, $\bar{\alpha}_V$ = 7 occurs if $Pe \approx 15$. In the other cases considered, the limiting value of *Pe* is less than 10, which indicates a generally weak influence of the diffusion factor on the hydraulic characteristics.

REFERENCES

Berdanov, V.M., Bogolyubov, K.S., & Molozhavaya, E.I. (1978). *Artificial replenishment of groundwater in systems of water management and drinking water supply*. Stroyizdat. (in Russian)

Bochever, F. M. (1977). About estimation of filtration intensity from basins in systems of artificial groundwater recharge. *Proceedings of VNII VODGEO*, 63, 3–10. (in Russian)

Burchak, T. V. (1978). *Infiltration basins*. Budivelnik. (in Russian)

Dunin, F. Kh. (1980). Modeling of infiltration in approximation to field conditions. In *Frontiers of Hydrology* (pp. 241–275). Gidrometeoizdat. (in Russian)

Geiger, S. L., & Durnford, D. S. (2000). Infiltration in homogeneous sands and a mechanistic model of unstable flow. *Soil Science Society of America Journal, 64*(2), 460–469. DOI: 10.2136/sssaj2000.642460x

Gusev, E. M. (1981). Experimental studies of pressure seepage. In *Physics of Soil Waters* (pp. 195–205). Nauka. (in Russian)

Guyonnet, D., Amraoui, N., & Kara, R. (2000). Analysis of transient data from infiltrometer tests in fine – grained soils. *Groundwater, 38*(3), 396–402. DOI: 10.1111/j.1745-6584.2000.tb00225.x

Kadlec, R. H., & Knight, R. L. (1996). *Treatment wetlands. Boca Ration*. CRC Press.

Kommunar, G. M., & Bogolyubov, K. S. (1977). Some features of calculation of infiltration basins under conditions of colmatization and clogging. *Proceedings of VNII VODGEO*, 63, 10–16. (in Russian)

Kosterin, A. V., & Skvortsov, E. V. (2002). Liquid infiltration from the ground surface into the fractured-porous aeration zone. *Izv.* [in Russian]. *RAS MZhG*, (5), 93–98.

Latifi, H., Prasad, S. N., & Helveg, O. J. (1994). Air entrapment and water infiltration in two-layered soil column. *Journal of Irrigation and Drainage Engineering, 120*(5), 871–891. DOI: 10.1061/(ASCE)0733-9437(1994)120:5(871)

Mein, R. G., & Farrel, D. A. (1974). Determination of wetting front suction in the Green-Ampt equation. *Soil Science Society of America Journal, 38*(6), 872–876. DOI: 10.2136/sssaj1974.03615995003800060014x

Neuman, S. P. (1976). Hydraulic head at the wetting front in the Green-Ampt model. *Water Resources Research, 12*, 564–566. DOI: 10.1029/WR012i003p00564

Oleynik, A. Ya., & Tkach, V. N. (1972). Water flow calculations of rice systems under two-layer aquifer structure. [in Russian]. *Land Reclamation and Water Management, 24*, 88–99.

Parlange, J.-Y., Haverkamp, R., & Touma, J. (1985). Infiltration under ponded conditions: 1. Optimal analytic solution and comparison with experimental observations. *Soil Science, 139*(4), 305–311. DOI: 10.1097/00010694-198504000-00003

Plotnikov, N.A. (1983). *Design of systems of artificial replenishment of underground waters for water supply*. Stroyizdat. (in Russian)

Poliakov, V.L. (2008). Intensive wetting of multilayer soils. *Applied Hydromechanics*, 10(82), 1, 69–79. (in Russian)

Poliakov, V. L. (2014). *Filtration flow deformations in drained soils: theory and applications*. Agrar Media Group. (in Russian)

Polubarinova-Kochina, P. Ya. (1977). *Theory of groundwater movement*. Nauka. (in Russian)

Recommendations on design and operation of artificial groundwater recharge systems. (1976). VNII VODGEO. (in Russian)

Romashchenko, M. I., & Balyuk, S. F. (2000). *Irrigation of lands in Ukraine. State and ways of improvement*. Svit. (in Ukrainian)

Shtykov, V. I., Shevelev, Ya. Z., & Koshevoy, O. Yu. (1987). *Use of livestock waste complexes on specialized systems*. Rosselhozizizdat. (in Russian)

Tkachuk, O., Yaruta, Y., & Shevchuk, O. (2022). Assessment of application conditions of infiltration basins for regulation of urban rainwater drainage. *Journal of Ecological Engineering*, *23*(2), 191–195. DOI: 10.12911/22998993/144718

Usenko, V. S. (1972). *Artificial recharge and infiltration water intakes of underground waters*. Science and Technology. (in Russian)

Verigin, N. N., & Vasiliev, S. V. (1987). Soil and ground wetting of the aeration zone. [in Russian]. *Appl. Techn. Phys*, (1), 133–137.

Watson, K., & Awadalla, S. A. (1986). Comparative study of the Green and Ampt analysis for a falling water table in a stratified profile. *Water Resources Research*, *22*(5), 723–730. DOI: 10.1029/WR022i005p00723

Yakovenko, Y. P. (2009). Mathematical modeling and research of infiltration basins operation. [in Ukrainian]. *Land Reclamation and Water Management*, *97*, 166–179.

Chapter 9
Concluding Remarks and Future Trends

ABSTRACT

This chapter briefly outlines the prerequisites for writing the monograph, its conception, the most important theoretical results, and related suggestions for practical applications to improve technology and the design of filtration facilities. It describes opportunities and prospects for future research on purifying contaminated waters of different origins using the specific properties of porous and pseudo-porous media, particularly when implementing biological treatment methods and membrane technologies.

9.1. SUMMARY OF RESULTS AND PROPOSALS

The monograph is devoted to theoretical studies of the removal of dispersed contaminants from natural and wastewater by rapid filtration, which is implemented in practice in special artificial (filters) and natural (ground bed) porous media, as well as the development of a methodology for technological and design analysis based on these studies.

The rationale for writing it is due to the deterioration of water supply for the economy and population in Ukraine as a result of irrational use of limited water resources, martial law and occupation of significant territories in the east and south of the country. The state of its drinking component is of particular concern, namely, the gradual reduction of reserves and increasing contamination of surface water and groundwater suitable for people.

Working media made of porous material (porous structures) in suspension and solution separation technologies are fundamentally capable of providing intensive mass exchange between the mobile (liquid, gaseous) and immobile (skeleton, biological) phases for a prolonged period. On the one hand, a multitude of regularly arranged structure-forming elements creates a material basis for active absorption and fixation of impurities. On the other hand, due to the relatively large free space (porosity) and its regular configuration, it is possible to arrange intensive filtration flow and thus pass a very significant volume of contaminated liquid through the filter media in a short time.

When writing the monograph, two main goals were pursued:

DOI: 10.4018/979-8-3693-8211-0.ch009

- systematic development of engineering methods for predicting changes over time and within the volume of a well-sorbing filtering material under controlled conditions of operational characteristics (concentrations, pressures, flow rates, levels)
- based on the proposed calculation methods, creation of a methodological basis for technological and design modeling of the clarifying effect of filtration facilities (rapid filters) for the purpose of rational selection of control algorithm, their technical arrangement.

The theoretical basis of the monograph consists of exact and approximate solutions to a class of mathematical problems of aqueous suspensions filtration and the resulting calculation dependencies. They have become a reliable tool, primarily for analytical calculations of the development of the physical-chemical conditions in facilities designed to remove dispersed impurities of various origins, mainly rapid filters.

Since water treatment filters are engineering structures for multiple use, with one change of filtering material capable of effectively purifying a significant volume of contaminated water over many operating periods, and in real time over many months and even several years, the absorption and flow capacities of the filtering material volume (filter media) becomes important. Therefore, when arranging the filter media, preference is generally given to non-cohesive materials with individual elements of almost spherical shape (grains). As a result, its pore space is formed only by flow-through pores and it is relatively easy to wash due to fluidization. Nevertheless, in water clarification practice by filtration, cohesive (fibrous) materials with increased strength, absorption capacity and permeability are increasingly being used. Their regeneration is significantly more complex and expensive, so the final decision on the choice of (non)cohesive material should be made based on technical and economic calculations.

Since the main function of filtration facilities in water treatment technologies is precisely the removal of harmful impurities from contaminated water, the focus of experimental and theoretical studies of their operation has always been the study of patterns of processes of different nature that jointly ensure its implementation. It is possible to operatively and reliably monitor the current physicochemical state of the working layer and the operability of the entire facility, evaluate its effectiveness based on water quality requirements (sanitary and hygienic standards) and purification efficiency (mechanical energy consumption, performance) exclusively thanks to, first of all, the formalization of the complex technological process using conservation laws and phenomenological relationships, and then the formulation and solution of corresponding mathematical problems, that is, thanks to mathematical modeling.

Mathematical modeling methods are a powerful tool in research on the problems of (pre)treatment of natural and wastewater in working media made of special porous materials. Thanks to their widespread application, it is possible to reliably monitor the deterioration of water treatment filters performance, to develop effective measures in advance to intensify the technological process, to improve design and management, and to prevent the occurrence of critical situations.

The results of theoretical studies using these methods, primarily through solving a class of mathematical problems of detachment and non-detachment, plane and radial, depth and surface, regulated and unregulated, multi-flow and single-flow filtration of aqueous suspensions, allow establishing general patterns in the behavior of all sought filtration characteristics; assessing their sensitivity in relation to model coefficients for better planning of experimental research; serve as a basis for technological and design modeling.

Due to the complexity of the original mathematical models of suspension filtration (complex of interrelated compartments, large number of dependent and independent variables, nonlinearity of basic equations and phenomenological relations, layered structure and curvilinear form of the filter media, mobility of physical domain boundaries), approximate solution methods and accuracy assessment of derived calculation dependencies acquire special importance when developing engineering methods for calculating its characteristics and justifying substantiating technological and structural parameters. Errors in approximate calculations usually were 1-3%, sometimes less than 1% and only in exceptional cases exceeded 10%. Such methods make it possible to promptly predict the development of the physicochemical environment in the working medium and at its output, to conduct a relatively simple but thorough analysis for the applied optimization of technological and design parameters.

A necessary condition for bringing the calculated pattern in the filtering material layer closer to the actual pattern, for the reliability of technological modeling results, is providing the basic mathematical model with initial empirical information in full, including the values of all coefficients and criterial parameters contained in the original mathematical model. A reliable source of such information can only be data from own research on experimental and industrial installations, as well as regulatory documents. Data for test calculations are suitable from reference literature and available publications suitable to similar filtration conditions.

A key role in modeling the clarification of aqueous suspensions through (non)detachment filtration is played by the kinetic coefficient, which generally characterizes the capability of the filter media material to absorb dispersed impurities. Significant difficulties with the form of representation of this coefficient are primarily due to the presence of two stages in the transition of suspended particles from the liquid phase to the solid (approaching the solid surface of the grains and fixing on it). It also reflects the significant influence of filtration conditions on the absorption rate, mainly the characteristic sizes of the filtering material and contamination particles, filtration flow rate, and medium temperature. Naturally, its reliable values can only be established experimentally. Obviously, optimizing the technological process is only possible with reliable information about kinetic coefficients. Their approximate values, borrowed from reference literature, fundamentally do not allow achieving maximum performance from filtration facilities.

The limited mechanical energy resource of filtration facilities is spent mainly on overcoming the hydraulic resistance of the contaminated filter media layer. When suspension particles are deposited, a disproportionately larger amount of water is bound (hydration), so that their volume constitutes only a small part of the total deposit volume (several percent). Thus, the magnitude of this resistance is determined precisely by the amount of bound water. Therefore, calculations of head losses in the filter media and technological times, especially the duration of the filter run, can be reliable only if the composition of the deposit and its most important characteristic (bulk factor) are reliably determined. This characteristic is a variable, and throughout the filter run, as deposit accumulates, it can change twofold or more. Taking into account the scarcity of information about the bulk factor in the literature and the complexity of its experimental determination, it is acceptable to use its constant (average) value in mathematical and technological modeling.

Filtration facilities operate throughout the year and clarify aqueous suspensions at ambient temperature. Therefore, the operating temperature constantly changes, reaching a maximum in warm weather and a minimum in cold weather, with a maximum difference between them of about 20 . Temperature fluctuations, due to its close connection with liquid viscosity, significantly affect the absorption of impurities and filtration of the suspension. When conducting applied analysis in predictive and technological calculations, the values of the main kinetic coefficient and hydraulic conductivity should be taken in accordance with the current season. Indeed, the duration of a representative filter run due to seasonal changes in the temperature factor can differ by many tens or even hundreds of percent.

The regularly performed sensitivity analysis of technological times, mainly the duration of the filter run, in relation to the main mass transfer coefficient and design parameter (height of the packed bed) demonstrated the presence of a pronounced maximum in the sought value (time function) within the calculated interval (covering normal, overestimated and underestimated values of the argument). Thus, there are usually unique (optimal) values of the specified coefficient and height, at which the duration of the filter run is maximal. Significantly, even a slight deviation of the argument value from the optimal one most often leads to a noticeable reduction in the effective operating time of the filter, and consequently, a proportionate increase in the cost of purified water. Therefore, taking into account the duration and cost of filtration facilities operation, from an economic point of view, despite additional resources and efforts, it is advisable to substantiate and implement in practice the procedure of applied optimization in relation to their design and implemented technology.

Water suspensions in modern rapid filters are clarified when implementing the filtration flow process in two rate modes. Modeling filtration at unregulated rate is significantly complicated due to the variable nature and uncertainty of mass transfer coefficients. In practice, this type of depth detachment filtration, especially when equipping the filter with a storage tank, contributes to a more complete use of the sorption resource of the filtering material. At the same time, the operation of filters becomes simpler and more reliable.

One of the effective ways to intensify the filtration of suspensions is to create a stable uneven rate field in the main working body of water treatment filters. Giving the filter media volume an axisymmetric shape and supplying water in the radial direction (plane-radial filtration) allows, firstly, to significantly reduce mechanical energy expenditure, and secondly, to use well-absorbing filtering materials. As a result, with plane-radial filtration compared to traditional plane filtration, it is possible to significantly increase the duration of the filter run (with intensive absorption by 2 times or more), to purify large volumes of water with increased content of dispersed impurities in a short time, and to noticeably reduce operational costs.

With enhanced clogging of the filtering material due to its high absorption capacity and severe initial water contamination, it is possible to noticeably slow down the increase in the output concentration of impurities and head losses in the filter media during filtration at a constant rate by, firstly, supplying water directly to the filter by non-traditional methods; secondly, arranging heterogeneous filter media with a special layered structure; thirdly, using stepped filters with a prefilter.

Reorientation of the filtration flow without interrupting the technological process is advisable when the deposit is firmly fixed (non-detachment filtration). An important factor in this is choosing the right time to change the direction of filtration. The effect achieved in this way is expressed in an increase in the duration of the representative filter run by approximately 25%. With detachment filtration, changing the direction of the filtration flow causes a sharp decrease in the quality of the filtrate due to partial immobilization of a large amount of deposit now in the outlet section of the bed.

A significant clarifying effect, measured by several tens of percent, can be realistically achieved by organizing the suspension input into a well-absorbing filter media with two (two-flow filter) or more (multi-flow filter) flows. To maximize the duration of the filter run in this way, it is necessary to substantiate the selection of the supply location and flow rates of all flows (with unchanged filter performance).

The use of surface filtration is promising due to the possible and sometimes significant reduction in operational costs. It is important to theoretically substantiate the duration of the critical period for the operation of a precoat filter, when the surface active (dynamic) layer is formed and, as a result, the quality of water treatment is unsatisfactory. The concentration of structure-forming particles in the impurity (coarse dispersion fraction) has a significant impact on the formation of the dynamic layer. By purposefully changing it, it is possible to achieve a substantial reduction in head losses and the duration of the preparatory period.

With high demand for clarified water and large volumes of contaminated water, intensification of rapid filters operation by increasing the hydraulic load appears promising in the case of non-consolidated deposit. Then, when modeling the filtration process, it is necessary to additionally take into account inertial forces, and in the forced rate mode (ultra-rapid filters) also flow turbulization (non-linear filtration). An increase in the hydraulic resistance of the filter media can lead to a significant reduction in technological times. The question of whether it is (in)expedient to significantly increase the filter performance is preferably solved taking into account practical needs through mathematical modeling with mandatory availability of reliable information about the dependence of the hydraulic resistance of the contaminated filtering material and the rate of interphase mass transfer on the filtration rate during forced filtration.

For multiple use of natural porous media as a protective barrier during infiltration of contaminated surface waters for their subsequent conservation in underground storage or use for water supply to the population and economy, optimization of technological parameters and primarily hydraulic load and initial impurity content becomes of great importance. In this case, it is necessary to take into account the peculiarities of mass transfer and filtration flow processes in porous media of natural composition with reduced throughput and absorption capacities.

In conclusion, it should be emphasized that due to the formalization of technological processes based on universal conservation laws and universal phenomenological relations, the fundamental mathematical models have a generalized character and are able to adequately reflect also specific conditions that develop, for example, during the separation of suspensions for the purpose of isolating and accumulating useful chemicals. As a result, the theoretical results presented in the monograph, which, incidentally, are often obtained using standard simplification techniques, can be widely and successfully (in some cases after minor correction) used not only for improving existing physicochemical and biochemical technologies and facilities, but also for developing technologies and facilities of a new generation.

9.2. ON THE PROSPECTS OF MODELING WATER TREATMENT USING POROUS MEDIA

The progressive pollution of surface and underground waters in Ukraine is explained not only by the development of chemical production but also by their extensive use in industrial facilities, agriculture, and communal services; insufficient measures for environmental protection due to limited financial capabilities, dishonesty of major water users, and uncontrolled consumption of water resources in zones of occupation and military operations. As a result, pollutants are often present in significant quantities not only in wastewater but also in natural waters, have a complex composition, and are more toxic and resistant to external influences.

Therefore, when modeling the operation of water treatment plants and, in particular, filtration facilities, it is justified to conditionally divide complex pollutants into a number of components that differ significantly in mechanical, physicochemical, and (micro)biological properties. It is logical that the scientific justification for removing such pollutants from waters of different origins should be based on the results of in-depth research using experimental and theoretical methods on the behavior of each component. At the current stage of research on water treatment problems, first of all, it is advisable to separately consider the behavior of dispersed and dissolved forms of pollutants. When necessary, one has to resort to greater detailing of the nature of pollution based, in essence, on the characteristic particle sizes as the dominant distinguishing feature of the impurity (coarsely dispersed, finely dispersed, microfiltration, ultrafiltration, etc.). Obviously, for effective extraction of impurities from solutions, special purification methods and apparatus that implement them should be used. Moreover, the designs of such apparatus may have fundamental (filter with biporous filtering material, wetland, aerotank, bio-conveyor, membrane module) and non-fundamental (adsorber, bioreactor-filter) features.

The information basis of technological modeling consists of a set of values of criteria parameters, and among them, first of all, the maximum permissible concentrations of all controlled pollutants. In the future, due to emerging trends in healthcare and the global economy, we should expect stricter requirements for water treatment quality and an expansion of the range of monitored pollutants. In the first case, the methodology of technological calculations developed in this monograph does not need improvement and can be successfully applied after appropriate adjustment of standards. In the second case, when it is necessary to additionally predict spatial and temporal changes in the content of a previously non-relevant toxic substance, mathematical models should be developed that would reliably describe its transport, transformations, interphase exchange, and possibly reflect the influence of other pollutants.

Some guidelines in the development of water treatment technologies using porous media are discussed below. Materials from research in these areas with extensive use of mathematical modeling methods are planned to be presented in a separate monograph. Theoretical analysis of the effectiveness of modern practical methods for removing dissolved impurities is primarily valuable due to recommendations for rational selection of technological and design parameters, which are developed based on solutions to correctly formulated mathematical problems of water treatment. The basic mathematical models should combine simplicity, which will facilitate the application of new calculation methods in engineering practice, with a complete formal description of the technological process under consideration, taking into account all factors that actually affect it. And certainly, they must be provided with reliable specific initial information in full.

Of significant practical interest is the adsorption method, which contributes to the effective removal of dissolved impurities through physical adsorption. Thanks to physical adsorption, it is possible to promptly extract from aqueous solutions various impurities - organic compounds, heavy metal ions, biogenic elements, radionuclides, residual chlorine, and other chemical and biological substances. The physical affinity of the interaction of contamination particles of different sizes with the solid phase predetermines the similarity of the designs of the rapid filter-clarifier and adsorber. A key role in assessing the purification potential of adsorption apparatuses is played by the absorption capacity of the adsorbing material, which can be in a suspended state and in small quantities (static adsorption) or, conversely, in a stationary and densely packed state (dynamic adsorption). Formally, such capacity is usually characterized by significantly nonlinear adsorption isotherms (Langmuir, Freundlich and others). If the adsorbent grains are porous (granular activated carbon), then in the transition of the adsorbate from the liquid phase to the solid, three characteristic stages must be distinguished. It appears on the surface of the grain, and therefore, at the entrance to the network of micropores at the end of the first stage. The adsorbate molecules move inside due to molecular and surface diffusion in the second stage. And they firmly attach to the walls of the micropores only in the third stage. A reliable forecast of the development of the adsorption process in a (bi)porous medium, which is realistic only due to mathematical modeling of the three-stage transition process, allows for improving the design of adsorption reactors and filters, selecting technological parameters close to optimal (size of porous grains, adsorbent dose, etc.).

The inevitable contradiction in developing the methodology of engineering calculations for adsorption water treatment between the completeness of its formal description in this case, the action of adsorbents in adsorption apparatus, the significantly limited capabilities of analytical methods, and the request for flexible control of the technological process in a simple and reliable method can be substantially mitigated through a reasonable compromise. On the one hand, simplification of nonlinear non-stationary models of water treatment through non-activated adsorbent allows approaching an exact solution of the corresponding approximate analog, and in rare cases even obtaining it. On the other hand, the discrepancy between the calculated (approximate) pattern and the true (mathematically exact) pattern may increase significantly in this case. Here, the (in)adequacy of the initial mathematical model to operating conditions is not discussed, and a corresponding conclusion can only be made after comparing the calculated and actual states of the working medium. Obviously, estimating the computational error due to the use of a simplifying technique is in principle possible only by comparing approximate and exact results. Nevertheless, even relying on simpler, for example linear, models, it is possible to obtain a correct understanding of trends in the behavior of the sought characteristics during the implementation of the technological process in adsorbers. Incidentally, the corresponding calculated data can be useful in solving and verifying tasks of static and dynamic adsorption, but in a more general formulation (at two levels).

Groundwater is of particular importance for drinking water supply in Ukraine. More than 50% of such waters contain elevated amounts of iron and manganese and therefore they pose a danger to human health in their original state. In conditions of increasing deficit of clean water, they are an important reserve, and at the same time are already being used without alternatives in rural areas of various regions of the country. To use this most important resource of quality water with maximum benefit, it is necessary to develop a methodological basis for technological and design modeling of long-term operation of specialized filters (physicochemical iron removal). And first of all, based on existing concepts and a conceptual model, it is necessary to formulate a fundamental mathematical model according to the compartment principle. Such a model should jointly describe the processes of oxidation, transformation of two forms of iron (ferrous and ferric) with subsequent formation of aging deposit based on its hydroxide. It is necessary to conduct comprehensive experimental studies under laboratory and production conditions to collect complete initial data for modeling the technological process. Of fundamental importance is the analytical solution of the corresponding complex nonlinear non-stationary problem with minimal assumptions, which will allow operational monitoring of the deterioration of filtrate quality and the increase in hydraulic resistance of the filter media. The forecast of the state and dynamics of the most important filtration characteristics logically concludes with proposals for the applied optimization of the standard duration of the filter run and the service life of the next change of filter media in the operating bed based on the known age of the deposit and its hydrodynamic stability. Based on technical and economic calculations of reduced costs, it is planned to perform a detailed technological analysis of a number of algorithms for controlling a specialized filter, establish the duration of its effective operation until the next washing and the overall service life of the new filter media, taking into account the gradual consolidation of the deposit. From a practical point of view, it is especially important to study the possibilities of optimizing the operation of iron removal filters and reducing the cost of purified groundwater through the rational choice of the bed height and (effective) diameter of its grains.

Biofiltration is a highly effective method for removing dissolved organic compounds and biogenic elements from natural and wastewater. It is successfully carried out in submerged bioreactor-filters during biological treatment of domestic wastewater containing predominantly small quantities of easily oxidizable substances. The porous layer of media in such filters is formed by multiple densely packed elements of various configurations. Due to the significant contact surface of profiled elements (grains) of the media with the liquid phase, a large active biomass of specific microorganisms (in accordance with the nutrient medium) accumulates and is effectively evenly distributed on it. A specially selected population of microorganisms, together with their metabolic products, forms biofilms that firmly attach to the elements of the porous medium. The growth of the hydrobiocenosis, which includes in granular media along with microorganisms-destructors also protozoa and predatory microflora, is significantly constrained due to unfavorable conditions for its existence - limited free space, nutrition, oxygen; force impact of the filtration flow; low temperature; presence of toxins, inhibitors, etc. As a result, experimental studies of biofiltration and generalization of their results, as well as theoretical analysis of the oxidative action of biofilters, become extremely complicated. Their effective stable operation is ensured by adopting rational design and technological parameters. It is much easier and more reliable to choose them if mathematical modeling methods are used, based on modern models of biofiltration of weakly concentrated organic solutions. Two standard approaches are successfully used in developing such models. From a formal point of view, the first approach is significantly simpler, in which the media is considered as a physical-biological continuum with properties and characteristics that change continuously and smoothly over time and height. However, the coefficients in the corresponding models depend on a multitude of external (filter volume) and internal (representative biofilm) factors, which explains the significant dispersion of experimental values and complicates their systematization and generalization. When formulating mathematical problems of biofiltration with the aim of developing engineering methods for calculating physical and chemical characteristics, and subsequently substantiating technological and design parameters, it is necessary to resort to a number of simplifications of the complex formal description of the removal of dissolved substances from wastewater. It is advisable to use a system of assumptions, the composition of which would correspond to specific operating conditions and be selected taking into account the accumulated extensive experience in the operation of biofiltration facilities, the results of previously conducted theoretical and experimental studies on pilot plants. Taking into account the practical importance and cost of removing organic pollutants from wastewater at special filtration facilities, the applied optimization of their operation acquires great importance. It is possible to operatively perform the multivariate calculations necessary for it only by theoretical means. The hydraulic component of the complex technological process in biofilters turns out to be less significant than in filters-clarifiers.

The bioconveyor technology is promising, which is implemented in artificial pseudo-porous media formed by a multitude of vertically fixed synthetic threads. Its high efficiency is explained by the active involvement in the technological process of various hydrobionts - from microorganisms to filter feeders and predators. An important component of this technology is biofiltration in the absence of oxygen. Anaerobic biofiltration, which often precedes and complements aerobic biofiltration, is intended to be analyzed predominantly by analytical methods, taking into account the peculiarities of formation and interaction of intermediate and final products of biochemical transformation of dissolved organic matter in oxygen-free conditions. A separate study of biodegradation of organic substrate in a representative anaerobic biofilm (as a two-stage process) and its convective transfer in the operating volume is forthcoming. When developing an engineering method, it is planned to implement a calculation scheme reflecting possible significant differences in the state of the operating medium (maximum and partial saturation with biomass).

Special attention should be paid to studying the consequences of dispersed contamination formed during the purification of aqueous solutions, i.e., patterns of secondary (bacterial) contamination. As a rule, their source is the biological phase, namely, soluble metabolic products released from biofilms, as well as fragments detached from them in various ways. The direct cause becomes conditions unfavorable for the vital activity of microorganisms. In such cases, it is advisable to jointly investigate the purifying effect of the biological and solid phases.

The subject of in-depth consideration by theoretical methods should be trickling biofiltration and the operation of high-loaded trickling biofilters. Much attention should be paid to joint gas filtration and moisture transfer.

The method of biological wastewater treatment through biofiltration, including aerobic, anaerobic, trickling varieties, as well as special-purpose biofilters that implement it in practice, require significant capital investments and operational costs. Such filters are also difficult to manage. Therefore, for water supply, primarily for small settlements (up to 2,000 thousand inhabitants), special biological treatment facilities, namely, wetlands are increasingly being used. They are economical, environmentally safe, have a simple design, are easily managed, quickly adapt to fluctuations in hydraulic load, and are capable of treating even concentrated wastewater, withstanding shock loads. In addition, they require little energy for stable operation. Their main disadvantages should be considered the need for a large area and dependence of efficiency on the season. Wetlands allow effective treatment of various contaminated waters - domestic, industrial, agricultural effluents, drainage, and stormwater. At the same time, it is possible to simultaneously remove a significant part of the main types of pollutants, namely, organic and suspended substances, biogenic elements, heavy metals, and pathogens. A key role in the technological process, as in biofilters, is played by microorganisms. It is microbial biocenoses that directly ensure the mineralization of organic substances, transformations of mineral nitrogen, using their products and released energy for biosynthesis and respiration. As a result, most of the removed pollutants are usually spent on the needs of the microbiocenosis. It is important to note that its vital activity is significantly facilitated by growing vegetation. Progressive clogging of the wetland media is an inevitable negative consequence of wastewater treatment. It has physical and chemical causes, with the main causes being the deposition of dispersed impurities and the growth of microbial biomass. The active and diverse participation of higher plants in the operation of wetland facilities causes a significant complication in the understanding of physical and microbiological processes, which together provide removal of the main types of pollutants from wastewater, including nitrogen compounds, and, as a consequence, their formal description. Therefore, the obvious discrepancy between the current level and correctness of mathematical modeling and practical needs is natural. At present, in essence, one can rely only on one's own experience of their operation, since finding information and practical recommendations in the literature that would correspond to unique conditions in each specific case is unrealistic. Nevertheless, certain successes have been achieved at different levels of modeling water treatment by infiltration systems. In general mathematical models of wetland, claiming a broad coverage of physical, chemical, and biological processes, it is necessary to resort to significant complications of their structure (increasing the number of compartments) and structural elements. So far, it has been possible to achieve only a satisfactory correspondence between operational and calculated data. Wetland engineering structures are essentially specialized bioreactor-filters, in which microbiota is supplemented by higher plants. Due to the release of oxygen and chemical substances by roots into the environment, as well as an increase in the available surface, more favorable conditions for its vital activity are created. As a result, it is possible to improve generalized treatment indicators, namely, to reduce outlet values of COD and BOD, to achieve a significant reduction in the content of organic and suspended substances, biogenic elements, and heavy metals simultaneously.

It is relevant to develop a simple calculation tool that would allow separately assessing the contribution to the removal of excess $NH_4^+ - N$ from wastewater on a vertical infiltration system with a two-layer structure of microbial biocenoses (nitrifiers and denitrifiers) and plants with a developing root system. Through the rational choice of the ratio between the thicknesses of both layers, it is realistic to achieve the removal of a significantly greater amount of nitrogen on the aforementioned infiltration system.

Membrane technologies have been rapidly developing over the past decades due to their great advantages over traditional methods of separating suspensions, solutions, mixtures, and setting new practical tasks for chemical technologies. Their main advantages include low energy costs, wide possibilities for control, combination with other separation processes, and automation. A relatively new field of application is membrane water treatment, which allows achieving drinking water standards by eliminating hardness, color, microorganisms, toxins, mutagens, carcinogens, etc. It is planned to conduct meaningful theoretical studies of membrane water treatment. Dead-end filtration will be studied by exact methods applied to membranes with regular and irregular structure, taking into account progressively uneven clogging of micropores by height up to their complete blockage, as well as surface contamination with the formation of a deposit layer (secondary membrane) based on non-stationary nonlinear models of continuous filtration. Of great practical interest is the modeling of cross-flow filtration after stabilization of hydrodynamic processes and the physical and mechanical state of the pressure channel and adjacent membranes, taking into account the phenomenon of concentration polarization in critical conditions. Special attention should be paid to the effect of (de)hydration of microparticles and quantitative assessment of its possible consequences for membrane selectivity.

Basic Symbols

C– (volumetric) concentration of suspended solids

C_e– (outlet) concentration of contaminants in filtrate

C_0– inlet (initial) concentration of contaminants

C_*– maximum allowable (volumetric) concentration of contaminants

D– convective diffusion coefficient

d_g– (equivalent) diameter of structural particles (grains) of porous medium (filter bed, ground)

d_p– (equivalent) diameter of suspended solids

f_k– symbol of normalized function of hydraulic conductivity (permeability)

f_s – symbol of function of deposit hydrodynamic resistance

f_α– symbol of function characterizing rate of suspension particles attachment to solid phase

f_β– symbol of function characterizing rate of deposited particles detachment

g– gravity acceleration

h– piezometric head

H_{out}– head at filtrate collector

H_d– head at filter bed outlet

H_{in}– head at the initial suspension (raw water) collector

H_W– depth of surface suspension layer (above the filter or ground bed, in storage tank)

H_*– mark of the edge of the chute discharging the washing water

k– actual current (saturated) hydraulic conductivity

L– height of filter (bed) operating layer

L_f– height of cylindrical packed bed

n– current actual filter bed porosity

n_0– clean filter bed porosity

Q_f– filter productivity

Q_{in}– volumetric flow rate in the supply pipeline (communications)

R_{f1}– total resistance of supply communications (pipeline)

R_{f2}– total resistance of offtake communications (pipeline)

S– (volumetric) concentration of deposited suspension particles

S_d– dynamic solids take-up capacity with respect to deposited particles

S_m– true solids take-up capacity with respect to deposited particles

S_S– volumetric concentration of newly formed solid matter (deposit, biomass)

T_t– (thermodynamic) temperature of porous medium

t_f– filter run duration

t_H– time of suspension level rise in storage tank to maximum mark

t_h – time of reaching maximum allowable head losses in filter bed

t_p– time of protective effect of filter bed

t_V– time to reduce filtration rate to economically feasible minimum value

V– filtration (flow) rate

V_*– the economically feasible minimum value of the filtration rate

W_f– operating bed volume

α– rate coefficient of suspension particles attachment (sticking)

β– rate coefficient of deposited particles detachment

χ_S– symbol of adhesion function

Δh– head losses in operating layer

Δh_*– maximum allowable head losses in filter bed

γ– symbol of deposit composition function (bulk factor)

γ_a– (representative) average value of deposit composition function

κ_f – shape factor

λ– filtration coefficient

λ_0 – reduced filtration coefficient

μ– dynamic viscosity

ν– kinematic viscosity

θ– autocatalytic coefficient

ω– inlet surface area of filter bed

Compilation of References

Abboud, N. M., & Corapcioglu, M. Y. (1993). Modeling of compressible cake filtration. *J. Colloid. Int. Sci., 160*, 304–316.

Aberg, B. (1996). Void sizes in granular soils. *Journal of Geotechnical Engineering, 122*(3), 236–239. DOI: 10.1061/(ASCE)0733-9410(1996)122:3(236)

Abramson, A., & Brown, D. B. (2007). Influence of solution ionic strength on the collision efficiency distribution and predicted transport distance of a Sphingomonas sp. flowing through porous media. *Water Research, 41*(19), 4435–4445. DOI: 10.1016/j.watres.2007.06.005 PMID: 17644150

Adelman, M. J., Monroe, L., Weber-Shirk, M. L., Cordero, A. N., Coffey, S. L., Maher, W. J., Dylan Guelig, D., Jeffrey, C., Will, J. C., Stodter, S. C., Hurst, M. W., & Lion, L. W. (2012). Stacked filters: Novel approach to rapid sand filtration. *Journal of Environmental Engineering, 138*(10), 999–1008. DOI: 10.1061/(ASCE)EE.1943-7870.0000562

Adin, A., & Elimelech, M. (1989). Particle filtration for wastewater irrigation. *Journal of Irrigation and Drainage Engineering, 115*(3), 474–487. DOI: 10.1061/(ASCE)0733-9437(1989)115:3(474)

Adin, A., & Rebhun, M. (1977). A model to predict concentration and head loss profiles in granular bed filtration. *Journal - American Water Works Association, 69*(8), 444–453. DOI: 10.1002/j.1551-8833.1977.tb06787.x

Adin, A., & Rebhun, M. (1987). Deep-bed filtration: Accumulation-detachment model parameters. *Chemical Engineering Science, 42*(5), 1213–1219. DOI: 10.1016/0009-2509(87)80072-6

Aerov, M. E., Todes, O. M., & Narinsky, D. A. (1979). *Apparatuses with stationary granular layer*. Khimiya. Leningrad Branch. (in Russian)

Ahsan, S., Kaneco, S., Ohta, K., Mizuno, T., & Kani, K. (2001). Use of some natural and waste materials for waste water treatment. *Water Research, 35*(15), 3738–3742. DOI: 10.1016/S0043-1354(01)00047-1 PMID: 11561637

Akgiray, O., & Saatci, A. M. (1998). An algorithm for bank operation of declining rate filters. *Water Research, 32*(7), 2095–2105. DOI: 10.1016/S0043-1354(97)00459-4

Aksogan, S., Basturk, A., Yuksel, E., & Akgiray, O. (2003). On the use of crushed shells of apricot stones as the upper layer in duel media filters. *Water Science and Technology, 48*(11-12), 497–503. DOI: 10.2166/wst.2004.0899 PMID: 14753573

Alekseev, V.S., Kommunar, G.M., & Sherzhukov, B.S. (1989). *Mass transfer in water-saturated rocks. Results of Science and Technology. V.11*. Hydrogeology, Engineering Geology. (in Russian)

Alekseev, V. S., & Kommunar, G. M. (1974). Clogging of filters and prefilter zones of water intake wells. [in Russian]. *Water Resources*, (5), 170–177.

Alonso, C., Zhu, X., Suidan, M. T., Kim, B. R., & Kim, B. J. (2000). Parameter estimation in biofilter systems. *Environmental Science & Technology*, *34*(11), 2318–2323. DOI: 10.1021/es990329o

Altshul, A. H., Usenko, V. S., & Chaban, M. O. (1977). *Regulation of groundwater reserves*. Kolos. (in Russian)

Amburgey, J. E., & Amirtharajah, A. (2005). Strategic filter backwashing techniques and resulting particle passage. *Journal of Environmental Engineering*, *131*(4), 535–547. DOI: 10.1061/(ASCE)0733-9372(2005)131:4(535)

American Water Works Assosiation. (1999). *Water quality and treatment, a handbook of community water supply* (5th ed.). Mc Graw–Hill.

Anderson, A., Laurent, P., Kihn, A., Prévost, M., & Servais, P. (2001). Impact of temperature on nitrification in biological activated carbon (BAC) filters used for drinking water treatment. *Water Research*, *35*(12), 2923–2934. DOI: 10.1016/S0043-1354(00)00579-0 PMID: 11471692

Anderson, E., Schilling, K. E., Just, C. L., & Seo, B. C. (2024). Quantifying the impact of a constructed wetland on downstream nitrate concentrations and loads in the US Midwest. *Frontiers in Environmental Science*, *12*, 1416018. DOI: 10.3389/fenvs.2024.1416018

Aravin, V. I., & Kochina, P. Ya. (Eds.). (1969). *Development of research on the theory of filtration flow in the USSR (1917–1967)*. Nauka. (in Russian)

Aravin, V. I., & Nosova, O. N. (1969). *Natural studies of water flow (theoretical bases)*. Energia. (in Russian)

Aronino, R., Dludy, C., Arkhangelsky, E., Shandalov, S., Oron, G., Brenner, A., & Gitis, V. (2009). Removal of viruses from surface water and secondary effluents by sand filtration. *Water Research*, *43*(1), 87–96. DOI: 10.1016/j.watres.2008.10.036 PMID: 19013631

Arora, S., Dhaliwal, S. S., & Kukreja, V. K. (2006). Simulation of washing of packed bed of porous particles by orthogonal collocation on finite elements. *Computers & Chemical Engineering*, *30*(6-7), 1054–1060. DOI: 10.1016/j.compchemeng.2006.02.002

Arun, N., Abdul, A., Razack, B., & Sivasubramanian, V. (2013). Recent progress in hydrodynamics of inverse fluidized bed reactors: A review. *Chemical Engineering Communications*, *200*(9), 1260–1277. DOI: 10.1080/00986445.2012.744747

Ayukaev, R. I., & Meltser, V. Z. (1985). *Production and application of filtering materials for water purification. Reference manual*. Stroyizdat. (in Russian)

Ayukaev, R. R. (2003). Slow filters in household and drinking water supply of small water consumers in Karelia. [in Russian]. *Water and Ecology*, (3), 16–28.

Babenkov, E. D. (1977). *Water purification by coagulants*. Nauka. (in Russian)

Badriyha, B. N., Ravindran, V., Den, W., & Pirbazari, M. (2003). Bioadsorber efficiency, design and performance forecasting for alachlor removal. *Water Research, 37*(17), 4051–4072. DOI: 10.1016/S0043-1354(03)00266-5 PMID: 12946887

Bai, R., & Mackie, R. I. (1995). Modeling the transition between deposition modes in deep bed filtration. *Water Research, 29*(1), 2601–2604. DOI: 10.1016/0043-1354(95)00078-Y

Bai, R., & Tien, C. (1997). Particle detachment in deep bed filtration. *Journal of Colloid and Interface Science, 186*(2), 307–317. DOI: 10.1006/jcis.1996.4663 PMID: 9056359

Bai, R., & Tien, C. (2000). Effect of deposition in deep-bed filtration determination and search of rate parameters. *Journal of Colloid and Interface Science, 231*(2), 299–311. DOI: 10.1006/jcis.2000.7130 PMID: 11049680

Barbusiński, K., Szeląg, B., Parzentna-Gabor, A., Kasperczyk, D., & Rene, E. R. (2024). Application of a generalized hybrid machine learning model for the prediction of H2S and VOCs removal in a compact trickle bed bioreactor (CTBB). *Chemosphere, 360*, 142181. DOI: 10.1016/j.chemosphere.2024.142181 PMID: 38685329

Barenblatt, G. I. (1987). Nonlinear flow filtration – past, present and future. In *Problems of flow filtration theory and mechanics of oil recovery enhancement processes* (pp. 15–27). Nauka. (in Russian)

Barenblatt, G. I., Yentov, V. M., & Ryzhik, V. M. (1972). *Theory of non-stationary flow and gas*. Nedra. (in Russian)

Barry, D. A., & Sposito, P. (1989). Analytical solution of a convection-dispersion model with time-dependent transport coefficients. *Water Resources Research, 25*(12), 2407–2416. DOI: 10.1029/WR025i012p02407

Bass, G.M., Grabovskiy, P.P., & Larkina, G.M. (1978). Optimal duration of the filter run of rapid filters operating with falling rate. *Science and Technology in Municipal Economy. Collected papers*, 38, 37–39. (in Russian)

Bear, J. (1988). *Dynamics of fluids in porous media*. Mineola.

Belyaev, N. N., & Kozachina, V. A. (2014). *Mathematical modeling of mass transfer in horizontal settling tanks*. Accent PP. (in Russian)

Ben Aim, R., Vigneswaran, S., Prasanthi, H., & Jegatheesan, V. (1997). Influence of particle size and size distribution in granular bed filtration and dynamic microfiltration. *Water Science and Technology, 36*(4), 207–215. DOI: 10.2166/wst.1997.0121

Benefield, L., & Molz, F. (1985). Mathematical simulation of a biofilm process. *Biotechnology and Bioengineering, 27*(7), 921–931. DOI: 10.1002/bit.260270702 PMID: 18553761

Benkli, Y. E., Can, M. F., Turan, M., & Celik, M. S. (2005). Modification of organo-zeolite surface for the removal of reactive azo dyes in fixed-bed reactors. *Water Research, 39*(2-3), 487–493. DOI: 10.1016/j.watres.2004.10.008 PMID: 15644257

Berdanov, V.M., Bogolyubov, K.S., & Molozhavaya, E.I. (1978). *Artificial replenishment of groundwater in systems of water management and drinking water supply.* Stroyizdat. (in Russian)

Bergendahi, J. A., & Grasso, D. (2003). Mechanistic basis for particle detachment from granular media. *Environmental Science & Technology*, *37*(10), 2317–2322. DOI: 10.1021/es0209316 PMID: 12785543

Berman, A. S. (1953). Laminar flow in channels with porous walls. *Journal of Applied Physics*, *24*(9), 1232–1235. DOI: 10.1063/1.1721476

Bespamyatnov, G. P., & Krotov, Y. A. (1985). *Maximum permissible concentrations of chemical substances in the environment*. Khimiya. (in Russian)

Beuling, E. E., van den Heuvel, J. C., & Ottengraf, S. P. P. (2000). Diffusion coefficients of metabolites in active biofilms. *Biotechnology and Bioengineering*, *67*(1), 53–60. DOI: 10.1002/(SICI)1097-0290(20000105)67:1<53::AID-BIT6>3.0.CO;2-H PMID: 10581435

Biserov, R. A. (1967). *Research of mud capacity of sand beds at clarification of preliminary coagulated water by the method of superrapid filtration*. Ph.D. dissertation abstract. LECI Press. (in Russian)

Bochever, F. M. (1977). About estimation of filtration intensity from basins in systems of artificial groundwater recharge. *Proceedings of VNII VODGEO*, 63, 3–10. (in Russian)

Boller, M. A., & Kavanauch, M. C. (1995). Particle characteristics and head loss increase in granular media filtration. *Water Research*, *29*(4), 1139–1149. DOI: 10.1016/0043-1354(94)00256-7

Bolshakov, V. A., Konstantinov, Y. M., Popov, V. N., & Dadenkov, V. Y. (1979). *Handbook on Hydraulics*. Vishcha Shkola. (in Russian)

Bolton, G. T., Hooper, C. W., Mann, R., & Stitt, T. H. (2004). Flow distribution and velocity measurement in a radial fixed bed reactor using electrical resistance tomography. *Chemical Engineering Science*, *59*(10), 1989–1997. DOI: 10.1016/j.ces.2004.01.049

Bradford, S. A., & Bettahar, M. (2005). Straining, attachment and detachment of Cryptosporidium Oocysts in saturated porous media. *Journal of Environmental Quality*, *34*(2), 469–478. DOI: 10.2134/jeq2005.0469 PMID: 15758099

Bradford, S. A., Yates, S. R., Bettahar, M., & Simunek, J. (2005). Physical factors affecting the transport and fate of colloids in saturated porous media. *Water Resources Research*, *38*(12), 1327–1339.

Bresler, E., McNeil, B. L., & Carter, D. L. (1987). *Solonchaks and salt marshes. Principles. Dynamics. Modeling*. Gidrometeoizdat. (in Russian)

Brook, O. L. (1989). Theory and calculation of industrial filtration of suspensions with formation of highly compressible precipitates. [in Russian]. *Theoretical Foundations of Chemical Engineering*, *23*(4), 495–498.

Brook, O. L. (1992). To calculation of parameters of industrial filtration processes. [in Russian]. *Theoretical Foundations of Chemical Engineering*, *26*(6), 867–871.

Bugay, N.G., Krivonog, A.I., & Krivonog, V.V. (2004). Reagentless water purification by filters from fiber-porous polyethylene. *Applied Hydromechanics*, 6(78), 2, 27–43. (in Russian)

Bugay, N. G., Krivonog, A. I., & Krivonog, V. V. (2007). Contact reagent treatment of filtration water from fiber-porous material. [in Russian]. *Applied Hydromechanics*, *9*(1), 8–22.

Burchak, T. V. (1978). *Infiltration basins*. Budivelnik. (in Russian)

Caliskaner, O., & Tchobanoglous, G. (2005). Modeling depth filtration of activated sludge effluent using a compressible medium filter water. *Environmental Research, 77*(7), 3080–3091. PMID: 16381157

Calvin, R. M. (1992). Ripening of silica sand used for filtration. *Water Research, 26*(5), 683–688. DOI: 10.1016/0043-1354(92)90245-Y

Cameron, D. A., & Klute, A. (1977). Convective-dispersive solute transport with a combined equilibrium and kinetic adsorption model. *Water Resources Research, 13*(1), 183–188. DOI: 10.1029/WR013i001p00183

Caputo, A. C., Pelagagge, P. M., & Scacchia, F. (2001). Software tool for environmental plant modeling. *Environmental Progress, 20*(4), 229–239. DOI: 10.1002/ep.670200411

Carrier, W. D. (2000). III. Compressibility of a compacted sand. *Journal of Geotechnical and Geoenvironmental Engineering, 126*(3), 273–275. DOI: 10.1061/(ASCE)1090-0241(2000)126:3(273)

Chang, H. T., & Rittmann, B. E. (1987). Verification of the model of biofilm on activated carbon. *Environmental Science & Technology, 21*(3), 280–288. DOI: 10.1021/es00157a009 PMID: 22185107

Chang, Y.-I., Cheng, W.-Y., & Chan, H.-C. (2009). A proposed correlation equation for predicting filter coefficient under unfavorable deposition conditions. Short communication. *Separation and Purification Technology, 65*(3), 248–250. DOI: 10.1016/j.seppur.2008.10.031

Chaundry, F.H. (1987), Theory of declining rate filtration. I. Continuous operation. *J. Environ. Eng. Div., ASCE*, 113(4), 834–851.

Cheng, N.-S. (2006). Application of Ergun equation to computation of critical shear velocity subject to seepage. *Journal of Irrigation and Drainage Engineering, 129*(4), 278–283. DOI: 10.1061/(ASCE)0733-9437(2003)129:4(278)

Chen, J., Truesdall, S., Zu, F., Zhan, G., Belvin, C., Koopmah, B., Farrah, S., & Shan, D. (1998). Long-term evaluation of aluminium hydroxide-coated sand for removal of bacteria from wastewater. *Water Research, 32*(7), 2171–2179. DOI: 10.1016/S0043-1354(97)00427-2

Chern, J.-M., & Wu, C.-Y. (2001). Desorption of dye from activated carbon beds: Effects of temperature, pH and alcohol. *Water Research, 35*(17), 4159–4165. DOI: 10.1016/S0043-1354(01)00127-0 PMID: 11791845

Chiang, H.-W., & Tien, C. (1985). Dynamics of deep-bed filtration. I. Analysis of two limiting situations. II. Experiment. *AIChE Journal. American Institute of Chemical Engineers, 31*(8), 1349–1371. DOI: 10.1002/aic.690310814

Chou, H., Wu, L., Zeng, L., & Chang, A. (2012). Evaluation of solute diffusion tortuosity factor models for variously saturated soils. Water Resour. Res., 48, 1–11. W10539, .DOI: 10.1029/2011WR011653

Chrysikopoulos, C. V., & Katzourakis, V. E. (2015). Colloid particle size-dependent dispersivity. *Water Resources Research, 51*(6), 4668–4683. DOI: 10.1002/2014WR016094

Chuang, C.-J., & Li, K.-Y. (1997). Effect of coagluant dosage and grain size on the performance of direct filtration. *Separation and Purification Technology, 12*(3), 229–241. DOI: 10.1016/S1383-5866(97)00048-8

Cleasby, J. L. (1993). Status of declining rate filtration design. *Water Science and Technology*, *27*(10), 151–164. DOI: 10.2166/wst.1993.0224

Colton, J. F., Hillis, P., & Fitpatrick, C. S. B. (1996). Filter backwash and start-up strategies for enhanced particulate removal. *Water Research*, *30*(10), 2502–2507. DOI: 10.1016/0043-1354(96)00085-1

Corapcioglu, M. Y., & Haridas, U. (1985). Microbial transport in soils and ground water: A numerical model. *Advances in Water Resources*, *8*(4), 188–200. DOI: 10.1016/0309-1708(85)90063-6

Corapciooglu, M. Y., & Jiang, S. (1993). Colloid-facilitated groundwater contaminant transport. *Water Resources Research*, *29*(7), 2215–2226. DOI: 10.1029/93WR00404

Corapciooglu, M. Y., & Wang, S. (1999). Dual porosity groundwater contaminant transport in the presence of colloids. *Water Resources Research*, *35*(11), 3261–3273. DOI: 10.1029/1999WR900122

Crank, J. (1975). *The mathematics of diffusion* (2nd ed.). Clarendon Press.

Crittenden, J. C., Hutzler, N. J., Geyer, D. G., Oravitz, J. L., & Friedman, G. (1986). Transport of organic compounds with saturated groundwater flow: Model development and parameter sensitivity. *Water Resources Research*, *22*(3), 271–284. DOI: 10.1029/WR022i003p00271

Crohn, D. M., & Valenzuela-Solano, C. (2003). Modeling temperature effects on decomposition. *Journal of Environmental Engineering*, *12*(12), 1149–1156. DOI: 10.1061/(ASCE)0733-9372(2003)129:12(1149)

Dagan, G. (1984). Solute transport in heterogeneous porous formations. *Journal of Fluid Mechanics*, *145*(-1), 151–177. DOI: 10.1017/S0022112084002858

Daley, J., & Harleman, D. (1971). Fluid Mechanics. [in Russian]. *Energy*.

Dalias, P., Anderson, J. M., Bottner, P., & Couteaux, M.-M. (2001). Long-term effects of temperature on carbon mineralization processes. *Soil Biology & Biochemistry*, *33*(7-8), 1049–1057. DOI: 10.1016/S0038-0717(01)00009-8

Dantas, A. D. B., & Bernardo, L. D. (2006). Performance of two-stage filtration system for treating high turbidity water. *J. Wat. Suppl.: Res. and Technol. Aqua (Neu-Isenburg, Germany)*, *55*(7-8), 499–515.

Darby, J. L., & Lawler, D. F. (1990). Ripening in depth filtration: Effect of particle size on removal and head loss. *Environmental Science & Technology*, *24*(7), 1069–1079. DOI: 10.1021/es00077a018

Darby, J. L., Lawler, D. F., & Wilshusen, T. P. (1991). Depth filtration of wastewater: Particle size and ripening. *Research Journal of the Water Pollution Control Federation*, *63*(3), 228–238.

Davidson, I., & Harrison, D. (Eds.). (1974). *Fluidization*. Khimiya. (in Russian)

De Beer, D., Stoodley, P., & Lewandowski, Z. (1997). Measurement of local diffusion coefficients in biofilms by microinjection and confocal microscopy. *Biotechnology and Bioengineering*, *53*(2), 151–158. DOI: 10.1002/(SICI)1097-0290(19970120)53:2<151::AID-BIT4>3.0.CO;2-N PMID: 18633959

De Jong, J. T., & Christoph, G. C. (2009). Influence of particle properties and initial specimen state on one-dimensional compression and hydraulic conductivity. *Journal of Geotechnical and Geoenvironmental Engineering*, *135*(3), 449–454. DOI: 10.1061/(ASCE)1090-0241(2009)135:3(449)

Degremont (Ed,). (2007). *Water Treatment Handbook* (7th ed.). 2 Volume Set. Editions Tec & Doc (Lavoisier).

Demchin, I. I., Orlov, V. O., & Venitsianov, E. V. (1994). Calculation of technological parameters of filters with floating filter beds. [in Russian]. *Journal of Water Chemistry and Technology*, *16*(2), 115–118.

Dharmappa, H. B., Prasanthi, H., Krishna, M. U., & Xiao, Y. (1997). Estimation of empirical coefficients of a granular filtration model. *Water Research*, *31*(5), 1083–1091. DOI: 10.1016/S0043-1354(96)00329-6

Dharmappa, H. B., Verink, J., Fujiwara, O., & Vigneswaran, S. (1992). Optimization of granular bed filtration treating polydispersed suspension. *Water Research*, *26*(10), 1307–1318. DOI: 10.1016/0043-1354(92)90125-N

Dickinson, C. (1992). *Filters and filtration handbook*. Elsevier Advanced Technology.

Diliunas, I. P., Kaminskas, M. M., & Yagminas, E. A. (1984). Evaluation of hydrodynamic parameters for artificial groundwater recharge from open flow structures. In *Methods of hydrogeological data analysis and processing for groundwater resources forecasting* (pp. 190–202). Academy of Sciences of the ESSR. (in Russian)

Dmitrenko, G. N., & Gvozdyak, P. I. (2002). Biotechnology of treatment of highly concentrated wastewater from organic solvents. [in Russian]. *Journal of Water Chemistry and Technology*, *24*(2), 185–190.

Droste, R. L., & Kennedy, K. J. (1988). Dynamic anaerobic fixed-film reactor. *Journal of Environmental Engineering*, *114*(3), 606–620. DOI: 10.1061/(ASCE)0733-9372(1988)114:3(606)

du Plessis, J. P., & Woudberg, S. (2008). Pore-scale derivation of the Ergun equation to enhance its adaptability and generalization. *Chemical Engineering Science*, *63*(9), 2576–2586. DOI: 10.1016/j.ces.2008.02.017

Duck, J., Zvetanov, E., & Neebe, T. (2000). Porosity prediction for fine-grained filter cake. *Chemical Engineering & Technology*, *22*(1), 18–22. DOI: 10.1002/(SICI)1521-4125(200001)23:1<18::AID-CEAT18>3.0.CO;2-P

Dunin, F. Kh. (1980). Modeling of infiltration in approximation to field conditions. In *Frontiers of Hydrology* (pp. 241–275). Gidrometeoizdat. (in Russian)

Dutournie, P., Limousy, L., Deon, S., & Bourseau, P. (2011). Unsteady transport of divalent salt through a mineral membrane of ultrafiltration: Numeracal estimation of physical parameters. *Desalination*, *265*(1-3), 184–189. DOI: 10.1016/j.desal.2010.07.050

Dytnersky, Y. I. (1986). *Baromembrane processes. Theory and calculation*. Khimiya. (in Russian)

Dzyubo, V.V. (2006). Radial filters of groundwater iron removal. Engineering and technological solutions. *Sanitary Engineering*, 16–19. (in Russian)

Dzyubo, V. V., & Alferova, L. I. (2007). Natural water filtration in the mode of non-uniform rates. [in Russian]. *Bulletin of TSASU*, (2), 180–190.

Elimelech, M., & O'Melia, C. R. (1990). Kinetics of deposition of colloidal particle in porous media. *Environmental Science & Technology*, *24*(10), 1528–1536. DOI: 10.1021/es00080a012

Emelko, M. B. (2003). Removal of viable and inactivated Cryptosporidium by dual and tri-media filtration. *Water Research, 37*(12), 2998–3008. DOI: 10.1016/S0043-1354(03)00113-1 PMID: 12767303

Epoyan, S., Karagiaur, A., Volkov, V., & Yarkin, V. (2016). Improvement of filtration facilities operation efficiency at application of drainage systems from porous polymeric materials. *MOTROL. Commision of Motorization and Energetic on Agriculture, 18*(6), 102–109.

Ergun, S. (1952). Fluid flow through packed columns. *Chemical Engineering Progress, 48*(2), 89–94.

Evtushenko, A. I., & Kotelkin, V. D. (1991). Modeling of transfer processes in a fixed granular layer. [in Russian]. *Theoretical Foundations of Chemical Engineering, 25*(4), 511–523.

FAO. (2006). *Guidelines for soil description* (4th ed.).

FAO-UNESCO. (1974). *Soil map of the world* (Vol. 1). Legend.

Faroog, S., & Al-Yousel, A. K. (1993). Slow sand filtration of secondary effluent. *Journal of Environmental Engineering, 119*(4), 615–630. DOI: 10.1061/(ASCE)0733-9372(1993)119:4(615)

Fayollea, Y., Cockxb, A., Gillota, S., Roustanb, M., & Héduit, A. (2007). Oxygen transfer prediction in aeration tanks using CFD. *Chemical Engineering Science, 62*(24), 7163–7171. DOI: 10.1016/j.ces.2007.08.082

Fedorov, A.A., Larionov, A.I., & Ushakov, V.A. (1984). *Porous filtering materials.* SRITEhim. (in Russian)

Finno, R. J., & Rechenmacher, A. L. (2003). Effects of consolidation history on critical state of sand. *Journal of Geotechnical and Geoenvironmental Engineering, 129*(4), 350–360. DOI: 10.1061/(ASCE)1090-0241(2003)129:4(350)

Fiori, A., & Dagan, G. (1999). Concentration fluctuations in transport by groundwater. Comparison between theory and field experiments. *Water Resources Research, 35*(1), 105–112. DOI: 10.1029/98WR01862

Fogden, A., Landman, K. A., & White, L. R. (1988). Contaminant transport in fractured porous media steady state solutions by a boundary integral method. *Water Resources Research, 24*(8), 1384–1396. DOI: 10.1029/WR024i008p01384

Fominykh, A. M. (1980). Mathematical model of water purification process by filtration and its practical application. [in Russian]. *Water Resources,* (3), 194–199.

Foppen, J. W., Liem, Y., & Schijven, I. (2008). Effect of humic acid on the attachment of Escherichia coli in columns of goethite-water sand. *Water Research, 42*(1-2), 211–219. DOI: 10.1016/j.watres.2007.06.064 PMID: 17825871

Foppen, J. W., & Schijven, J. F. (2005). Transport of E. coli in columns of geochemically heterogeneous sediment. *Water Research, 39*(13), 3082–3088. DOI: 10.1016/j.watres.2005.05.023 PMID: 15996706

Foppen, J. W., & Schijven, J. F. (2006). Evaluation of data from the literature on the transport and survival of Escherichia coli and thermotolerant coliforms in aquifers under saturated conditions. [Review]. *Water Research, 40*(4), 401–426. DOI: 10.1016/j.watres.2005.11.018 PMID: 16434075

Frank, J., Ruhl, A. S., & Jekel, M. (2015). Impacts of backwashing on granular activated carbon filters for advanced wastewater treatment. *Water Research, 87*, 166–174. DOI: 10.1016/j.watres.2015.09.020 PMID: 26405842

Fried, J. J., & Combarnous, M. A. (1971). Dispersion in porous media. *Adv. Hydrosci, 7*, 169–282. DOI: 10.1016/B978-0-12-021807-3.50008-4

Gal, E., Tardos, G., & Pfeffer, R. (1971). A study of inertial effects in granular bed filtration. *AIChE Journal. American Institute of Chemical Engineers, 31*(7), 1093–1104. DOI: 10.1002/aic.690310707

Gao, M., Lv, M., Gong, X., Song, W., Qi, Y., & Zhen, L. (2017). Effect of temperature and aging on the adsorption and desorption of menadione in soil. *Environmental Progress & Sustainable Energy, 36*(4), 997–1004. DOI: 10.1002/ep.12540

Gao, P., Xue, G., Song, X., & Lin, Z. (2012). Depth filtration using novel fiber-ball filter media for the treatment of high-turbicity surface water. *Separation and Purification Technology, 95*, 32–38. DOI: 10.1016/j.seppur.2012.04.010

Gao, X., da Costa, J. C. D., & Bhatia, S. K. (2014). Understanding the diffusional tortuosity of porous materials: An effective medium theory perspective. *Chemical Engineering Science, 110*, 55–71. DOI: 10.1016/j.ces.2013.09.050

Gąszczak, A., Bartelmus, G., Burghardt, A., Rotkegel, A., & Sarzyński, R. (2018). Experiments and modelling of a biotrickling filter (BTF) for removal of styrene from airstreams. *Journal of Chemical Technology and Biotechnology, 93*(9), 2659–2670. DOI: 10.1002/jctb.5620

Geiger, S. L., & Durnford, D. S. (2000). Infiltration in homogeneous sands and a mechanistic model of unstable flow. *Soil Science Society of America Journal, 64*(2), 460–469. DOI: 10.2136/sssaj2000.642460x

Girole, N. N., Zhurba, M. G., Semchuk, G. M., & Yakimchuk, B. N. (1998). Secondary treatment of waste water on granular filters (special ed.) SPOOO, "Levoberezhnaya". (in Russian)

Glatstein, D. A., & Francisca, F. M. (2014). Hydraulic conductivity of compacted soils controlled by microbial activity. *Environmental Technology, 35*(15), 1886–1892. DOI: 10.1080/09593330.2014.885583 PMID: 24956782

Goncharuk, V. V., Shevchuk, E. A., & Galinker, E. V. (2001). Theoretical studies of regularities of the process of filtration flow of low-concentration suspensions through granular beds. [in Russian]. *Journal of Water Chemistry and Technology*, (6), 159–161.

Gordin, I.V., & Markov, P.P. (1991). *Closed systems of agrarian-industrial water use.* VO "Agropromizdat". (in Russian)

Grabovsky, P.O., & Gurinchyk, N.A. (2005). Numerical realization of the mathematical model of filtration. *Problems of Water Supply,* Sewage *and Hydraulics. Collected papers*, 6, 4–13. (in Russian)

Grabovsky, P.O., & Gurinchyk, N.A. (2007). Numerical modeling of filtration structures operation. *Bulletin of NUWEE*, 4(40), part 2, 429–434. (in Russian)

Grabovsky, P. A. (2016). Water filtration through a granular layer with decreasing rate. [in Russian]. *Rep. NAS of Ukraine*, (8), 40–45.

Grabovsky, P. A., Larkina, G. M., & Progulny, V. I. (2012). Washing of water treatment filters. [in Russian]. *Optimum (Paris)*.

Grabovsky, P. O., & Gurinchyk, N. A. (2008). Water filtration through a granular layer with decreasing rate in time. *Water and Ecology*. [in Russian]. *Problems and Solutions*, *37*(4), 3–11.

Grant, S., & Salehzadeh, A. (1996). Calculation of temperature effects on wetting coefficients of porous solids and their capillary pressure functions. *Water Resources Research*, *32*(2), 261–270. DOI: 10.1029/95WR02915

Gray, W. G., & Miller, C. T. (2004). Examination of Darcy's law for flow in porous media with variable porosity. *Environmental Science & Technology*, *38*(22), 5895–5901. DOI: 10.1021/es049728w PMID: 15573587

Gremyachkin, V. M., & Dementiev, V. V. (1994). Inhomogeneities of flow rate distribution in a model reactor with a fixed granular bed. [in Russian]. *Theoretical Foundations of Chemical Engineering*, *28*(2), 99–103.

Guo, S., Zhang, F., Song, X., & Wang, B. (2015). Deposited sediment settlement and consolidation mechanisms. *Water Science and Engineering*, *8*(4), 335–344. DOI: 10.1016/j.wse.2015.10.002

Gupta, V., Johnson, W. P., Shafieian, P., Ryu, H., Alum, A., Abbaszadegan, M., Hubbs, S. A., & Rauch-Williams, T. (2009). Riverbank filtration: Comparison of pilot scale transport with theory. *Environmental Science & Technology*, *43*(3), 669–676. DOI: 10.1021/es8016396 PMID: 19245000

Gurinchik, N. A. (2010). Determination of the duration of operation of filters for different modes of filtration. [in Russian]. *Bulletin of ONACA*, *42*, 56–62.

Gursch, J., Giner, R., & Krammer, G. (2016). Influence of wall effects on filter upscaling during solid-liquid filter tests. *Chemical Engineering & Technology*, *39*(3), 417–424. DOI: 10.1002/ceat.201500312

Gusev, E. M. (1981). Experimental studies of pressure seepage. In *Physics of soil waters* (pp. 195–205). Nauka. (in Russian)

Gusev, E. M. (1981). Experimental studies of pressure seepage. In *Physics of Soil Waters* (pp. 195–205). Nauka. (in Russian)

Guyonnet, D., Amraoui, N., & Kara, R. (2000). Analysis of transient data from infiltrometer tests in fine – grained soils. *Groundwater*, *38*(3), 396–402. DOI: 10.1111/j.1745-6584.2000.tb00225.x

Gvozdyak, P. I. (2003). On the principle of bioprocessing (Biotechnology of environmental protection). [in Russian]. *Bulletin of NAS of Ukraine*, (3), 29–36.

Haarstrick, A., Mora-Naranjo, N., Meima, J., & Hempel, D. C. (2004). Modeling anaerobic degradation in municipal landfills. *Environmental Engineering Science*, *21*(4), 471–484. DOI: 10.1089/1092875041358494

Hall, D., & Fitzpatrick, C. S. B. (1998). A mathematical filter backwash model. *Water Science and Technology*, *37*(12), 371–379. DOI: 10.2166/wst.1998.0563

Han, C., Yang, J., Dong, S., Ma, L., Dai, Q., & Guo, J. (2024). *Zeolite preparation from industrial solid waste: Current status, applications, and prospects*. Sep. and Purif. Tech.

Happel, J. (1958). Viscous flow in multiparticle systems: Slow motion of fluids relative to beds of spherical particles. *AIChE Journal. American Institute of Chemical Engineers*, *4*(2), 197–201. DOI: 10.1002/aic.690040214

Hassan, A. E., Cushman, J. H., & Delleur, J. W. (1998). Significance of porosity variability to transport in heterogeneous porous media. *Water Resources Research*, *34*(9), 2249–2259. DOI: 10.1029/98WR01470

Hazen, H. (1911). Dams on sand formation [discussion]. *Trans. ASCE*, *73*, 199–203.

Hekmat, D., Stephan, M., Bauer, R., Feuchtinger, A., & Vortmeyer, D. (2006). Modelling of multispecies biofilm population dynamics in a trickle-bed bioreactor used for waste gas treatment. *Proc. Biochem.*, 41, 1409–1416. Doi:, 2006.02.003.DOI: 10.1016/j.procbio

Hekmat, D., & Vortmeyer, D. (1994). Modelling of biodegradation processes in trickle-bed bioreactors. *Chemical Engineering Science*, *49*(24), 4327–4345. DOI: 10.1016/S0009-2509(05)80025-9

Henderson, N., Brettas, J. C., & Wagner, F. S. (2010). A three-parameter Kozeny-Carman generalized equation for fractal porous media. *Chemical Engineering Science*, *65*(15), 4432–4442. DOI: 10.1016/j.ces.2010.04.006

Henze, M., Armoes, P., La-Coeur-Jansen, I., & Arvan, E. (2006). *Treatment of sewage waters*. Mir. (in Russian)

Herzig, J. P., Leclerc, D. M., & Le Goff, P. (1979). Flow of suspension through porous media. Application to deep filtration. *Industrial & Engineering Chemistry*, *62*(5), 8–35. DOI: 10.1021/ie50725a003

Ho, C.-C., & Zydnev, A. L. (2000). A combined pore blockage and cake filtration model for protein fouling during microfiltration. *Journal of Colloid and Interface Science*, *232*(2), 389–399. DOI: 10.1006/jcis.2000.7231 PMID: 11097775

Hønning, J., Broholm, M. M., & Bjerg, P. L. (2007). Role of diffusion in chemical oxidation of PCE in a dual permeability system. *Environmental Science & Technology*, *41*(24), 8426–8432. DOI: 10.1021/es0708417 PMID: 18200874

Hopper-Jones, C., Oldham, G., & Drewes, J. E. (2010). Attenuation of total organic carbon and unregulated trace organic chemicals in US riverbank filtration systems. *Water Research*, *44*(15), 4643–4659. DOI: 10.1016/j.watres.2010.06.022 PMID: 20599244

Horan, N. I., & Lowe, M. (2007). Full-scale trials of recycled glass as tertiary filter medium for wastewater treatment. *Water Research*, *41*(1), 253–259. DOI: 10.1016/j.watres.2006.08.028 PMID: 17070893

Huang, J. C., Rothstein, D., & Madey, R. (1984). Analytical solution for a first-order reaction considered in a packed bed with diffusion. *AIChE Journal. American Institute of Chemical Engineers*, *30*(4), 660–662. DOI: 10.1002/aic.690300423

Huang, Q.-Z., Hsu, S.-Y., & Syu, C.-H. (2024). Empirical formulas for dynamic effective suction head in the modified Green-Ampt model. *Journal of Hydrology (Amsterdam)*, *642*(131820), 131820. Advance online publication. DOI: 10.1016/j.jhydrol.2024.131820

Huisman, L. (1979). *Rapid filtration. Part 1*. Delft University of Technology Press.

Hutchison, W. R. (1976). High-rate direct filtration. *Journal - American Water Works Association*, *68*(6), 292–298. DOI: 10.1002/j.1551-8833.1976.tb02412.x

Hutt, B. E., Fletcher, T. D., & Deletic, A. (2007). Treatment performance of gravel filter media: Implications for design and application of storm water infiltration system. *Water Research*, *41*(12), 2513–2524. DOI: 10.1016/j.watres.2007.03.014 PMID: 17475303

Hu, X., Do, D. D., & Rao, G. N. (1994). Experimental cjncentration dependence of surface diffusivity of hydrocarbons in activated carbon. *Chemical Engineering Science*, *49*(13), 2145–2152. DOI: 10.1016/S0009-2509(94)85059-3

Ibaraki, M., & Sudicky, E. A. (1995). Colloid-facilitated contaminant transport in discretely fractured porous media. 1. Numerical formulation and sensitivity analysis. *Water Resources Research*, *31*(12), 2945–2960. DOI: 10.1029/95WR02180

Iliuta, I., & Larachi, F. (2004). Biomass accumulation and clogging in trickle-bed bioreactors. *AIChE Journal. American Institute of Chemical Engineers*, *50*(10), 2541–2551. DOI: 10.1002/aic.10201

In: *Proceedings of 9 th International Scientific Conference* "Topical Problems of Fluid Mechanics", 42–47.

Indraratna, B. F., Basack, S., & Rujikiatkamjorn, C. (2013). Numerical solution of stone column–improved soft soil considering arching, clogging, and smear effects. *Journal of Geotechnical and Geoenvironmental Engineering*, *139*(3), 377–394. DOI: 10.1061/(ASCE)GT.1943-5606.0000789

Iritani, E., Katagiri, N., Sugiyama, Y., & Yagishita, K. (2007). Analysis of flux decline behaviors in filtration of very dilute suspensions. *AIChE Journal. American Institute of Chemical Engineers*, *53*(9), 2275–2283. DOI: 10.1002/aic.11271

Isaji, C. R. (2003). Integrated water quality management for drinking water of good quality. *Water Science and Technology*, *47*(9), 15–23. DOI: 10.2166/wst.2003.0482 PMID: 12830935

Ishiyama, E. M., Coletti, F., Macchietto, S., Paterson, W. R., & Wilson, D. I. (2010). Impact of deposit ageing on thermal fouling: Lumped parameter model. *AIChE Journal. American Institute of Chemical Engineers*, *56*(2), 531–545. DOI: 10.1002/aic.11978

Ishiyama, E. M., Paterson, W. R., & Wilson, D. I. (2011). Exploration of alternative models for the ageing of fouling deposits. *AIChE Journal. American Institute of Chemical Engineers*, *57*(11), 3199–3209. DOI: 10.1002/aic.12514

Ison, C. R., & Ives, K. J. (1969). Removal mechanism in deep bed filtration. *Chemical Engineering Science*, *24*(4), 717–729. DOI: 10.1016/0009-2509(69)80064-3

Ivanchenko, O., Sindhwani, N., & Linninger, A. A. (2012). Exact solution of the diffusion-convection equation in cylindrical geometry. *AIChE Journal. American Institute of Chemical Engineers*, *58*(4), 1299–1302. DOI: 10.1002/aic.12663

Ives, K. J. (1981). Optimization model of rapid filtration at tertiary treatment. In A. James (Ed.), *Mathematical models of water pollution control.* (pp.160 – 205). Mir. (in Russian)

Ives, K.J. (1979). The basis for the application of multiple layer filters to water treatment. *Wasser und Abwasser – Forchung*, 12 (3), 106–110.

Ives, K. J. (1970). Rapid filtration. *Water Research*, *4*(3), 201–223. DOI: 10.1016/0043-1354(70)90068-0

Iwasaki, I. (1937). Some notes on sand filtration. *Journal - American Water Works Association*, *29*(10), 1591–1602. DOI: 10.1002/j.1551-8833.1937.tb14014.x

James, A. (Ed.). (1981). *Mathematical models of water pollution control.* Mir. (in Russian)

James, S. C., & Chrysikopoulos, C. (2003). Effective velocity and effective dispersion coefficient for finite-sized particles flowing in a uniform fracture. *Journal of Colloid and Interface Science*, *263*(1), 288–295. DOI: 10.1016/S0021-9797(03)00254-6 PMID: 12804914

Jegatheesan, V., & Vigneswaran, S. (2000). Transient stage deposition of submicron particles in deep bed filtration under unfavorable conditions. *Water Research*, *34*(7), 2119–2131. DOI: 10.1016/S0043-1354(99)00381-4

Jegatheesan, V., & Vigneswaran, S. (2005). Deep bed filtration: Mathematical models and observations. *Critical Reviews in Environmental Science and Technology*, *35*(6), 515–569. DOI: 10.1080/10643380500326432

Jia, H.-Y. (2005). Bioplato for wastewater treatment. *Problems of Water Supply,* Sewage *and Hydraulics. Collected papers*, 4, 114–124. (in Russian)

Jimenez, B., & Buitron, G. (1990). Comparison between three secondary effluents in tertiary high rate filtration. *Environmental Technology*, *117*(9), 987–995. DOI: 10.1080/09593331708616467

Jimenez, B., Chavez, A., Leyva, A., & Tchobanoglous, G. (2000). Sand and synthetic medium filtration of advanced primary treatment effluent from Mexico-city. *Water Research*, *34*(2), 473–480. DOI: 10.1016/S0043-1354(99)00180-3

Jönsson, K. A.-S., & Jönsson, B. T. L. (1992). Fluid flow in compressible porous media. I. Steady-state conditions. II. Dynamic behavior. *AIChE Journal. American Institute of Chemical Engineers*, *38*(9), 1340–1356. DOI: 10.1002/aic.690380904

Kadlec, R. H., & Knight, R. L. (1996). *Treatment wetlands*. Boca Ration. FL.: CRC Press.

Kadlec, R. H., & Knight, R. L. (1996). *Treatment wetlands. Boca Ration*. CRC Press.

Kamke, E. (1976). *Handbook on ordinary differential equations*. Nauka. (in Russian)

Karagyaur, A. S., Epoyan, S. M., & Pabolkov, V. V. (2013). Choice of filter media for multilayer filters with horizontal filtration direction. [in Russian]. *MOTROL Commission of Motorization and Energetics in Agricultural*, *15*(6), 11–18.

Kau, S. M., & Lawler, D. F. (1995). Dynamics of deep-bed filtration: Velocity, depth and media. *Journal of Environmental Engineering, 121*(12), 850–859. DOI: 10.1061/(ASCE)0733-9372(1995)121:12(850)

Kawamura, S. (1985). Two-stage filtration. *Journal - American Water Works Association, 77*(12), 42–47. DOI: 10.1002/j.1551-8833.1985.tb05656.x

Kessler, J. H., & Hunt, J. R. (1994). Dissolved and colloidal contaminant transport. *Water Resources Research, 30*(4), 1195–1206. DOI: 10.1029/93WR03555

Khapun, T. F., & Miltreiger, I. M. (1981). About mathematical models and engineering methods of calculating the process of natural water purification by filtration. [in Russian]. *Water Resources*, (3), 189–195.

Khilar, K. C., & Fogler, H. S. (1998). *Migration of bines in porous media*. Kluwer Academic Publishers. DOI: 10.1007/978-94-015-9074-7

Kim, A. N., Kolodkin, I. V., & Sharavii, Ch. O. (2006). Study of physicochemical and technological properties of filtering and sorption material of zeolite of Kholinsky deposit of Buryatia. [in Russian]. *Water and Ecology*, (1), 9–15.

Kim, J., & Lawler, D. F. (2012). The influence of hydraulic loads on depth filtration. *Water Research, 46*(2), 433–441. DOI: 10.1016/j.watres.2011.10.059 PMID: 22112925

Kim, J., & Tobiasson, J. E. (2004). Origins of filter effluent particles: Experimental study of particle deposition and detachment. *Water Science and Technology, 50*(12), 215–222. DOI: 10.2166/wst.2004.0716 PMID: 15686024

Kim, J., & Tobiasson, J. E. (2004). Particles in filter effluent: The roles of deposition and detachment. *Environmental Science & Technology, 38*(22), 6132–6138. DOI: 10.1021/es0352698 PMID: 15573617

Kiranmai, D., Jyothirmai, A., & Murty, C. V. S. (2005). Determination of kinetic parameters in fixed-film bio-reactors: An inverse problem approach. *Biochemical Engineering Journal, 23*(1), 73–83. DOI: 10.1016/j.bej.2004.10.005

Kiselev, S.K. (1997). Change of filtration properties of filter bed at water deferrization. *Land Reclamation and Water Management. Collected papers*, 84, 190–194. (in Russian)

Knapp, R. B., Chiaropa, M. L., & Durham, W. B. (2000). An experimental exploration of the transport and capture of abiotic colloids in a single fracture. *Water Resources Research, 36*(11), 3139–3149. DOI: 10.1029/2000WR900145

Ko, D. C. K., Porter, J. F. P., & McKay, G. (2001). Film-pore diffusion model for the fixed-bed sorption of copper and cadmium ions onto bone char. *Water Research, 35*(16), 3876–3886. DOI: 10.1016/S0043-1354(01)00114-2 PMID: 12230170

Koganovskiy, A. M., Levchenko, T. M., Roda, I. G., & Marotovskiy, R. M. (1981). *Adsorption technology of wastewater treatment*. Technics. (in Russian)

Komarova, E. A., Nakorchevskaya, V. F., & Kulsky, L. A. (•••). Application of stepped filtration in water treatment technology. [in Russian]. *Journal of Water Chemistry and Technology, 4*(3), 240–244.

Kommunar, G. M., & Bogolyubov, K. S. (1977). Some features of calculation of infiltration basins under conditions of colmatization and clogging. *Proceedings of VNII VODGEO*, 63, 10–16. (in Russian)

Kosterin, A. V., & Skvortsov, E. V. (2002). Liquid infiltration from the ground surface into the fractured-porous aeration zone. *Izv.* [in Russian]. *RAS MZhG*, (5), 93–98.

Kovalchuk, V.A. (2002). *Wastewater treatment*. OJSC "Rivne Printing House". (in Ukrainian)

Kozeny, J. (1931). Grundwasserbewegung bei freiem Spiegel, Fluss- und Kanal – versickerung. *Wasserkraft und Wasserwirts chaft*, 4.3, 28–31.

Kozhinov, V. F. (2008). *Purification of drinking and technical water. Textbook for universities* (4th ed.). BASTET. (in Russian)

Kravchuk, A.M. (2000). *Contaminant transport and phase transfer in the structures of water supply and sewage systems*. KNUCA Press. (in Russian)

Kretzsehmar, R., Barmettler, K., & Grolimund, D. (1997). Experimental determination of colloid deposition rates and collision efficiencies in natural porous media. *Water Resources Research*, *33*(5), 1129–1137. DOI: 10.1029/97WR00298

Kryuchkov, Yu. N. (1998). Determination of porous structure parameters and permeability of porous materials. [in Russian]. *Theoretical Foundations of Chemical Engineering*, *32*(5), 515–523.

Kryvonoh, O.I., & Kryvonoh, V.V. (1997). Study of the process of water clarification by filters made of fiber-porous polyethylene. *Land Reclamation and Water Management. Collected papers*, 84, 170–174. (in Russian)

Kuberkar, V. T., & Davis, R. H. (2000). Modeling of fouling reduction by secondary membranes. *Journal of Membrane Science*, *168*(1-2), 243–258. DOI: 10.1016/S0376-7388(99)00324-5

Kuhn, M., & Briesen, H. (2016). Dynamic modeling of filter-aid filtration including surface and filtration effects. *Chemical Engineering & Technology*, *39*(3), 425–434. DOI: 10.1002/ceat.201500347

Kulikov, N.I., Naimanov, A.Y., Omelchenko, N.P., & Chernyshev, V.N. (2009). *Theoretical bases of water purification*. Publishing House "Znanie", (Donetsk branch). (in Russian)

Kulkarni, P., Dutari, G., Weingeis, D., Adin, A., Haught, R., & Biswas, P. (2005). Capture of water-borne colloids in granular beds using external electric fields: Improving removal of Cryptosporidium parvum. *Water Research*, *39*(6), 1047–1060. DOI: 10.1016/j.watres.2004.12.026 PMID: 15766959

Kulskiy, L. A., Goronovskiy, I. T., Koganovskiy, A. M., & Shevchenko, M. A. (1980). *Reference book on properties, methods of analysis and purification of water: in 2 parts*. Naukova dumka. (in Russian)

Kulskiy, L. A., & Strokach, P. P. (1986). *Technology of natural water purification*. Vishcha Shkola. (in Russian)

Kulsky, L. A., Yaroshevskaya, N. V., & Shevchuk, E. A. (1991). Increasing the retention capacity of filters in two-stage filtration schemes. [in Russian]. *Journal of Water Chemistry and Technology*, *13*(3), 239–241.

Kumari, M., & Kazmi, A. A. (2014). Effect of physical property of supporting media and variable hydraulic loading on hydraulic characteristics of advanced onsite wastewater treatment system. *Environmental Technology, 36*(11), 1414–1422. DOI: 10.1080/09593330.2014.992480 PMID: 25428652

Kuwabara, S. (1959). The forces experienced by randomly distributed parallel circular cylinders or spheres in a viscous flow at small Reynolds numbers. *Journal of the Physical Society of Japan, 14*(4), 527–532. DOI: 10.1143/JPSJ.14.527

Kwapinski, W., Winterberg, M., Tsotsas, E., & Mewes, D. (2004). Modeling of the wall effect in packed bed adsorption. *Chemical Engineering & Technology, 27*(11), 1176–1186. DOI: 10.1002/ceat.200407001

Landkamer, L. L., Harvey, R. W., Scheibe, T. D., & Ryan, J. N. (2013). Colloid transport in saturated porous media: Elimination of attachment efficiency in a new colloid transport model. *Water Resources Research, 49*(5), 2952–2965. DOI: 10.1002/wrcr.20195

Landman, K., & White, L. (1994). Solid/liquid separation of flocculated suspensions. *Advances in Colloid and Interface Science, 51*, 175–246. DOI: 10.1016/0001-8686(94)80036-7

Last, W., & Stichimair, J. (2002). Influence of stagnant liquid elements on the effective interfacial area during chemical adsorption in packed columns. *Chemical Engineering & Technology, 25*(11), 1075–1079. DOI: 10.1002/1521-4125(20021105)25:11<1075::AID-CEAT1075>3.0.CO;2-C

Latifi, H., Prasad, S. N., & Helveg, O. J. (1994). Air entrapment and water infiltration in two-layered soil column. *Journal of Irrigation and Drainage Engineering, 120*(5), 871–891. DOI: 10.1061/(ASCE)0733-9437(1994)120:5(871)

Lau, B. L., Harrington, G. W., Anderson, M. A., & Tejedor, I. (2004). Removal of nano and microparticles by granular dilter media coated with nanoporous aluminium oxide. *Water Science and Technology, 50*(12), 223–228. DOI: 10.2166/wst.2004.0717 PMID: 15686025

Lebedev, I. A., Somin, V. A., Kondratyuk, E. V., & Komarova, L. F. (2006). Ecological technologies: Intensification of water purification from suspended solids by filtration method. [in Russian]. *Engineering Ecology*, (2), 20–27.

Lee, J., Fox, P. J., & Lenhart, J. J. (2009). Investigation of consolidation-Induced solute transport. I. Effect of consolidation on transport parameters. *Journal of Geotechnical and Geoenvironmental Engineering, 135*(9), 1228–1238. DOI: 10.1061/(ASCE)GT.1943-5606.0000047

Lee, Y., & Clark, M. M. (1998). Modeling of flux decline during crossflow ultrafiltration of colloidal suspension. *Journal of Membrane Science, 149*(2), 181–202. DOI: 10.1016/S0376-7388(98)00177-X

Leichkis, I. M. (1975). *Filtering with the precoating of auxiliary substances*. Tehnika. (in Russian)

Leichkis, I. M. (1975). *Filtering with the use of auxiliary substances*. Tehnika. (in Russian)

Leichkis, I. M., & Ventzel, A. D. (1987). Filtering with flush of a thin layer of auxiliary filtering agent. *Theoretical Foundations of Chemical Engineering, 25*(6), 832.

Lerch, A., Nahrstedt, A., Gimbel, R., Gray, S., & Booker, N. (2002). On the development of a high-rate deep bed filtration unit. *Water Science and Technology: Water Supply*, *2*(1), 259–269. DOI: 10.2166/ws.2002.0031

Levenspiel, O. (1999). *Chemical reaction engineering* (3rd ed.). Wiley.

Lewandowska, J., Ngoc, T. D. T., Vauclin, M., & Bertin, H. (2008). Water drainage in double-porosity soils: Experiments and micro–macro modeling. *Journal of Geotechnical and Geoenvironmental Engineering*, *134*(2), 231–243. DOI: 10.1061/(ASCE)1090-0241(2008)134:2(231)

Leyva-Ramos, R., & Geankoplis, C. J. (1985). Model simulation and analysis of surface diffusion of liquids in porous solids. *Chemical Engineering Science*, *40*(5), 799–807. DOI: 10.1016/0009-2509(85)85032-6

Liao, Q., Tian, X., Chen, R., & Zhu, X. (2007). Mathematical model for gas-liquid two-phase flow and biodegradation of a low concentration volatile organic compound (VOC) in a trickling biofilter. Internal. J. Heat Mass Transfer, 1–13.

Limbach, K. W., & Wei, J. (1990). Restricted diffusion through granular materials. *AIChE Journal. American Institute of Chemical Engineers*, *36*(2), 242–248. DOI: 10.1002/aic.690360210

Lindstrem, F. T. (1992). A mathematical model for the one-dimensional transport and fate of oxygen and substrate in a water-saturated sorbing homogeneous porous medium. *Water Resources Research*, *28*(9), 2499–2511. DOI: 10.1029/92WR00898

Liu, C., & Zachara, J. M. (2001). Uncertainties of Monod kinetic parameters nonlinearly estimated from batch experiments. *Environmental Science & Technology*, *35*(1), 133–141. DOI: 10.1021/es001261b PMID: 11351998

Liu, L., Xue, Q., Wan, Y., & Tian, Y. (2016). Evaluation of dual permeability of gas flow in municipal solid waste: Experiment and modeling. *Environmental Progress & Sustainable Energy*, *35*(1), 1–7. DOI: 10.1002/ep.12184

Liu, Q., Chen, L., Minglei, Z., Ying, L., Yangyang, Y., Yuxuan, L., & Siyuan, M. (2024). Performance of a double-filter-medium tandem membrane bioreactor with low operating costs in domestic wastewater treatment. *Water (Basel)*, *16*(2), 361. DOI: 10.3390/w16020361

Liu, S., Gao, P. F., Li, S., Fu, H., Wang, L., Dai, Y., & Fu, M. (2023). A review of the recent progress in biotrickling filters: Packing materials, gases, micro-organisms, and CFD. *Environmental Science and Pollution Research International*, *30*(60), 125398–125416. DOI: 10.1007/s11356-023-31004-7 PMID: 38012483

Li, Y.-C., & Cleall, P. J. (2010). Analytical solutions for contaminant diffusion in double-layered porous media. *Journal of Geotechnical and Geoenvironmental Engineering*, *136*(11), 1542–1554. DOI: 10.1061/(ASCE)GT.1943-5606.0000365

Logan, B. E., Jewett, D. G., Arnold, D. G., Bouwer, E. J., & O'Melia, C. R. (1995). Clarification of clean-bed filtration models. *Journal of Environmental Engineering*, *12*(12), 869–873. DOI: 10.1061/(ASCE)0733-9372(1995)121:12(869)

Lopato, L., Galaj, Z., Delpont, S., Binning, P. J., & Arvin, E. (2011). Heterogeneity of rapid sand filters and its effect on contaminant transport and nitrification performance. *Journal of Environmental Engineering*, *137*(4), 248–257. DOI: 10.1061/(ASCE)EE.1943-7870.0000321

Lozada, P. T., & Cruz-Velez, C. H. (2014). Influence of the flow control (declining – constant rate) on the filtration process in drinking water treatment. *Inginieria y Universidad*, (July), 217–234.

Lu, C., Chang, K., & Hsu, S. (2004). A model for treating isopropyl alcohol and acetone mixtures in a trickle-bed air biofilter. *Process Biochemistry (Barking, London, England)*, *39*(12), 1849–1858. DOI: 10.1016/j.procbio.2003.09.019

Lukinykh, N. A., Lipman, B. L., & Krishtul, V. P. (1978). *Methods of additional treatment of sewage waters*. Stroyizdat. (in Russian)

Lyn, D. A. (1996). Asymptotic analysis of intraparticle diffusion in GAC batch reactors. *Journal of Environmental Engineering*, *122*(11), 1013–1022. DOI: 10.1061/(ASCE)0733-9372(1996)122:11(1013)

MacDonald, M. J., Chu, C.-F., Guilloit, P. P., & Ng Ka, M. (1991). A generalized Blake-Kozeny equation for multisized spherical particles. *AIChE Journal. American Institute of Chemical Engineers*, *37*(10), 1583–1588. DOI: 10.1002/aic.690371016

MacDonald, T. R., Kitanidis, P. K., McCarty, P. L., & Roberts, P. V. (1999). Effects of shear detachment on biomass growth and in situ bioremediation. *Groundwater*, *37*(4), 555–562. DOI: 10.1111/j.1745-6584.1999.tb01142.x

Mackie, R. I., Horne, R. M. W., & Jarvis, R. J. (1987). Dynamic modeling of deep-bed filtration. *AIChE Journal. American Institute of Chemical Engineers*, *33*(11), 1761–1775. DOI: 10.1002/aic.690331102

Mackie, R. I., & Zhao, Q. (1999). A framework for modeling removal in the filtration of polydisperse suspensions. *Water Research*, *33*(3), 794–806. DOI: 10.1016/S0043-1354(98)00267-X

Martin, R.E., Bouwer, E.J., & Hanna, L.H. (1992). Application of clean-bed filtration theory to bacterial deposition in porous media. *Environ. Scl. Technol.*, 1053-1058.

Martynov, S. Yu. (2014). Modeling the operation of polystyrene filters with variable filtration rate. [in Ukrainian]. *Bulletin of NUWEE*, *2*(66), 262–269.

Martynov, S. Yu., & Orlova, A. M. (2022). *Design of water clarification and decolorization stations. Study guide*. NUWEE Press. (in Ukrainian)

Martynov, S. Yu., & Poliakov, V. L. (2022). Experimental studies of iron transformations kinetics and autocatalysis during its physicochemical removal from underground water. *Water Science and Technology: Water Supply*, *22*(3), 2883–2895. DOI: 10.2166/ws.2021.428

Martynov, S. Yu., & Poliakov, V. L. (2022). Experimental studies on the hydrodynamic properties of a deposit in rapid filters during physicochemical removal of iron from groundwater. *Water Science and Technology: Water Supply*, *22*(11), 7603–7617. DOI: 10.2166/ws.2022.305

Maslov, B. S., Minaev, I. V., & Guber, K. V. (1989). *Reference book on land reclamation*. Rosagropromizdat. (in Russian)

Mayer, K. U., Frind, E. O., & Blowes, D. W. (2001). Multicomponent reactive transport modeling in variably saturated porous media using a generalized formulation for kinetically controlled reactions. *Water Resources Research, 38*(9), 1174–1183. DOI: 10.1029/2001WR000862

Mays, D. C. (2010). Contrasting clogging in granular media filters, soils, and dead-end membranes. *Journal of Environmental Engineering, 136*(5), 475–480. DOI: 10.1061/(ASCE)EE.1943-7870.0000173

Mays, D. C., & Hunt, J. R. (2005). Hydrodynamic aspects of particle clogging in porous media. *Environmental Science & Technology, 39*(2), 577–584. DOI: 10.1021/es049367k PMID: 15707058

McDowell, L. M., Hunt, J. R., & Sitar, N. (1986). Particle transport through porous media. *Water Resources Research, 22*(13), 1901–1921. DOI: 10.1029/WR022i013p01901

Mein, R. G., & Farrel, D. A. (1974). Determination of wetting front suction in the Green-Ampt equation. *Soil Science Society of America Journal, 38*(6), 872–876. DOI: 10.2136/sssaj1974.03615995003800060014x

Meltser, V.Z. (1995). *Filtration structures in municipal water supply*. Stroyizdat.

Meric, S., Tunay, O., & Ali San, H. (2002). A new approach to modelling substrate inhibition. *Environmental Technology, 23*(2), 163–177. DOI: 10.1080/09593332508618428 PMID: 11950069

Mints, D. M. (1964). *Theoretical bases of water purification technology*. Publishing House on Construction. (in Russian)

Mints, D. M., & Meltser, V. Z. (1970). Hydraulic resistance of granular porous medium in the process of clogging. *Rep.* [in Russian]. *Academy of Sciences of USSR, 192*(2), 304–306.

Mints, D. M., & Meltser, V. Z. (1970). Hydraulic resistance of granular porous medium in the process of clogging. *Rep.* [in Russian]. *USSR Academy of Sciences, 192*(2), 304–306.

Mintz, D. M., & Meltser, V. Z. (1973). Simplified method of technological modeling of the filtration process. *Scientific Proceedings of AKH*, 98, 57–62. (in Russian)

Miralles-Wilhelm, F., & Gelhar, L. W. (1996). Stohastic analysis of sorption macrokinetics in heterogeneous aquifers. *Water Resources Research, 32*(6), 1541–1549. DOI: 10.1029/96WR00791

Mitchell, C., & McNevin, D. (2001). Alternative analysis of BOD removal in subsurface flow constructed wetlands employing Monod kinetics. *Water Research, 35*(5), 1295–1303. DOI: 10.1016/S0043-1354(00)00373-0 PMID: 11268849

Mitchell, J. K., & Santamarina, J. C. (2005). Biological considerations in geotechnical engineering. *Journal of Geotechnical and Geoenvironmental Engineering, 131*(10), 1222–1233. DOI: 10.1061/(ASCE)1090-0241(2005)131:10(1222)

Mitin, B. A. (1969). Research of influence of structural and mechanical properties of deposit on operation of granular clarifying filters. Ph.D. dissertation abstract. LECI Press. (in Russian)

Mitin, B. A. *Research of influence of structural and mechanical properties of deposit on operation of granular clarifying filters.* Ph.D. dissertation abstract. LECI Press. (in Russian)

Mohanka, S. S. (1969). Theory of multilayer filtration. *Journal of the Sanitary Engineering Division, 98*(6), 1079–1095. DOI: 10.1061/JSEDAI.0001033

Mokrik, R. V., & Strimaitite, R. I. (1984). About numerical solution of the inverse problem of stationary flow in multilayer strata. In *Methods of analysis and processing of hydrogeological data for groundwater resources forecasting* (pp. 175–178). Academy of Sciences of the ESSR. (in Russian)

Molchak, Y. A. (1984). *Rainfall runoff of the over-humidification zone of Ukraine (studies and calculations).* "Vishcha Shkola", Lviv State University Press. (in Russian)

Mollah, A. H., & Robinson, C. W. (1996). Pentachlorophenol adsorption and desorption characteristics of granular activated carbon. I. Isotherms. II. Kinetics. *Water Research, 30*(2), 2901–2913. DOI: 10.1016/S0043-1354(96)00131-5

Molnar, I. L., Gerhard, J. I., Willson, C. S., & O'Carroll, D. M. (2015). The impact of immobile zones on the transport and retention of nanoparticles in porous media. *Water Resources Research, 51*(11), 8973–8994. DOI: 10.1002/2015WR017167

Moran, M. C., Moran, D. C., Cushing, R. S., & Lawler, D. F. (1993). Particle behavior in deep bed filtration. 2. Particle detachment. *Journal - American Water Works Association, 85*(12), 82–93. DOI: 10.1002/j.1551-8833.1993.tb06121.x

Myint, M., Nirmalakhandan, N., & Speece, R. E. (2007). Anaerobic fermentation of cattle manure: Modeling of hydrolysis and acidogenesis. *Water Research, 41*(2), 323–332. DOI: 10.1016/j.watres.2006.10.026 PMID: 17141822

Naumenko, I. I. (2005). *Hydraulics.* NUWEE Press. (in Russian)

Ncube, P., Pidou, M., & Jarvis, P. (2018). The impact of filter bed depth and solids loading using a multimedia filter. *Separation Science and Technology, 53*(14), 2249–2258. DOI: 10.1080/01496395.2018.1439961

Nelson, K. E., & Ginn, T.R. (2011). New collector efficiency equation for colloid filtration in both natural and engineered flow conditions. *Water Resour. Res.*, 47, 1-17. W05543, .DOI: 10.1029/2010WR009587

Neuman, S. P. (1976). Hydraulic head at the wetting front in the Green-Ampt model. *Water Resources Research, 12*, 564–566. DOI: 10.1029/WR012i003p00564

Nielsen, D. R., van Genuchten, M. T., & Biggar, J. W. (1986). Water flow and solute transport processes in the unsaturated zone. *Water Resources Research, 22*, 89–108. DOI: 10.1029/WR022i09Sp0089S

Niu, S., Park, K., Yu, J., & Kim, Y. (2016). Operation and performance evaluation of high-speed filter using porous non-woven filamentous fibre for the treatment of turbid water. *Environmental Technology, 37*(5), 577–589. DOI: 10.1080/09593330.2015.1075596 PMID: 26209268

Nivala, J., Knowles, P., Dotro, G., Garcı, J., & Wallace, S. (2012). Clogging in subsurface-flow treatment wetlands: Measurement, modeling and management. [Review]. *Water Research, 46*(6), 1625–1640. DOI: 10.1016/j.watres.2011.12.051 PMID: 22284912

Noskov, M. D., Zaitseva, M. S., Istomin, A. D., & Lukashevich, O. D. (2008). Mathematical modeling of the rapid filters operation. [in Russian]. *Bulletin of TSASU*, (2), 126–137.

Odegaard, H., Liao, Z., & Hansen, A. T. (2003). Coarse media filtration – an alternative to settling in wastewater treatment. *Water Science and Technology*, *47*(12), 81–88. DOI: 10.2166/wst.2003.0631 PMID: 12926673

Ojha, C. S. P., & Graham, N. J. D. (1992). Appropriate use of deep-bed filtration models. *Journal of Environmental Engineering*, *118*(6), 964–980. DOI: 10.1061/(ASCE)0733-9372(1992)118:6(964)

Ojha, C. S., & Graham, N. J. D. (1993). Theoretical estimates of bulk specific deposit in deep-bed filters. *Water Research*, *27*(3), 377–387. DOI: 10.1016/0043-1354(93)90037-I

Ojha, C. S., & Graham, N. J. D. (1994). Computer-aided simulation of slow sand filter performance. *Water Research*, *28*(5), 1025–1030. DOI: 10.1016/0043-1354(94)90187-2

Oleynik, A. Ya. (1978). *Flow calculations of vertical drainage*. Naukova dumka. (in Russian)

Oleynik, A. Ya., & Kolpakova, O. A. (2012). Theoretical substantiation of wastewater treatment on trickling filters. [in Russian]. *Rep. NAS of Ukraine*, (3), 120–126.

Oleynik, A. Ya., & Tkach, V. N. (1972). Water flow calculations of rice systems under two-layer aquifer structure. [in Russian]. *Land Reclamation and Water Management*, *24*, 88–99.

Oleynik, A. Ya., & Tugay, A. M. (2001). Modeling of the processes of clogging and suffusion in the prefilter zone of the well. [in Russian]. *Rep. NAS of Ukraine*, (9), 190–194.

Oliynyk, O.Ya., & Kalugin, Y.I. (2005). Some results of theoretical studies of mass transfer processes in porous media. *Problems of Water Supply, Sewage and Hydraulic. Collected papers,* 5, 100–112. (in Ukrainian)

Orlov, V. O. (2005). *Water purification filters with granular filter media*. NUWEE Press. (in Russian)

Orlov, V. O. (2008). *Deferrization of groundwater by simplified aeration and filtration*. NUWEE Press. (in Ukrainian)

Orlov, V. O. (2008). *Iron removal from groundwater by simplified aeration and filtration*. NUWEE Press. (in Ukrainian)

Orlov, V. O., & Martynov, S. Y. (2011). Aeration methods of water deferrization. [in Ukrainian]. *Water and Water Treatment Technologies*, *2*(4), 42–52.

Orlov, V. O., Martynov, S. Yu., Orlova, A. M., Zoshchuk, V. O., & Kunytskyi, S. O. (2017). *Water treatment on polystyrene foam filters*. NUWEE Press.

Ornatsky, N. S., Sergeev, E. M., & Shehtman, Y. M. (1955). *Investigation of the process of sand clogging*. Moscow State University Press. (in Russian)

Oyanader, M., & Arce, P. (2007). An effective analytical–algorithmic approach to solve diffusion-type differential equations with non-linear sources: Unbounded domains. *Computers & Chemical Engineering*, *31*(9), 1174–1180. DOI: 10.1016/j.compchemeng.2006.10.021

Paal, L. L., Karu, Y. A., Melder, H. A., & Repin, B. N. (1994). *Reference book on purification of natural and sewage waters*. Vishchya Shkola. (in Russian)

Pachepsky, J. O. (1990). *Mathematical models of physicochemical processes in soils*. Nauka. (in Russian)

Panfilov, M., & Fourar, M. (2006). Physical splitting of non-linear effects in high-velocity stable flow through porous media. *Advances in Water Resources*, *29*(1), 30–41. DOI: 10.1016/j.advwatres.2005.05.002

Pang, L., Goltz, M., & Close, M. (2003). Application of the method of temporal moments to interpret solute transport with sorption and degradation. *Journal of Contaminant Hydrology*, *60*(1-2), 123–134. DOI: 10.1016/S0169-7722(02)00061-X PMID: 12498577

Parker, E. A., Rippy, M. A., Mehring, A. S., Winfrey, B. K., Ambrose, R. F., Levin, L. A., & Grant, S. B. (2017). Predictive power of clean bed filtration theory for fecal indicator bacteria removal in stormwater biofilters. *Environmental Science & Technology*, *51*(10), 5703–5712. DOI: 10.1021/acs.est.7b00752 PMID: 28445642

Parlange, J.-Y., Haverkamp, R., & Touma, J. (1985). Infiltration under ponded conditions: 1. Optimal analytic solution and comparison with experimental observations. *Soil Science*, *139*(4), 305–311. DOI: 10.1097/00010694-198504000-00003

Payatakes, A. C., Tien, C., & Turian, R. M. (1973). A new model for granular porous media. I. Model formulation. *AIChE Journal. American Institute of Chemical Engineers*, *19*(1), 58–67. DOI: 10.1002/aic.690190110

Perkins, T. K., & Johnston, O. C. (1963). A review of diffusion and dispersion in porous media. *Society of Petroleum Engineers Journal*, *3*(1), 70–84. DOI: 10.2118/480-PA

Philip, J. R. (1994). Some exact solutions of convection-diffusion and diffusion equations. *Water Resources Research*, *30*(12), 3545–3551. DOI: 10.1029/94WR01329

Piron, E., Rene, F., & Latrille, E. (1995). A cross-flow microfiltration model based on integration of the mass transport equation. *Journal of Membrane Science*, *108*(1-2), 57–70. DOI: 10.1016/0376-7388(95)00141-4

Pittoors, E., Guo, Y., & Van Hulle, S. W. H. (2014). Modeling dissolved oxygen concentration for optimizing aeration systems and reducing oxygen consumption in activated sludge processes: A review. *Chemical Engineering Communications*, *201*, 983–1002. DOI: 10.1080/00986445.2014.883974

Pivovar, N.G., Bugay, N.G., Friedrichson, V.L., Krivonog, A.I., & Krivonog, V.V. (2000). *Drainage with fiber filters for protection of territories from waterlogging*. NASU. Institute of Hydromechanics. (in Russian)

Plotnikov, N.A. (1983). *Design of systems of artificial replenishment of underground waters for water supply*. Stroyizdat. (in Russian)

Plusnin, I. I., & Golovanov, A. I. (1983). *Reclamation soil science*. Kolos. (in Russian)

Poliakov V.L. (2013). About one strict solution of the problem of plane-radial filtration of a suspension. *Problems of Water Supply, Sewage and Hydraulics. Collected papers*, 22, 124–131. (in Russian)

Poliakov V.L. (2015). On the theory of plane-radial filtration and its some applications. *Problems of Water Supply, Sewage and Hydraulics. Collected papers*, 25, 217–222. (in Russian)

Poliakov, V. L. (2019). Theoretical substantiation of filtration intensification by changing the direction of the suspension flow. In: *Proceedings of* III International Scientific and Technical Conference "Water supply and drainage, design, construction, operation, monitoring", 194–196.

Poliakov, V.L. (2006). Calculation of suspension separation by filtration through a three-layer bed at linear mass transfer: *Problems of Water Supply, Sewage and Hydraulics. Collected papers*, 7, 122–131. (in Russian)

Poliakov, V.L. (2007). Modeling of suspensions clarification in granular beds at nonlinear filtration flow. *Municipal economy of cities. Collected papers*, 74, 77– 86. (in Russian)

Poliakov, V.L. (2007). On the choice of the rate coefficient of suspended solids removal from suspensions during filtration. *Problems of Water Supply, Sewage and Hydraulics. Collected papers*, 8, 120–131. (in Russian)

Poliakov, V.L. (2007). On the choice of the rate coefficient of suspended solids removal from suspensions during filtration. *Problems of Water Supply, Sewage and Hydraulics. Collected papers*, 8, 120–131. (in Russian).

Poliakov, V.L. (2008). Filtration of suspensions through granular beds at variable suspended solids concentration. *Problems of Water Supply, Sewage and Hydraulics. Collected papers*, 11, 103–111. (in Russian)

Poliakov, V.L. (2008). Intensive wetting of multilayer soils. *Applied Hydromechanics*, 10(82), 1, 69–79. (in Russian)

Poliakov, V.L. (2009). About mathematical modeling and engineering calculations of suspension clarification by filtration taking into account the deposition structure. *Problems of Water Supply, Sewage and Hydraulics. Collected papers*, 13, 91–102. (in Russian)

Poliakov, V.L. (2009). New analytical solutions of mathematical problems of suspension filtration at linear mass-exchange kinetics. *Bulletin of NUWEE*, 3(47), part 2, 264-271.

Poliakov, V.L. (2009). On the influence of grain sorting during filter bed operation on its efficiency. *Problems of Water Supply, Sewage and Hydraulics. Collected papers*, 12, 103–118. (in Russian)

Poliakov, V.L. (2010). About influence of bed height and filtration rate on suspension clarification. *Problems of Water Supply, Sewage and Hydraulics. Collected papers*, 15, 131–141. (in Russian)

Poliakov, V.L. (2010). Modeling the dynamics of secondary pollution in bioreactors-filters. *Problems of Water Supply, Sewage and Hydraulics. Collected papers*, 14, 152–162. (in Russian)

Poliakov, V.L. (2012). Filtration of suspension through a granular bed at limiting saturation of its part with deposit. *Applied hydromechanics*, 14(86), 3, 55–63. (in Russian)

Poliakov, V.L. (2012). Filtration of suspension through a granular bed at limiting saturation of its part with depositt. *Applied hydromechanics*, 14(86), 3, 55–63. (in Russian)

Poliakov, V.L. (2013). About rational choice of design parameters of cylindrical bed of water purification filter. *Problems of Water Supply, Sewage and Hydraulics. Collected papers*, 22, 113–123. (in Russian)

Poliakov, V.L. (2014). Theoretical substantiation of plane-radial filtration of suspension. *Applied hydromechanics*, 16(88), 3, 52–62. (in Russian)

Poliakov, V.L. (2014). To the calculation of suspension filtration at nonlinear mass transfer kinetics. *Problems of Water Supply, Sewage and Hydraulics. Collected papers*, 23, 145–153. (in Russian)

Poliakov, V.L. (2023). Modeling of the surface filtration of an aqueous suspension.

Poliakov, V.L., & Shevchuk, E.A. (2007). Calculation of stepped filtration of suspensions through granular filter beds. Bulletin of NUWEE, 4(40), part 2, 544–553. (in Russian)

Poliakov, V. L. (2006). About filtration of suspension at initial contamination of filter bed (linear kinetics of mass exchange). [in Russian]. *Rep. NAS of Ukraine, 10*, 65–71.

Poliakov, V. L. (2006). About filtration of suspension at initial contamination of filter bed (linear kinetics of mass transfer). [in Russian]. *Rep. NAS of Ukraine, 10*, 65–71.

Poliakov, V. L. (2008). Calculation of suspension filtration through multilayer filter bed under linear mass transfer kinetics. [in Russian]. *Journal of Water Chemistry and Technology, 30*(1), 3–31.

Poliakov, V. L. (2009). Theoretical analysis of filter run duration. [in Russian]. *Journal of Water Chemistry and Technology, 31*(6), 605–618.

Poliakov, V. L. (2010). Clarification of suspension by filtration with declining flow rate. [in Russian]. *Rep. NAS of Ukraine*, (5), 57–63.

Poliakov, V. L. (2010). Clarification of suspension by filtration with decreasing flow rate. [in Russian]. *Rep. NAS of Ukraine*, (5), 57–63.

Poliakov, V. L. (2010). On the forecast of head losses in the filter bed. [in Russian]. *Rep. NAS of Ukraine*, (3), 70–76.

Poliakov, V. L. (2011). Engineering calculation of suspension filtration through a two-layer filter bed under linear mass transfer kinetics. [in Russian]. *Journal of Water Chemistry and Technology, 33*(4), 367–380.

Poliakov, V. L. (2011). Filtration of suspensions through multilayer filter bed at nonlinear mass exchange kinetics. 1. Theory. [in Russian]. *Journal of Water Chemistry and Technology, 33*(1), 3–14.

Poliakov, V. L. (2011). Filtration of suspensions through multilayer packed bed at nonlinear kinetics of mass exchange. 1. Theory. [in Russian]. *Journal of Water Chemistry and Technology, 33*(1), 3–14.

Poliakov, V. L. (2011). Filtration of suspensions through multilayer packed bed at nonlinear kinetics of mass exchange. 2. Quantitative analysis. [in Russian]. *Journal of Water Chemistry and Technology, 33*(2), 111–127.

Poliakov, V. L. (2012). Filtration of suspensions with decreasing rate at linear kinetics of mass transfer. [in Russian]. *Journal of Water Chemistry and Technology, 34*(2), 107–130.

Poliakov, V. L. (2012). Modeling of the water treatment process in the adsorber bed. [in Russian]. *Rep. NAS of Ukraine*, (5), 63–71.

Poliakov, V. L. (2012). Theoretical analysis of the adsorbent grain action. [in Russian]. *Rep. NAS of Ukraine*, (4), 64–69.

Poliakov, V. L. (2013). Theoretical analysis of the plane-radial filtration of suspension. [in Russian]. *Rep. NAS of Ukraine*, (12), 52–59.

Poliakov, V. L. (2014). *Filtration flow deformations in drained soils: theory and applications*. Agrar Media Group. (in Russian)

Poliakov, V. L. (2015). About rational choice of the form and sizes of bed of a water treatment filter. *Vodoochistka, Vodopodgotovka* [in Russian]. *Vodosnabzheniya*, *9*(93), 8–16.

Poliakov, V. L. (2015). Calculation of plane-radial filtration at suspension feeding into the filter bed from inside. *Vodoochistka, Vodopodgotovka* [in Russian]. *Vodosnabzheniy*, *5*(89), 16–21.

Poliakov, V. L. (2015). Theoretical analysis of plain-radial filtration. *Water and Ecology.* [in Russian]. *Problems and Solutions*, *2*(62), 14–24.

Poliakov, V. L. (2019). Mathematical modeling of suspension clarification in dynamic layer. [in Russian]. *Rep. NAS of Ukraine*, (12), 33–40.

Poliakov, V. L. (2020). Mathematical modeling of the non-detachment filtration of water suspension with change in direction. [in Russian]. *Rep. NAS of Ukraine*, (5), 31–41.

Poliakov, V. L. (2020). Theoretical analysis of the influence of the seasonal temperature changes on aqueous suspension filtration. In: *Proceedings of International Scientific and Practical Conference "Energoeffektivne misto XXI stolittya"*, 42–47. (in Russian)

Poliakov, V. L. (2022). Theoretical analysis of reagent filtration of aqueous suspensions with unregulated filtration rate. *Theoretical Foundations of Chemical Engineering*, *56*(4), 477–496. DOI: 10.1134/S0040579522030101

Poliakov, V., Kravchuk, A., Kochetov, G., & Kravchuk, O. (2019). Clarification of aqueous suspensions with a high content of suspended solids in rapid sand filters. *Eureka: Physics and Engineering*, *1*(1), 28–45. DOI: 10.21303/2461-4262.2019.00827

Poliakov, V., & Martynov, S. (2021). Mathematical modeling of physicochemical iron removal from groundwater at rapid filters. *Chem. Eng. Sci.*, 231, Saiers, J.E., Hornberger, G.M., & Liang, Z. (1994). First- and second-order kinetics approaches for modeling the transport of colloidal particles in porous media. *Water Resources Research*, *30*(9), 2499–2506.

Polubarinova-Kochina, P. Ya. (1977). *Theory of groundwater movement*. Nauka. (in Russian)

Polubarinova-Kochina, P. Ya., Pryazhinskaya, V. G., & Emikh, V. N. (1969). *Mathematical methods in irrigation issues*. Nauka. (in Russian)

Polyakov, Y. S., & Dilman, V. V. (2006). Approximate method for nonlinear differential and integro-differential equations. *AIChE Journal. American Institute of Chemical Engineers*, *52*(11), 3813–3824. DOI: 10.1002/aic.10995

Polyanin, A. D., & Zaitsev, V. F. (2002). *Reference book on nonlinear equations of mathematical physics: exact solutions*. Fizmatlit. (in Russian)

Popkovich, G. S., & Ropin, B. N. (1986). *Systems of aeration of sewage waters*. Stroyizdat. (in Russian)

Pritzker, M. D. (2003). Model for parallel surface and pore diffusion of an adsorbate in a spherical adsorbent particle. *Chemical Engineering Science, 58*(2), 473–478. DOI: 10.1016/S0009-2509(02)00554-7

Progulny, V. I., & Ryabkov, M. V. (2019). Porous polymer concrete in drains of pressure polystyrene foam filters. [in Russian]. *Bulletin of ONACA, 74*, 147–153.

Quintana-Solorzano, R., Che-Galicia, G., Trejo-Reyes, M., & Armendariz-Herrera, H. (2018). On the influence of particle shape and process conditions in the pressure drop and hydrodynamics in a wall-effect fixed bed. *Chemical Engineering Communications, 205*(10), 1323–1341. DOI: 10.1080/00986445.2018.1448274

Rahli, O., Tadrist, L., & Miscevic, M. (1996). Experimental analysis of fibrous porous media permeability. *AIChE Journal. American Institute of Chemical Engineers, 42*(12), 3547–3549. DOI: 10.1002/aic.690421224

Rajagopalan, R., & Tien, C. (1976). Trajectory analysis of deep bed filtration with sphere in cell porous media model. *AIChE Journal. American Institute of Chemical Engineers, 22*(3), 523–533. DOI: 10.1002/aic.690220316

Rastvorova, O. G. (1983). *Soil Physics. Practical guide*. Leningrad State University Press. (in Russian)

Recommendations on design and operation of artificial groundwater recharge systems. (1976). VNII VODGEO. (in Russian)

Reddy, L. N., Ming, X., Hajra, M. G., & Lee, I. M. (2000). Permeability reduction of soil filters due to physical clogging. *Journal of Geotechnical and Geoenvironmental Engineering, 126*(3), 236–246. DOI: 10.1061/(ASCE)1090-0241(2000)126:3(236)

Reddy, L. N., Xiao, M., Hajra, M. C., & Lee, I. M. (2005). Physical clogging of soil filters under constant flow rate versus constant head. *Canadian Geotechnical Journal, 42*(3), 804–811. DOI: 10.1139/t05-018

Ren, J., Packman, A. J., & Welty, C. (2000). Correlation of colloid efficiency with hydraulic conductivity of silica sands. *Water Resources Research, 36*(9), 2493–2500. DOI: 10.1029/2000WR900163

Reznikov, G. D. (1999). About the particle absorption in filters of water treatment facilities. [in Russian]. *Engineering and Physical Journal, 72*(1), 38–43.

Richard, L. (2004). Filter media surface modification technology: State of art. *Filtration & Separation, 41*(9), 20–21. DOI: 10.1016/S0015-1882(04)00343-X

Riekert, L. (1985). The relative contribution of pore volume diffusion and surface diffusion to mass transfer in capillaries and porous media. *AIChE Journal. American Institute of Chemical Engineers, 31*(5), 863–864. DOI: 10.1002/aic.690310525

Rittmann, B. E. (1993). The significance of biofilms in porous media. *Water Resources Research, 27*(7), 2195–2202. DOI: 10.1029/93WR00611

Rittmann, B. E., & Stiwell, D. S. (2002). Modeling biological processes in drinking water treatment: An integral biofiltration model. [in Russian]. *Water and Ecology*, *3*, 13–29.

Rode, A. A. (1965). *Fundamentals of the doctrine of soil moisture* (Vol. 1). Gidrometeoizdat. (in Russian)

Rodgers, M., Healy, M. G., & Mulqueen, J. (2005). Organic removal and nitrification of high strength wastewater using stratified sand filters. *Water Research*, *39*(14), 3279–3286. DOI: 10.1016/j.watres.2005.05.035 PMID: 16005491

Romankov, P. G., & Kurochkina, M. I. (1982). *Hydromechanical processes of chemical technology*. Khimiya. (in Russian)

Romashchenko, M. I., & Balyuk, S. F. (2000). *Irrigation of lands in Ukraine. State and ways of improvement*. Svit. (in Ukrainian)

Rudnytsky, E.A. (2008). Hydraulic characteristics of the upward flow through the floating filter bed. *Land Reclamation and Water Management. Collected papers*, 96, 213–222. (in Russian)

Rudzskiy, G. G., & Kim, A. N. (1983). Application of superfast filtration method for additional treatment of waste water. [in Russian]. *Chemistry and Water Technology*, *5*(1), 65–67.

Saatci, A.M. (1990). Application of the declining rate filtration theory. – Continuous operation. *J. Environ. Eng. Div., ASCE*, 116(1), 87–105.

Sabliy, L. A. (2013). *Physical, chemical and biological treatment of highly concentrated wastewater*. NUWEE Press. (in Russian)

Saiers, J. E., Hornberger, G. M., & Liang, Z. (1994). First- and second-order kinetics approaches for modeling the transport of colloidal particles in porous media. *Water Resources Research*, *30*(9), 2499–2506. DOI: 10.1029/94WR01046

Sakkas, N. D., & Lekkas, T. D. (1989). Hydrodynamic characteristics of the solid-liquid fluidized bed developing during the backwash of filter media. *Environmental Technology Letters*, *10*(2), 151–166. DOI: 10.1080/09593338909384728

Salthes, T., Eikenbrokk, B., & Odegard, H. (2003). Effect of filtration rate in a filter with two-layer expanded clay (filtralite) in comparison with a sand-anthracite filter. [in Russian]. *Water and Ecology*, (2), 23–32.

Samani, J. M. V., Fathi, P., & Homaee, M. (2007). Simultaneous prediction of saturated hydraulic conductivity and drainable porosity using the inverse problem technique. *Journal of Irrigation and Drainage Engineering*, *133*(2), 110–115. DOI: 10.1061/(ASCE)0733-9437(2007)133:2(110)

Samuel, M. P., Senthilvel, S., Tamilmani, D., & Mathew, A. C. (2012). Performance evaluation and modeling studies of gravel-coir fibre-sand multimedia stormwater filter. *Environmental Technology*, *33*(17), 2057–2069. DOI: 10.1080/09593330.2012.660642 PMID: 23240200

Sanda, A., Nopolaa, T., Hjelt, T., & Toivakka, M. (2009). A particle motion model for the study of consolidation phenomena. *Computers & Chemical Engineering*, *33*(7), 1227–1239. DOI: 10.1016/j.compchemeng.2009.03.001

SBC V.2.5-74:2013. (2013). *Water supply. External networks and structures. Basic design provisions.* MRRBZHKG of Ukraine. (in Ukainian)

SBC V.2.5-75:2013. (2013). *Sewerage. Design of external networks and structures. Basic provisions of design.* UkrNDIvodokanalproekt. (in Ukainian)

Schneider, O. D., Nickols, D., VandeVenter, L., & Leggiero, S. (2002). An evaluation of high rate filtration using particle counting and turbidimetry for low turbidity waters. *Water Science and Technology: Water Supply*, *2*(1), 249–258. DOI: 10.2166/ws.2002.0030

Schuberth, J., Mauschitz, G., & Höflinger, W. (2011). Clogging mechanisms involved in the aging process of cleanable filter media. *Separation and Purification Technology*, *77*(2), 196–201. DOI: 10.1016/j.seppur.2010.12.004

Schulze-Makuch, D., Carlson, D. A., Cherkauer, D. S., & Malik, P. (1999). Scale dependency of hydraulic conductivity in heterogeneous media. *Groundwater*, *37*(6), 904–919. DOI: 10.1111/j.1745-6584.1999.tb01190.x

Seifert, D., & Engesgaard, P. (2012). Sand box experiments with bioclogging of porous media: Hydraulic conductivity reductions. *Journal of Contaminant Hydrology*, *136–137*, 1–9. DOI: 10.1016/j.jconhyd.2012.04.007 PMID: 22647500

Sen, T. K., Shanbhag, S., & Khilar, K. C. (2004). Subsurface colloids in groundwater contamination: A mathematical model. *Colloids and Surfaces. A, Physicochemical and Engineering Aspects*, *232*(1), 29–38. DOI: 10.1016/j.colsurfa.2003.10.004

Senyavin, M. M., Venitsianov, E. V., & Ayukaev, R. I. (1997). About mathematical models and engineering methods for calculating the process of natural water purification by filtration. [in Russian]. *Water Resources*, (2), 157–170.

Serrano, S. E. (2001). Solute transport under non-linear sorption and decay. *Water Research*, *35*(6), 1525–1533. DOI: 10.1016/S0043-1354(00)00390-0 PMID: 11317900

Shandyba, A. B., & Vainshtein, I. A. (1988). To calculation of washing of dispersed materials. [in Russian]. *Hydromechanics*, *57*, 72–76.

Shani, C., Weisbrod, N., & Yakirevich, A. (2008). Colloid transport through saturated sand columns: Influence of physical and chemical properties on deposition. *Colloids and Surfaces. A, Physicochemical and Engineering Aspects*, *316*(1-3), 142–150. DOI: 10.1016/j.colsurfa.2007.08.047

Shaoging, Q. (1998). Expression for critical suspended particle size in granular media filtration processes. *Water Research*, *32*(9), 2856–2859. DOI: 10.1016/S0043-1354(98)00028-1

Sharma, M. M., Chaumon, H., Sarma, D. S., & Schecher, R. S. (1992). Factors controlling the hydrodynamic detachment of particles from surfaces. *Journal of Colloid and Interface Science*, *149*(1), 121–134. DOI: 10.1016/0021-9797(92)90398-6

Sharma, M. M., & Yortos, Y. C. (1987). Transport of particulate suspensions in porous media: Model formulation. *AIChE Journal. American Institute of Chemical Engineers*, *33*(10), 1636–1643. DOI: 10.1002/aic.690331007

Shehtman, Yu. M. (1961). *Filtration of low–concentration suspensions*. AS USSR Publishing House. (in Russian)

Shein, E. V. (Ed.). (2001). *Field and laboratory methods of research of physical properties and soil regimes: Methodical guide*. Moscow State University Press. (in Russian)

Shen, C., Huang, Y., Li B., & Yan Jin, Y. (2010). Predicting attachment efficiency of colloid deposition under unfavorable attachment conditions. *Water Resour. Res.*, 46, 1-12. W11526, .DOI: 10.1029/2010WR009218

Shena, L., & Chen, Z. (2007). Critical review of the impact of tortuosity on diffusion. *Chemical Engineering Science*, *62*(14), 3748–3755. DOI: 10.1016/j.ces.2007.03.041

Shen, W., Yang, G.-H., Liu, P., Yao, Z., Shi, H., Cui, Y., & Li, Z. (2018). Effects of temperature on the filtration characteristics of a dual–layer granular bed filter. *Chemical Engineering & Technology*, *41*(9), 1759–1766. DOI: 10.1002/ceat.201800071

Sherard, J. L., Dunnigan, L. P., & Talbot, J. R. (1984). Basic properties of sand and gravel filters. *Journal of Geotechnical Engineering*, *110*(6), 684–700. DOI: 10.1061/(ASCE)0733-9410(1984)110:6(684)

Sherwood, T., Pigford, R., & Wilkie, C. (1982). *Mass Transfer*. Khimiya. (in Russian)

Shevchuk, E. A., Mamchenko, A. V., & Goncharuk, V. V. (2005). Technology of direct flow filtration of natural and waste water through granular loads. [in Russian]. *Journal of Water Chemistry and Technology*, *27*(4), 369–383.

Shevtsov, N. M. (1988). *Intra-soil purification and utilization of sewage waters*. Agropromizdat. (in Russian)

Shirasaki, N., Matsushita, T., Matsui, Y., Oshiba, A., & Ohno, K. (2010). Estimation of norovirus removal performance in a coagulation-rapid sand filtration process by using recombinant norovirus VLPs. *Water Research*, *44*(5), 1307–1316. DOI: 10.1016/j.watres.2009.10.038 PMID: 19922975

Shtykov, V.I., Shevelev, Ya, Z., & Koshevoy, O.Yu. (1987). *Use of livestock waste complexes on specialized systems*. Rosselhozizizdat. (in Russian)

Shtykov, V. I., Shevelev, Ya. Z., & Koshevoy, O. Yu. (1987). *Use of livestock waste complexes on specialized systems*. Rosselhozizizdat. (in Russian)

Shvidler, M. I. (1987). Stochastic filtration flow transfer. In *Problems of filtration flow theory and mechanics of oil recovery enhancement* (pp. 203–211). Nauka. (in Russian)

Šimůnek, J., Jarvisb, N. J., van Genuchten, M. Th., & Gärdenäs, A. (2003). Review and comparison of models for describing non-equilibrium and preferential flow and transport in the vadose zone. *Journal of Hydrology (Amsterdam)*, *272*(1-4), 14–35. DOI: 10.1016/S0022-1694(02)00252-4

Skouras, E. D., Burganos, V. N., Paraskeva, C. A., & Payatakes, A. C. (2011). Phenomenological description and network simulation of horizontal filtration dynamics. *Separation and Purification Technology*, *80*(1), 105–118. DOI: 10.1016/j.seppur.2011.04.016

Sokovnin, O. T., & Zagoskina, N. V. (2002). Selection and determination of characteristic particle size. [in Russian]. *Theoretical Foundations of Chemical Engineering, 36*(6), 658–669. DOI: 10.1023/A:1021274019713

Song, L. (1998). Flux decline in crossflow microfiltration and ultrafiltration: Mechanisms and modeling of membrane fouling. *Journal of Membrane Science, 139*(2), 183–200. DOI: 10.1016/S0376-7388(97)00263-9

Sposito, G., Jury, W. A., & Gupta, V. K. (1986). Fundamental problems in the stochastic convection-dispersion model of solute transport in aquifers and field soils. *Water Resources Research, 22*(1), 77–88. DOI: 10.1029/WR022i001p00077

Stechkina, I. B., & Kirsh, V. A. (2003). Optimization of filter parameters in a multistage system of fine gas purification. [in Russian]. *Theoretical Foundations of Chemical Engineering, 37*(3), 236–245. DOI: 10.1023/A:1024071202844

Stevenson, D. G. (1995). Process conditions for the backwashing of filter with simultaneous air and water. *Water Research, 29*(11), 2594–2597. DOI: 10.1016/0043-1354(95)00041-I

Stevenson, D. G. (1997). Flow and filtration through granular media. The effect of grain and particle size dispersion. *Water Research, 31*(2), 310–322. DOI: 10.1016/S0043-1354(96)00271-0

Stevik, T. K., Aa, K., Ausland, G., & Hansen, J. F. (2004). Retention and removal of pathogenic bacteria in wastewater percolating through porous media: A review. *Water Research, 38*(6), 1355–1367. DOI: 10.1016/j.watres.2003.12.024 PMID: 15016513

StSanRandN 2.2.4-171-10. (2010). *Hygienic requirements for drinking water intended for human consumption*. Ministry of Health of Ukraine. (in Ukrainian)

Sutherland, K. (2008). Filtration overview: A closer look at depth filtration. *Filtration & Separation, 45*(October), 25–27. DOI: 10.1016/S0015-1882(08)70296-9

Sutzkover-Gutman, I., Hasson, D., & Semiat, R. (2010). Humic substances fouling in ultrafiltration processes. *Desalination, 261*(3), 218–231. DOI: 10.1016/j.desal.2010.05.008

Swanson, W. J., & Loeh, R. C. (1997). Biofiltration: Fundamentals, design and operations principles, and applications. *Journal of Environmental Engineering, 123*(6), 538–546. DOI: 10.1061/(ASCE)0733-9372(1997)123:6(538)

Tanaka, Y., Miyajuma, K., Funakosi, T., & Chida, S. (1995). Filtration of municipal sewage by ring shaped floating plastic net media. *Water Research, 29*(5), 1387–1392. DOI: 10.1016/0043-1354(94)00288-I

Tang, Y., & Aral, M. M. (1992). Contaminant transport in layered porous media. 1. General solution. *Water Resources Research, 28*(5), 1389–1397. DOI: 10.1029/92WR00292

Tan, H. K. S. (1994). Analysis of cyclic fixed bed sorption processes for second-order irreversible kinetics. *Chemical Engineering Science, 49*(8), 1277–1285. DOI: 10.1016/0009-2509(94)85096-8

Tan, X.-H., Jianh, L., Li, X.-P., Li, Y.-Y., & Zhang, K. (2017). A complex model for the permeability and porosity of porous media. *Chemical Engineering Science, 172*, 230–238. DOI: 10.1016/j.ces.2017.06.041

Tarnapolskaya, M. G. (2005). Filtering materials for water purification from oil products and criterion of their choice. [in Russian]. *Water and Ecology*, (3), 74–79.

Thomasa, D., Penicota, P., Contala, P., Leclerca, D., & Vende, J. (2001). Clogging of fibrous filters by solid aerosol particles. Experimental and modelling study. *Chemical Engineering Science*, *56*(11), 3549–3561. DOI: 10.1016/S0009-2509(01)00041-0

Tiller, F. M., Kirby, J. M., & Nguyen, H. L. (1996). Approximate theory for radial filtration / consolidation. *Journal of Geotechnical Engineering*, *122*(10), 797–805. DOI: 10.1061/(ASCE)0733-9410(1996)122:10(797)

Tiller, F. M., & Yeh, C. S. (1985). The role of porosity infiltration. Part X: Deposition of compressible cakes on external radial surfaces. *AIChE Journal. American Institute of Chemical Engineers*, *31*(8), 1241–1248. DOI: 10.1002/aic.690310803

Tkachuk, O. A. (2022). *Hydraulic calculations of pipeline systems of water supply and drainage: monograph*. NUWEE Press. (in Russian)

Tkachuk, O., Yaruta, Y., & Shevchuk, O. (2022). Assessment of application conditions of infiltration basins for regulation of urban rainwater drainage. *Journal of Ecological Engineering*, *23*(2), 191–195. DOI: 10.12911/22998993/144718

Tobiason, J. E., Johnson, G. S., Westerhoff, P. K., & Vigneswaran, B. (1993). Particle size and chemical effects on contact filtration performance. *Journal of Environmental Engineering*, *119*(3), 520–533. DOI: 10.1061/(ASCE)0733-9372(1993)119:3(520)

Tobiason, J. E., & Vigneswaran, B. (1999). Evaluation of a modified model for deep bed filtration. *Water Research*, *28*(2), 335–342. DOI: 10.1016/0043-1354(94)90271-2

Tom, C. S., Zimbron, J. A., & Dandy, D. S. (2008). Effects of reduced contaminant loading on downgradient water quality in an idealized two-layer granular porous media. *Journal of Contaminant Hydrology*, *102*(1-2), 72–85. DOI: 10.1016/j.jconhyd.2008.08.002 PMID: 18930336

Trussell, R. R., & Chang, M. (1999). Review of flow through porous media as applied to head loss in water filters. *Journal of Environmental Engineering*, *125*(11), 998–1006. DOI: 10.1061/(ASCE)0733-9372(1999)125:11(998)

Tsitsishvili, G. V., & Andronikashvili, T. G. (1985). *Natural zeolites*. Khimiya. (in Russian)

Tufenkji, N., Dixon, D. R., Considine, R., & Drummond, C. J. (2006). Multi-scale Cryptosporidium sand interactions in water treatment. [Review]. *Water Research*, *40*(18), 3315–3330. DOI: 10.1016/j.watres.2006.07.036 PMID: 16979211

Tufenkji, N., & Elimelech, M. (2004). Correlation equation for predicting single-collector efficiency in physicochemical filtration in saturated porous media. *Environmental Science & Technology*, *38*(2), 529–536. DOI: 10.1021/es034049r PMID: 14750730

Tugay, A.M., & Orlov, V.O. (2009). *Water supply*. Znannya. (in Ukrainian)

Tugay, A. M., Oleynyk, A. Ya., & Tugay, Y. A. (2004). *Productivity of water intake wells under conditions of clogging*. KHNAME Press. (in Russian)

Twort, A. C., Ratnayaka, D. D., & Brandt, M. J. (2006). *Water Supply* (5th ed.). IWA Publishing.

Usenko, V. S. (1972). *Artificial recharge and infiltration water intakes of underground waters*. Science and Technology. (in Russian)

Vadyunina, A.F., & Korchagina, Z.A. (1961). *Methods of determination of physical properties of soils and grounds in the field and laboratory*. State Publishing House "Higher School". (in Russian)

Van Genuchten, M. T. (1980). A closed-form equation for predicting the hydraulic conductivity of unsaturated soils. *Soil Science Society of America Journal, 44*(5), 892–898. DOI: 10.2136/sssaj1980.03615995004400050002x

Vandevivere, P., & Haveye, P. (1992). Saturated hydraulic conductivity reduction caused by aerobic bacteria in sand columns. *Soil Science Society of America Journal, 56*(1), 1–13. DOI: 10.2136/sssaj1992.03615995005600010001x

Vasiliev, S. V., Verigin, N. V., & Glazer, B. A.. (1970). *Methods of water flow calculations of hydroreclamation systems*. Kolos.

Veerapaneni, S., & Wiesner, M. R. (1993). Role of suspension polydispersivity in granular media filtration. *Journal of Environmental Engineering, 119*(1), 172–190. DOI: 10.1061/(ASCE)0733-9372(1993)119:1(172)

Venitsianov, E. V. (1998). Some features of heavy metals sorption by a layer of bottom sediments and soil. [in Russian]. *Water Resources, 25*(4), 462–466.

Venitsianov, E. V., & Rubinstein, R. N. (1983). *Dynamics of sorption from liquid media*. Nauka. (in Russian)

Verigin, N. N. (1971). *Clogging of radial filters with activated carbon*. Stroyizdat. (in Russian)

Verigin, N. N., & Kommunar, G. M. (1978). Filtration of suspensions through porous media. *Proceedings of VNII VODGEO*, 74, 120–124. (in Russian)

Verigin, N. N., & Vasiliev, S. V. (1987). Soil and ground wetting of the aeration zone. [in Russian]. *Appl. Techn. Phys*, (1), 133–137.

Vigneswaran, S., & Chang, J. S. (1986). Mathematical modeling of the entire cycle of deep bed filtration. *Water, Air, and Soil Pollution, 29*(2), 155–164. DOI: 10.1007/BF00208406

Vigneswaran, S., Thiyagaram, M., & Ben Ami, R. (1985). Mathematical modeling of radial flow filtration and its application to groundwater recharge problems. *AIChE Journal. American Institute of Chemical Engineers, 31*(11), 1929–1933. DOI: 10.1002/aic.690311121

Vigneswaran, S., & Tulachan, R. K. (1988). Mathematical modeling of transient behavior of deep bed filtration. *Water Research, 22*(9), 1093–1100. DOI: 10.1016/0043-1354(88)90003-6

Vorobyev, E. I., & Tarasenko, A. P. (1987). Influence of sludge compressibility on the regularities of the process of suspension filtration. [in Russian]. *Theoretical Foundations of Chemical Engineering, 21*(4), 507–514.

Voytov, E. L. (1974). *Pretreatment of biologically treated urban sewage on radial filters.* Ph.D. dissertation abstract. Novosibirsk State University Press. (in Russian)

Voznyuk, S. T., Kuzmich, P. K., Krysztof, V. G., Kirienko, T. N., & Olinevich, V. A. (1984). *Reclamation soil science with the basics of hydrology*. "Vishcha Shkola", Lviv State University Press. (in Russian)

Vyakhirev, G. I., Zagnitiko, A. V., Burbasov, A. N., Pushko, A. I., Pushko, G. I., & Chaplygin, Yu. O. (2002). Gas-dynamic resistance at high-speed purification of gases by multilayer metal-ceramic filters. [in Russian]. *Theoretical Foundations of Chemical Engineering, 36*(4), 353–357. DOI: 10.1023/A:1019887330128

Wakeman, R. (2007). The influence of particles properties on filtration. *Separation and Purification Technology, 58*(2), 234–241. DOI: 10.1016/j.seppur.2007.03.018

Wang, Y., Kim, J.-H., Baek, J.-B., Miller, G. W., & Pennell, K. D. (2012). Transport behavior of functionalized multi-wall carbon nanotubes in water-saturated quartz sand as a function of tube length. *Water Research, 46*(14), 4521–4531. DOI: 10.1016/j.watres.2012.05.036 PMID: 22704927

Watson, K., & Awadalla, S. A. (1986). Comparative study of the Green and Ampt analysis for a falling water table in a stratified profile. *Water Resources Research, 22*(5), 723–730. DOI: 10.1029/WR022i005p00723

Weber-Shirk, M. L., & Chan, K. L. (2007). The role of aluminum in slow sand filtration. *Water Research, 41*(6), 1350–1354. DOI: 10.1016/j.watres.2006.12.002 PMID: 17276479

Whitaker, S. (1988). Diffusion in packed beds of porous particles. *AIChE Journal. American Institute of Chemical Engineers, 34*(4), 679–683. DOI: 10.1002/aic.690340419

Widdowson, M. A., Molz, F. J., & Benefield, L. D. (1988). A numerical transport model for oxygen- and nitrate-based respiration linked to substrate and nutrient availability in porous media. *Water Resources Research, 24*(9), 1553–1565. DOI: 10.1029/WR024i009p01553

Williams, G. J., Sheikh, B., Holden, R. B., Kouretas, T. J., & Nelson, K. L. (2007). The impact of increased loading rate on granular media, rapid depth filtration of wastewater. *Water Research, 41*(19), 4535–4545. DOI: 10.1016/j.watres.2007.06.018 PMID: 17632202

Wolska, J., & Smolinska–Kempisty, K. (2017). Polypropylene prefilters with surface imprinted layer. *Separation and Purification Technology, 174*, 89–96. DOI: 10.1016/j.seppur.2016.10.007

Xue, B., & Sun, Y. (2003). Modeling of sedimentation of polydisperse spherical beads with a broadsize distribution. *Chemical Engineering Science, 58*(8), 1531–1543. DOI: 10.1016/S0009-2509(02)00656-5

Xu, S., Gao, B., & Saiers, J. E. (2006). Straining of colloidal particles in saturated porous media. *Water Resources Research, 42*(12), 2006WR004948. Advance online publication. DOI: 10.1029/2006WR004948

Yakovenko, Y. P. (2009). Mathematical modeling and research of infiltration basins operation. [in Ukrainian]. *Land Reclamation and Water Management, 97*, 166–179.

Yakovlev, S. V., Krasnoborodko, I. G., & Rogov, V. M. (1987). *Technology of electrochemical water purification*. Stroyizdat. (in Russian)

Yang, Z., & Langdon, A. G. (2004). The use of magnetic bed conditioning and pH control to enhance filtration by natural titanomagnetite. *Water Research, 38*(14-15), 3304–3312. DOI: 10.1016/j.watres.2004.04.017 PMID: 15276747

Yaroshevskaya, N. V., Andrievskaya, M. D., Shvidenko, O. G., & Kulskiy, L. A. (1989). Water purification by filtration with non-stationary change in the rate. [in Russian]. *Journal of Water Chemistry and Technology*, *11*(3), 251–253.

Yaroshevskaya, N. V., & Kulskiy, L. A. (1985). Calculation method for filtration facilities operating with decreasing rate. [in Russian]. *Journal of Water Chemistry and Technology*, *7*(2), 3–5.

Yaroshevskaya, N. V., & Muravyev, V. R. (2002). Comparative evaluation of water purification filters loading. [in Russian]. *Journal of Water Chemistry and Technology*, *24*(1), 71–79.

Yaroshevskaya, N. V., & Sotkova, G. Z. (1996). Calculation of parameters of a block of filters with a drainage of a large resistance, operating with decreasing rate. [in Russian]. *Journal of Water Chemistry and Technology*, *18*(4), 356–366.

Yavorsky, B. M., & Detlaf, A. A. (1977). *Reference book on physics*. Nauka. (in Russian)

Yeh, W. W.-G. (1986). Review of parameter identification procedures in groundwater hydrology. The inverse problem. *Water Resources Research*, *22*(2), 95–108. DOI: 10.1029/WR022i002p00095

Yenikeeva, S. N. (1985). Application of nonlinear optimization method for determination of formation migration parameters from field experiments. In *Modeling of hydrogeochemical processes and scientific bases of hydrogeochemical forecasts* (pp. 110–117). Nauka. (in Russian)

Yiantsios, S. G., & Karabelas, A. J. (2003). Deposition of micron-sized particles on flat surfaces: Effects of hydrodynamic and physicochemical conditions on particle attachment efficiency. *Chemical Engineering Science*, *58*(14), 3105–3113. DOI: 10.1016/S0009-2509(03)00169-6

Yin, C. Y., Aroua, M. K., & Daud, W. M. A. W. (2007). Review of modifications of activated carbon for enhancing contaminant uptakes from aqueous solutions. [Review]. *Separation and Purification Technology*, *52*(3), 403–415. DOI: 10.1016/j.seppur.2006.06.009

Young, J. C., & Dahab, M. F. (1983). Effect of media design on the performance of fixed–bed anaerobic filters. *Water Science and Technology*, *15*(8-9), 369–383. DOI: 10.2166/wst.1983.0179

Yuksel, E., Akgiray, O., & Soyer, E. (2003). Direct filtration with preozonation for small water treatment systems. *Water Science and Technology*, *48*(11-12), 473–479. DOI: 10.2166/wst.2004.0896 PMID: 14753570

Yu, Y., & Rowe, R. K. (2011). Improved solutions for porosity and specific surface of a uniform porous medium with attached film. *J. Environ. Eng*, *137*, 248–257.

Zakariae, N., & Nore, S. M. M. (2024). Harvesting solutions: Discover the evolution o" agriculture wastewater treatment through comprehensive bibliometric analysis using scopus database 1971-2023. *Desalination and Water Treatment*, *317*, 100291. DOI: 10.1016/j.dwt.2024.100291

Zapolsky, A. K., & Baran, A. A. (1987). *Coagulants and flocculants in water purification processes*. Khimiya. (in Russian)

Zapolsky, A. K., Mishkova-Klimenko, N. A., Astrelin, I. M., Bryk, M. G., Gvozdyak, P. I., & Knyazkova, T. V. (2000). *Physical and chemical bases of wastewater treatment technologies: Textbook*. Libra. (in Ukrainian)

Zhang, L. M., & Li, X. (2010). Microporosity structure of coarse granular soils. *Journal of Geotechnical and Geoenvironmental Engineering*, *136*(10), 1425–1436. DOI: 10.1061/(ASCE)GT.1943-5606.0000348

Zhernov, I. E. (1982). *Dynamics of underground waters*. Vishcha Shkola. (in Russian)

Zhurba, M. G. (1980). *Water purification on granular filters*. Vyshchaya shkola, Lviv State University Press. (in Russian)

Zhurba, M.G. (1993). Application of filtration theory in engineering calculations. *WSE. Water supply and sanitary engineering*, N°7, 2–6. (in Russian)

Zhurba, M.G. (2011). *Water treatment filters with floating filtering media. Scientific edition.* RNO VoSTU. (in Russian)

Zhurba, M.G., & Myakishev, V.A. (1992). Technology of water purification subjected to anthropogenic impact. *WSE. Water supply and sanitary engineering*, N°8, 3–5. (in Russian)

Zhurba, M.G., Sokolov, L.I., & Govorova, J.M. (2010). *Water Supply. Design of systems and structures: Textbook in 3 vol.* ASV Publishing House. (in Russian)

Zielina, M. (2011). Particle shapes in the drinking water filtration process. *Clean (Weinheim)*, *39*(11), 941–946. DOI: 10.1002/clen.201000432

Zosin, A. L., Priymak, T. I., Schall, E. E., Aleev, N. G., & Sulimenko, L. P. (2003). Filtering systems based on complex mineral loadings for treatment of mine water. [in Russian]. *Water and Ecology*, *3*(16), 58–65.

Zou, R. P., Gan, M. L., & Yu, A. B. (2011). Prediction of the porosity of multi-component mixtures of cohesive and non-cohesive particles. *Chemical Engineering Science*, *66*(20), 4711–4717. DOI: 10.1016/j.ces.2011.06.037

About the Author

Vadym Leontiiovych Poliakov graduated from Moscow (diploma), Kiev State Universities. For fifty years I have been working as (leading) researcher at the Institute of Hydromechanics (NASU, Kiev); defended dissertations devoted to nonlinear theory of waterlogged lands drainage, for twenty years taught at the National University of Construction, Architecture. I am also the chief researcher (IWPLR, NAASU). The specific areas of research concern studying theoretical issues related to purification of natural and waste waters by physical-chemical and biological methods, regulation of water and nutrient conditions of agricultural lands under climate change. The results are presented in more than three hundred publications, including three monographs and textbook.

Index

H

I

L

M

N

O

P

Q

R

S

T

W

www.ingramcontent.com/pod-product-compliance
Lightning Source LLC
LaVergne TN
LVHW080621300926
849162LV00005B/1223